D1325441

THE ESTIMATION OF ANIMAL ABUNDANCE
and related parameters

THE ESTIMATION OF ANIMAL ABUNDANCE

and related parameters

G. A. F. SEBER

M.Sc. (Auckland), Ph.D. (Manchester)

Professor of Statistics, Auckland University

New Zealand

SECOND EDITION

Edward Arnold

A member of the Hodder Headline Group

LONDON SYDNEY AUCKLAND

Edward Arnold is a division of Hodder Headline PLC
338 Euston Road, London NW1 3BH

First published in the United Kingdom 1973 by
Charles Griffin & Co Ltd
Second edition published 1982

2 4 6 5 3
95 97 99 98 96

British Library Cataloguing in Publication Data
Seber, G A F
Estimation of animal abundance and related parameters. –2nd ed.
1. Animal populations – Statistical metholds
I. Title
591.52'48'072 QH352

ISBN 0-85264-262-8

Typeset by The Alden Press Ltd, Oxford, London and Northampton
Printed and bound in the United Kingdom by
The Athenæum Press Ltd, Gateshead, Tyne and Wear

PREFACE TO SECOND EDITION

Since the manuscript was submitted for the first edition there has been a steadily growing interest in problems of estimating animal abundance and related parameters such as survival rate, and research articles on the subject are now widely scattered throughout the journals. In particular there have been several important Ph.D. theses which have led to a number of useful developments, including more complex parametric models on the one hand and simple but more robust non-parametric methods on the other. Although the time is not ripe for a complete rewriting of the first edition, these new developments are important enough to warrant a detailed review. I have therefore decided in the second edition to make minor alterations to the text, to add two extra chapters (Chapters 12 and 13) giving recent developments from about 1971 to 1977, and, with minor changes, renumber Chapter 12 as Chapter 14. Also the number of references has more than doubled.

Since writing the first edition it has become very clear to me that accurate and precise estimates of population parameters can only be achieved by careful and extensive experimentation. Frequently samples are too small or probabilities of capture are too low for the detection of significant population changes, a necessary requirement for environmental impact assessment. I have also noticed that inappropriate methods of analysis are still being used by some authors, particularly in bird-banding and fishery experiments. Clearly a major problem at present is that of communication: ecologists need to be made aware of the many techniques that are now available. Unfortunately many of the techniques, particularly mark–recapture methods, require from the ecologist a certain level of mathematical maturity and the ability to handle complex notation. Fortunately more ecologists are receiving statistical training these days, so that most of the methods in this book should become readily available to those interested in quantitative ecology. However, it is clear that some techniques will need to be presented in more appealing wrappers if they are to achieve widespread acceptance.

I would like to express my sincere thanks to the many authors who have sent me pre-publication copies of their articles. In particular my thanks go to Drs D.R. Anderson, K.P. Burnham, L.L. Eberhardt, G.G. Koch, J.K. Ord, A.R. Sen and K.H. Pollock for their interest in my work.

ACKNOWLEDGEMENTS

Several tables and figures have been added to the first edition and for permission to reproduce these my thanks go to the Editors of *Biometrical Journal* (Table 13.1), *Journal of the Fisheries Research Board of Canada* (Fig. 13.2 and 13.3) and the *Journal of Wildlife Management* (Table 12.1).

May 1980 G. A. F. Seber

PREFACE TO THE FIRST EDITION

Today ecology is one of the "in" subjects. With the steady growth of technology and its resulting problems of pollution, more and more attention is being directed towards our environment and our natural resources. This upsurge of interest in ecological problems is being stimulated by the use of statistical methods in much the same way that the physical sciences have been stimulated by theoretical developments in mathematics. Although mathematical theory tends to leap way ahead of mathematical practice, it is now widely recognised that many branches of mathematics have a vital role to play in the study of biological systems, no matter how complex these systems may be. I believe that the interaction between mathematics and biology is of benefit to the mathematician as well as to the biologist.

In ecological studies the first problem that is usually encountered is a census one. We do not know the sizes of the various wildlife populations in our study area, nor do we know how these populations are changing with time. However, over the past forty years a large body of techniques have been developed for estimating animal population numbers and related parameters such as the mortality and birth rates, and this book is an attempt to bring this material together. Originally I had intended to write a short monograph, but as I searched deeper into the literature my working bibliography began to grow at an alarming rate; the monograph eventually became a book.

In organising this large body of research material my biggest problem has been to classify the various techniques in such a way that an appropriate technique could be readily located in the text. Eventually I decided to classify the techniques according to the type of population studied and the nature of the sampling information available from the population. Populations are divided into two categories called "closed" and "open", depending on whether the population remains unchanged during the period of investigation, or changes through such processes as mortality, migration, etc. For the two types of population, one can obtain sample information from plot studies, capture-tag-recapture data, catch-effort statistics, and demographic data such as the sex ratio or age structure.

Before using a statistical technique it is important that the user be fully aware of the assumptions that must hold for the technique to be valid. For this reason I have paid considerable attention to underlying assumptions and, where possible, have given methods for testing their validity. Regression techniques are widely used throughout as they are fairly robust with regard to departures from the underlying assumptions, and can be examined graphically. Sometimes a less efficient but more robust method is to be preferred to a highly efficient method requiring strong assumptions.

In an endeavour to make this book more useful for the field worker I have included at least one numerical example demonstrating each technique, and, apart from one or two isolated cases, these examples are based on "real-life" data. To some extent the choice of examples has been rather arbitrary in that the one chosen for a particular technique has sometimes been the first one I encountered in my reading. However, in general, I have tried to use examples in which the authors have made some attempt to discuss the underlying assumptions from a field point of view. I hope that future users of these techniques will adopt the same critical attitude.

I am very grateful to a number of authors who kindly sent me prepublication copies of their articles. This has, to some extent, compensated for various delays caused by typing difficulties, inaccessibility of certain journals, etc. I would also express my debt to Dr E.G. White of Lincoln College, Christchurch, N.Z., not only for reading the manuscript, but also for helping me to update some of the references prior to publication. Finally I would like to thank Mr John Whale of Auckland University for writing several computer programs and calculating Table 7.6 on page 321.

G.A.F. SEBER

ACKNOWLEDGMENTS

For permission to reproduce certain published tables, thanks are due to the Editors of the *Annals of the Institute of Statistical Mathematics* (**A4**), *Annals of Mathematical Statistics* (**A5**), *Biometrics* (Tables 7.5, 9.3, 9.4; Fig. 7.3, 7.4), *Biometrika* (**A3**; Tables 4.18, 9.1, 12.2), *Journal of the Fisheries Research Board of Canada* (**A2**; Fig. 6.2), *Journal of Wildlife Management* (Table 7.3; Fig. 9.1–9.3), and *Transactions of the American Fisheries Society* (**A6**; Tables 3.1, 3.2, 10.9; Fig. 3.1–3.6).

CONTENTS

CONTENTS

CONTENTS

CONTENTS

CONTENTS

CONTENTS

CONTENTS

CONTENTS

CONTENTS

CHAPTER 1

PRELIMINARIES

1.1 INTRODUCTION

During the past thirty years there has been a growing realisation of the importance of sound statistical technique in the analysis of ecological data. This change of emphasis from qualitative to quantitative methods is as reflected for example in the bibliographies of Schultz [1961], Schultz *et al.* [1976] and Gilbert *et al.* [1976], and recent issues of journals in ecology and wildlife management.

The ecologist has also recognised the importance of obtaining data in the field from "natural" or free-ranging populations as opposed to data from "artificial" or laboratory populations. So often the population changes that occur in the laboratory give little indication as to what happens in the natural state, particularly when the animal under investigation is capable of exhibiting certain social tendencies in captivity. But the study of natural populations is not easy, as the experimenter is faced with the conflicting requirements of finding out as much about the population as possible and, at the same time, leaving the population undisturbed. Because of this lack of control over natural populations, statistical models should be used with caution, as they all depend on the validity of certain underlying assumptions which, in some cases, may be difficult to investigate. Some models are very sensitive to departures from their underlying assumptions, and it is therefore essential that with the development of a new model, there should be a corresponding development of procedures for testing the assumptions and for investigating the "robustness" of the model with regard to departures from the assumptions. Generally, the stronger the assumptions underlying a particular population model, the more "powerful" is the model *when valid* (e.g. estimates have smaller variances). However, a low-powered but robust model requiring few assumptions is often more useful than a complex high-powered model requiring many assumptions.

As a first step in understanding the structure and dynamics of a natural population it is essential to know something about the population size and related parameters such as the birth- and death-rates, etc. at given points in time. Since 1930, following Lincoln's [1930] work on banded waterfowl and the series of papers by Jackson [1933, 1937, 1939, 1940, 1944, 1948] on the tsetse fly, there has been a growing interest in methods of estimating population parameters, and today the literature is extensive. Much of the pioneering work has been done by fishery scientists, culminating in the monumental volumes of Beverton and Holt [1957] and Ricker [1958]. The importance of such methods, as for example in pest control and wildlife management, has led research workers of widely differing interests to enter the fray, and today the whole subject is expanding rapidly.

This book is an attempt to systematise the growing body of literature according to types of statistical models used and, where possible, to discuss in some detail the assumptions underlying the models. In checking my coverage of the literature I have been helped considerably by a number of reviews, especially De Lury [1954], Scattergood [1954], Ricker [1958: fish], Giles [1969: wildlife management], Southwood [1966: insects], and in particular Cormack [1968, 1979].

The remainder of this chapter is devoted to introducing the notation and outlining a number of statistical methods which are used throughout this book but are not usually dealt with in statistics textbooks.

1.2 NOTATION AND TERMINOLOGY

1.2.1 Notation

One of the difficulties in writing this book has been the choice of a suitable notation which could be maintained fairly consistently throughout. The international notation adopted by the F.A.O. for fishery research was found to be unsuitable and I finally opted for a "mnemonic" notation. For example, N and n denote the *number* of individuals in population and sample; M and m refer to the number of *marked* (or tagged members) of the population and sample; s represents the number of *samples*, etc. Where there has been a choice of symbols available I have endeavoured to choose those symbols which would help the reader in consulting the original articles. Occasionally, for the sake of clarity, I have found a change of notation necessary, and to avoid ambiguity I have made each chapter self-contained as far as the notation is concerned. In most cases the convention of distinguishing between a random variable and its observed value has not been followed because of notational difficulties.

Some statistical symbols are required: $E[y]$, $\sigma[y]$, $V[y]$ $(= \sigma^2[y])$, $C[y]$ $(= \sigma[y]/E[y])$ will represent the mean, standard deviation, variance, and coefficient of variation, respectively, of the random variable y; $\mathrm{cov}[x, y]$ is the covariance of two random variables x and y, and $E[x|y]$, $V[x|y]$ are the mean and variance of x conditional on fixed y. The normal distribution with mean θ and variance σ^2 is represented by $\mathfrak{N}[\theta, \sigma^2]$, and the random variable with a *unit normal* distribution $(\theta = 0, \sigma^2 = 1)$ will usually be denoted by z. The symbols z_α and $t_k[\alpha]$ will represent the 100α per cent upper tail values for z and the t-distribution with k degrees of freedom, respectively; thus

$$\Pr[z \geqslant z_\alpha] = \Pr[t_k \geqslant t_k[\alpha]] = \alpha,$$

where Pr stands for Probability. The chi-squared distribution with k degrees of freedom is denoted by χ^2_k.

Occasionally the symbols $O[N]$ and $o[N]$ will be used: if g is a function of N, then $g[N] = O[N]$ if there exists an integer N_1 and a positive number A such that, for $N > N_1$, $|g(N)/N| < A$; $g(N) = o[N]$ if $\lim_{N\to\infty} \{g(N)/N] = 0$. Roughly speaking, $O[N]$ means "of the same order of magnitude as N when N is large", while $o[N]$ means "of smaller order of magnitude than N when N is large".

In one or two chapters (especially Chapter 11) it is necessary to introduce vectors and matrices; these will be denoted by bold-face roman type. For example,

$$\mathbf{x} = [(x_i)] \quad \text{and} \quad \mathbf{\Sigma} = [(\sigma_{ij})]$$

represent respectively the column vector $\mathbf{x}$ with ith element x_i and the matrix $\mathbf{\Sigma}$ with i, jth element σ_{ij}.

If the vector $\mathbf{x}$ has a multivariate normal distribution with mean vector $\boldsymbol{\theta}$ and variance–covariance (dispersion) matrix $\mathbf{\Sigma}$, we shall write $\mathbf{x}$ is $\mathfrak{N}[\boldsymbol{\theta}, \mathbf{\Sigma}]$.

All logarithms, written $\log x$, will be to the base e unless otherwise stated.

1.2.2 Terminology

The size of an animal population in a given area will be determined by the processes of *immigration* (or movement into the area), *emigration* (or movement out of the area), *total mortality*, and *recruitment*.

TOTAL MORTALITY. In dealing with exploited populations we shall usually subdivide total mortality into mortality due to exploitation and natural mortality, i.e. mortality due to natural processes such as predation, disease, climatic conditions: contrary to some authors, emigration is not included here under "mortality". We shall also distinguish between *mortality rate* and *instantaneous mortality rate* as follows:

Let ϕ_t be the probability that an animal survives for the period of time $[0, t]$, then if N_0 animals are alive at time zero we would expect $N_0\phi_t$ ($= N_t$ say) to be alive at time t. The proportion ϕ_t, sometimes expressed as a percentage, is called the *survival rate* over period t, and $1 - \phi_t$ is called the *mortality rate* over period t. If, however, the mortality may be regarded as a Poisson process with parameter μ – that is, the probability that an animal dies in the time-interval $(t, t+\delta t)$ is $\mu\delta t + o(\delta t)$ – then (e.g. Feller [1957]: Chapter 17),

$$\phi_t = e^{-\mu t}, \tag{1.1}$$

$$\frac{dN_t}{dt} = -\mu N_t, \tag{1.2}$$

and the parameter μ is called the *instantaneous mortality rate*.

MEAN LIFE EXPECTANCY. Let Y be the time at which a member of N_0 dies. Then

$$\begin{aligned} F[y] &= \Pr[Y \leqslant y] \\ &= 1 - \Pr[Y > y] \\ &= 1 - \Pr[\text{animal survives until time } y] \\ &= 1 - \exp(-\mu y), \end{aligned}$$

and Y has probability density function

$$f(y) = F'(y) = \mu e^{-\mu y}, \quad (y \geqslant 0).$$

Therefore the *mean life expectancy* is

$$\begin{aligned} E[Y] &= \int_0^\infty \mu y e^{-\mu y}\, dy \\ &= 1/\mu \\ &= -1/\log \phi_1. \end{aligned} \tag{1.3}$$

RECRUITMENT. By recruitment we shall refer to those animals born into the population or, where applicable, those animals which grow into the catchable part of the population. In fishery research, recruitment sometimes denotes those fish which grow into the class of *legally* catchable fish. Thus we do not treat immigration as a component of recruitment.

OPEN AND CLOSED POPULATIONS. A population which remains unchanged during the period of investigation (i.e. the effects of migration, mortality and recruitment are negligible) is called a *closed* population. If the population is changing due to one or more of the above processes operating, then the population is said to be *open*.

1.3 SOME STATISTICAL METHODS

1.3.1 Maximum-likelihood estimation*

Let $x_1, x_2, \ldots, x_n$ be a random sample from the probability (or probability density) function $f(x, \theta)$, and let

$$L(\theta) = \prod_{i=1}^{n} f(x_i, \theta)$$

be the likelihood function. Then, provided the function f has certain reasonable properties, $\hat{\theta}$, the maximum-likelihood estimate of θ, is a solution of

$$\frac{\partial \log L(\theta)}{\partial \theta} = 0,$$

and as $n \to \infty$, $\hat{\theta}$ is asymptotically distributed as $\mathfrak{N}[\theta, \sigma_{\hat{\theta}}^2]$, where

$$\sigma_{\hat{\theta}}^2 = \left\{ -E\left[\frac{\partial^2 \log L(\theta)}{\partial \theta^2}\right]\right\}^{-1} \tag{1.4}$$

(*)For a helpful discussion and review see Norden [1972, 1973].

Replacing θ by $\hat{\theta}$ leads to the estimate $\hat{\sigma}$, say, of σ_θ, and an approximate large-sample $100(1-\alpha)$ per cent confidence interval for θ is given by

$$\hat{\theta} \pm z_{\alpha/2}\hat{\sigma}. \tag{1.5}$$

In a number of standard situations (e.g. when x is binomial or Poisson) a more accurate confidence interval is (θ_1, θ_2), where θ_1 and θ_2 are the appropriate roots of the following equation in θ:

$$(\hat{\theta}-\theta)^2 = \sigma_\theta^2 z_{\alpha/2}^2. \tag{1.6}$$

For a 95 per cent confidence interval, $\alpha = 0{\cdot}05$ and $z_{0\cdot025} = 1{\cdot}96$.

COEFFICIENT OF VARIATION. The coefficient of variation of $\hat{\theta}$ is asymptotically

$$C[\hat{\theta}] = \sigma_\theta/\theta$$

which can be estimated by

$$\hat{C} = \hat{\sigma}/\hat{\theta}.$$

Here $\hat{C}$ is related to the proportional width of the interval (1.5) and is therefore a useful measure of the "accuracy" of $\hat{\theta}$.

BIAS. Occasionally an expression for the bias can be calculated, so that

$$E[\hat{\theta}] = \theta + b_\theta.$$

The estimate $\hat{b}$ of the bias term should be included in the interval $\hat{\theta} - \hat{b} \pm 1{\cdot}96\hat{\sigma}$ if it is more than 10 per cent of the magnitude of $\hat{\sigma}$ (Cochran [1977: 12–15]). We shall call b_θ/θ the *proportional bias* of $\hat{\theta}$.

SEVERAL PARAMETERS. Let $x_1, x_2, \ldots, x_n$ be a random sample from $f(x, \boldsymbol{\theta})$, where $\boldsymbol{\theta}$ is now a vector of parameters $\theta_1, \theta_2, \ldots, \theta_r$. Then, when f has certain reasonable properties, $\hat{\boldsymbol{\theta}}$, the vector of maximum-likelihood estimates, is a solution of the r equations

$$\frac{\partial \log L(\boldsymbol{\theta})}{\partial \theta_i} = 0, \quad (i = 1, 2, \ldots, r),$$

and is asymptotically distributed as the multivariate normal distribution $\mathfrak{N}[\boldsymbol{\theta}, \mathbf{V}_\theta^{-1}]$, where $\mathbf{V}_\theta$ is the r-by-r matrix with i, jth element

$$-E\left[\frac{\partial^2 \log L(\boldsymbol{\theta})}{\partial \theta_i \partial \theta_j}\right]. \tag{1.7}$$

The matrix $\mathbf{V}_\theta$ is sometimes called the *information matrix*.

MOMENT ESTIMATES. The above maximum-likelihood theory is applicable to more general situations than those stated above. For example, the x_i may have different distributions or the x_i may not be independent but have a joint multinomial distribution. In these cases, if the number of random variables equals the number of unknown parameters, then the maximum-likelihood estimates can usually be obtained by equating each random variable to its expected value and solving the resulting equations for the unknown parameters. This method of estimation is called *moment estimation* and the estimates *moment estimates*.

1.3.2 Estimating a mean

WEIGHTED MEAN. Let x_i $(i = 1, 2, \ldots, n)$ be n independent random variables with known variances σ_i^2 and common mean θ. For the class of unbiased estimates of θ of the form

$$\bar{x}_w = (\sum_{i=1}^{n} w_i x_i)/(\sum_{i=1}^{n} w_i),$$

it is readily shown that $\bar{x}_w$ has minimum variance when w_i is proportional to $1/\sigma_i^2$. In particular, if $w_i\sigma_i^2 = a$, say, for $i = 1, 2, \ldots, n$, then

$$V[\bar{x}_w] = a/\sum w_i \tag{1.8}$$

and, by setting $y_i = x_i - \theta$, it is readily shown that

$$v[\bar{x}_w] = \frac{\sum w_i(x_i - \bar{x}_w)^2}{(n-1)\sum w_i}$$

is an unbiased estimate of this minimum variance.

UNWEIGHTED MEAN. If the variances σ_i^2 are unknown then we can simply use the sample mean

$$\bar{x} = \sum x_i/n$$

as our estimate of θ. In this case it transpires that

$$v[\bar{x}] = \frac{\sum (x_i - \bar{x})^2}{n(n-1)} \tag{1.9}$$

is an unbiased estimate of $V[\bar{x}]$.

A similar estimate of $V[\bar{x}]$ can also be obtained when the x_i's are not independent but correlated. Suppose that

$$\text{cov}[x_i, x_j] = \begin{cases} \sigma_{ij} & j = i+1 \\ 0 & j > i+1 \end{cases}$$

so that the (unknown) non-zero covariances are $\sigma_{12}, \sigma_{23}, \ldots, \sigma_{n-1,n}$. Then

$$V[\bar{x}] = \frac{1}{n^2}\{\sum_{i=1}^{n} \sigma_i^2 + 2\sum_{i=1}^{n-1} \sigma_{i,i+1}\}$$

$$= \frac{1}{n^2}\{A + 2B\}, \text{ say,}$$

and the problem reduces to finding estimates of A and B. Let

$$S_1^2 = \sum_{i=1}^{n} (x_i - \bar{x})^2$$

and

$$S_2^2 = \sum_{i=1}^{n} (x_{i+1} - x_i)^2,$$

where $x_{n+1} = x_1$, then

$$E[S_1^2] = \frac{(n-1)}{n} A - \frac{2B}{n},$$

$$E[S_2^2] = 2A - 2B$$

and, using moment estimation, unbiased estimates of A and B are

$$\hat{A} = \frac{nS_1^2 - S_2^2}{n-3}$$

and

$$\hat{B} = \hat{A} - \tfrac{1}{2}S_2^2.$$

Therefore an unbiased estimate of $V[\bar{x}]$ is given by

$$v[\bar{x}] = \frac{\hat{A} + 2\hat{B}}{n^2} = \frac{3S_1^2 - S_2^2}{n(n-3)}. \qquad (1.10)$$

We note in passing that $\hat{A} > 0$ since

$$nS_1^2 = \tfrac{1}{2} \sum_i \sum_j (x_i - x_j)^2 > S_2^2.$$

If the x_i's actually have different means θ_i then

$$E[v[\bar{x}]] = V[\bar{x}] + (3C_1 - C_2)/[n(n-3)],$$

where $C_1 = \sum_{i=1}^{n} (\theta_i - \bar{\theta})^2$ and $C_2 = \sum_{i=1}^{n} (\theta_{i+1} - \theta_i)^2$.

1.3.3 The delta method

A useful method used repeatedly in this book for finding approximate means, variances and covariances is demonstrated by the following examples.

MEAN. Let x_i be a random variable with mean θ_i ($i = 1, 2, \dots, n$) and suppose we wish to find the mean of some function $g(x_1, x_2, \dots, x_n)$ ($= g(\mathbf{x})$ say). Then using the first few terms of a Taylor expansion about $\boldsymbol{\theta}$, we have

$$g(\mathbf{x}) \approx g(\boldsymbol{\theta}) + \sum_{i=1}^{n} (x_i - \theta_i) \frac{\partial g}{\partial x_i} + \sum_{i=1}^{n} \sum_{j=1}^{n} \frac{(x_i - \theta_i)(x_j - \theta_j)}{2!} \frac{\partial^2 g}{\partial x_i \partial x_j},$$

where all partial derivatives are evaluated at $\mathbf{x} = \boldsymbol{\theta}$. Therefore, taking expected values,

$$E[g(\mathbf{x})] \approx g(\boldsymbol{\theta}) + b,$$

where
$$b = \sum_{i=1}^{n} \sum_{j=1}^{n} \frac{1}{2} \operatorname{cov}[x_i, x_j] \frac{\partial^2 g}{\partial x_i \partial x_j}$$

$$= \sum_{i=1}^{n} \frac{1}{2} V[x_i] \frac{\partial^2 g}{\partial x_i^2} + \sum_{i<j}\sum \operatorname{cov}[x_i, x_j] \frac{\partial^2 g}{\partial x_i \partial x_j}.$$

VARIANCE. If we ignore the bias b and neglect quadratic terms in the above Taylor expansion, then

$$V[g(\mathbf{x})] \approx E[\{g(\mathbf{x}) - g(\boldsymbol{\theta})\}^2]$$
$$\approx \sum_{i=1}^{n} V[x_i]\left(\frac{\partial g}{\partial x_i}\right)^2 + 2\sum\sum_{i<j}\text{cov}[x_i, x_j]\left(\frac{\partial g}{\partial x_i}\right)\left(\frac{\partial g}{\partial x_j}\right).$$

Example: Let $x_1, x_2, \ldots, x_n$ have a multinomial distribution

$$f(x_1, x_2, \ldots, x_n) = \frac{N!}{\prod_{i=1}^{n} x_i!}\prod_{i=1}^{n} p_i^{x_i} \quad (\sum_{i=1}^{n} p_i = 1);$$

then, defining $q_i = 1 - p_i$,

$$E[x_i] = Np_i \quad (= \theta_i),$$
$$V[x_i] = Np_i q_i$$

and

$$\text{cov}[x_i, x_j] = -Np_i p_j \quad (i \neq j).$$

Suppose we wish to find the asymptotic variance of

$$g(\mathbf{x}) = \frac{x_1 x_2 \ldots x_r}{x_{r+1} x_{r+2} \ldots x_s} \quad (s \leqslant n).$$

Then, using the Taylor approximation,

$$g(\mathbf{x}) - g(\boldsymbol{\theta}) \approx \sum_i (x_i - \theta_i)\left[\frac{\partial g}{\partial x_i}\right]_{\mathbf{x}=\boldsymbol{\theta}},$$

and hence

$$\frac{g(\mathbf{x}) - g(\boldsymbol{\theta})}{g(\boldsymbol{\theta})} \approx \sum_{i=1}^{r}\left(\frac{x_i - \theta_i}{\theta_i}\right) - \sum_{i=r+1}^{s}\left(\frac{x_i - \theta_i}{\theta_i}\right)$$

Squaring both sides and taking expected values, we have

$$V[g(\mathbf{x})] \approx E[\{g(\mathbf{x}) - g(\boldsymbol{\theta})\}^2]$$
$$\approx [g(\boldsymbol{\theta})]^2\left\{\sum_{i=1}^{s}\left(\frac{V[x_i]}{\theta_i^2}\right) + 2\frac{\text{cov}[x_1, x_2]}{\theta_1\theta_2} + \ldots + \right.$$
$$\left. + 2\frac{\text{cov}[x_{r+1}, x_{r+2}]}{\theta_{r+1}\theta_{r+2}} + \ldots - 2\frac{\text{cov}[x_1, x_{r+1}]}{\theta_1\theta_{r+1}} - \ldots\right\}.$$

Now

$$\frac{V[x_i]}{\theta_i^2} = \frac{Np_i(1-p_i)}{N^2 p_i^2} = \frac{1}{\theta_i} - \frac{1}{N},$$
$$\frac{\text{cov}[x_i, x_j]}{\theta_i\theta_j} = -\frac{Np_i p_j}{N^2 p_i p_j} = -\frac{1}{N},$$

and therefore

$$V[g(\mathbf{x})] \approx [g(\boldsymbol{\theta})]^2 \left\{ \sum_{i=1}^{s} \theta_i^{-1} - \frac{s}{N} - \frac{2}{N}\left[\binom{r}{2} + \binom{s-r}{2} - r(s-r)\right]\right\}$$

$$= \frac{[g(\boldsymbol{\theta})]^2}{N} \{\sum_{i=1}^{s} p_i^{-1} - (s-2r)^2\},$$

a result given by Seber [1967]. We note that the second term in the above expression vanishes when $s = 2r$.

COVARIANCE. The delta method can also be used for finding an approximate formula for the covariance of two functions $g(\mathbf{x})$ and $h(\mathbf{x})$. Thus

$$\begin{aligned} \operatorname{cov}[g(\mathbf{x}), h(\mathbf{x})] &\approx E[\{g(\mathbf{x}) - g(\boldsymbol{\theta})\}\{h(\mathbf{x}) - h(\boldsymbol{\theta})\}] \\ &\approx E\left[\left\{\sum_i (x_i - \theta_i) \frac{\partial g}{\partial x_i}\right\}\left\{\sum_j (x_j - \theta_j) \frac{\partial h}{\partial x_j}\right\}\right] \\ &= \sum_i \sum_j \operatorname{cov}[x_i, x_j] \frac{\partial g}{\partial x_i} \frac{\partial h}{\partial x_j}. \end{aligned}$$

AN EXACT FORMULA. If x and y are independent random variables then we have the exact relation (Goodman [1960])

$$V[xy] = (E[x])^2 V[y] + (E[y])^2 V[x] + V[x]V[y]. \tag{1.11}$$

Some generalisations of this result are given by Bohrnstedt and Goldberger [1969].

1.3.4 Conditional variances

Let x and y be a pair of random variables. Then from Kendall and Stuart [1968: 191] we have

$$E[x] = \underset{y}{E}\{E[x \mid y]\}$$

and

$$V[x] = \underset{y}{E}\{V[x \mid y]\} + \underset{y}{V}\{E[x \mid y]\}, \tag{1.12}$$

where $\underset{y}{E}$, etc. denotes taking the expected value with respect to the distribution of y. We note that if $E[x \mid y]$ does not depend on y then the second term of (1.12) is zero and, by the delta method of **1.3.3**,

$$\begin{aligned} V[x] &= \underset{y}{E}\{V[x \mid y]\} \\ &= \underset{y}{E}\{g(y)\}, \text{ say} \\ &\approx g(\theta) \\ &= \{V[x \mid y]\}_{y=\theta} \end{aligned} \tag{1.13}$$

where $\theta = E[y]$.

1.3.5 Regression models

1 Weighted linear regression

Consider the regression line

$$Y_i = \beta_0 + \beta x_i + e_i \quad (i = 1, 2, \ldots, n), \tag{1.14}$$

where the x_i are constants, the e_i are random variables independently distributed as $\mathfrak{N}[0, \sigma^2/w_i]$, the weights w_i are known, and β_0, β and σ^2 are unknown parameters. Then the weighted least-squares estimates $\hat{\beta}_0$ and $\hat{\beta}$ are found by minimising $\Sigma\, w_i e_i^2$ with respect to β_0 and β. Thus

$$\hat{\beta} = \Sigma\, w_i(Y_i - \bar{Y})(x_i - \bar{x})/\Sigma\, w_i(x_i - \bar{x})^2$$

and

$$\hat{\beta}_0 = \bar{Y} - \hat{\beta}\bar{x},$$

where $\bar{Y} = \Sigma\, w_i Y_i/\Sigma\, w_i$ and $\bar{x} = \Sigma\, w_i x_i/\Sigma\, w_i$.

Also,

$$V[\hat{\beta}] = \sigma^2/\Sigma\, w_i(x_i - \bar{x})^2$$

and an unbiased estimate of σ^2 is

$$\hat{\sigma}^2 = \Sigma\, w_i(Y_i - \bar{Y} - \hat{\beta}(x_i - \bar{x}))^2/(n-2).$$

A $100(1-\alpha)$ per cent confidence interval for β can be obtained in the usual manner from the t-distribution, namely

$$\hat{\beta} \pm t_{n-2}[\alpha/2](\hat{\sigma}^2/\Sigma\, w_i(x_i - \bar{x})^2)^{\frac{1}{2}}.$$

The assumptions underlying the above model can be investigated by examining the weighted residuals $\sqrt{w_i}(Y_i - \bar{Y} - \hat{\beta}(x_i - \bar{x}))$, using the methods of Seber [1977: Chapter 6].

CASE 1. When $\beta_0 = 0$ in (1.14) the least-squares estimate of β is now

$$\tilde{\beta} = \Sigma\, w_i Y_i x_i/\Sigma\, w_i x_i^2.$$

The corresponding confidence interval for β is

$$\tilde{\beta} \pm t_{n-1}[\alpha/2](\tilde{\sigma}^2/\Sigma\, w_i x_i^2)^{\frac{1}{2}}$$

where

$$\begin{aligned}(n-1)\tilde{\sigma}^2 &= \Sigma\, w_i(Y_i - \tilde{\beta}x_i)^2 \\ &= \Sigma\, w_i Y_i^2 - (\Sigma\, w_i Y_i x_i)^2/(\Sigma\, w_i x_i^2)\end{aligned}$$

CASE 2. When $\beta = 0$ in (1.14) the least-squares estimate of β_0 is

$$\tilde{\beta}_0 = \Sigma\, w_i Y_i/\Sigma\, w_i = \bar{Y}$$

with confidence interval

$$\tilde{\beta}_0 \pm t_{n-1}[\alpha/2](\tilde{\sigma}_0^2/\Sigma\, w_i)^{\frac{1}{2}},$$

where

$$\begin{aligned}(n-1)\tilde{\sigma}_0^2 &= \Sigma\, w_i(Y_i - \tilde{\beta}_0)^2 \\ &= \Sigma\, w_i Y_i^2 - (\Sigma\, w_i Y_i)^2/(\Sigma\, w_i).\end{aligned}$$

2 Weighted multiple linear regression

A generalisation of (1.14) is the multiple linear regression model

$$\mathbf{Y} = \mathbf{X}\boldsymbol{\beta} + \mathbf{e},$$

where $\mathbf{e}$ has the multivariate normal distribution $\mathcal{N}[\mathbf{0}, \sigma^2 \mathbf{B}]$, $\mathbf{X}$ is a known n-by-r matrix of rank r, $\mathbf{B}$ is a known n-by-n positive-definite matrix, and $\boldsymbol{\beta}$ and σ^2 are unknown parameters. From Seber [1977: 60–2] the weighted least squares estimate of $\boldsymbol{\beta}$, obtained by minimising

$$(\mathbf{Y} - \mathbf{X}\boldsymbol{\beta})'\mathbf{B}^{-1}(\mathbf{Y} - \mathbf{X}\boldsymbol{\beta})$$

with respect to $\boldsymbol{\beta}$, is

$$\hat{\boldsymbol{\beta}} = (\mathbf{X}'\mathbf{B}^{-1}\mathbf{X})^{-1}\mathbf{X}'\mathbf{B}^{-1}\mathbf{Y}.$$

The variance–covariance matrix of this estimate is

$$\mathbf{V}[\hat{\boldsymbol{\beta}}] = \sigma^2(\mathbf{X}'\mathbf{B}^{-1}\mathbf{X})^{-1}$$

and σ^2 is estimated by

$$\hat{\sigma}^2 = (\mathbf{Y}'\mathbf{B}^{-1}\mathbf{Y} - \mathbf{Y}'\mathbf{B}^{-1}\mathbf{X}\hat{\boldsymbol{\beta}})/(n - r).$$

Multiple confidence intervals for linear combinations of the form $\mathbf{h}'\boldsymbol{\beta}$ can be obtained in the usual manner (Seber [1977: Chapter 5]).

3 A special model

Consider the linear regression model

$$Y_i = \gamma(\theta - x_i) + e_i, \quad (i = 1, 2, \dots, n), \tag{1.15}$$

where the e_i and x_i are defined as for (1.14) above (though for most applications in this book $w_i = 1$). Then, using the notation of section *1* above, the weighted least-squares estimates of γ and θ are

$$\hat{\gamma} = -\hat{\beta},$$
$$\hat{\theta} = \hat{\beta}_0/\hat{\gamma} = \bar{x} + (\bar{Y}/\hat{\gamma}),$$

and a $100(1-\alpha)$ per cent confidence interval for γ is

$$\hat{\gamma} \pm t_{n-2}[\alpha/2]\,(\hat{\sigma}^2/\Sigma\, w_i(x_i - \bar{x})^2)^{\frac{1}{2}}. \tag{1.16}$$

A confidence interval for θ can be obtained using a technique due to Fieller [1940] as follows. Now

$$\frac{E[\bar{Y}]}{E[\hat{\gamma}]} = \theta - \bar{x} \quad (= \delta \text{ say})$$

and, since $\text{cov}[\bar{Y}, Y_i - \bar{Y}] = 0$,

$$\text{cov}[\bar{Y}, \hat{\gamma}] = 0,$$
$$V[(\bar{Y} - \delta\hat{\gamma})] = V[\bar{Y}] + \delta^2 V[\hat{\gamma}]$$
$$= \sigma^2\left[\frac{1}{\Sigma\, w_i} + \frac{\delta^2}{\Sigma\, w_i(x_i - \bar{x})^2}\right]$$
$$= \sigma^2 v, \text{ say,}$$

and $\bar{Y} - \delta\hat{\gamma}$ is $\mathfrak{N}[0, \sigma^2 v]$. Also $(n-2)\hat{\sigma}^2/\sigma^2$ is distributed as χ^2_{n-2} and is independent of both $\bar{Y}$ and $\hat{\gamma}$. Therefore

$$t = \frac{\bar{Y} - \delta\hat{\gamma}}{\hat{\sigma}\sqrt{v}}$$

has the t-distribution with $n-2$ degrees of freedom, and a $100(1-\alpha)$ per cent confidence interval for δ is given by

$$t^2 \leqslant t^2_{n-2}[\alpha/2], \quad \text{or } d_1 \leqslant \delta \leqslant d_2,$$

where d_1 and d_2 are the roots of the quadratic

$$d^2\left(\hat{\gamma}^2 - \frac{\hat{\sigma}^2 t^2_{n-2}[\alpha/2]}{\sum w_i(x_i - \bar{x})^2}\right) - 2d\bar{Y}\hat{\gamma} + \left(\bar{Y}^2 - \frac{\hat{\sigma}^2 t^2_{n-2}[\alpha/2]}{\sum w_i}\right) = 0. \qquad (1.17)$$

The corresponding confidence interval for θ is $(d_1 + \bar{x}, d_2 + \bar{x})$.

Writing (1.17) in the form

$$d^2(\hat{\gamma}^2 - A) - 2d\bar{Y}\hat{\gamma} + \bar{Y}^2 - B = 0,$$

we see that it has real unequal roots when

$$A\bar{Y}^2 + B\hat{\gamma}^2 - AB > 0.$$

Sufficient conditions for this are $\hat{\gamma}^2 \geqslant A$ or $\bar{Y}^2 \geqslant B$.

When n is large, an approximate expression for the variance of $\hat{\theta}$ can be derived using the delta method, thus

$$\begin{aligned} V[\hat{\theta}] &= V[\hat{\theta} - \bar{x}] \\ &= V[\bar{Y}/\hat{\gamma}] \\ &\approx \frac{\sigma^2}{\gamma^2}\left[\frac{1}{\sum w_i} + \frac{(\theta - \bar{x})^2}{\sum w_i(x_i - \bar{x})^2}\right]. \end{aligned} \qquad (1.18)$$

Assuming $\hat{\theta}$ to be approximately normally distributed, an approximate 95 per cent confidence interval for θ is

$$\hat{\theta} \pm 1{\cdot}96\, \hat{V}^{\frac{1}{2}}, \qquad (1.19)$$

where $\hat{V}$ is $V[\hat{\theta}]$ with θ, γ and σ^2 replaced by their estimates. However, for the applications considered in this book n will usually be small (< 10), so the assumption that $\hat{\theta}$ is approximately normal is open to question. Therefore (1.17) will be used in all applications.

1.3.6 Goodness-of-fit tests

1 Binomial distribution

Let $x_1, x_2, \ldots, x_n$ be a random sample from the binomial distribution

$$f(x) = \binom{N}{x} p^x q^{N-x}, \quad (q = 1-p, \quad x = 0, 1, \ldots, N).$$

Suppose that x_i takes a value x with frequency f_x ($\sum_x f_x = n$), then it is readily shown that the maximum-likelihood estimate of p is

$$\hat{p} = \bar{x}/N = \sum_{x=0}^{n} x f_x/(nN)$$

and the expected frequencies, E_x, are given by

$$E_x = n \binom{N}{x} \hat{p}^x \hat{q}^{N-x}, \quad x = 0, 1, \ldots, N.$$

Since the joint distribution of the random variables f_x is multinomial with $N + 1$ categories, the goodness-of-fit statistic for testing the appropriateness of the binomial model is

$$T_1 = \sum_{x=0}^{N} (f_x - E_x)^2/E_x,$$

which is approximately distributed as χ^2_{N-1} when n is large.

An alternative test statistic can be obtained by putting the data in the form of a contingency table, namely

x_1	x_2	...	x_n	$\sum x_i$
$N - x_1$	$N - x_2$	...	$N - x_n$	$nN - \sum x_i$
N	N	...	N	nN

and carrying out a test for homogeneity. The test statistic is then

$$T_2 = \sum_{i=1}^{n} \frac{(x_i - N\hat{p})^2}{N\hat{p}\hat{q}}$$

$$= \frac{\sum_{i=1}^{n} (x_i - \bar{x})^2}{\bar{x}[1 - \bar{x}/N]}$$

$$= \frac{\sum_{x=0}^{N} f_x (x - \bar{x})^2}{\bar{x}[1 - \bar{x}/N]}$$

the so-called *Binomial Dispersion Test*, and it is asymptotically distributed as χ^2_{n-1}. We note that $T_2/(n-1)$ is effectively based on comparing the observed variance estimate $\sum (x_i - \bar{x})^2/(n-1)$ with $\hat{N}\hat{p}\hat{q}$, an estimate of the expected variance under a binomial model.

2 Poisson distribution

Let $x_1, x_2, \ldots, x_n$ be a random sample from a Poisson distribution

$$f(x) = e^{-\lambda} \frac{\lambda^x}{x!} \quad x = 0, 1, 2, \ldots .$$

Then the maximum-likelihood estimate of λ is $\hat{\lambda} = \bar{x}$ and the expected frequencies are

$$E_x = n e^{-\hat{\lambda}} \cdot \frac{\hat{\lambda}^x}{x!}.$$

Usually the expected frequencies are pooled for $x \geqslant X$ so as to ensure that $n - \sum_{x=0}^{X-1} E_x$ ($= E_{X+}$ say) is greater than about 5 (though a value as small as 1 can usually be tolerated if $X \geqslant 4$), and the goodness-of-fit statistic

$$T_1 = \sum_{x=0}^{X-1} \left\{ \frac{(f_x - E_x)^2}{E_x} \right\} + \frac{(f_{X+} - E_{X+})^2}{E_{X+}}$$

is then approximately distributed as χ^2_{X-1}.

Alternatively, we can use the *Poisson Dispersion Test* (Perry and Mead [1979][(*)])

$$T_2 = \sum_{i=0}^{n} (x_i - \bar{x})^2/\bar{x} = \sum_x f_x(x - \bar{x})^2/\bar{x},$$

which is asymptotically distributed as χ^2_{n-1}. Since the mean of a Poisson variable equals its variance, $T_2/(n-1)$ can be regarded as a statistic for comparing the observed variance estimate with the estimate, $\bar{x}$, of the expected variance under a Poisson model.

In general, T_2 will provide a more sensitive test than T_1 (Kendall and Stuart [1973: 599]), though when the underlying distribution is not Poisson a comparison of the f_x and E_x may give some idea as to the form of departure from Poisson. Also T_2 can be used for quite small values of n ($n > 20$; or if $\bar{x} > 1$, $n > 6$: Kathirgamatamby [1953]), while T_1 requires a much larger sample size n in order to ensure that $E_x \geqslant 5$ for several values of x. The statistic T_2 is further discussed in Rao and Chakravarti [1956] who suggest that the chi-squared approximation to the distribution of T_2 is a good one when $\bar{x} > 3$; they also give a useful table for carrying out an exact small-sample test based on the statistic $\sum x_i^2$.

3 *Multinomial distribution with N unknown*

Let $y_1, y_2, \ldots, y_k$ have a multinomial distribution

$$f_1(y_1, y_2, \ldots, y_k) = \frac{N!}{\left(\prod_{i=1}^{k} y_i!\right)(N-r)!} \left(\prod_{i=1}^{k} p_i^{y_i}\right) Q^{N-r}, \qquad (1.20)$$

where $r = \sum_{i=1}^{k} y_i$, $Q = 1 - \sum_{i=1}^{k} p_i$.

We wish to test the hypothesis H that $p_i = p_i(\boldsymbol{\theta})$ ($i = 1, 2, \ldots, k$), where $p_i(\boldsymbol{\theta})$ is a function of t unknown parameters $\theta_1, \theta_2, \ldots, \theta_t$. When N is known, we can test H using the standard multinomial goodness-of-fit statistic

$$T_1 = \sum_{i=1}^{k} \left\{ \frac{(y_i - N\tilde{p}_i)^2}{N\tilde{p}_i} \right\} + \frac{(N - r - N\tilde{Q})^2}{N\tilde{Q}},$$

(*) In Additional References.

where $\tilde{p}_i = p_i(\tilde{\boldsymbol{\theta}})$, $\tilde{Q} = 1 - \sum_{i=1}^{k} \tilde{p}_i$, and $\tilde{\boldsymbol{\theta}}$ is the maximum-likelihood estimate of $\boldsymbol{\theta}$ for (1.20).

When N is unknown, it transpires that we can work with the conditional multinomial distribution (cf. **1.3.7**(*2*) for the method of derivation)

$$f_2(y_1, y_2, \ldots, y_k \mid r) = \frac{r!}{\prod_{i=1}^{k} y_i!} \prod_{i=1}^{k} \left(\frac{p_i}{1-Q}\right)^{y_i} \tag{1.21}$$

and use

$$T_2 = \sum_{i=1}^{k} \frac{[y_i - r\hat{p}_i/(1-\hat{Q})]^2}{r\hat{p}_i/(1-\hat{Q})}$$

$$= \sum_{i=1}^{k} \frac{(y_i - \hat{N}\hat{p}_i)^2}{\hat{N}\hat{p}_i},$$

where $\hat{N} = r/(1-\hat{Q})$, $\hat{p}_i = p_i(\hat{\boldsymbol{\theta}})$, and $\hat{\boldsymbol{\theta}}$ is the maximum-likelihood estimate of $\boldsymbol{\theta}$ for (1.21). It can be shown, using, for example, the methods of Birch [1964] or Darroch [1959: 347], that when H is true, T_2 is asymptotically distributed as χ^2_{k-t-1} as $N \to \infty$.

By solving the equations $\partial \log f_1/\partial\theta_j = 0$ $(j = 1, 2, \ldots, t)$ and $\nabla \log f_1 = 0$ (where ∇ denotes the first backward difference with respect to N), we find that when N is unknown, $\hat{\boldsymbol{\theta}}$ and $\hat{N}$ are close to the maximum-likelihood estimates of $\boldsymbol{\theta}$ and N for the model (1.20). This topic is discussed further in **12.8.1**.

1.3.7 Some conditional distributions

1 Poisson variables

If x_1 and x_2 are independent Poisson random variables with means θ_1 and θ_2 respectively, then it is readily shown that the distribution of x_1 conditional on $y = x_1 + x_2$ is binomial, namely

$$f(x_1 \mid y) = \binom{y}{x_1} p^{x_1} q^{x_2}$$

where $p = \theta_1/(\theta_1 + \theta_2)$. Conversely, if x_1 and y are a pair of random variables such that the conditional distribution of x_1 given y is binomial with parameters y and P, and y is Poisson with mean λ, then the unconditional distribution of x_1 is Poisson with mean λP (cf. Feller [1957: 160, Example 27]).

2 Multinomial variables

Let $x_1, x_2, \ldots, x_k$ have a multinomial distribution

$$f(x_1, x_2, \ldots, x_k) = \frac{n!}{\prod_{i=1}^{k} x_i!} \prod_{i=1}^{k} p_i^{x_i}, \quad (\sum_i p_i = 1);$$

then the joint marginal distribution of x_1 and x_2 is

$$f_1(x_1, x_2) = \frac{n!}{x_1!\, x_2!\, (n-x_1-x_2)!} p_1^{x_1} p_2^{x_2} (1-p_1-p_2)^{n-x_1-x_2}.$$

If $y = x_1 + x_2$, then y has probability function

$$f_2(y) = \binom{n}{y} (p_1+p_2)^y (1-p_1-p_2)^{n-y}$$

and

$$\begin{aligned} f(x_1 \mid y = x_1+x_2) &= \Pr[X_1 = x_1 \mid Y = y] \\ &= \frac{\Pr[X_1 = x_1 \text{ and } Y = y]}{\Pr[Y = y]} \\ &= \frac{\Pr[X_1 = x_1 \text{ and } X_2 = x_2]}{\Pr[Y = y]} \\ &= \frac{f_1(x_1, x_2)}{f_2(y)} \\ &= \binom{y}{x_1}\left(\frac{p_1}{p_1+p_2}\right)^{x_1}\left(\frac{p_2}{p_1+p_2}\right)^{x_2}. \end{aligned} \tag{1.22}$$

1.3.8 Iterative solution of equations

1 One unknown

Suppose we wish to find a root θ_0 of the equation $h(\theta) = 0$. Then, starting with a trial solution $\theta_{(1)}$, a number of iterative procedures are available in which the ith step of the iteration takes the form

$$\theta_{(i+1)} = \theta_{(i)} - h(\theta_{(i)})/k_i. \tag{1.23}$$

For example, if $\theta_{(i)} - \theta_0$ is small, we have the Taylor approximation

$$\begin{aligned} 0 &= h(\theta_0) \\ &\approx h(\theta_{(i)}) + (\theta_0 - \theta_{(i)})h'(\theta_{(i)}), \end{aligned}$$

where $h'(\theta_{(i)})$ is the first derivative of $h(\theta)$ evaluated at $\theta = \theta_{(i)}$. Rearranging this last equation, we have

$$\theta_0 \approx \theta_{(i)} - h(\theta_{(i)})/h'(\theta_{(i)}), \tag{1.24}$$

so that one method of iteration is to set $k_i = h'(\theta_{(i)})$ in (1.23), the so-called Newton–Raphson method. Another possibility is to choose

$$k_i = \frac{h(\theta_{(i)}) - h(\theta_{(i-1)})}{\theta_{(i)} - \theta_{(i-1)}};$$

this is known as the method of "false positions". When θ_0 is only required to the nearest integer, a useful discrete analogue of the Newton–Raphson method, used for example by Robson and Regier [1968], is to set

$$k_i = \nabla h(\theta_{(i)}) = h(\theta_{(i)}) - h(\theta_{(i)} - 1).$$

Conditions for the convergence of such procedures are discussed in standard textbooks on numerical analysis (see also Kale [1961, 1962]).

2 Several unknowns

Suppose that we have r equations

$$h_i(\boldsymbol{\theta}) = 0 \quad (i = 1, 2, \ldots, r)$$

in r unknowns $\theta_1, \theta_2, \ldots, \theta_r$. Then the multivariate analogue of the Newton–Raphson method (1.24) is

$$\boldsymbol{\theta}_{(i+1)} = \boldsymbol{\theta}_{(i)} - \mathbf{H}_{(i)}^{-1}\mathbf{h}(\boldsymbol{\theta}_{(i)}), \tag{1.25}$$

where

$$\mathbf{H}_{(i)} = \left[\left(\frac{\partial h_i(\boldsymbol{\theta})}{\partial \theta_j}\right)\right]_{\boldsymbol{\theta}=\boldsymbol{\theta}_{(i)}}$$

and

$$\mathbf{h}(\boldsymbol{\theta}_{(i)}) = \begin{bmatrix} h_1(\boldsymbol{\theta}_{(i)}) \\ h_2(\boldsymbol{\theta}_{(i)}) \\ \cdots \\ h_r(\boldsymbol{\theta}_{(i)}) \end{bmatrix}.$$

3 Maximum-likelihood equations

ONE PARAMETER. Let $x_1, x_2, \ldots, x_n$ be a random sample from a probability (or probability density) function $f(x, \theta)$ and let

$$L(\theta) = \prod_{i=1}^{n} f(x_i, \theta)$$

be the likelihood function. Then $\hat{\theta}$, the maximum-likelihood estimate of θ, is in general a solution of

$$h(\theta) = \partial \log L(\theta)/\partial\theta = 0, \tag{1.26}$$

which can be solved by the methods of section *1* above.

If f has certain reasonable properties, then $\hat{\theta}$ is asymptotically $\mathfrak{N}[\theta, \sigma_{\hat{\theta}}^2]$, where

$$\sigma_{\hat{\theta}}^2 = \left\{-E\left[\frac{\partial^2 \log L(\theta)}{\partial\theta^2}\right]\right\}^{-1}$$

can be estimated by

$$v = \left\{-\left[\frac{\partial^2 \log L(\theta)}{\partial\theta^2}\right]\right\}^{-1}_{\theta=\hat{\theta}}. \tag{1.27}$$

Since $-1/h'(\theta_{(i)})$ tends to v as $\theta_{(i)}$ approaches $\hat{\theta}$, we find that v can be calculated from the last iteration of the Newton–Raphson method.

Some of the problems involved in solving (1.26) are discussed by Barnett [1966]. He shows that the method of false positions is preferred when (1.26) has multiple roots.

SEVERAL PARAMETERS. If θ is now replaced by a vector $\boldsymbol{\theta} = [(\theta_i)]$ of r parameters, then $\hat{\boldsymbol{\theta}}$, the maximum-likelihood estimate of $\boldsymbol{\theta}$, is a solution of the r equations

$$h_i(\boldsymbol{\theta}) = \partial \log L(\boldsymbol{\theta})/\partial\theta_i = 0 \quad (i = 1, 2, \ldots, r).$$

Using the Newton–Raphson method (equation (1.25)) we have that

$$\mathbf{H}_{(i)} = [(\partial^2 \log L(\boldsymbol{\theta})/\partial\theta_i\partial\theta_j)]_{\boldsymbol{\theta}=\boldsymbol{\theta}_{(i)}}$$

and, as in the case of one parameter, the term $-\mathbf{H}^{-1}$ in the last step of the iteration provides an estimate of

$$-[(E[\partial^2 \log L(\boldsymbol{\theta})/\partial\theta_i\partial\theta_j])]^{-1}, \tag{1.28}$$

the asymptotic variance–covariance matrix of $\hat{\boldsymbol{\theta}}$.

1.3.9 The jackknife method

A technique of potential use in ecology is the so-called jackknife method, proposed originally by Quenouile [1956] and Tukey [1958]. This method, which is described briefly in **13.1.2**, is reviewed by Miller [1974: mainly theory] and Bissel and Ferguson [1975: practical aspects], and investigated further by Hinkley [1977] and Thorburn [1977]. Further generalisations, including the so-called "generalized" jacknife, are given by Gray and Schucany [1972], Gray *et al.* [1975] and Sharot [1976a, b]. The method has been applied to population studies by Burnham (cf. Burnham and Overton [1978, 1979]), Manly [1977b, c], Zahl [1977b] and Seber and Pemberton [1979]. It can be used for finding variances of "awkward" estimates, and for reducing the bias of $f(\hat{\theta})$ as an estimate of $f(\theta)$ when $\hat{\theta}$ is an unbiased estimate of θ, e.g. $f(\hat{\theta}) = 1/\hat{\theta}$.

A related method, also of potential use in ecology, is the so-called bootstrap method (Efron [1979]).

1.3.10 Monte Carlo test

Another technique of considerable potential in population studies is the so-called Monte Carlo test. This technique was suggested by Barnard [1963], investigated by Hope [1968], and Marriott [1979] and applied to spatial problems by Cliff and Ord [1973: ch. 2 and 4; 1975: 313], Ripley [1977: 181], Besag and Diggle [1977], and Diggle [1979a]. The method may be described briefly as follows.

Suppose we are given a simple null hypothesis H_0 and a set of data from which we calculate the value u_1 of a test statistic u. Suppose further that, using $(n-1)$ simulated data sets, values $u_2, u_3, \ldots, u_n$ ($n = 100$, or 200 for a two-sided test) are generated as a random sample from the null distribution of u and that $u_{(1)} < u_{(2)} < \ldots < u_{(n)}$ are the corresponding order-statistics. Under H_0, it is then exactly true that

$$\Pr[u_1 = u_{(j)}] = n^{-1}, \quad j = 1, 2, \ldots, n, \tag{1.29}$$

so that the rank of u_1 may be used to construct an exact test of H_0. If the null distribution of u is discrete, and ties occur, we choose the most conservative rank possible, for example the lowest rank if we are interested in the upper tail of (1.29), and the highest rank for the lower tail. The advantage of this approach is that a test is readily available even when the null distribution of u is unknown, provided the distribution can be simulated. Also the choice of the test statistic need not be constrained by known distribution theory.

CHAPTER 2

DENSITY ESTIMATES FOR CLOSED POPULATIONS

2.1 ABSOLUTE DENSITY

2.1.1 Total count over whole area

Occasionally it is possible to count all the animals of a particular species in a given area by systematically traversing the whole area on foot or, in the case of large animals, by air. For example, territorial species such as birds can be counted in this manner as the animals tend to stay in their territories when observed, thus reducing the possibility of duplication in counting through movement in and out of adjacent areas. Animals which congregate in groups can often be photographed and counted later (Davis [1963]), while fish which must migrate through rivers during part of their life cycles can be counted individually using traps or weirs (Scattergood [1954]). Recently radar has been used for estimating bird densities (Dyer [1967], Myres [1969]) and echo-sounding for counting fish (see **12.2.2** for further references).

The accuracy of the total count will depend on the visibility of the animals; this in turn depends on a number of factors such as animal size, vegetation cover available, and activity of the animal at the particular time of the year. During the censusing the population should be disturbed as little as possible, to avoid both the duplication in counting mentioned above and the possibility of frightened animals moving out of the area altogether.

In an effort to carry out a total count some attempts at killing or live-trapping all the members of a given population have been recorded in the literature. For example, fish populations have been counted by draining ponds in order to check on the accuracy of other methods of estimation (Buck and Thoits [1965]). Also with some populations of small mammals it is possible to tag and release animals until only tagged animals are being caught. In this way virtually the whole population can be counted, provided all the animals are of trappable age and there is no evidence of trap avoidance. When such counts are to be converted to density estimates of so many per unit area, the problem arises as to just what area has been trapped.* One can add to the area enclosed by the traps a strip of width one-half the home range of the animal (Dice [1938]). Some authors subtract from the total population area all regions which are definitely uninhabitable, as far as the given species is concerned, to obtain a "maximal" density estimate. A method resembling direct enumeration, called the "calendar of captures" method, has been widely used in

(*)This problem is discussed further in **12.1.1**.

Poland (cf. Petrusewicz and Andrzejewski [1962], Andrzejewski [1963], Ryszkowski *et al.* [1966], Andrzejewski [1967]).

The home range, if it exists, can be estimated from the trap records of individuals (Southwood [1966: 261], Mohr and Stumpf [1966], Sanderson [1966], Kikkawa [1964: 276], Jennrich and Turner [1969]) or by using various tracking devices such as radio transmitters (Giles [1963], Adams and Davis [1967]), radioactive tracers (Miller [1957], Kaye [1961], Kikkawa [1964], Harvey and Barbour [1965]), smoked kymograph paper (Justice [1961], Sheppe [1965], Bailey [1968]), dyes (New [1958]), and even spools of thread (Dole [1965]); for further comments see **12.1.1**. Techniques for the simultaneous estimation of home range and population density are described in **2.1.5** and **12.1.1** (*4*). A cluster method for bird territories is given by North [1977] (see Additional References).

2.1.2 Total counts on sample plots

With some populations it is impracticable or impossible to count the animals over the whole area because of the disturbance caused or the number of personnel required. In this case a sampling scheme is required whereby total counts are made on randomly chosen sample areas. An estimate of total population size is then obtained by multiplying the average density per unit area, estimated from the sample areas, by the total area of the population. The main steps in devising such a sampling scheme are as follows:

(1) The size and shape of the sample area or "plot" should be determined and one's choice will be guided by such factors as the species under consideration, the terrain, the distribution of vegetation cover and the distribution of food supply. One can use either quadrats, a term often loosely applied to plots such as squares, rectangles, circles etc. (though strictly it applies to square plots), or transects, the term for strips running the length of the population area. Line transects, which are discussed in **2.1.3**., are also being widely used in ecological research.

(2) The number of sample plots should be determined in advance, unless of course sequential methods are used (e.g. Southwood [1966: 43]). This number will depend on the precision required for the estimate of population density, and such factors as staff and time available for the sampling.

(3) The sample plots should be located by some random procedure so that valid sampling errors can be calculated. If there are marked differences in, for example, vegetation cover, one can divide up the total area into domains or "strata" of uniform habitat and then use stratified random sampling (Cochran [1963]). This stratification can be achieved using any factor related to population density: for example, in the study of ruffed grouse populations Ammann and Ryel [1963] used a stratification based on deer density. Another useful application of stratification is in sampling for soil animals. Here, very small plots are used, so that choosing the plots at random over the whole area may lead to an undesirable bunching of the plots (Healy [1962]). However, by using stratification and randomisation within each stratum a certain degree of coverage is achieved.

We note that for mobile species the number counted on a sample plot may not be the actual number present before the arrival of the investigator. Some of the animals will hide while others may move away, so that the counts are more of an index of population density than a measure of absolute density. In big game studies aerial sampling can be used, but once again the observer may not be able to see all the animals present. For example, the animals may tend to stay under cover in certain weather conditions or at certain times of the day. Sometimes the terrain is so rugged that planes cannot fly safely at the low levels necessary for adequate observation. Statistical methods for aerial censusing are considered in **12.3**.

We shall now discuss the above three steps in some detail with regard to quadrats and strip transects: line transects will be dealt with in a separate section.

1 Quadrats: simple random sampling

Quadrat sampling is particularly useful for dealing with ant and termite mounds (Waloff and Blackith [1962], Sands [1965]) and animals which are fairly stationary, e.g. soil animals (Murphy [1962]), some species of insects (Southwood [1966: 108]), and birds which are readily detected (Kendeigh [1944], Einarsen [1945: pheasants], Ammann and Baldwin [1960: woodpeckers]). It has also been used for mobile animals such as small mammals (Bole [1939], Turček [1958], Pelikán *et al.* [1964]) and deer (Trippensee [1948]); in this case the size of the quadrat appears to be crucial as small quadrats are more affected by movements across the quadrat boundary.

SHAPE. Although circular and even hexagonal plots (e.g. Lloyd [1967]) have been used, squares and rectangles are generally the easiest shapes to mark out or identify, particularly if large plots are required. Also the sampling procedures for selecting such plots are usually simpler than those for plots with curved sides. In comparing squares and rectangles, sometimes long narrow rectangles are more efficient than squares of the same area (e.g. Clapham [1932]). However, the longer the rectangle, the greater the length of boundary per unit area, so that there is a greater chance of error per unit area in determining whether individuals on the edges of the plot are inside or outside the boundary, and whether or not the plot population is closed during the counting. Although this error per unit area decreases as the shape of the plot approaches a circle, circular plots do not fit together without leaving spaces. This complicates the sampling procedure unless the plot areas are very small compared to the total area, as in soil sampling. Another possibility is the hexagonal, a compromise between a square and a circle as far as edge effect is concerned. However, although hexagonals can be fitted together, the time taken to mark them out precludes their use in most situations. Therefore, as a general recommendation, square sample plots should be used for most populations.

SIZE. In determining the size of a quadrat we note that a density estimate from a large number of small quadrats will usually have a smaller variance than that from a few large quadrats of the same total area (though the reverse can occur when the population is not randomly distributed). It is therefore recommended that the quadrat be reasonably small, bearing in mind that the edge effect mentioned above increases as the plot size decreases. For example, in populations of small mammals where a quadrat count is obtained by trapping out the quadrat area over several trap-nights, the smaller quadrats are more affected by animals drifting into them from adjacent areas (Bole [1939]). Also the quadrat should not be so small that the majority of quadrats are found to be empty. One working rule suggested by Greig-Smith [1964] is that there should not be more empty quadrats than quadrats with just one individual.

NUMBER OF PLOTS. Let

N = total population size,
A = area occupied by the population,
D = N/A (the average population density),
pA = area sampled,
and n = number of animals found in the sampling area.

We shall assume that the population is randomly distributed,* i.e. the probability of finding an individual at any particular point in the population area is the same for all points, and the presence of one individual does not influence the position occupied by another. Then n is binomially distributed with probability function

$$f(n) = \binom{N}{n} p^n q^{N-n} \quad (q = 1-p)$$

and a natural estimate of N is

$$\hat{N} = n/p. \tag{2.1}$$

This is unbiased with variance

$$V[\hat{N}] = Nq/p$$

and coefficient of variation

$$C = \sqrt{\left(\frac{q}{Np}\right)}.$$

Rearranging this last equation, we have

$$p = 1/(1 + NC^2), \tag{2.2}$$

so that given a rough lower bound for N and a prescribed value of C, p can be determined. For example, one general rule often stated (e.g. Dice [1952: 33]) is that about 5–10 per cent of the population area should be sampled. This would mean that for $p = 0{\cdot}1$, say, we could only obtain a precision of $C \leqslant 0{\cdot}2$ if $N \geqslant 225$. Therefore it is not possible to obtain reasonably precise estimates for small populations without sampling a very high proportion

(*) This assumption is not required in practice (cf. **12.1.2 (1)**).

of the population area. In fact, for $p < 0{\cdot}2$ we have $\sqrt{q} \approx 1$, and since n is an estimate of Np, C is estimated by $1/\sqrt{n}$. This means that the precision depends only on the total number of animals seen when less than 20 per cent of the population area is sampled.

When the population is not randomly distributed it is difficult to determine the precise effect of non-randomness on (2.2), though one or two comments may be made. If the animals tend to cluster (i.e. the population is "contagious"), as is often the case, $V[\hat{N}]$ will generally be greater than Nq/p and (2.2) will underestimate p. However, this effect will be small if the size of the quadrat is very much smaller than the average cluster size, or if the quadrat is large enough so that it generally includes one or more clusters (Greig-Smith [1964: 56–57]). In general, (2.2) will be satisfactory if the total area sampled is sufficiently large, so that the clustering tends to affect only the distribution and not the number of animals in the sampled area.

When p has been determined, the next step is to choose the quadrat area a and the number of sample plots s, subject to $sa = pA$. Normally one would endeavour to make s large as as possible, bearing in mind the time taken to map out a quadrat in the field and the possible edge effects of small quadrats.

SELECTING THE QUADRATS. Once s and a are determined, the quadrats must be chosen by some random procedure if the statistical error in our estimate of N is to be calculated. If a is very small compared with A, as in soil sampling where circular plots of less than 10 in. (260 cm) diameter are used, the simplest method for locating the centre of the plot (or one corner if square plots are used) is to read off a pair of numbers from a table of random numbers and use them as coordinates on a grid system (cf. Sampford [1962: Chapter 4]). However, if a is an appreciable fraction of A then the above method of random selection will lead to repeated overlapping of the sample quadrats. In this case the simplest method is to divide up the whole area into numbered quadrats of area a (ignoring incomplete quadrats on the boundary), and then use a table of random numbers to obtain a selection of s of these plots.

ESTIMATION. Let x_i be the number of animals seen in the ith quadrat $(i = 1, 2, \ldots, s)$; then $n = \sum x_i$ and, since $sa = pA$,

$$\hat{N} = n/p = \bar{x}A/a = \bar{x}S, \quad \text{say.} \tag{2.3}$$

This is the usual estimate of a total based on a simple random sample without replacement of s quadrats from a population of S $(= A/a)$ quadrats (Cochran [1977; Chapter 2]). Therefore $\hat{N}$ will still be an unbiased estimate of N, even when the population is not randomly distributed, though the formula for $V[\hat{N}]$ above will no longer be valid. In this case an unbiased estimate of the true variance is given by (Cochran [1977: 25]).

$$\hat{\sigma}_{\hat{N}}^2 = S^2 \frac{v}{s}\left(1 - \frac{s}{S}\right),$$

where

$$v = \sum_{i=1}^{s} (x_i - \overline{x})^2/(s-1).$$

Also an approximate $100(1-\alpha)$ per cent confidence interval for N is given by $\hat{N} \pm t_{s-1}[\alpha/2]\hat{\sigma}_N$: dividing this interval by A gives a confidence interval for the population density D.

PILOT SURVEY. In designing an experiment, if a rough estimate of N is not available, an alternative approach would be to choose a so that S is large and then carry out a pilot sample using s_0 quadrats, say. Calculating $\overline{x}$ and v from this sample, it is readily shown that $s - s_0$ further quadrats are required, where s is the smallest integer satisfying

$$\frac{1}{s} \leqslant \frac{1}{S} + \frac{C^2\overline{x}^2}{v}, \tag{2.4}$$

to give a prescribed coefficient of variation C.

VARIABLE PLOT SIZE. In some populations the total area may already be divided up into T plots of varying shapes and sizes by natural or man-made barriers such as watercourses, fences, etc. We then find that, provided the s plots are chosen at random, the above theory still holds but with S replaced by T. Obviously, when the population density per unit area is constant, so that the number of animals in a plot is proportional to plot size, the greater the variation in plot size the larger is v. This means that for a constant density and a given value of T, the most efficient method of sampling is to use plots of equal area.

TESTS OF RANDOMNESS. When the population is randomly distributed then x, the number of individuals in a sample plot of area δA, has a binomial distribution with parameters M and θ, where M is the maximum number of individuals the sample area can contain and θ is the probability that any one of the M possible "places" in δA is occupied. When M is large, θ small and $M\theta$ moderate, the distribution of x is best approximated by the Poisson distribution with mean $M\theta$, where $M\theta$, the expected number found in δA, is $D\delta A$ (cf. Greig-Smith [1964: 11–14] or, for a rigorous derivation, Dacey [1964a]). For animal populations which are randomly distributed, this Poisson model would normally be more appropriate than the binomial, as δA is usually very large compared with the area taken up by an individual, and D is usually small, so that the actual number of places occupied by individuals is only a small fraction of the M possible places. Therefore, if $x_1, x_2, \ldots, x_s$ are the numbers of animals found in s sample plots of equal area, a test of randomness amounts to testing the hypothesis that these observations come from a Poisson distribution. One can use either the standard chi-squared goodness-of-fit test or the Poisson Dispersion Test (cf. **1.3.6** (*2*)). However, it is generally advisable to use both tests, as the rare situation can arise in which one of the tests may detect evident non-randomness when the other fails to do so (Greig-Smith [1964: 68–70], Hughes [1970]). We note that the degree of non-randomness detected will depend on the size of the sample plot used (cf. Greig-Smith [1964: Chapter 3] and Kershaw [1964: Chapter 6]).

It seems that, in general, natural plant or animal populations are rarely distributed at random and that organisms are usually found to be clustered together more than a Poisson distribution would predict. For example, small mammals will tend to congregate in regions where there is shelter, thus avoiding open areas. This means that there will be more empty plots and more plots with many individuals than expected, so that the variance of x will be greater than its mean. When the Poisson distribution is not applicable a number of alternative distributions have been put forward to account for the spatial patterns observed (cf. Greig-Smith [1964], Southwood [1966], Pielou [1969], and King [1969] for references). In particular the negative binomial distribution

$$f(x) = \frac{k(k+1)\ldots(k+x-1)}{x!} P^x Q^{-x-k} \qquad x = 0, 1, 2, \ldots$$

obtained by expanding $(Q-P)^{-k}$, where $Q = 1 + P$ and $k > 0$, has been found to be the most useful because of its flexibility. Since

$$E[x] = kP = \mu, \quad \text{say,}$$

and

$$V[x] = kPQ = \mu + \mu^2/k,$$

we see that this distribution can be fitted to a wide range of distributions, ranging from the Poisson with variance equal to the mean ($k = \infty$) to those in which the variance is much greater than the mean. For example, in insect populations k is often found to be about 2 (Southwood [1966: 25]). A general discussion on the negative binomial is given in Southwood, and some interesting examples of fitting data to Poisson and negative binomial distributions are given in Andrewartha [1961], Debauche [1962] and South [1965];* further applications to fish populations are given by Moyle and Lound [1960], Roessler [1965], and Houser and Dunn [1967]. A study on goodness-of-fit tests for the negative binomial is given in Pahl [1969].

If it is known from past experience that a population is scattered according to a negative binomial then from (2.3) we have

$$N = E[\hat{N}] = SE[\bar{x}] = SkP,$$

$$V[\hat{N}] = \frac{S^2}{s} V[x] = \frac{S^2}{s} kP(1+P) = \frac{N}{s}\left(S + \frac{N}{k}\right)$$

and

$$(C[\hat{N}])^2 = V[\hat{N}]/N^2 = \frac{1}{s}\left(\frac{S}{N} + \frac{1}{k}\right).$$

Rearranging this last equation yields

$$s = \frac{1}{C^2}\left(\frac{S}{N} + \frac{1}{k}\right),$$

so that if approximate estimates of N and k are available, s, the number of

(*) See also Siniff *et al.* [1970: seals].

sample plots of area a, can be calculated to give a prescribed C. If a pilot sample is used then N/S is estimated by $\bar{x}$, and k can be estimated using one of the methods described in Southwood [1966: 26] (see also **12.1.2** (*2*)).

2 *Quadrats: stratified random sampling*

NUMBER OF PLOTS. Suppose now that the total population area is divided into J domains (or strata), each of area A_j ($j = 1, 2, \ldots, J$; $\Sigma A_j = A$), and let N_j be the number of animals in the jth domain ($\Sigma N_j = N$). If an area $p_j A_j$ is sampled in the jth domain and found to contain n_j animals, then if the animals are randomly distributed within each domain, an unbiased estimate of N is given by (cf. (2.1))

$$\tilde{N} = \sum_j \hat{N}_j = \sum_j (n_j/p_j)$$

with variance

$$V[\tilde{N}] = \sum_j N_j q_j/p_j .$$

To determine the optimum allocation of sampling effort we minimise $V[\tilde{N}]$, subject to

$$\Sigma p_j A_j = \text{constant} (= pA, \text{ say})$$

and obtain

$$p_j = K\sqrt{(N_j/A_j)} = K\sqrt{D_j}, \quad \text{say,}$$

where $K = pA/\sum_j \sqrt{(N_j A_j)}$.

If these values of p_j are used, we find that

$$V[\tilde{N}] = \frac{\{\Sigma \sqrt{(N_j A_j)}\}^2}{pA} - N$$

and for a given coefficient of variation C we must have

$$p = \frac{\{\Sigma \sqrt{(N_j A_j)}\}^2}{AN(1+NC^2)} .$$

When rough estimates of the N_j are available, then for a given C, p — and hence the p_j — can be determined by the above equations.

We note that for optimum allocation we must choose p_j proportional to the square root of the population density D_j in the jth domain. In fact, even just relative density estimates (cf. **2.2**) can be used for determining the p_j. For example, if $p_j = K_1\sqrt{R_j}$, where R_j is an estimate of relative density (the same units being used for each domain), and p is determined in advance (by the resources available for sampling), then

$$K_1 = pA/\Sigma A_j\sqrt{R_j}$$

and the p_j can be calculated.

If a proportional allocation of area is used, so that $p_j = p$, then

$$\tilde{N} = \sum_j n_j/p$$

and p can be determined from (cf. (2.2))

$$p = 1/(1+NC^2).$$

ESTIMATION. Suppose that s_j plots are chosen at random from the T_j plots available in the jth domain and let x_{ij} be the count on the ith sample plot ($i = 1, 2, \ldots, s_j$; $\Sigma s_j = s$, $\Sigma T_j = T$). We shall assume, for completeness, the general case where the plots may have different sizes. Then if

$$\bar{x}_j = \sum_{i=1}^{s_j} x_{ij}/s_j$$

is the average sample count for the jth domain, N is estimated by (cf. (2.3))

$$\tilde{N} = \sum_j \hat{N}_j = \sum_j \bar{x}_j T_j$$

and an unbiased estimate of $V[\tilde{N}]$ is given by

$$\tilde{\sigma}_{\tilde{N}}^2 = \sum_j \hat{\sigma}_{\hat{N}_j}^2 = \sum_j T_j^2 \frac{v_j}{s_j}\left(1-\frac{s_j}{T_j}\right),$$

where $v_j = \sum_i (x_{ij} - \bar{x}_j)^2/(s_j - 1)$.

A $100(1 - \alpha)$ per cent confidence interval for N is given by $\tilde{N} \pm t_k[\alpha/2]\tilde{\sigma}_N$, where k, the number of degrees of freedom, is given by (Cochran [1977: 95])

$$k = \frac{\tilde{\sigma}_{\tilde{N}}^4}{\sum_j \left(\dfrac{\hat{\sigma}_{\hat{N}_j}^4}{s_j - 1}\right)}.$$

We note that the above theory does not depend on the assumption that the population is randomly distributed within each domain.

PILOT SURVEY. When estimates $\tilde{N}$ and v_j are available from a pilot sample then, for a given s, the optimum value of s_j which minimises $V[\tilde{N}]$ is estimated by (Cochran [1977: 97])

$$s_j = \frac{sT_j\sqrt{v_j}}{\Sigma T_j\sqrt{v_j}} = sr_j, \quad \text{say}.$$

With this choice of s_j it is readily shown that, for a given coefficient of variation C, we must have

$$s = \frac{\Sigma T_j^2 v_j r_j^{-1}}{\tilde{N}^2 C^2 + \Sigma T_j v_j}. \tag{2.5}$$

If, however, proportional sampling is used, so that $s_j/T_j = s/T$, then we simply set $r_j = T_j/T$ in (2.5).

In some situations it may not be possible or practicable to use either optimum or proportional sampling. However, data from a pilot sample can still be used to design an experiment in which $\tilde{N}$ has a coefficient of variation no greater than C. Calculating $\bar{x}_j$, v_j and $\hat{N}_j$ from the pilot sample, we choose s_j so that the coefficient of variation of the estimate of N_j will be no greater than C, i.e. from (2.4) we have

$$\frac{1}{s_j} \leqslant \frac{1}{T_j} + \frac{C^2 \bar{x}_j^2}{v_j}.$$

Then, defining

$$b_j = \left\{ T_j^2 \frac{v_j}{s_j} \left(1 - \frac{s_j}{T_j}\right)\right\}^{\frac{1}{2}}$$

we have $b_j/\hat{N}_j \leqslant C$, and

$$\hat{C}[\tilde{N}] = \frac{(\Sigma\, b_j^2)^{\frac{1}{2}}}{\Sigma\, \hat{N}_j} < \frac{\Sigma\, b_j}{\Sigma\, \hat{N}_j} \leqslant \max_j \left\{\frac{b_j}{\hat{N}_j}\right\} \leqslant C.$$

Some examples of stratified sampling applied to animal populations are given in Foote *et al.* [1958], Wiegert [1964], Siniff and Skoog [1964], Evans *et al.* [1966], and Sen [1970] (see also Schultz *et al.* [1976]).

3 Strip transects

In comparing quadrat and strip sampling we note that strip sampling is usually easier to carry out, so that the time saved can be devoted to covering a greater area. In particular, by simply working from a base-line running across the population area, a strip is more readily located than a quadrat. This is a considerable advantage when large quadrats are needed, as in the case of sparsely scattered populations. Also, quadrat sampling may be impracticable where the terrain is difficult or where the animals are very mobile (e.g. Brock [1954: reef fish populations], Pielowski [1969: European hare]). Strip sampling is generally preferred in aerial surveys (cf. **12.3**).

Usually population areas are irregular in shape, so that strips running the full length of the area will vary in size. This means that as far as the estimation of N is concerned, the general theory of the previous section as applied to sampling with a variable plot size can be used here. However, more efficient sampling schemes are available (cf. **12.3**). A correction for the lack of visibility in immobile populations can be made (cf. **12.4**).

2.1.3 Line transect methods: parametric models*

1 General theory

In the line transect method, an observer walks a distance L across the population in non-intersecting and non-overlapping lines, counting n, the number of animals sighted, and recording one or more of the following statistics at the time of first sighting:

(a) radial distance r_i $(i = 1, 2, \ldots, n)$ from observer to animal,
(b) right-angle distance y_i from the animal sighted to the path of the observer, and
(c) angle of sighting θ_i from the observer's path to the point at which the animal was first sighted.

(*) This section should be read in conjunction with **12.4.3**.

This technique seems to be most useful for populations in which the animals can only be seen when they are disturbed and flushed into the open; in this case r_i is called the "flushing" distance.

A number of different methods have been proposed for estimating the population density $D(=N/A)$ from the above data, and for a summary of the methods to date the reader is referred to Eberhardt [1968] and Gates [1969]. All the formulae are of the form $\hat{D} = n/(2Lw)$, where w is some measure of one-half the "effective width" of the strip covered by the observer as he moves down the line transect. The basic assumptions underlying the various models are:

(1) The animals are randomly and independently distributed over the population area, i.e. the probability of a given animal being in a particular region of area δA is $\delta A/A$.
(2) The sighting of one animal is independent of the sighting of another.
(3) No animal is counted more than once.
(4) When animals are seen through being flushed into the open, each animal is seen at the exact position it occupied when startled.
(5) The response behaviour of the population as a whole does not substantially change in the course of running a transect.
(6) The individuals are homogeneous with regard to their response behaviour, regardless of sex, age, etc.
(7) The probability of an animal being seen, given that it is a right-angle distance y from the line transect path (irrespective of which side of the path it is on), is a simple function $g(y)$, say, of y, such that $g(0) = 1$ (i.e. probability 1 of seeing an animal on the path).

Given the above assumptions, and using a slight generalisation of a method given by Gates *et al.* [1968], we shall now derive the joint distribution of the observed distances $y_1, y_2, \ldots, y_n$ and n, the number of sightings.

If an animal is on a particular line at right angles to the transect, then by assumptions (1) and (7),

$$\Pr[\text{animal in } (y, y+dy)] = 2L\,dy/A,$$
$$\Pr[\text{animal seen} \mid \text{animal in } (y, y+dy)] = g(y),$$

so that

$$\Pr[\text{animal seen in } (y, y+dy)] = 2Lg(y)dy/A$$

and the probability of sighting an animal from the transect is

$$P = \frac{2L}{A}\int_0^\infty g(y)dy = \frac{2Lc}{A}, \quad \text{say.} \qquad (2.6)$$

Since

$$\Pr[\text{animal in } (y, y+dy) \mid \text{animal is seen}]$$

$$= \Pr[\text{animal seen in } (y, y+dy)]/\Pr[\text{animal is seen}]$$
$$= 2Lg(y)dy/(AP)$$
$$= c^{-1}g(y)dy,$$

the probability density function (p.d.f.) for y is

$$f(y) = c^{-1}g(y), \tag{2.7}$$

and the joint p.d.f. of $y_1, y_2, \ldots, y_n$ conditional on n is

$$f(y_1, y_2, \ldots, y_n \mid n) = c^{-n} \prod_{i=1}^{n} g(y_i). \tag{2.8}$$

From assumption (2), the probability function of n is binomial, namely

$$f(n) = \binom{N}{n} P^n Q^{N-n}, \tag{2.9}$$

where $Q = 1 - P$, so that the joint p.d.f. of the y_i's and n is

$$f(y_1, y_2, \ldots, y_n, n) = \binom{N}{n} P^n Q^{N-n} c^{-n} \prod_{i=1}^{n} g(y_i). \tag{2.10}$$

Once $g(y)$ is specified as a function of y and of certain unknown parameters, c and P can be derived in terms of these parameters and then (2.10) can be maximised to obtain maximum-likelihood estimates of N and the unknown parameters. It transpires that maximising (2.10) is equivalent to maximising (2.8) with respect to the unknown parameters and then, using these estimates in P, maximising (2.9) with respect to N, i.e. $\hat{N} = n/\hat{P}$. We shall now consider three different models for the function g.

2 *Exponential law for probability of detection*

Gates *et al.* [1968] choose

$$g(y) = \exp(-\lambda_1 y)$$

so that $c = 1/\lambda_1$, $P = 2L/(A\lambda_1)$, and show that the maximum-likelihood estimate of λ_1 (corrected for bias) is

$$\hat{\lambda}_1 = (n-1)/\Sigma\, y_i.$$

Hence $\hat{N}$, the maximum-likelihood estimate of N, is given by

$$\hat{N}_5 = n/\hat{P} = nA\hat{\lambda}_1/(2L)$$

and the population density is estimated by

$$\hat{D}_5 = (n-1)/(2\bar{y}L).$$

(The suffix 5 is introduced in preparation for a later comparison of density estimates.) Setting $\hat{c} = 1/\hat{\lambda}_1$, we also have

$$\hat{D}_5 = n/(2L\hat{c}). \tag{2.11}$$

We note in passing that the error in using infinity as the upper limit in (2.6) when the population area is strictly finite will in most cases be negligible. This is because A is usually large, so that the line transect is well away from the boundaries, i.e. $g(y)$ is approximately zero for y, the distance to the nearest boundary.

Gates *et al.* [1968] prove that

$$E[\hat{N}_5] = N(1-Q^{N-1})$$

and give an approximate expression for the variance, namely

$$V[\hat{N}_5] \approx \frac{N}{P}\left\{Q + NPE\left[\frac{1}{n-2}\right]\right\}. \tag{2.12}$$

This means that for moderate N, $\hat{N}_5$ is almost unbiased, and a consistent estimate of the above variance approximation is

$$v_5 = \frac{n}{\hat{P}^2}\left[1 - \hat{P} + \frac{n}{n-2}\right]. \tag{2.13}$$

We note that in applying the above theory y_i, L and A should be measured in the same units (e.g. miles (km) and square miles (sq km) respectively).

EXPERIMENTAL DESIGN. As in quadrat sampling, the first problem in designing a line transect survey is to determine in advance what L or n should be in order to obtain an estimate of N with a prescribed coefficient of variation C. From (2.13) C^2 is estimated by

$$\hat{C}^2 = \frac{1-\hat{P}}{n} + \frac{1}{n-2},$$

and neglecting $\hat{P}$ (which is usually small), this quadratic in n can be solved for the larger root. For example, if $C = 0{\cdot}25$ then $n = 33$. Therefore, choosing $C < 0{\cdot}25$, $C^2 \approx 2/n$ which leads to a value of n approximately twice that required for quadrat sampling (p. 23: see also **12.1.2** (*1*) for a cost comparison). This means that line transect estimates are not very precise unless n is large. We note that if $\hat{P}$ is not negligible then ignoring it will lead to a conservative value of n, i.e. n will be larger than necessary.

If a preliminary pilot survey is carried out with length L, then Gates *et al.* show that the total length, L_T say, required to give a prescribed C is given by

$$L_T = \frac{\hat{\lambda}_1 A(n-1)}{(C^2\hat{N}_5 + 1)(n-2)},$$

where $\hat{\lambda}_1$, $\hat{N}_5$ and n are obtained from the pilot survey. This expression can derived by noting that if n_T is the number of animals that would be observed for length L_T then, since $n_T > n$,

$$n/(n-2) > n_T/(n_T-2)$$

and $n/(n-2)$ can be used as a conservative estimate of $NPE[1/(n_T-2)]$ in (2.12).

UNDERLYING ASSUMPTIONS. Assumptions (1), (2) and (3) (p. 29) can be tested by dividing L into a large number of smaller segments of length l and noting n_l, the number seen in each segment. If P_l is the probability of sighting an animal from such a segment, then P_l will be very small, so that when assumptions (1)–(3) hold, n_l will have a Poisson rather than a binomial distribution with mean NP_l. Therefore one can test the goodness of fit of the values n_l to the Poisson distribution by using a standard chi-squared test or the Poisson Dispersion Test (**1.3.6** (*2*)).

Assumption (2) is violated if the animals tend to flush in pairs, for example. However, according to Gates [1969], the estimate $\hat{N}_5$ will not be biased by this tendency, though the variance of $\hat{N}_5$ will be increased.

Assumption (4) can be relaxed where it is possible to assume that a certain constant proportion of the animals flee and are not seen on the approach of the observer. For example, some game birds run along the ground instead of flushing and may not be seen in heavy cover. Where such a constant proportion is known to exist and can be determined independently, one may substitute $P' = kP$ for P in the general theory and still obtain an approximately unbiased estimate of N.

Assumption (6) can be investigated by analysing the data from the different subgroups in the population separately. However, when the other assumptions hold, we find that even in the extreme case where every individual is from a different subgroup with regard to behaviour pattern, the true variance of $\hat{N}_5$ will be *smaller* than that given by (2.12) (Kendall and Stuart [1969: 127]).

To test whether $g(y)$ is exponential, we note from (2.7) that y is distributed as $\lambda_1 \exp(-\lambda_1 y)$. If n is large enough, we can therefore carry out a standard chi-squared goodness-of-fit test to see whether the observations y_i come from this negative exponential distribution (cf. p. 46).

Any inaccuracies in the estimation of the right-angle distances y_i will not bias $\hat{N}_5$, provided the errors are not biased; though the variance of $\hat{N}_5$ will be increased. Therefore, in view of the fact that some of the departures mentioned above do not bias $\hat{N}_5$ but tend to increase the variance, it is recommended that one should also calculate the less efficient estimate $\bar{N}_5$, the average of the estimates $\hat{N}_{5j}$, say ($j = 1, 2, \ldots, s$), for s transects of length L/s. Then, from (1.9), a robust estimate of $V[\bar{N}_5]$ is given by

$$\frac{1}{s(s-1)} \sum_{j=1}^{s} (\hat{N}_{5j} - \bar{N}_5)^2 .$$

Example 2.1 Ruffed grouse (*Bonasa umbellus*): Gates *et al.* [1968].

Estimated densities for ruffed grouse at the Cloquet Forest Research Centre, Minnesota, in the line transect surveys conducted there during the years 1950–58 are presented in Table 2.1. The King, Hayne and Webb estimates of density are also included for comparison (see section 5 for definitions).

TABLE 2.1

Estimated densities for a ruffed grouse population at Cloquet Forest Research Centre, Minnesota, 1950–58: from Gates *et al.* [1968: Table 1].

Date of survey	L (miles)	n (number seen)	Density estimate (birds per square mile)			
			$\hat{D}_1$ (King)	$\hat{D}_2$ (Hayne)	$\hat{D}_3$ (Webb)	$\hat{D}_4$ ± standard deviation (Gates *et al.*)
1950A	28·00	45	98·5	188·7	155·7	148·8 ± 30·6
1950B	30·75	35	79·1	167·8	84·0	121·6 ± 29·1
1951A	26·50	41	85·3	132·2	183·3	195·2 ± 43·2
1951B	31·40	27	59·8	97·2	114·4	122·4 ± 33·0
1954	34·25	8	10·9	13·3	21·3	17·1 ± 9·1
1955	34·25	8	17·2	50·5	30·3	28·8 ± 15·4
1956A	31·50	12	20·9	42·5	29·0	36·2 ± 15·2
1956B	31·50	8	12·6	19·4	36·9	41·5 ± 22·2
1957A	36·50	4	6·4	7·2	18·3	14·5 ± 12·4
1957B	34·30	9	15·9	25·1	23·7	30·6 ± 15·2
1958A	32·00	12	17·0	23·2	26·2	27·3 ± 11·4
1958B	31·65	17	21·4	37·3	36·0	48·5 ± 16·9
All surveys: 226						

A goodness-of-fit test of the y_i observations to a negative exponential distribution was carried out using the data of all 7 years, and the observed and expected frequencies are given in Table 2.2 (y_i was not recorded for 2 birds, giving a total of 224 observations). Class intervals of 5 yards were used and, because of bias in reporting, those observations recording at exactly 5 yards were divided equally between the class above and below. A value $\hat{\lambda}_1 = 0{\cdot}126$ was calculated from the ungrouped data and used to derive the expected frequencies. The chi-squared goodness-of-fit test yielded a value of 2·295 which, with 6 degrees of freedom, indicates an excellent fit.

TABLE 2.2

Frequency distribution of 224 right-angle flushing distances (y) fitted to a negative exponential distribution: from Gates *et al.* [1968: Table 2].

y (yards)	Frequencies	
	Observed	Expected
0–5	103·5	104·6
5–10	61·5	55·8
10–15	28·5	29·8
15–20	11·5	15·9
20–25	9·0	8·3
25–30	3·5	4·5
> 30	6·5	5·2
Total	224·0	224·1

To test assumptions (1)–(3), the numbers n_l were recorded for each l = quarter-mile and a test of fit for the Poisson distribution was carried out. The observed and expected frequencies are given in Table 2.3, and pooling

TABLE 2.3

Distribution of ruffed grouse flushings per quarter-mile segment of line transect fitted to a Poisson distribution: from Gates *et al.* [1968: Table 3].

n_l	Frequencies	
	Observed	Expected
0	2450	2410·8
1	292	361·7
2	52	27·1
3	6	1·4
4	0	0·0
5	0	0·0
6	0	0·0
7	1	0·0
Total	2801	2801·0

the last 5 classes into one class gives a chi-squared value of 59·4 with 2 degrees of freedom. This is highly significant, thus suggesting a breakdown in the underlying assumptions. An examination of Table 2.3 indicates that there are either too many one-quarter mile segments with two or more birds flushing at the expense of single-bird flushes, or alternatively an excess of zeros. The authors suggest two possible explanations, not mutually exclusive, for the former and one possible explanation for the latter failure to follow a Poisson distribution, namely (i) the grouse have a tendency to flush in pairs (assumption (2) false), (ii) some grouse may fly ahead of the observer and be flushed twice in the same one-quarter mile segment (assumption (3) false), and (iii) some parts of the area may not be suitable for occupation (assumption (1) false).

3 Power law for probability of detection

Eberhardt [1968] suggests using the function

$$g(y) = \begin{cases} 1 - \left(\dfrac{y}{W}\right)^a & 0 \leq y \leq W \\ 0 & y > W \end{cases}$$

which leads to $c = Wa/(a+1)$ and the density function (cf. (2.7))

$$f(y) = \frac{(a+1)}{aW}\left(1 - \left(\frac{y}{W}\right)^a\right). \tag{2.14}$$

If a is known, then from

$$E[\bar{y}](=E[y]) = \frac{W(a+1)}{2(a+2)} \tag{2.15}$$

we have the moment estimate $\hat{W} = 2\bar{y}(a+2)/(a+1)$, which leads to the estimate (cf. (2.11))

$$\hat{D}_8 = \frac{n(a+1)^2}{4L\bar{y}a(a+2)}.$$

Both $\hat{D}_5$ (the biased version $n/(2\bar{y}L)$) and $\hat{D}_8$ are derived as moment estimates by Eberhardt [1968], using a different method. However, his estimates differ from ours by a factor of two as he deals with only one side of the transect line (setting $\nu = 2DL$ in his paper leads to $\hat{D}_5$, $\hat{D}_8$ and $\hat{D}_9$ given below).

When a is unknown the maximum-likelihood estimates of W and a are y_M, the maximum observed value of y, and the solution of

$$\frac{n}{a(a+1)} + \sum_i \left\{ \left(\frac{y_i}{y_M}\right)^a . \log\left(\frac{y_i}{y_M}\right) . \frac{1}{1-(y_i/y_M)^a} \right\} = 0.$$

As this equation is rather intractable it would seem preferable to use a simpler, though less efficient, estimate of a. For example, setting $W = y_M$ one can use (2.15) to obtain such an estimate.

4 *All animals detected over a constant unknown distance*

The simplest, and probably the least realistic, assumption one can make about $g(y)$ is that *all* animals are sighted for some unknown distance W out from the transect and none are sighted beyond this distance, i.e.

$$g(y) = 1, \quad 0 \leqslant y \leqslant W,$$

$c = W$, and y has the uniform distribution (cf. (2.7))

$$f(y) = 1/W, \quad 0 \leqslant y \leqslant W.$$

Since $E[y] = W/2$, c can be estimated by $2\bar{y}$, which leads to the density estimate (cf. (2.11))

$$\hat{D}_9 = n/(4L\bar{y}).$$

Alternatively c can be estimated by y_M, leading to

$$\hat{D}_{10} = n/(2Ly_M).$$

This estimate, suggested by Amman and Baldwin [1960], was found to be useful in the study of woodpecker populations, as virtually all the birds could be detected from feeding sounds.

5 *Density estimates based on flushing distances*

All the previous models can be used either when the animals are stationary and visibility is limited (e.g. mortality surveys of dead deer: Eberhardt [1968: 86–7]), or when the animals are mobile and have to be flushed into the open. In the latter situation one can also make use of the flushing distances r_i which are generally easier to measure than right-angle distances. For example, setting $\hat{D}_i = n/(2Lw_i)$ we have, as possibilities: $w_1 = \bar{r}$, the so-called "King method" (Leopold [1933]); $w_2 = n/\Sigma\,(1/r_i)$ (Hayne [1949a]); $w_3 = \bar{r}\sin\bar{\theta}$, where $\bar{\theta}$ is the average angle of sighting (Webb [1942]); and $w_4 = \bar{G}$, the geometric mean of the r_i (Gates [1969]).

Hayne examined the King method $\hat{D}_1$ and pointed out that it is based on three main assumptions: (i) in a population animals vary with regard to the distance at which they will flush upon the approach of an observer; (ii) the various classifications of animals (with regard to flushing distance) are scattered about over the study area in a random fashion, or at least random with respect to the path of the observer; and (iii) the average flushing distance $\bar{r}$ observed by the investigator is a good estimate of the average flushing distance for the whole population. There is also a fourth assumption implicit in his discussion, namely (iv) all animals flushed are actually seen by the observer. The first two assumptions are generally satisfied if there is adequate randomisation in the choice of line transect, though (ii) will not hold if there is any directional movement of animals in relation to the observer's path. For example, certain animals appear to avoid an observer so successfully that the flushing distance in this case must be considered greater than the range of visibility of the observer. Also any tendency for the animals to "drive", that is move down the line of travel ahead of the observer after being flushed, will increase the apparent population density.

The third assumption appears to be false, and Hayne demonstrated mathematically that w_1 will generally lead to an overestimate of effective strip width with a consequent underestimate of population density. This is readily seen by comparing the observed average flushing distance with the true average flushing distance for a strip centred on the line transect. Obviously animals in the strip at a perpendicular distance from the line greater than their flushing distance will not be flushed, so that the observed average will be based on fewer short flushing distances than the true average. To allow for this bias, Hayne suggested the modification w_2 which will generally give unbiased results, provided assumptions (i), (ii) and (iv) are satisfied (see also Overton and Davis [1969: 420–23]).

The King method was initially devised for censusing ruffed grouse populations. Webb [1942], however, felt that the method could not be applied to snowshoe hares because of the difference in flushing behaviour. When the grouse are flushed they make a considerable noise with their wings, so that the birds flushed are less likely to be overlooked. In flushing hares, however, Webb argued that the observer may see the hares that move ahead of him and those which remain motionless until he is very close, but will tend to miss those that move directly to the right or left. This means that the effective strip width w should be reduced and on intuitive grounds Webb proposed the modification w_3.

We note that w_1 and w_2 are the arithmetic mean and the harmonic mean respectively of the flushing distances, so that from Kendall and Stuart

[1969: 36] $w_2 \leqslant w_4 \leqslant w_1$, and $\hat{D}_1 \leqslant \hat{D}_4 \leqslant \hat{D}_2$.

Although Webb's estimate allows for the possibility of flushed animals not being seen, a general method for dealing with this problem is given by Gates [1969] as follows. We shall make assumptions (1)–(7) of p. 29 above, together with the following:

(8) $g(y) = \exp(-\lambda_1 y)$, i.e. the p.d.f. of y is

$$f(y) = \lambda_1 \exp(-\lambda_1 y). \tag{2.16}$$

(9) The conditional probability density function of the radial distance r, given the right-angle distance y, is

$$f(r \mid y) = \lambda_2 \exp[-\lambda_2(r-y)], \quad y \leqslant r < \infty, \lambda_2 > 0. \tag{2.17}$$

Then, using the transformation $y = r \sin\theta$, Gates shows that the joint distribution of r and θ (the angle of flushing), given that the animal is seen, is

$$f(r, \theta) = \lambda_1\lambda_2 (\cos\theta)\, r \exp(-\alpha r), \quad 0 \leqslant \theta \leqslant \pi/2, \tag{2.18}$$

where $\alpha = (\lambda_1 - \lambda_2)\sin\theta + \lambda_2$.

CASE 1: $\lambda_1 = \lambda_2 = \lambda$. It follows from (2.18) that

$$f(r, \theta) = \lambda^2 r (\cos\theta) \exp(-\lambda r)$$

with marginal distributions

$$f_1(r) = \lambda^2 r \exp(-\lambda r) \tag{2.19}$$

and

$$f_2(\theta) = \cos\theta, \quad 0 \leqslant \theta \leqslant \pi/2. \tag{2.20}$$

By the same argument which led to equation (2.10) we find that the joint distribution of $r_1, r_2, \ldots, r_n$ and n is

$$f(r_1, r_2, \ldots, r_n, n) = \binom{N}{n} P^n Q^{N-n} \prod_{i=1}^{n} \{\lambda^2 r_i \exp(-\lambda r_i)\}. \tag{2.21}$$

Gates shows that the maximum-likelihood estimates of λ and N, after correction for bias, are

$$\hat{\lambda} = (2n-1)/\Sigma r_i$$

and

$$\hat{N}_7 = n/\hat{P} = nA\hat{\lambda}/(2L),$$

so that the estimate of population density is given by

$$\hat{D}_7 = (2n-1)/(2L\bar{r}).$$

Also, using the delta method, Gates proves that

$$V[\hat{N}_7] \approx N^2 E\left[\frac{1}{2n-2}\right] + \frac{N(1-P)}{P},$$

and a consistent estimate of this expression is

$$v_7 = \frac{n}{\hat{P}^2}\left[\frac{3n-2}{2n-2} - \hat{P}\right].$$

It can be shown that $v_7 < v_5$ (cf. equation (2.13)), so that radial distances are not only easier to measure, but when the above model is applicable, they also give a more efficient estimate of population size.

Using (2.19), it is readily shown that

$$\hat{\lambda}' = \frac{1}{n}\Sigma\frac{1}{r_i}$$

is also unbiased for λ, and the corresponding unbiased estimate of N is

$$\hat{N}_2 = \frac{An\hat{\lambda}'}{2L} = \frac{A}{2L}\Sigma\frac{1}{r_i},$$

which, expressed in terms of density, is Hayne's [1949a] estimate $\hat{D}_2$. However, $V[\hat{D}_2]$ does not exist for the present model, as $E[1/r^2]$ does not exist.

From (2.16), (2.17) and (2.19) we find that

$$f(y \mid r) = 1/r, \quad 0 \leqslant y \leqslant r$$

so that for given r, y has a uniform distribution. Katz, in an appendix to Hayne's 1949a paper, assumed this distribution to start with, and by writing $y = r\sin\theta$, proved that $\sin\theta$ is uniformly distributed on [0, 1] and that θ has the probability density function (2.20). These two distributional properties of θ have been used as a check on the validity of the above model.* For example, from (2.20) we have

$$\begin{aligned} E[\theta] &= \frac{\pi}{2} - 1 \quad \text{radians} \\ &= 32{\cdot}704°, \end{aligned}$$

and we can test whether the sample mean $\bar{\theta}$ differs significantly from 32·704°. One can also carry out a goodness-of-fit test to see whether the sample values $\sin\theta_i$ $(i = 1, 2, \ldots, n)$ come from a uniform distribution. A variety of tests of fit are available, and these are listed in many textbooks (e.g. Kendall and Stuart [1973], Keeping [1962]). When n is sufficiently large, the usual Pearson chi-squared goodness-of-fit statistic can be used, though the Kolmogorov statistic gives a more sensitive test and its exact distribution has been tabled. The validity of (2.19) can also be tested using a chi-squared goodness-of-fit test, and an example of this is given in Gates [1969].

CASE 2: $\lambda_1 \neq \lambda_2$. From (2.18) the marginal distributions of r and θ are

$$f_1(r) = \frac{\lambda_1\lambda_2}{\lambda_2 - \lambda_1}\{\exp(-r\lambda_1) - \exp(-r\lambda_2)\}$$

and

$$f_2(\theta) = \lambda_1\lambda_2(\cos\theta)/\alpha^2. \tag{2.22}$$

Using the same argument that led to (2.10) and (2.21), the maximum-likelihood estimates $\hat{N}$, $\hat{\lambda}_1$ and $\hat{\lambda}_2$ are given by

$$\hat{N} = nA\hat{\lambda}_1/(2L),$$

where $\hat{\lambda}_1$, $\hat{\lambda}_2$ are the solutions of

(*) But see 12.4.3 (3)

$$\frac{n\lambda_1}{\lambda_2(\lambda_2-\lambda_1)} = \sum_i \left\{\frac{r_i \exp(-r_i\lambda_2)}{\exp(-r_i\lambda_1)-\exp(-r_i\lambda_2)}\right\}$$

$$\frac{n\lambda_2}{\lambda_1(\lambda_2-\lambda_1)} = \sum_i \left\{\frac{r_i \exp(-r_i\lambda_1)}{\exp(-r_i\lambda_1)-\exp(-r_i\lambda_2)}\right\}$$

Unfortunately these equations, obtained from $\prod_i f_1(r_i)$, do not have simple solutions and must be solved iteratively (cf. **1.3.8**). Gates suggests obtaining initial values for the iterative process by using the moment estimates

$$\bar{r} = E[r] = (\lambda_1+\lambda_2)/(\lambda_1\lambda_2), \tag{2.23}$$

$$s^2 = V[r] = (\lambda_1^2+\lambda_2^2)/(\lambda_1^2\lambda_2^2), \tag{2.24}$$

where $s^2 = \Sigma(r_i-\bar{r})^2/(n-1)$.

Solving these two equations we find the estimates, if they exist, to be

$$\frac{\bar{r} \pm \sqrt{(2s^2-\bar{r}^2)}}{\bar{r}^2-s^2}.$$

Because of the symmetry of λ_1 and λ_2 in (2.23) and (2.24) this result has the unfortunate feature that the two estimates of λ are not distinguishable. However, one possible method for identifying these estimates is described below.

From (2.22) Gates shows that the average angle of flushing depends on the relative magnitudes of λ_1 and λ_2; thus

$$E[\theta] = \begin{cases} k\left[\dfrac{-\pi}{2\lambda_1} + \dfrac{2}{\sqrt{c}}\tan^{-1}\left\{\dfrac{2\lambda_2-\lambda_1}{\sqrt{c}}\right\}\right], & \text{for } \lambda_2 > \tfrac{1}{2}\lambda_1 \\ k\left[\dfrac{-\pi}{2\lambda_1} + \dfrac{1}{d}\log\left\{\dfrac{(\lambda_1-d)(\lambda_1-\lambda_2+d)}{(\lambda_1+d)(\lambda_1-\lambda_2-d)}\right\}\right], & \text{for } \lambda_2 < \tfrac{1}{2}\lambda_1 \end{cases}$$

where $c = 2\lambda_1\lambda_2 - \lambda_1^2$, $d = \sqrt{(-c)}$, $k = \lambda_1\lambda_2/(\lambda_1-\lambda_2)$ and $\lambda_1 \neq \lambda_2$. When λ_2 is greater or less than λ_1 it can be shown that $E[\theta]$ is respectively greater or less than 32·704°. Therefore any significant departure of $\bar{\theta}$ from 32·704° would suggest that $\lambda_1 \neq \lambda_2$, and the direction of the departure would indicate which λ is the greater.

6 *Comparison of estimators*

Using computer simulation, Gates [1969] compared the estimates $\hat{D}_i$ ($i = 1, 2, \ldots, 5$) (cf. pp. 36 and 30 for definitions) with respect to bias, variance, and robustness towards certain departures from the underlying assumptions. The estimate $\hat{D}_7$ was derived after most of the computer calculations had been carried out, so that it was compared with the other estimators in a few cases only. His conclusions are summarised below.

BIAS. In this section we shall say that an estimator is "biased" or "unbiased" according to whether or not it is significantly different from the true population density. Gates found that when assumptions (1)–(9) were satisfied and $\lambda_1 = \lambda_2 = \lambda$, $\hat{D}_2$, $\hat{D}_5$ and $\hat{D}_7$ were generally unbiased while $\hat{D}_1$ and $\hat{D}_4$ were negatively biased (the differences usually being significant at the 1 per cent level of significance). The Webb estimator $\hat{D}_3$, although frequently biased, appeared to be less biased than $\hat{D}_1$ or $\hat{D}_4$. It was found that the relative bias of the various estimators did not depend on λ, though the the variances were directly proportional to λ.

VARIANCE. The Hayne estimator $\hat{D}_2$ consistently had a larger variance than the other estimators. When the estimators are ranked according to increasing size of variance we have the order $\hat{D}_1$, $\hat{D}_4$, $\hat{D}_5$, $\hat{D}_3$ and $\hat{D}_2$.

ROBUSTNESS. When there are unbiased errors in measuring distances it was found that the relative bias in all the estimators appeared to be unaffected, though the variances were increased as expected. $\hat{D}_2$ appeared to be particularly susceptible to measurement error, while the other estimators appeared to be less so.

Sensitivity to departure from the negative exponential distribution of assumption (8) was investigated by considering two alternative distributions: (i) a right-triangle type of distribution (cf. (2.14) with $a = 1$), and (ii) a half-normal distribution. It was found that all the estimators $\hat{D}_1$ to $\hat{D}_5$ appeared sensitive to such departures, and $\hat{D}_5$ consistently overestimated the true population density in both cases.

In some populations, as for example grouse, there is often the tendency for animals to flush in pairs, thus violating assumption (2) that animals act independently. Using simulated pairing, Gates found that such pairing did not change the relative bias of the estimators, and in particular $\hat{D}_2$, $\hat{D}_5$ and $\hat{D}_7$ were unbiased for the four computer runs based on 10, 20 and 30 per cent pairing. It was also found that the variances of the estimators increased as the percentage pairing increased.

As pointed out above, $E[\theta]$ is greater or less than 32·704° when λ_2/λ_1 is greater or less than unity, respectively. Gates found that the estimators $\hat{D}_i$ ($i = 1, 2, 3, 4$), and in particular $\hat{D}_2$, were sensitive to departures from the ratio $\lambda_2/\lambda_1 = 1$, while $\hat{D}_5$ was apparently insensitive. His results also verified the findings of Robinette *et al.* [1956] who concluded that when $E[\theta] > 40°$, the Hayne method is positively biased.

OTHER ESTIMATORS. Several density estimates have been derived by Yapp [1956] and Skellam [1958] for the more complex situation where the population is on the move. Unfortunately the requirement of knowing the velocities of the animals encountered rather precludes the use of these methods on practical grounds. However Gates *et al.* [1968] state that under certain reasonable conditions the two formulae proposed by Yapp reduce to the King estimator $\hat{D}_1$.

In conclusion it should be stressed that no matter which estimator is used, the data should be recorded separately for each segment of length l ($= L/s$) as described on p. 32. In this way the estimate and its theoretical variance estimate can be compared with the sample mean and variance estimate based on s repeated samples.

2.1.4 Density estimates based on distances*

1 Introduction

When a population is randomly distributed over the population area, one can use the distances of individuals from randomly chosen points (closest-individual techniques) or the distances between neighbours (nearest neighbour techniques) to calculate estimates of population density. Such techniques, although most suitable for stationary populations such as plants or trees (Greig-Smith [1964]), can be applied to conspicuous and relatively stationary animal populations (e.g. snails: Keuls *et al.* [1963]) or to well-marked colonies such as ant or termite mounds (Waloff and Blackith [1962], Blackith *et al.* [1963], Sands [1965]). They have also been applied to more mobile populations such as grasshoppers (Blackith [1958]), cricket frogs (Turner [1960a]), and quail coveys (Ellis *et al.* [1969]).

Before discussing these distance estimates of density in detail it should be emphasised that most of the estimates depend on the assumption of random distribution, so that tests of randomness must be carried out. However, it is hoped that further research will eventually show that some of the methods are robust with regard to certain departures from randomness: some useful methods for handling non-randomly distributed populations have been given by Dacey [1964b, 1965, 1966]. If the density is known (or estimated by other techniques), these distance methods can be used for detecting non-randomness (e.g. Holgate [1965b], McLaren [1967], Pielou [1969; 115]).

Finally it should be pointed out that some of the published work using the above distance methods reveals a lack of understanding with regard to the background theory (Kendall and Moran [1963: 38]): this is particularly the case in the sampling methods used for choosing animals at random (see section *4* below).

2 Closest-individual techniques: Poisson model

UNBIASED ESTIMATION. We shall assume that:

(i) The population area is infinite with a constant population density D (restriction to a finite area A is considered later).

(ii) The number of individuals found in an area δA, chosen at random, has a Poisson distribution with parameter $D\delta A$ (cf. p. 24).

Suppose that a sample point is chosen at random and let $X_{(1)}$ be the distance to the nearest animal. Then from assumption (ii) we have (Eberhardt [1967])

$$F_1(x) = \Pr[X_{(1)} \leqslant x]$$

$$= \Pr[\text{finding at least one individual in circular plot of radius } x]$$

(*) See also **12.5** for more recent work.

$$= 1 - \Pr[\text{no individuals in plot of area } \pi x^2]$$

$$= 1 - \exp(-D\pi x^2), \tag{2.25}$$

and the probability density function (p.d.f.) for $X_{(1)}$ is

$$f_1(x) = F_1'(x) = 2D\pi x \exp(-D\pi x^2). \tag{2.26}$$

If $X_{(r)}$ is the distance to the rth nearest animal, then, from Eberhardt [1967],

$$F_r(x) = \Pr[X_{(r)} \leqslant x]$$

$$= \Pr[r \text{ or more animals in circular plot of area } \pi x^2]$$

$$= \sum_{i=r}^{\infty} \exp(-D\pi x^2) \frac{(D\pi x^2)^i}{i!}$$

and

$$f_r(x) = F_r'(x) = \frac{2(D\pi)^r}{(r-1)!} x^{2r-1} \exp(-D\pi x^2). \tag{2.27}$$

This probability density was derived independently by Morisita [1954], Moore [1954], and Thompson [1956] (see Dacey [1964a] for a review) and generalised to the case of n-dimensional space by Dacey [1963].

Setting $Y = \pi X_{(r)}^2$, then the probability density of Y has the simplest form

$$g_r(y) = D^r y^{r-1} \exp(-Dy)/(r-1)!\,. \tag{2.28}$$

To estimate D, suppose that s such sample points are chosen at random and let Y_i $(i = 1, 2, \ldots, s)$ be the corresponding values of Y. If $Y_{\bullet} = \Sigma Y_i$, then from Moore [1954] and Holgate [1964a] an unbiased estimate of D is the modified maximum-likelihood estimate

$$\hat{D} = (sr - 1)/Y_{\bullet} \tag{2.29}$$

with variance

$$V[\hat{D}] = D^2/(sr-2). \tag{2.30}$$

Since $2DY$ is distributed as χ^2_{2r}, $2DY_{\bullet}$ is χ^2_{2rs}, and this statistic can be used for constructing confidence intervals for D. For example a 95 per cent confidence interval for D is given by $(c_1/(2Y_{\bullet}), c_2/(2Y_{\bullet}))$ where c_1 and c_2 are the lower and upper 2·5 per cent points of χ^2_{2rs} respectively. Generally $2rs$ will be greater than the degrees of freedom tabulated, and in this case a normal approximation is used: thus

$$z = \sqrt{(4DY_{\bullet})} - \sqrt{(4rs-1)}$$

is approximately unit normal and

$$0{\cdot}95 \approx \Pr\left[\frac{-1{\cdot}96 + \sqrt{(4rs-1)}}{\sqrt{(4Y_{\bullet})}} < \sqrt{D} < \frac{1{\cdot}96 + \sqrt{(4rs-1)}}{\sqrt{(4Y_{\bullet})}}\right].$$

Since, from (2.28),

$$E[Y^{-1}] = D/(r-1),$$

Eberhardt [1967] suggests the unbiased estimate

$$\tilde{D} = (r-1) \sum_{i=1}^{s} (Y_i^{-1}/s) = \sum_{i=1}^{s} \tilde{D}_i/s, \text{ say} \tag{2.31}$$

which is evidently equivalent to one proposed by Morisita [1957]. It can be shown that, for $r \geqslant 3$,

$$V[\tilde{D}] = V[\tilde{D}_i]/s = D^2/[s(r-2)], \tag{2.32}$$

which, although greater than $V[\hat{D}]$ of (2.30), has an unbiased replicated-sample estimate (cf. (1.9))

$$v[\tilde{D}] = \frac{\sum_{i=1}^{s} (\tilde{D}_i - \tilde{D})^2}{s(s-1)}. \tag{2.33}$$

FINITE POPULATION AREA. In practice the population area is finite, and to avoid the possibility of "edge effect" one should choose a subarea in the population away from the boundary and restrict sample points to this subarea only. If a sample point then falls on the edge of the subarea, one would expect the distribution of $X_{(r)}$ to be the same as that for points within the subarea, provided that the edge of the subarea was at a sufficient distance from the population boundary. As a rough guide, this distance, x say, should be such that for $y = \pi x^2$, $g_r(y)$ is sufficiently small for the appropriate truncation of (2.28) to have a negligible effect. We note in passing that a common method of choosing a sample point is to divide up the total area into quadrats of equal area (allowing a "buffer" strip around the boundary), choose a quadrat (with replacement) using random numbers, and then select a point in the quadrat by reading off a pair of coordinates from the random numbers (cf. Sampford [1962: Chapter 4]).

COMPARING DENSITIES. A simple test for comparing the densities D_1 and D_2 of two areas is available from Moore [1954]. If s_j sample points are located in the jth area ($j = 1, 2$) and $Y_{\cdot j}$ is the corresponding value of $Y_{\cdot}$, then

$$F = (r_2 s_2 D_1 Y_{\cdot 1})/(r_1 s_1 D_2 Y_{\cdot 2})$$

has an F-distribution with $2r_1 s_1$ and $2r_2 s_2$ degrees of freedom respectively. When $D_1 = D_2$, F reduces to a simple statistic which can be entered in the the tables of the F-distribution. As in the F-test for equal variances, one can observe the convention of putting the larger value of $Y_{\cdot j}/(r_i s_i)$ in the numerator.

EXPERIMENTAL DESIGN. We note that $V[\hat{D}]$ of (2.30) and the coefficient of variation of $\hat{D}$, $1/\sqrt{(sr-2)}$, depend on the product sr, so that we can, for example, use either sr samples of nearest individuals ($r = 1$) or s samples of rth nearest individuals. The choice of procedure would depend on a comparison of the time taken in the field to locate a random point and the time to search for the rth nearest animal. In mobile populations, however, it may only be possible to measure the distance to the nearest animal: this special case has been considered by Skellam [1952: 349], Clark and Evans [1954, 1955], Hopkins [1954], and Moore [1954].

When the main effort involved is in searching for the animals rather than the location of sample points or quadrat boundaries, Holgate [1964a] shows that the above method has approximately the same efficiency as quadrat sampling. For example, if quadrats of total area A_0 yield a count of n individuals, then, assuming n to have a Poisson distribution with mean DA_0 (cf. p. 24), D is estimated by $D' = n/A_0$ with variance D/A_0. To compare this with the closest-individual technique based on $X_{(r)}$, it is assumed that the experimenter searches in increasing circles until exactly r animals have been found. Then the expected total area covered for s sample points will be $E[Y_.]$ $(= sr/D)$. Therefore, setting $A_0 = E[Y_.]$, we find that

$$V[D'] = D/A_0 = D^2/(sr),$$

which is slightly less than $V[\hat{D}]$ of (2.30). However, this analysis does not apply to the common situation where animals are readily seen, and the main effort is involved in laying out the random quadrats or sample points, and making the measurements (for such a comparison see **14.1.2** (*1*)).

QUARTER METHOD. A modification of the above technique, called the "point-centred quarter method", has been discussed by Cottam *et al.* [1953], Cottam and Curtis [1956] and Morisita [1954]. The sample point is chosen at random, together with two perpendicular directions fixed in advance, and the distances Z_1, Z_2, Z_3, and Z_4 say, from the sample point to the nearest individual in each of the four quadrants are measured. Let

$$Y = \tfrac{1}{4}\pi(Z_1^2 + Z_2^2 + Z_3^2 + Z_4^2),$$

then, from Kendall and Moran [1963: 40], the $\frac{1}{2}\pi D Z_i^2$ are independently distributed as χ_2^2, so that $2DY$ is χ_8^2. If s sample points are used and $Y_. = \Sigma Y$, then $2DY_.$ is χ_{8s}^2. Therefore, setting $r = 4$ in the above general theory, we see that this quarter method is equivalent to looking for the fourth nearest individual. This accounts for the fact discovered empirically by Cottam and Curtis [1956: 457] that $4s$ sample points using closest-individual measurements ($r = 1$) are needed to give an estimate of D with approximately the same accuracy as s points using the quarter method (Cottam and Curtis actually use a less efficient method of estimation based on the mean distance $\bar{Z}$: it is readily shown that $E[\bar{Z}] = 1/\sqrt{D}$). We note that in the field it will usually be easier to locate the nearest individual in each of the four quadrants than to look for the fourth nearest individual, though either method could only be applied to relatively immobile animals.

APPROXIMATE METHODS. Approximate methods for obtaining point and interval estimates of D based on the untransformed distances X_r have also been given. From Thompson [1956] we have

$$E[X_{(r)}] = a_r/\sqrt{D},$$
$$V[X_{(r)}] = \sigma_r^2 = b_r/D,$$

where

$$a_r = \frac{(2r)!r}{(2^r r!)^2} \quad \text{and} \quad b_r = \frac{r}{\pi} - a_r^2.$$

If $\bar{X}_{(r)}$ is the mean of a sample of s values of $X_{(r)}$, then a moment estimate of D is

$$D^* = a_r^2/\bar{X}_{(r)}^2.$$

Although the distribution of $X_{(r)}$ is fairly skew for small values of r (cf. Thompson [1956]), $\bar{X}_{(r)}$ will be approximately normally distributed for large s by the Central Limit theorem. We can then obtain an approximate 95 per cent confidence interval for $\sqrt{D}$ as follows:

$$\begin{aligned} 0{\cdot}95 &\approx \Pr\left[-1{\cdot}96 < \frac{\bar{X}_{(r)} - E[\bar{X}_{(r)}]}{\sigma_r/\sqrt{s}} < 1{\cdot}96\right] \\ &= \Pr\left[\frac{-1{\cdot}96\sqrt{(b_r/s)} + a_r}{\bar{X}_{(r)}} < \sqrt{D} < \frac{1{\cdot}96\sqrt{(b_r/s)} + a_r}{\bar{X}_{(r)}}\right]. \end{aligned}$$

When $r = 1$ we have $a_1 = \frac{1}{2}$ and

$$D^* = 1/(4\bar{X}_{(1)}^2),$$

an estimate first given by Clark and Evans [1954].

INDEX OF NON-RANDOMNESS. If $C[X_{(r)}]$ is the coefficient of variation of $X_{(r)}$, then from the previous paragraph we can define

$$\begin{aligned} I_r &= (C[X_{(r)}])^2 + 1 \qquad (2.34) \\ &= E[X_{(r)}^2]/(E[X_{(r)}])^2 \\ &= \frac{r/(\pi D)}{a_r^2/D} \\ &= \frac{(2^r r!)^4}{(2r!)^2 \pi r}, \end{aligned}$$

which is independent of D: for example, when $r = 1$, $I_1 = 4/\pi = 1{\cdot}27$. Since I_r is readily estimated from sample data, and increases with increasing tendency for aggregation, Eberhardt [1967] suggests that it may be a useful measure or index of "non-randomness" (see Hines and Hines [1979]).

TESTING UNDERLYING ASSUMPTIONS. Several methods are available for testing the validity of the Poisson assumptions. For example, if one of the above methods of estimation is used along with quadrat sampling, then quadrat counts will provide a test of fit to the Poisson distribution. It was noted on p. 25 that the two standard tests of fit, and in fact all tests of randomness based on quadrat counts, are dependent on the size of the quadrat used. Quadrats which are too small or too large with respect to the average size of the "patches" of individuals may fail to detect any non-randomness (Greig-Smith [1964: 28, 56–7]). As a rough guide it is suggested that the quadrat size should be such that the average number of animals per quadrat is not less than unity. In this case it appears that the Poisson Dispersion Test (cf. **1.3.6** (*2*)) is satisfactory for as few as 6 sample quadrats (Kathirgamatamby [1953]).

When quadrat sampling is not used it is simpler to work with the transformed distances Y_i and calculate a standard chi-squared goodness of fit to the probability density (2.28). In particular, when $r = 1$, (2.28) reduces to the negative exponential distribution.

$$g_1(y) = D \exp(-Dy),$$

and the expected frequencies E_j, say, are readily calculated. For example, with the class-interval $[a_{j-1}, a_j)$, $(j = 1, 2, \ldots, k;\ a_0 = 0,\ a_k = \infty)$,

$$E_j = s \int_{a_{j-1}}^{a_j} \hat{D} \exp(-\hat{D}y)dy$$

$$= s\{\exp(-\hat{D}a_{j-1}) - \exp(-\hat{D}a_j)].$$

If f_j is the observed frequency of the y values in the jth class interval, then the goodness-of-fit statistic

$$\sum_{j=1}^{k} (f_j - E_j)^2/E_j$$

is approximately distributed as χ^2_{k-1} when s is large. We note that one degree of freedom is not subtracted for the estimation of D as $\hat{D}$ is calculated from the ungrouped data (Kendall and Stuart [1973: 447]).

Since the times between events for a Poisson time-process follow the negative exponential distribution (cf. p. 4), a number of other test procedures, summarised in Cox and Lewis [1966: 153 ff.], can be used here. For example, if we transform from the Y_i to

$$Z_i = \sum_{r=1}^{i} Y_r \Big/ \sum_{r=1}^{s} Y_r,$$

then $Z_1, Z_2, \ldots, Z_{s-1}$ represent an ordered sample from the uniform distribution on $[0, 1]$, and the usual tests of the goodness of fit of the Z_i to this distribution can be carried out (Seshadri *et al.* [1969]).

A good description of the various test methods is also given in Epstein [1960a, b].

3 *Closest-individual techniques: alternative models*

BINOMIAL MODEL. Consider a population of N animals randomly distributed in a region of area A. Then if a sampling point is chosen at random in the area, there is the possibility that the point may be closer to the boundary than to the nearest individual. However, if sampling points are restricted to an interior area sufficiently far from the boundaries to avoid the above difficulty, then, using the same argument that led to (2.26), we have (Eberhardt [1967])

$$F_1(x) = 1 - \left(1 - \frac{\pi x^2}{A}\right)^N$$

and

$$f_1(x) = \frac{2\pi N x}{A}\left(1 - \frac{\pi x^2}{A}\right)^{N-1}. \tag{2.35}$$

When $N \to \infty$, $A \to \infty$ in such a way that $D = N/A$ is constant, we find that (2.35) tends to (2.26) as expected.

Eberhardt shows that (2.27) now becomes

$$f_r(x) = \frac{2\pi N x\,(N-1)\,!}{A(N-r)\,!\,(r-1)\,!} \cdot \left(\frac{\pi x^2}{A}\right)^{r-1} \left(1 - \frac{\pi x^2}{A}\right)^{N-r} \tag{2.36}$$

or, setting $Y = \pi X_{(r)}^2$,

$$g_r(y) = \frac{N\,!}{A(N-r)\,!\,(r-1)\,!} \cdot \left(\frac{y}{A}\right)^{r-1} \left(1 - \frac{y}{A}\right)^{N-r}, \quad 0 \leqslant y \leqslant A,$$

which is beta-distribution. Once again $\tilde{D}$ of (2.31) is an unbiased estimate of D and, for $r \geqslant 3$,

$$V[\tilde{D}] = \frac{D(D-(r-1)/A)}{s(r-2)}.$$

For the case when a sampling point may be closer to the boundary than to the nearest animal we can use the following method, due to Craig [1953a]. Suppose that sampling is carried out in a quadrat of side a as shown in Fig. 2.1. Let Q be the point in the square with coordinates (u, v) with respect to the origin O and let $w = a - u$, $z = a - v$. Then the basis of

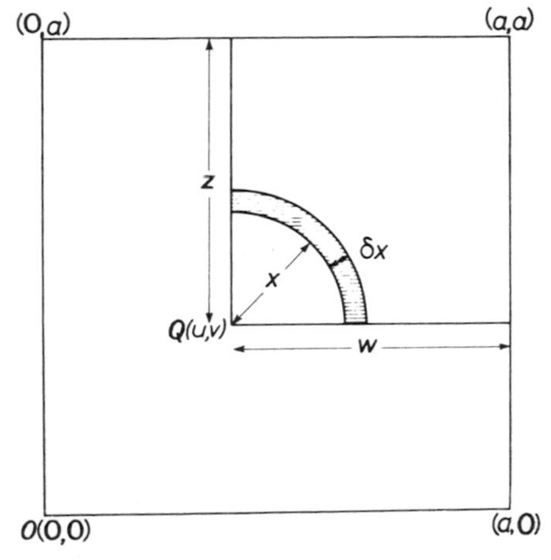

Fig. 2.1 Q is a point chosen at random in a quadrat of side a. The shaded circular strip of width δx contains the individual in the first quadrant with Q as centre which is nearest to Q.

Graig's method is to choose Q at random and, provided $X_{(1)} \leqslant d_0 =$ minimum (w, z), to measure $X_{(1)}$, the distance to the nearest animal in the first quadrant with Q as centre. Now if there are n animals in the square, $X_{(1)}$ is the distance to the nearest animal if one animal lies in the curved shaded area of Fig. 2.1 and $n-1$ animals lie outside the quarter circle with centre Q and radius $X_{(1)}$. Hence

$$\Pr[x < X_{(1)} < x + \delta x \mid w, z] = \binom{n}{1} \frac{\pi x \delta x}{2a^2}\left(1 - \frac{\pi x^2}{4a^2}\right)^{n-1} \quad (x \leqslant d_0)$$

and

$$\Pr[x < X_{(1)} < x + \delta x] = \int_x^a \int_x^a \frac{1}{a^2} \Pr[x < X_{(1)} < x + \delta x \mid w, z]\, dw\, dz$$

$$= \Pr[x < X_{(1)} < x + \delta x \mid w, z]\left(1 - \frac{x}{a}\right)^2 \quad (x \leqslant d_0),$$

since the above conditional probability does not depend functionally on w and z when $x \leqslant d_0$. Therefore the probability density function for $X_{(1)}$, given that Q is chosen at random and $x \leqslant d_0$, is given by

$$f(x) = \frac{n\pi x}{2a^2}\left(1 - \frac{\pi x^2}{4a^2}\right)^{n-1}\left(1 - \frac{x}{a}\right)^2.$$

If s points Q are chosen, k say of these points will have x's less than or equal to the distance from Q to the nearest of the right and upper boundaries of the square. For these k values of $X_{(1)}$ the likelihood function $\prod_i f(x_i)$ can be maximised to give a maximum-likelihood estimate of n, namely

$$\hat{n} = -k/\sum_{i=1}^{k} \log\left(1 - \frac{\pi x_i^2}{4a^2}\right) \tag{2.37}$$

with asymptotic variance (cf. (1.4))

$$V[\hat{n}] = n^2/k.$$

The corresponding estimate of population density is $\hat{n}/a^2$ and, from a random sample of quadrats of side a, an average density estimate can be calculated.

One advantage of Craig's method is that it takes into account the problem of boundary points, so that the square could have a side along or near the boundary of the population area under consideration. This convenience, however, is obtained at the expense of $s - k$ sampling points. Defining $P[n]$ as the probability that a sampling point has an x value satisfying $x \leqslant d_0$, Craig gives the following table of $P[n]$ for different values of n:

n:	10	20	30	50	100	200
$P[n]$:	0·5058	0·6216	0·6804	0·7442	0·8134	0·8652

If a is sufficiently large so that $n > 50$, we see from this table that less than about one quarter of the sampling points will be wasted.

Blackith [1958] discusses Craig's model and points out that if $x_i/a < 0{\cdot}1$ (which will often be the case for animal populations) we can approximate $-\log(1-\epsilon)$ by ϵ in (2.37) and obtain the simpler expression

$$\hat{n} = 4ka^2/(\pi \sum_i x_i^2).$$

However, he does not interpret the x_i correctly (Eberhardt [1967: 213]) which to some extent accounts for his inability to reconcile the models of Craig [1953a] and Clark and Evans [1954].

NEGATIVE BINOMIAL MODEL. Suppose that animals are not randomly distributed but are distributed according to a negative binomial law (cf. p. 25)

$$f(u) = \frac{k(k+1)\dots(k+u-1)}{u\,!} P^u(1+P)^{-u-k}, \tag{2.38}$$

where $f(u)$ is the probability of finding $U = u$ animals in a randomly chosen plot of area πx^2. Then, setting

$$D\pi x^2 = E[U]\ (= kP)$$

we have

$$P = D\pi x^2/k.$$

If we assume that k, the "heterogeneity parameter," is independent of x, then $X_{(1)}$, the distance from a randomly chosen point to the nearest animal, has distribution function:

$$\begin{aligned} F_1(x) &= \Pr[X_{(1)} \leqslant x] \\ &= 1 - \Pr[\text{no animals in } \pi x^2] \\ &= 1 - f(0) \\ &= 1 - (1+P)^{-k} \\ &= 1 - \left(1 + \frac{D\pi x^2}{k}\right)^{-k} \end{aligned}$$

and

$$f_1(x) = F_1'(x) = 2D\pi x \left(1 + \frac{D\pi x^2}{k}\right)^{-k-1}. \tag{2.39}$$

In deriving (2.39) Eberhardt [1967: 209] points out that k will tend to vary with x, so that the above derivation may not be realistic. However, (2.39) may serve as a more useful distribution than either (2.35) or (2.26) for data displaying heterogeneity.

Using the same argument which led to (2.27), Eberhardt shows that $X_{(r)}$ has probability density

$$f_r(x) = \frac{2D\pi x\, \Gamma(r+k)(D\pi x^2)^{r-1} k^k}{\Gamma(k)(r-1)!\ (k+D\pi x^2)^{r+k}}$$

which tends to (2.27) when $k \to \infty$. Setting $Y = \pi X_{(r)}^2$, we have

$$g_r(y) = \frac{\Gamma(r+k)(Dy)^{r-1} k^k D}{\Gamma(k)(r-1)!\ (k+Dy)^{r+k}}$$

which is related to the beta distribution. Using $g_r(y)$, it is readily shown that $\tilde{D}$ of (2.31) is still an unbiased estimate of D and, for $r \geqslant 3$,

$$V[\tilde{D}] = \frac{D^2(1 + (r-1)/k)}{s(r-2)},$$

which tends to (2.32) as $k \to \infty$.

COMPARISON OF MODELS. When $r > 1$, $\tilde{D}$ of (2.31) is an unbiased estimate of D for the Poisson, binomial and negative binomial models, so that it can be used for most animal populations along with the robust estimate of variance $v[\tilde{D}]$ (cf. (2.33)). When $r = 1$, some indications as to which model is appropriate may be given by Eberhardt's index I_1 of (2.34). For example, Eberhardt gives the following values:

Poisson:	I_1	=	1·27				
Negative binomial:	k	=	10	2	1·50	1·10	1·01
	I_1	=	1·31	1·62	2·00	5·19	42

and he evaluates the sample value of I_1 for published data from Clark and Evans [1954], Cottam and Curtis [1956], and Blackith [1958].

4 *Nearest-neighbour techniques*

If instead of choosing a sampling point at random, an animal is chosen at random and the distance $X'_{(r)}$ to its nearest neighbour is measured, then provided the animals are randomly distributed, $X'_{(r)}$ will have the same distribution as $X_{(r)}$, the distance from a sample point to the nearest animal. A number of methods for selecting an animal have been suggested, such as (a) numbering all the animals in the population and using a table of random numbers, (b) choosing a quadrat at random and then applying method (a) to the quadrat population, and (c) choosing a point at random and using the nearest animal. Method (a) is of course out of the question except in the situation when the population size is easily determined and one is primarily interested in spatial distribution (e.g. termite mounds). Even then, however, if the population is large, considerable labour is involved in identifying all the members of the population. Method (b), although not giving a proper random sample of nearest-neighbour distances, could perhaps be used along with quadrat sampling methods, provided the locating of all the animals in the quadrat did not disturb their pattern of distribution. Method (c) seems to be the most popular, no doubt because of its simplicity, but it also does not give a random sample (Cottam and Curtis [1956]). This method has been considered by Kendall and Moran [1963: 39] who show that if $\bar{X}'_{(r)}$ is the mean of a sample of rth nearest-neighbour measurements chosen by method (c) then $0{\cdot}8396\, \bar{X}'_{(r)}$ is an unbiased estimate of $1/\sqrt{D}$. A variation of the nearest-neighbour method, called the "random pairs" method, has been proposed (Cottam and Curtis [1949, 1955, 1956]), but this also uses method (c) for sampling. A helpful discussion on this sampling problem is given in Pielou [1969: 117].

In conclusion we see that on account of sampling difficulties and the present lack of theory for methods (b) and (c), the closest-individual techniques of the previous section are preferred to the above nearest-neighbour techniques for estimating population density. The same comments apply to a number of tests of randomness based on comparing closest-individual and nearest-neighbour measurements. These and other tests of randomness are discussed in detail in **12.5.1**.

2.1.5 Simultaneous estimation of density and home range

MacLulich [1951] has given a trapping method for the simultaneous estimation of population density and home range. It is based on the observation that, given a sufficient number of traps and a reasonable trapping time (e.g. four trap-nights, though a longer period may be necessary: Sanderson [1950]), a line or quadrat of traps will catch most of the animals whose ranges of movement overlap the line or area enclosed by the traps.

Suppose that R is the mean diameter of the home range or range of movement of the animals during the trapping period. Then the effective trapping area of a rectangular quadrat of length a and width b, obtained by adding a strip of width $\frac{1}{2}R$ to the quadrat boundary (Dice [1938]) as in Fig. 2.2, is given by

$$(a+R)(b+R) - (1-\tfrac{1}{4}\pi)R^2.$$

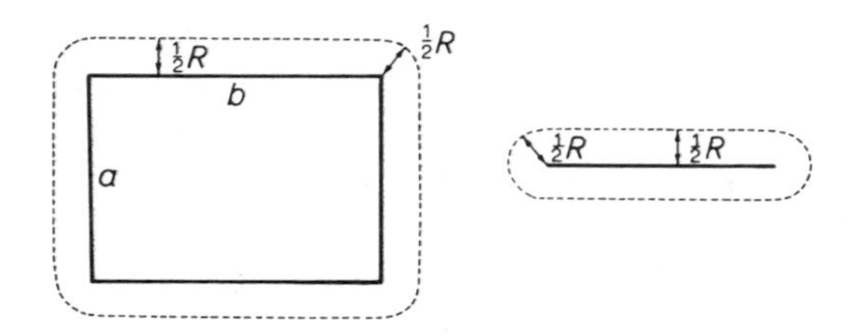

Fig 2.2 Effective trapping areas for a quadrat and a line of traps.

Assuming that the population density D is constant throughout the population area, and defining n_1, n_2 to be the number of animals resident in the effective trapping areas of two such quadrats of different areas, we have for $i = 1, 2$ the deterministic equation

$$n_i = D[(a_i+R)(b_i+R) - (1-\tfrac{1}{4}\pi)R^2] = DA_i, \quad \text{say.}$$

Therefore, setting $n_1/n_2 = A_1/A_2$ leads to the quadratic in R

$$\frac{\pi}{4}\left(1-\frac{n_1}{n_2}\right)R^2 + \left[(a_1+b_1) - \frac{n_1}{n_2}(a_2+b_2)\right]R + \left(a_1b_1 - \frac{n_1}{n_2}a_2b_2\right) = 0,$$

which can be solved for the positive root. Substituting this value of R in A_i, the population density is then given by

$$D = n_1/A_1 = n_2/A_2.$$

If live-traps and tagging are used, then the n_i can be estimated by the methods of Chapters 3 and 4 or by trapping until no new animals are caught. Alternatively, if breakback traps are used, n_i can be estimated by the removal methods of Chapter 7 or by trapping out the quadrat areas completely. Obviously the two quadrats should be of reasonably different areas so as to reduce the probability of the larger quadrat giving the smaller value of n_i through variations in population density.

Instead of using two quadrats the experimenter can use one quadrat and one straight line of traps of length L, say, for which the effective trapping area is (Fig. 2.2) $RL + \frac{1}{4}\pi R^2$. If the number of individuals in this area is n_1, then using such a line instead of the first quadrat above leads to the following quadratic in R:

$$\frac{\pi}{4}\left(1-\frac{n_1}{n_2}\right) R^2 + \left(L - \frac{n_1}{n_2}(a_2+b_2)\right) R - \frac{n_1}{n_2} a_2 b_2 = 0 .$$

Again, R is given by the positive root, and $D = n_2/A_2$. For three specified line–quadrat combinations MacLulich [1951] gives a graph for reading off the solution of the above quadratic for a given value of n_1/n_2. He also gives a number of practical recommendations for the spacing of the traps, and the reader is referred to his article for further details.

In conclusion we note that the two main assumptions underlying the above mathematical models are (i) the population has constant density, and (ii) the range of movement is not too variable, so that R is approximately constant over the population area. Obviously the best check on these assumptions is by replication, as any marked departures from (i) and (ii) will lead to wide fluctuations in the estimates of D and R. From replicate quadrat–quadrat or quadrat–line pairs, average estimates of D and R can be calculated along with the usual estimates of variance (cf. (1.9)). A useful generalisation of the above method is described in **12.1.1** *(4)*.

2.2 RELATIVE DENSITY(*)

2.2.1 Direct methods

Often it is easier, and sometimes more appropriate, to measure population density in units other than area. For example, in the study of insect populations a useful concept is that of population intensity (Southwood [1966]) which is the number of animals per unit of habitat, e.g. per leaf, per tree, per host. Other units which are commonly employed are distance (e.g. animals seen per mile while driving or walking through the area on a prescribed route), time (e.g. animals per hour crossing a given path), and trapping effort (e.g. mice per trap-night, insects per sweep, pheasants killed per hunter-hour); some examples are given in Dice [1952: 39–42]. The roadside census method, in which counting is done from a car usually driven at constant speed, has the advantage that large areas can be quickly covered, and in the U.S.A. it is one of the few methods applicable on a state-wide basis. It has

(*) See also **12.2** and **12.3.8**.

been applied for example to rabbits (Newman [1959], Wight [1959], Kline [1965]), small birds (Howell [1951]), woodcocks (Kozicky *et al.* [1954]), mourning doves (Foote *et al.* [1958]), pheasants (Fisher *et al.* [1947], Kozicky *et al.* [1952], Hartley *et al.* [1955]), blackbirds (Hewitt [1967]), and rhesus monkeys in India (Southwick and Siddiqi [1968]). Also some states of U.S.A. secure the cooperation of rural mail carriers to obtain relative estimates of their principal game species.

These relative estimates of population density are particularly useful in detecting changes in population density with time or in comparing populations in different areas. It should, however, be stressed that if any comparisons are to be made then the censusing should be carried out under as nearly identical conditions as possible. For example, an increase in roadside counts may be due to a genuine increase in population size or to an increase in activity of the animals. As so many factors such as time of day, weather, food supply, and vegetation cover can affect activity (Davis [1963]), a blind application of one of the above methods without a careful study of daily and seasonal activity of the species concerned could give a completely false picture. Newman [1959], for example, using regression analysis, examined the factors which could possibly influence a winter roadside count of cottontails. In comparing counts for different times of the year and different localities, regression and analysis-of-variance methods have proved useful (e.g. Hartley *et al.* [1955], Schultz and Byrd [1957], Schultz and Muncy [1957], Schultz and Brooks [1958], Sen [1970]): see Overton and Davis [1969] for a helpful discussion.

If absolute estimates of population density are also available from time to time from other methods, then the conversion factors thus provided can be used to convert the relative estimates to absolute estimates. Howell [1951], for example, calculated a measure of conspicuousness for various bird species by determining the ratio of those birds seen on the roadside count to the absolute number known to be in the area. These ratios could then be used to convert roadside counts to absolute densities. However, such conversion factors should be used with caution (Dice [1952: 42]) as they will vary with time and place, and even if applicable will give a density estimate with a large coefficient of variation. (See also Gates and Smith [1972].).

An ingenious method for converting roadside counts of territorial birds into absolute density estimates has been suggested by Hewitt [1967]. A census route is driven at approximately 15 miles per hour, and on the first trip the investigator tallies the birds seen and describes their locations in relation to roadside landmarks on a continuously operating tape recorder. The route is then rerun immediately, at similar speed and with a continuous playback of the recording, so the investigator can tally and distinguish "marked" birds seen on the first trip and new "unmarked" birds. The Petersen or Schnabel methods of Chapters 3 and 4 can then be used to estimate the population size. When suitable landmarks for locating the birds are absent, a sensitive resettable odometer can be used for reading off bird locations to the nearest 0·01 mile,say (Harke and Stickley [1968]).

2.2.2 Indirect methods*

Sometimes it is not easy to observe an animal, and its presence has to be inferred by some sign such as dens, burrows, nests, houses, tracks, faeces, dead individuals, songs, calls, shed antlers, etc. (Scattergood [1954: 279], Davis [1963]). Such measures of animal abundance, usually called population indices, are generally more unreliable than the direct methods described above. However, in studying population trends the more methods that can be used and compared the better. For secretive animals an index is sometimes all that is available.

Before a particular index is used it should be studied carefully to see how it is affected by changing conditions: here regression and analysis-of-variance techniques can often be used (e.g. Kozicky [1952], Carney and Petrides [1957], Cohen *et al.* [1960], Smith [1964], Progulske and Duerre [1964], Gates [1966]). Generally it is not possible to examine the whole population area for animal signs, so that sampling methods such as those described in **2.1.2** must be used. For example, Foote *et al.* [1958] used stratified random sampling in choosing routes for collecting data on the call counts of mourning doves.

A useful index for some populations is the number of dwellings per unit area (e.g. Leedy [1949: pheasants], Reid *et al.* [1966: mountain pocket gophers]). However, unless it is possible to distinguish between used and unused dwellings, the number of dwellings will not necessarily reflect the population size.

Large animals can often be detected by their tracks, and in snow these tracks may be distinguishable from the air. Small mammals can be detected by their tracks on smoked kymograph paper (Justice [1961], Sheppe [1965], Marten [1972a]), or by runways that they build up (Carroll and Getz [1976]).

Faecal pellets have been used as a basis for a population index in big game (Neff [1968]), pheasants (McClure [1945]), mice (Emlen *et al.* [1957]), ruffed grouse (Dorney [1958]), rabbits (Taylor and Williams [1956]), and snowshoe hares (Adams [1959]). Indices based on the call counts of birds are widely used, and the following references cover some of the problems involved in using such indices: woodcocks (Kozicky *et al.* [1954]); mourning doves (Lowe [1956], Foote *et al.* [1958], Cohen *et al.* [1960]); pheasants (Davis [1963: 101], Nelson *et al.* [1962], Gates [1966], Martinson and Grondahl [1966]); and ruffed grouse (Petraborg *et al.* [1953], Dorney *et al.* [1958], Gullion [1966]). In particular, Kozicky *et al.* [1954] use an analysis-of-variance technique for analysing the data with respect to years and routes, and they give a method for estimating the number of routes required to detect a given percentage change in the count index.

Sometimes an index can be converted to a measure of absolute density

---

*For a useful review see Overton and Davis [1969: 427–32].

if a reliable conversion factor of signs per animal is available, e.g. calls per bird (Gates and Smith [1972]), pellet groups per deer (Rogers *et al.* [1958]). A useful technique given by Davis [1963] for estimating absolute density is to calculate the index before and after a known number of animals are removed. For example, suppose that a population of size N occupying a given area yields an index I_1 for the area (e.g. I_1 is the total number of calls or pellet groups). If the index is I_2 after the removal of n animals from the area, and the number of signs per animal remains constant, we have the relation

$$\frac{I_1}{N} = \frac{I_2}{N-n} \left(= \frac{I_1 - I_2}{n}\right)$$

or

$$N = nI_1/(I_1 - I_2).$$

This method is further exploited in Chapter 9 (cf. (9.22)).

FREQUENCY INDEX. One other index worthy of consideration is the so-called "frequency index" (Dice [1952], Scattergood [1954: 279], Davis [1963]), which is the proportion, $\hat{p}$ say, of the sample units which contain at least one or more animals (or signs). This index is particularly useful when it is difficult to count the number of individuals in a unit (e.g. fleas per rat) but easy to determine the presence or absence of individuals. We note that $\hat{p}$ will depend on the size of the sampling unit, and the same size unit must be used if comparisons are to be made. If quadrat sampling is used and the quadrat is too big, $\hat{p}$ will be unity, and obviously no comparison can be made between two populations that are present in 100 per cent of the quadrats, even if the densities are widely different. To overcome this problem, quadrats of several sizes should be used and a practical way of doing this (Davis [1963]) is to divide a large quadrat into smaller-size quarters for recording data. The data can then be analysed at any time in the future by quarters, halves or wholes. If several species are to be compared it is recommended that the quadrat size should be such that $\hat{p}$ is about 0·8 for the more important species (Fisher [1954: 62]).

Suppose that the population area consists of S quadrats with a proportion p $(= 1-q)$ occupied by animals. If n quadrats from a sample of s quadrats are found to contain animals, then from Cochran [1963], n has a hypergeometric distribution, $\hat{p} = n/s$ is an unbiased estimate of p, and

$$V[\hat{p}] = \frac{pq\,(S-s)}{s\,(S-1)}$$

with unbiased estimate

$$v[\hat{p}] = \frac{\hat{p}\hat{q}}{s-1}\left(1-\frac{s}{S}\right).$$

If the sampling fraction s/S is less than 0·1 it can be ignored and a knowledge of S is then not required for $v[\hat{p}]$. The question of constructing a confidence interval for p is discussed in Cochran [1963: Chapter 3]. By redefining p, the above theory can be extended to the case of counting the number of plots with t or less animals (cf. Gerrard and Chiang [1970]).

When the animals are randomly distributed, the frequency index can be converted to an estimate of the absolute population density D. In this case the number of animals in a quadrat of area a follows a Poisson distribution with parameter Da (p. 24), and D can be estimated from

$$\hat{q} = \exp(-\hat{D}a). \tag{2.40}$$

From the delta method (**1.3.3**) we have

$$V[\hat{D}] \approx V[\hat{p}]/(qa)^2.$$

We note that if the population is not randomly distributed and the animals tend to group together more than as predicted by the Poisson distribution (which will usually be the case), then the probability of finding a given quadrat empty will be greater than $\exp(-Da)$ and $\hat{D}$ will underestimate the true density. When a non-random distribution is known or suspected, Dice [1952] suggests that the best procedure is to reduce the quadrat size or sampling unit until only rarely will more than one individual be found in any quadrat. In this case $\hat{D}a$ will be small, so that, expanding the exponential in (2.40),

$$\hat{p} = 1 - \hat{q} \approx \hat{D}a, \quad \text{or} \quad \hat{D} = n/(sa).$$

However, this is simply the same as saying that n is a good approximation to the total count on a region of area sa when the probability that a quadrat contains more than one individual is small. For a further discussion on the question of non-random distribution see Gerrard and Chiang [1970].

The frequency indexes of two or more populations can be compared by means of a contingency table, and details are given, for example, in Greig-Smith [1964: 37–41]. It should, however, be stressed that a difference in frequency indexes does not necessarily imply a difference in population densities. For example, two populations with the same average density will give different p values if the animals in one population tend to be clustered together more than in the other. Obviously the greater the clustering the greater the number of empty quadrats and the smaller the value of p.

Finally it is noted that if the presence or absence of several species is recorded for each quadrat, one can test for association between species (Greig-Smith [1964: Chapter 4], Pielou [1969: Chapter 13]).

Example 2.2 Grey fox (*Urocyon cinereoargenteus*): Wood [1959]

Two basic methods for obtaining information on the relative density of grey fox populations were considered, namely trapping with standardised trap-lines and track-counts. The trap-lines consisted of pairs of traps or "stations" spaced at equal intervals along the line. In looking for a suitable method of standardising the trap-line sampling, such questions as how to trap, where to trap, how long to trap, and how to space the stations were considered. For example, it was found that the highest catches occurred when the trap-lines were set along primitive roads, and a spacing of 0·2 mile between stations yielded the highest proportion of "positive" stations, i.e. stations which recorded at least one capture. A trapping period of 7 days was adopted as it was found that lines run for 8, 9 or 10 days caught over 90 per cent of their total catch by the 7th day, while lines run for 11–15 days caught over 80 per cent by the 7th day.

Two variations of the track-count method were used. The first procedure, called the scent-post method, was conducted in the same manner as the above trapping method except that no traps were set. At each trap site there was a raked area of about 3 feet in diameter with a clump of grass containing a scent lure in the centre or near one edge. Animals visiting the lure had to cross part of the raked area, and their visit was recorded by the track left in the soft soil. The second procedure, called the random track-count method, required no field preparation. Undisturbed sandy, dusty or muddy areas that would record the track of a passing animal were randomly selected and checked for the presence or absence of fox signs.

Several indices of relative density were considered for the trap-line data. However, it was found that the frequency index $\hat{p}$, the observed proportion of positive stations, gave the simplest means for comparing data from different areas. This index could also be applied to data from either of the track-count methods.

In Table 2.4 the observed numbers of stations catching 0, 1, 2, 3 or 4 foxes in a period of 7 days are compared with the expected number of stations

TABLE 2.4

Comparison of the number of observed and of expected stations that captured 0, 1, 2, 3 or 4 grey foxes: from Wood [1959: Table 5].

No. of foxes	Observed stations (O)	Expected stations (E)	$\frac{(O-E)^2}{E}$
0	3771	3702·6	1·264
1	524	647·0	23·383
2	97	55·2	31·652
3	15	3·2	
	17	3·4	54·400
4	2	0·2	
Total	4409	4408·2	110·699

calculated on the assumption that catches follow a Poisson distribution. The chi-squared goodness-of-fit statistic (cf. **1.3.6**(*2*)) yields the high value of 110·699 (2 degrees of freedom), thus indicating that the distribution of foxes was not random and that some form of social behaviour was affecting the numbers caught (Davis [1963: 98]). However, the observed number of stations catching zero foxes does not differ significantly from the expected number. We also find that the average catch of

$$[1(524) + 2(97) + 3(15) + 4(2)]/4409 = 0{\cdot}175$$

foxes per station is close to $\log(1/\hat{q}) = \log(4409/3771) = 0{\cdot}16$

(cf. equation (2.40) with a equivalent to one station). This suggests that $\hat{q}$ (and hence $\hat{p}$) is a suitable index for comparing population densities. For example, in Table 2.5 we have a 2-by-2 contingency table for comparing the

TABLE 2.5

Comparison of the number of positive stations (i.e. stations catching at least one fox) for summer and winter seasons.

	Summer	Winter	Total
Positive stations	16	14	30
Zero-catch stations	60	27	87
Total	76	41	117
$\hat{p}$	0·210	0·341	$\chi_1^2 = 1{\cdot}76$

frequency indexes for two seasons. The value of chi-squared is not significant, so that there is no evidence of a seasonal difference.

In conclusion it is noted that the two track-count methods were not developed early enough for extensive use in the above population study. However, they showed promise as a means of obtaining density indexes, and the random track-count could obviously be used in areas lacking a sufficient number of primitive roads.

BOUNDED COUNTS METHOD. Population units (whether animals or signs) are sometimes not easy to count because of mobility or lack of distinguishability, and a person counting units on an area or sample plot may not be sure that he has counted all the inits. If it is theoretically possible to count all the animals on a single occasion, no units are counted twice and repeated counts are possible, Regier and Robson [1967] suggest the following "bounded counts" method based on the theory of Robson and Whitlock [1964].

Let N be the true number of units and let N_m, N_{m-1} be the largest and second largest counts obtained, respectively. Then N can be estimated by

$$\hat{N} = N_m + (N_m - N_{m-1}) = 2N_m - N_{m-1}$$

and an approximate $100(1-\alpha)$ per cent confidence interval for N is $N_m < N < [N_m - (1-\alpha)N_{m-1}]/\alpha$. Regier and Robson give several possible applications of this method to fresh-water fish populations. For example, it can be used in stream censuses by a number of divers obtaining independent counts under conditions similar to those described by Northcote and Wilke [1963]. It could also be applied in counting migrating fish-runs from a number of vantage points by equally perceptive enumerators or mechanical devices, and in small ponds through which large seines may be drawn at least twice during an interval when the population is closed.* If s independent counts are made, then the bias of $\hat{N}$ is of order $1/s^2$. For cases when more than two counts are made, Robson and Whitlock [1964] derive further corrections to reduce the bias (c.f. Cook [1979], Additional References).

(*) Further examples are given by Overton and Davis [1969: 426] and Francis [1973].

CHAPTER 3

CLOSED POPULATION: SINGLE MARK RELEASE

3.1 ESTIMATION

3.1.1 Hypergeometric model

A simple method, which we shall call the "Petersen method", for estimating N, the number of animals in a closed population, is the following. A sample of n_1 animals is taken from the population, the animals are marked or tagged for future identification and then returned to the population. After allowing time for marked and unmarked to mix, a second sample of n_2 animals is then taken and it is found that m_2 are marked. Assuming that the proportion of marked in the second sample is a reasonable estimate of the unknown population proportion, we can equate the two and obtain an estimate of N. Thus:

$$\frac{m_2}{n_2} = \frac{n_1}{\hat{N}}, \quad \text{or} \quad \hat{N} = \frac{n_1 n_2}{m_2},$$

the so-called "Petersen estimate" (or "Lincoln Index": cf. Le Cren [1965]). As this estimate is widely used in ecological investigations we shall now discuss the above method in some detail.

The first step is to decide what assumptions must hold if $\hat{N}$ is to be a suitable estimate of N. These are usually listed as follows:

(a) The population is closed (cf. **1.2.2**), so that N is constant.
(b) All animals have the same probability of being caught in the first sample.
(c) Marking does not affect the catchability of an animal.
(d) The second sample is a simple random sample, i.e. each of the $\binom{N}{n_2}$ possible samples has an equal chance of being chosen.
(e) Animals do not lose their marks in the time between the two samples.
(f) All marks are reported on recovery in the second sample.

We note that these assumptions are not mutually exclusive. For example, (d) will depend on the validity of (b) and (c), as any variation in the catchability of the animals, whether natural or induced by the handling and marking, will lead to a non-random second sample. This point is discussed further in **3.2.2**.

When assumptions (a), (d), (e) and (f) are satisfied, the conditional distribution of m_2, given n_1 and n_2, is the hypergeometric distribution

$$f(m_2 \mid n_1, n_2) = \binom{n_1}{m_2}\binom{N-n_1}{n_2-m_2} \Big/ \binom{N}{n_2}, \tag{3.1}$$

where $m_2 = 0, 1, 2, \ldots,$ minimum (n_1, n_2). The properties of $\hat{N}$ with respect to this distribution have been discussed fully by Chapman [1951]. He shows that although $\hat{N}$ is a best asymptotically normal estimate of N as $N \to \infty$, it is biased, and the bias can be large for small samples. However, when $n_1 + n_2 \geqslant N$, his modified estimate

$$N^* = \frac{(n_1+1)(n_2+1)}{(m_2+1)} - 1$$

is exactly unbiased, while if $n_1 + n_2 < N$ we have, to a reasonable degree of approximation (Robson and Regier [1964]),

$$E[N^* \mid n_1, n_2] = N - Nb,$$

where $b = \exp\{-(n_1+1)(n_2+1)/N\}$. Defining

$$\mu = n_1 n_2 / N = E[m_2 \mid n_1, n_2],$$

Robson and Regier recommend that in designing a Petersen type experiment it is essential that $\mu > 4$, so that b is small (in this case less than 0·02). They also state that if $m_2 \geqslant 7$ in a given experiment, then we are 95 per cent confident that $\mu > 4$. This means that for 7 or more recaptures we can be 95 per cent confident that the bias of N^* is negligible.

Chapman [1951] shows that N^* not only has a smaller expected mean square error than $\hat{N}$ for values encountered in practice, but it also appears close to being a minimum variance unbiased estimate over the range of parameter values for which it is almost unbiased. Using what is essentially a Poisson approximation to (3.1), he shows that the variance of N^* is approximately given by

$$V[N^* \mid n_1, n_2] = N^2(\mu^{-1} + 2\mu^{-2} + 6\mu^{-3}). \tag{3.2}$$

If the expected number of recaptures μ is small, this variance is large, and Chapman [1951: 148] concludes that "sample census programmes in which the expected number of tagged members in the sample is much smaller than 10 may fail to give even the order of magnitude of the population correctly." For example, when $\mu = 10$, (3.2) yields a standard deviation of $0{\cdot}36N$.

An estimate, V^* say, of the variance of N^* is obtained by simply replacing N by N^* in (3.2). However, an approximately unbiased estimate has been given by Seber [1970a] and Wittes [1972], namely

$$v^* = \frac{(n_1+1)(n_2+1)(n_1-m_2)(n_2-m_2)}{(m_2+1)^2(m_2+2)},$$

which has a positive proportional bias of order $\mu^2 \exp(-\mu)$. It can be shown that v^* is exactly unbiased when $n_1 + n_2 \geqslant N$. (In fact, when $n_1 + n_2 \geqslant N$, N^* and v^* are both *unique* unbiased estimates because of the completeness of m_2.)

The coefficient of variation of N^* is approximately given by

$$C[N^*] = 1/\sqrt{\mu},$$

and if a rough estimate of N is available before the experiment, n_1 and n_2 can be chosen beforehand to give a desired value of C; this question of experimental design is discussed more fully in **3.1.5.** We note that an estimate of C is obtained by replacing μ by m_2, giving $C = 1/\sqrt{m_2}$. This means that the "accuracy" of N^* is almost solely dependent on the number of recaptures m_2. For further comments on the Petersen method see **12.7.1.**

3.1.2 Bailey's binomial model

Using a binomial approximation to the hypergeometric distribution (3.1), we have (Bailey [1951, 1952]):

$$f(m_2 \mid n_1, n_2) \approx \binom{n_2}{m_2}\left(\frac{n_1}{N}\right)^{m_2}\left(1 - \frac{n_1}{N}\right)^{n_2 - m_2}, \tag{3.3}$$

and the maximum-likelihood estimate of N is the Petersen estimate $\hat{N}$. Bailey shows that $\hat{N}$ is biased with respect to this binomial distribution and suggests the modification

$$\hat{N}_1 = n_1(n_2 + 1)/(m_2 + 1)$$

which has a proportional bias of order $\exp(-\mu)$. The variance of this estimate may be estimated, with a positive proportional bias of order $\mu^2 \exp(-\mu)$, by

$$v_1 = \frac{n_1^2(n_2 + 1)(n_2 - m_2)}{(m_2 + 1)^2(m_2 + 2)}.$$

If the sampling fraction n_2/N is sufficiently small for one to be able to ignore the complications of sampling without replacement then $\hat{N}_1$ can be used instead of N^*, though in practice there will often be little difference in the two estimates. Obviously (3.3) will hold exactly for the less common situation when random sampling *with* replacement is used (cf. Example 3.9, p. 110, where the animals are merely observed and not actually captured: this point is discussed further in **14.1.2** (*3*)). However, there is one other practical situation where (3.3) may be more appropriate than (3.1). We saw above that $\hat{N}$ is an intuitively reasonable estimate when the sample proportion of marked in the second sample faithfully reflects the population proportion of marked. This means that $\hat{N}$ can still be used even when assumption (d) is false and the second sample is a systematic rather than a random sample, provided that (i) there is uniform mixing of marked and unmarked so that the proportion (n_1/N) of marked throughout the population is constant, and (ii) given that a certain location in the population area is sampled, all animals at that location, whether marked or unmarked, have the same probability of being caught. When (i) and (ii) are satisfied, the probability that an animal is found to be marked, given that it is caught in the second sample, is n_1/N, and the binomial model (3.3) applies. The question of systematic sampling is discussed further on p. 82.

An alternative model to the above hypergeometric and binomial models, using a distribution derived by Skellam [1948] on the assumption that the probability of capture is not constant but follows a beta distribution, has been given by Eberhardt [1969a] (cf. (4.34) in **4.1.6** (*4*)).

3.1.3 Random sample size

In practice it is not always possible to fix n_2 in advance as the sample size may depend on the effort or time available for sampling. However, when n_2 is regarded as a random variable rather than a fixed parameter, N^* is still approximately unbiased, since

$$\begin{aligned} E[N^* \mid n_1] &= \underset{n_2}{E}\, E[N^* \mid n_1, n_2] \\ &\approx \underset{n_2}{E}\, [N] \\ &= N, \end{aligned}$$

and using a similar argument it is readily shown that v^* is an approximately unbiased estimate of $V[N^* \mid n_1]$. Also, from **1.3.4** we have

$$\begin{aligned} V[N^* \mid n_1] &= \underset{n_2}{E}\, \{V[N^* \mid n_1, n_2]\} + \underset{n_2}{V}\{E[N^* \mid n_1, n_2]\} \\ &\approx \underset{n_2}{E}\, \{V[N^* \mid n_1, n_2]\} \\ &\approx \{V[N^* \mid n_1, n_2]\}_{n_2 = E[n_2 \mid n_1]} \,. \end{aligned}$$

This means that for large N, the only difference between $V[N^* \mid n_1]$ and $V[N^* \mid n_1, n_2]$ is that in the former, n_2 is replaced by $E[n_2 \mid n_1]$. Since in practice one would estimate $E[n_2 \mid n_1]$ by n_2 in a variance formula, there is therefore little difference between treating n_2 as a fixed parameter or as a random variable as far as estimation is concerned. But it can be argued that once n_2 is known, we are only interested in the distribution of m_2 given n_1 and n_2, and that $f(m_2 \mid n_1, n_2)$ is then the appropriate distribution, irrespective of whether n_2 is fixed or random.

3.1.4 Confidence intervals

We now turn our attention to the problem of finding confidence intervals for N. As $N \to \infty$, N^* is asymptotically normally distributed, so that an approximate 95 per cent confidence interval for N is given by

$$N^* \pm 1{\cdot}96\sqrt{v^*}.$$

However, according to Ricker [1958], $1/N^*$ is more symmetrically distributed and more nearly normal than N^*, so that in general it is better to base confidence intervals on the probability distribution of m_2. This hypergeometric distribution (3.1) has been tabled, and exact confidence limits for $p = n_1/N$ when N is known and n_1 unknown are available (e.g. Chung and De Lury [1950]). Unfortunately no such tables are available for the case when N is unknown and n_1 known, so that approximate methods have to be used. For various values of n_1, n_2 and N the hypergeometric distribution can be satisfactorily approximated by the Poisson, binomial and normal distributions (Chapman [1948], Lieberman and Owen [1961], Molenaar [1973]). But the choice of which approximation to use when N is unknown still needs further investigation, so that the following recommendations should be regarded as a general guide only.

Let $\hat{p} = m_2/n_2$; then when $\hat{p} < 0{\cdot}1$ and $m_2/n_1 < 0{\cdot}1$, the Poisson approximation is recommended using m_2 as the "entering" variable in appropriate tables. For example, a confidence interval for μ $(= n_1 n_2/N)$ can be read off from tables such as Pearson and Hartley [1966: 227, $m_2 \leqslant 50$], Crow and Gardner [1959: $m_2 \leqslant 300$] or from a graph (Adams [1951: $m_2 \leqslant 50$]). For $m_2 \leqslant 50$, however, it is simpler to use a table specially prepared by Chapman [1948], giving the shortest 95 per cent confidence intervals for N/λ where $\lambda = n_1 n_2$ (for example, when $m_2 = 10$, Chapman's interval is 5 per cent shorter than the "equi-tail" interval of Pearson and Hartley). This table is reproduced in **A1**, and we demonstrate its use with the following example. Suppose $n_1 = 1000$, $n_2 = 500$, $m_2 = 20$, then $\hat{p} = 0{\cdot}04$, $m_2/n_1 = 0{\cdot}02$ and the Poisson approximation is appropriate. Using m_2 as the entering variable, a 95 per cent confidence interval for N/λ is (0·030 04, 0·0773), and multiplying these limits by $n_1 n_2$ gives the corresponding interval for N, namely (15 020, 38 650).

When $\hat{p} < 0{\cdot}1$ and $m_2 > 50$ we can use a normal approximation given by Cochran [1963: 87] to obtain a 95 per cent confidence interval for p, namely

$$\hat{p} \pm \{1{\cdot}96[(1-f)\hat{p}(1-\hat{p})/(n_2-1)]^{\frac{1}{2}} + 1/(2n_2)\} \tag{3.4}$$

which can be inverted to give a confidence interval for N. Here $f(= n_2/N)$, the unknown "sampling fraction", can be neglected if its estimate $\hat{f} = m_2/n_1$ is less than 0·1; also $1/(2n_2)$, the correction for continuity, will often be negligible. For example, if $n_1 = 2000$, $n_2 = 1000$, $m_2 = 80$, then $\hat{p} = 0{\cdot}08$, $\hat{f} = 0{\cdot}04$ and the interval for p is

$$0{\cdot}08 \pm 1{\cdot}96[(0{\cdot}08)(0{\cdot}92)/1000]^{\frac{1}{2}},$$

or (0·0632, 0·0968). The corresponding interval for N is (20 700, 31 600). Neglecting f when $f > 0{\cdot}1$ will lead to conservative confidence intervals, i.e. intervals which are overwide. This is not serious, however, as the assumptions given in **3.1.1** are never exactly true and variances have a habit of being larger than predicted by theory!

When $N > 150$, $n_1 > 50$ and $n_2 > 50$, m_2 is approximately normal (Robson and Regier [1964]), and a more accurate method than the one above is to solve the following cubic equation in N:

$$\frac{\left(m_2 - \dfrac{n_1 n_2}{N}\right)^2}{n_2 \cdot \dfrac{n_1}{N}\left(1 - \dfrac{n_1}{N}\right)\left(\dfrac{N - n_2}{N - 1}\right)} = 1{\cdot}96^2.$$

The two largest roots then give an approximate 95 per cent confidence interval for N (Chapman [1948]); a graphical method of solution is discussed in Schaefer [1951].

If $\hat{p} > 0{\cdot}1$ we can use either the binomial approximation (3.3) or the normal approximations mentioned above. A rough guide as to the smallest values of n_2 for which the normal approximation (3.4) is applicable is given by the following table reproduced from Cochran [1963: 57]:

$\hat{p}$ (or $1-\hat{p}$)	0·5	0·4	0·3	0·2	0·1
n_2	30	50	80	200	600

For example, if $\hat{p} = 0{\cdot}3$ (or 0·7), the normal approximation can be used if $n_2 > 80$. When the normal approximation is not applicable, a binomial confidence interval for p can be obtained from the Clopper–Pearson charts in Pearson and Hartley [1966: 228–229] and Adams [1951], or from extensive binomial tables such as those of the Harvard Computation Laboratory [1955]. If, for example, $m_2 = 18$, $n_2 = 60$ then $\hat{p} = 0{\cdot}3$, and the first Clopper–Pearson chart (with $n = 60$) gives (0·190, 0·433) as the 95 per cent confidence interval for p.

3.1.5 Choice of sample sizes

1 Prescribed accuracy

It was mentioned in **3.1.1** that given a rough estimate of N, the product $n_1 n_2$ can be determined to give a prescribed coefficient of variation $C[N^*]$; useful graphs for doing this are given in Davis [1964]. However, although accuracy is often measured in terms of coefficient of variation, a more appropriate definition of the accuracy A of an estimate is given by Robson and Regier [1964] as follows.

Let $(1-\alpha)$ be the probability that the Petersen estimate $\hat{N}$ will not differ from the true population size by more than $100A$ per cent. Then

$$1 - \alpha \leqslant \Pr\left[-A < \frac{\hat{N} - N}{N} < A\right],$$

where α and A are to be chosen by the experimenter. Three standard levels for α and A are suggested:

(i) $1 - \alpha = 0{\cdot}95$, $A = 0{\cdot}50$; recommended for preliminary studies or management surveys where only a rough idea of population size is needed.

(ii) $1 - \alpha = 0{\cdot}95$, $A = 0{\cdot}25$; recommended for more accurate management work.

(iii) $1 - \alpha = 0{\cdot}95$, $A = 0{\cdot}10$; recommended for careful research into population dynamics.

When $\hat{N}$ is multiplied or divided by some other estimate (e.g. multiplied by weight to obtain total biomass or divided by another population estimate to obtain probability of survival) a relatively high accuracy is required of each estimate if the variance of the product or ratio is to be reasonably small. In this situation, level (iii) would then be most appropriate.

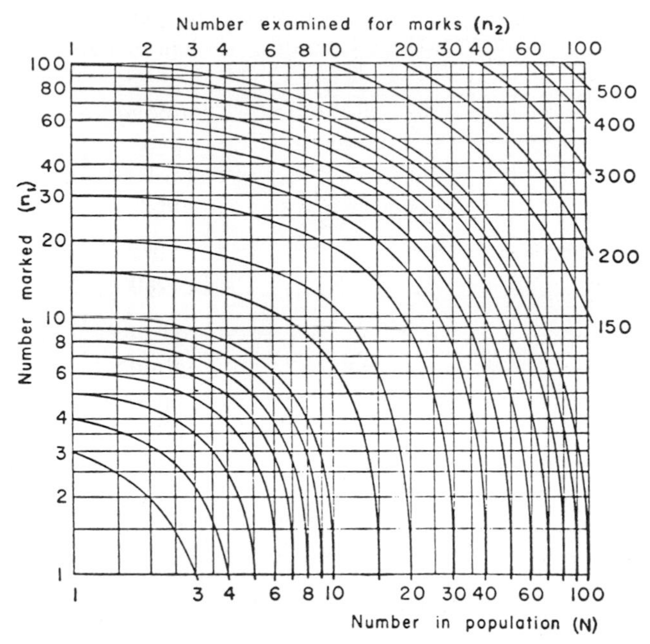

Fig. 3.1 Sample sizes when $1 - \alpha = 0{\cdot}95$, $A = 0{\cdot}5$ and $N \leqslant 500$; recommended for preliminary studies or management surveys. (From Robson and Regier [1964].)

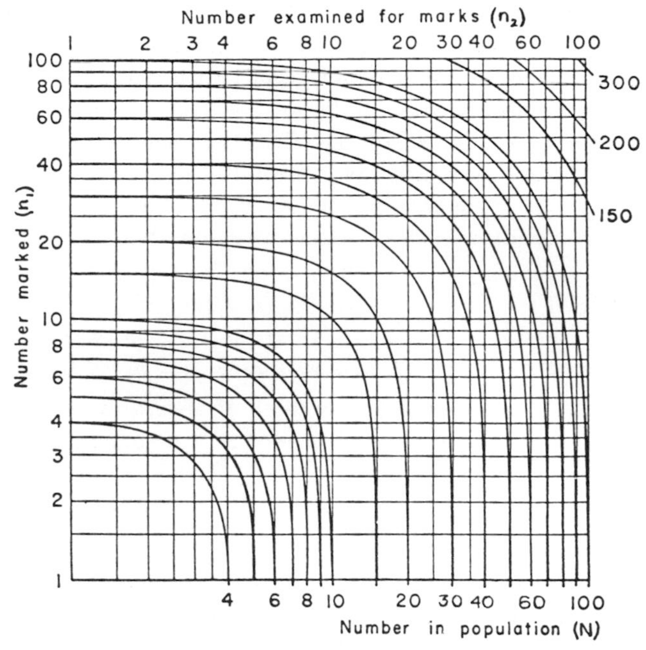

Fig. 3.2 Sample sizes when $1 - \alpha = 0{\cdot}95$, $A = 0{\cdot}25$ and $N \leqslant 300$; recommended for management studies. (From Robson and Regier [1964].)

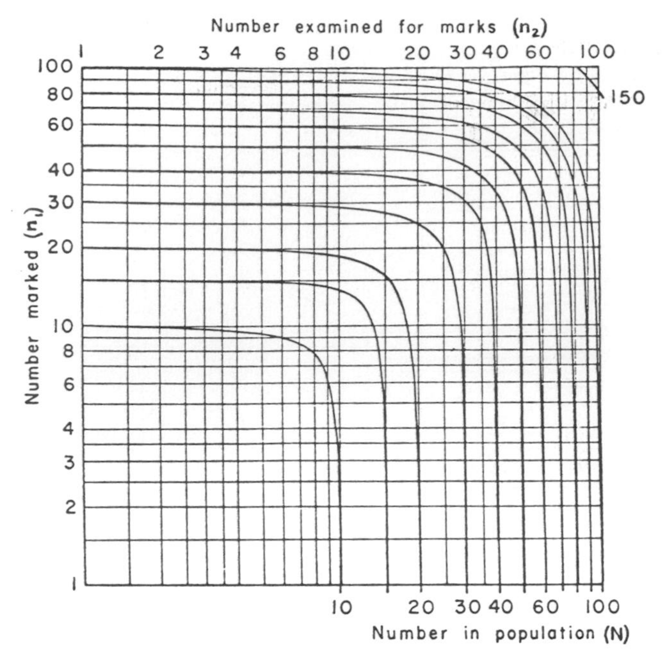

Fig. 3.3 Sample sizes when $1 - \alpha = 0{\cdot}95$, $A = 0{\cdot}10$ and $N \leqslant 150$; recommended for research. (From Robson and Regier [1964].)

Robson and Regier give charts (Fig. 3.1–3.6) for determining n_1 and n_2 from a rough estimate of N for the above three levels of α and A. They point out that for the values of n_1 and n_2 given by these charts the bias mentioned in **3.1.1** is negligible (less than about 1 per cent). The first three charts ($N \leqslant 100$) were calculated using tables of the hypergeometric distribution given by Lieberman and Owen [1961], while the second three charts ($N > 100$) were obtained using a normal approximation to the hypergeometric. In this latter case only part of the range of the combinations of n_1 and n_2 for each N are drawn, as the relationships determining the charts (equations (3.5) and (3.6)) are symmetric for n_1 and n_2. Thus, for a given α, sample sizes $n_1 = 300$, $n_2 = 600$ will yield an estimate $\hat{N}$ of the same accuracy A as sample sizes $n_1 = 600$, $n_2 = 300$. To illustrate the use of the tables suppose that N is estimated to be 10 000 and we require $1 - \alpha = 0{\cdot}95$, $A = 0{\cdot}25$. Then from Fig. 3.5 we obtain possible pairs such as (6000, 50), (4000, 100) and (1000, 600),

If, for $N > 100$, we wish to use values of α and A other than those mentioned above, charts for n_1 and n_2 can be drawn using the normal approximation as follows. Given α and A, we solve

$$1 - \alpha = \phi\left(\frac{A\sqrt{D}}{1 - A}\right) - \phi\left(\frac{-A\sqrt{D}}{1 + A}\right) \tag{3.5}$$

for D, where $\phi(z)$ is the cumulative unit normal distribution, and then, using an estimate of N, plot n_1 and n_2 subject to the constraint

TABLE 3.1

Values of D satisfying equation (3.5) for selected α and A: from Robson and Regier [1964: Table 2].

$1-\alpha$	A	D
0·75	0·50	4·75
0·90	0·50	14·8
0·90	0·25	45·5
0·95	0·50	24·4
0·95	0·25	69·9
0·95	0·10	392
0·99	0·10	695
0·99	0·01	66 300

TABLE 3.2

Combinations of A and $1-\alpha$ satisfying (3.5) for selected D: from Robson and Regier [1964: Table 3].

D	A	$1-\alpha$
24·4	0·88	0·99
24·4	0·50	0·95
24·4	0·25	0·79
24·4	0·10	0·38
24·4	0	0
69·9	0·39	0·99
69·9	0·25	0·95
69·9	0·10	0·60
69·9	0	0
392	0·14	0·99
392	0·10	0·95
392	0·05	0·75
392	0	0

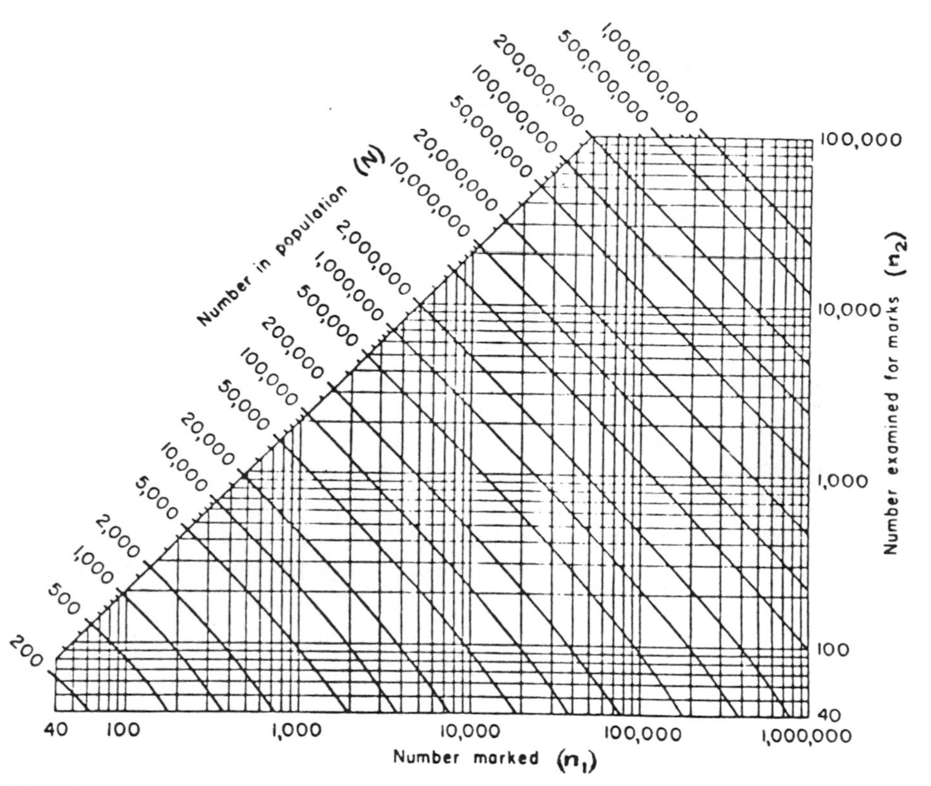

Fig. 3.4 Sample sizes when $1-\alpha = 0{\cdot}95$, $A = 0{\cdot}5$ and $200 \leqslant N \leqslant 10^9$: from Robson and Regier [1964].

$$\frac{n_1 n_2(N-1)}{(N-n_1)(N-n_2)} = D. \tag{3.6}$$

Robson and Regier give two useful tables (Tables 3.1 and 3.2) for handling (3.5); Table 3.1 gives the solution D for various combinations of $1-\alpha$ and A, and Table 3.2 gives various combinations of $1-\alpha$ and A satisfying (3.5)

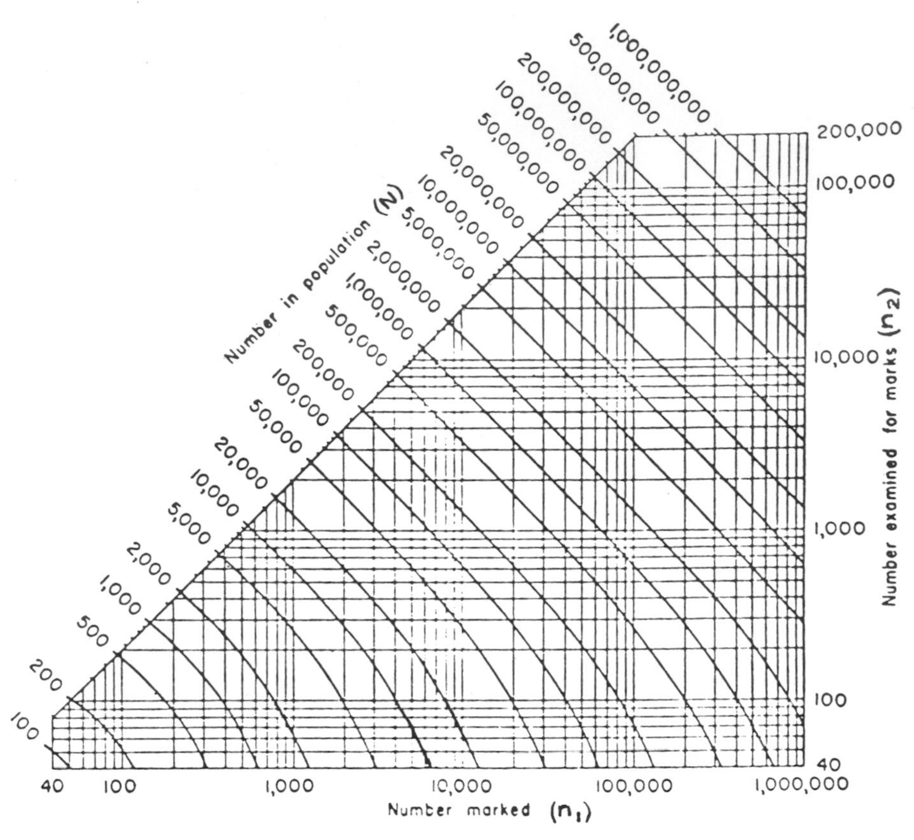

Fig. 3.5 Sample sizes when $1-\alpha = 0{\cdot}95$, $A = 0{\cdot}25$ and $100 \leqslant N \leqslant 10^9$: from Robson and Regier [1964].

for the values of D used in the construction of Fig. 3.4–3.6, namely $D = 24{\cdot}4$, $69{\cdot}9$, 392. For example, given $1-\alpha = 0{\cdot}99$, $A = 0{\cdot}10$, then from Table 3.1, $D = 695$. If N is estimated to be 1000, n_1 and n_2 are the solutions of

$$\frac{n_1 n_2(1000-1)}{(1000-n_1)(1000-n_2)} = 695;$$

a particular solution is $n_1 = n_2 = 455$.

In conclusion it is noted that sample sizes less than 40 are not graphed in Fig. 3.4–3.6 as the normal approximation is not sufficiently accurate in this case. The lines for smaller sample sizes could have been derived using the Poisson or binomial approximations to the hypergeometric, but the authors did not consider it worth the effort. As demonstrated below, experiments utilising markedly unequal sample sizes (either large n_1 and very small n_2 or vice versa) would almost certainly be much more costly than if more nearly equal-sized samples were used. For a further discussion on the accuracy of the Petersen estimate see **14.1.2** (*3*)

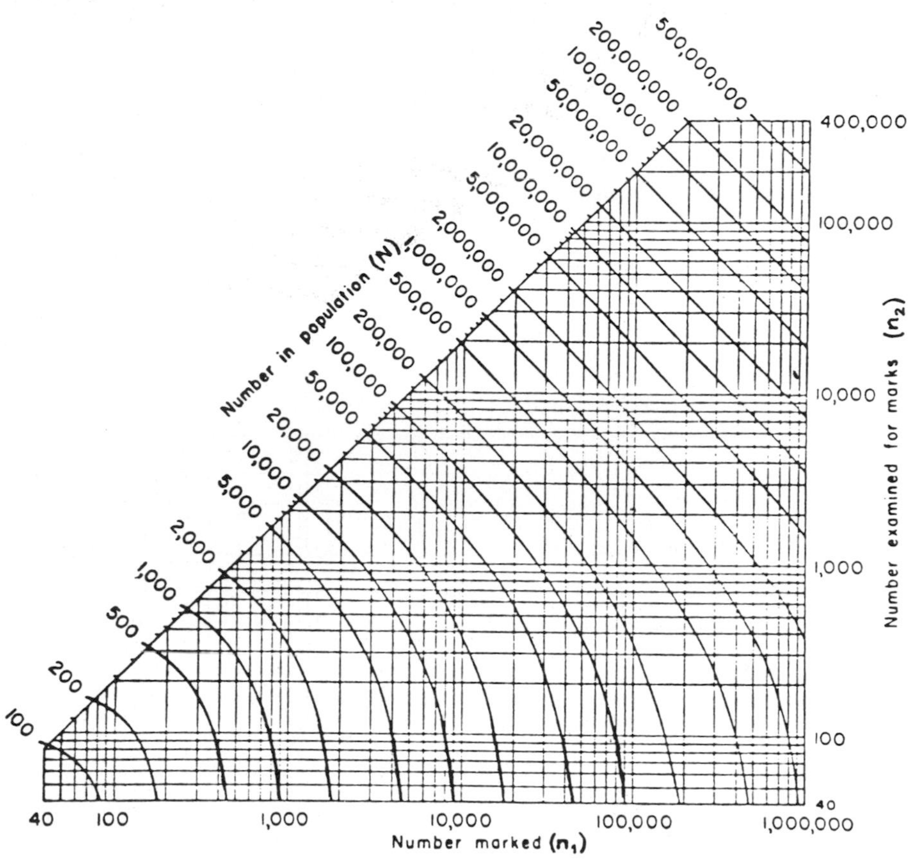

Fig. 3.6 Sample sizes when $1 - \alpha = 0{\cdot}95$, $A = 0{\cdot}10$ and $100 \leqslant N \leqslant 5 \times 10^8$: from Robson and Regier [1964].

2 *Optimum allocation of resources*

MINIMUM α WITH LIMITED FUNDS. Suppose c_1 and c_2 are the costs per animal of catching–marking and catching–examining and let f represent the known fixed or overhead costs, so that the total cost of the experiment is

$$c_T = c_1 n_1 + c_2 n_2 + f. \tag{3.7}$$

If funds are limited and the accuracy A is predetermined, then an optimum allocation would be to choose n_1 and n_2 which maximise $1 - \alpha$, subject to c_T being fixed. Robson and Regier [1964] show that the solution of this is

$$\frac{c_1 n_1}{c_2 n_2} = \frac{N - n_2}{N - n_1}, \tag{3.8}$$

and given estimates of c_1, c_2 and N, equations (3.7) and (3.8) can be solved for n_1 and n_2. When n_1 and n_2 are small relative to N, then (3.8) becomes $c_1 n_1 \approx c_2 n_2$, so that an equal division of resources between catching–marking and catching–examining is optimal.

MINIMUM COSTS FOR FIXED α AND A. If funds are not restricted, so that both α and A can be fixed in advance, then one would wish to choose n_1 and n_2 so that c_T is minimised. Given rough estimates of N and c_1/c_2, Robson and Regier show that the solution is obtained by consulting the appropriate curve in Fig. 3.1–3.6 and choosing the pair (n_1, n_2) satisfying (3.8). If a combination of α and A other than one of the three given above is required, then one must solve equations (3.6) and (3.8), where D is calculated from (3.5).

ALTERNATIVE METHODS OF CAPTURE. If two or more alternative methods of capture are available and α and A are predetermined, then the problem of choosing the least costly method arises. Suppose we compare two methods with sample sizes n_{i1}, n_{i2} and costs c_{i1}, c_{i2} ($i = 1, 2$), where, for the ith method, n_{i1} and n_{i2} are optimal in the sense of the previous paragraph. Then, assuming the same fixed overhead cost for each method, Robson and Regier show that

$$\frac{\text{cost for 2nd method}}{\text{cost for 1st method}} \approx \sqrt{\left(\frac{c_{21}c_{22}}{c_{11}c_{12}}\right)} = r, \quad \text{say.}$$

For example, if the second method of capture costs twice as much per captured animal as the first method, and both methods utilise the same marking technique, then we can write

$$c_{11} = a + b, \; c_{12} = a; \quad c_{21} = 2a + b, \; c_{22} = 2a;$$

and

$$r = \sqrt{\left(\frac{4a + 2b}{a + b}\right)}.$$

3.2 VALIDITY OF ASSUMPTIONS

3.2.1 Assumption (a): closed population

If the assumption of constant population size is to hold, the experiment should be carried out over a short period of time, in fact, ideally at a single point in time. For this reason the Petersen method is often called a "point census".

Departures from assumption (a) can occur in a number of ways, and we shall now discuss these in detail.

1 Accidental deaths

If there are d accidental deaths through the process of catching and marking the first sample, the general theory of **3.1** remains unchanged, provided that n_1 now refers to the number returned *alive* to the population. $\hat{N}$ and N^* are estimates of $N - d$.

2 Natural mortality

Suppose that mortality is taking place in the time between the two samples, and let N be the size of the population when the first sample is released. Robson [1969] points out that when assumptions (d), (e) and (f) of **3.1.1** are true and the only departure from (a) is due to mortality, the hypergeometric model (3.1) still holds, provided the mortality process is such that the deaths constitute a *simple random sample* of unknown size. This follows from the fact that the survivors will also constitute a simple random sample of which the second sample is a random subsample (assumption (d)). Since a random subsample of a random sample is itself a random sample, the second sample will still represent a simple random sample from the original population. Chapman [1952: 300, 1954: 5] also demonstrates this feature of the Petersen method by using a binomial model for the mortality process.

When the deaths do not constitute a simple random sample, the Petersen estimate can still be used, provided the marked and unmarked have the same average probability ϕ of surviving up till the time of the second sample. This can be seen intuitively from the equation

$$E\left[\frac{m_2}{n_2}\middle| n_1\right] \approx \frac{\phi n_1}{\phi N} = \frac{n_1}{N}.$$

We note that mortality is often selective with regard to size or age of the animal. However, if the first sample is a simple random sample, the more "vulnerable" individuals will be proportionately represented in both marked and unmarked populations, thus ensuring that marked and unmarked have the same average survival probabilities. This point is discussed further on p. 87

To examine the effect of variable mortality in the marked portion, suppose that there are various subcategories in the population with numbers $X, Y, \ldots, W$, where $N = X + Y + \ldots + W$. For the *marked* members in these categories let $\phi_x, \phi_y, \ldots, \phi_w$ be the respective survival probabilities and let $p_{2x}, p_{2y}, \ldots,$ be the probabilities of recapture in the second sample. Then, using the suffix x to denote membership of category X, we have

$$E[m_{2x}/n_{1x}] = \phi_x p_{2x}$$

and we can test the hypothesis $H_0 : \phi_x p_{2x} = \ldots = \phi_w p_{2w}$ using a standard chi-squared test based on the contingency Table 3.3 (Dunnet [1963: 92],

TABLE 3.3

	Subcategory				
	X	Y	...	W	Total
Recaptured	m_{2x}	m_{2y}	...	m_{2w}	m_2
Not recaptured	$n_{1x} - m_{2x}$	$n_{1y} - m_{2y}$	...	$n_{1w} - m_{2w}$	$n_1 - m_2$
Number released	n_{1x}	n_{1y}	...	n_{1w}	n_1

Robson and Regier [1968]). When $\phi_x = \phi_y = \ldots = \phi_w$, a test of H_0 is a test that the second sample is random with respect to the marked individuals in the various categories. Conversely, when the second sample is random so that $p_{2x} = p_{2y} = \ldots = p_{2w}$, this test is a test of constant survival probability for marked members.

3 Catchable population

It should be noted that N may sometimes refer to the catchable portion of the population only, and not to the whole population. For example, in sampling foraging ants, Ayre [1962] obtained a Petersen estimate of 109 for the population of an anthill known to be in the region of 3000. This enormous bias was explained by the fact that only a small proportion of the ants ever go foraging, so that the majority remain uncatchable.

If an approximately unbiased estimate $(\hat{p}_c)$ of the catchable proportion (p_c) of the population is available then we can estimate the total population N_T by

$$\hat{N}_T = N^*/\hat{p}_c.$$

If N^* and $\hat{p}_c$ are based on separate sampling experiments, as will usually be the case, they are statistically independent. Therefore, using the delta method (**1.3.3**), we have

$$\begin{aligned} E[\hat{N}_T] &= E[N^*]E[1/\hat{p}_c] \\ &\approx N\left(\frac{1}{p_c} + \frac{V[\hat{p}_c]}{p_c^3}\right) \\ &= N_T\left(1 + \frac{V[\hat{p}_c]}{p_c^2}\right) \end{aligned}$$

and

$$V[\hat{N}_T] \approx \frac{V[N^*]}{p_c^2} + \frac{N_T^2}{p_c^2}\, V[\hat{p}_c].$$

If binomial sampling is used to obtain $\hat{p}_c$ then $V[\hat{p}_c] = p_c q_c/n$, where n is the number of animals investigated for catchability.

Assuming $\hat{N}_T$ to be approximately normally distributed, an approximate confidence interval for N_T can be calculated in the usual manner. However, unless $\hat{p}_c$ and N^* are accurate estimates, this interval may be too wide to be of much practical use.

4 Recruitment

Sometimes the time-lapse between the two samples is sufficient to allow the recruitment of younger animals into the catchable part of the population. These new recruits would tend to decrease the proportion of marked in the second sample, and the Petersen estimate $\hat{N}$ would overestimate the initial population size. In the situation where there is recruitment but no mortality, $\hat{N}$ will be a valid estimate of the population number at the time when the second sample is taken. However, as pointed out by Robson and Regier [1968], when both recruitment and mortality occur, $\hat{N}$ will overestimate both the initial and final population size. This is seen mathematically by noting that if there are r recruits in the population at the time of the second sample, we have the approximate relation

$$E\left[\frac{m_2}{n_2} \mid n_1\right] \approx \frac{\phi n_1}{\phi N + r},$$

or

$$E[\hat{N} \mid n_1] \approx N + r\phi^{-1},$$

where $N + r\phi^{-1}$ is greater than N and $\phi N + r$ $(0 < \phi < 1)$. By enlarging the definition of r to include *permanent* immigrants, and redefining ϕ as the average probability that an animal in the population at the release of the first sample is alive and still in the population at the time of the second sample, then, provided ϕ is the same for marked and unmarked, the above comments apply to a population in which there is immigration and emigration also.

If an animal becomes immediately catchable as soon as it reaches a certain age, then an age analysis of the second sample would provide an estimate of the ratio of recruits to non-recruits. Using this ratio and a Petersen estimate of non-recruits (from the proportion of marked in the non-recruited members of the second sample) we could then obtain an estimate of total recruits. But the process of age determination is usually time-consuming and the "threshold" age for recruitment is not usually well defined, so that partial recruitment may occur over a range of younger ages. Usually the more readily available information such as length or weight is used to classify the individual, and such data can be used for carrying out the following two tests of recruitment.

CHI-SQUARED TEST. If the individual measurements are not actually recorded and the animals are simply allotted to particular size-classes, we can test for recruitment as follows. Let $X, Y, \ldots, W$ denote both the classes and the numbers in the classes at the beginning of the experiment, and suppose that X increases to $X + r_x$, etc. through recruitment. Then if the second sample is random with respect to within (but not necessarily between) classes, we have

$$E\left[\frac{m_{2x}}{n_{2x}}\right] = \underset{n_{1x}}{E}\, E\left[\frac{m_{2x}}{n_{2x}} \mid n_{1x}\right]$$

$$= \underset{n_{1x}}{E}\left[\frac{n_{1x}}{X + r_x}\right] = p_{1x}, \quad \text{say},$$

and the hypothesis $H_0 : p_{1x} = p_{1y} = \ldots = p_{1w}$ can be tested using a standard chi-squared statistic based on the contingency Table 3.4 (Robson and Regier [1968]). When there is no recruitment, so that $r_x = 0$ for each class, then this test will be a test that the first sample is random with respect to size-class. Conversely, if the first sample is a simple random sample then $E[n_{1x}/X]$ will

TABLE 3.4

	Size-class				
	X	Y	...	W	Total
Marked	m_{2x}	m_{2y}	...	m_{2w}	m_2
Unmarked	$n_{2x} - m_{2x}$	$n_{2y} - m_{2y}$	...	$n_{2w} - m_{2w}$	$n_2 - m_2$
Total	n_{2x}	n_{2y}	...	n_{2w}	n_2

be the same for each class and, since there is no recruitment in the classes with larger animals, a test of H_0 will then amount to a test of $r_x = 0$ for *all* the classes. In this latter situation the test will be unaffected by mortality, provided that the survival probabilities are the same for marked and unmarked. This follows from the simple relationship

$$E\left[\frac{m_{2x}}{n_{2x}}\right] = E\left[\frac{\phi_x n_{1x}}{\phi_x X + r_x}\right]$$

$$= E\left[\frac{n_{1x}}{X + r_x \phi_x^{-1}}\right],$$

where ϕ_x is the average survival probability for class X. Finally it is noted that the above test can still be used, even when animals grow from one class into another with unknown overlap during the course of the experiment. The reason for this is that although size-classes are usually determined from the second sample, X may be regarded as the "conceptual" population, existing at the time of the first sample, which grows into the required class: n_{1x} will then be unknown.

NON-PARAMETRIC TEST. When individual size measurements are recorded, Robson and Flick [1965] suggest the following non-parametric method for detecting and eliminating recruits. To simplify the notation we shall drop the suffix 2 from m_2 and n_2 and define $u = n_2 - m_2$. In the following discussion the word "length" will denote some readily available measurement of size.

We shall assume that n_1 is sufficiently large and the first sample sufficiently random for the length distribution of animals to be the same for both marked and unmarked. Suppose that the lengths of the m recaptures are $L_1 < L_2 < \ldots < L_m$ and let u_i $(i = 1, 2, \ldots, m + 1)$ be the number of unmarked animals caught in the second sample with lengths L in the interval $L_{i-1} \leqslant L < L_i$ $(L_0 = 0, L_{m+1} = \infty)$. If the second sample is random with respect to mark status and length, the probability that the length of an unmarked animal falls into any one of the above $m + 1$ length-classes is $1/(m + 1)$, and the expected value of each u_i will be $u/(m + 1)$. However, if recruitment has occurred in the shorter size range, the observed u_i for the intervals in this range will be greater than expected. Thus, if recruitment in the length-class $[0, L_1)$ has occurred, u_1 will be significantly larger than $u/(m+1)$. To determine the significance of u_1 we calculate the tail probability

$$\Pr[U_1 \geqslant u_1] = \binom{u+m-u_1}{m} \Big/ \binom{u+m}{m}, \tag{3.9}$$

where U_1 is the random variable taking the values u_1, and compare this with the significance level α (usually $\alpha = 0{\cdot}05$). For example, if $u = 110$, $m = 50$ and $u_1 = 8$, then

$$u_1 > u/(m+1) = 2{\cdot}2$$

and

$$\Pr[U_1 \geqslant 8] = \binom{160-8}{50} \Big/ \binom{160}{50}$$
$$= \frac{152!\ 110!}{102!\ 160!}.$$

Using Stirling's approximation for large factorials, namely

$$\log K! \approx \tfrac{1}{2}\log(2\pi) + (K+\tfrac{1}{2})\log K - K,$$

we have, after some simplification,

$$\log \Pr[U_1 \geqslant 8] \approx (152{\cdot}5)\log 1{\cdot}52 + (110{\cdot}5)\log 1{\cdot}10 - (102{\cdot}5)\log 1{\cdot}02 - (160{\cdot}5)\log 1{\cdot}60$$
$$= -3{\cdot}0821,$$

or

$$\Pr[U_1 \geqslant 8] \approx 0{\cdot}046.$$

Since this probability is just less than $0{\cdot}05$, we reject the hypothesis of no recruitment in the length-class $[0, L_1)$ at the 5 per cent level of significance.

The same procedure can now be applied to the second length-class $[L_1, L_2)$ by eliminating L_1 and the class of u_1 animals from the data. Thus we compare u_2 with its expected value $(u-u_1)/m$ and if this difference is significant we compare u_3 with $(u-u_1-u_2)/(m-1)$, and so on. Proceeding in this stepwise fashion through the larger classes, the recruits, if any, will dwindle in number until the rth step, say, is reached when the recruits no longer make a significant contribution. Thus u_r will not be significantly greater than $(u-a_{r-1})/(m-r+2)$, where $a_r = u_1 + u_2 + \ldots + u_r$; for this step the tail probability is

$$\Pr[U_r \geqslant u_r \mid u_1, u_2, \ldots, u_{r-1}] = \frac{\binom{n-a_r-r+1}{m-r+1}}{\binom{n-a_{r-1}-r+1}{m-r+1}}.$$

This would suggest that the remaining sample of $u_{r+1} + u_{r+2} + \ldots + u_{m+1}$ unmarked animals is free of recruits, and that the average

$$\bar{u}_{r+1} = (u_{r+1} + \ldots + u_{m+1})/(m+1-r)$$

is therefore an estimate of the number of unmarked non-recruits that should occur between every adjacent pair of marked recaptures. Hence the estimated number of unmarked non-recruits in the second sample is $(m+1)\bar{u}_{r+1}$ and the modified Petersen estimate of N becomes

$$N^* = \frac{(n_1+1)[\bar{u}_{r+1}(m+1)+m+1]}{m+1} - 1$$

$$= (n_1+1)(\bar{u}_{r+1}+1) - 1.$$

In evaluating the mean and variance of N^* we run into the difficulty of r being a random variable. This same problem arises, for example, in fitting a polynomial regression where the degree of the final polynomial obtained is strictly a random variable. However, as with the regression problem, treating r as though it were a constant would not seem unreasonable and would perhaps lead to a slight underestimate of $V[N^*]$. Now under the assumption of non-recruitment after the rth class, we have

$$E[\bar{u}_{r+1} \mid m, r, a_r] = \frac{u - a_r}{m+1-r}$$

leading to

$$E[N^* \mid m, r, a_r, n_1] = \frac{(n_1+1)(n-r-a_r+1)}{(m-r+1)} - 1$$

$$= \frac{(n_1+1)(n'+1)}{(m'+1)} - 1, \text{ say,}$$

where m' and n' are simply the values of m and n obtained by truncating the second sample at length L_r. If this truncation successfully eliminates recruits we would then expect

$$E[N^* \mid n_1] \approx N.$$

Also

$$V[N^* \mid n_1] = (n_1+1)^2 \, V[\bar{u}_{r+1}],$$

where the variance of $\bar{u}_{r+1}$ can be estimated robustly from the replicated u's (cf. (1.9)), namely from

$$v[u_{r+1}] = \frac{1}{(m+1-r)(m-r)} \sum_{i=1}^{m+1-r} (u_{r+i} - \bar{u}_{r+1})^2.$$

Robson and Flick [1965] point out that as recruitment will generally tend to decrease with increasing body-length, there will be a decrease in the probability of detecting these recruits. Also this decrease in detectability is further accentuated by a decrease in the length interval between marked animals as the test progresses from the lower tail toward the centre of the length distribution, and also by the decrease in sample size resulting from the successive removal of the intervals tested. To overcome this difficulty we require some method of pooling intervals as the number of recruits falls off.

The authors mention that a further need for combining intervals arises when length measurements are sufficiently crude to permit ties to occur. In particular, if several marked animals are recorded as having the same body-length, then the resulting degenerate intervals must be combined to include all unmarked animals having that same recorded length.

It transpires that the optimal pooling procedure is simply to combine adjoining intervals giving a new total interval and a new total number of unmarked. If the first k intervals are combined, then a test for recruitment in this total interval has a tail probability of

$$\begin{aligned} T[s_k] &= \Pr[S_k \geq s_k] \\ &= \sum_{r=0}^{k-1} \binom{s_k+r-1}{r}\binom{u+m-s_k-r}{m-r} \Big/ \binom{u+m}{m} \end{aligned} \tag{3.10}$$

where $s_k(=u_1+u_2+\ldots+u_k)$ is the number of unmarked in the total interval. As k gets large, (3.10) becomes computationally awkward, and for small values of s_k the recursive relation

$$\Pr[S_k = s_k] = \frac{(s_k+k-1)(u-s_k+1)}{s_k(u+m-s_k-k+1)} \Pr[S_k = s_k-1], \tag{3.11}$$

where $\Pr[S_k = 0] = \binom{u+m-k}{u} \Big/ \binom{u+m}{u}$,

is useful ($u-s_k-1$ should be replaced by $u-s_k+1$ in equation (3) of Robson and Flick [1965]; the correct equation has been used in their calculations). When u is much greater than m, so that

$$\frac{m+1-k}{kn} \approx 0 \quad \text{and} \quad \frac{u(n+1)}{n^2} \approx 1,$$

Robson and Flick suggest the incomplete beta approximation

$$\begin{aligned} \Pr[S_k \geq s_k] &= \frac{\Gamma(m+1)}{\Gamma(k)\,\Gamma(m+1-k)} \int_p^1 t^{k-1}(1-t)^{m-k}\,dt \\ &= \sum_{i=0}^{k-1} \binom{m}{i} p^i(1-p)^{m-i}, \end{aligned} \tag{3.12}$$

where $p = (s_k+k-1)/n$, and they give a useful table indicating the accuracy of the approximation. We note that (3.12) can be evaluated from standard binomial tables.

Suppose now that the first k_1 intervals, the second k_2 intervals ... are combined to give new length-classes $[0, L_{k_1})$, $[L_{k_1}, L_{k_2})$, etc. with unmarked numbers s_{k_1}, s_{k_2}, etc. Then if

$$\Pr[S_{k_1} \geq s_{k_1}] < \alpha,$$

we reject the hypothesis of no recruitment in the length-class $[0, L_{k_1})$ at the α level of significance and proceed to consider s_{k_2}. Dropping the data in the first interval from the sample, we now evaluate the tail probability (cf. (3.10))

$$T[s_{k_2};\, s_{k_1}] = \Pr[S_{k_2} \geqslant s_{k_2} \mid S_{k_1} = s_{k_1}]$$

$$= \sum_{r=0}^{k_2-1} \binom{s_{k_2}+r-1}{r}\binom{u'+m'-s_{k_2}-r}{m'-r}\bigg/\binom{u'+m'}{m'}, \quad (3.13)$$

where $m' = m - k_1$ and $u' = u - s_{k_1}$ are the "new" values of m and u. We again reject the hypothesis of no recruitment in $[L_{k_1}, L_{k_2})$ if the above probability is less than α. This process can then be repeated for s_{k_3}, s_{k_4}, etc. until non-significance is achieved; the number of non-recruits in the second sample can then be estimated as before.

One of the problems in combining adjacent intervals is to determine the best sequence $k_1, k_2, \ldots$. Since there is a practical possibility that all the recruits are shorter than the shortest non-recruit, it would seem reasonable to use $k_1 = 1$. Also, because of the steady reduction in numbers of recruits between successively larger marked animals, the k-sequence should be increasing, so that $1 = k_1 \leqslant k_2 \leqslant \ldots$. Unfortunately the optimum sequence can only be determined if the length–frequency distributions are known for both recruits and non-recruits, although as the statistics S_{k_1}, S_{k_2}, etc. are virtually independent, the k sequence could perhaps be determined sequentially by regression methods. For example, Robson and Flick suggest extrapolating the regression of u_i on i for $i = 2, 3, \ldots, k_2$ to indicate the number k_3 of intervals which must be next combined in order to achieve the relation

$$u_{k_1+k_2+k_3} \approx \frac{u - s_{k_1} - s_{k_2} - s_{k_3}}{m - k_1 - k_2 - k_3 + 1}.$$

Further research needs to be done on such methods of finding a suitable k-sequence.

If one wishes to combine intervals still further (e.g. the first $k_1 + k_2$ intervals), then, as mentioned above, the optimal procedure is simply to use the sum $S_{k_1} + S_{k_2}$ $(= S_{k_1+k_2})$ and evaluate the tail probability

$$T[s_{k_1+k_2}] = \Pr[S_{k_1+k_2} \geqslant s_{k_1+k_2}].$$

However, to avoid this additional computation Robson and Flick suggest a number of approximate procedures such as using $T[s_{k_1}] + T[s_{k_2};\, s_{k_1}]$ with a significance level of $(2\alpha)^{1/2}$, or

$$T[s_{k_1}] + T[s_{k_2};\, s_{k_1}] + T[s_{k_3};\, s_{k_2}, s_{k_1}]$$

with a significance level of $(6\alpha)^{1/3}$ if three groups are pooled.

Example 3.1 Brook trout (*Salvelinus fontinalis*): Robson and Flick [1965]

In 1962 a number of brook trout were captured with trap nets from an experimental pond and were returned with jaw tags. The following year 70 marked and 165 unmarked trout were caught from the same pond, again using trap nets, and Table 3.5 gives the length-frequency distribution of these fish. The catchable trout in 1962 were presumed to be of age I+ and

TABLE 3.5

Length–frequency distributions of marked and unmarked brook trout in a trap-netted sample: from Robson and Flick [1965: Table 2].

Length-class (in.)	Marked	Unmarked	
< 9·7	0	54	*
9·7 (L_1)	1	1	
9·8	0	4	
9·9	0	7	
10·0	0	8	
10·1 (L_2)	1	1	
10·2	0	2	*
10·3 (L_3)	1	4	
10·4 (L_4, L_5)	2	7	
10·5 (L_6, L_7)	2	4	
10·6 (L_8)	1	2	
10·7 (L_9, L_{10})	2	4	*
10·8 (L_{11})	1	5	
10·9 (L_{12})	1	5	
11·0 ($L_{13}, \dots, L_{15}$)	3	1	*
11·1 ($L_{16}, \dots, L_{19}$)	4	3	
11·2 (L_{20})	1	4	
11·3 ($L_{21}, \dots, L_{23}$)	3	4	*
11·4	2	6	
11·5	5	8	*
11·6	3	2	
11·7	3	3	
11·8	4	2	*
11·9	7	1	
12·0	6	2	*
12·1	4	3	
12·2	4	2	
12·3	2	1	*
12·4	1	0	
12·5	4	1	
12·6	0	2	
12·7	2	0	
> 12·7	0	12	*
Total	70	165	

* Intervals used in testing for recruits.

the 1963 sample was then expected to include these same fish now at age II+ plus the new recruits of age I+.

The first few rows of Table 3.5 indicate that recruitment clearly occurred; also the number with length greater than 12·7 in. seems significantly high, but, for simplicity of exposition, we shall neglect this fact for the moment.

The k-sequence was arbitrarily chosen as $k_i = i$, subject to any necessary modifications due to ties. Thus the length-intervals are $[0, L_1)$, $[L_1, L_3)$, $[L_3, L_6)$, $[L_6, L_{11})$, etc.; the interval $[L_6, L_{11})$ includes 5 rather than 4 lengths because 2 marked fish have the same length of 10·7 in. The k-sequence is now 1, 2, 3, 5, 5, 8, 7, 10, 13 and 10. We recall that u_i is the number of unmarked animals with lengths in $[L_{i-1}, L_i)$ and the various tests are as follows.

(i) $u = 165$, $m = 70$, $u_1 = 54$. From equation (3.9)

$$\Pr[U_1 \geqslant 54] = \binom{165+70-54}{70} \Big/ \binom{165+70}{70}$$

$$= 0{\cdot}2098 \times 10^{-9}.$$

(ii) $s_2 = u_2 + u_3 = 20 + 3 = 23$.
$u' = 165 - 54 = 111$, $m' = 70 - 1 = 69$.
From equation (3.13) we have

$$\Pr[S_2 \geqslant 23] = \left[\binom{22}{0}\binom{69+111-23}{69} + \binom{23}{1}\binom{69+111-24}{68}\right] \Big/ \binom{69+111}{69}$$

$$= \frac{156\,!\ 111\,!}{88\,!\ 180\,!}[157 + 23(69)]$$

$$= 0{\cdot}6160 \times 10^{-4}.$$

(iii) $s_3 = u_4 + (u_5 + u_6) = 4 + 7 = 11$.
$u' = 111 - 23 = 88$, $m' = 69 - 2 = 67$.

$$\Pr[S_3 \geqslant 11] = \left[\binom{10}{0}\binom{144}{67} + \binom{11}{1}\binom{143}{66} + \binom{12}{2}\binom{142}{65}\right] \Big/ \binom{155}{67}$$

$$= 0{\cdot}030\ 04.$$

(iv) $s_5 = (u_7 + u_8) + u_9 + (u_{10} + u_{11}) = 4 + 2 + 4 = 10$, $u' = 77$, $m' = 64$.
$\Pr[S_5 \geqslant 10] = 0{\cdot}147\ 72$.

(v) $s_5 = u_{12} + u_{13} + (u_{14} + u_{15} + u_{16}) = 5 + 5 + 1 = 11$, $u' = 67$, $m' = 59$.
$\Pr[S_5 \geqslant 11] = 0{\cdot}086$.

(vi) $s_8 = u_{17} + \dots + u_{24} = 11$, $u' = 56$, $m' = 54$.
$\Pr[S_8 \geqslant 11] = 0{\cdot}246\ 05$.

(vii) $s_7 = u_{25} + \dots + u_{31} = 14$, $u' = 45$, $m' = 46$.
$\Pr[S_7 \geqslant 14] = 0{\cdot}033\ 06$.

(viii) $s_{10} = u_{32} + \dots + u_{41} = 7$, $u' = 31$, $m' = 39$.
To calculate $\Pr[S_{10} \geqslant 7]$ it is simpler to use the recursive relation (3.11) rather than (3.13). Thus

$$\Pr[S_{10} = 0] = \binom{31+39-10}{31} \Big/ \binom{31+39}{31} = \binom{60}{31} \Big/ \binom{70}{31},$$

$$\Pr[S_{10} = 1] = \frac{(1+10-1)(31-1+1)}{1(31+39-1-10+1)} \Pr[S_{10} = 0]$$

$$= \frac{10(31)}{60} \Pr[S_{10} = 0], \text{ etc.}$$

Hence

$$\Pr[S_{10} \geqslant 7] = 1 - \sum_{i=0}^{6} \Pr[S_{10} = i] = 0{\cdot}628\,96.$$

All the higher length-intervals fail to show a significant recruitment until the last interval is reached, for which the present procedure provides no test. This last interval has been ignored for purposes of exposition, but, as the authors point out, there is obviously an excessive number of unmarked trout longer than 12·7 in. (322 mm). If it had been suspected before the experiment that recruitment was possible at both ends of the length distribution then the appropriate test procedure would have been to alternate from one tail to the other, progressing toward the centre as far as possible from each end.

It is noted that the probabilities of the above tests fluctuate considerably, with (i), (ii), (iii) and (vii) being significant at the 5 per cent level of significance. These erratic results could be due to the arbitrary choice of $k_i = i$, and further pooling of the intervals is suggested. Thus, starting with the interval [9·7, 10·2), we again follow the rule of combining one, then two, then three intervals, etc., and using the approximate method of adding tail probabilities described above we obtain

$$0{\cdot}6160 \times 10^{-4} < 0{\cdot}05$$

$$0{\cdot}030\,04 + 0{\cdot}147\,72 < 0{\cdot}3162 \; (= [2(0{\cdot}05)]^{1/2})$$

$$0{\cdot}0806 + 0{\cdot}2461 + 0{\cdot}0331 < 0{\cdot}6694 \; (= [6(0{\cdot}05)]^{1/3}),$$

thus obtaining significant recruitment up to 11·5 inches (292 mm).

3.2.2 Assumptions (b), (c) and (d)

1 Practical considerations

VARIABLE CATCHABILITY.* One of the crucial assumptions underlying the theory of **3.1.1** is assumption (d) that the second sample is a simple random sample (though we shall see below in section *2* that the theory still holds under alternative assumptions). Strictly speaking, such a random sample can only be obtained by numbering the animals 1, 2, ... , N and using a table of random numbers to select n_2 animals. However, in practice, if all the animals have the same catchability, we can approximate to a random sample by arranging that every point of the population area has the same probability of being sampled (e.g. using random number pairs as coordinates) and that all points selected are sampled with the same effort. If a more even coverage is required one can use stratified random sampling, whereby the population area is divided into equal subareas and one or more points are allotted at random within each subarea (e.g. Leslie *et al.* [1953: 139]). Unfortunately the requirement of constant probability of capture may not hold, either because of an inherent variation in catchability or because catching and handling in the first sample affect future catchability. Very often the probability of capture will vary between various subgroups defined by age, sex, species, etc. (Kikkawa [1964: 260, 290], Pucek [1969]). For example,

(*) See also **12.7.3** (*1*).

certain subgroups may be more mobile and have a different habitat preference (e.g. Corbet [1952: male and female dragonflies]) while others may have certain bait and trap preferences (e.g. Chitty and Shorten [1946: bait indifference shown by some rats], Dobzhansky *et al.* [1956: differential attraction of species of Drosophila to different kinds of yeast]). Strandgaard [1967] found age and sex variations in the sighting and trapping of roe deer.

In fisheries, catchability usually varies with the size of the fish, and considerable research has been carried out on such problems as "gear selectivity" and "length-selection" curves (cf. International Commission for Northwest Atlantic Fisheries [1963], Hamley [1975]*). When recruitment and mortality are negligible (or any mortality is random with respect to size, cf. p. 71) and the marked members are individually identifiable, Robson [1969] suggests testing the randomness of the second sample with respect to size by partitioning the first sample into that portion which is ultimately recaptured and the portion which is not. A Mann–Whitney (Wilcoxon) rank–sum test (e.g. Keeping [1962]) comparing these two subsamples with respect to body size will then be a test for a monotonic relation between body size and probability of capture in the second sample. Alternatively, if the marked members are allotted to size-classes one can carry out the goodness-of-fit test described on p. 71 (Table 3.3).

When there is variation in the inherent catchabilities of the individuals and the first sample is not random, assumption (b) will be false, and the more catchable individuals will be caught in the first sample. This means that for the second sample the marked will in general be more catchable than the unmarked, and assumption (d) will be false. Unfortunately, apart from a careful choice of catching method and preliminary studies – for example, on activity, feeding habits, "length-selection" curves, etc. – little can be done to overcome this problem of variable catchability in the first sample. However, we shall see in section *2* below that the bias in the Petersen estimate $\hat{N}$ due to variation in catchability can be reduced by using different trapping methods for the two samples.

If the catchability is constant within certain well-defined subgroups and there are sufficient recaptures from each subgroup, the numbers in each subgroup should be estimated separately (e.g. Gunderson [1968]). The question of pooling data from different subgroups is discussed in **3.2.5** and **11.1.4**.

SYSTEMATIC SAMPLING. It was noted in **3.1.2** that the Petersen estimate can still be used, even when assumption (d) is false and a systematic rather than a random sample is taken, provided there is uniform mixing of marked and unmarked and all animals are equicatchable in the second sample. But in many populations uniform mixing is unlikely because of territorial behaviour and the presence of well-defined home ranges. Another situation where it is difficult to obtain a uniform mixing is when the population is not randomly distributed throughout the population area and the animals are relatively immobile. For example, Hancock [1963] suggested that the

(*) See also Hamley and Regier [1973] and Jester [1977].

excessive variability in the monthly returns of marked whelks may have been due to the random distribution of marked individuals among essentially non-randomly distributed unmarked individuals. Banks and Brown [1962] found the same lack of dispersal of the marked individuals in the study of Sunn Pest populations (cf. Example 3.10, p. 115).

It would seem that where possible the experimenter should aim for a random sample rather than rely on the assumption of uniform mixing. However, in many population studies it is helpful to arrange the release of the first sample so that mixing can take place as much as possible. For example, one can divide up the total area into subareas, sample each subarea with the same effort, and then release the marked animals back into the same area from which they were taken. If the catchability is independent of subarea one would hope that this method produces roughly the same proportion of marked in each subarea. To check this, one can carry out a standard goodness-of-fit test to see whether the proportions of marks recorded in the second sample from the various subareas are significantly different (cf. **3.4.1** (*1*)).

CATCHING AND HANDLING. Departures from assumption (c) that trapping and marking do not affect catchability can be minimised if the following points are observed by the experimenter:

(i) *Type of trap*. It is essential that a trapping method be used which will not harm the animal in any way. For example, in small mammal populations such expedients as placing the traps under cover of vegetation, drugging the bait, and visiting traps frequently can reduce trap mortality (Buckner [1957]). But in spite of careful trapping technique the shock of actually being trapped may have considerable effect on the animal (e.g. Guthrie *et al.* [1967: squirrels], Keith *et al.* [1968: snowshoe hares], Bouck and Ball [1966: rainbow trout]).

If several types of trap are available one would endeavour to choose the type which is most efficient, as the accuracy of the Petersen estimate increases with n_1 and n_2 (cf. Pucek [1969]). Also to increase trap efficiency, the bait or lure should be selective for the species under investigation and some consideration should be given to the spacing and distribution of the traps (cf. Andrzejewski *et al.* [1966]).

(ii) *Method of handling*. Care is needed in handling the captured animals so that they quickly recover on their return to the population. For example, a major problem in much fishery work is the high mortality of tagged fish immediately after release, and recent evidence indicates that tagging and marking can have a greater effect on fish than is apparent (Ricker [1958: Chapter 3], North Atlantic Fish Marking Symposium [1963: 7–13], Paulik [1963a: 41–2], Clancy [1963], Shetter [1967]): see also **12.7.3** (*2*)

Another problem arises with small mammals and birds where "trap addiction" or "trap shyness" can alter an animal's pattern of behaviour after it has been caught for the first time. Numerous examples of this are given in the literature, and the following are a selection: rats and voles (Chitty and

Shorten [1946], Tanaka [1951], Bailey [1968]), house mice (Young *et al.* [1952]), squirrels (Evans [1951], Flyger [1959], Nixon *et al.* [1967]), rabbits (Geis [1955], Huber [1962], Eberhardt *et al.* [1963], Edwards and Eberhardt [1967]), small mammals (Morris [1955], Tanaka [1956, 1963a], Getz [1961], Pucek [1969: 421]) and bluetits (Taylor [1966]). Obviously there is a need for more research into methods of measuring the response of individuals to traps (e.g. Sealander *et al.* [1958], Balph [1968], Bailey [1968], Bailey [1969]). One way of detecting trap response, which is discussed later in Chapter 4, is by analysing the frequencies of recaptures of individuals from multiple-recapture experiments (cf. **4.1.5**(*2*) and **4.1.6**); another method, which can be used when the second sample is taken in stages, is discussed at the end of this chapter on p. 129.

The effect of trap shyness can be minimised by prebaiting the traps for a suitable period of time before the census, thus allowing the animals to get used to the presence of the traps (Chitty and Kempson [1949], Tanaka [1970]). This, however, does not always work (e.g. Young *et al.* [1952], Crowcroft and Jeffers [1961], Balph [1968]),* and of course trap addiction is not necessarily helped by prebaiting. Sometimes trap addiction can be reduced by altering the trap positions throughout the trapping period, thus, for example, preventing individuals from building their runaways up to the mouth of the trap (Brown [1954]). Another way of minimising the effect of trap response is to use a different trapping method for taking the second sample. For instance, animals can be live-trapped for marking and then shot for recapture (Robel *et al.* [1972], Nixon *et al.* [1975]) or, if the mark is conspicuous, sight of the mark itself could be the means of "recapture" (e.g. Dunnet [1957: rabbits], Flyger [1959: squirrels], Strandgaard [1967: deer]). In the latter case, observing the animals and noting the proportion of marked amounts to sampling with replacement, so that Bailey's binomial model (**3.1.2**) is appropriate.

In some circumstances the tag itself may affect the longevity and behaviour of the animals. For example, jaw tags on fish can interfere with feeding and thus affect growth rate, while Petersen disc tags can make the fish more vulnerable to gill nets through the net catching under the disc (Ricker [1958]). Newly emerged insects may be more sensitive to the toxic substances used in paint markers than older insects, and labels attached to their wings may interfere with blood circulation (Southwood [1966: 58]).

Another aspect of marking particularly relevant to insects is that the presence of conspicuous marks may well destroy an animal's natural camouflage and make it more or less liable to predation. Also when animals are sampled by a method which relies on the sight of the collector then the marked may tend to be collected more than the unmarked (Southwood [1966: 58]). On the other hand, if the tags are not conspicuous enough they may be overlooked, particularly if one relies on huntsmen, fishermen, farmers, etc., to return the tags.

In general the effect of a particular marking method should be checked

(*)See also Smith *et al.* [1975: 28], Tanaka and Murakami [1977].

by both laboratory and field experiments where possible. A comprehensive review of methods of catching and handling wild animals is given in Taber and Cowan [1969].

(iii) *Method of release.* Animals often show a high level of activity immediately on release, and efforts should be made to minimise this. For example, birds and insects could perhaps be restrained from flying immediately by covering them with small cages until the effects of handling wear off, while tagged fish could be held in tanks so as to reduce and measure initial tagging mortality (e.g. North Atlantic Fish Marking Symposium [1963: 8]). If the animals have a rhythm of activity during the day they could perhaps be released during an inactive period (Southwood [1966: 74]). Where an animal is released may have some effect on its future behaviour, particularly if released in strange surroundings (Flyger [1960: 369]).

2 *Theoretical analysis of catchability*

For the jth member of the population ($j = 1, 2, \ldots, N$), let x_j be the probability that it is caught in the first sample, let y_j be the conditional probability that it is caught in the second sample given that it is caught in the first, and let z_j be the conditional probability that it is caught in the second, given that it is not caught in the first. Assuming that the population represents a random sample of N triples (x_j, y_j, z_j) with regard to the species as a whole and the particular trapping methods used, then (x_j, y_j, z_j) may be regarded as a random observation from a trivariate probability density function, $f(x, y, z)$ say. Then from Seber [1970a] we have that the conditional distribution of m_2 given n_1 and n_2 is approximately binomial, i.e.

$$f(m_2 \mid n_1, n_2) \approx \binom{n_2}{m_2} P^{m_2}(1-P)^{n_2-m_2}, \tag{3.14}$$

where $$P = \frac{n_1}{(N-n_1)k + n_1},$$

$$k = \frac{(p_3 - p_{13})p_1}{p_{12}(1-p_1)},$$

and $p_1 = E[x]$, $p_3 = E[z]$, $p_{13} = E[xz]$ and $p_{12} = E[xy]$; all expectations being with respect to $f(x, y, z)$. From (3.14) it can be shown that

$$\begin{aligned} E[N^* \mid n_1, n_2] &= \frac{n_1+1}{P}\{1-(1-P)^{n_2+1}\} - 1 \\ &\approx n_1/P \\ &= (N-n_1)k + n_1, \end{aligned}$$

or

$$E[N^* - n_1 \mid n_1, n_2] \approx (N-n_1)k, \tag{3.15}$$

so that N^* is an approximately unbiased estimate of N if and only if $k = 1$. Seber [1970a] also shows that V^* and v^* of **3.1.1** are still satisfactory estimates of the "true" variance (i.e. the variance of N^* with respect to (3.14)), even when $k \neq 1$.

ASSUMPTION (c) TRUE. In studying this special case it is helpful to make the transformation

$$k = (B-p_1)/(1-p_1), \tag{3.16}$$

where $B = p_1(p_3-p_{13}+p_{12})/p_{12}$.

Obviously $k = 1$ if and only if $B = 1$ and $k < B$ when $B < 1$. Now when assumption (c) is true, i.e. marking does not affect catchability, we have $y_j = z_j$ $(j = 1, 2, \ldots, N)$, so that $p_{13} = p_{12}$, $p_3 = p_2$ and

$$\begin{aligned} 1 - B &= 1 - (p_1 p_2/p_{12}) \\ &= \text{cov}\,[x, y]/E[xy]. \end{aligned} \tag{3.17}$$

This equation was first considered by Junge [1963] and its implications discussed in some detail by him. He pointed out that $B = 1$ if and only if x and y are uncorrelated; a positive correlation will lead to an underestimate of N (as $k < B < 1$) and a negative correlation to an overestimate. We note that if a correlation exists we would generally expect it to be positive. This may account for the persistent underestimation observed by Buck and Thoits [1965] who checked on several estimates of fish population numbers by draining the ponds containing the populations.

We conclude from the above that variation in catchability due to, say, trapping selectivity could exist for both samples without introducing bias if the sources of selectivity in the two samples were independent. This leads support to the statement by a number of authors that bias due to difference in catchability can be reduced by using a different sampling method for each sample (Buck and Thoits [1965], Waters [1960]). For example, Ricker [1958: 96] argues that if fish for marking are trapped while the second sample is obtained by angling, it would seem unlikely for a similar sampling bias to be present in both gears. Junge [1963] gives an interesting example on river tagging of migrating salmon. In this situation it may be possible to tag non-selectively with respect to size but not with respect to time, e.g. sampling effort may vary with time. On the other hand, the recovery process on the spawning grounds may be selective with respect to fish size and spawning area, so that if the time of passage past the tagging site is uncorrelated with fish size and area of spawning, no bias is introduced.

Junge also discusses the sensitivity of B to variation in catchability by studying the special case of $y_j = bx_j$ $(b > 0)$ when the correlation is unity. We find that

$$B = 1 - V[x]/(E[x])^2 \leqslant 1$$

with equality if and only if $V[x] = 0$ or x is constant (i.e. assumption (b) is true). He shows that if the range of x is $[c, d]$ $(0 \leqslant c < d \leqslant 1)$ then, provided $w = c/d$ is not too small, B will be near unity and insensitive to the shape of $f(x)$, the probability density function of x. For example, if $f(x)$ is the uniform distribution, then

$$B = 3(1+w)^2/4(1+w+w^2)$$

and $B = 1, \frac{27}{28}, \frac{27}{31}, \frac{3}{4}$ when $w = 1, \frac{1}{2}, \frac{1}{5}$ and 0. On the other hand, if $c = 0$, B may be significantly less than unity; if in addition $f(x)$ is concentrated near $x = 0$ (i.e. probabilities of capture are near zero for a substantial proportion of the population) Junge shows that B could be much smaller still (see also **4.1.6**(*4*) and Table 4.20). Since $k < B$ when $B < 1$ these effects will be accentuated in the value of k, e.g. for the extreme case when $f(x)$ is the uniform distribution on [0, 1] we have $p_1 = \frac{1}{2}$, $B = \frac{3}{4}$ and $k = \frac{1}{2}$.

It was noted by Junge that x and y are uncorrelated if either x or y is constant, i.e. $B = 1$ if at least one of the two samples is a random sample. In particular, if the first sample is random, the second sample need not be random and in fact could be highly selective, provided the selectivity was independent of mark status (assumption (c)). This fact is also noted by Robson [1969] who points out that when assumption (c) is true, the hypergeometric model given by (3.1) still holds if either sample is random because of the symmetry of n_1 and n_2 in the formula.

When mortality is taking place and just the second sample is random we saw in **3.2.1**(*2*) that (3.1) is valid only when the mortality process is random. However, if just the first sample is random, (3.1) will still hold even if the mortality is non-random, provided that it is independent of mark status. This follows from the fact that when the first sample is random, the tags are distributed randomly through every possible subgroup or category existing in the population and therefore throughout any portion of the population subsequently removed for investigation. Since most mortality tends to be selective, Robson therefore suggests that "the most effective plan for the two-sample experiment consists of a determined effort to obtain a random sample for marking and then exploiting the habits of the creature to obtain a large, if selective, sample in the recapture stage."

If the first sample is random with respect to body size and there is no recruitment, we would expect the marked and unmarked portions of the second sample to represent random subsamples from the same size-distribution. Therefore, as suggested by Robson [1969], a Mann–Whitney (Wilcoxon) rank sum test (e.g. Keeping [1962]) would provide a test of randomness with respect to body size. Alternatively the goodness-of-fit test given on p. 74 (Table 3.4) could be used to test the randomness of the first sample with respect to size or any other category.

ASSUMPTION (b) TRUE. When x is constant we find that $p_{13} = p_1 p_3$, $p_{12} = p_1 p_2$ and $k = p_3/p_2$. This means that $k = 1$ if and only if the average probability of capture of the marked in the second sample is the same as the average for the unmarked.

CONCLUSION. We see from the above discussion that if assumption (c) is true and $y_j = bx_j$, then $k = 1$ if and only if assumption (b) is true; a test for this case based on taking two samples from a known (e.g. marked) population is given below in section *3*. Conversely, if assumption (b) is true then $k = 1$ if and only if assumption (c) is true. In practice there may be departures from both assumptions, and a general method for testing $k = 1$ based on multiple recaptures is given in **4.1.3**(*3*). A model due to Skellam [1948], based on a beta distribution for the probability of capture, is used by Eberhardt [1969] to estimate N: this is discussed in **4.1.6**(*4*).

3 Test for constant catchability

We shall now consider the problem of testing assumption (b), given that assumptions (a), (c), (e) and (f) are true, by taking three samples and using the first sample as an identifiable population (or else by taking two samples from a population of known size).

Suppose that the m_2 tagged animals in the second sample are given another tag (if individually numbered tags are not used for the first sample) and the second sample then returned to the population. If a third sample of size n_3 is now taken, then on the basis of the tagging information obtained and assuming that catching and tagging do not affect future catchability, Cormack [1966] gives two procedures, which we discuss below, for testing the hypothesis of constant catchability. The choice of procedure depends on which of two methods is used for catching n_1. In the first method, sample one is captured by a different technique from that used for samples two and three. We then require an additional assumption that the first sample is a random sample of probabilities of capture with respect to the *latter* catching technique. This will imply that the distribution of probability of capture over the first sample with regard to the latter catching technique is the same as over the entire population. For the second method the same catching technique is used for all three samples and no further assumptions are necessary.

In the following discussion we note that n_1, n_2 and n_3 play the same role as N, n_1, n_2 in the theoretical analysis of the previous section above.

METHOD 1. The probability p_j ($j = 1, 2, \ldots, n_1$) that the jth member of the first sample is captured in a later sample will be proportional to its inherent catchability and to the intensity of sampling or sampling effort. Assuming the sampling effort f to be the same for each individual, we therefore define $c_j = p_j/f$ as the catchability of the jth individual for the particular catching method used. We shall assume the c_j's to be a random sample of size n_1 from a probability density function $g(c)$ with moments μ'_r about the origin and moments μ_r about the mean. If we standardise c_j so that the domain of g is $[0, 1]$ then f will be uniquely determined and $0 \leqslant f \leqslant 1$. Cormack points out that the catchability of an animal may be regarded as the probability with which it places itself in a position where the experimenter is able to catch it, and the sampling intensity is then the probability that an animal in this position will be caught. Alternatively, if c_j is not standardised we can regard it as the probability that one unit of sampling effort catches the ith individual. Then, considering f as the number of units of effort expended on the sample ($f > 1$), and assuming units of effort to be additive, $p_j = fc_j$ as before.

Let f_2 and f_3 be the sampling efforts for samples two and three respectively. Then if x_j and y_j are the probabilities that the jth member of sample one is caught in samples two and three respectively, and assumption (c) is true, we have $x_j = f_2 c_j$, $y_j = f_3 c_j$, and hence $y_j = b x_j$. Let m_{10} be those individuals caught in the first sample only, m_{12} those caught in both samples one and two only, m_{13} those caught in both samples one and three only, and m_{123} those caught in all three samples. Then Cormack shows that the joint probability function of m_{12}, m_{13} and m_{123} is given by

$$\frac{n_1!}{m_{12}!\, m_{13}!\, m_{123}!\, m_{10}!} [\alpha_2(1-\alpha_3\lambda)]^{m_{12}} [\alpha_3(1-\alpha_2\lambda)]^{m_{13}} [\alpha_2\alpha_3\lambda]^{m_{123}} \times [1-\alpha_2-\alpha_3+\alpha_2\alpha_3\lambda]^{m_{10}}, \tag{3.18}$$

where $\alpha_2 = f_2\mu'_1$, $\alpha_3 = f_3\mu'_1$ and $\lambda = \mu'_2/(\mu'_1)^2$. To test the hypothesis of constant catchability it is sufficient to test whether the variance of c is zero. This is equivalent to testing $H_0 : d = 0$, where d, the square of the coefficient of variation, is given by

$$\begin{aligned} d &= \mu_2/(\mu'_1)^2 \\ &= \lambda - 1. \end{aligned} \tag{3.19}$$

Now, for the multinomial distribution (3.18) the maximum-likelihood estimates of α_2, α_3 and λ are simply the moment estimates

$$\hat{\alpha}_1 = (m_{12}+m_{123})/n_1 ;$$

$$\hat{\alpha}_2 = (m_{13}+m_{123})/n_1 ,$$

and

$$\hat{\lambda} = n_1 m_{123}/[(m_{12}+m_{123})(m_{13}+m_{123})].$$

Therefore, writing $\hat{d} = \hat{\lambda} - 1$, we have, from the delta method, the asymptotic expressions

$$E[\hat{d}] = d + d(1+d)/n_1 \tag{3.20}$$

and

$$V[\hat{d}] = \frac{d+1}{n_1\alpha_2\alpha_3}[1-(\alpha_2+\alpha_3)(d+1) + \alpha_2\alpha_3(d+1)(2d+1)]. \tag{3.21}$$

The asymptotic bias and variance of $\hat{d}$ can be estimated as usual by replacing each unknown parameter by its estimate; thus

$$\hat{V}[\hat{d}] = \frac{m_{123}^2 n_1^2}{(m_{12}+m_{123})^2 (m_{13}+m_{123})^2}\left(\frac{1}{m_{123}} - \frac{1}{n_1} - \frac{m_{12}+m_{13}}{(m_{12}+m_{123})(m_{13}+m_{123})}\right). \tag{3.22}$$

Under the null hypothesis H_0 we have, assuming approximate normality, that $\hat{d}$ is $\mathfrak{N}(0, (1-\alpha_2)(1-\alpha_3)/(n_1\alpha_2\alpha_3))$. Therefore a one-sided test of H_0 (one-sided since $d \geqslant 0$) is given by the statistic

$$z = \hat{d}\sqrt{\{n_1\hat{\alpha}_2\hat{\alpha}_3/[(1-\hat{\alpha}_2)(1-\hat{\alpha}_3)]\}}, \tag{3.23}$$

which is approximately distributed as the unit normal when H_0 is true: if z is negative we accept $d = 0$ as the most reasonable hypothesis. It is readily seen that the power of this test will be maximised when f_2, f_3 and n_1 are as large as possible.

In conclusion we note from Cormack that the above experiment does not have to be carried out on a free-ranging population. As the interest is in the physiological or psychological behaviour of the animal with regard to one particular sampling technique, it is possible that the assumptions necessary for the method could still be satisfied by a controlled population of size n_1 living within fixed boundaries.

METHOD 2. Suppose now that the sampling technique is the same for all three samples and let c_j be the catchability of the jth individual in the *total* population ($j = 1, 2, \ldots, N$). If the sampling intensity for the first sample is f_1 then the probability that the jth individual is caught in the first sample is $f_1 c_j$. Hence

$$\begin{aligned} \Pr[\text{in } n_1 \text{ and } c_j < c < c_j + \delta c_j] &= \Pr[\text{in } n_1 \mid c_j < c < c_j + \delta c_j] \\ &\quad \times \Pr[c_j < c < c_j + \delta c_j] \\ &\approx f_1 c_j\, g(c_j)\, \delta c_j\,, \\ \Pr[\text{in } n_1] &= \int_0^1 f_1 c\, g(c)\, dc = f_1 \mu_1' \end{aligned}$$

and

$$\Pr[c_j < c < c_j + \delta c_j \mid \text{in } n_1] \approx (1/\mu_1')\, c_j\, g(c_j)\, \delta c_j\,.$$

Therefore, in effect, subsequent sampling for n_2 and n_3 is from a marked population of size n_1 with catchability density function

$$h(c) = (1/\mu_1') c\, g(c)\,.$$

Hence $\hat{d}$ defined in Method 1 is now an estimate of D, the square of the coefficient of variation for the density $h(c)$. This means that d must be replaced by D in (3.20) and (3.21); the statistic z (3.23) now provides a test for $D = 0$ rather than for $d = 0$.

From Cormack [1966] it can be shown that

$$\begin{aligned} D &= [\mu_3' \mu_1' / (\mu_2')^2] - 1 \\ &= \frac{1}{(1+d)^2}\left[\frac{\mu_3}{(\mu_1')^3} + d(1-d)\right], \end{aligned} \qquad (3.24)$$

so that $\hat{d}$ cannot be used for estimating d without some knowledge of μ_3 and μ_1'. However, if we can make the additional assumption that the original density $g(c)$ is symmetrical, then $\mu_3 = 0$, $\mu_1' = \frac{1}{2}$, and solving (3.24), d can be estimated by

$$\hat{d}_1 = \frac{1 - 2\hat{d} - (1 - 8\hat{d})^{\frac{1}{2}}}{2(1 + \hat{d})},$$

where $\hat{d}$ is the estimate of D. Using the delta method, the asymptotic variance of $\hat{d}_1$ is given by

$$V[\hat{d}_1] = \frac{(1+d)^6}{(1-3d)^2}\, V[\hat{d}],$$

where d is estimated by $\hat{d}_1$ and $V[\hat{d}]$ by (3.22). Now $d = 0$ implies both $D = 0$ and $V[\hat{d}_1] = V[\hat{d}]$, so that the statistic for testing $d = 0$ is the same as before, namely (3.23) but with $\hat{d}$ replaced by $\hat{d}_1$. Since from (3.24) $D < d$ when $g(c)$ is symmetrical, we have $\hat{d} < \hat{d}_1$ and a situation could arise in which $\hat{d}_1$ is significantly different from zero but $\hat{d}$ is not. This means that we must be clear which of the two methods is appropriate to a given experimental situation and choose the correct statistic, $\hat{d}$ or $\hat{d}_1$, for testing $d = 0$.

COMPARING THE METHODS. For a certain class of alternatives (namely $g(c)$ is a beta distribution) Cormack compares the relative powers of the two methods given above and notes that for moderate deviations from the null hypothesis of constant catchability the first method is the more powerful. However, as the alternative hypothesis comes closer to the null hypothesis, the second method eventually becomes the more powerful. Unfortunately, as with the Petersen method, both tests will be rather insensitive unless n_1 is large and a significant proportion of the first sample is captured in the second and third samples. For example, Table 3.6 gives the size n_1 required for different intensities f_1 $(= f_2 = f_3)$, to obtain a 90 per cent probability of disproving the null hypothesis at the 5 per cent level of significance when $g(c)$ is actually the uniform distribution on $[0, 1]$. Thus, for a recapture rate of $12\frac{1}{2}$ per cent we require $n_1 = 4133$ for the first method.

TABLE 3.6

The size (n_1) of the first sample required to discover a uniform distribution for different sampling intensities (f_1): from Cormack [1966: Table 1].

f_1	0·25	0·5	1
% recaptured ($100\mu_1' f_1$)	$12\frac{1}{2}$%	25%	50%
n_1 (method 1)	4133	730	84
n_1 (method 2)	8485	1559	174

MORTALITY PRESENT. Suppose we relax the assumption of a closed population to the extent of allowing mortality. Let ϕ_1, ϕ_2 be the probabilities of survival for a tagged animal between the first two and the second two samples respectively. To allow for the estimation of the ϕ_i, we require additional information provided by releasing a further r_2 tagged animals (in addition to the m_2) into the population after the second sample and noting the recaptures, m_{23} say, from this group in the third sample.

Under the assumptions of method 1, and assuming the members of r_2 to represent a random sample of catchabilities from $g(c)$, the joint probability of m_{12}, m_{13}, m_{23}, m_{123} is given by

$$\frac{n_1!}{m_{12}!\, m_{13}!\, m_{123}!\, m_{10}!} [\phi_1\alpha_2(1-\phi_2\alpha_3\lambda)]^{m_{12}}[\phi_1(1-\alpha_2\lambda)\phi_2\alpha_3]^{m_{13}}$$

$$\times (\phi_1\alpha_2\phi_2\alpha_3\lambda)^{m_{123}}\, \theta^{m_{10}} \binom{r_2}{m_{23}} (\phi_2\alpha_3)^{m_{23}}(1-\phi_2\alpha_3)^{r_2-m_{23}},$$

where $\theta = 1 - \phi_1\alpha_2 - \phi_1\phi_2\alpha_3(1-\alpha_2\lambda)$.
In the above probability function we have four independent observations but five parameters, and we find that ϕ_2 and α_3 cannot be estimated separately; only the product $\phi_2\alpha_3$ is estimable.

The maximum-likelihood estimates (which are also the moment estimates in this case) are

$$\hat{\phi}_1 = (m_{13}+m_{123})r_2/(m_{23}n_1),$$

$$\hat{\alpha}_2 = m_{23}(m_{12}+m_{123})/[r_2(m_{13}+m_{123})],$$

$$\hat{\phi}_2\hat{\alpha}_3 = m_{23}/r_2,$$

and

$$\hat{\lambda} = m_{123}r_2/[m_{23}(m_{12}+m_{123})].$$

Setting $\hat{d} = \hat{\lambda} - 1$ as before, we have approximately

$$E[\hat{d}] = d + \frac{\lambda}{r_2}\left(\frac{1}{\phi_2\alpha_3} - 1\right)$$

and

$$V[\hat{d}] = \frac{\lambda}{n_1\phi_1\alpha_2\phi_2\alpha_3}[1 + \lambda(\phi_1\alpha_2 n_1 r_2^{-1} - \phi_2\alpha_3 - \phi_1\alpha_2\phi_2\alpha_3 n_1 r_2^{-1})].$$

To test the hypothesis $d = 0$ we again use a one-tailed test based on the statistic $z = (\hat{d} - \hat{b})/\hat{\sigma}$, where

$$\hat{b} = \hat{\lambda}\left(\frac{1}{m_{23}} - \frac{1}{r_2}\right)$$

and

$$\hat{\sigma}^2 = \frac{r_2^2 m_{123}^2}{m_{23}^2(m_{12}+m_{123})^2}\left\{\frac{1}{m_{123}} + \frac{1}{m_{23}} - \frac{1}{(m_{12}+m_{123})} - \frac{1}{r_2}\right\}.$$

In testing $d = 0$ there is unfortunately a very considerable loss in power through having to estimate ϕ_1. Cormack [1966] states that even when the death-rate is actually zero ($\phi_1 = \phi_2 = 1$), five to ten times (depending on the sampling intensity) the number of tagged animals $n_1 + r_2$ are required to give the same discriminatory power as the test for a closed population. Because of this lack of sensitivity and the need for such large numbers of tagged animals Cormack suggests that the experiment should be arranged so that the possibility of death can be neglected.

If information on ϕ_1 is required we can use $\hat{\phi}_1$ and the approximate variance formula

$$V[\hat{\phi}_1] \approx \frac{\phi_1}{n_1 r_2\phi_2\alpha_3}[r_2(1-\phi_1\phi_2\alpha_3) + n_1\phi_1(1-\phi_2\alpha_3)]$$

to obtain an approximate confidence interval for ϕ_1. We note that if f_3/f_2 ($= \alpha_3/\alpha_2$) is known, we can obtain the estimates

$$\hat{\alpha}_3 = \hat{\alpha}_2 f_3/f_2$$

and

$$\hat{\phi}_2 = m_{23}/(r_2\hat{\alpha}_3).$$

When $\phi_1 = 1$ and f_3/f_2 is known, the second release of r_2 animals is unnecessary as the parameters α_2, ϕ_2 and λ can be estimated from the joint multinomial distribution of m_{12}, m_{13}, m_{123} and m_{10}. In this case the maximum-likelihood estimates are again the moment-estimates, and their asymptotic means and variances can be derived using the delta method. Also the hypothesis $\phi_2 = 1$ can be tested using a standard goodness-of-fit statistic.

3.2.3 Assumption (e): no loss of tags

If animals lose their marks or tags, the observed recaptures will be smaller than expected and N^* will overestimate N. Therefore considerable thought should be given to the choice of tag, and some experiments should be carried out either before or during the census period to check for tag losses or tag deterioration. The type of tag chosen will depend on such factors as the species studied, the information required by the tag, time and personnel available for tagging, and the method of tag return – whether by hunter, fisherman or research worker. Obviously tags should be durable so that they are not lost through the effects of weather or physical changes in the animal, such as moulting.

1 Types of marks and tags

There are a number of useful review articles on marking methods, e.g. mammals (Taber [1956]), birds (Cottam [1956]), amphibians and reptiles (Woodbury [1956]), fish (Ricker [1956], North Atlantic Fish Marking Symposium [1963], Stott [1968]), insects (Dobson [1962], Gangwere *et al.* [1964], Southwood [1966] and White [1970]) and for general references Taber and Cowan [1969] and Southwood [1966]. The following is a brief list of the main methods of marking and tagging; see also **12.7.3** (*2*).

PAINTS. Slow-drying oil paints, quick-drying cellulose paints and lacquers, and reflecting paints for night detection, are particularly useful for the study of insect populations.

DYES. Dyes in solution or powder form have been used for a wide variety of populations. They can be applied externally, and in this respect automatic marking without capturing is sometimes possible (Taber and Cowan [1969]), or they can be applied internally by injecting into the tissues. For relative estimates of population density, dyes can also be introduced into food, giving marked faeces (New [1958, 1959]).

TAGS. Various tags, bands and rings can be attached externally, and small internal tags can sometimes be used, particularly when the recapturing is done by trained personnel.

MUTILATION. Mutilation methods have been widely used for vertebrates, e.g. fin clipping or punching for fish, toe clipping and fur clipping for small mammals.

MUTATIONS. Occasionally mutants which are readily distinguished from the normal species can be introduced into the population to act as the marked sample n_1. This method has been used for studying insect populations (Dobson [1962]).

FLUORESCENT SOLUTIONS. These can for example be sprayed onto insects and later detected with an ultraviolet lamp.

RADIOACTIVE ISOTOPES. Despite the obvious problems of cost and danger to personnel, radioactive isotopes are being widely used for population studies; see Giles [1969], Southwood [1966], Peterle [1969] and Kemp and Keith [1970].

RADIO TRACKING. With the development of miniature transistorised radio transmitters, radio tracking using transmitter "tags" is now being used for many species (Giles [1969]).(*) Transmitters can be used not only for locating animals irrespective of cover, weather or time of day, but they can also be used for relaying physiological data such as body temperature, pulse-rate etc. (e.g. Slater [1963], Stoddart [1970]). Recently transmitters have been used for studying prey–predator movements (Mech [1967]). This approach is a step towards solving the problem of obtaining biological data from free-ranging animals rather than from animals in captivity where behaviour patterns may be very different. Some recent examples of radio tracking, together with further references, are given in Sanderson [1966], *Journal of Wildlife Management* [1967], Adams and Davis [1967], Heezen and Tester [1967], Knowlton *et al.* [1968], and Hessler *et al.* [1970]; ultrasonic transmitters have also been useful for studying fish movements (e.g. Henderson *et al.* [1966]). More recent references are given in **12.1.1** (*2*).

PARASITES. These can sometimes be used as "natural" tags (e.g. Sindermann [1961]).

2 Estimating tag loss

One simple method of detecting tag loss is to give all the n_1 animals in the first sample two types of tags and then to note those recaptures with just one tag and those with both tags intact (Beverton and Holt [1957], Gulland [1963]). Denoting the two types of tag by A and B, we define:

π_x = probability that a tag of type x is lost by the time of the second sample ($x = A, B$),
π_{AB} = probability that both tags are lost,
m_x = number of tagged animals in the second sample, with tag x only ($x = A, B$),
m_{AB} = number of tagged animals in the second sample with both tags,
and m_2 = members of n_1 caught in n_2.

Assuming that the tags are independent of each other ($\pi_{AB} = \pi_A \pi_B$), the

(*)See also Brander and Cochran [1969].

joint probability function of m_A, m_B, m_{AB} and m_2 is given by

$$f(m_A, m_B, m_{AB}, m_2 \mid n_1, n_2) = f(m_A, m_B, m_{AB} \mid m_2) f(m_2 \mid n_1, n_2)$$

where

$$f(m_A, m_B, m_{AB} \mid m_2) = \frac{m_2!}{m_A!\, m_B!\, m_{AB}!\, m_0!} [(1-\pi_A)\pi_B]^{m_A}$$
$$\times [\pi_A(1-\pi_B)]^{m_B} [(1-\pi_A)(1-\pi_B)]^{m_{AB}} [\pi_A \pi_B]^{m_0},$$
$$m_0 = m_2 - m_A - m_B - m_{AB},$$

and $f(m_2 \mid n_1, n_2)$ is given by equation (3.1). The maximum-likelihood estimates of N, m_2, π_A and π_B (which are also the moment estimates) are given by

$$\begin{aligned} \hat{N}_{AB} &= n_1 n_2 / \hat{m}_2, \\ m_A &= \hat{m}_2(1-\hat{\pi}_A)\hat{\pi}_B, \\ m_B &= \hat{m}_2 \hat{\pi}_A (1-\hat{\pi}_B), \\ m_{AB} &= \hat{m}_2(1-\hat{\pi}_A)(1-\hat{\pi}_B), \end{aligned}$$

which have solutions

$$\begin{aligned} \hat{\pi}_A &= m_B/(m_B + m_{AB}), \\ \hat{\pi}_B &= m_A/(m_A + m_{AB}), \end{aligned}$$

and

$$\begin{aligned} \hat{m}_2 &= (m_A + m_{AB})(m_B + m_{AB})/m_{AB} \\ &= c(m_A + m_B + m_{AB}), \quad \text{say.} \end{aligned}$$

This means that the observed recaptures $m_A + m_B + m_{AB}$ must be corrected by a factor

$$c = \left[1 - \frac{m_A m_B}{(m_A + m_{AB})(m_B + m_{AB})}\right]^{-1} = \frac{1}{(1-\hat{\pi}_A)(1-\hat{\pi}_B)}$$

to give an estimate of the actual number of recaptures m_2. For large samples, $\hat{N}_{AB}$ is an approximately unbiased estimate of N, and defining $\hat{N} = n_1 n_2 / m_2$, we have from **1.3.4**

$$\begin{aligned} V[\hat{N}_{AB} \mid n_1, n_2] &= \underset{m_2}{E} \{V[\hat{N}_{AB} \mid n_1, n_2, m_2]\} + \underset{m_2}{V} \{E[\hat{N}_{AB} \mid n_1, n_2, m_2]\} \\ &\approx \underset{m_2}{E} \{V[\hat{N}_{AB} \mid n_1, n_2, m_2]\} + V\{\hat{N} \mid n_1, n_2\}, \end{aligned}$$

which, by the delta method (and equation 3.2), is approximately equal to

$$\frac{N^3}{n_1 n_2} \pi_A \pi_B \left[\frac{1}{(1-\pi_A)(1-\pi_B)}\right] + \frac{N^3}{n_1 n_2}\left[1 + \frac{2N}{n_1 n_2} + 6\left(\frac{N}{n_1 n_2}\right)^2\right]. \tag{3.25}$$

In some situations the only information recorded is the number of tags for each tagged individual, so that just the numbers m_{AB} and m_C $(= m_A + m_B)$ are available. For this case we can still estimate m_2 if we can assume that $\pi_A = \pi_B$ $(= \pi$ say). We then have

$$f(m_C, m_{AB} \mid m_2) = \frac{m_2!}{m_C!\, m_{AB}!\, m_0!}[2\pi(1-\pi)]^{m_C}[(1-\pi)^2]^{m_{AB}}[\pi^2]^{m_0},$$

and the maximum-likelihood estimates (and moment estimates) of m_2 and π are

$$\tilde{m}_2 = (m_C + 2m_{AB})^2/4m_{AB}$$

and

$$\tilde{\pi} = m_C/(m_C + 2m_{AB}),$$

an estimate obtained independently by several authors (**12.7.3** (*3*)). Setting

$$\tilde{N}_{AB} = n_1 n_2/\tilde{m}_2,$$

this estimate is asymptotically unbiased, and its asymptotic variance is the same as (3.25) but for the first expression which is replaced by

$$\frac{N^3\pi^2}{n_1 n_2(1-\pi)^2}.$$

When information on m_A and m_B is available for the whole or perhaps a part of the second sample, we can test the hypothesis $\pi_A = \pi_B$ as follows. Let $\pi_B = k\pi_A$; then, from (**1.3.7**(*2*)), the conditional probability function of m_A given $m_A + m_B$ is given by

$$f(m_A \mid m_A + m_B) = \binom{m_A + m_B}{m_A} p^{m_A} q^{m_B}, \tag{3.26}$$

where $p = \dfrac{k(1-\pi_A)\pi_A}{k(1-\pi_A)\pi_A + \pi_A(1-k\pi_A)}$ and $q = 1 - p$.

Testing $k = 1$ is therefore equivalent to testing $p = \frac{1}{2}$ for the above binomial distribution (3.26).

We see from **3.2.1**(*2*) that the above theory will still hold when there is mortality, provided that the deaths constitute a random sample and the second sample is random. It is also noted that when tag losses are small, which will be the case for many populations, the above variance terms involving π_A, π_B and π will be negligible, so that the effect of tag loss on the variance can be neglected. For example, when π_A and π_B are both less than 0·1 the contribution of the first expression to (3.25) is less than about $3\frac{1}{2}$ per cent. In such cases, therefore, the general theory of **3.1** can still be used but with m_2 replaced by $\hat{m}_2$ or $\tilde{m}_2$.

In conclusion we note that the above theory can be extended to the case of more than two marks by defining π_A as the probability that a particular mark is lost and π_B as the probability of losing at least one of the other marks. Whatever the marking method used, it is recommended (Southwood [1966: 72]) that one should follow the policy of Michener *et al.* [1955], and use a system in which all individuals bear the same number of marks (e.g. if colour marking is used one can allocate a colour to zero in the units, tens, hundreds position). An interesting coding technique for insects based on this system is given by White [1970].

3.2.4 Assumption (f): all tags reported

When there are incomplete tag returns the observed value of m_2 will be too small and N^* will overestimate N. This problem arises when tags are returned by hunters, commercial fishermen, local inhabitants, etc. who may or may not be interested in the experiment.(*) It is found that the percentage return is usually related to such factors as the training of observers, size of reward (Bellrose [1955], Atwood and Geis [1960], Paulik [1961]), publicity given to the experiment, and ease of visibility of the tag. However, if the second sample can be classified into two categories, one which has a known reported ratio of unity or nearly so (e.g. through inspection by special observers), and the other with an unknown reported ratio, then Paulik's [1961] method described below can be used to test whether the unknown ratio is significantly less than unity.

Let $n_2 = n_{2a} + n_{2b}$, where the suffixes a and b denote the two categories respectively. Let m_{2a}, m_{2b} be the number of recaptures in the two groups and let r_{2a}, r_{2b} be the number of recaptures actually reported (i.e. $r_{2a} = m_{2a}$). Then if n_2 is large, the tag ratio n_1/N small, and the tag ratio the same for both groups, we can use the Poisson approximation to the hypergeometric distribution (cf. **3.1.4**) and assume that the recaptures m_{2i} in each group have independent Poisson distributions with parameters $n_1 n_{2i}/N$ $(i = a, b)$. If ρ is the (constant) probability that a member of m_{2b} is reported, then the conditional probability function of r_{2b} given m_{2b} is

$$f(r_{2b} \mid m_{2b}) = \binom{m_{2b}}{r_{2b}} \rho^{r_{2b}} (1-\rho)^{m_{2b}-r_{2b}} \tag{3.27}$$

and, from **1.3.7** (*1*), the unconditional probability function of r_{2b} is Poisson with parameter $n_1 n_{2b}\rho/N$. As r_{2a} and r_{2b} are independent Poisson variables it then follows from **1.3.7** (*1*) that

$$f(r_{2a} \mid r) = \binom{r}{r_{2a}} p^{r_{2a}} q^{r_{2b}}, \tag{3.28}$$

where $r = r_{2a} + r_{2b}$ and $p = n_{2a}/(n_{2a} + \rho n_{2b})$. Therefore an estimate of ρ is given by

$$r_{2a}/r = n_{2a}/(n_{2a} + \hat{\rho} n_{2b}),$$

or

$$\hat{\rho} = \left(\frac{r_{2b}}{n_{2b}}\right) \Big/ \left(\frac{r_{2a}}{n_{2a}}\right),$$

and a test of $H_0 : \rho = 1$ against the one-sided alternative $\rho < 1$ is equivalent to testing $p = n_{2a}/n_2$ $(= p_0$ say) against $p > p_0$ for the binomial model (3.28). We note that an estimate of m_2, the actual number of tagged individuals recaptured, is given by

$$\hat{m}_2 = m_{2a} + (r_{2b}/\hat{\rho}) = m_{2a} n_2/n_{2a}$$

(*)See also p. 360 and **12.7.3** (*4*).

and

$$\hat{N} = n_1 n_2/\hat{m}_2 = n_1 n_{2a}/m_{2a}.$$

This means that we base the Petersen estimate on the recapture data for which we have a 100 per cent reporting rate.

In fishery research, if the tag ratio remains constant during the entire season, the above scheme is very flexible. For example, all the fish landed during the first part of the season may be inspected and none of those during the latter part of the season, or vice versa. However, if the tag ratio varies, suppose that the season can be divided into intervals for which the tag ratio is constant in each interval. Then, provided a constant fraction of the catch in each interval is inspected, i.e. $n_{2b}/n_{2a}\ (=\gamma)$ is constant, Paulik shows that (3.28) still holds but with $p = (1+\rho\gamma)^{-1}$; r and r_{2a} now refer to the season's total.

If the tag ratio n_1/N is not small, the m_{2i} may be more appropriately represented by a binomial law (cf. **3.1.4**), namely

$$f_1(m_{2i}) = \binom{n_{2i}}{m_{2i}} \left(\frac{n_1}{N}\right)^{m_{2i}} \left(1-\frac{n_1}{N}\right)^{n_{2i}-m_{2i}}.$$

Assuming (3.27), the above equation then leads to

$$f_2(r_{2i}) = \binom{n_{2i}}{r_{2i}} \left(\frac{\rho_i n_1}{N}\right)^{r_{2i}} \left(1-\frac{\rho_i n_1}{N}\right)^{n_{2i}-r_{2i}}, \quad (i = a,\ b), \qquad (3.29)$$

where $\rho_a = 1$ and $\rho_b = \rho$. Therefore testing H_0 is now equivalent to a test for homogeneity in the 2-by-2 table:

r_{2a}	r_{2b}	r
$n_{2a} - r_{2a}$	$n_{2b} - r_{2b}$	$n_2 - r$
n_{2a}	n_{2b}	n_2

As the alternative hypothesis is one-sided, the usual test is modified slightly by only rejecting H_0 when $r_{2a}/n_{2a} > r_{2b}/n_{2b}$, and the chi-squared statistic is significant at the 2α level of significance, where α is the size of the test. When the r_{2i} and n_{2i} are small, Fisher's exact test (Keeping [1962]) should be used.

It is important not only to detect incomplete reporting after the experiment, but also to decide before the experiment how many tags should be released and how much of the sample should be inspected to be reasonably sure of detecting non-reporting of a given magnitude. Paulik shows that such information can be obtained by examining the power of the test of H_0 for particular alternatives $\rho < 1$. When the smaller of rp and rq is greater than 5 (preferably greater than 10: Raff [1956: 296]), and the correction for continuity is used, the normal approximation to the binomial distribution (3.28) can be used. In this case the most powerful one-sided test of H_0 is to reject H_0 when $r_{2a} \geqslant d$, where

$$d = z_\alpha \sqrt{(rp_0 q_0)} + \tfrac{1}{2} + rp_0$$

(or the nearest integer greater than the right-hand side). Here z_α satisfies $\Pr[z \geqslant z_\alpha] = \alpha$, where z has a unit normal distribution. If β is the Type II error for a given alternative ρ, i.e. $1 - \beta = \Pr[r_{2a} \geqslant d \mid \rho]$, then the number of recoveries needed to test H_0 at pre-set values of α and β for different values of ρ can be expressed as

$$r = (z_\beta\sqrt{(pq)} + z_\alpha\sqrt{(p_0 q_0)})^2/(p-p_0)^2 ,$$

where $$p = \frac{n_{2a}}{n_{2a} + \rho n_{2b}} = \frac{p_0}{p_0 + \rho(1-p_0)} .$$

Paulik has tabulated r as a function of β, ρ and p_0 for $\alpha = 0{\cdot}10,\ 0{\cdot}05,\ 0{\cdot}01$, and these tables are reproduced in **A2**. He points out that if the binomial model (3.28) is more appropriate than the Poisson model which led to (3.29), then the value of r obtained from the tables is conservative in that the true power of the test exceeds $1 - \beta$.

To demonstrate the use of Paulik's table, suppose that a biologist wants to be 99 per cent sure of detecting a non-reporting of 20 per cent or more of the tags recovered by a commercial fishery. From past experience he knows that at least 30 per cent of the tags released will be recovered. If $\alpha = 0{\cdot}10$ and 25 per cent of the total catch is to be inspected, how many tags should be released? Entering Table **A2** for $\alpha = 0{\cdot}10$, $\rho = 0{\cdot}80$, $p_0 = 0{\cdot}25$ and $1 - \beta = 0{\cdot}99$, we find that $r = 1340$. Now

$$\begin{aligned} E[r] &= E[m_{2a}] + E[r_{2b}] \\ &= \frac{n_1 n_{2a}}{N} + \frac{n_1 n_{2b}\rho}{N} \\ &= \frac{n_2}{N}[p_0 + \rho q_0]n_1 \end{aligned}$$

and substituting the above values (with $n_2/N \approx m_2/n_1$),

$$1340 = 0{\cdot}30[0{\cdot}25 + (0{\cdot}80)(0{\cdot}75)]n_1$$

or $n_1 = 5255$ tags liberated. If, however, the 30 per cent recovery estimate had been based on the percentage of tags actually reported then $n_2/N \approx 0{\cdot}30/\rho$ and $n_1 = 4204$.

The above method of Paulik's is a very flexible one and can be applied to a number of different experimental situations. In particular it can be used for the case when p_0 refers to the proportion of the total *effort* inspected rather than the proportion of the total catch. For example, in the above numerical application p_0 could refer to the percentage of traps to be inspected, or even to the percentage of fishermen who agree to attend a special course of instruction and cooperate fully in the experiment. A theoretical justification for this extension of the above theory, along with several examples, is given in **8.3.1**.

We note that incomplete reporting can arise either through (i) tags being accidentally overlooked or (ii) tags being deliberately withheld. For larger animals a tag can usually be designed so that the chance of it being overlooked is negligible. In fisheries, however, where large numbers are rapidly handled and tags are small, this source of error can be a real problem. The second source of error can usually be minimised by extensive advertising and the offering of an "adequate" reward for a returned tag. One obvious method of separating the two types of incomplete reporting is by planting tagged fish in the catches (Margetts [1963]).

In conclusion we see that the above methods of Paulik will still apply even if natural mortality is operating, provided that tagged and untagged have the same mortality rate, so that the tag ratio remains constant throughout.

3.2.5 Populations with sub-categories

1 Pooling

It will frequently happen that the population being investigated can be divided into several sub-categories according to sex, age, size, species, etc. If the Petersen method is used and the appropriate assumptions are satisfied for the population as a whole, the question arises as to whether it is better to estimate N by pooling the data or by estimating the size of each subgroup separately and then summing the estimates. This question is particularly crucial when the number of recaptures in each subgroup is small. By considering just two subgroups, Chapman [1951: 151] shows that the former method, namely pooling, is to be preferred. That is, if $N = X + Y$ (where X and Y are the subgroup sizes), $n_1 = n_{1x} + n_{1y}$, etc., and the assumptions of **3.1.1** are satisfied for the *whole* population, then

$$N^* = \frac{(n_1 + 1)(n_2 + 1)}{(m_2 + 1)} - 1$$

has smaller variance than

$$X^* + Y^* = \frac{(n_{1x} + 1)(n_{2x} + 1)}{(m_{2x} + 1)} + \frac{(n_{1y} + 1)(n_{2y} + 1)}{(m_{2y} + 1)} - 2.$$

However, if we wish to estimate X, say, Chapman suggests the estimate

$$\begin{aligned}\tilde{X} &= \frac{n_{1x} + n_{2x} - m_{2x}}{n_1 + n_2 - m_2} . N^* \\ &= \tilde{P}_x N^*, \quad \text{say,}\end{aligned}$$

where $P_x = X/N$. When both samples are random, so that $E[n_{1x} \mid n_1] = n_1 P_x$, $E[n_{2x} \mid n_2] = n_2 P_x$, etc. then

$$\begin{aligned}E[\tilde{X} \mid n_1, n_2] &= \underset{m_2}{E}\, E[\tilde{X} \mid n_1, n_2, m_2] \\ &= P_x \underset{m_2}{E}\, [N^* \mid n_1, n_2] \\ &\approx P_x N \\ &= X\end{aligned}$$

and $\tilde{X}$ is almost unbiased (exactly unbiased when $n_1 + n_2 \geqslant N$). The asymptotic variance of $\tilde{X}$ is given by

$$V[\tilde{X}\,|\,n_1, n_2] = P_x^2 N^2 \left[\left(\frac{N}{n_1 n_2}\right) + 2\left(\frac{N}{n_1 n_2}\right)^2 + 6\left(\frac{N}{n_1 n_2}\right)^3\right] + \frac{P_x(1-P_x)N^2}{n_1 + n_2 + 1}$$

and, estimating N by the Petersen estimate $\hat{N}$ in the above square bracket, we have the variance estimate

$$v[\tilde{X}] = \tilde{X}^2\left[\frac{1}{m_2} + \frac{2}{m_2^2} + \frac{6}{m_2^3}\right] + \frac{\tilde{X}(N^* - \tilde{X})}{n_1 + n_2 + 1}.$$

To compare the efficiencies of $\tilde{X}$ and X^* we note that the variance of X^* can be estimated by (cf. (3.2))

$$v[X^*] = X^{*2}\left[\frac{1}{m_{2x}} + \frac{2}{m_{2x}^2} + \frac{6}{m_{2x}^3}\right].$$

Since $m_2 > m_{2x}$ and the second term of $v[\tilde{X}]$ is relatively small for large n_1 and n_2, a comparison of the above two estimates would indicate that $\tilde{X}$ will generally have a smaller variance than X^*. In fact, m_{2x} may be so small (e.g. less than 7) that X^* not only has a comparatively large variance but also has considerable bias.

We note that all the above formulae still apply when there are more than two subgroups. The more subgroups there are, the smaller the ratios m_{2x}/m_2 and $v[\tilde{X}]/v[X^*]$ will be.

2 *Variable catchability*

As pointed out in **3.2.2**, the catchability of individuals may vary between subgroups because of such factors as mobility, bait or trap preference, etc. It is therefore of interest to investigate the effect of this variation on our estimates for the special case when the probability of capture is constant within each subgroup for both samples. Let p_{ix} be the probability that a member of X is caught in the ith sample ($i = 1, 2$) and let $X, Y, Z, \ldots$ denote both the respective subgroups and their sizes. Then since, for a given subgroup, the probability of capture in the second sample is constant irrespective of mark status, assumption (c) that marking does not affect catchability is satisfied. Therefore, in the notation of **3.2.2**(2), the pair of random variables (x_j, y_j) takes values (p_{1x}, p_{2x}), (p_{1y}, p_{2y}), etc. with probabilities X/N, Y/N, etc., so that

$$p_1 = E[x_j] = (p_{1x}X + p_{1y}Y + \ldots)/N$$

and

$$p_{12} = E[x_j y_j] = (p_{1x}p_{2x}X + p_{1y}p_{2y}Y + \ldots)/N.$$

Hence from (3.14), (3.15) and (3.16):

$$E[(N^* - n_1)\,|\,n_1, n_2] \approx \frac{(B - p_1)(N - n_1)}{(1 - p_1)},$$

where
$$1 - B = 1 - (p_1 p_2 / p_{12})$$
$$= 1 - \frac{(p_{1x}X + p_{1y}Y + \ldots)(p_{2x}X + p_{2y}Y + \ldots)}{N(p_{1x}p_{2x}X + p_{1y}p_{2y}Y + \ldots)}$$
$$= \frac{\Sigma(p_{1x} - p_{1y})(p_{2x} - p_{2y})XY}{N(p_{1x}p_{2x}X + p_{1y}p_{2y}Y + \ldots)}$$

and Σ denotes summation over all pairs of subgroups (X, Y), (X, Z), (Y, Z), etc. Now for any pair (X, Y) we would expect $(p_{1x} - p_{1y})$ and $(p_{2x} - p_{2y})$ to have the same sign. Therefore N^* will generally underestimate N, though the bias will be small if the p_{1x}, p_{1y}, etc. are not too different. This of course follows directly from the general theory of p. 86, where it was noted that a positive correlation between the probabilities of capture in the two samples leads to underestimation.

When $p_{ix} \neq p_{iy}$ $(i = 1, 2)$, N^* is biased and $\widetilde{P}_x$ is no longer a satisfactory estimate of P_x; in this case X^* should be used instead of $\widetilde{X}$. However, when the probabilities of capture are the same for the first sample (assumption (b) true) and different for the second, we have $p_{1x} = p_{1y} = \ldots$, $B = 1$, and N^* is approximately unbiased. This means that $\widetilde{X}$ can now be used but with P_x estimated by n_{1x}/n_1. The only change in $V[\widetilde{X} \mid n_1, n_2]$ above is to replace the second expression by $P_x(1 - P_x)N^2/n_1$. For the opposite situation, when the probabilities of capture are the same for the second sample only, we simply estimate P_x by n_{2x}/n_2, and the corresponding variance term is $P_x(1 - P_x)N^2/n_2$. We recall that when mortality and recruitment are negligible, and tagging does not affect catchability, the goodness-of-fit tests mentioned on pp. 71 and 73 are tests of the hypotheses $p_{2x} = p_{2y} = \ldots = p_{2w}$ and $p_{1x} = p_{1y} = \ldots = p_{1w}$ respectively. In conclusion we note that populations can also be stratified geographically and temporally; a detailed discussion of this problem is given in Chapter 11.

3.2.6 Independent estimates of migration, mortality and recruitment

Suppose that a_m and a_u are the net additions to the numbers of marked and unmarked between the two samples and let $a = a_m + a_u$ ($a_m \leqslant 0$, though a_u may be positive or negative). Then, if $\hat{a}_m$ and $\hat{a}_u$ are unbiased estimates of a_m and a_u respectively, the Petersen estimate of N, the initial population size, is

$$\hat{N} = \frac{n_1 + \hat{a}_m}{\hat{p}} - \hat{a}, \tag{3.30}$$

where $\hat{a} = \hat{a}_m + \hat{a}_u$ and $\hat{p} = m_2/n_2$. This estimate is asymptotically unbiased, and when $\hat{a}_m$, $\hat{a}_u$ and $\hat{p}$ are independent estimates (as will usually be the case) we have, using the delta method,

$$V[\hat{N} \mid n_1, n_2] \approx p^{-2}\{(N + a)^2 V[\hat{p} \mid n_1, n_2] + (1 - p)^2 V[\hat{a}_m] + p^2 V[\hat{a}_u]\}, \tag{3.31}$$

where $p = E[\hat{p}]$.

If the second sample is drawn randomly without replacement so that the hypergeometric model (3.1) is applicable, then

$$v[\hat{p}] = \frac{\hat{p}\hat{q}}{n_2}\left(1 - \frac{n_2}{\hat{N}+\hat{a}}\right)$$

is an approximately unbiased estimate of $V[\hat{p}]$. In this situation we also find that the modified estimate

$$N^* = \frac{(n_1 + \hat{a}_m + 1)(n_2 + 1)}{m_2 + 1} - \hat{a} - 1$$

is still approximately unbiased (exactly unbiased when $n_1 + a_m + n_2 \geqslant N + a$).

When subsampling is used, p may be estimated by the methods of **3.4**.

Example 3.2 Large-mouth bass (*Micropterus salmoides*): Paulik and Robson [1969]

In a mark-recapture experiment to estimate the number of large-mouth bass in a reservoir, 3000 bass were seined from the population for marking a month before the fishing season. A subsample of 500 marked bass were retained in a live-pen for observation and the remaining 2500 bass were returned to the population. It was found that 100 of the marked bass in the pen died in the first three weeks and none died in the fourth week, while in a control group of unmarked bass held in the same live-pen there were no mortalities. This suggests that the actual marking process was responsible for the mortality among the marked, and that marking mortalities were complete before the season started: we shall assume this to be true for the whole population. During the following fishing season 2300 bass were inspected and 400 were found to be marked.

Let θ be the proportion of marked bass surviving until the fishing season; then θ is estimated by $\hat{\theta} = 400/500 = 0{\cdot}80$. Now $n_1 = 3000 - 500 = 2500$, $\hat{a}_m = -n_1(1-\hat{\theta}) = -500$, $a_u = -500$ (those marked fish kept in the live-pen), $\hat{p} = 400/2300 = 0{\cdot}1739$, $n_1 + \hat{a}_m = n_1\hat{\theta} = 2000$, and from (3.30)

$$\hat{N} = \frac{2000(2300)}{400} + (500+500) = 12\,500.$$

As a first step in estimating the variance of $\hat{N}$ we note that

$$\frac{n_2}{\hat{N} + \hat{a}} = \frac{m_2}{n_1 + \hat{a}_m} = \frac{400}{2000} = \frac{1}{5}$$

and

$$v[\hat{p}] = \frac{0{\cdot}1739(0{\cdot}8261)}{2300} \cdot \frac{4}{5} = 0{\cdot}000\,500.$$

To derive $V[\hat{a}_m]$ we assume that $\hat{\theta}$ is a binomial random variable with parameters 500 and θ, i.e.

$$\begin{aligned} V[\hat{a}_m] &= n_1^2 V[1-\hat{\theta}] \\ &= n_1^2 V[\hat{\theta}] \\ &= n_1^2\theta(1-\theta)/500, \end{aligned}$$

which is estimated by

$$v[\hat{a}_m] = 2500^2\,\hat{\theta}(1-\hat{\theta})/500 = 2000.$$

Since a_u is known, $V[a_u] = 0$, so that from (3.31) $V[\hat{N} \mid n_1, n_2]$ is finally estimated by

$$v[\hat{N}] = \left(\frac{2300}{400}\right)^2 \{(11\,500)^2(0{\cdot}000\,050\,0) + (0{\cdot}8261)^2(2000)\}$$

$$= 263\,830.$$

Hence

$$\sqrt{v\,[\hat{N}]} = 514$$

and the approximate 95 per cent confidence interval $\hat{N} \pm 1.96\sqrt{v[\hat{N}]}$ is (11 500, 13 500).

3.3 EXAMPLES OF THE PETERSEN METHOD

In this section we shall consider seven applications of the Petersen method to biological populations; the first two examples are included for their historical interest.

Example 3.3 Population of France: Laplace [1783].

Laplace used the Petersen method for estimating the total population size N from a register of births for the whole country (the "marked" individuals n_1). His second "sample" consisted of a number of parishes of known total population size n_2, and m_2 was the number of births recorded for these parishes.

Example 3.4 North American Ducks: Lincoln [1930].

Lincoln trapped and banded large numbers of ducks before the annual dispersal from the breeding grounds. At every shooting season after a release he consistently received from the shooters about 12 per cent of the bands. From this he inferred that if 12 per cent of the banded had been shot, 12 per cent of the unbanded ducks also met the same fate. Estimating the total kill for a particular year at 5 million ducks, he used the Petersen method as follows to obtain an estimate of the total duck population N for the North American continent.

Let n_1 = number banded, n_2 = total kill, and m_2 = number of returned bands. Then $m_2/n_1 = 0{\cdot}12$ and

$$\hat{N} = n_2 \Big/ \left(\frac{m_2}{n_1}\right)$$

$$= 5 \times 10^6 \times 100/12$$

$$\approx 42 \text{ million birds.}$$

We note that this estimate depends on the assumption of a 100 per cent band return from the shooters.

Example 3.5 Underground ant (*Lasius flavus*): Odum and Pontin [1961].

It was found that radioactive tagging with P^{32} is a satisfactory method of tagging large numbers of ants in a colony of *Lasius flavus*. The population was effectively constant in size throughout the experiment as (i) the workers were relatively long-lived, (ii) the individuals newly emerged from the pupae could be recognised by their lighter colour and therefore eliminated from the counts, and (iii) there was little or no movement of individuals from one colony to another. The method of sampling on both occasions was to place a flat stone on top of the mound and then carefully lift the stone after a suitable period of time. It was found that the ants construct extensive galleries directly under the stone where warmth from the sun is favourable for the development of the pupae. During the summer the workers move through these chambers in a continuous stream to and from underground tunnels leading to the aphids clustered on roots of nearby plants. By gently lifting the stone, large numbers of individuals could therefore be removed for tagging and then conveniently returned to the mound with a minimum disturbance of the colony. It was stressed that a sufficient period of time should be allowed (5–10 days in this case) for the proper mixing of tagged and untagged before the stones are again lifted for the second sample. But this period should not be too long (no greater than 10 days) as the short half-life of P^{32} and the rapid biological turnover leads to the eventual "loss" of the tag.

For mass tagging, a dipping device was used to soak the ants thoroughly in a P^{32} solution. As soon as they had dried out on filter-paper and were beginning to move about, the paper was placed next to the hill and the ants allowed to crawl back under the stone on their own accord. Any individuals which appeared to be abnormal or injured were removed, so that only active, uninjured ants were returned. It was found that mortality during the tagging process was less than 1 per cent. A number of preliminary laboratory experiments carried out over 6 days seemed to indicate that tagging did not affect subsequent behaviour of the ants. Although in some cases untagged ants picked up a small amount of radioactivity, this "secondary" tagging was so much less than the "primary" tagging that there was never any doubt as to which individuals had received the primary tag.

For illustrative purposes only, I have arbitrarily selected the results of just six experiments from a table given by the authors and renumbered the colonies studied in order of magnitude of the number of recaptures, m_2. The data are set out in Table 3.7. We recall from **3.1.1** that

$$N^* = \frac{(n_1+1)(n_2+1)}{(m_2+1)} - 1,$$

$$v^* = \frac{(n_1+1)(n_2+1)(n_1-m_2)(n_2-m_2)}{(m_2+1)^2(m_2+2)}$$

is an almost unbiased estimate of $V[N^*]$, and that

TABLE 3.7

Recapture data for six ant colonies: modified from Odum and Pontin [1961: Table 1].

Colony	n_1	n_2	m_2	N^*	$\sqrt{v^*}$	$100\ \hat{C}$
1	500	149	7	9 393	3 022	38
2	500	159	11	6 679	1 761	30
3	500	189	17	5 287	1 133	24
4	1 000	243	21	11 101	2 184	22
5	500	437	68	3 179	324	12
6	600	321	89	2 149	176	11

$$\hat{C} = 1/\sqrt{m_2}$$

is an estimate of the coefficient of variation of N^*. To obtain confidence intervals for N we calculate $\hat{p} = m_2/n_2$ and $\hat{f} = m_2/n_1$ as a guide to the choice of approximation mentioned in **3.1.4.** Using the rules outlined there, we find that the Poisson approximation is appropriate to the first four colonies ($\hat{p}$ and $\hat{f}$ both less than 0·1) and the normal approximation (3.4) for the last two colonies. The 95 per cent confidence limits thus calculated are set out in Table 3.8, and comparing these intervals with those based on $N^* \pm 1{\cdot}96\sqrt{v^*}$ we notice that the latter are too narrow for small $\hat{p}$. In particular, the upper limits are too low, thus reflecting the skewness of the distribution of N^* when p is small.

TABLE 3.8

95 per cent confidence intervals for the sizes of six ant colonies (workers only): data from Table 3.7.

$\hat{p}$	$\hat{f}$	Confidence interval	$N^* \pm 1{\cdot}96\sqrt{v^*}$
0·047	0·014	(4 179, 23 021)	(3 470, 15 316)
0·069	0·022	(3 522, 13 118)	(3 228, 10 130)
0·089	0·034	(3 177, 8 930)	(3 066, 7 508)
0·086	0·021	(7 049, 17 715)	(6 820, 15 382)
0·156	0·136	(2 671, 4 032)	(2 545, 3 815)
0·277	0·148	(1 860, 2 591)	(1 803, 2 495)

It should be emphasised that the point and interval estimates of N are for the number of workers only (the "catchable" population) and not for the whole population in the colony. For further studies of the above marking technique cf. Stradling [1970] and Kruk-de Bruin *et al.* [1977].

Example 3.6 Snowshoe hares (*Lepus americanus*): Green and Evans [1940]

The aim of this investigation was to study the fluctuations in abundance of snowshoe hares on the Lake Alexander area, Minnesota, over a period of years from 1932 to 1939, using a series of Petersen estimates. The trapping season extended from October or November to April or May of the following spring, and trapping activities were discontinued during the warmer months of the year as hares would not then enter traps in appreciable numbers, probably because desirable food was abundant. In each year the major part of the trapping season was utilised to band as many new hares as possible, and the final two and a half weeks of the season were used to retrap the entire area.

The animals were tagged with numbered metal ear-bands which were clipped on with special pliers. Although the bands made only a minute perforation they seemed to stay on throughout the animal's entire life and caused no irritation. The sampling method consisted of dividing up the area into "stations" and then operating each station for six trap-days, either by using six traps for one day or more commonly three traps for two days. Some of the experimental results for the whole Lake Alexander Area are given in Table 3.9: the normal approximation (3.4) in **3.1.4** was used to calculate the confidence intervals for N.

TABLE 3.9

95 per cent confidence intervals for the size of a snowshoe hare population, Lake Alexander, Minnesota: data from Green and Evans [1940].

Year	n_1	n_2	m_2	N^*	$\hat{p}$	$\hat{f}$	Confidence interval
1932–3	948	421	167	2383	0·3976	0·176	(2153, 2685)
1933–4	876	329	154	1866	0·4681	0·176	(1689, 2098)
1934–5	659	272	105	1699	0·3860	0·159	(1494, 1991)

The population of hares was not strictly closed as there was a natural mortality, particularly over the winter months. However, as mentioned on page 71 this would not affect the Petersen estimate if the mortality rate was the same for tagged and untagged. As the hares lived in a markedly restricted range, seldom travelling more than $\frac{1}{8}$ mile from the point where they were first trapped, migration to and from the area was considered to be negligible. Usually a hare trapped at one station was recaptured at the same station or an adjoining one, even when a year or two had elapsed between the two trappings. As already mentioned, tag losses were considered negligible, and trapping did not seem to affect an animal's behaviour with regard to future trapping.

Systematic rather than random sampling was used, so that for the Petersen estimate to apply, the tagged proportion must be constant over the population area before the second sample is taken (cf. **3.1.2**). Since the hares had restricted home ranges, then, provided that all the animals were equicatchable, this uniformity would have been achieved by systematically trapping the whole area with a uniform trapping effort. The uniform effort would ensure that a fixed proportion of animals in any subarea would be caught for tagging. But in this particular investigation this was not done as there were more traps and more trapping effort used on a particular subarea which the authors called the Mile Area. The authors felt that this departure from the assumptions of the Petersen method led to an underestimate of the total population by about 5 to 10 per cent.

Example 3.7 Climbing cutworms: Wood [1963]

Population estimates of climbing cutworms are difficult to obtain as the larvae feed only at night and then conceal themselves in the litter layer during the day. It was found that the sweep net method (cf. Southwood [1966]) provided a satisfactory measure of the relative abundance of cutworms between fields or from year to year. The Petersen method was also used to estimate the number of insects per unit area in a small field ($\frac{1}{4}-\frac{1}{2}$ acre) of blueberry plants. The field was swept systematically at night and all larvae captured were taken to the laboratory for examination, where they were painted or sprayed with a solution of a fluorescent compound and then redistributed systematically over the field on the same night; any specimen injured in collecting or handling was discarded. On the following two or three evenings the field was again swept and marked specimens were detected by passing an ultraviolet lamp over the collected larvae. The results of the experiment are given in Table 3.10, and the confidence limits were obtained using the Poisson approximation (cf. **3.1.4**). We note that for $m_2 < 10$ the intervals are very wide.

TABLE 3.10

95 per cent confidence intervals for the density of a climbing cutworm population: data from Wood [1963: Table 1].

Year	n_1	n_2	m_2	N^*	$\hat{p}$	Confidence interval	Density per square foot
1956	125	107	8	15 119	0·074	(7 035, 34 240)	(0·06, 0·29)
1958	75	143	6	15 633	0·042	(6 435, 41 613)	(0·08, 0·52)
1959	93	114	3	27 024	0·026	(7 803, 130 405)	(0·07, 1·30)
1961	200	122	14	16 481	0·115	(9 321, 29 524)	(0·09, 0·30)
1962	1 000	1 755	41	41 850	0·023	(30 414, 57 915)	(1·73, 3·29)

Wood felt that by and large the assumptions underlying the Petersen model were satisfied and in particular the method of marking had no adverse effect on the insects. Although the fluorescent marker did not wash off, it was found that the proportion of marked larvae decreased with each successive night of collecting. This apparent loss of marked insects was attributed to loss of markings through moulting. Since the longer the interval the greater the possibility of moulting, it was decided to limit the counts to those taken on the first recapture date (24 hours after marking).

Example 3.8 Redpolls (*Acanthis linaria*): Nunneley [1964]

In 1962 a flock of redpolls frequented a banding station in Massachusetts, U.S.A., from March 3 to April 5, and 122 of them were banded. Three categories of information about the flock were sought: (i) the lengths of stay for individuals and the flock, (ii) the changing population size produced by the invasion movement from day to day, and (iii) information on whether the number of birds passing through the station was much larger than the number entering traps. I shall concern myself with just (ii) as a means of illustrating the Petersen method; the recapture data in Table 3.11 are taken from Table 4 of Nunneley's article.

TABLE 3.11

95 per cent confidence intervals for the size of a redpoll population: data from Nunneley [1964: Table 4].

Date (March)	n_1	n_2	m_2	N^*	$\hat{p}$	Confidence interval
3	–	7	0	–	–	–
6	6	12	5	14	0·417	(8, 40)
8	11	16	9	19	0·563	(14, 37)
13	13	45	9	63	0·200	(37, 133)
19	33	39	23	56	0·590	(45, 78)
22	24	43	18	57	0·419	(42, 88)
25	25	28	17	41	0·607	(32, 62)
27	13	6	3	24	0·500	(15, 108)
28	12	17	11	19	0·647	(14, 31)

Of the 122 banded there were 37 males, 74 females and 11 of unknown sex. As there appeared to be negligible differences in the arrival, average length of stay and recapture behaviour of the two sexes, the data for the two sexes were pooled and the redpolls treated as a single population. The assumption of a closed population (assumption (a)) seemed reasonable for two reasons. First, by confining tagging and recapturing to a single day, natural mortality and new additions were considered negligible over this short period. Secondly, banded birds appeared to be confined to the vicinity of the banding station. This meant that they had a very narrow feeding range, so that over a day, losses through emigration would be small. Tag losses would be negligible, and comparing captures per day with both length of stay and number of repeats suggested that trap shyness or trap addiction was negligible. The 95 per cent confidence intervals in Table 3.11 were calculated using the binomial approximation (**3.1.4**).

Example 3.9 Roe-deer (*Capreolus capreolus* L.): Andersen [1962]

In most European countries, the roe-deer is a widely distributed and popular animal, even though it causes considerable damage to forests. For example, in Denmark about 25 000 roe-deer are killed each year by more than 100 000 licensed sportsmen. With such a high annual kill it is imperative that deer management and shooting policy should be based on a sound knowledge of roe-deer biology. In particular, Andersen mentions the danger of using purely subjective methods in estimating roe-deer density as roe-deer are extremely good at escaping detection.

The aim of Andersen's work was to analyse both the size and composition of a herd which filled to carrying capacity the two woods of an experimental game farm (Kalø estate), using the Petersen method. The tag used was a leather collar studded with different coloured plastic buttons and bearing a copper plate giving a serial number and address. As a special precaution, a metal ear-tag supplying the relevant information was attached at the same time so as to take care of the rare cases when the collar was lost. From experience it was felt that the collar did not hamper the deer or expose them to danger, and there was no evidence to indicate that tagged animals behaved differently from untagged.

The traps were fenced enclosures and there was a period of prebaiting to allow the deer time to become used to the traps. Trapping took place in January and February, and the trapped animals were released immediately after tagging. Many entered the traps again, sometimes on the next day, in which case the collars were read and the animals released. The total number n_1 of animals tagged formed the first sample. During the following months, March, April and May, one man had the job of combing the woods and their immediate surroundings with binoculars, and noting all the individuals observed; this total n_2 formed the "recaptured" sample. Recording was not equally easy at all times as light-coloured buttons showed up better during dusk hours or in dense woodland, and it is here that observer bias could possibly have affected the data as doubtful records were excluded. For example, if the observer sees a group of, say, five deer together he must be able to "read" all five; if that is not possible Andersen recommends that they should all be disregarded. To achieve adequate sampling of the whole area, a number of observation posts were established at strategic points. In addition the observer stalked the deer, creating the least possible disturbance, and took special precautions to cover the entire woodland and to vary the time of his visit to different parts of the wood.

Concerning the assumption of constant N which underlies the Petersen method the following observations were made. Although the estate was fenced, public roads crossed the perimeter and the animals used them as routes to pass in and out of the area. However, it was felt that during the trapping and observation periods (autumn and winter respectively) migration took place to a very limited extent. As far as emigration out of the area was concerned, the marked deer were easy to spot, and all forest owners and sportsmen in the neighbourhood knew of the experiments. In particular a reward of approximately $1·50 was paid to sportsmen who handed in a collar. As far as mortality is concerned, a few died through trapping; although the natural mortality during the experimental period was not estimated, it would not seriously affect the Petersen estimate if it was small and proportionately the same for tagged and untagged (cf. p. 71).

During the trapping two age-classes were distinguished, i.e. fawns born during the preceding summer and adults. It appeared that the fawns were more easily trapped than the adults and that about 75 per cent of the population had to be trapped before there was stability in the estimate of the fawn–adult ratio. The data for fawns and adults were then pooled as the percentage of adults in both n_1 and m_2 was approximately the same (57 per cent and 55 per cent respectively).

It is noted that members of the population could be "recaptured" (observed) several times, so that n_2 was much larger than N. Here the sampling was effectively with replacement, so that Bailey's binomial model (**3.1.2**) could be used, provided that the method of observing was random, or at least representative, at all times. For example, if a particular subarea contained a higher proportion than average of marked deer and that subarea was observed more often than other areas, then n_2 would contain too many tagged animals and N would consequently be underestimated.

The following numbers were obtained: $n_1 = 74$, $n_2 = 462$ and $m_2 = 340$, so that from **3.1.2**

$$\hat{N}_1 = \frac{74(463)}{341} = 100 \quad \text{and} \quad \hat{p} = \frac{340}{462} = 0{\cdot}736.$$

Using the normal approximation to the binomial distribution, an approximate 95 per cent confidence interval for $p = n_1/N$ is

$$\hat{p} \pm 1{\cdot}96\,[\hat{p}(1-\hat{p})/n_2]^{\frac{1}{2}}.$$

Inverting this gives the corresponding confidence interval for N, namely (95, 106).

For a further study of this deer population the reader is referred to Strandgaard [1967].

3.4 ESTIMATION FROM SEVERAL SAMPLES

Let n_1 animals be captured, marked, and released throughout the population and let $p = n_1/N$, where N is the total population size. Suppose that the total population area can be divided up into K subareas of which k are selected at random for further sampling. Let

N_i = number of animals in ith subarea ($i = 1, 2, \ldots, K$),

$\bar{N} = N/K$,

n_{1i} = number of marked animals in the ith subarea,

$\{n_{1i}\}$ = set of n_{1i} $(i = 1, 2, \ldots, K)$,

$\{n_{1i}\}_k$ = sample set of n_{1i} $(i = 1, 2, \ldots, k)$,

n_{2i} = number of animals caught in the ith subarea $(i = 1, 2, \ldots, k)$,

m_{2i} = number of marked in n_{2i},

p_i = n_{1i}/N_i,

$\hat{p}_i$ = m_{2i}/n_{2i},

f_1 = k/K,

and f_{2i} = n_{2i}/N_i.

We note that

$$p = n_1/N = \left(\sum_{i=1}^{K} N_i p_i\right) \Big/ \sum_{i=1}^{K} N_i.$$

3.4.1 Ratio estimate

With the above experimental set-up we effectively have two-stage sampling, in which we choose k unequal size-units or clusters at the first stage and then subsample from each unit, noting the proportion $\hat{p}_i$ of marked in each subsample. Therefore, if the cluster sizes N_i were known, p could be estimated by the ratio-type estimator (cf. Cochran [1963: 300])

$$\left(\sum_{i=1}^{k} N_i \hat{p}_i\right) \Big/ \left(\sum_{i=1}^{k} N_i\right).$$

However, as the N_i are unknown, one possibility is to use weights proportional to the sample sizes n_{2i}, so that p is estimated by

$$\begin{aligned} \hat{p} &= \left(\sum_{i=1}^{k} n_{2i} \hat{p}_i\right) \Big/ \left(\sum_{i=1}^{k} n_{2i}\right) \\ &= \left(\sum_{i=1}^{k} m_{2i}\right) \Big/ \left(\sum_{i=1}^{k} n_{2i}\right) \\ &= m_2/n_2 \quad \text{say.} \end{aligned}$$

and N by $n_1/\hat{p}$, the Petersen estimate once again. It is noted that

$$E[\hat{p} \mid \{n_{2i}\}, \{n_{1i}\}_k] = \left(\sum_{i=1}^{k} n_{2i} p_i\right) / n_2 \tag{3.32}$$

and we now consider two cases where $\hat{p}$ is unbiased and approximately unbiased, respectively.

1 Constant mark ratio

When $p_i = p$ $(i = 1, 2, \ldots, k)$ the right-hand side of (3.32) reduces to p and $\hat{p}$ is unbiased. In this case $V[\hat{p} \mid \{n_{1i}\}]$, the variance of $\hat{p}$, can be estimated by (ignoring finite population corrections due to sampling without replacement) –

$$v_1[\hat{p}] = \hat{p}(1-\hat{p})/(n_2 - 1).$$

This estimate was used, for example, by Welch [1960]. The hypothesis H that $p_i = p$ can be tested using the standard goodness-of-fit statistic

$$\sum_{i=1}^{k} \frac{(m_{2i} - n_{2i}\hat{p})^2}{n_{2i}\,\hat{p}(1-\hat{p})},$$

which, when H is true, is approximately distributed as chi-squared with $k - 1$ degrees of freedom. When H is true and n_2 is large (say greater than 30), $\hat{p}$ is asymptotically normal with mean p and variance estimate $v_1[\hat{p}]$. An approximate confidence interval for p (and hence for N) can then be calculated in the usual manner. Also $V[\hat{N}]$ is given by (3.35).

2 *Constant sampling effort*

In general, if H is not true $\hat{p}$ will be biased. If, however, the same sampling effort is used within each subarea and the expected fraction caught (θ say) is proportional to the sampling effort, then $E[n_{2i} \mid n_{1i}] = E[n_{2i}] = N_i\theta$ and $E[m_{2i} \mid n_{1i}] = n_{1i}\theta$. Hence, for large n_{2i},

$$E[\hat{p} \mid \{n_{1i}\}_k] \approx \frac{\sum_{i=1}^{k} E[m_{2i} \mid n_{1i}]}{\sum_{i=1}^{k} E[n_{2i} \mid n_{1i}]}$$

$$= \left(\sum_{i=1}^{k} n_{1i}\right) \Big/ \left(\sum_{i=1}^{k} N_i\right). \qquad (3.33)$$

Since the k subareas are chosen at random, (3.33) represents a ratio estimate of p with respect to the first stage in the sampling (cf. Cochran [1963: 29]). Therefore, taking expectations with respect to this first stage, we have, for large k,

$$E[\hat{p} \mid \{n_{1i}\}] = \underset{k}{E}\, E[\hat{p} \mid \{n_{1i}\}_k]$$

$$\approx \left(\sum_{i=1}^{K} n_{1i}\right) \Big/ \left(\sum_{i=1}^{K} N_i\right)$$

$$= n_1/N = p.$$

Thus $\hat{p}$ is approximately unbiased for large n_{2i} and large k.

Noting that

$$\theta = \frac{E[n_{2i}]}{N_i} = \frac{E[n_2]}{\sum_{i=1}^{k} N_i} \approx \frac{E[n_2]}{k\bar{N}} = \frac{E[\bar{n}_2]}{\bar{N}},$$

where $\bar{n}_2 = n_2/k$, $N_i/\bar{N}$ can be estimated by $n_{2i}/\bar{n}_2$. Therefore, from Cochran [1963: 300–2 and particularly 324, Ex. 11.5, with $M_i = N_i$] an approximate estimate of $V[\hat{p} \mid \{n_{1i}\}]$ is given by

$$v_2[\hat{p}] = \frac{1-f_1}{k(k-1)} \sum_{i=1}^{k} \left(\frac{n_{2i}}{\bar{n}_2}\right)^2 (\hat{p}_i - \hat{p})^2 +$$

$$+ \frac{f_1(1-f_2)}{kn_2} \sum_{i=1}^{k} \left(\frac{n_{2i}}{\bar{n}_2}\right) \cdot \frac{n_{2i}}{n_{2i}-1} \cdot \hat{p}_i(1-\hat{p}_i), \qquad (3.34)$$

where f_2 ($\approx \bar{n}_2/\bar{N}$) can either be ignored, or estimated using $n_1/\hat{p}$ as an estimate of N.

When $k = K$, i.e. $f_1 = 1$, we have the special case of stratified random sampling. The expression (3.33) then reduces to p and the first term in (3.34) is zero.

When f_1 is small the second term in (3.34) can be neglected, and $V[\hat{p} \mid \{n_{1i}\}]$ is now estimated by

$$v_3[\hat{p}] = \frac{k}{(k-1)n_2^2} \sum_{i=1}^{k} (m_{2i} - n_{2i}\hat{p})^2$$

Using the delta method, the variance of the Petersen estimate $\hat{N} = n_1/\hat{p}$ is approximately given by

$$n_1^2 p^{-4} V[\hat{p} \mid \{n_{1i}\}], \tag{3.35}$$

which can then be estimated by

$$\begin{aligned} v[\hat{N}] &= n_1^2 \hat{p}^{-4} v_3[\hat{p}] \\ &= \frac{\hat{N}^2 k}{m_2^2(k-1)} \sum_{i=1}^{k} (m_{2i} - n_{2i}\hat{p})^2. \end{aligned} \tag{3.36}$$

From the relationship

$$\Sigma(m_{2i} - n_{2i}\hat{p})^2 = m_2^2 \left[\frac{\Sigma(m_{2i} - \bar{m}_2)^2}{m_2^2} - \frac{2\Sigma(m_{2i} - \bar{m}_2)(n_{2i} - \bar{n}_2)}{m_2 n_2} + \frac{\Sigma(n_{2i} - \bar{n}_2)^2}{n_2^2} \right],$$

it can be shown that (3.36) is identical with a formula used by Banks and Brown [1962] in Example 3.11 below.

3.4.2 Mean estimates

An alternative estimate of p is the average

$$\hat{\bar{p}} = \sum_{i=1}^{k} \hat{p}_i/k$$

and from the general theory of Cochran [1963: 272–8, setting $y_i' = \hat{p}_i/k$, $Y_i' = p_i/k$, $\pi_i = k/K$ etc.] an unbiased estimate of the variance is given by

$$v[\hat{\bar{p}}] = \frac{(1-f_1)}{k(k-1)} \sum_{i=1}^{k} (\hat{p}_i - \hat{\bar{p}})^2 + \frac{f_1}{k^2} \sum_{i=1}^{k} (1-f_{2i}) \frac{\hat{p}_i \hat{q}_i}{(n_{2i}-1)},$$

which reduces to the usual sample estimate of variance (cf. (1.9)) when f_1 is ignored: for an application see Best and Rand [1975]. As

$$\begin{aligned} E[\hat{\bar{p}} \mid \{n_{1i}\}] &= \underset{k}{E}\, E[\hat{\bar{p}} \mid \{n_{1i}\}_k] \\ &= \underset{k}{E} \left[\sum_{i=1}^{k} p_i/k \right] \\ &= \left(\sum_{i=1}^{K} p_i \right) /K \\ &= \bar{p}, \text{ say,} \end{aligned}$$

then $\bar{\hat{p}}$ is unbiased when either H ($p_i = p$ for all i) is true or when $\bar{p} = p$. If H is true, $\bar{\hat{p}}$ seems preferable to $\hat{p}$ because of the general robustness of a mean with regard to normality and because $v[\bar{\hat{p}}]$ is more robust than $v_1[\hat{p}]$ with regard to departures from H. On the other hand, if H is rejected by the goodness-of-fit test and the sampling effort is uniform, then $\hat{p}$ can be used with $v_2[\hat{p}]$ of (3.34).

In some experiments the numbers n_{1i} of marked animals in the individual subareas are known and N can be estimated by (cf. **3.1.1**)

$$N' = K \sum_{i=1}^{k} N_i^*/k$$

$$= K \sum_{i=1}^{k} \left\{ \frac{(n_{1i}+1)(n_{2i}+1)}{(m_{2i}+1)} - 1 \right\} / k .$$

Then

$$E[N'] = \underset{k}{E}\, E[N_i' \mid k]$$

$$\approx K \underset{k}{E} \left[\sum_{i=1}^{k} N_i/k \right]$$

$$= K\bar{N}$$

$$= N,$$

and using the general theory of Cochran [1963: 272–4, setting $y_i' = N_i^*/k$, $Y_i' = N_i/k$, $\pi_i = k/K$, etc.] it can be shown that an approximately unbiased estimate of the variance of N' is

$$v[N'] = \frac{K(K-k)}{k(k-1)} \sum_{i=1}^{k} (N_i^* - N')^2 + \frac{K}{k} \sum_{i=1}^{k} v[N_i^*],$$

where

$$v[N_i^*] = \frac{(n_{1i}+1)(n_{2i}+1)(n_{1i}-m_{2i})(n_{2i}-m_{2i})}{(m_{2i}+1)^2(m_{2i}+2)} .$$

When $n_{1i} + n_{2i} \geqslant N_i$ for each i, then N' and $v[N']$ are *exactly* unbiased.

In comparing N' with the Petersen estimate $\hat{N}$, we note from **3.2.5** that when $k = K$ (i.e. stratified sampling) the pooled estimate $\hat{N}$ will have smaller variance than N', the sum of the individual estimates. We would also expect this to be true when $k < K$. However, N' has several advantages. First of all, it is based on a mean, so that for large k it is approximately normally distributed by the Central Limit theorem. Secondly, if the assumptions underlying the Petersen method hold within each subarea then, in general, N' and $v[N']$ are almost unbiased estimates, irrespective of whether H is true or whether the same sampling effort is used in each subarea. This means that confidence intervals based on N' will often be more reliable than those based on $\hat{p}$ or $\bar{\hat{p}}$, particularly when these estimates of p are biased.

Example 3.10 Adult Sunn Pest (*Eurygaster integriceps* Put.): Banks and Brown [1962]

This pest seriously damages wheat and other cereals in various countries of the Middle East. In attempting to investigate chemical and biological methods of control, accurate estimates of the insect population density are required for assessing the effects of insecticides on the mortality of the insects; while quick, but less accurate, estimates are needed for deciding the best time to apply the insecticides to the overwintered insects newly arrived on the wheat fields. Banks and Brown considered three methods of estimation: sweeping with a hand net, quadrat sampling, and the Petersen recapture method.

A wheat plot of about 1000 square metres was studied within each of two fields (called A and B). Each plot was staked out into 36 equal sectors (of 28 sq.m. for A and 23 sq.m. for B) and each sector in turn was swept systematically with a hand net by the same person. The insects caught were counted and marked with a spot of quick-drying paint. As soon as the paint was dry, the marked insects were scattered evenly over the sector from which they were caught and the team then moved into the next sector. Sixteen hours were allowed for the insects to return to the plants and emerge from cracks in the soil after being disturbed, and for the marked to mix with the unmarked. Assuming uniform mixing of marked and unmarked, systematic sampling could then be used for taking the second sample (cf. **3.1.2**). The two systematic methods used here were the sweepnet method, again over the whole plot, and quadrat sampling. In the latter case fifty quadrats of one square metre were marked out systematically over the whole plot and each quadrat was carefully examined. All bugs visible on the plants were first picked off and then the vegetation and ground were thoroughly searched.

Applying the quadrat method to plot A, it was found that $K = 28(36) = 1008$, $k = 50$, $n_1 = 3538$, $n_2 = 980$, $m_2 = 106$, $\Sigma\, n_{2i}^2 = 30\,380$, $\Sigma\, m_{2i}^2 = 450$ and $\Sigma\, m_{2i}n_{2i} = 2700$. The authors effectively used the estimates $\hat{p}$ and $v_3[\hat{p}]$, and these are calculated as follows:

$$\hat{p} = 106/980 = 0\cdot108\,163,$$

$$\begin{aligned}\Sigma(m_i - n_i\hat{p})^2 &= 450 - 2(2700)(0\cdot108\,163) + 30\,380(0\cdot108\,163)^2 \\ &= 221\cdot34,\end{aligned}$$

and

$$v_3[\hat{p}] = \frac{50(221\cdot34)}{49(980)^2} = 0\cdot000\,235\,16.$$

Assuming $\hat{p}$ to be asymptotically normal, the 95 per cent confidence interval for p is (0·078 45, 0·1380) and the corresponding interval for N is (25 640, 45 100). Also $\hat{N} = 3538(980)/106 = 32\,710$ and, from (3.36),

$$\begin{aligned}\{v[\hat{N}]\}^{\frac{1}{2}} &= \frac{32\,710}{106}\left\{\frac{50}{49}(221\cdot34)\right\}^{\frac{1}{2}} \\ &= 4638.\end{aligned}$$

We shall now briefly consider the assumptions underlying the above use of the Petersen method. To begin with, the population was expected to be fairly constant from one day to the next, as the deaths overnight were considered to be few and the dispersal was negligible owing to lack of horizontal movement shown by the insects. We can presume that there were no tag losses or unreported tags, and since no mention is made of mortality through handling it was no doubt negligible in both samples. It was assumed that a period of 16 hours was sufficient for the marked insects to get over the effects of handling and that the paint spot would not affect their movements. However, as far as the validity of the Petersen estimate is concerned, the main assumption that was difficult to satisfy was the assumption of uniform mixing of marked and unmarked which is needed when systematic rather than random sampling is carried out. The method used to achieve this mixing was to scatter the marked insects evenly over the sector from which they were caught; this would be satisfactory if the density of the population was uniform over the sector to start with and the marked insects dispersed in a random manner. To check on the mixing, we can calculate (cf. **3.1.2**)

$$\sqrt{v_1} = \frac{n_1}{m_2+1}\left\{\frac{(n_2+1)(n_2-m_2)}{(m_2+2)}\right\}^{\frac{1}{2}} = 2946.$$

Since this is much less than $\{v[\hat{N}]\}^{\frac{1}{2}}$ (= 4638) it would seem that the binomial model is not valid and that the dispersal of the marked bugs was not complete at the time of recapture. This lack of dispersal, and in fact the lack of mobility, indicates that the Petersen method is not suitable as a standard method of studying sunn pest populations. Another disadvantage of the Petersen method is that it is laborious, requiring a team of several people to obtain satisfactory results.

The two other methods of census suggested, namely quadrat counting and sweeping, are easier to carry out and would therefore be used more often. The main disadvantage of direct counting using quadrats is that a certain proportion of the population, though small in this case, would hide in soil crevices and would need to be smoked out for more accurate counts. On the credit side, however, the accuracy can be increased by increasing the number of quadrats: the same is true with the sweeping method, though in this case the maximum obtainable accuracy depends on the sweeping efficiency.

The quickest method of the three, namely sweeping, relies heavily on the assumption of uniform sweeping efficiency, so that this assumption needs to be checked from time to time. The authors conclude that the Petersen recapture method "may be useful as an occasional check on other sampling methods and, in particular, to provide a measure of the efficiency of routine sampling by sweeping".

3.4.3 Interpenetrating subsamples

Chapman and Johnson [1968] suggest one other estimate of N based on interpenetrating subsamples (cf. Cochran [1963: 383]). Here the total sample of size n_2 is subdivided randomly into r subsamples and an estimate N_i'' of N is obtained from each subsample, using one of the above methods. Then

N is estimated by the subsample mean $N'' = \sum_{i=1}^{r} N_i''/r$, and $V[N'']$ is estimated by

$$v[N''] = \sum_{i=1}^{r} (N_i'' - N'')^2/[r(r-1)].$$

Chapman and Johnson apply this method to a fur seal pup population.

If $N''_{(1)}$ and $N''_{(r)}$ are the smallest and largest of the N_i'' then (Overton and Davis [1969]) $\Pr[N''_{(1)} < N < N''_{(r)}] = 1 - (\frac{1}{2})^{r-1}$ provides exact limits for the median of the distribution of N_i'': these are quite acceptable as limits for N.

3.5 INVERSE SAMPLING METHODS*

We now consider an inverse sampling method for the second sample which, in contrast to the "direct" Petersen method considered so far, provides an unbiased estimate of N with an *exact* (rather than a large-sample) expression for the variance, and a coefficient of variation which is almost independent of N. The method is to tag or mark n_1 animals as before and then continue taking the second sample until a prescribed number m_2 of marked animals have been recovered. This means that n_1 and m_2 are now considered as fixed parameters and n_2 is the random variable. As in the direct Petersen method, the second sample can be taken with or without replacement, and we shall now consider these two cases separately.

3.5.1 Sampling without replacement

When the assumptions (a), (d), (e) and (f) of **3.1.1** are satisfied, Bailey [1951] shows that the probability function of n_2, conditional on n_1 and m_2, is the negative hypergeometric

$$f(n_2 \mid n_1, m_2) = \frac{\binom{n_1}{m_2 - 1}\binom{N - n_1}{n_2 - m_2}}{\binom{N}{n_2 - 1}} \cdot \frac{n_1 - m_2 + 1}{N - n_2 + 1},$$

where $n_2 = m_2, m_2 + 1, \ldots, N + m_2 - n_1$, and suggests the modified maximum-likelihood estimate of N

$$\hat{N}_2 = \{n_2(n_1 + 1)/m_2\} - 1 \tag{3.37}$$

This estimate is unbiased and has exact variance

$$V[\hat{N}_2 \mid n_1, m_2] = \{(n_1 - m_2 + 1)(N + 1)(N - n_1)\}/\{m_2(n_1 + 2)\} \approx N^2/m_2.$$

Assuming $N + 1$ and $N - n_1$ to be approximately equal to N, the coefficient of variation of $\hat{N}_2$ is close to

$$C[\hat{N}_2] = [(n_1 - m_2 + 1)/\{m_2(n_1 + 2)\}]^{\frac{1}{2}},$$

(*) See also 12.7.2.

and since n_1 is known, m_2 can be chosen beforehand, so that this coefficient has a prescribed value. For example, if $n_1 = 100$ and $C = 0{\cdot}1$, then sampling must continue until $m_2 = 50$ marked individuals have been caught.

Chapman [1952] mentions several useful properties of $\hat{N}_2$. As m_2 tends to infinity, n_2 is asymptotically normally distributed, so that for large m_2, $\hat{N}_2$ is approximately normal. Also, on the average, the inverse method is slightly more efficient than the direct Petersen method. By this we mean that for a given coefficient of variation, the inverse method provides an estimate of N with an expected sample size $E[n_2 \mid n_1, m_2] (= (N+1)m_2/(n_1+1))$ which is smaller than the sample size needed for the direct method though usually the difference is small (Robson and Regier [1964: 218]). The inverse method also shares the same advantage as the direct method in that it is still applicable when mortality is taking place, provided that ϕ, the probability of survival between the two samples, is the same for marked and unmarked (cf. page 71). In this case $\hat{N}_2$ remains approximately unbiased, and its variance is now (Chapman [1952: 301])

$$\frac{N^2}{m_2} + \frac{N(1-\phi)}{\phi}\left(1 + \frac{1}{m_2}\right),$$

which is still approximately N^2/m_2 unless ϕ is very small.

However, an undesirable feature of the inverse method is that if the experimenter knows absolutely nothing about N, the expected sample size may be very large with an improper choice of n_1 and m_2. Also the variance of n_2 (approximately $m_2 N^2/n_1^2$) is large, and this increases the unpredictability of what n_2 will actually turn out to be. These difficulties may be partly overcome by considering an inverse sampling scheme of Chapman's in which u_2 $(= n_2 - m_2)$, the number of unmarked individuals caught in n_2, is fixed instead of m_2; m_2 and n_2 are now both random variables, though completely dependent. For this scheme

$$E[n_2 \mid n_1, u_2] = u_2(N+1)/(N-n_1+1)$$

and

$$V[n_2 \mid n_1, u_2] = \frac{u_2 n_1 (N+1)(N-u_2-n_1+1)}{(N-n_1+1)^2(N-n_1+2)}$$

$$\approx u_2 n_1/(N-n_1) \quad \text{(for large } N\text{)},$$

so that in contrast with the method above, the variance of n_2 is small and the expected sample size does not depend so critically on N. We now find that an approximately unbiased estimate of N is given by

$$\hat{N}_3 = \frac{n_2(n_1+1)}{(m_2+1)} - 1 \tag{3.38}$$

with approximate variance

$$V[\hat{N}_3 \mid n_1, u_2] = (n_1+1)^2[q_{22} + q_{23} + 2q_{24} - q_{12} - q_{13} - 2q_{14}] - N^2 - 2N - 2, \tag{3.39}$$

where

$$q_{ij} = \frac{(N+i)_i(N-n_1)_k}{(n_1+j)_j(u_2-1)_k} \quad (i \leqslant j, \quad k = j-i)$$

and $(a)_i = a(a-1)\ldots(a-i+1)$, etc.

Writing $p_0 = n_1/(N+1)$, Chapman [1952] shows that when n_2 is large,

$$z = \frac{n_2 - u_2/(1-p_0)}{\{u_2p_0/(1-p_0)\}^{\frac{1}{2}}}$$

is approximately unit normal, so that approximate $100(1-\alpha)$ per cent confidence limits for p_0 are the roots of the quadratic

$$\{n_2(1-p_0) - u_2\}^2 = z_{\alpha/2}^2\, u_2p_0(1-p_0).$$

Inverting these limits leads to a confidence interval for N. In conclusion, we note that this method can be applied to the situation where n_1 is also unknown and an estimate of n_1/N ($\approx p_0$) is required, e.g. the estimation of sex or age ratios.

3.5.2 Sampling with replacement

We shall now consider the less common situation in which members of the second sample are caught one at a time and returned immediately to the population. For example, this model would apply when the animals are merely observed and not actually captured. In the inverse method, sampling is continued until a prescribed number m_2 of marked animals have been caught and released, so that the probability function of n is now the negative binomial*

$$f(n_2 \mid n_1, m_2) = \binom{n_2-1}{m_2-1} p^{m_2}(1-p)^{n_2-m_2}$$

where $n_2 = m_2, m_2+1, \ldots,$ and $p = n_1/N$. As $E[n_2 \mid n_1, m_2] = Nm_2/n_1$, the obvious estimate for N is the Petersen estimate $\hat{N} = n_1n_2/m_2$. This maximum likelihood estimate is unbiased with variance

$$V[\hat{N} \mid n_1, m_2] = (N^2 - Nn_1)/m_2,$$

which is estimated unbiasedly by (Chapman [1952: 287])

$$v[\hat{N}] = \frac{n_2n_1^2(n_2-m_2)}{m_2^2(m_2+1)}.$$

The coefficient of variation of $\hat{N}$ is close to $1/\sqrt{m_2}$, so that for C to be no greater than 20 per cent, for example, sampling must continue until $m_2 = 25$ marked individuals have been captured. We note that, as in sampling without replacement, this inverse method is more efficient than Bailey's direct method of **3.1.2** (Goodman [1953: 67]). However, as the mean and variance of n_2 are almost the same as for sampling without replacement, the inverse method above suffers from the same disadvantages, namely that the variance of n_2 is large and the expected sample size may also be large with an improper choice of n_1 and m_2.

(*)This model is compared with (3.3) by Robson [1979].

When m_2 is large, Chapman [1952] shows that

$$z = \frac{\hat{N}-N}{\{N(N-n_1)/m_2\}^{\frac{1}{2}}} \tag{3.40}$$

is approximately distributed as the unit normal distribution, so that $100(1-\alpha)$ per cent confidence limits for N are the roots of

$$(\hat{N}-N)^2 = z_{\alpha/2}^2 N(N-n_1)/m_2.$$

When m_2 and p are both small, Chapman shows that $2n_2p$ is approximately distributed as chi-squared with $2m_2$ degrees of freedom. Therefore an approximate $100(1-\alpha)$ per cent confidence interval for N is $(2n_1n_2/c_2, 2n_1n_2/c_1)$ where c_1 and c_2 are the lower and upper $\alpha/2$ significance points respectively for $\chi^2_{2m_2}$.

If n_1 is also unknown, confidence intervals for p can be obtained by the above methods. For example, using the normal approximation (3.40), $100(1-\alpha)$ per cent limits are the roots of

$$(n_2p-m_2)^2 = z_{\alpha/2}^2 m_2(1-p),$$

while the chi-squared approximation leads to $(c_1/(2n_2), c_2/(2n_2))$. Haldane [1945] gives an unbiased estimate of p, namely

$$\hat{p} = (m_2-1)/(n_2-1)$$

with variance estimated by $\hat{p}^2(1-\hat{p})/(m_2-2)$ (Mikulski and Smith [1975]).

In conclusion we note that several authors have recently considered the problem of finding a minimax estimate of N, given that N is known to be greater than a certain integer (Czen Pin [1962], Zubrzycki [1963, 1966]).

3.6 COMPARING TWO POPULATIONS

3.6.1 Goodness-of-fit method

Suppose we have two closed populations of unknown sizes N_a and N_b, and on the basis of a single Petersen experiment in each population we wish to test the null hypothesis H_0 that $N_a = N_b$. Such a situation could arise, for example, where a control area and an experimental area are under observation and one wishes to test for any difference in population size due to experimental management practice. Alternatively, N_a and N_b could refer to the same population area but at different times. A third possibility is when N_a and N_b refer to the same population at the same time but two different sampling methods are used, e.g. seine and gill-net fishing. A test of H_0 would then indirectly provide a test for the hypothesis that marked animals are equally vulnerable to both methods of sampling.

In the notation of **3.1.1**, we can test H_0 by assuming that

$$z = \frac{N_a^* - N_b^*}{\sqrt{(v_a^* + v_b^*)}}$$

is approximately unit normal when H_0 is true. A more sensitive test is given by Chapman [1951] as follows. Using suffixes a and b to denote the two populations, let

$$\tilde{N} = \frac{\lambda_a^3 m_{2b} u_{1b} u_{2b} + \lambda_b^3 m_{2a} u_{1a} u_{2a}}{m_{2a} m_{2b} [\lambda_a^2 u_{1b} u_{2b} + \lambda_b^2 u_{1a} u_{2a}]},$$

where $\lambda_a = n_{1a} n_{2a}$, $u_{1a} = n_{1a} - m_{2a}$, $u_{2a} = n_{2a} - m_{2a}$, etc., and define

$$T_1 = \sum \frac{(m_{2c} - \lambda_c/\tilde{N})^2}{\dfrac{\lambda_c}{\tilde{N}}\left(1 - \dfrac{n_{1c}}{\tilde{N}}\right)\left(1 - \dfrac{n_{2c}}{\tilde{N}}\right)},$$

$$T_2 = \sum \frac{\lambda_c (m_{2c} - \lambda_c/\tilde{N})^2}{m_{2c} u_{1c} u_{2c}},$$

and

$$T_3 = \sum \frac{2(m_{2c} - \lambda_c/\tilde{N})^2}{\dfrac{\lambda_c}{\tilde{N}}\left(1 - \dfrac{n_{1c}}{\tilde{N}}\right)\left(1 - \dfrac{n_{2c}}{\tilde{N}}\right) + \dfrac{m_{2c} u_{1c} u_{2c}}{\lambda_c}},$$

where Σ denotes summation over the two values $c = a, b$. Then, when H_0 is true, T_1, T_2 and T_3 are each approximately distributed as chi-squared with one degree of freedom when N is large, and all three statistics are candidates for testing H_0. Chapman suggests using T_2 when $\lambda_a = \lambda_b$, T_3 when these quantities differ moderately, and T_1 when λ_a is widely different from λ_b. To test H_0 against the two-sided alternative $N_a \neq N_b$, the criterion is to reject H_0 at the 100α per cent level of significance when T_i is greater than the α critical value of χ_1^2.

3.6.2 Using Poisson approximations

When experimental circumstances are such that the hypergeometric distributions of m_{2a} and m_{2b} can be approximated by Poisson distributions (cf. **3.1.4**), the following technique of Chapman and Overton [1966] can be used for testing H_0. Let $m_2 = m_{2a} + m_{2b}$ and $N_b = kN_a$; then from **1.3.7**(*1*) the conditional probability function of m_{2a} given m_2 is

$$f(m_{2a} \mid m_2) = \binom{m_2}{m_{2a}} p^{m_{2a}} q^{m_{2b}},$$

where $p = 1 - q = k\lambda_a/(k\lambda_a + \lambda_b)$. Setting

$$p_0 = \lambda_a/(\lambda_a + \lambda_b),$$

we note that p is greater or less than p_0 if and only if k is greater or less than unity, so that testing H_0 against the two-sided alternative $N_a \neq N_b$ is equivalent to testing $p = p_0$ against $p \neq p_0$. Therefore, given m_2, m_{2a} and p_0, we can test H_0 by evaluating the exact tail probabilities, using such tables as those of the Harvard Computation Laboratory [1955], or we can obtain a confidence interval for p and reject H_0 if p_0 lies outside this interval. Confidence intervals for p can be determined from the Clopper–Pearson charts of Pearson and Hartley [1966: 228–9], using $\hat{p} = m_{2a}/m_2$ as the entering variable, or from tables such as Owen [1962: 273–85]. A bibliography of "equi-tail" confidence intervals is given by Owen [1962: 273], and for $m_2 \leqslant 30$, shorter intervals are given by Crow [1956]. When $\hat{p} < 0{\cdot}1$, and for any m_2 (Raff [1956]), the Poisson approximation to the binomial can be used to find a confidence interval for $m_2 p$, using m_{2a} as the entering variable in Poisson tables (e.g. Crow and Gardner [1959], Pearson and Hartley [1966: 227]). When $0{\cdot}1 \leqslant \hat{p} \leqslant 0{\cdot}9$, $m_2\hat{p}$ and $m_2\hat{q}$ are both greater than 5, and a correction for continuity is used, the normal approximation is applicable (Raff [1956: 296]) and we can use the following statistic for testing H_0.

$$z = \frac{|m_{2a} - m_2 p_0| - \frac{1}{2}}{\sqrt{(m_2 p_0 q_0)}}$$

When $\lambda_a = \lambda_b$ ($> 2{\cdot}5$) another procedure is given by Sichel [1973].

We note that the above methods can be used to obtain a confidence interval for k. In particular, when N_a and N_b refer to the same population but at different times, and the population is closed except for mortality, k can be interpreted as the proportion of N_a surviving, i.e. as a survival probability. If we are interested in just a one-sided alternative, say $N_b > N_a$, then we would reject H_0 if m_{2a} lay in the upper tail of the binomial distribution with $p = p_0$; the appropriate tail probability could be evaluated using binomial tables or the normal approximation.

To illustrate the above theory we shall now consider a number of examples which have been adapted from Chapman and Overton.

Example 3.11

Suppose we wish to test H_0 against the two-sided alternative $N_a \neq N_b$ for the following data:

$$\lambda_a = 430, \; m_{2a} = 11; \quad \lambda_b = 1040, \; m_{2b} = 7.$$

Then $m_2 = 11 + 7 = 18$ and $p_2 = 430/1470 = 0{\cdot}2925$. For illustrative purposes we shall consider four methods of testing H_0.

(i) From Pearson and Hartley [1966], the 99 per cent confidence interval for p, namely ($0{\cdot}29$, $0{\cdot}87$) does not contain p_0, so we reject H_0 at the 1 per cent level of significance.

(ii) Working with m_{2b} instead of m_{2a}, we can use Owen's tables with $x = 7$, $n - x = 11$ to obtain a 99 per cent confidence interval ($0{\cdot}1284$, $0{\cdot}7068$) for $1 - p$. As this interval does not contain $1 - p_0$, we reject H_0 at the 1 per cent level of significance.

(iii) Using the normal approximation, we have

$$z = \frac{11 - 18(0{\cdot}2925) - \frac{1}{2}}{\sqrt{[18(0{\cdot}2925)(0{\cdot}7075)]}} = 2{\cdot}71,$$

which is greater than 2·58, the two-tailed 1 per cent significant point.

(iv) If we wish to carry out a more accurate test of H_0, the exact probabilities can be calculated with appropriate binomial tables. In this case the required probability for a two-tailed test is the sum of all terms $\Pr[m_{2a} = d \mid p = p_0]$ in both tails, such that

$$\Pr[m_{2a} = d \mid p = p_0] \leqslant \Pr[m_{2a} = 11 \mid p = p_0].$$

From binomial tables (interpolating between 0·29 and 0·30 for p_0) we obtain

$$\begin{aligned} \Pr[m_{2a} = 0 \mid p_0] &= 0{\cdot}0020 \\ \Pr[m_{2a} = 11 \mid p_0] &= 0{\cdot}0038 \\ \Pr[m_{2a} \geqslant 12 \mid p_0] &= 0{\cdot}0011 \\ \text{Sum} &= 0{\cdot}0069 \end{aligned}$$

Since Sum < 0·01 we reject H_0 at the 1 per cent level of significance.

Example 3.12

Suppose that for a given fish population we wish to test the hypothesis that marked fish are more vulnerable to seining than to gill-net fishing. If the suffix a represents seining then this is equivalent to testing the hypothesis $p = p_0$ against the one-sided alternative $p > p_0$. For the following data

$$\lambda_a = 85\,042, \quad m_{2a} = 73, \quad \lambda_b = 43\,818, \quad m_{2b} = 20$$

we have $m_2 = 93$ and $p_0 = 0{\cdot}660$. The normal approximation is justified here, so that the test statistic is

$$z = \frac{73 - 93(0{\cdot}660) - \frac{1}{2}}{\sqrt{[93(0{\cdot}660)(0{\cdot}340)]}} = 2{\cdot}43.$$

For a one-sided test this corresponds to a significance level of 0·0075, so that we reject the null hypothesis at the 1 per cent level of significance. There is therefore evidence that the seine samples or gill-net samples are not random with respect to marked fish.

It should be noted that if m_{2a} and m_{2b} are too small, the test of H_0 will have low power and be very insensitive. To illustrate this point, Chapman and Overton [1966] discuss the following example.

Example 3.13

Suppose we wish to test H_0 against the one-sided alternative $N_b > N_a$ with $\lambda_a = \lambda_b = 1000$ (i.e. $p_0 = \frac{1}{2}$). If in actual fact $N_a = 100$, $N_b = 200$, then from the theory of the Petersen method (cf. **3.1.1**) the expected recaptures in the two populations (λ/N) will be 10 and 5, respectively. Hence from the binomial tables we have

$$\begin{aligned} \Pr[m_{2a} \geqslant 12 \mid m_2 = 15,\ p_0 = \tfrac{1}{2}] &= 0{\cdot}018 \\ \Pr[m_{2a} \geqslant 11 \mid m_2 = 15,\ p_0 = \tfrac{1}{2}] &= 0{\cdot}059 \end{aligned}$$

so that H_0 will be rejected at the 0·05 level of significance for $m_{2a} \geqslant 12$. The power of this test for the alternative $N_b = 2N_a$ (i.e. $p = \frac{2}{3}$) is then given approximately by

$$\Pr[m_{2a} \geqslant 12 \mid m_2 = 15, p = \tfrac{2}{3}] = 1 - \Pr[m_{2a} \geqslant 4 \mid m_2 = 15, p = \tfrac{1}{3}]$$
$$= 0{\cdot}209.$$

This means that the above program has only about 1 chance in 5 of detecting an effect as great as doubling the population. The main reason for this insensitivity is the small numbers of recaptures. We recall from **3.1.1** that the Petersen estimate may not even give the right order of magnitude when there are less than 10 recaptures. For example, if $\lambda_a = \lambda_b = 4000$ in the above experiment then the expected recaptures are 40 and 20 respectively, and the approximate power for $p = \frac{2}{3}$ is now 0·83, a considerable improvement over 0·209. We note that although such power calculations are only approximate, m_2 being strictly a random variable, they will be useful in providing some guidance on the choice of λ_a and λ_b.

In concluding this section we mention briefly two other problems considered in the literature. Chapman and Overton [1966] describe a paired comparison technique, based on a number of pairs of areas, for testing whether a certain management technique has any effect on population size. Chapman [1951: 156] briefly considers the problem of comparing ratios of populations, say N_a/N_b and N_c/N_d, using four separate Petersen type experiments and a chi-squared goodness-of-fit test similar to those discussed in **3.6.1.** Such an experimental situation could arise, for example, in comparing two populations before and after immigration or in detecting changes in sex or age ratios for populations which tend to segregate.

3.7 ESTIMATION BY LEAST SQUARES

In commercially exploited populations the second sample in the Petersen method may consist of a sequence of samples, each sample being permanently removed from the population. For this situation N can be estimated by the following least-squares method due to Paloheimo [1963].

It is convenient to have a change in notation. Let

N_0 = initial size of the total population,
M_0 = initial size of the marked population,
$U_0 = N_0 - M_0$,
n_i = size of the ith sample removed from the population ($i = 1, 2, \ldots s$),
m_i = number of marked individuals in the ith sample,
$u_i = n_i - m_i$,
$y_i = m_i/n_i$,

$$M_i = M_0 - \sum_{j=1}^{i-1} m_j, \text{ and}$$

$$N_i = N_0 - \sum_{j=1}^{i-1} n_j.$$

Then, if the assumptions underlying the Petersen method (**3.1.1**) hold for each sample,

$$E[y_i \mid M_i, N_i] = M_i/N_i \quad (i = 1, 2, \dots, s).$$

Paloheimo suggests estimating N_0 by minimising $\sum w_i(y_i - M_i/N_i)^2$ with respect to N_0, where the w_i's are appropriate weights, customarily taken to be proportional to the inverse of the variances of the y_i. When the sampling is random, or the marked and unmarked are randomly mixed, these variances may be calculated by assuming Poisson or binomial sampling. For example, assuming Poisson sampling, the variance of y_i equals its expected value M_i/N_i and the weights would have to be estimated iteratively as they contain the unknown N_0. Not only are such weights awkward to compute, but very often, in practice, the y_i vary more than expected on the assumption of random fluctuations. Under these circumstances De Lury [1958] argues that one should preferably choose weights equal to the sample sizes. Also, if the marked and unmarked are removed at the same rate, we have approximately $M_i/N_i = M_0/N_0$, so that

$$E[y_i \mid M_i, N_i] = M_0/N_0 = \beta_0, \quad \text{say.}$$

Therefore, assuming the y_i to be approximately independently and normally distributed with variances σ^2/n_i, and setting $w_i = n_i$, we can use the general theory of **1.3.5** (*1*) to obtain an estimate and $100(1-\alpha)$ per cent confidence interval for β_0, namely

$$\tilde{\beta}_0 = \sum w_i y_i / \sum w_i$$
$$= \sum m_i / \sum n_i$$

and

$$\tilde{\beta}_0 \pm t_{s-1}[\alpha/2](\tilde{\sigma}_0^2/\sum n_i)^{\frac{1}{2}},$$

where $$(s-1)\tilde{\sigma}_0^2 = \sum_i \frac{m_i^2}{n_i} - \frac{(\sum m_i)^2}{\sum n_i}.$$

The least-squares estimate of N_0 is then

$$\tilde{N}_0 = M_0/\tilde{\beta}_0 = M_0 \sum n_i / \sum m_i,$$

and Paloheimo notes that this is simply the usual Petersen estimate based on pooling the data from all the catches. Inverting the above interval for $\tilde{\beta}_0$ gives the following $100(1-\alpha)$ per cent confidence interval for N_0, namely

$$\frac{M_0 \sum n_i}{\sum m_i \pm t_{s-1}[\alpha/2](\tilde{\sigma}_0^2 \sum n_i)^{\frac{1}{2}}}.$$

In the same way we can obtain an estimate and confidence interval for $\sum n_i/N_0$, the rate of exploitation.

As a first step in examining the underlying assumptions of the above least-squares method we can plot y_i against i as a visual check on the constancy of M_i/N_i. If necessary, a test of $\beta = 0$ for the model $E[y_i] = \beta_0 + \beta i$ could be carried out using the theory of **1.3.5** (*1*). Also, by drawing the line

$y = \tilde{\beta}_0$, an examination of the deviation of each y_i from this line would provide a rough check on the reliability of the weights $w_i = n_i$. When there is mortality taking place the above method can still be used, provided that the mortality rates for marked and unmarked are the same, so that M_i/N_i remains approximately constant. However, if recruitment and immigration into the population are appreciable then the more complex models of Chapter 6 are required. Sometimes the effect of recruitment can be eliminated, and Ricker [1958: 86] gives several methods for fish populations based on growth data. For example, if the population can be divided into age-groups which overlap only slightly in length (or some other suitable measurement), then by choosing the minimum length of fish to be marked (L, say) at the trough between two age-groups, a boundary can be established whose position will advance as the season progresses and the fish grow larger. In this way recruitment into the marked length-range can be eliminated, and M_i/N_i will remain constant for this particular section of the population, provided the marked grow as much as the unmarked and suffer the same mortality rates. Thus N_0, the initial size of the population of all fish longer than L, can be estimated as above. Alternatively, if suitable length boundaries are not available, the recruits can be eliminated from each sample by the method of Robson and Flick [1965] described on pp. 74–78.

Example 3.14 Lobsters: Paloheimo [1963]

A full discussion of the lobster fishery is given in Paloheimo's paper and the reader is referred there for details. The data exhibited in Table 3.12 represent just one season's tagging experiments (1953–4) based on a release of 1000 tagged lobsters in an area centred at Port Maitland on the Atlantic coast of Canada. As lobsters are relatively non-migratory, the population of lobsters in this study area can be treated as an isolated unit.

TABLE 3.12

Recapture data for lobsters at Port Maitland (1953–4): from Paloheimo [1963: extracts from Tables 2 and 3].

i	m_i	n_i ('00)	y_i ($\times 10^2$)	$\frac{m_i^2}{n_i}$	i	m_i	n_i ('00)	y_i ($\times 10^2$)	$\frac{m_i^2}{n_i}$
1	95	323	0·29	0·279 412	8	30	77	0·39	0·116 883
2	28	173	0·16	0·045 318	9	36	193	0·19	0·067 150
3	53	202	0·26	0·139 059	10	42	253	0·16	0·069 723
4	17	155	0·11	0·018 645	11	34	336	0·10	0·034 405
5	7	61	0·11	0·008 033	12	53	286	0·18	0·098 217
6	3	58	0·05	0·001 552	13	45	135	0·33	0·150 000
7	12	92	0·13	0·015 652	Sum	455	2344		1·044 049

Since $M_0 = 1000$, $s = 13$, we have from Table 3.12:

$$\tilde{N}_0 = 1000(234\,400)/455 = 515\,200,$$
$$12\tilde{\sigma}_0^2 = 1{\cdot}044\,049 - (455)^2/234\,400 = 0{\cdot}160\,837,$$

and a 95 per cent confidence interval for N_0 is given by

$$\frac{1000\,(234\,400)}{455 \pm 2{\cdot}179(3141{\cdot}682)^{1/2}} \quad \text{or } (406\,100,\ 704\,200).$$

A plot of y_i versus i shows little trend, so that the assumption of constant M_i/N_i seems reasonable. However, for other study areas considered by Paloheimo there is often a strong trend in the y_i, particularly at the beginning of an experiment. This trend is put down to the lack of mixing of tagged and untagged rather than to the effects of tagging on the lobsters. The tagged lobsters when released are distributed more or less uniformly over the study area, while general observations indicate that the resident population on the other hand would be more concentrated on ledges and on rocky parts of the sea-bed close to their potential hiding places. This concentration of residents seems to be greater when the water is cold, and less when the water is warm and the lobsters more active. The differences, therefore, between the uniform distribution of tagged lobsters and the non-uniform distribution of the resident population at the start of the season would be more pronounced in the cold- than in warm-water areas. These differences would presumably disappear as the tagged lobsters gradually establish themselves on the bottom. Other sources of trend in the y_i would be immigration into the study area through the depletion of stock by the fishery, and recruitment to legal size during the season by the moulting and growth of undersized lobsters.

MARKED AND UNMARKED EQUICATCHABLE. Let p_i be the average probability of catching an unmarked individual in the ith sample: then

$$\begin{aligned} E[u_i \mid M_i, N_i] &= (N_i - M_i)p_i \\ &= \left(U_0 - \sum_{j=1}^{i-1} u_j\right)p_i \\ &= (U_0 - x_i)p_i, \quad \text{say.} \end{aligned}$$

If we assume that the average probability of catching a marked individual in the ith sample is p_i/k (where k is constant from sample to sample), then, estimating this probability by m_i/M_i, we are led to consider the regression model (Marten [1970a])

$$E\left[\frac{u_i M_i}{m_i} \,\middle|\, x_i\right] \approx k(U_0 - x_i)$$

or, adjusting for bias,

$$E\left[\frac{u_i(M_i+1)}{m_i+1} \mid x_i\right] \approx k(U_0 - x_i).$$

The constancy of k can be checked visually by plotting $u_i(M_i+1)/(m_i+1)$ against x_i, and both k and U_0 can be estimated using the method of **1.3.5**(*3*). We can also test the hypothesis $k = 1$ using the usual t-test for the slope of a linear regression.

3.8 REPEATED SAMPLES

From 3.1.5 we find that large sample sizes, n_1 and n_2, are generally required for an estimate of N of reasonable accuracy. As tagging can be very time-consuming, the value of n_1 determined by the charts on pages 65–69 may be prohibitively large. One way round this problem is to repeat the second sample k times and thus effectively increase the size of n_2. A good example of this approach is given by Rice and Harder [1977] who use a helicopter for taking the "recapture" samples. They also give a number of useful graphs which plot n_1/N against k for different n_2/N and a given accuracy of $A = 0{\cdot}1$ (cf. p. 64).

CHAPTER 4

CLOSED POPULATION : MULTIPLE MARKING

4.1 SCHNABEL CENSUS[*]

4.1.1 Notation

A simple extension of the Petersen method to a series of s samples of sizes $n_1, n_2, \ldots, n_s$ is the so-called Schnabel census (Schnabel [1938]). In this method each sample captured (except the first) is examined for marked members and then every member of the sample is given another mark before the sample is returned to the population. If different marks or tags are used for different samples, then the capture–recapture history of any animal caught during the experiment is known. In particular, if individual numbered tags are used then the same information is available if just the untagged members in each sample are tagged.

For the closed population (i.e. one in which immigration, death, etc., are negligible) a variety of theoretical models have been suggested, but before we discuss these we shall need some notation. Let

N = total population size,
s = number of samples,
n_i = size of the ith sample ($i = 1, 2, \ldots, s$),
m_i = number of marked individuals in n_i,
u_i = $n_i - m_i$,

$$M_i = \sum_{j=1}^{i-1} u_j \quad (i = 1, 2, \ldots, s+1)$$

= number of marked individuals in the population just before the ith sample is taken.

Since there are no marked animals in the first sample, we have $m_1 = 0$, $M_1 = 0$, $M_2 = u_1 = n_1$ and we define M_{s+1} ($= r$ say) as the total number of marked animals in the population at the end of the experiment, i.e. the total number of *different* animals caught throughout the experiment.

4.1.2 Fixed sample sizes

1 The generalised hypergeometric model

Let a_w be the number of animals with a particular capture history w, where w is a non-empty subset of the integers $\{1, 2, \ldots, s\}$: thus a_{124} represents those animals caught in the first, second and fourth samples only, also $r = \sum_w a_w$. If P_w, the probability that an animal chosen at random from the population has history w, is the same for each animal, and animals act

[*]This section should be read in conjunction with **12.8**.

independently, the animals may be regarded as N independent "trials" from a multinomial experiment. Therefore the joint probability function of the random variables $\{a_w\}$ is

$$f(\{a_w\}) = \frac{N!}{\prod_w a_w!\,(N-r)!}\, Q^{N-r} \prod_w P_w^{a_w}, \tag{4.1}$$

where $Q = 1 - \sum_w P_w$. We shall assume that:

(i) all individuals have the same probability $p_i(= 1 - q_i)$ of being caught in the ith sample, and

(ii) for any individual the events "caught in the ith sample ($i = 1, 2, \ldots, s$)" are independent. Then

$$Q = \prod_{i=1}^{s} q_i, \quad P_{124} = p_1 p_2 q_3 p_4 \ldots q_s = \frac{p_1 p_2 p_4 Q}{q_1 q_2 q_4}, \quad \text{etc.},$$

and Darroch [1958] shows that (4.1) reduces to

$$f(\{a_w\}) = \frac{N!}{\prod_w a_w!\,(N-r)!} \prod_{i=1}^{s} p_i^{n_i} q_i^{N-n_i}. \tag{4.2}$$

From the above assumptions the $\{n_i\}$ are independent binomial variables, so that

$$f(\{n_i\}) = \prod_{i=1}^{s} \binom{N}{n_i} p_i^{n_i} q_i^{N-n_i}$$

and the joint probability function of the $\{a_w\}$, conditional on fixed sample sizes $\{n_i\}$ (i.e. the sample sizes are chosen in advance), is

$$f(\{a_w\} \mid \{n_i\}) = \frac{N!}{\prod_w a_w!\,(N-r)!} \prod_{i=1}^{s} \binom{N}{n_i}^{-1}. \tag{4.3}$$

By setting $\nabla \log f(\{a_w\}|\{n_i\}) = 0$ and using the fact that $\nabla \log N! = \log N$, etc., Darroch shows that the maximum-likelihood estimate $\hat{N}$ of N for the model (4.3) is the *unique* root (cf. **A7**), greater than r, of the $(s-1)$th degree polynomial given by

$$\left(1 - \frac{r}{N}\right) = \prod_{i=1}^{s} \left(1 - \frac{n_i}{N}\right). \tag{4.4}$$

This equation has a very simple interpretation when it is noted that the left-hand side is equal to $\prod_i (1 - u_i/(N - M_i))$, the product of the probabilities that an unmarked individual is not caught in the ith sample. The right-hand side represents the same product of probabilities, but now with respect to all individuals rather than just the unmarked.

The above equation (4.4) was first obtained by Chapman [1952], using a slightly different model; summing on the $\{a_w\}$ in each sample, the joint probability function of the $\{m_i\}$ is

$$f(m_2, \ldots, m_s \mid \{n_i\}) = \prod_{i=2}^{s} \binom{M_i}{m_i}\binom{N-M_i}{u_i}\Big/\binom{N}{n_i}$$

$$= \frac{\prod_{i=2}^{s} \binom{M}{m_i}}{\prod_{i=1}^{s} u_i!} \cdot \frac{N!}{(N-r)!} \prod_{i=1}^{s} \binom{N}{n_i}^{-1}. \tag{4.5}$$

This product of hypergeometric distributions, derived directly by Chapman on the assumption that each sample is a simple random sample, leads to the same maximum-likelihood equation (4.4).

When $s = 2$, (4.4) is of first degree and we find that $\hat{N} = n_1 n_2 / m_2$, the Petersen estimate. For $s = 3$ we have the quadratic

$$N^2(m_2 + m_3) - N(n_1 n_2 + n_1 n_3 + n_2 n_3) + n_1 n_2 n_3 = 0,$$

which can be readily solved for the larger root $\hat{N}$. When $s > 3$ we require some iterative method of solution, and three useful techniques are described in **1.3.8**(*1*). However, since we are only interested in finding $\hat{N}$ to the nearest integer, Robson and Regier's technique will be used as it is the easiest method for a desk calculator. Let

$$g(N) = \prod_{i=1}^{s} (1 - n_i/N)$$

and let $h(N) = N - r - Ng(N)$; then the ith step of the iteration is given by

$$N_{(i+1)} = N_{(i)} - h(N_{(i)})/\nabla h(N_{(i)}),$$

where

$$\nabla h(N_{(i)}) = h(N_{(i)}) - h(N_{(i)} - 1)$$
$$= 1 - N_{(i)} g(N_{(i)}) + (N_{(i)} - 1) g(N_{(i)} - 1).$$

To begin the iterations we require first of all a trial solution $N_{(1)}$ where $N_{(1)} > r$. If N is large, then expanding $g(N)$ in powers of $1/N$, neglecting powers greater than the second and using

$$\sum_{i=1}^{s} n_i - r = \sum_{i=2}^{s} m_i,$$

we find that (4.4) yields the approximate solution

$$N_B = \left(\sum_{i=1}^{s} \sum_{j=i+1}^{s} n_i n_j\right)\Big/\left(\sum_{i=2}^{s} m_i\right)$$
$$= R_2/m, \text{ say,}$$

where R_2 can be expressed in the form

$$R_2 = \tfrac{1}{2}\left[\left(\sum_{i=1}^{s} n_i\right)^2 - \sum_{i=1}^{s} n_i^2\right].$$

However, if the cubic terms are retained, (4.4) reduces to the quadratic

$$N^2 m - NR_2 + R_3 = 0, \tag{4.6}$$

$$\text{where} \quad R_3 = \sum_{i=1}^{s} \sum_{j=i+1}^{s} \sum_{k=j+1}^{s} n_i n_j n_k$$

$$= \tfrac{1}{3}\left[\left(\sum_{i=1}^{s} n_i\right)\left(R_2 - \sum_{i=1}^{s} n_i^2\right) + \sum_{i=1}^{s} n_i^3\right]$$

and the desired solution of the quadratic is the larger root, N_A say. Chapman [1952] shows that under certain conditions, which are often satisfied when N is much greater than $\Sigma\, n_i$, $N_A < \hat{N} < N_B$. However his proof requires modification (Sandvik [1978]). Chapman also gives another pair of numbers

$$N_C = \text{maximum}\,\{r, \underset{2 \leq i \leq s}{\text{minimum}}\,(n_i M_i/m_i)\},$$

$$N_D = \underset{2 \leq i \leq s}{\text{maximum}}\,(n_i M_i/m_i)$$

and in general this pair will be satisfactory, provided that no m_i (except m_1) is zero. We note that the methods discussed later in this chapter can also be used for providing a first approximation (cf. Example 4.7 on p. 144).

In solving (4.4) we see that the only recapture information required is r, the number of different animals caught during the experiment. This follows from the fact that r is a sufficient statistic for N and means that as far as the estimation of N is concerned, distinguishing marks are not needed for each sample. In fact, at each stage, we need only mark the unmarked members of the sample. However, as pointed out in **4.1.1**, if the tags have sufficient information (e.g. are numbered) then we can record all the recapture histories: this information is useful for testing some of the underlying assumptions (cf. **4.1.5**).

Using the model (4.3), Darroch [1958] proves that asymptotically (i.e. $N \to \infty$, $n_i \to \infty$ such that n_i/N remains constant)

$$E[\hat{N}] = N + b,$$

where b, the bias, is estimated by

$$\hat{b} = \frac{\left[\dfrac{s-1}{\hat{N}} - \Sigma\left(\dfrac{1}{\hat{N}-n_i}\right)\right]^2 + \left[\dfrac{s-1}{\hat{N}^2} - \Sigma\left(\dfrac{1}{\hat{N}-n_i}\right)^2\right]}{2\left[\dfrac{1}{\hat{N}-r} + \dfrac{s-1}{\hat{N}} - \Sigma\left(\dfrac{1}{\hat{N}-n_i}\right)\right]^2},$$

and the asymptotic variance of $\hat{N}$ is estimated by

$$v[\hat{N}] = (\hat{N}-r)/h'(\hat{N})$$

$$= \left[\frac{1}{\hat{N}-r} + \frac{s-1}{\hat{N}} - \Sigma\left(\frac{1}{\hat{N}-n_i}\right)\right]^{-1},$$

where all summations are for $i = 1, 2, \ldots, s$. Obviously the last step in the Newton–Raphson method (which requires $h'(\hat{N})$, cf. **1.3.8**(*1*)) can be used for evaluating $v[\hat{N}]$. Also $\nabla h(\hat{N}) \approx h'(\hat{N})$, so that the last step of Robson and Regier's method will provide a reasonable approximation for $v[\hat{N}]$. It can be shown that

$$\hat{b} = -\tfrac{1}{2}(\hat{N}-r)h''(\hat{N})/[h'(\hat{N})]^2$$

so that $\hat{b}$ can be approximated by

$$-\tfrac{1}{2}(\hat{N}-r)\nabla^2 h(\hat{N})/[\nabla h(\hat{N})]^2.$$

Although this approximation may not be very accurate it does at least indicate the order of magnitude of $\hat{b}$.

Assuming $\hat{N}$ to be asymptotically normal, we have the approximate 95 per cent confidence interval for N, namely

$$\hat{N} - \hat{b} \pm 1{\cdot}96\sqrt{v[\hat{N}]}, \tag{4.7}$$

where $\hat{b}$ can be neglected if it is less than one-tenth of $\sqrt{v}$ (Cochran [1963: 12]). However, the statistic r is more nearly normally distributed than $\hat{N}$, and Darroch [1958: 348] shows that we can use r as a basis for a confidence interval as follows.

The expected value of r, regarded as a function of N, is

$$\rho(N) = N - \left(\prod_{i=1}^{s} (N-n_i)\right)/N^{s-1},$$

and equation (4.4) is simply $\rho(\hat{N}) = r$. The variance of r, expressed as a function of N, is (Lee [1972]).

$$\sigma^2(N) = [N-\rho(N)][\rho(N)-\rho(N-1)]$$

and we have

$$\begin{aligned} 0{\cdot}95 &\approx \Pr[r - 1{\cdot}96\,\sigma(N) < \rho(N) < r + 1{\cdot}96\,\sigma(N)] & (4.8)\\ &\approx \Pr[r - 1{\cdot}96\,\sigma(\hat{N}) < \rho(N) < r + 1{\cdot}96\,\sigma(\hat{N})] \\ &= \Pr[r_1 < \rho(N) < r_2] \\ &= \Pr[\rho^{-1}(r_1) < N < \rho^{-1}(r_2)] \\ &= \Pr[N_1 < N < N_2], \quad \text{say}, & (4.9) \end{aligned}$$

since $\rho(N)$ is a monotonic increasing function of N. The confidence limits N_1 and N_2 can be calculated by setting $h(N) = \rho(N) - r_i$ and solving $h(N) = 0$ iteratively as above. Alternatively we could deal with the interval (4.8) directly by solving the equations $r \pm 1{\cdot}96\,\sigma(N) = \rho(N)$ iteratively on a computer.

RANDOM SAMPLE SIZES. We now mention briefly the more common situation in which the sample sizes n_i are random variables rather than fixed parameters. Darroch [1958] has investigated this model (namely 4.2) in some detail and shows that, as far as the point and interval estimation of N is concerned, there is no difference (asymptotically) between the two cases of fixed and random sample size. The reason for this is that the maximum-likelihood estimate $\hat{N}$ is almost the same in both cases, and in estimating the variance of $\hat{N}$, one effectively replaces $E[n_i]$ by n_i when n_i is random (cf. **3.1.3**). For a further discussion see **12.8.2**.

Example 4.1 Cricket frog (*Acris gryllus*): Turner [1960a]

From the data in Table 4.1 we have $r = 87 + (41 - 36) = 92$ and $(N_C, N_D) = (92, 99)$. We note in passing that $N_B = 150$, but N_A is not applicable as (4.6) does not have real roots. Choosing $N_{(1)} = 97$ as our first approximation

TABLE 4.1
Capture–recapture data for a population of cricket frogs: from Turner [1960a: Table 3].

Sample	n_i	m_i	M_i	$n_i M_i / m_i$
1	32	–	–	–
2	54	18	32	96
3	37	31	68	81
4	60	47	74	94
5	41	36	87	99

and using Robson and Regier's iterative method, we have

$$\begin{aligned} N_{(1)} - r &= 5 \\ N_{(1)}g(N_{(1)}) &= (97-32)(97-54)(97-37)(97-60)(97-41)/97^4 \\ &= 3{\cdot}924\ 965\ 0, \\ (N_{(1)}-1)g(N_{(1)}-1) &= 3{\cdot}697\ 103\ 0, \\ \nabla h(N_{(1)}) &= 1 + 3{\cdot}697\ 103 - 3{\cdot}924\ 965\ 0 = 0{\cdot}772\ 138\ 0 \\ h(N_{(1)}) &= 5 - 3{\cdot}924\ 965\ 0 = 1{\cdot}075\ 035\ 0 \end{aligned}$$

and the correction is

$$-h(N_{(1)})/\nabla h(N_{(1)}) = -1{\cdot}075\ 035\ 0/0{\cdot}772\ 138\ 0 = -1{\cdot}4.$$

Therefore our next approximation is

$$N_{(2)} = N_{(1)} - 1 = 96$$

from which we calculate

$$\begin{aligned} N_{(2)} - r &= 4, \\ N_{(2)}g(N_{(2)}) &= 3{\cdot}697\ 103\ 0, \\ (N_{(2)}-1)g(N_{(2)}-1) &= 3{\cdot}476\ 320\ 4, \\ \nabla h(N_{(2)}) &= 1 + 3{\cdot}476\ 320\ 4 - 3{\cdot}697\ 103\ 0 = 0{\cdot}779\ 217\ 4, \\ h(N_{(2)}) &= 4 - 3{\cdot}697\ 103\ 0 = 0{\cdot}302\ 897 \end{aligned}$$

and the correction is

$$-0{\cdot}302\ 897/0{\cdot}779\ 217\ 4 = -0{\cdot}39.$$

Therefore, to the nearest integer, $N_{(3)} = N_{(2)}$ and $\hat{N} = 96$. Also

$$(N_{(2)} - r)/\nabla h(N_{(2)}) = 4/0{\cdot}779\ 217\ 4 = 5{\cdot}13$$

which is close to

$$v[\hat{N}] = 1/0{\cdot}189\ 32 = 5{\cdot}282.$$

Since

$$\begin{aligned} \nabla^2 h(\hat{N}) &\approx \nabla^2 h(\hat{N}+1) \\ &= \nabla h(N_{(1)}) - \nabla h(N_{(2)}) \\ &= 0{\cdot}772\ 138\ 0 - 0{\cdot}779\ 217\ 4 \\ &= -0{\cdot}007\ 079, \end{aligned}$$

$$\hat{b} \approx \tfrac{1}{2} \times 4 \times 0{\cdot}007\ 079/(0{\cdot}779\ 217\ 4)^2 = 0{\cdot}02,$$

indicating that the bias of $\hat{N}$ is negligible. The asymptotic confidence interval (4.7) is then found to be (91·5, 100·5).

We note that

$$1{\cdot}96\,\sigma(\hat{N}) = 1{\cdot}96 \times 4(0{\cdot}18932)^{\frac{1}{2}} = 3{\cdot}412,$$

so that $r_1 = 88{\cdot}41$ and $r_2 = 95{\cdot}41$. Since $\rho(91) = 89{\cdot}3$, $\rho(90) = 88{\cdot}5$, $\rho(102) = 96{\cdot}29$, $\rho(101) = 95{\cdot}8$ and $\rho(100) = 95{\cdot}35$, we have, from (4.8), $(N_1, N_2) = (90, 100)$.

2 *Samples of size one*

The special case of sampling one at a time was considered by Craig [1953b] for the study of butterfly populations. Putting $n_i = 1$ in equation (4.4), we find that $\hat{N}$ is the solution of

$$\left(1 - \frac{r}{N}\right) = \left(1 - \frac{1}{N}\right)^s, \tag{4.10}$$

and this equation can be solved in much the same way as (4.4). However, since s is generally large, Craig suggests taking logarithms and solving

$$H(N) = (s-1)\log_{10} N + \log_{10}(N-r) - s\log_{10}(N-1) = 0,$$

using a good table of logarithms (e.g. Spenceley *et al.* [1952]) and a suitable first approximation N_0 such as the following. Let f_x be the frequency of cases in which the same individual is caught x times ($x = 1, 2, \ldots$) and let $s_2 = \sum x^2 f_x$; then Craig suggests

$$N_0 = s^2/(s_2 - s).$$

Alternatively, following Darroch [1958: 349] and letting $N \to \infty$, $s \to \infty$, subject to s/N ($= D$, say) remaining constant, (4.10) becomes

$$1 - (r/N) = e^{-D}.$$

A first approximation to $\hat{N}$ (in fact an upper bound) is then $N_0' = s/D_0'$ where D_0', the solution of

$$(1 - e^{-D})/D = r/s \quad (= a \text{ say}) \tag{4.11}$$

is obtained by linear interpolation in Table **A3**. Samuel [1969] suggests a further approximation $N_0'' = s/D_0''$, where D_0'' is the solution of (4.11) with $a = r/(s + D_0')$.(*)

For the limiting process mentioned above, Darroch [1958: 34] shows that asymptotically

$$E[\hat{N}] = N + b,$$

where $b = \frac{1}{2}D^2(e^D - 1 - D)^{-2}$,

and

$$V[\hat{N}] = N(e^D - 1 - D)^{-1}.$$

Both b and $V[\hat{N}]$ can be estimated by replacing D by $\hat{D} = s/\hat{N}$: extensive tables of e^D are given, for example, in Becker and Van Orstrand [1924], and Comrie [1959].

(*) For a useful approximation cf. Darroch and Ratcliff [1980], Additional References.

Confidence limits for N based on r can be calculated as in (4.8) above, using

$$\rho(N) = N(1-e^{-D})$$

and $$\sigma^2(N) = Ne^{-2D}(e^D - 1 - D).$$

If the interval (4.9) is used, we have to solve two equations of the form $\rho(N) = r_i$ or

$$(1-e^{-D})/D = r_i/s$$

which, as for the case $r_i = r$ above (equation 4.11), can be solved for D by interpolating linearly in Table **A3.**

Example 4.2 Butterflies (*Colias eurytheme*): Craig [1953b]

From the data in Table 4.2 we have $r = 341$, $s = 435$, $s_2 = 645$ and $N_0 = (435)^2/(645-435) = 901$. Using $a = r/s = 0{\cdot}7839$ and interpolating linearly in

TABLE 4.2
Capture–recapture data for samples of size one from a population of butterflies: from Craig [1953b].

x	1	2	3	$\geqslant 4$	Total
f_x	258	72	11	0	341
xf_x	258	144	33	0	435
$x^2 f_x$	258	288	99	0	645

Table **A3**, we find that $D_0' = 0{\cdot}5084$ and $\hat{N} < N_0' = 435/(0{\cdot}5084) = 856$. Also $r/(s+D_0') = 0{\cdot}7830$ so that $D_0'' = 0{\cdot}5110$ and $N_0'' = 851$. Since

$$H(851) = -0{\cdot}000\,222 < 0,$$

where $$H(N) = 434 \log_{10} N + \log_{10}(N-341) - 435 \log_{10}(N-1),$$

we have that $851 < \hat{N} < 856$.

To solve $H(N) = 0$ for $\hat{N}$ we shall use Robson and Regier's iterative method (cf. **1.3.8** (*1*)), starting with the trial solution $N_{(1)} = 851$. To get the next approximation we calculate

$$H(850) = -0{\cdot}000\,329,$$
$$\nabla H(851) = H(851) - H(850) = 0{\cdot}000\,107,$$

and the correction is

$$-H(851)/\nabla H(851) = 0{\cdot}000\,222/0{\cdot}000\,107 \approx 2,$$

so that $N_{(2)} = 851 + 2 = 853$. Successive iterations yield $N_{(3)} = N_{(4)} = 854$, so that to the nearest integer $\hat{N} = 854$. If we use Craig's first approximation $N_{(1)} = 901$, the successive approximations are $N_{(2)} = 867$, $N_{(3)} = 853$, $N_{(4)} = N_{(5)} = 854$.

Now $$\hat{D} = 435/854 = 0{\cdot}509\,36,$$
$$\exp(\hat{D}) - 1 - \hat{D} = 0{\cdot}1549,$$

$$\hat{b} = \frac{1}{2}\left(\frac{0{\cdot}509\ 36}{0{\cdot}1549}\right)^2 = 5,$$

$$\hat{V}[\hat{N}] = 854/0{\cdot}1549 = 5513,$$

and the approximate 95 per cent confidence interval (4.7) is 849 ± 146 or (703, 995). If we use (4.9) then we require

$$\sigma^2(\hat{N}) = 854(0{\cdot}600\ 88)^2\,(0{\cdot}1549) = 47{\cdot}76,$$

$$r_1 = r - 1{\cdot}96\sigma(\hat{N}) = 327{\cdot}5,$$

$$r_2 = r + 1{\cdot}96\sigma(\hat{N}) = 354{\cdot}5,$$

$$r_1/s = 0{\cdot}7529, \quad r_2/s = 0{\cdot}8149,$$

and $(D_1, D_2) = (0{\cdot}597, 0{\cdot}424)$ where D_i is the solution of (4.11) when $r = r_i$. Hence the confidence interval (4.9) is

$$\left(\frac{435}{0{\cdot}597}, \frac{435}{0{\cdot}424}\right) \quad \text{or } (729, 1026).$$

3 Mean Petersen estimate

At each stage of the sampling a modified Petersen estimate of N can be calculated, namely (cf. **3.1.1**)

$$N_i^* = \frac{(M_i+1)(n_i+1)}{(m_i+1)} - 1 \quad (i = 2, 3, \ldots, s)$$

with variance estimate

$$v_i^* = \frac{(M_i+1)(n_i+1)(M_i-m_i)(n_i-m_i)}{(m_i+1)^2(m_i+2)}.$$

Therefore a natural estimate of N, suggested by Chapman [1952: 293], is the average

$$\bar{N} = \Sigma N_i^*/(s-1).$$

(All summations throughout this section are for $i = 2, 3, \ldots, s$). Since the covariances of the N_i^* are asymptotically negligible compared with their variances, we have approximately

$$V[\bar{N} \mid \{n_i, M_i\}] = \Sigma V[N_i^* \mid n_i, M_i]/(s-1)^2.$$

This can be estimated by either

$$v^* = \Sigma v_i^*/(s-1)^2$$

which is approximately unbiased if and only if each v_i^* is almost unbiased, or by (cf. 1.9)

$$v[\bar{N}] = \Sigma (N_i^* - \bar{N})^2/(s-1)(s-2),$$

which is almost unbiased when the N_i^* have the same mean. When these conditions for unbiasedness are not satisfied, both estimates are conservative in that they tend to overestimate the true variance. For a numerical example using the above approach the reader is referred to p. 144 (Example 4.7).

A more efficient way of averaging the N_i^* is to take a weighted average with weights inversely proportional to the variances (cf. **1.3.2**). This produces an estimate with the same variance as the maximum-likelihood estimate $\hat{N}$ (Darroch [1958: 354] proves this for the case when the n_i are random variables). However, as the weights themselves have to be estimated, bias would be introduced, which $\bar{N}$ aims to avoid. Therefore, although $\bar{N}$ is less efficient than $\hat{N}$, it is almost unbiased (under certain reasonable conditions, cf. **3.1.1**), and we would expect $\bar{N}$ to have some degree of robustness with regard to departures from the assumptions underlying the Schnabel census.

4 Schnabel's binomial model

An alternative approach to the Schnabel census can be made by assuming that the M_i are fixed parameters (which they are, conditionally, at each sampling) and then using the binomial approximation of **3.1.2.** This leads to Schnabel's [1938] model (cf. (4.5))

$$f(m_2, \ldots, m_s \mid \{n_i, M_i\}) = \prod_{i=2}^{s} \binom{n_i}{m_i} \left(\frac{M_i}{N}\right)^{m_i} \left(1 - \frac{M_i}{N}\right)^{n_i - m_i}, \qquad (4.12)$$

and the maximum-likelihood estimate of N is now the appropriate root of

$$\sum_{i=2}^{s} \left\{\frac{(n_i - m_i)M_i}{N - M_i}\right\} = \sum_{i=2}^{s} m_i. \qquad (4.13)$$

This model assumes that at each stage, n_i/N is sufficiently small (say less than 0·1) for one to ignore the complications of sampling without replacement. If each M_i/N is also small, a first approximation to the solution of (4.13) is

$$N' = \left(\sum_{i=2}^{s} n_i M_i\right) \Big/ \left(\sum_{i=2}^{s} m_i\right) = \lambda/m, \quad \text{say.}$$

We note that, irrespective of any assumptions concerning the probability function of the m_i or the magnitudes of the various parameters, N' has a certain intuitive appeal, being simply a weighted average of the Petersen estimates $n_i M_i/m_i$.

When n_i/N and M_i/N are both less than, say, 0·1 for each i, a modification of N' which is almost unbiased is

$$N'' = \lambda/(m+1).$$

In fact, from Chapman [1952: 293] we have

$$E[N'' \mid \{n_i, M_i\}] = N(1 - \exp(-\lambda/N))$$

and (correcting a misprint in Chapman's equation)

$$V[N'' \mid \{n_i, M_i\}] \approx N^2 \left(\frac{N}{\lambda} + 2\frac{N^2}{\lambda^2} + 6\frac{N^3}{\lambda^3}\right).$$

These formulae were derived by Chapman on the basis that, when the above conditions hold, m_i is approximately distributed as Poisson with parameter $M_i n_i/N$. Hence the sum, m, of independent Poisson variables is also Poisson with parameter λ/N. Actually a study of Raff [1956: Table 4] would suggest that the Poisson approximation still applies, even if $0{\cdot}1 < M_i/N < 0{\cdot}2$, provided that n_i/N is much less than 0·1, so that the hypergeometric distribution of m_i is well approximated by the binomial. But the errors in these approximations have an accumulative effect on the sum m, so that for m to be approximately Poisson we could not have more than one or two samples with M_i/N greater than 0·1.

Assuming N'' to be asymptotically normal, we can calculate a confidence interval for N in the usual manner. However, as in the Petersen method (cf. **3.1.4**), it is recommended to base confidence intervals on the distribution of m. For $m \leqslant 50$ we can use Chapman's Poisson table (Table **A1**) to obtain the shortest interval for N/λ, and hence for N. When $m > 50$ we can use the normal approximation to the Poisson, and a 95 per cent confidence interval for N is given by the roots of the quadratic

$$\frac{(m-\lambda/N)^2}{\lambda/N} = 1{\cdot}96^2,$$

namely

$$(N/\lambda) = \frac{2m + 1{\cdot}96^2 \pm 1{\cdot}96\sqrt{(4m+1{\cdot}96^2)}}{2m^2}. \tag{4.14}$$

If the Poisson approximation to the binomial is not satisfactory, the variance of m will be less than λ/N, and the above normal confidence interval will be too wide. In this case λ/N is replaced by a sum of binomial variances, and (4.14) becomes

$$(N/\lambda) = \frac{2m + 1{\cdot}96^2(1-\delta) \pm 1{\cdot}96\sqrt{(1-\delta)[4m+1{\cdot}96^2(1-\delta)]}}{2m^2} \tag{4.15}$$

where $\delta = \sum_{i=2}^{s} n_i M_i^2/(\lambda N')$.

Example 4.3

Consider the data in the first half of Table 4.3, opposite. Here $m = 15$, $\lambda = 22\,000$ and $N'' = 22\,000/16 = 1375$. Entering Table **A1** with $m = 15$, a 95 per cent confidence interval for N/λ is (0·0365, 0·111) and the corresponding interval for N is (803, 2442). This interval is very wide because of the small number of recaptures.

COMPARING TWO POPULATIONS. Chapman and Overton [1966] consider the problem of comparing two populations of sizes N_a and N_b respectively. Suppose a Schnabel census is carried out in each population, yielding λ_a, m_a, λ_b and m_b as the respective values of λ and m. Then, provided m_a and m_b can be regarded as Poisson random variables, the methods described in **3.6.2** can be used here to test the hypothesis H_0 that $N_a = N_b$.

Example 4.4

Artificial data for two hypothetical populations are set out in Table 4.3. We have $m_a = 15$, $m_b = 14$, $m_a + m_b = 29$, $\lambda_a = 22\,000$, $\lambda_b = 27\,765$ and $p_0 = 22\,000/49\,765 = 0{\cdot}4421$. To test H_0 against the two-sided alternative

$N_a \neq N_b$ we can use the unit normal statistic

$$z = \frac{|15 - 29(0{\cdot}4421)| - \frac{1}{2}}{\sqrt{29(0{\cdot}4421)(0{\cdot}5579)}} = 0{\cdot}63,$$

which is not significant at the 5 per cent level of significance.

4.1.3 Regression methods

The maximum-likelihood method described in **4.1.2** (*1*) will give the most efficient estimate $\hat{N}$ of N, provided the assumptions underlying the model are

TABLE 4.3
Artificial capture–recapture data for two hypothetical populations.

N_a				N_b			
n_i	m_i	M_i	n_iM_i	n_i	m_i	M_i	n_iM_i
70	–	–	–	204	–	–	–
43	3	70	3 010	90	9	204	18 360
75	6	110	8 250	33	5	285	9 405
60	6	179	10 740				
Total	15		22 000	Total	14		27 765

satisfied. $\hat{N}$ will, however, tend to be sensitive to departures from the underlying assumptions, particularly those relating to constant N and the random behaviour of marked animals. Therefore, in practice, $\hat{N}$ may sometimes be an inefficient estimate and $v[\hat{N}]$ may be unreliable. For this reason less efficient but more robust estimates of N, like $\bar{N}$ of **4.1.2** (*3*), are desirable. In particular a useful regression method has been suggested by Schumacher and Eschmeyer [1943] and we now discuss this technique in detail.

1 Schumacher and Eschmeyer's method

In Schnabel's model (4.12) each m_i is assumed to be binomially distributed, so that $y_i = m_i/n_i$ has mean M_i/N and variance

$$\sigma_i^2 = \frac{M_i}{N}\left(1 - \frac{M_i}{N}\right) \cdot \frac{1}{n_i}.$$

We may therefore write

$$y_i = \beta M_i + e_i \quad (i = 2, 3, \ldots, s),$$

where $\beta = 1/N$ and the "error" e_i has mean zero and variance σ_i^2. If we plot y_i against M_i, the plotted points should lie approximately on a straight line of slope β passing through the origin. Since the variance of e_i is not constant, the least-squares fitting of a straight line should be done using weights w_i, say, as in **1.3.5** (*1*). Thus $\tilde{N}$, the least-squares estimate of N, is given by

$$1/\tilde{N} = \tilde{\beta} = (\Sigma\, w_i y_i M_i)/(\Sigma\, w_i M_i^2),$$

where all summations throughout this section are for $i = 2, 3, \ldots, s$. If the weights are chosen in the usual manner, namely proportional to the reciprocal of the variances, then the above equation becomes the maximum-likelihood equation (4.13). However, although these weights will give the most efficient estimate of N when sampling is truly random, we are computationally no better off than before, as these unknown weights have to be estimated iteratively. De Lury [1958] also points out that "owing to the tendency of fishes to stratify and for other reasons that lead to similar effects, the proportion of marked individuals available to the sampling at any one time is likely to differ widely from the 'true' proportion, and the weights are therefore likely to be seriously wrong. In these circumstances, weighting by sample size alone is preferable to weighting according to proportions tagged." In support of this last statement we note that $(M_i/N)[1-(M_i/N)]$ does not vary much as M_i/N varies from 0·2 to 0·8. Therefore, putting $w_i = n_i$, $\tilde{\beta}$ is now given by

$$\tilde{\beta} = (\Sigma\, m_i M_i)/(\Sigma\, n_i M_i^2), \tag{4.16}$$

which is equivalent to the formula given by Schumacher and Eschmeyer [1943]. This formula was also given by Hayne [1949b], so that the method is sometimes called Hayne's method. The mean and variance of $\tilde{N}$ could be calculated using the delta method (**1.3.3**) and an approximate confidence interval for N obtained in the usual manner. However, following De Lury [1958] it seems preferable to assume that the e_i's are independently normally distributed with variances σ^2/n_i, and to invert the confidence interval for β given in **1.3.5**(*1*). Hence a $100(1-\alpha)$ per cent confidence interval for N is given by

$$\frac{\Sigma\, n_i M_i^2}{\Sigma\, m_i M_i \pm t_{s-2}[\alpha/2](\tilde{\sigma}^2 \Sigma\, n_i M_i^2)^{\frac{1}{2}}}, \tag{4.17}$$

where $$(s-2)\tilde{\sigma}^2 = \sum \frac{m_i^2}{n_i} - \frac{(\Sigma\, m_i M_i)^2}{\Sigma\, n_i M_i^2}.$$

From bead sampling experiments, De Lury showed that the above confidence interval compared favourably with the confidence interval based on the more efficient binomial weights. We would also expect (4.17) to be robust with regard to departures from the underlying assumptions, and this model should therefore be used in conjunction with the other methods mentioned so far in this chapter. In particular a graph is always a useful indicator of any marked departures from the assumptions underlying the model.

In conclusion we mention a small point concerning the number of degrees of freedom used in the above theory. Some authors (e.g. Hayne [1949b], Ricker [1958]) include the first sample in the theory, so that the point (0, 0) is used in the regression analysis. In this case the number of degrees of freedom should be $s-1$ rather than $s-2$. However, as y_1 is always zero when $M_1 = 0$, y_1 is not strictly a random observation and for this reason is not included in the above theory.

Example 4.5 Red-ear sunfish (*Lepomis microlophus*): Ricker [1958: 103]

The data and calculations are set out in Table 4.4. We have

$$\tilde{N} = 970\ 296/2294 = 423,$$

$$\tilde{\sigma}^2 = \tfrac{1}{12}\{7{\cdot}7452 - (2294)^2/970\ 296\} = 0{\cdot}1935,$$

and (4.17) becomes

$$\frac{970\ 296}{2294 \pm 2{\cdot}179[0{\cdot}1935(970\ 296)]^{\frac{1}{2}}}$$

TABLE 4.4

Capture–recapture data for a population of sunfish: from Ricker [1958: 103].

i	n_i	m_i	y_i	M_i	n_iM_i	m_iM_i	$n_iM_i^2$	m_i^2/n_i
1	10	–	–	–	–	–	–	–
2	27	0	0	10	270	0	2 700	0
3	17	0	0	37	629	0	23 273	0
4	7	0	0	54	378	0	20 412	0
5	1	0	0	61	61	0	3 721	0
6	5	0	0	62	310	0	19 220	0
7	6	2	0·33	67	402	134	26 934	0·6667
8	15	1	0·07	71	1 065	71	75 615	0·0667
9	9	5	0·56	85	765	425	65 025	2·7778
10	18	5	0·28	89	1 602	445	142 578	1·3889
11	16[2]	4	0·25	102	1 632	408	166 464	1·0000
12	5	2	0·40	112	560	224	62 720	0·8000
13	7[1]	2	0·29	115	805	230	92 575	0·5714
14	19	3	0·16	119	2 261	357	269 059	0·4737
Total		24			10 740	2294	970 296	7·7452

[2, 1] Number of deaths in this sample.

or (300, 719). A cursory glance at Table 4.4 shows that there is no obvious linear relation between y_i and M_i, but any trend could have been masked by the wide fluctuations of the values of y_i due to the small numbers of recaptures m_i.

For comparison, the Schnabel estimates are given by (cf. **4.1.2**(*4*))

$$N' = 10\ 740/24 = 448,$$

and

$$N'' = 10\ 740/25 = 430.$$

Although $m < 50$, $M_i/\tilde{N} > 0{\cdot}1$ for most samples, so that m will not have a Poisson distribution. It would seem, therefore, that the normal approximation (4.15) is appropriate here, and this yields a 95 per cent confidence interval for N of

(314, 639). We note that the confidence intervals are wide as the number of recaptures in each sample is small.

Example 4.6 Cricket frog (*Acris gryllus*): Turner [1960a]

Using the data of Example 4.1 (p. 135) we find that the plot of y_i versus M_i is approximately linear and the regression estimate is $\tilde{N} = 93$ with 95 per cent confidence interval (82, 108). This may be compared with the Schnabel method: $N' = 93$, (85, 101), and, from Example 4.1, the maximum-likelihood method: $\hat{N} = 96$, (90, 100).

Example 4.7 Cricket frog (*Acris crepitans*): Pyburn [1958]

A population of cricket frogs was studied to determine the population size and to obtain information about the movements of the frogs. Cricket frogs commonly occur about the margins of permanent and semi-permanent areas of water and are usually found in large numbers where conditions are favourable. The study site was centred round a pond $112' \times 79'$ which was staked at intervals of 10 feet around its circumference so that the points of capture could be located. The sampling method was to proceed around the pond from a given point, attempting to take frogs at random. Each frog was marked by toe clipping, located according to the nearest stake, and released at the point of capture. Six times round the pond constituted a "sample", and the data from six consecutive "samples" are given in Table 4.5.

TABLE 4.5

Capture–recapture data for a population of cricket frogs: from Pyburn [1958: Table 1].

Date of capture	n_i	m_i	M_i	N_i^*	v_i^*
Sept. 19–Sept. 21	109	–	–	–	–
Sept. 22–Sept. 25	133	15	109	920·3	37 568
Sept. 28–Oct. 2	138	30	227	1021·3	21 927
Oct. 7–Oct. 9	72	23	335	1021·0	26 041
Oct. 11–Oct. 14	134	47	384	1081·8	13 498
Oct. 16–Oct. 21	72	33	471	1012·4	14 548
Total		148		5056·8	113 582

The assumptions underlying the Schnabel census seemed to be satisfied throughout the experiment. For example, the pond was completely isolated from any other source of water, and the fact that only 15 of the 510 caught were 15 mm or less indicated that the recruitment of young frogs was small. The method of marking by toe clipping did not seem to affect the behaviour of the frogs, and the good fit of the data to a straight line (Fig. 4.1) would suggest that marking did not affect their catchability.

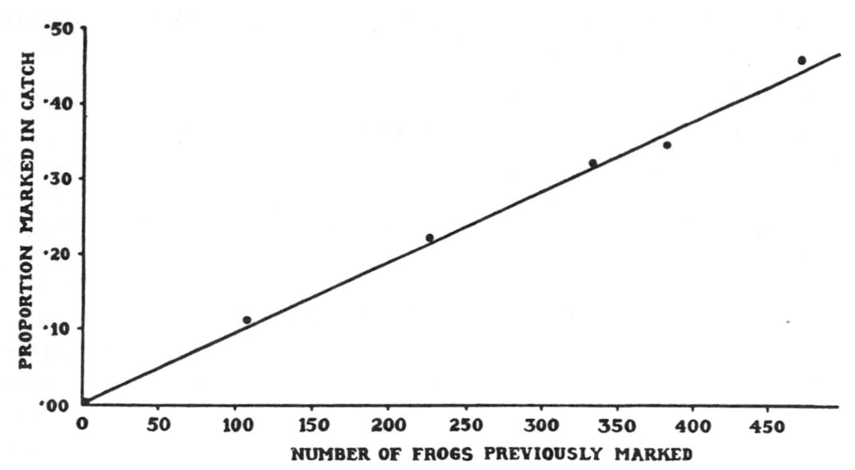

Fig 4.1 Plot of the proportion marked in the sample (y_i) versus the number of frogs previously marked (M_i): from Pyburn [1958].

From Table 4.5 we find that $r = 471 + (72-33) = 510$, $\tilde{N} = 1056$ and the 95 per cent regression confidence interval for N is (1012, 1104). The Schnabel estimate is $N' = 1049$, and (4.15) leads to the confidence interval (957, 1149). Using $\tilde{N}$ as a first approximation, (4.4) can be solved iteratively using the method of Robson and Regier to give $\hat{N} = 1052$ and a confidence interval (cf. 4.7) of (919, 1181). From **4.1.2**(*3*) we have the mean estimate

$$\bar{N} = 5056{\cdot}8/5 = 1011,$$

with variance estimate

$$v^* = 113\,582/25 = 4543,$$

and $\bar{N} \pm 1{\cdot}96\sqrt{v^*}$ gives the interval (879, 1143). Also

$$\sum_{i=2}^{s} (N_i^* - \bar{N})^2 = \Sigma N_i^{*2} - (\Sigma N_i^*)^2/(s-1) = 13\,347,$$

so that

$$v[\bar{N}] = 13\,347/20 = 6673$$

and $\bar{N} \pm 1{\cdot}96\sqrt{(v[\bar{N}])}$ leads to (851, 1171).

The general agreement of the above confidence intervals indicates that the assumptions underlying the various models are approximately satisfied.

2 Tanaka's model

Sometimes a plot of y_i versus M_i, as in the previous section, yields a graph which is definitely curved. For this situation Tanaka [1951, 1952] has proposed a non-linear relationship of the form $y = (M/N)^\gamma$ or, taking logarithms, the linear regression model

$$E[-\log_{10} y_i] \approx \gamma(\log_{10} N - \log_{10} M_i), \quad (i = 2, 3, \ldots, s).$$

Least-squares estimates and confidence intervals for γ and $\theta = \log_{10} N$ can be obtained by setting $Y_i = -\log_{10} y_i = \log_{10}(n_i/m_i)$, $x_i = \log_{10} M_i$ and using the methods of **1.3.5**(*3*). A visual estimate of θ can also be obtained by drawing the regression line by eye and extending this line to meet the x-axis. However, before actually looking up the logarithms it is simpler to plot y_i versus M_i on log–log paper first.

The parameter γ can be interpreted as an index of trap response (e.g. Tanaka and Teramura [1953]). For example, if $\gamma < 1$, $E[m_i/n_i] > M_i/N$ or, rearranging, $E[m_i/M_i] > E[n_i/N]$, and the marked individuals have a higher probability of capture than the unmarked. However, care should be exercised in interpreting the graph of y_i versus M_i as several interpretations are possible (e.g. Tanaka [1951: 452], Hayne [1949b: 407] and Davis [1963: 111]). If, for example, the graph curves downwards, Hayne argues that the fall-off in the proportion of marked in the sample could be due to the immigration of unmarked animals into the trapping area. But if the graph is interpreted in the light of Tanaka's model we have $\gamma < 1$ and the curvature is due to the marked animals having a higher probability of capture than the unmarked. In this case the fall-off is simply due to the curve settling down to its "correct" position instead of dropping away from its "correct" position as suggested by Hayne. Obviously both interpretations are possible, and one could perhaps distinguish between the two by an analysis of the recaptures to see whether any individuals were being recaptured more often than expected (cf. **4.1.5**). Alternatively Marten's regression model discussed below may be applicable.

Example 4.8 Red-backed vole (*Clethrionomys smithi*): Tanaka [1951]

Tanaka studied populations and home ranges of voles and mice in a bushy area near the summit of Mt Ishizuchi in central Shikoku. Thirty-two small live-traps were set in a grid pattern spaced 15 metres apart, and the traps were baited with peanuts and sweet potatoes. Unfortunately some individuals escaped because of a deficiency in the door apparatus, and a number of animals were found dead in the traps. The voles were numbered using toe clipping, and in Table 4.6 we have the capture–recapture data for a red-backed vole population.

TABLE 4.6
Capture–recapture data for a population of red-backed voles: from Tanaka [1951: Table 2].

Day (i)	$n_i - m_i$	m_i	n_i	M_i	y_i (m_i/n_i)	Y_i $\log_{10}(n_i/m_i)$	x_i $\log_{10} M_i$
1	5[1]	–	5	–	–	–	–
2	5[1]	2	7	4	0·29	0·544 07	0·6021
3	7[1]	3	10	8	0·30	0·522 84	0·9031
4	5	7	12	14	0·58	0·234 01	1·1461
5	5[1]	11	16	19	0·69	0·162 86	1·2788
6	5[1]	12	17	23	0·71	0·151 37	1·3617
7	4	14	18	27	0·78	0·109 24	1·4314

[1]One death.

From (4.16) we find that the Schumacher–Eschmeyer regression estimate of N is $\widetilde{N} = 21$ with a 95 per cent confidence interval (cf. (4.17)) of (27, 37). This method is obviously unsatisfactory both graphically (Fig. 4.2) and

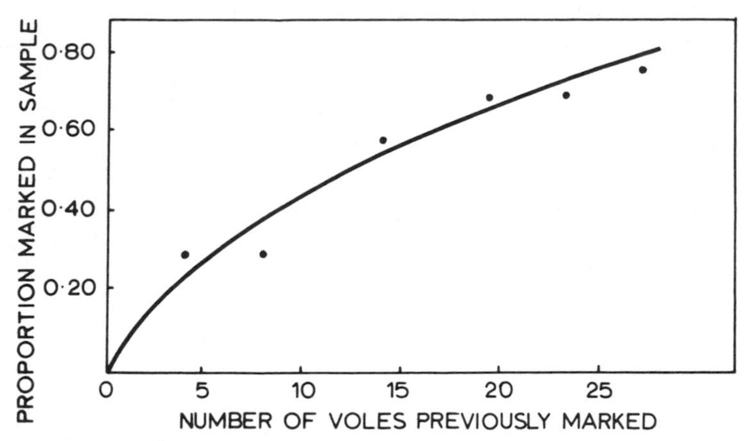

Fig 4.2 Plot of the proportion marked in the sample (y_i) versus the number of voles previously marked (M_i): from Tanaka [1951].

mathematically, as the initial population size is at least $r = 36$. However, a plot of Y_i versus x_i (Fig. 4.3) is fairly linear, thus indicating that Tanaka's method can be used. The steps in the calculations are as follows:

$$\bar{Y} = 0{\cdot}287\,40, \qquad \bar{x} = 1{\cdot}120\,52,$$

$$\Sigma\,(Y_i - \bar{Y})^2 = 0{\cdot}189\,888, \qquad \Sigma\,(x_i - \bar{x})^2 = 0{\cdot}496\,467,$$

$$\Sigma\,(Y_i - \bar{Y})(x_i - \bar{x}) = \Sigma\,Y_i\,(x_i - \bar{x}) = -0{\cdot}293\,474.$$

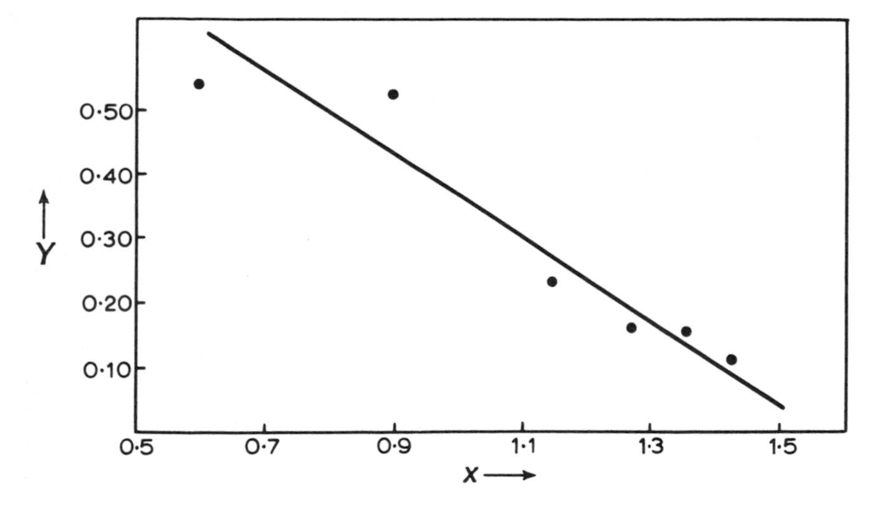

Fig 4.3 Plot of Y_i $(= \log_{10}(n_i/m_i))$ versus x_i $(= \log_{10} M_i)$ for a population of red-backed voles: redrawn from Tanaka [1951].

Setting $w_i = 1$ in **1.3.5** (3), we have

$$\begin{aligned}
\hat{\gamma} &= 0{\cdot}293\,474/0{\cdot}496\,467 = 0{\cdot}591\,125, \\
\hat{\theta} &= \log_{10}\hat{N} = \bar{x} + (\bar{Y}/\hat{\gamma}) = 1{\cdot}606\,71, \\
\hat{N} &= 40, \\
\hat{\sigma}^2 &= \tfrac{1}{4}\{\Sigma\,(Y_i - \bar{Y})^2 - [\Sigma\, Y_i(x_i - \bar{x})]^2/\Sigma\,(x_i - \bar{x})^2\} \\
&= 0{\cdot}004\,101\,9, \\
t_4[0{\cdot}025] &= 2{\cdot}776, \\
\hat{\gamma}^2 - A &= \hat{\gamma}^2 - \hat{\sigma}^2(2{\cdot}776)^2/\Sigma\,(x_i - \bar{x})^2 = 0{\cdot}285\,769, \\
2\bar{Y}\hat{\gamma} &= 0{\cdot}339\,776, \\
\bar{Y}^2 - B &= \bar{Y}^2 - \hat{\sigma}^2(2{\cdot}776)^2/6 = 0{\cdot}077\,329\,4,
\end{aligned}$$

and the quadratic to be solved, namely (1.17), is

$$0{\cdot}285\,769\, d^2 - 0{\cdot}339\,776\, d + 0{\cdot}077\,329\,4 = 0.$$

This has roots 0·8823, 0·3067, and adding $\bar{x}$ to each root gives (1·4272, 2·0028) as the 95 per cent confidence interval for θ, or (27, 100) for N.

We note from (1.18) that

$$\begin{aligned}
\hat{V}[\hat{\theta}] &\approx \frac{\hat{\sigma}^2}{\hat{\gamma}^2}\left\{\frac{1}{s} + \frac{(\hat{\theta} - \bar{x})^2}{\Sigma\,(x_i - \bar{x})^2}\right\} \\
&= 0{\cdot}007\,546,
\end{aligned}$$

and $\hat{\theta} \pm 1{\cdot}96\sqrt{\hat{V}[\hat{\theta}]}$ is (1·436, 1·777), which leads to the confidence interval (27, 60) for N. However, as s is small, this confidence interval is of doubtful validity and the wider interval based on $t_4[0{\cdot}025] = 2{\cdot}776$ is probably more reliable. One should expect a wide interval as the recaptures m_i are small.

The 95 per cent confidence interval for γ, namely

$$\hat{\gamma} \pm t_{s-2}[\alpha/2]\sqrt{\hat{\sigma}^2/\Sigma\,(x_i - \bar{x})^2},$$

is (0·3388, 0·8434). As this interval does not contain the value 1, γ is significantly less than unity, which, interpreted in the light of Tanaka's model, suggests the presence of trap addiction among the marked animals.

There were five deaths during the experiment through trapping, which would be expected to have some effect on the Schumacher–Eschmeyer regression model. In particular, with decreasing N the expected value of y_i would increase from M_i/N to $M_i/(N - D_i)$, where D_i is the total number of deaths before the ith sample. This increase, ranging from about 3 per cent for $i = 2$ to about 14 per cent for $i = 7$, would have the effect of curving the regression line upwards. As the reverse is true, it would seem that the degree of trap addiction is even greater than that indicated by the above analysis.

Example 4.9 Meadow vole (*Microtus pennsylvanicus*): Hayne [1949b]

Table 4.7 gives the capture–recapture data for a population of adult female meadow voles trapped in a field in East Lansing, Michigan. It is found

TABLE 4.7

Capture–recapture data for a population of meadow voles: from Hayne [1949b: Table 2].

Date of capture	n_i	m_i	M_i	y_i (m_i/n_i)	Y_i $\log_{10}(n_i/m_i)$	x_i $\log_{10} M_i$
July 19 p.m.	8	–	–	–	–	–
20 a.m.	19	0	8	0	–	–
20 p.m.	10	2[1]	27	0·20	0·698 97	1·431 36
21 a.m.	23	8	34	0·35	0·458 64	1·531 48
21 p.m.	9	0	49	0·00	–	–
22 a.m.	14	9	58	0·64	0·192 01	1·763 43
22 p.m.	9	7[1]	63	0·78	0·109 24	1·799 34
23 a.m.	21	13	64	0·62	0·208 17	1·806 18

[1]One death

that the Schumacher–Eschmeyer regression estimate of N is $\widetilde{N} = 106$ with a 95 per cent confidence interval of (78, 165). However, the graph of y_i versus M_i (Fig. 4.4) seems to curve upwards. Therefore, ignoring the points for which $y_i = 0$ and replotting the remaining five points on log–log paper, we obtain approximate linearity (Fig. 4.5), thus indicating that Tanaka's model can be

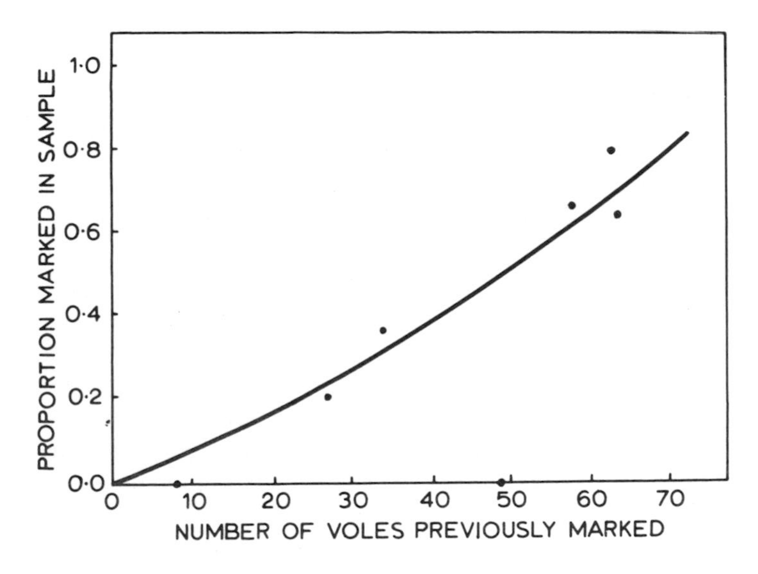

Fig 4.4 Plot of the proportion marked in the sample (y_i) versus the number of voles previously marked (M_i): from Hayne [1949b]. The above freehand curve is drawn to demonstrate the upward trend in the data.

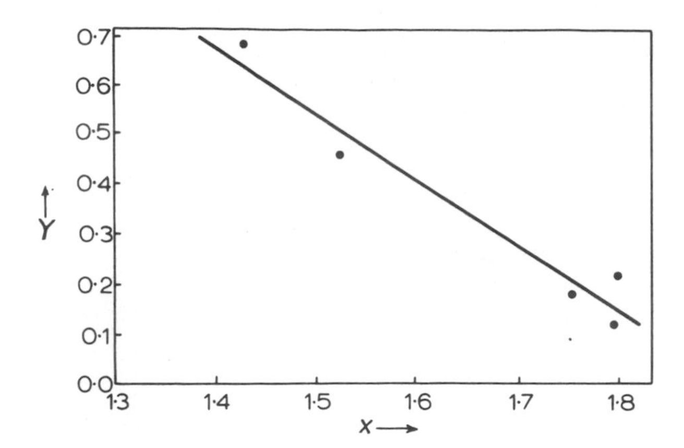

Fig 4.5 Plot of Y_i (= $\log_{10}(n_i/m_i)$) versus x_i(= $\log_{10}M_i$) for a population of meadow voles: data from Hayne [1949b].

used. Using the same technique as outlined in the previous example, we find that N is now estimated as 81 with a confidence interval of (66, 126). Also $\hat{\gamma} = 1{\cdot}36$ with a confidence interval for γ of (0·79, 1·94). This indicates that γ is not significantly different from unity.

A further application of Tanaka's method is given in Pearson [1955: 251, spadefoot toad].

3 Marten's model

One of the difficulties in using Tanaka's regression model above is the interpretation of the parameter γ. Although one may obtain a good straight-line fit to the graph of Y_i versus x_i, the model lacks a simple "physical" interpretation and other regression curves may give just as good a fit. For example, suppose that the average catchability of the m_i marked individuals in each sample bears a constant ratio to the average catchability of the u_i unmarked members, then we have approximately

$$\frac{m_i}{M_i} = \frac{u_i}{k(N-M_i)} \left(= \frac{n_i}{k(N-M_i)+M_i} \right)$$

and

$$E[y_i \mid M_i] \approx \frac{M_i}{k(N-M_i)+M_i}.$$

This means that the plot of y_i versus M_i will be curved upwards or downwards, depending on whether k is less than or greater than unity. Instead of fitting Tanaka's model we can rearrange the above equation, apply a bias correction, and obtain the linear regression model

$$E[Y_i \mid M_i,\ n_i] \approx k(N-M_i), \tag{4.18}$$

where $Y_i = u_i(M_i+1)/(m_i+1)$. This model, first suggested by Marten [1970a], can be analysed using the methods of **1.3.5**(*3*) (with $\theta = N$, $\gamma = k$, $x_i = M_i$) to obtain point and interval estimates of N, and to test $k = 1$. Marten points out that although the plot may not be linear, there may be a subclass M_i', say, whose average catchability relative to the unmarked animals remains constant. If m_i' members of this subclass are caught in the ith sample and

$$Y_i' = u_i(M_i'+1)/(m_i'+1),$$

we can plot Y_i' versus M_i (not M_i'): an example of this is given below.

Seber [1970a] derives (4.18) from the general theory of **3.2.2** (*2*), and such a derivation throws some light on the nature of k. Using the notation of **3.2.2** (*2*), we define x_j as the probability that the jth member of the population ($j = 1, 2, \ldots, N$) is caught at least once before the ith sample, and y_j, z_j now refer to the conditional probabilities of the jth member being caught in the ith sample, given that it is caught or not caught, respectively, before the ith sample, i.e., in terms of **3.2.2** (*2*), the first $i-1$ samples are regarded as the "first" sample and the ith sample as the "second". Therefore, setting $n_1 = M_i$, $u_2 = u_i$, $m_2 = m_i$, and defining Y_i as above, we have from (3.15)

$$E[Y_i \mid M_i, n_i] \approx k_i(N-M_i),$$

which reduces to (4.18) if $k_i = k$. The plot of Y_i versus M_i will indicate whether k_i can be regarded as approximately constant and will also show up any marked heterogeneity of variance. In many cases the variances of the Y_i will not be too different. For example, since the variance of Y_i depends on the product $M_i n_i$ (cf. (3.2)), the variance of Y_i will be constant if, at each stage, n_i is chosen so that $M_i n_i$ has some predetermined constant value; methods for doing this are discussed in **3.1.5**.

If $k \neq 1$ we see from the discussion in **3.2.2** (*2*) that we cannot be sure whether this departure is due to marking affecting future catchability, or to variation in the inherent catchability of individuals, or both. When there is a variable catchability the more catchable individuals are caught first, resulting in marked animals becoming, on the average, more catchable than the unmarked. If marking does not affect catchability, then excluding the unlikely case when the catchability of an individual before every sample is uncorrelated with its catchability after the sample, a test of $k = 1$ is a test of constant catchability. On the other hand, if the inherent catchability is constant over the population, it will not remain constant (except for the unmarked) if marking affects catchability. In this case one way of separating the two effects is to consider a regression using the subclass $M_i' = u_{i-1}$ of individuals first caught in the $(i-1)$th sample. Defining x_j to be the probability that the jth member of the population is first caught in the $(i-1)$th sample, then, if the unmarked are equicatchable, we would expect x_j to be constant over the population. This implies from p. 87 that a test of $k = 1$ is then a test that, for each sample, the average catchability of the marked members from u_{i-1} in the ith sample is the same as the (constant) catchability of the unmarked in the ith sample.

Example 4.10 Tide-pool snail (*Polinices duplicatus*): Marten [1970a]

Marten used data from Hunter and Grant [1966] on tide-pool snails to illustrate the above regression technique. From Table 4.8 a plot of Y_i versus

TABLE 4.8

Capture–recapture data for tide-pool snails: from Marten [1970a: Table 1].

i	u_i	m_i	M_i	Y_i	m_i'	M_i'	Y_i'
1	142	–	–	–	–	–	–
2	129	3	142	4611	–	–	–
3	122	23	271	1383	19	139	854
4	99	58	393	661	35	248	685
5	94	67	492	682	63	335	494
6	147	273	586	315	219	425	285

M_i showed considerable scatter, thus indicating that the entire class of marked snails was not suitable for applying the regression method. However, Hunter and Grant found that after handling, the snails tended to burrow into the ground and remain immobile for about 24 hours. This meant that in any particular sample there were fewer individuals than expected from those caught in the previous sample. Therefore, choosing the subgroup M_i' consisting of all marked individuals except those caught in the $(i-1)$th sample, Marten found the regression of Y_i' on M_i to be closely linear (Fig. 4.6). Hence, using Fieller's technique of **1.3.5** (*3*), N is estimated to be 756 with a 95 per cent confidence interval of (670, 923). Also $\hat{k} = 1{\cdot}81$ with a 95 per cent confidence interval of (1·21, 2·41). As this interval does not contain unity, we reject the hypothesis that $k = 1$ and conclude that marked snails are under-represented in the samples.(*)

4.1.4 Allowing for known removals

1 The hypergeometric model

In many population experiments there are accidental deaths due to trapping and handling, and some animals may be deliberately removed for further study. If the percentage of such removals is appreciable, some allowance must be made for them in the particular model used. For example, the removal could form a major part of the sample as in commercially exploited populations, with the remainder of the sample being tagged (or retagged) and returned to the population. Suppose, then, that d_i members of the ith sample are not returned to the population and let M_i be the number of marked animals alive in the population before the ith sample is taken. Then, assuming the n_i to be fixed parameters, Chapman's model (4.5) now becomes

$$f(m_2, \ldots, m_s \mid \{n_i\}) = \prod_{i=2}^{s} \binom{M_i}{m_i}\binom{N-M_i-D_i}{n_i-m_i} \Big/ \binom{N-D_i}{n_i},$$

(*)For a similar experiment see Eisenberg [1972].

where $D_i = \sum_{j=1}^{i-1} d_j$ (the total removal up to but not including the ith sample), and N is now the *initial* population size. It is readily shown that N_D, the maximum-likelihood estimate of N, is the unique root greater than r of the polynomial

$$\frac{N - r}{N} = \prod_{i=1}^{s} \left\{\frac{N - D_i - n_i}{N - D_i}\right\}, \tag{4.19}$$

where r is the total number of different animals caught during the whole experiment, *including* the ones not returned.

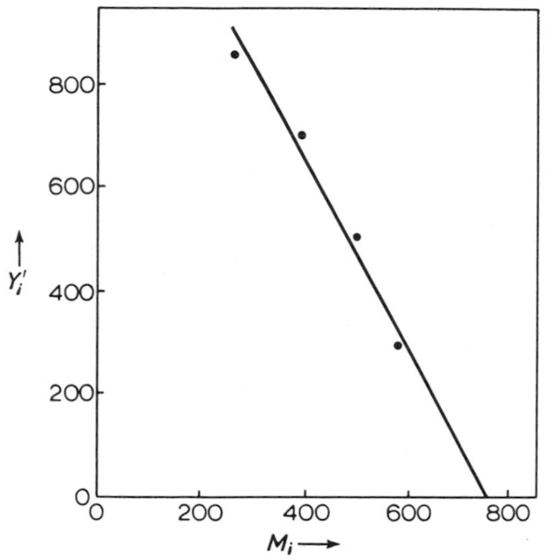

Fig 4.6 Application of Marten's regression model to a population of tide-pool snails: modified from Marten [1970a].

By setting $g(N)$ equal to the right-hand side of (4.19) and defining $h(N) = N - r - Ng(N)$, (4.19) can be solved in exactly the same way as (4.4). If N is large, so that each n_i/N is less than about 0·1, a reasonable first approximation of (4.19) is (Robson and Regier [1968: 138])

$$N_{(1)} = \frac{[\sum (n_i + D_i)]^2 - \sum (n_i + D_i)^2 - (r + \sum D_i)^2 + r^2 + \sum D_i^2}{2 \sum m_i}, \tag{4.20}$$

where all summations are for $i = 1, 2, \ldots, s$ ($m_1 = 0$, $D_1 = 0$). This approximation is obtained by cross-multiplying in (4.19), dividing both sides by N^s, expanding the products, and neglecting powers of $1/N$ greater than the second. Another possible first approximation is the mean estimate

$$\bar{N}_D = \sum_{i=2}^{s} (N_i^* + D_i)/(s-1), \tag{4.21}$$

where N_i^* is the modified Petersen estimate $[(n_i+1)(M_i+1)/(m_i+1)] - 1$.

The mean and variance of $\hat{N}_D$ can be evaluated using the method outlined in Darroch [1958: 346–7]. It transpires that, asymptotically,

$$E[\hat{N}_D] = N + b,$$

where b, the bias, is estimated by

$$\hat{b} = \frac{\left[\frac{1}{\hat{N}_D} + \sum_{i=1}^{s}\left(\frac{1}{\hat{N}_D - D_i - n_i} - \frac{1}{\hat{N}_D - D_i}\right)\right]^2 - \left[\frac{1}{\hat{N}_D^2} + \sum_{i=1}^{s}\left(\frac{1}{(\hat{N}_D - D_i - n_i)^2} - \frac{1}{(\hat{N}_D - D_i)^2}\right)\right]}{2\left[\frac{1}{\hat{N}_D - r} - \frac{1}{\hat{N}_D} - \sum_{i=1}^{s}\left(\frac{n_i}{(\hat{N}_D - D_i - n_i)(\hat{N}_D - D_i)}\right)\right]^2}$$

and the asymptotic variance of $\hat{N}_D$ is estimated by

$$\begin{aligned} v[\hat{N}_D] &= (\hat{N}_D - r)/h'(\hat{N}_D) \\ &= \left[\frac{1}{\hat{N}_D - r} - \frac{1}{\hat{N}_D} - \sum_{i=1}^{s} \frac{n_i}{(\hat{N}_D - D_i - n_i)(\hat{N}_D - D_i)}\right]^{-1}. \end{aligned}$$

Since

$$\hat{b} = -\tfrac{1}{2}(\hat{N}_D - r)h''(\hat{N}_D)/[h'(\hat{N}_D)]^2,$$

the bias and the variance of $\hat{N}_D$ can be estimated from the last steps in the Robson–Regier iterative procedure as for the case of no removals (cf. **4.1.2** (*1*)); a numerical example is given in Robson and Regier [1968: 134–138].

2 Overton's method

Overton [1965] has given the following method for modifying Schnabel's estimate N' (p. 139) to allow for known removals. Since

$$E[m_i | M_i, n_i, D_i] = M_i n_i/(N - D_i), \quad (i = 2, 3, \ldots, s),$$

then summing this equation for $i = 2, 3, \ldots, s$ and setting $m = \sum_{i=2}^{s} m_i$ leads to

$$\begin{aligned} E[m \mid \{M_i, n_i, D_i\}] &= \frac{1}{N} \frac{\sum n_i M_i (N - D_i + D_i)}{N - D_i} \\ &= \frac{\sum n_i M_i}{N} + \frac{\sum n_i M_i D_i}{N(N - D_i)}. \end{aligned}$$

Equating m to its expected value leads to an estimate N_D' of N given by

$$\begin{aligned} N_D' &= \frac{\sum n_i M_i}{m} + \frac{\sum n_i M_i D_i}{(N_D' - D_i)m} \qquad (4.22) \\ &= N' + A, \quad \text{say,} \end{aligned}$$

where A is to be added to the usual Schnabel estimate N'. Equation (4.22) must be solved iteratively for N_D', and Overton suggests the following first approximation which will usually be close unless the removal is heavy, namely

$$N_{(1)} = N' + A_{(1)},$$

where $$A_{(1)} = \Sigma\, n_i M_i D_i/(mN') = (\Sigma\, n_i M_i D_i)/(\Sigma\, n_i M_i),$$

so that $N_{(1)} < N'_D$. Another first approximation, suggested by Robson and Regier [1968: 143, $D_s < 0{\cdot}1\,N$], is obtained from (4.22) directly by neglecting D_i in the denominator of the right-hand side and solving for N'_D, namely

$$\tfrac{1}{2}\{N' + \sqrt{N'^2 + 4\,\Sigma\, n_i M_i D_i/m}\}.$$

Whichever first approximation is used, however, subsequent approximations are

$$N_{(j)} = N' + A_{(j)},$$

where $$A_{(j)} = \Sigma\, n_i M_i D_i/(N^*_{(j)} - D_i)m \tag{4.23}$$

and $N^*_{(j)}$ is to be determined by $N_{(j-1)}$. In determining a suitable method for choosing $N^*_{(j)}$, Overton points out that the iterative process is not necessarily convergent if we set $N^*_{(j)} = N_{(j-1)}$. But if $N^*_{(j)} < N'_D$, then $N_{(j)} > N'_D$ and vice versa, so that N'_D will be between any pair $N^*_{(j)}, N_{(j)}$. Overton therefore suggests the reasonable procedure of choosing

$$N^*_{(j+1)} = \tfrac{1}{2}(N^*_{(j)} + N_{(j)})$$

and taking $N^*_{(2)}$ as the integer nearest to $N_{(1)}$.

Confidence limits for N can be obtained as in **4.1.2** (*4*) on p. 140, using $\lambda = \Sigma\, n_i M_i + \Sigma\, n_i M_i D_i/(N'_D - D_i)$ (which can be obtained from the last step of (4.23)) and either Chapman's table (Table **A1**, $m \leqslant 50$) or the normal approximation (4.14); (4.15) can also be used with

$$\delta = \frac{N'_D}{\lambda} \sum_{i=2}^{s} \left\{ \frac{n_i M_i^2}{(N'_D - D_i)^2} \right\}. \tag{4.24}$$

Example 4.11 Hypothetical data: Overton [1965]

In Table 4.9, d_i consists of the members of n_i permanently removed from the population, a_i $(= n_i - m_i - d_i)$ is the *net* increase in the marked population after the ith sample,

$$D_i = \sum_{j=1}^{i-1} d_j \quad \text{and} \quad M_i = \sum_{j=1}^{i-1} a_j.$$

Now

$$N' = 6231/16 = 389{\cdot}4$$

and

$$N_{(1)} = N' + A_{(1)} = \frac{6231}{16} + \frac{362\,781}{6231} = 447{\cdot}6.$$

Taking the nearest integer $N^*_{(2)} = 448$, and substituting in (4.23), leads to

TABLE 4.9

Hypothetical data for demonstrating Overton's iterative procedure for estimating population size when there are known removals: from Overton [1966: Table 1].

i	n_i	m_i	d_i	a_i	M_i	D_i	n_iM_i	$n_iM_iD_i$	$N^*_{(2)}-D_i$	$\frac{n_iM_iD_i}{N^*_{(2)}-D_i}$
1	20	–	10	10	–	0	–	–	448	0
2	22	1	11	10	10	10	220	2 200	438	5
3	18	0	10	8	20	21	360	7 560	427	18
4	21	1	11	9	28	31	588	18 228	417	44
5	16	2	7	7	37	42	592	24 864	406	61
6	18	1	11	6	44	49	792	38 808	399	97
7	17	3	9	5	50	60	850	51 000	388	131
8	14	2	8	4	55	69	770	53 130	379	140
9	17	2	8	7	59	77	1003	77 231	371	208
10	16	4	7	5	66	85	1056	89 760	363	247
					(71)	(92)				
Total		16	92				6231	362 781		951

$$N_{(2)} = \frac{6231}{16} + \frac{951}{16} = 448{\cdot}875.$$

From the discussion above we now have $448 < N'_D < 448{\cdot}875$ and normally it would be sufficient to choose N'_D equal to 448 or 449. Overton, however, takes the calculations a step further to determine which integer N'_D is nearer. Thus, choosing $N^*_{(3)} = 448{\cdot}5$, it transpires that $N_{(3)} = 448{\cdot}86$ and $448{\cdot}5 < N'_D < 448{\cdot}86$, so that, to the nearest integer, $N'_D = 449$.

To calculate an approximate confidence interval for N we require

$$\begin{aligned}\lambda &= \Sigma\, n_iM_i + \Sigma\, n_iM_iD_i/(N'_D - D_i)\\ &\approx 6231 + 951 = 7182.\end{aligned}$$

Entering $m = 16$ in Chapman's table (Table **A1**), we obtain 95 per cent limits (0·0350, 0·1020) for N/λ, or (251, 733) for N. However, as five of the values $M_i/(N^*_{(2)} - D_i)$ are greater than 0·1 (cf. p. 140), m will no longer be strictly Poisson. Therefore calculating $\delta = 0{\cdot}1271$ from (4.24), (4.15) leads to the approximate 95 per cent confidence interval for N (285, 707): if δ is neglected, (4.14) leads to the interval (276, 729).

3 Regression methods

All the regression methods discussed in **4.1.3** depend heavily on the assumption of N remaining constant, and these methods cannot be used unless the removal is negligible. However, the mean estimate $\bar{N}_D$ given above by (4.21) will provide a robust estimate of N, provided the number of recaptures in each sample is not too small (cf. **3.1.1**). The variance of $\bar{N}_D$ can be estimated by (cf. (1.9))

$$v[\bar{N}_D] = \sum_{i=2}^{s} (N_i^* + D_i - \bar{N}_D)/(s-1)(s-2),$$

or, when the assumptions underlying the Petersen method hold for each sample, more efficiently by

$$v_D^* = \sum_{i=2}^{s} \left\{ \frac{(M_i+1)(n_i+1)(M_i-m_i)(n_i-m_i)}{(m_i+1)^2(m_i+2)(s-1)^2} \right\}.$$

4.1.5 Testing the underlying assumptions

1 Validity of the models

MULTINOMIAL MODEL. Using the notation of **4.1.2**(*1*) we wish to test the hypothesis H that the probabilities P_w in (4.1) can be written as products of the $\{p_i\}$ and $\{q_i\}$ so that (4.1) reduces to Darroch's multinomial model (4.2). From the theory of **1.3.6**(*3*), an appropriate goodness-of-fit test statistic for H is therefore given by

$$T = \sum_{w} (a_w - \hat{N}\hat{P}_w)^2/(\hat{N}\hat{P}_w),$$

where, for example, $\hat{P}_{124} = \hat{p}_1\hat{p}_2\hat{p}_4\hat{Q}/(\hat{q}_1\hat{q}_2\hat{q}_4)$ and $\hat{p}_i = 1 - \hat{q}_i = n_i/\hat{N}$. When H is true, T is asymptotically distributed as chi-squared with $d - s - 1$ degrees of freedom, where d is the number of different recapture histories w. If any of the groups a_w are too small (some of them may be zero) they can be pooled in the usual manner.

HYPERGEOMETRIC MODEL. Chapman [1952: 295] has suggested a non-parametric test for the validity of the model (4.3) using the random variables $\{b_{ij}\}$, where b_{ij} $(i < j)$ is the number of marked in the jth sample which were first caught and marked in the ith. When the sampling is random we have

$$E[b_{ij}/n_j \mid n_j, u_i] = u_i/N \; (= \theta_i, \text{ say}), \tag{4.25}$$

and an array

$$\begin{matrix} \frac{b_{12}}{u_1 n_2}, & \frac{b_{13}}{u_1 n_3}, & \frac{b_{14}}{u_1 n_4}, \ldots, & \frac{b_{1s}}{u_1 n_s} \\ & \frac{b_{23}}{u_2 n_3}, & \frac{b_{24}}{u_2 n_4}, \ldots, & \frac{b_{2s}}{u_2 n_s} \\ & & \ldots & \ldots \\ & & & \frac{b_{s-1,s}}{u_{s-1} n_s} \end{matrix}$$

may be formed, in which each element is a random variable with expectation $1/N$. These random variables are independent within each row, but are dependent between rows as $b_{i_1 j}$, $b_{i_2 j}$ belong to the same sample and are therefore correlated. For large N, however, the correlation is small and Chapman suggests testing for the validity of the underlying model by testing whether the $t = \frac{1}{2}s(s-1)$ elements formed by putting the rows one after another is a sequence of random observations from a common distribution. The test suggested is the sign test of Moore and Wallis [1943] based on D, the number of negative signs in the sequence of successive differences of observations (i.e. first observation minus the second, etc.). When the hypothesis of a common distribution is true, $E[D] = \frac{1}{2}(t-1)$, $\sigma^2[D] = \frac{1}{12}(t+1)$, and $(D-E[D])/\sigma[D]$ is approximately unit normal for $t \geqslant 12$; Moore and Wallis have tabled the exact probability distribution of D for small values of t.

Chapman points out that in many cases the alternatives to randomness are essentially one-sided. For example, possible alternatives are:

(a) some marked individuals die off more rapidly or disappear, so that they are not available for sampling;
(b) the marked individuals disperse from the tagging location slowly and are more likely to be recaptured in the samples taken soon after marking than later; and
(c) the population size N is increasing through recruitment.

If any of these alternatives is true, the numbers in each row of the array will tend to decrease from left to right. In this case a test based on the whole array as a single sequence has the following defect: in each row the probability of a negative difference between successive elements is less than $\frac{1}{2}$, but the probability of a negative difference between the last element of any row and the first element of the next row will be much greater than $\frac{1}{2}$. Also another disadvantage of considering the whole array as a single sequence is that the variances of the elements will vary from row to row. However, if the sample sizes n_i are approximately the same, the elements within a given row will have approximately the same distribution when the underlying model is valid. Therefore, to overcome the above objections, Chapman recommends treating each row separately so that the array may be considered as $(s-1)$ sequences of observations decreasing in length from $(s-1)$ to 1. A test of randomness may then be made using the statistic

$$X = D_1 + D_2 + \ldots + D_{s-2},$$

where D_i is the number of negative differences in row i (no difference is obtained from the last row). Then

$$E[X] = \tfrac{1}{4}(s-1)(s-2), \quad \sigma^2[X] = \tfrac{1}{4}(s+3)(s-2),$$

and X is asymptotically normal. A partial tabulation of the distribution of X (reproduced from Chapman [1952]) for $s = 5, 6, 7, 8$ is given in Table 4.10.

An alternative test can be carried out using Table 4.11. Neglecting the complications of sampling without replacement, the columns of Table 4.11 represent independent multinomial distributions. Let p_{ij} be the probability of being in the class containing b_{ij} individuals, then from (4.25), $p_{ij} = \theta_i$ $(j = i+1, \ldots, s)$ and the likelihood function for Table 4.11 is proportional to

$$\prod_{j=2}^{s}\left\{\left(\prod_{i=1}^{j-1}\theta_i^{b_{ij}}\right)(1-\theta_1-\theta_2-\ldots-\theta_{j-1})^{u_j}\right\}$$
$$= \theta_1^{b_{1.}}\,\theta_2^{b_{2.}}\ldots(1-\theta_1)^{u_2}(1-\theta_1-\theta_2)^{u_3}\ldots.$$

The maximum-likelihood estimates $\hat{\theta}_i$ of θ_i are then solutions of the equations

$$\frac{b_{1.}}{\theta_1}-\frac{u_2}{(1-\theta_1)}-\frac{u_3}{(1-\theta_1-\theta_2)}-\ldots-\frac{u_s}{(1-\theta_1-\theta_2-\ldots-\theta_{s-1})}=0$$

$$\frac{b_{2.}}{\theta_2}-\frac{u_3}{(1-\theta_1-\theta_2)}-\ldots-\frac{u_s}{(1-\theta_1-\theta_2-\ldots-\theta_{s-1})}=0$$

$$\ldots\ldots\ldots\ldots\ldots\ldots$$

$$\frac{b_{s-1.}}{\theta_{s-1}}-\frac{u_s}{(1-\theta_1-\theta_2-\ldots-\theta_{s-1})}=0$$

TABLE 4.10

Cumulative distribution of X: from Chapman [1952: Table 2].

$s \backslash x$	0	1	2	3	4	5	6	7
				$\Pr[X \leqslant x]$				
5	0·0035	0·0590	0·3056	0·6944	0·9410	0·9965	1·0000	1·0000
6	–	0·0012	0·0172	0·1052	0·3392	0·6608	0·8948	0·9828
7	–	–	0·0001	0·0020	0·0166	0·0627	0·2010	–
8	–	–	–	–	0·0001	0·0009	0·0323	0·1103

TABLE 4.11

Contingency table for carrying out Leslie's test for dilution.

						Total
	b_{12}	b_{13}	b_{14}	...	b_{1s}	$b_{1.}$
		b_{23}	b_{24}	...	b_{2s}	$b_{2.}$
			b_{34}	...	b_{3s}	$b_{3.}$
				...	...	...
					$b_{s-1,s}$	b_{s-1}
	u_2	u_3	u_4	...	u_s	$u.$
Total	n_2	n_3	n_4	...	n_s	$n.$

and the expected frequencies corresponding to the observed frequencies b_{ij} are $n_j\hat{\theta}_i$; the expected frequencies for the u_j are obtained by subtraction. The goodness-of-fit statistic based on comparing the observed frequencies with the expected frequencies in Table 4.11 is chi-squared with $(s-1)(s-2)/2$ degrees of freedom.

Unfortunately the above maximum-likelihood equations do not seem to have explicit solutions, and alternative estimates of the θ_i are desirable. For example, if there were no blanks in Table 4.11 the estimate of θ_i would be $b_{i.}/n_.$; Leslie

[1952: 385] uses this estimate in his so-called test for "dilution" (immigration and recruitment). Another problem that arises in the use of the above test is tha the expected frequencies are often small and pooling may be needed. Leslie *et al.* [1953] suggest pooling the b_{ij} (and their expected frequencies) in each column, thus reducing the table to $s - 1$ pairs of frequencies (m_j, u_j).

The above method can still be used when mortality is taking place, provide that all subclasses of marked and unmarked have the same mortality rates between successive samples. In this case the proportion of the population first marked at the ith sample will remain constant and equal to u_i/N, so that (4.25) is still satisfied.

Examples of Leslie's test are given in Leslie *et al.* [1953], Turner [1960a], Dunnet [1963: 92], Krebs [1966: 243], Delong [1966] and Otis *et al.* [1978].

REGRESSION MODEL. As far as the Schumacher–Eschmeyer regression method (cf. **4.1.3**(*1*)) is concerned, the best evidence for its validity is obviously the linearity of the graph. Any change in N through recruitment, mortality, etc., or any variation in catchability will affect the basic equation $E[m_i/n_i] = M_i/N$, and this in turn will show up in the graph, provided the m_i are not too small (say greater than 10). In practice, point and interval estimates of N should be obtained using as many different methods as possible, as any departures from underlying assumptions will usually affect different models in different ways. A substantial agreement among the estimates would then give qualitative support for the validity of the models concerned. If mortality, recruitment, etc. are definitely affecting the population then the recapture data should be analysed using the general methods of Chapter 5.

2 *Tests for random sampling*

Apart from poor experimentation and inadequate experimental design, there are three basic sources of non-randomness:

(a) There may be subcategories in the population due to size, sex, species, etc. for which the sampling is random within each subcategory but not between the subcategories (e.g. Takahasi [1961]). In this case, if there is no mortality, the chi-squared goodness-of-fit test based on the contingency table 3.3 on page 71 can be applied to each sample (except the first) using the pairs m_{ix}, $M_{ix} - m_{ix}$. As pointed out by Robson and Regier [1968], these $(s-1)$ chi-squares are independent and can be added together to give a total chi-square.

(b) Catching and handling may affect catchability, so that marked and unmarked have different probabilities of capture in a given sample. However, in some populations, once an animal has been caught its catchability remains fairly constant irrespective of future recaptures (e.g. Young *et al.* [1952]). In this case the ratio k of the average probability of capture of an unmarked animal to the average probability of capture of a marked may remain approximately constant from sample to sample, so that Marten's regression method

(**4.1.3**(*3*)) can be used for testing $k = 1$.

(c) If the catching and handling go on affecting the catchability of marked individuals after their first capture, then the sampling will not be random within the marked population. Such non-randomness can be detected using the following technique of Leslie [1958], based on the frequency of recapture of individuals. This technique is discussed further on p. 226 and in **13.1.7**.

Suppose a multiple-recapture experiment consisting of t samples is carried out in a closed population containing an identifiable group of animals, and let G denote both the group and the number in the group. If g_j of this group are caught in the jth sample ($j = 1, 2, \ldots, t$), then, on the assumption of simple random sampling, the probability P_j $(= 1-Q_j)$ that an individual member of G bears the recovery mark j is g_j/G. Suppose a particular member of G is caught x times, then from Kendall and Stuart [1969: 126–7] we have

$$E[x \mid \{g_j\}] = \sum_{i=1}^{t} P_j = \mu, \quad \text{say},$$

$$V[x \mid \{g_j\}] = \sum_{i=1}^{t} P_j Q_j$$

$$= \mu - \sum_{j=1}^{t} g_i^2/G^2$$

$$= \sigma^2, \quad \text{say}.$$

If f_x members of G are caught x times, then

$$\mu = \sum g_i/G$$

$$= \sum x f_x / \sum f_x = \bar{x},$$

and

$$T = \sum_{x=0}^{t} f_x(x-\mu)^2/\sigma^2$$

is approximately distributed as chi-squared with $G - 1$ degrees of freedom when the sampling is random. Leslie suggests that the approximation is satisfactory when $G > 20$ and $t \geqslant 3$. We note that any samples for which $g_j = 0$ are ignored in the above analysis.

To apply this test to a Schnabel census of s samples we first of all define $G = n_1$ $(= u_1)$, the animals tagged in the first sample. In this case g_1, g_2, etc. are the members of this group caught in the second and third samples, etc., and $t = s - 1$. We can then apply this procedure to the newly tagged individuals in each sample, so that G successively represents u_2, $u_3, \ldots, u_{s-3}$ with corresponding t values $s - 2, s - 3, \ldots, 3$ respectively. If there are accidental deaths through catching and handling, then G refers to the members of u_i which are still alive at the end of the experiment. Since the test statistics thus obtained are based on different individuals they are independent and can be combined to give a total chi-square. In practice, G will often be greater than the degrees of freedom tabled, so that one must use the usual normal approximation

$$z = \sqrt{(2T)} - \sqrt{(2G-3)},$$

which is approximately distributed as the unit normal distribution when sampling is random (cf. **13.1.7** for a generalisation by Carothers [1971]).

We note that any group of identifiable animals can be used for G. In particular, if the population size is known (e.g. Crowcroft and Jeffers [1961], Huber [1962]) we can put $G = N$, and T is now a test that sampling is random with respect to the whole population and not just the marked population. One or two authors (e.g. Geis [1955]) have put G equal to the group of animals caught at least once throughout the experiment and then applied one of the approximate methods described below, with x now representing the number of *re*captures rather than captures (i.e. $x = 0, 1, 2, \ldots, s-1$). However, I cannot see how such a model can be justified in terms of the above theory.

One advantage of the above method is that it can be adapted to "open" populations in which there is natural death and recruitment. In this case the group G consists of the members of u_i known to be alive over a certain sequence of samples, say samples $i+1, i+2, \ldots, i+t$, through having been caught after sample $i+t$. This method, however, does not apply if there is migration, as some of the marked animals may be out of the sampling area for several sampling occasions.

Example 4.12 Hypothetical data

In a Schnabel census of 5 samples the following recapture histories were recorded:

a_1	a_{12}	a_{13}	a_{14}	a_{15}	a_{123}	a_{124}	a_{125}	a_{134}	a_{135}	a_{145}	a_{1235}	other	n_1
84	20	14	15	12	2	1	3	3	1	4	1	0	160

a_2	a_{23}	a_{24}	a_{25}	a_{234}	a_{235}	other	u_2
83	8	7	10	1	1	0	110

From this set of data Table 4.12 was constructed. For the group n_1 we have

TABLE 4.12

Data for demonstrating Leslie's test for simple random sampling.

$G = 160$						$G = 110$					
Sample	g_i	g_i^2	x	f_x	$x^2 f_x$	Sample	g_i	g_i^2	x	f_x	$x^2 f_x$
2	27	729	0	84	0	3	10	100	0	83	0
3	21	441	1	61	61	4	8	64	1	25	25
4	23	529	2	14	56	5	11	121	2	2	8
5	21	441	3	1	9		29	285	3	0	0
	92	2140	4	0	0					110	33
				160	126						

$$G = 160$$

$$g_1 = a_{12} + a_{123} + a_{124} + a_{125} + a_{1235} = 27$$

$$g_2 = a_{13} + a_{123} + a_{134} + a_{135} + a_{1235} = 21, \text{ etc.}$$

$$\mu = 92/160 = 0{\cdot}575$$

$$\sigma^2 = 0{\cdot}575 - 2140/160^2 = 0{\cdot}4914$$

$$\Sigma f_x(x-\mu)^2 = 126 - 92^2/160 = 73{\cdot}1$$

$$T = 73{\cdot}1/(0{\cdot}4914) = 148{\cdot}7$$

and

$$z = \sqrt{297{\cdot}4} - \sqrt{317} = -0{\cdot}6,$$

which is not significant. Similarly for the group u_2 we find that $z = -0{\cdot}5$, which again is not significant. Therefore, on the basis of these two tests we do not reject the hypothesis of simple random sampling as far as the marked population is concerned.

APPROXIMATE TESTS. If $g_j = g$, say $(j = 1, 2, \ldots, t)$, we have

$$P_j = \Sigma P_j/t = \bar{P} \; (= \bar{x}/t)$$

and x is the outcome of t binomial trials. The test for randomness is then a test that the G values of x constitute a random sample of size G from a binomial distribution with parameters t and $\bar{P}$. In this case $\sigma^2 = t\bar{P}\bar{Q}$ and T reduces to the standard Binomial Index of Dispersion (**1.3.6** (*1*))

$$T' = \frac{\Sigma f_x(x-\bar{x})^2}{\bar{x}(1-\bar{x}/t)}$$

which may be regarded as an approximation for T when the g_j are not too different. In fact

$$\begin{aligned} t\bar{P}\bar{Q} - \Sigma P_j Q_j &= -\Sigma (P_i - \bar{P})(Q_i - \bar{Q}) \\ &= \Sigma (P_i - \bar{P})^2 \\ &\geqslant 0, \end{aligned}$$

so that $T' \leqslant T$ with equality only in the unlikely event of the g_j being equal. Therefore, if T' is significant then T will be significant, and T' is a "conservative" approximation.

When P is small ($< 0{\cdot}05$ say), we can use the Poisson approximation to the binomial with $\sigma^2 = \mu = t\bar{P}$. The statistic T then becomes the Poisson Index of Dispersion, $\Sigma f_x(x-\bar{x})^2/\bar{x}$, and since $\bar{P} > \bar{P}\bar{Q}$ this statistic will be smaller than T'. The Poisson approximation is particularly relevant to the situation where the sampling is a continuous process and the animals are caught one at a time (i.e. for the $\Sigma x f_x$ samples in which a member of G is caught we have $P_j = 1/G$).

COMPARING OBSERVED AND EXPECTED FREQUENCIES. We note that the statistic T is based on comparing the observed variance of x with the theoretical variance, calculated on the assumption of random sampling. In general this test will be more sensitive than a goodness-of-fit test based on comparing the observed frequencies f_x with the expected frequencies (Cochran [1954]). However, if T is significant, a comparison of the observed and expected frequencies can be helpful in detecting where departures from random sampling occur. Unfortunately, using Leslie's method, the expected frequencies require lengthy calculations, particularly for large values of x. For example,

$$\Pr[x = 0] = Q_1 Q_2 \ldots Q_t,$$

$$\Pr[x=1] = Q_1 Q_2 \ldots Q_t \sum_i (P_i/Q_i),$$

$$\Pr[x=2] = Q_1 Q_2 \ldots Q_t \sum_{i<j}\sum (P_i P_j/Q_i Q_j), \text{ etc.}$$

Using the binomial approximation, however, the expected frequencies are more readily calculated, namely

$$E_x = G\binom{t}{x}\bar{P}^x \bar{Q}^{t-x}.$$

4.1.6 Models based on constant probability of capture

1 General theory

Suppose that the trapping effort is the same for each sample, so that p_i, the probability of capture in the ith sample ($i = 1, 2, \ldots, s$), is constant ($= p$, say). Then (4.2) reduces to

$$f(\{a_w\}) = \frac{N!}{\prod_w a_w!\,(N-r)!}\, p^{\Sigma n_i} q^{sN-\Sigma n_i}$$

and $\hat{N}_p$, the maximum-likelihood estimate of N, is now the unique root greater than r of

$$\left(1-\frac{r}{N}\right) = \left(1-\frac{\Sigma n_i}{sN}\right)^s. \tag{4.26}$$

Darroch [1958: 355] shows that as $N \to \infty$ the asymptotic variance of $\hat{N}_p$ is

$$V[\hat{N}_p] = N\left[\frac{1}{q^s} + s - 1 - \frac{s}{q}\right]^{-1}$$

which may be compared with

$$V[\hat{N}] = N\left[\frac{1}{q_1 q_2 \ldots q_s} + s - 1 - \Sigma\frac{1}{q_i}\right]^{-1},$$

the corresponding expression when the p_i are unequal. Since $V[\hat{N}_p]$ follows from $V[\hat{N}]$ by simply putting $p_i = p$, Darroch concludes that, asymptotically, no information is gained by using the knowledge that p_i is constant. It is therefore recommended that the methods of **4.1.2** be used irrespective of whether one suspects p_i to be constant or not, except possibly for small samples.

2 Frequency of capture(*)

Several models based on the frequency of capture have been developed recently, and these have been used mainly for detecting any variation in trap response. For example, if p is constant, the probability that an animal is caught x times ($x = 0, 1, 2, \ldots, s$) is given by the binomial probability function

$$f(x) = \binom{s}{x} p^x q^{s-x}. \tag{4.27}$$

If N is known, we can regard the animals as representing N independent observations from (4.27) and carry out a standard goodness-of-fit test (**1.3.6** (*1*)) to test for constant p. When p is small ($< 0{\cdot}05$), the Poisson

(*)See also **12.8.7**.

approximation to (4.27) can be used. Examples of such tests are given in Crowcroft and Jeffers [1961] and in Example 4.14 below.

Some authors have also used these tests for populations in which N has been estimated by independent methods not affected by trap response. For example, Geis [1955], in his study of a rabbit population, estimated N by the Petersen method using data from a hunting kill. He assumed that in spite of a variable trap response, shooting would provide an unbiased sample from the population. Keith and Meslow [1968] also used the Petersen estimate for a population of snowshoe hares. Three independent methods of estimation were used, based on the ratios of (i) colour-marked to unmarked hares observed after winter and spring trapping periods, (ii) marked to unmarked hares taken in late summer live-snaring, and (iii) radioactive to non-radioactive young in summer resulting from implantation of adult females with calcium-45 in spring. Edwards and Eberhardt [1967] carried out two studies: one based on a known population of rabbits which had been released in an enclosure and the other using an unconfined population. In the second study, N was estimated using the Petersen method: the tag ratio was obtained from a drive census, which would not be biased by any differential trap response.

It should be noted that the above theory is not just a repetition of the approximate method given on p. 163. In the above theory, p_i is the probability of catching *any* individual in the ith sample, while P_i in Leslie's method is the conditional probability that an individual from an identifiable group is caught in the ith sample, *given* that at least one member of this group is caught in the sample.

Example 4.13 House mice (*Mus musculus* L.): Young *et al.* [1952]

A program of live-trapping was carried out in two populations of house mice in an endeavour to detect any heterogeneity in trap response. Both populations inhabited essentially the same kind of environment, namely heated buildings with an ample supply of food (stored grain), water and cover. Since there appeared to be no differences in the trap reactions of the two populations, they were treated as one population and their records pooled.

Mice were caught in a fixed number of live-traps which operated for three consecutive nights each week followed by a four-day rest period. Each mouse, when first caught, was marked with an individual pattern of toe-clipping, and the termination of any mouse's exposure to trapping was indicated either by its being found dead in the course of the study, or by its capture during an extensive snap-trapping programme at the conclusion. Mice not finally recovered were treated as residents of the population only up to the dates on which they last appeared in live-traps. By interpreting the recapture records in terms of exposure to trapping, the authors eliminated some of the effects of mortality and migration. An examination of local movements in the two populations indicated that "there were essentially no irregular wandering movements and shifts of base which might have affected the calculations of exposure to trapping (Young *et al.* [1950])".

In order to detect any tendencies for the pattern of trap reaction to change with time, the mice were separated into groups known to have survived through successive periods of ten nights of trapping following the date of their original capture. The frequencies of capture within each period were then compared with the frequencies expected under a Poisson model. In each case the agreement was poor as the observed frequencies were too large at either end of the distribution, thus indicating a heterogeneity of trap response. From the histograms, however, there did not appear to be any substantial change in the form of the capture distribution with respect to time, indicating perhaps that the tendency towards trap addiction or avoidance is a relatively permanent characteristic of any one mouse, i.e. once supplied with an attitude towards traps, a mouse "maintains" this attitude in spite of an accumulation of experiences with traps. In support of this statement the authors presented Table 4.13 in which the trapping history of a cohort of 56

TABLE 4.13

Frequencies of captures as a function of time for 56 mice surviving 50 nights of trapping: from Young *et al.* [1952: Table 2].

	Number of captures (x)						
Period	0	1	2	3	4	5	Total
Nights 1–10	41	12	2	0	1	0	56
Nights 11–20	44	9	3	0	0	0	56
Nights 21–30	44	9	3	0	0	0	56
Nights 31–40	43	11	1	0	1	0	56
Nights 41–50	48	6	1	0	1	0	56

mice that survived through at least 50 nights of trapping was broken down into five periods of ten nights. The capture distributions for each of these periods are remarkably similar.

The large observed frequencies at both ends of the distributions can be explained by the presence of trap-shy individuals boosting the "zero" class and by trap-addicted animals affecting the higher frequencies. However, an alternative explanation is available. In a previous study (Young *et al.* [1950]) there were indications that mice had a tendency to be recaptured in traps from which they were most recently released. This is an important factor when the normal foraging distances of the mice are small in relation to trap distances, for then some of the marked mice might be expected to have moved temporarily out of range of any trap, while others may tend to establish themselves close to a trap in which they would then be caught frequently. But the aim of the study was not to determine whether the departures from randomness were due to trap distribution or to innate trap vulnerability, but rather to show that such departures actually occur and to demonstrate how they may be detected.

Example 4.14 Cottontail rabbit (*Sylvilagus floridanus mearnsii*): Geis [1955]

The trap response of rabbits was evaluated by considering live-trapping data collected during Nov.–Dec. 1951 and Oct.–Nov. 1952 from a population of cottontails at the Kellogg Station of Michigan State College. Fifty wooden traps were used, and to ensure complete coverage, the main study area of 160 acres, largely surrounded by open, cultivated land, was divided into halves which were trapped separately. An irregular spacing of traps was used because an irregularly shaped lake in the centre of the study area made the operation of a grid or a straight trap-line impractical.

Rabbits were marked with numbered metal tags inserted near the centre of each ear. There was no evidence of tags being lost except that an occasional one was torn out by shot. The trap location, age, sex, weight and other data were recorded each time a rabbit was handled. After each of the trapping periods, closely supervised hunting took place and the location at which each rabbit was shot was recorded.

The capture frequencies for one particular trap-line (called trap-line A) are given in Table 4.14, and f_0, the number not captured, is obtained from

TABLE 4.14

Expected distribution of captures for a population of rabbits made up of subpopulations with three different probabilities of capture, trap-line A: from Geis [1955: Table 4].

Times captured x	Observed frequency f_x	Expected frequency E_x	Expected frequencies			
			(1) (Poisson) $\hat{p} = 0{\cdot}027$	(2) (Binomial) $\hat{p} = 0{\cdot}314$	(3) (Binomial) $\hat{p} = 0{\cdot}571$	Total (4) =(1)+(2)+(3)
0	121	65·8	118·2	0·2	–	118·4
1	35	77·2	41·5	1·1	–	42·6
2	12	40·8	7·3	3·0	0·0	10·3
3	6	12·8	0·8	5·1	0·1	6·0
4	11	3·0	0·2	5·8	0·2	6·2
5	8	0·4	0·0	4·8	0·6	5·4
6	3	0·0	–	2·9	1·1	4·0
7	0	–	–	1·4	1·5	2·9
8	3	–	–	0·6	1·5	2·1
9	0	–	–	0·1	1·1	1·2
10	1	–	–	0·0	0·6	0·6
11	0	–	–	–	0·2	0·2
12	0	–	–	–	0·1	0·1
13	0	–	–	–	0·0	0·0
Total	200	200·0	168·0	25·0	7·0	200·0

a Petersen estimate of total population size using the marked–unmarked ratio in the hunting kill. From this table we have that p, the probability of capture in a sample, is estimated by

$$\hat{p} = \frac{\Sigma\, x f_x}{\Sigma\, f_x s} = \frac{213}{200(13)} = 0{\cdot}081\ 92$$

and the expected frequencies E_x are given by

$$E_x = 200 \binom{13}{x} \hat{p}^x \hat{q}^{13-x}.$$

Pooling for $x > 3$, the chi-squared statistic $\Sigma\,(f_x - E_x)^2/E_x = 244$ which at 3 degrees of freedom is highly significant, thus indicating that the assumption of constant p is not true. In fact, from Table 4.14 we see that there are too many animals in the no-capture and many-capture categories. This could be due to the fact that the data for adults and juveniles, males and females are pooled, whereas the four groupings of sex and age may have different probabilities of capture. To investigate this possibility we can calculate the expected frequencies for each class separately as in Table 4.15, add the

TABLE 4.15

Comparison to show the effect of combining data from four age and sex combinations of rabbits: adapted from Geis [1955: Table 3].

Sex and Age	Number captured	$\hat{p}$	Expected frequencies $x=$ 0	1	2	3	4	5	6	Total
Adult–Male	5	0·084	1·60	1·91	1·05	0·35	0·08	0·02	0·00	5·01
Adult–Female	9	0·131	1·45	2·84	2·57	1·42	0·54	0·14	0·05	9·01
Juv.–Male	28	0·084	8·96	10·67	5·88	1·96	0·45	0·08	0·00	27·99
Juv.–Female	27	0·071	10·37	10·29	4·73	1·32	0·24	0·03	0·00	26·98
Total	69	0·082	22·38	25·71	14·23	5·05	1·31	0·27	0·05	69·00
E_x	69	0·082	22·70	26·63	14·08	4·42	1·04	0·14	0·00	69·01

expected frequencies together, and compare these with the expected frequencies based on the pooled data (Geis uses 69 of the 79 animals captured). As these two groups of expected frequencies are close together, it is clear that the differences observed in Table 4.14 cannot be accounted for by the pooling of the four categories. Geis suggests, however, that the observed frequencies may still be the product of pooling segments of the population with different probabilities of capture, and to demonstrate this possibility he divides the data into three groups: 168 caught 0–2 times, 25 caught 3–5 times and 7 caught 6–10 times. The corresponding values of $\hat{p}$ for each of these groups are 0·027, 0·314 and 0·571 respectively, and three separate expected distributions are computed as in Table 4.14 (columns (1), (2) and (3)). A combined distribution (column (4), formed by adding the corresponding frequencies from the separate distributions, is compared with the distribution of observed frequencies f_x, and the difference is found to be not significant. But there is a highly significant difference between the compound distribution and the distribution of expected frequencies E_x.

This analysis shows that the observed frequencies of capture can be explained in terms of a compound distribution which reflects a variable probability of capture. According to this view, most animals had a low probability of capture and relatively few had higher probabilities. Geis offers three explanations as to why some rabbits entered traps more readily than others: first, some animals may have a small home range within which a trap was favourably located and therefore frequently encountered; second, rabbits vary in an innate tendency to enter traps; and, third, once a rabbit has had some experience with traps its behaviour is altered so that it is more likely to be recaptured. The first explanation is ruled out by the fact that frequently captured rabbits turned up in many trap locations. The third explanation is supported by the fact that the mean time-interval between first and second captures was significantly longer than the interval between later captures. Additional evidence for trap conditioning comes from the fact that rabbit tracks were observed in the snow around sprung, unbaited traps after the trapping period had been completed.

3 Truncated models: constant probability of capture

BINOMIAL. When N is unknown we can test whether sampling is random with respect to just the marked population by using Leslie's method of **4.1.5**(2). When p_i is constant, an alternative approach is to truncate (4.27) by ignoring the group of $N - r$ animals not captured during the experiment. Thus x, the number of times an animal is captured *given* that it is captured at least once, has probability function

$$f(x) = \binom{s}{x} p^x q^{s-x}/(1-q^s), \quad x = 1, 2, \ldots, s.$$

For this model the maximum-likelihood estimate $\hat{q}$ of q is the unique root of

$$0 = h_s(q) = \frac{1-q^s}{1-q} - \frac{s}{\bar{x}} = 1 + q + \ldots + q^{s-1} - (s/\bar{x}),$$

where $\bar{x}$ $(= \Sigma\, x_i/r)$ is now the mean number of captures per animal for the r animals actually captured. For $s > 3$ this equation can be solved iteratively using the Newton–Raphson method; the ith step is given by

$$q_{i+1} = q_i - h_s(q_i)/h_s'(q_i),$$

and a possible first approximation q_1 is given by the positive root of the quadratic $h_3(q) = 0$. Alternatively we can use the following general technique given by Hartley [1958] for handling truncated distributions. Beginning with a first approximation $N_{(1)}$ of N, we carry out the chain of iterations

$$p_{(i)} = \frac{\bar{x}}{s} \cdot \frac{r}{N_{(i)}}$$

and

$$N_{(i+1)} = \frac{r}{(1-q_{(i)}^s)}.$$

This procedure not only gives us $\hat{p}$ but, as a bonus, we also get $\hat{N}_p$, the solution of (4.26): this follows from the fact that $r\bar{x} = \Sigma\, n_i$. Once $\hat{p}$ is calculated, a standard goodness-of-fit test for the above truncated binomial model can be carried out.

POISSON. When p is small we can use the Poisson approximation to the binomial and consider the truncated distribution (Craig [1953b])

$$f(x) = \frac{e^{-\lambda}}{(1-e^{-\lambda})} \cdot \frac{\lambda^x}{x!}, \quad x = 1, 2, \ldots . \qquad (4.28)$$

This distribution has been investigated by David and Johnson [1952], who show that the maximum-likelihood estimate $\hat{\lambda}$ of λ is the solution of

$$(1-e^{-\lambda})/\lambda = 1/\bar{x},$$

which can be solved by interpolating in Table **A3**. One can then carry out a chi-squared goodness-of-fit test by comparing observed and expected frequencies in the usual manner (examples of this are given in Keith and Meslow [1968]). David and Johnson suggest the alternative procedure of using the usual Poisson Dispersion Test, but with the class of zero captures left out, i.e.

$$T = \sum_{x=1}^{X} f_x(x-\bar{x})^2/\bar{x},$$

where $\bar{x} = \sum_{x=1}^{X} xf_x / \sum_{x=1}^{X} f_x$

and X is the largest observed value of x. They show that treating T as chi-squared with $r-1$ degrees of freedom leads to a conservative test, for if T is significant then the Poisson Dispersion Test derived from the complete data is also significant. Rao and Chakravarti [1956] and Chakravarti and Rao [1959] suggest using $\bar{x}(1+\hat{\lambda}-\bar{x})$ in the denominator of T: they also give an exact small-sample test based on $\Sigma\, x_i^2$.

We note that, strictly speaking, (4.28) should be truncated on the right at $x = s$ as no more than s recaptures are possible. However, if s is sufficiently large for $\Pr[X \leqslant s]$ to be almost 1, the effect of truncation on the data will be negligible.

An estimate of N, the population size, is given by $r/[1-\exp(-\hat{\lambda})]$ or $r\bar{x}/\hat{\lambda}$.

4 Truncated models: allowing for trap response(*)

GEOMETRIC. Eberhardt *et al.* [1963] found that the capture frequencies for a rabbit population are well fitted by the geometric distribution

$$f(x) = PQ^x, \quad x = 0, 1, 2, \ldots, (0 < P < 1;\ Q = 1-P),$$

and we now outline one of the two derivations that they give for this model. Suppose that conditional on λ, the "average capture rate", x has a Poisson distribution

(*)This section should be read in conjunction with **12.8.7** (*1*).

$$f(x \mid \lambda) = e^{-\lambda}\lambda^x/x!$$

Then, assuming a circular home range of radius R, we would expect the average capture rate to be proportional to the area of the home range, i.e. $\lambda = d\pi R^2$, where d is a constant depending on such factors as the density of traps and the probability of capture, given that there is "contact" with one or more traps. Following Calhoun and Casby [1958], it is assumed that R has density function

$$f_1(R) = \frac{R}{\sigma^2} e^{-R^2/(2\sigma^2)}$$

so that if $c = 2d\pi\sigma^2$,

$$f_2(\lambda) = c^{-1} e^{-\lambda/c}, \quad (\lambda \geqslant 0).$$

Hence

$$f(x) = \int_0^\infty f(x \mid \lambda) f_2(\lambda)\, d\lambda = PQ^x,$$

where $P = 1/(1+c)$.

When the size of the zero class is unknown, this distribution can be truncated at the origin as in the previous models. If trapping is carried out on s occasions then the distribution should also be truncated on the right, so that we are led to consider

$$f(x) = PQ^{x-1}/(1-Q^s), \quad x = 1, 2, \ldots, s. \tag{4.29}$$

For a sample of r observations from this distribution, the maximum-likelihood estimate $\hat{P}$ $(= 1-\hat{Q})$ is the unique solution of

$$\bar{x} = \frac{sQ^{s+1} - (s+1)Q^s + 1}{Q^{s+1} - Q^s - Q + 1}$$

which can be solved by interpolating linearly in Table **A4** (reproduced from Thomasson and Kapadia [1968]). For example, if $s = 10$ and $\bar{x} = 3{\cdot}644$,

$$\hat{P} = 0{\cdot}20 + \frac{3{\cdot}797 - 3{\cdot}644}{3{\cdot}797 - 3{\cdot}043}(0{\cdot}10) = 0{\cdot}2203.$$

When s is large, the effect of truncation on the right is negligible and $\hat{P} = 1/\bar{x}$; or allowing for bias (Eberhardt [1969a]), $\hat{P} = (r-1)/(r\bar{x}-1)$ which, from Chapman and Robson [1960], is the minimum variance unbiased estimate of P. In this case the total population size can be estimated by $\hat{N} = r/\hat{Q}$ (Edwards and Eberhardt [1967]). To determine when the truncation can be neglected, we enter $1/\bar{x}$ at the top of Table **A4** and in the nearest column we note when the entry becomes independent of s.

If f_x animals are caught x times then, truncating the distribution of x on the right only,

$$E[f_x] = NPQ^x/(1-Q^{s+1}), \quad x = 0, 1, 2, \ldots, s,$$

and taking logarithms we have

$$\begin{aligned} E[\log f_x] &\approx \log[NP/(1-Q^{s+1})] + x \log Q \\ &= \beta_0 + \beta x, \text{ say}, \quad (x = 1, 2, \ldots, s), \end{aligned}$$

which, neglecting Q^{s+1}, is the regression model suggested by Edwards and Eberhardt [1967] (see also Kelley and Barker [1963]). We can then estimate N and P from the usual least-squares estimates of β_0 and β. One method of obtaining a confidence interval for N is to calculate the confidence limits for $NP/(1-Q^{s+1})$, the expected number of animals not caught, and add r to both limits.

In applying these methods to squirrel populations Edwards and Eberhardt make the empirical suggestion that reliable estimates of N are obtained when about 50 per cent of the population is captured and $\bar{x}$ is about $1\frac{1}{2}$ to 2.

Example 4.15 Squirrel (*Sciurus carolinensis*, etc.): Nixon *et al.* [1967]

The aim of the study was to investigate the accuracy of the Schnabel (**4.1.2**(*4*)) and Schumacher–Eschmeyer (**4.1.3**(*1*)) estimates of population size for squirrel populations, and to consider the possible application of the above geometric model. The study area occupied 237 acres of continuous forest habitat in the 1250 acre Waterloo Wildlife Experiment Station, Athens County, Ohio. Both fox and grey squirrels occurred in the area, with the grey squirrels comprising about 85 per cent of the squirrel population. The area was gridded on a 3 × 3 chain interval, with a trap placed at the discretion of the trapper within a $\frac{1}{5}$-acre plot surrounding each point of intersection. This yielded a trap density of about one (0·96) trap per acre.

Prebaiting was used for 10 days before the experiment and trapping was carried out for 11 consecutive days just before the hunting season. All squirrels captured were ear-tagged and released at their points of capture. Squirrels killed by the hunters on the study area during the hunting season provided an estimate of the proportion tagged, from which N could be estimated using the Petersen method.

Recapture data for the year 1962 are given in Table 4.16, and a plot of m_i/n_i versus M_i is not linear (cf. p. 141), thus suggesting that the Schnabel estimate and its modifications will not give reliable estimates of N. This was borne out from a more detailed analysis by the authors, who felt that the Schnabel and Schumacher–Eschmeyer methods led to an underestimation of population size. This is in agreement with Flyger [1959: 221] who reported considerably lower estimates using Schnabel's method than those derived from sight records of colour-marked squirrels, and it also agrees with the data obtained on cottontail rabbits by Edwards and Eberhardt [1967].

Using the capture frequencies f_x as given in Table 4.17, a goodness-of-fit test for (4.29) can be carried out. We have $1/\bar{x} = 72/223 = 0{\cdot}323$, and from Table **A4** we find that the truncation of the distribution at $s = 11$ must be taken into account. Therefore, entering the table with $s = 11$ and $\bar{x} = 3{\cdot}097$ we find that $\hat{P} = 0{\cdot}300$. The expected frequencies are then given by

$$E_x = 72\,\hat{P}\hat{Q}^{x-1}/(1-\hat{Q}^{11})$$

and

$$\Sigma\,(f_x - E_x)^2/E_x = 2{\cdot}3\,,$$

TABLE 4.16
Capture–recapture data from a Schnabel census: from Nixon *et al.* [1967: Table 1, 1963].

Trap day i	Sample size n_i	Marked m_i	M_i	m_i/n_i
1	38	–	–	–
2	29	19	38	0·66
3	31	23	48	0·74
4	16	13	56	0·81
5	20	19	59	0·95
6	18	17	60	0·94
7	17	14	61	0·82
8	19	13	64	0·68
9	16	14	70	0·88
10	14	14	72	1·00
11	5	5	72	1·00
Total	223	151		

TABLE 4.17
Observed capture frequencies fitted to zero-truncated geometric and Poisson distributions: data from Nixon *et al.* [1967: Table 3].

Number of captures x	Observed frequencies f_x		$x\ f_x$	Expected frequencies Geometric ($s = 11$) E_x	Geometric ($s = \infty$)	Poisson ($s = \infty$)
1	23		23	22·0	23·3	11·9
2	14		28	15·4	15·8	17·4
3	9		27	10·8	10·7	17·2
4	6		24	7·6	7·2	12·6
5	8		40	5·3	4·9	7·4
6	7	12	42	10·9	10·1	5·5
7	3		21			
8	0		0			
9	2		18			
10	0		0			
11	0		0			
Total	72		223	72·0	72·0	72·0

which at 4 degrees of freedom, indicates a close fit. The authors actually ignored the effect of truncation in their analysis ($s = \infty$) and used $\hat{P} = 0{\cdot}323$. However, we see from Table 4.17 that this makes little difference to the analysis. They also fitted a truncated Poisson which gave a very poor fit to the observed frequencies ($\chi_4^2 = 26{\cdot}1$).

The authors concluded that the probability of capture did not seem to be the same for all individuals and that the geometric model gave a reasonable to fit to the observed frequencies.[*] However, as pointed out by Eberhardt *et al.* [1963], a good fit to this model provides no conclusive evidence that the assumptions used in deriving the model actually occur in Nature: different sets of assumptions can give rise to the same model. For example, suppose that radio-tracking data indicate that the number of visits y of an animal to some small regular area around the trap follows a geometric distribution (e.g. Tester and Siniff [1965])

$$f_1(y) = \theta(1-\theta)^y, \quad y = 0, 1, 2, \ldots .$$

If x, the number of captures, is conditionally binomial with parameters (y, p), then we find that the unconditional distribution is once again geometric (Eberhardt [1969a]), namely

$$f(x) = w(1-w)^x, \quad x = 0, 1, 2, \ldots , \tag{4.30}$$

where $w = \theta/[\theta + (1-\theta)p]$.

Further examples comparing the merits of the above Poisson and geometric models are given in Eberhardt [1969a].

NEGATIVE BINOMIAL. The above derivation of (4.30) applies to the situation where a single trap is randomly located within a given animal's home range. If, however, the traps are closer together, so that k traps fall within the home range, and assuming that traps act independently, then it is readily shown (Eberhardt [1969a]) that the sum of k random variables independently sampled from the geometric distribution (4.30) has a negative binomial distribution

$$\frac{k(k+1)\ldots(k+x-1)}{x!} w^k(1-w)^x, \quad x = 0, 1, 2, \ldots \tag{4.31}$$

(which reduces to (4.30) when $k = 1$). Alternatively this distribution can also be derived by assuming that the Poisson model with parameter λ is appropriate, but with λ varying according to a Pearson Type III distribution (Kendall and Stuart [1969: 129]). In this case x has probability function

$$\frac{k(k+1)\ldots(k+x-1)}{x!} \cdot \frac{a^x}{(1+a)^{k+x}}, \quad x = 0, 1, 2, \ldots , \tag{4.32}$$

which reduces to (4.31) by putting $1 - w = a/(1+a)$. Thus, whichever method of derivation is used, the distribution of x, truncated at $x = 0$, is given by

$$f(x) = \frac{k(k+1)\ldots(k+x-1)}{x!} \cdot \frac{w^k(1-w)^x}{(1-w^k)} \quad x = 1, 2, \ldots , \tag{4.33}$$

where k may or may not be an integer.

This model seems to have been first applied to recapture data by Tanton [1965: 14]: it has also been used by Taylor [1966: 125] and by Tanton [1969: 515]. From Sampford [1955] the maximum-likelihood estimates of w and k

---

[*] For a similar study see Nixon *et al.* [1975].

for a sample of r observations from this distribution are the solutions of

$$\frac{rk}{w(1-w^k)} - \frac{r\bar{x}}{1-w} = 0$$

and

$$\frac{r \log w}{(1-w^k)} + \sum_{x=1}^{X} \left(\frac{1}{k} + \frac{1}{k+1} + \ldots + \frac{1}{(k+x-1)}\right) f_x = 0,$$

where X is the maximum observed value of x; and Sampford describes several methods for solving these equations iteratively. Alternatively we can use the iterative technique of Hartley [1958], which consists basically of estimating $N - r$, the size of the "zero" class, and then using Bliss and Fisher's [1953] technique for estimating k and a from (4.32) and the "completed" data. However, a simpler method of obtaining estimates for k and w has been proposed by Brass [1958: method A] as follows.

Let $\pi_1 = \Pr[x = 1] = kw^k(1-w)/(1-w^k)$; then if μ and σ^2 are the mean and variance of x for (4.33), Brass shows that

$$w = \mu(1-\pi_1)/\sigma^2$$

and

$$k = (w\mu - \pi_1)/(1-w).$$

Therefore, replacing μ, σ^2 and π_1 by their sample estimates

$$\bar{x} = \sum_{x=1}^{X} xf_x/r,$$

$$s^2 = \sum_{x=1}^{X} f_x(x-\bar{x})^2/(r-1),$$

and

$$\hat{\pi}_1 = f_1/r,$$

respectively, we have the simple estimates

$$\tilde{w} = \bar{x}(1-\hat{\pi}_1)/s^2$$

and

$$\tilde{k} = (\tilde{w}\bar{x}-\hat{\pi}_1)/(1-\tilde{w}).$$

Table 4.18, taken from Brass, gives the efficiency (based on the determinant of the variance–covariance matrix) of the above estimation procedure as

TABLE 4.18

Percentage efficiency of estimation by Method A: from Brass [1958: Table 1].

M \ k	0·5	1	2	3	4	5	10	∞
0·5	93·4	97·1	98·9	99·3	99·4	99·5	99·3	98·6
1	88·0	93·6	97·1	98·1	98·4	98·6	98·0	97·4
2	81·0	88·2	93·7	95·7	96·6	97·0	97·4	96·1
5	70·9	78·8	86·5	90·4	92·7	94·2	97·2	98·5
10	63·6	70·9	78·9	83·7	87·0	89·3	95·1	100·0
∞	22·7	38·8	57·5	67·6	73·9	78·2	88·0	100·0

compared to the maximum-likelihood method for different values of k and $M = k(1-w)/w$ (the mean of the complete distribution (4.32)). We see that for small M or large k, Brass's procedure is remarkably efficient.

Example 4.16 Wood mouse (*Apodemus sylvaticus* (L.)): Tanton [1965]

Using Brass's method, Tanton estimates w and k for a population of wood mice and carries out a goodness-of-fit test for (4.33) as set out in

TABLE 4.19

Capture frequencies of a male population of wood mice fitted to a zero-truncated negative-binomial distribution: from Tanton [1965: Table 5].

x	f_x		$x\,f_x$	$x^2\,f_x$	E_x	
1	71		71	71	68·10	
2	59		118	236	56·82	
3	41		123	369	45·90	
4	39		156	624	36·46	
5	20		100	500	28·68	
6	26		156	936	22·40	
7	19		133	931	17·41	
8	12		96	768	13·45	
9	9		81	729	10·38	
10	5		50	500	8·00	
11	8		88	968	6·15	
12	4		48	576	4·72	
13	9	11	117	1 521	3·62	6·39
14	2		28	392	2·77	
15	1	10	15	225	2·12	9·13
16	3		48	768	1·62	
17	3		51	867	1·24	
18	3		54	972	0·94	
19	0		0	0	0·72	
20	0		0	0	0·55	
21	0		0	0	0·42	
> 21	0		0	0	1·52	
Total	334		1 533	11 953	333·99	

Table 4.19. From this table we have $r = 334$,

$$\bar{x} = 1533/334 = 4{\cdot}5898\,,$$

$$s^2 = [11\,953 - 1533^2/334]/333 = 14{\cdot}765\,,$$

and $\hat{\pi}_1 = 71/334 = 0{\cdot}212\,57$

so that

$$\tilde{w} = 4{\cdot}5898\,(0{\cdot}787\,43)/14{\cdot}765 = 0{\cdot}2448\,.$$

Actually Tanton uses a divisor of 334 in calculating s^2, and his estimates $\tilde{w} = 0{\cdot}2455$ and $\tilde{k} = 1{\cdot}2118$ are used in calculating the column of expected frequencies E_x. He obtains a value of 1·10 for his chi-squared goodness-of-fit statistic which, at 11 degrees of freedom, indicates a good fit of the observed to the expected frequencies. He also suggests estimating the population size by

$$\tilde{N} = r/(1-\tilde{w}^{\tilde{k}}) = 408{\cdot}5,$$

though no variance formula for this estimate is given.

Finally, since $\tilde{M} = \tilde{k}(1-\tilde{w})/\tilde{w} = 3{\cdot}7$, we see from Table 4.18 that Brass's method has an efficiency relative to the maximum-likelihood method of approximately 80 per cent.

SKELLAM'S MODEL. Suppose that for a given animal the frequency of capture follows the binomial distribution (4.27) with parameters s and p. In many experimental situations p may not be the same for each animal but will vary according to some distribution $f_1(p)$ (cf. **3.2.2**(*2*)). For example, Skellam [1948] used the beta-distribution

$$f_1(p) = \frac{1}{B[\alpha,\beta]}\, p^{\alpha-1}(1-p)^{\beta-1}; \quad 0 \leqslant p \leqslant 1$$

where $B[\alpha,\beta] = \Gamma(\alpha)\Gamma(\beta)/\Gamma(\alpha+\beta)$,

and hence showed that

$$\begin{aligned} f(x) &= \int_0^1 f(x\,|\,p)f_1(p)dp \\ &= \binom{s}{x}\frac{B[\alpha+x,\ \beta+s-x]}{B[\alpha,\ \beta]}, \quad x = 0, 1, 2, \ldots, s. \end{aligned} \tag{4.34}$$

If $s \to \infty$, $\beta \to \infty$, $\beta/s \to c$ $(c \neq 0)$ and $p \to 0$ in such a way that sp remains finite, then Skellam showed that sp tends to a Pearson Type III distribution, and the limit of $f(x)$ is the negative binomial (4.32) with $k = \alpha$, $a = 1/c$.

The truncated version of (4.34), which is appropriate when the zero class is not observed, is given by

$$\binom{s}{x}\frac{B[\alpha+x,\ \beta+s-x]}{B[\alpha,\beta] - B[\alpha,\beta+s]}, \quad x = 1, 2, \ldots, s.$$

Unfortunately estimation for this distribution is not easy, which rather precludes its use in practice. However, when $s = 2$, (4.34) can be used for investigating the Petersen estimate. For example, if n_i is the size of the ith sample $(i = 1, 2)$, m_2 the number of recaptures in the second sample, and $\hat{N} = n_1 n_2/m_2$, then

$$E[n_i] = NE[p] = N\alpha/(\alpha+\beta), \tag{4.35}$$

$$E[m_2] = NE[p^2] = N\alpha(\alpha+1)/[(\alpha+\beta)(\alpha+\beta+1)], \tag{4.36}$$

and asymptotically

$$\begin{aligned} E[\hat{N}] &= E[n_1]E[n_2]/E[m_2] \\ &= \frac{N\alpha^2}{(\alpha+\beta)^2}\cdot\frac{(\alpha+\beta)(\alpha+\beta+1)}{\alpha(\alpha+1)} \\ &= \frac{N\alpha(\alpha+\beta+1)}{(\alpha+1)(\alpha+\beta)} \\ &= NB, \text{ say,} \end{aligned}$$

where B is tabulated in Table 4.20 for selected values of α and β. We note that the value of B could have also been derived directly from the general theory of **3.2.2**(*2*) by setting $x = y = z = p$, so that

$$B = p_1p_2/p_{12} = (E[p])^2/E[p^2].$$

TABLE 4.20

Table of $B = \alpha(\alpha+\beta+1)/[(\alpha+1)(\alpha+\beta)]$ for selected values of α and β.

β \ α	1	2	3	5	10	∞
1	0·75	0·89	0·94	0·97	0·99	1·00
2	0·67	0·83	0·90	0·95	0·98	1
3	0·63	0·80	0·88	0·94	0·98	1
5	0·58	0·76	0·84	0·92	0·97	1
10	0·55	0·72	0·81	0·89	0·94	1
∞	0·50	0·67	0·75	0·83	0·91	1

It was also mentioned there that B is small when a large proportion of the population has a low probability of capture (i.e. α small).

For the special case $\alpha = 1$ the limiting form of (4.34) is geometric rather than negative-binomial, and Eberhardt [1969a] uses this special case to derive a new estimate of N when $s = 2$. Thus, setting $\alpha = 1$ in (4.35) and (4.36), we have

$$\begin{aligned} E[n_1+n_2] &= 2N/(\beta+1) \\ E[m_2] &= 2N/[(\beta+1)(\beta+2)], \end{aligned}$$

and solving, we have the moment estimates

$$\hat{\beta} + 2 = (n_1+n_2)/m_2$$

and

$$\begin{aligned} \hat{N}_\beta &= (n_1+n_2)(\hat{\beta}+1)/2 \\ &= (n_1+n_2)(n_1+n_2-m_2)/(2m_2). \end{aligned}$$

If in fact p is actually constant, then $\hat{N}$ is asymptotically unbiased (since $B = 1$) and, asymptotically,

$$\begin{aligned} E[\hat{N}_\beta] &= E[n_1+n_2]E[n_1+n_2-m_2]/E[2m_2] \\ &= 2Np(2Np-Np^2)/(2Np^2) \\ &= N(2-p) \end{aligned}$$

which lies between N and $2N$.

To find the asymptotic variance of $\hat{N}_\beta$, let y_i be the number of animals caught i times, i.e. $y_1 = n_1 + n_2 - 2m_2$ and $y_2 = m_2$. Then the joint distribution of y_1 and y_2 is multinomial, namely

$$f(y_1, y_2)$$

$$= \frac{N!}{y_1!\, y_2!\, (N-y_1-y_2)!} P_1^{y_1} P_2^{y_2} P_0^{N-y_1-y_2}$$

$$= \frac{N!}{y_1!\, y_2!\, (N-y_1-y_2)!} \left[\frac{2\beta}{(\beta+1)(\beta+2)}\right]^{y_1} \left[\frac{2}{(\beta+1)(\beta+2)}\right]^{y_2} \left[\frac{\beta}{\beta+2}\right]^{N-y_1-y_2}$$

since, from (4.34) with $s = 2$, $P_i = \Pr[x = i] = f(i)$. Then the maximum-likelihood estimate of N is once again $\hat{N}_\beta$, which now takes the form

$$\hat{N}_\beta, = (y_1 + 2y_2)(y_1 + y_2)/(2y_2).$$

Hence, using the delta method, we find, after some algebra, that

$$V[\hat{N}_\beta] \approx \frac{N\beta}{2(\beta+1)^2(\beta+2)^2}(\beta^5 + 7\beta^4 + 20\beta^3 + 29\beta^2 + 21\beta + 6).$$

5 *Models based on waiting times between captures*(*)

TIME TO FIRST RECAPTURE. The probability that an animal is caught for the first time in the yth sample, *given* that it is caught at least once in s samples, is given by the truncated geometric probability function

$$f(y) = q^{y-1}p/(1-q^s), \quad y = 1, 2, \ldots, s,$$

where p is the probability of capture in a sample. This model has already been discussed above (p. 171), though in a different context. However, a slightly different model has been suggested by Young *et al.* [1952] which can still be used when there is migration and mortality.

Suppose that an animal is captured for the second time in sample number $y + z$ ($z = 1, 2, \ldots, s-y$). Then, given y, and given that an animal is caught at least twice, z has probability function

$$f(z \mid y) = q^{z-1}p/(1-q^{s-y}), \quad z = 1, 2, \ldots, s-y.$$

If we consider only those animals for which $s - y$ is large, then the truncation at $z = s - y$ can be neglected, and we are led to consider the model

$$f(z) = q^{z-1}p, \quad z = 1, 2, \ldots.$$

This model has the simple maximum-likelihood estimate $\hat{p} = 1/\bar{z}$. Young *et al.* [1952] point out that once an animal has been recaptured, we are not interested in its subsequent fate, so that we do not need to "correct" the data for those dying or disappearing before the end of the experiment. Also, if the tendency to die or emigrate is not related to trap vulnerability, then those animals which die or emigrate before being recaptured at all will be distributed randomly over the groups that would have been recaptured after

(*)See also **12.8.7.** (*3*)

1, 2, 3,... samples, and the disappearance of such animals will therefore not bias $\hat{p}$ and the associated goodness-of-fit test.

The above model has been used by several authors in bird banding studies (e.g. Young [1958], Swinebroad [1964]). Young also utilises 3rd, 4th, etc. captures in the estimation of p. This means that one animal can give rise to several values of z, representing the intervals between 1st and 2nd, 2nd and 3rd captures, etc. Obviously this estimate of p is more sensitive to the presence of any trap addiction among the animals, as the "repeaters" are counted several times.

TIME OF RESIDENCE. Suppose that the population under study is such that animals move into the population area, stay for a random length of time, and move out and stay out of the area for the remainder of the investigation. If the trapping is carried out at equally spaced intervals of time with constant trapping effort (i.e. p constant), then it is not unreasonable to assume that θ, the probability that an animal does not leave the trapping area some time between two successive trappings, is the same for all animals in the area and for all successive pairs of trappings. On the basis of these assumptions, Holgate [1964b] gives the following method for estimating θ and p from the observed values of Y, the recorded period of residence, i.e. the interval between the first and last occasions when it is actually captured.

Let Z denote the true period of residence of an individual in the study area, i.e. the interval between the first and last occasions when it is exposed to capture; then (ignoring truncation on the right)

$$\Pr[Z = z] = (1-\theta)\theta^z, \quad z = 0, 1, 2, \ldots. \tag{4.37}$$

Now an animal that remains in the area for z complete intervals is exposed to capture on $z+1$ occasions, so that the probability of its not being caught at all (i.e. Y undefined) is

$$\Pr[Y \text{ undefined} \mid Z = z] = q^{z+1}.$$

Also

$$\Pr[Y = 0 \mid Z = z] = (z+1)pq^z$$

and, noting that as far as Y is concerned, it does not matter how often an animal is recaptured between its first and last capture,

$$\Pr[Y = y \mid Z = z] = (z-y+1)p^2q^{z-y}, \quad y = 1, 2, \ldots, z.$$

Now

$$\begin{aligned}\Pr[Y \text{ undefined}] &= \sum_{z=0}^{\infty} \Pr[Y \text{ undefined} \mid Z = z]\Pr[Z = z] \\ &= \frac{(1-\theta)q}{1-q\theta},\end{aligned} \tag{4.38}$$

and in a similar fashion it is readily shown that

$$\Pr[Y = 0] = \frac{(1-\theta)p}{(1-q\theta)^2} \tag{4.39}$$

and

$$\Pr[Y = y] = \frac{p^2(1-\theta)\theta^y}{(1-q\theta)^2} \qquad y = 1, 2, \ldots . \tag{4.40}$$

Finally we have (dividing (4.39) and (4.40) by one minus the probability (4.38)) the zero modified geometric distribution

$$\Pr[Y = 0 \mid Y \text{ defined}] = \frac{1-\theta}{1-q\theta} \tag{4.41}$$

and

$$\Pr[Y = y \mid Y \text{ defined}] = \frac{(1-q)(1-\theta)\theta^y}{(1-q\theta)}, \quad y = 1, 2, \ldots . \tag{4.42}$$

If a sample of r observations is taken from the above distribution then the maximum-likelihood estimates of θ and q are

$$\hat{\theta} = 1 - (u/\bar{y})$$

and

$$\hat{q} = \frac{\hat{\theta} - u}{\hat{\theta}(1-u)}$$

where u is the proportion of individuals caught more than once (i.e. with $y > 0$). Holgate shows that as $r \to \infty$ the asymptotic variances and co-variances of these estimates are given by

$$V[\hat{\theta} \mid r] = \frac{(1-\theta)^2(1-q\theta)}{r(1-q)},$$

$$V[\hat{q} \mid r] = \frac{(1-q)(1-q\theta^2)(1-q\theta)}{r\theta^2(1-\theta)},$$

and

$$\text{cov}[\hat{\theta}, \hat{q} \mid r] = \frac{(1-\theta)(1-q\theta)}{r\theta}.$$

It is noted that, strictly speaking, (4.37) should be truncated on the right since the number of trappings is finite. Otherwise, at the end of the trapping series, the animals still in the area will be ascribed a duration of residence which is too short. Unfortunately, since the time of the first capture varies, each animal will have a different truncation point, thus leading to a complicated likelihood function for the estimation of θ and q. However, if the study period is long compared with the average time of residence, this effect will be negligible and the truncation can be ignored.

Andrzejewski and Wierzbowska [1961] and Wierzbowska and Petrusewicz [1963] consider using an exponential distribution for the true residence period Z. They find that the values of y (treated as grouped data since the observed period of residence is measured in integral units) are not well fitted by an exponential distribution as there are too many observations in the zero class. However, if this class of animals, which they call "ephemeral", is omitted, it is found that the remaining data are well fitted by an exponential distribution with its zero class omitted (which is again an exponential distribution). From this the authors conclude that "residents" have a constant disappearance rate, while the greater rate for the ephemeral group is caused by the capture of "migrants" passing through the area. Therefore by extrapolating the successfully fitted exponential distribution the number of residents can be estimated; subtracting this number from the ephemerals gives an estimate of the migrants. But the authors effectively assume that $Y = Z$, i.e. the observed period of residence is the actual period of residence, which is not true as not all the animals in the trapping area are captured on every trapping occasion ($p < 1$). Holgate [1964b] points out that this feature of the model could account for the excess of animals in the zero class. In carrying out a goodness-of-fit test for Holgate's model the estimates of p and θ are such that the zero class is fitted exactly.

Holgate [1966] has considered two generalisations of the above theory. In the first generalisation he assumes that θ varies from animal to animal according to a beta-distribution, and he derives a new distribution for Y which he calls the "zero modified Appell distribution". However, this distribution involves hypergeometric functions and is therefore rather difficult to fit to data. In the second generalisation θ is constant but p is assumed to vary according to a beta-distribution. This time the distribution of Y is again a zero modified geometric distribution like (4.41) and (4.42); although θ has the same estimate, α and β are not identifiable.

BIVARIATE DISTRIBUTION. Holgate [1966] has utilised the joint distribution of Y (the recorded period of residence) and W (the number of captures during the *intervening* period, i.e. between the first and last capture) to obtain more efficient estimates of q and θ as follows (in Holgate's notation $Y = X$ and $W = Y$).

Since $Y = y$ implies $y+1$ possible captures, the range of W is 0 to $y-1$ and

$$\Pr[W = w \mid Y = y] = \binom{y-1}{w}(1-q)^w q^{y-w-1} \quad (y > 0).$$

Hence, from (4.41) and (4.42), the joint probability function is given by

$$\Pr[Y = 0 \mid Y \text{ defined}] = (1-\theta)/(1-q\theta)$$

and

$$\Pr[Y = y, W = w \mid Y \text{ defined}] = \binom{y-1}{w}(1-q)^{w+1} q^{y-w-1} \frac{(1-\theta)\theta^y}{(1-q\theta)},$$

where $w = 0, 1, \ldots, y-1$ and $y = 1, 2, \ldots$. (For $Y = 0$, W is not defined.) Let n_0 and n_{yw} denote the corresponding sample frequencies and let r be the total number in the sample ($= n_0 + \sum_y \sum_w n_{yw}$). Then the likelihood function for the sample is proportional to

$$\left(\frac{1-\theta}{1-q\theta}\right)^{n_0} \prod_{y,w} \left\{\frac{(1-q)^{w+1} q^{y-w-1}(1-\theta)\theta^y}{(1-q\theta)}\right\}^{n_{yw}}$$

and it transpires that the maximum-likelihood estimates $\tilde{q}$ and $\tilde{\theta}$ are given by

$$\tilde{q} = (\bar{y} - \bar{w} - u)(1 + \bar{w} + u)/[\bar{y} - (\bar{w} + u)(1 - \bar{y})]$$

and

$$\tilde{\theta} = [\bar{y}-(\bar{w}+u)(1-\bar{y})]/[\bar{y}(1+\bar{w}+u)],$$

where $\bar{w} = \sum_y \sum_w w n_{yw}/r = \sum w n_{.w}/r$, say

$\bar{y} = \sum_y \sum_w y n_{yw}/r = \sum y n_{y.}/r$, say

and

$$u = 1-(n_0/r).$$

Holgate shows that the asymptotic variance–covariance matrix of $\tilde{q}$ and $\tilde{\theta}$ is given by

$$\frac{1}{r}\begin{bmatrix} \dfrac{q(1-q)(1-q\theta^2)(1-\theta)}{\theta^2(1-q\theta)}, & \dfrac{q(1-\theta)^3}{\theta(1-q\theta)} \\ \dfrac{q(1-\theta)^3}{\theta(1-q\theta)}, & \dfrac{(q+\theta+q^2\theta^2-3q\theta)(1-\theta)^2}{(1-q)(1-q\theta)} \end{bmatrix}$$

where the (1, 1) element is the asymptotic variance of $\tilde{q}$.

Using the determinant of the variance–covariance matrix as a measure of asymptotic efficiency, Holgate shows that the relative efficiency of the above method with respect to the previous method based on the marginal distribution of Y only is

$$e = q(1-\theta)^2/(1-q\theta)^2.$$

Example 4.17 Wood mouse (*Apodemus sylvaticus* (L.)): Holgate [1966]

Holgate applied the above theory to the data in Table 4.21. From this table we have

$$n_0 = 65, \qquad r = 65+47 = 112,$$

$$u = 47/112, \quad \bar{y} = 255/112 \quad \text{and} \quad \bar{w} = 26/112.$$

Multiplying through by r^2, we have

$$\tilde{q} = \frac{(255-36-47)(112+36+47)}{[255(112)-(36+47)(112-225)]} = 0{\cdot}8298,$$

and similarly

$$\hat{\theta} = 0{\cdot}8131.$$

Using these estimates in the variance formulae, we find that

$$\tilde{\sigma}[\tilde{q}] = 0{\cdot}0215 \quad \text{and} \quad \tilde{\sigma}[\tilde{\theta}] = 0{\cdot}0203.$$

If only the marginal distribution of Y is used, we have

$$u = 47/112, \qquad \bar{y} = 255/112,$$

$$\hat{\theta} = 1-(u/\bar{y}) = 1-(47/255) = 0{\cdot}8157,$$

and

$$\hat{q} = \frac{\bar{y}-u(\bar{y}+1)}{(\bar{y}-u)(1-u)} = \frac{255(112)-47(255+112)}{(255-47)(112-47)} = 0{\cdot}8366.$$

TABLE 4.21

Bivariate frequency distribution (n_{yw}) of length of residence and number of times caught in the intervening period for a population of wood mice: from Holgate [1966: Table 3].

Length of residence in weeks, y	Frequency of capture, w 0	1	2	3	4	5	6	$n_{y.}$	$yn_{y.}$
0	65	–	–	–	–	–	–	65	
1	6	0	0	0	0	0	0	6	6
2	5	1	0	0	0	0	0	6	12
3	6	0	0	0	0	0	0	6	18
4	9	0	0	0	0	0	0	9	36
5	1	2	0	0	0	0	0	3	15
6	0	1	1	0	0	0	0	2	12
7	0	2	0	0	0	0	0	2	14
8	0	0	0	1	1	0	0	2	16
9	3	0	1	0	0	0	0	4	36
10	0	1	0	0	0	0	0	1	10
11	0	0	1	1	0	0	0	2	22
13	0	0	1	0	1	0	0	2	26
16	0	1	0	0	0	0	1	2	32
$n_{.w}$	30	8	4	2	2	0	1	47	255
$wn_{.w}$	0	8	8	6	8	0	6	36	

Also $\hat{e} = 0{\cdot}28$, indicating that the method using the joint distribution is much more efficient.

Unfortunately the numbers are too small for an adequate test of fit to the bivariate distribution. However, using the estimates $\hat{q}$ and $\hat{\theta}$, Holgate carried out a goodness-of-fit test of the marginal distribution of Y to the modified geometric distribution given by (4.41) and (4.42). Here the observed frequencies are given by $f_0 = n_0$ and $f_y = n_{y.}$; the expected frequencies are

$$E_0 = r(1-\hat{\theta})/(1-\hat{q}\hat{\theta})$$

and

$$E_y = r\frac{(1-\hat{q})(1-\hat{\theta})\hat{\theta}^y}{(1-\hat{q}\hat{\theta})} \quad (y > 0).$$

These are set out in Table 4.22, and (pooling as indicated)

$$\Sigma\,(f_y - E_y)^2/E_y = 7{\cdot}65\,,$$

which, at 6 degrees of freedom, indicates a good fit. For further applications of the above methods see Tanton [1969: 519] and Van Den Avyle [1976].

6 *A runs test for randomness*

When the probability of capture is the same for each sample, Young [1961] gives a runs test for testing the randomness of the captures. He

TABLE 4.22

Observed length of residence compared with expected length of residence, fitted by Holgate's modified geometric distribution for a population of wood mice: from Holgate [1966: Table 1].

Length of residence in weeks, y	Observed frequency f_y	Expected frequency E_y	$\frac{(f_y - E_y)^2}{E_y}$
0	65	65·00	0
1	6	8·66	0·82
2	6	7·07	0·16
3	6	5·76	0·01
4	9	4·70	3·93
5	3	3·83	0·18
6	2	3·13	0·41
7	2	2·55	0·09
8	2	2·08	
9	4	1·70	1·85
10	1	1·38	
11	2	1·13	
12	0	0·92	0·20
13	2	0·75	
14	0	0·61	
15	0	0·50	
16	2	0·41	
⩾17	0	1·82	
Total	112	112·00	7·65

demonstrates the test with the following example.

From January 29 to April 21, 1958, a ten-trap banding station was in operation in a local cemetery. The Slate-coloured Junco (*Junco hyemalis*) was one of the species captured, and the data were first organised by designating as A any day on which at least one junco was caught, and as B any day on which no juncos were captured. These were then arranged in their natural chronological sequence, giving the pattern shown below:

BB A B AA B AA B AAA B A BBB A BB A B A BBBBBB AAAA BBB A BB A BBBB A BBBBB A

From this pattern we have $n_A = 20$ A's, $n_B = 32$ B's and $x = 26$ runs of either A or B (a single letter counts as a run). If the sampling is random, then (Mood [1940])

$$E[x \mid n_A, n_B] = 2np_A q_A + 1 = \mu, \quad \text{say,}$$

and

$$V[x \mid n_A, n_B] = 2np_A q_A(2np_A q_A - 1)/(n-1) = \sigma^2, \quad \text{say},$$

where $n = n_A + n_B$ and $p_A = 1 - q_A = n_A/n$.

Provided n_A and n_B are both greater than 10, x is approximately normal, so that we can test for randomness using the statistic $z = (x - \mu)/\sigma$. Therefore, from the above data we have $z = (26 - 25{\cdot}62)/11{\cdot}40$, which, for the unit normal is not significant, so that the hypothesis of randomness is not rejected.

Strictly speaking, the above conditional test is not correct, as n_A and n_B are random variables rather than fixed parameters chosen in advance (cf. Brunk [1965: 354] for an excellent discussion). From the delta method and **1.3.4**, equation (1.12), we have approximately

$$E[x] = 2npq + 1$$

and

$$V[x] = 2npq(2npq - 1)/(n-1) + V[2np_A q_A],$$

where p is the probability of capture on a given day. Since p can be estimated by p_A, we see that $V[x]$ exceeds σ^2 by approximately $V[2np_A q_A]$. This means that the above conditional test may reject the hypothesis of randomness when in fact the trapping is random. For a proper application of the conditional test one chooses the numbers n_A and n_B in advance, and if, for example, $n_A + k$ A's actually turn up, k of the A's are chosen at random (e.g. using tables of random numbers) and changed to B's. A correction for continuity and tables for carrying out an exact test are given in Swed and Eisenhart [1943].

In conclusion we note that the above test can be applied to the trapping record of a single marked individual.

4.2 INVERSE SCHNABEL CENSUS

Consider a Schnabel census in which, for each sample n_i, the sampling is now continued until a *predetermined* number of marked animals m_i are captured. This modification is a generalisation of **3.5** and for obvious reasons can be suitably called the "inverse" Schnabel census. Using the same notation as in **4.1.1**, we have fixed parameters N, s, n_1 $(= M_2)$, m_2, $m_3, \ldots, m_s$, random variables $M_3, M_4, \ldots, M_s$, r, $n_2, \ldots, n_s$, and the joint probability function of the random variables is a straightforward generalisation of (3.36), namely

$$\prod_{i=2}^{s} \left\{ \frac{\binom{M_i}{m_i - 1}\binom{N - M_i}{u_i}}{\binom{N}{n_i - 1}} \cdot \frac{M_i - m_i + 1}{N - n_i + 1} \right\}.$$

We find that the maximum-likelihood estimate $\hat{N}$ is still given by the appropriate root of (4.4), but the asymptotic bias and variance of $\hat{N}$ are not so readily found as in the "direct" Schnabel census. Alternative methods are therefore desirable, and by analogy with (3.37) Chapman [1952: 297] suggests the mean

$$\bar{N}_2 = \sum_{i=2}^{s}\left\{\frac{n_i(M_i+1)}{m_i} - 1\right\}\bigg/(s-1),$$

which is unbiased with approximate variance

$$V[\bar{N}_2] = \frac{N^2}{(s-1)^2}\sum_{i=2}^{s}\frac{1}{m_i}.$$

Here the coefficient of variation

$$C[\bar{N}_2] = \left\{\sum_{i=2}^{s}\frac{1}{m_i}\right\}^{\frac{1}{2}}\bigg/(s-1)$$

can be used for choosing the fixed parameters to give a predetermined precision. However, the correct choice of the m_i is also important from another point of view, for in **3.5.1** it was pointed out that a wrong choice of m_i, coupled with an unfavourable M_i, could give rise to a large n_i. Therefore a reasonable criterion, suggested by Chapman, for choosing these parameters is to minimise $E[\Sigma\, n_i]$, subject to $C[\bar{N}_2]$ being held constant; unfortunately this does not have a simple solution (a few values are tabulated in Chapman [1952: 306]). As pointed out by Cormack [1968: 448], an increasing sequence of the m_i spreads the sampling effort most evenly, and an examination of Chapman's table seems to indicate that a constant m_i provides the maximum precision for approximately the same expected sample size $E[\Sigma\, n_i]$.

To avoid the possibility of large n_i we can modify the above model as in **3.5.1** and continue the sampling until a predetermined number u_i of unmarked individuals is taken in the ith sample ($i = 2, 3, \ldots, s$). This means that our fixed parameters are now N, s, n_1, $u_2, \ldots, u_s$, $M_3, \ldots, M_s$, r and the random variables $n_2, n_3, \ldots, n_s$, $m_2, \ldots, m_s$. By analogy with (3.38) an approximately unbiased estimate of N is the mean

$$\bar{N}_3 = \sum_{i=2}^{s}\hat{N}_{3i}/(s-1),$$

where $$\hat{N}_{3i} = \frac{n_i(M_i+1)}{(m_i+1)} - 1,$$

and, asymptotically,

$$V[\bar{N}_3] = \sum_{i=2}^{s}V[\hat{N}_{3i}]/(s-1)^2,$$

where $V[\hat{N}_{3i}]$ can be evaluated using (3.39). A simpler estimate of variance is given by (cf. (1.9))

$$v[\bar{N}_3] = \sum_{i=2}^{s}(\hat{N}_{3i} - \bar{N}_3)^2/(s-1)(s-2).$$

4.3 SEQUENTIAL SCHNABEL CENSUS

4.3.1 General methods

In the inverse Schnabel census we fixed the m_i or u_i in advance and took a *fixed* number of samples. A more flexible, sequential-type method suitable for small recaptures m_i would be to fix the sequence $n_1, n_2, \ldots$ in advance and continue this sequence until a predetermined number of recaptures $\Sigma\, m_i$ have been taken, s being now a random variable. Thus we have the fixed parameters $N, n_1, n_2, \ldots, L$, and the random variables $m_2, m_3, \ldots, M_3, M_4, \ldots, r, s, n\ (= \sum_{i=1}^{s} n_i)$, and the procedure is to stop sampling at the completion of the sth sample, s being defined by

$$\sum_{i=2}^{s-1} m_i < L, \quad \sum_{i=2}^{s} m_i \geqslant L .$$

1 Goodman's model(*)

Goodman [1953] showed that r is sufficient for N and he deduced the existence of a minimum variance unbiased estimate (MVUE) as the ratio of two determinants of order $n - n_1 + 1$. In general, these determinants will be of high order, so that a fair amount of calculation is involved. However, an asymptotic MVUE is available in the form of

$$\hat{N}_4 = n^2/(2L)$$

with asymptotic variance

$$V[\hat{N}_4] = N^2/L .$$

Confidence intervals for $1/N$ (and hence for N) can then be obtained from n^2/N, which is asymptotically distributed as chi-squared with $2L$ degrees of freedom. The coefficient of variation of $\hat{N}_4$ takes the simple form $1/\sqrt{L}$, which can be used for choosing L to give a predetermined precision.

In looking up the chi-squared table we find that for $2L > 30$, which is generally the case for a reasonable coefficient of variation, we use

$$z = \sqrt{(2n^2/N)} - \sqrt{(4L-1)},$$

which is asymptotically distributed as $\mathfrak{N}(0, 1)$. Thus an approximate 95 per cent confidence interval for N is given by

$$\frac{2n^2}{[1{\cdot}96+\sqrt{(4L-1)}]^2} < N < \frac{2n^2}{[-1{\cdot}96+\sqrt{(4L-1)}]^2} . \quad (4.43)$$

When $2L > 30$ we have $E[n] \approx \sqrt{(2NL)}$ which, for the special case $n_1 = n_2 = \ldots (= n_0$ say), leads to $E[s] \approx \sqrt{(2NL)}/n_0$. Therefore, if L has been chosen and a rough estimate of N is available before the experiment, we can either calculate the expected duration $E[s]$ of the experiment for a given n_0, or determine n_0 for a given expected duration.

Example 4.18 Hypothetical data: Goodman [1953: Table 2]

Seven samples of 100 ($n_1 = n_2 = \ldots = n_7 = 100$), followed by samples of size 1 ($n_8 = n_9 = \ldots = 1$), were drawn from a population of $N = 10\ 000$

(*)See 12.8.5. for further comments.

random numbers. Sampling ceased when a total of $L = 25$ "recaptures" were made, giving the following data:

Sample	1	2	3	4	5	6	7	8, 9, . . . , 72	Total
n_i	100	100	100	100	100	100	100	65	765
m_i	–	4	4	1	3	2	7	4	25

From the above table we have $n = 765$, $\hat{N}_4 = (765)^2/50 = 11\,704$ and the confidence interval (4.43) is (8251, 18 334).

2 Chapman's model

Assuming the M_i to be fixed parameters as in (4.12) and approximating the distribution of m_i by the Poisson distribution with mean n_iM_i/N (which is satisfactory if n_i/N and M_i/N are both less than 0·1), Chapman [1954: 6] derives an alternative (MVUE) for N, namely

$$\hat{N}_5 = \sum_{i=2}^{s} n_iM_i/L = \lambda/L, \quad \text{say},$$

with variance N^2/L. Confidence intervals for N can be based on $2\lambda/N$, which is approximately distributed as chi-squared with $2L$ degrees of freedom. For $2L > 30$ an approximate 95 per cent confidence interval for N is given by

$$\frac{4\lambda}{[1{\cdot}96+\sqrt{(4L-1)}]^2} < N < \frac{4\lambda}{[-1{\cdot}96+\sqrt{(4L-1)}]^2}.$$

We note that Chapman's model would still be applicable even when there is a small natural mortality, provided that marked and unmarked have the same mortality rate, so that M_i/N remains constant between the $(i-1)$th and ith samples. An additional advantage of Chapman's model is that the marking can be done independently of the sampling.

4.3.2 Samples of size one

1 Fixed $s - r$

If we put $n_i = 1$ in Goodman's model above, the sampling procedure amounts to sampling one at a time with replacement until $s - r = L$ marked animals have been caught. The general theory of **4.3.1**(*1*) still applies, with $n = s$, but now the exact MVUE mentioned there has a simpler form, namely (Goodman [1953: 61])

$$\hat{N}_6 = K(r, L)/K(r, L-1),$$

where $K(r, 0) = r$ and $K(r, L) = r\sum_{i=1}^{r} K(i, L-1)$.

Darroch [1958: 351] has also shown that

$$\hat{N}_6 = \sigma_{r+L}^{r}/\sigma_{r+L-1}^{r},$$

where $\sigma_y^x = \Delta^x(0^y)/x!$, a Stirling number of the second kind. As far as maximum-likelihood estimation is concerned, it transpires that (cf. Darroch [1958: 350], Samuel [1968]) the maximum-likelihood estimate $\hat{N}$ of N is the same as for the direct census and is therefore given by (4.10).

A different approach to the problem is given by Boguslavsky [1955] as follows. He shows that

$$g(N, j) = \Pr[S \leqslant j \mid N] = \binom{N}{N-j+L} \sum_{i=0}^{j-L} \left\{ (-1)^i \binom{j-L}{i} \left(\frac{j-L-i}{N} \right)^j \left(\frac{N-j+L}{N-j+L+i} \right) \right\}, \quad (4.44)$$

where S is the random variable taking values s, and uses the above function as a basis for constructing uniformly most powerful (UMP) one-sided tests. Thus a UMP test of size α for the null hypothesis $H_0 : N \geqslant N_0$ against the alternative $H_1 : N < N_0$ consists of rejecting H_0 at the 100α per cent level of significance if $s \leqslant a$, where s is the observed value of S and a is a function of N_0 determined by $g(N_0, a) = \alpha$. Similarly the rejection rule for a UMP test of $H_0 : N \leqslant N_0$ against $H_1 : N > N_0$ is given by $s \geqslant b$, where $g(N_0, b-1) = 1 - \alpha$. However, for testing $H_0 : N = N_0$ against the two-sided alternative $H_1 : N \neq N_0$ we find that a UMP test does not exist. Therefore an approximation to the best unbiased test is to accept H_0 if $c + 1 \leqslant s \leqslant d - 1$, where c and d are functions of N_0 such that

$$g(N_0, c) = \alpha/2 \quad \text{and} \quad g(N_0, d-1) = 1 - (\alpha/2).$$

Such a method can also be used to obtain a $100(1-\alpha)$ per cent two-sided confidence interval for N as the set of all N_0 such that the interval $[c+1, d-1]$ contains the observed value s. Therefore, if this interval is $N_c \leqslant N \leqslant N_d$ then it can be shown that N_c and N_d are given by

$$g(N_c, s) = \alpha/2 \quad \text{and} \quad g(N_d, s-1) = 1 - (\alpha/2).$$

As N_c and N_d are integers, these equations, in general, will not have exact solutions, and we would choose the nearest integers so that

$$g(N_c, s) \leqslant \alpha/2 \quad \text{and} \quad g(N_d, s) \geqslant 1 - (\alpha/2).$$

Boguslavsky gives a chart for reading off N_c and N_d for $\alpha = 0{\cdot}2$, $L = 1(1)5$ and $s = 1$ to 80 (in his notation $s = n_k$, $L = k$, $N_c = N_L$ and $N_d = N_U$). However, it should be noted that, as in **4.3.1**(*1*), $1/\sqrt{L}$ is still a rough measure of the coefficient of variation and, in particular, of half the proportional width of the confidence interval. This means that the interval (N_c, N_d) will be wide if L is small. Boguslavsky also gives a method for testing the adequacy of the above model (4.44), but unfortunately the calculations involved are considerable unless L is small.

2 *Fixed r*

A variation on the above theme is to continue sampling until a fixed number r of unmarked individuals is caught. Darroch [1958: 350] shows that the maximum-likelihood estimate of N is still $\hat{N}$ and points out that (4.10) has no solution $\hat{N} \geqslant r$ when

$$s > r\left(1 + \frac{1}{2} + \dots + \frac{1}{r}\right).$$

He adds that this situation is unlikely to happen in practice and proves that asymptotically ($N \to \infty$, $r \to \infty$, r/N constant)

$$E[\hat{N}] = N\left\{1 + \sum_{k=1}^{r-1} \frac{k}{(N-k)^3}\left[\sum_{k=1}^{r-1} \frac{k}{(N-k)^2}\right]^{-2}\right\}$$

and

$$V[\hat{N}] = N\left[\sum_{k=1}^{r-1} \frac{k}{(N-k)^2}\right]^{-1}.$$

The above sums can be simplified using the following integral approximations:

$$\sum_{k=1}^{r-1} \frac{k}{(N-k)^2} \approx \int_0^1 \frac{x}{\left(\frac{N}{r} - x\right)^2} dx$$

and

$$\sum_{k=1}^{r-1} \frac{k}{(N-k)^3} \approx \frac{1}{r}\int_0^1 \frac{x}{\left(\frac{N}{r} - x\right)^3} dx.$$

Darroch also mentions how s, which is sufficient for N, can be used for constructing a confidence interval for N in much the same way as r was used on p. 134.

3 *Variable stopping rules*

Boguslavsky [1956], Darling and Robbins [1967] and Samuel [1968] have given a number of methods in which the stopping rule at any particular instant depends on the number of unmarked individuals already caught. The method given in the first two articles amounts to sampling until the absence of further unmarked individuals in the samples leads the experimenter to believe that he has marked the whole population. Mathematically this means that we require a stopping rule for s such that if r is the number of different (unmarked) individuals caught throughout the entire experiment, then

$$\Pr[r = N] \geqslant 1 - \alpha, \tag{4.45}$$

i.e. we have at least $100(1-\alpha)$ per cent confidence that the whole population has been marked. Since $r \leqslant N$, (4.45) can be rewritten as

$$\Pr[r = 1] + \Pr[r = 2] + \ldots + \Pr[r = N-1] \leqslant \alpha,$$

which was used by Boguslavsky to give the following stepwise procedure for calculating a suitable stopping rule.

The simplest situation that one can have is when the same individual keeps turning up, i.e. $r = 1$, so that at the end of $s = s_1$ samples the observer may wish to terminate the experiment with the statement that $N = 1$. In making this decision, the observer rejects the most likely alternative hypothesis that $N = 2$ with a Type II error of $2(\frac{1}{2})^s$ or $(\frac{1}{2})^{s-1}$. If we set this probability equal to α (say 0·10) and solve for s, we find that $s = 4{\cdot}3$. Therefore a suitable stopping rule would be to stop sampling if there are $s_1 = 5$ consecutive drawings of the same individual and accept $N = 1$ with, using an obvious notation,

$$\Pr[r = 1 \mid N = 2] = \Pr[s = s_1 \mid N = 2] = \left(\frac{1}{2}\right)^4 < 0{\cdot}10.$$

However, if a second individual turns up within the 5 samples, our stopping rule changes and we now look for an $s_2 > s_1$ such that we accept $N = 2$ if no new individual turns up by sample number s_2. This amounts to finding s_2 such that

$$\Pr[r = 1 \mid N = 3] + \Pr[r = 2 \mid N = 3] = \Pr[s = s_1 \mid N = 3] + \Pr[s = s_2 \mid N = 3] \leqslant 0{\cdot}10.$$

Thus, proceeding stepwise in this manner it is possible to calculate the following sequence $s_1, s_2, \ldots$, which we reproduce from Boguslavsky:

i	1	2	3	4	5	6	7	8	9	10	11	12	13	14	15
s_i	5	9	14	18	23	28	33	39	44	50	55	61	67	73	79

In general one would use this method for small populations only, as s is much greater than N. This means that the above table ($N \leqslant 15$, $\alpha = 0{\cdot}10$) is probably satisfactory for most practical purposes. To illustrate the use of the table, suppose that unmarked animals are caught at samples 1, 3, 6, 11, 17 and no new individuals are captured from samples 18 to 23 inclusive. Then our sampling would stop at $s = s_5$, and we would accept $N = 5$ with 90 per cent confidence.

Darling and Robbins [1967] also derive a stopping rule for which (4.45) is true, but use a different sequence $b_1, b_2, \ldots$, where b_i is the longest "waiting time" allowed between the appearances of the ith and $(i + 1)$th unmarked individuals. Suppose the ith unmarked animal appears on the (Y_i)th sample ($Y_1 = 1$, $Y_{N+1} = \infty$) and let $X_i = Y_{i+1} - Y_i$ ($i = 1, 2, \ldots, N$) be the ith "waiting time". Then, if the stopping rule is such that one stops sampling the first time the event $X_i > b_i$ occurs for some i, it transpires that

$$\Pr[r = N] = \prod_{i=1}^{N-1} \{1 - (i/N)^{b_i}\}. \tag{4.46}$$

Darling and Robbins suggest choosing the sequence (b_i) in such a way that the right-hand side of (4.46) is greater than or equal to α for $N = 2, 3, \ldots$. (i.e. essentially Boguslavsky's stepwise approach). Their solution is the sequence (c_i), where $c_1, c_2, c_3, \ldots$ are the smallest integers satisfying

$$1 - \left(\frac{1}{2}\right)^{c_1} \geqslant 1 - \alpha, \quad \left[1 - \left(\frac{1}{3}\right)^{c_1}\right]\left[1 - \left(\frac{2}{3}\right)^{c_2}\right] \geqslant 1 - \alpha,$$

$$\left[1 - \left(\frac{1}{4}\right)^{c_1}\right]\left[1 - \left(\frac{2}{4}\right)^{c_2}\right]\left[1 - \left(\frac{3}{4}\right)^{c_3}\right] \geqslant 1 - \alpha, \text{ etc.}$$

Therefore, if $\alpha = 0{\cdot}10$, we have, using logarithms, $c_1 = 4$, $c_2 = 9$, etc., so that if, for example, the second unmarked animal does not turn up after $1 + 4 = 5$ samples, then $X_1 > 4$ and we stop sampling. If, however, the second unmarked animal turns up at sample 3, then we stop sampling if a third unmarked animal has not been captured by the $3 + 9 = 12$th sample.

Darling and Robbins also consider a stopping rule based on the "cumulative waiting times" Y_i, but this turns out to be the same as that given by Boguslavsky above. They also suggest an asymptotic solution to the problem of finding a suitable sequence (s_i), namely (cf. Samuel [1968: rule D with $c = D + 1$])

$$s_i \geqslant \max\,(i+1,\ i \log i + ci), \quad -\infty < c < \infty,$$

where $\lim \Pr[r = N] = \exp\,(-\exp(-c)) \geqslant 1 - \alpha, \quad N \to \infty.$

But this stopping rule needs further investigation for small and moderate N as s may be impracticably large for large N: an upper bound for s is $N \log N + cN$.

One further stopping rule is considered in some detail by Samuel [1969: rule C]. Let $C > 0$ be a fixed constant, then one samples until the ratio of the number of samples which are marked $(s - r)$ to the number (r) which are unmarked is at least C, i.e. until $(s-r)/s \geqslant C$. This rule is investigated numerically by Samuel for different values of C, and the reader is referred to her article for further details.

Samuel also shows that for all the stopping rules considered in this section, $\hat{N}$, the solution of (4.10), is still the maximum-likelihood estimate of N.

4.4 THE MULTI-SAMPLE SINGLE-RECAPTURE CENSUS

We shall now consider a census technique which is specially suited to commercially exploited populations such as fisheries, where the samples are permanently removed from the population. The method is as follows. The experimenter, using differentiated marking, releases batches of marked individuals containing $R_1, R_2, \ldots, R_s$ members respectively into the population, and after each batch R_i is released, a sample of size n_i is removed permanently from the population, thus giving the sequence R_1 added, n_1 removed, R_2 added, n_2 removed, etc. Ideally the marked individuals should either be caught before the experiment and stored, or perhaps taken from a similar population not connected with the one under investigation. In actual practice, however, the experimenter will usually take each batch from the population during the experiment because the R_i, although large, will generally be much smaller than the n_i, so that the recaptures in the successive batches will be negligible. Also the overall reduction in the number of unmarked individuals due to the marking of $\sum R_i$ individuals will be small compared with the total population size. Let

N = initial population size,

m_j = number of marked in n_j $(j = 1, 2, \ldots, s)$,

m_{ij} = the number of individuals from R_i caught in n_j ($j = i, i+1, \ldots, s$),

$u_j = n_j - m_j$,

and $r = \sum_{j=1}^{s} u_j$.

Then, regarding the $\{n_j\}$ as fixed parameters and assuming that each sample is a simple random sample, the joint probability function of $\{m_{ij}, u_j\}$ is the multi-hypergeometric distribution (cf. Seber [1962: 347])

$$f(\{m_{ij}, u_j\} \mid \{R_i, n_j\}) = \prod_{i=1}^{s} \left\{ \frac{\binom{R_1 - \sum_{j=1}^{i-1} m_{1j}}{m_{1i}} \binom{R_2 - \sum_{j=2}^{i-1} m_{2j}}{m_{2i}} \cdots \binom{R_i}{m_{ii}} \binom{N - \sum_{j=1}^{i-1} u_j}{u_i}}{\binom{N + \sum_{j=1}^{i} R_j - \sum_{j=1}^{i-1} n_j}{n_i}} \right\}$$

Setting $\Delta \log f = 0$, the maximum-likelihood estimate $\hat{N}$ of N is the unique root, greater than r, of $h(N) = 0$, where

$$h(N) = N - r - N \prod_{i=1}^{s} \left(1 - \frac{n_i}{N + \sum_{j=1}^{i} R_j - \sum_{j=1}^{i-1} n_j}\right).$$

Seber shows that $\hat{N}$ is asymptotically unbiased and that the asymptotic variance of $\hat{N}$ is estimated by

$$v[\hat{N}] = (\hat{N} - r)/h'(\hat{N})$$

$$= \left[\frac{r}{\hat{N}(\hat{N} - r)} - \sum_{i=1}^{s} \frac{n_i}{\left(\hat{N} + \sum_{j=1}^{i} R_j - \sum_{j=1}^{i} n_j\right)\left(\hat{N} + \sum_{j=1}^{i} R_i - \sum_{j=1}^{i-1} n_j\right)}\right]^{-1}$$

As in **4.1.2** (*1*), $h(N) = 0$ can be solved using Robson and Regier's iterative method, and once again the last iteration provides an approximation for $v[\hat{N}]$. **By setting** $D_i = \sum_{j=1}^{i-1} n_j - \sum_{j=1}^{i} R_j$, **a first approximation to** $\hat{N}$ **is given by** (4.20) on page 154: alternatively one can use the mean estimate (4.21) with $M_i = \sum_{j=1}^{i} R_j - \sum_{j=1}^{i-1} m_j$ in N_i^*.

For the more realistic situation where the mark releases R_i are obtained from the population during the course of the experiment, we have

$$f(\{m_{ij}, u_j\} \mid \{R_i, n_j\}) = \prod_{i=1}^{s} \left\{ \frac{\binom{R_1 - \sum_{j=1}^{i-1} m_{1j}}{m_{1i}} \cdots \binom{R_i}{m_{ii}} \binom{N - \sum_{j=1}^{i} R_j - \sum_{j=1}^{i-1} u_j}{u_i}}{\binom{N - \sum_{j=1}^{i-1} n_j}{n_i}} \right\}$$

In this case R_i refers to the *newly* marked individuals released; any recaptures are not retagged but simply returned to the population. The maximum-likelihood estimate $\tilde{N}$ is the largest root of

$$\prod_{i=1}^{s}\left\{\frac{N-\sum_{j=1}^{i}R_j-\sum_{j=1}^{i}u_j}{N-\sum_{j=1}^{i}R_j-\sum_{j=1}^{i-1}u_j}\right\}=\prod_{i=1}^{s}\left\{\frac{N-\sum_{j=1}^{i}n_j}{N-\sum_{j=1}^{i-1}n_j}\right\}$$

and we note that this equation has the same simple interpretation as (4.4).

In actual practice the $\{n_j\}$ will be random variables, and to set up a probability model for this situation we make the usual assumption that p_i $(= 1 - q_i)$, the probability of capture of a given animal in the ith sample, given that the animal is in the population at the time of the sample, is the same for all animals, irrespective of whether they are marked or unmarked. Under this assumption the random variables $\{m_{ij}\}$ and $\{u_j\}$ are independent and

$$f(\{m_{ij}\}\mid\{R_i\})$$

$$=\prod_{i=1}^{s}\left\{\frac{R_i!}{\prod_{j=i}^{s}m_{ij}!\;(R_i-r_i)!}\,p_i^{m_{ii}}\,(q_ip_{i+1})^{m_{i,i+1}}\cdots(q_iq_{i+1}\cdots q_s)^{R_i-r_i}\right\}, \tag{4.47}$$

where $r_i = \sum_{j=1}^{s} m_{ij}$. As in the Schnabel census of **4.1.2**(*1*), $\hat{N}$ and $v[\hat{N}]$ are the same, irrespective of whether the n_i are regarded as fixed parameters or random variables. However, (4.47) can be used to provide a goodness-of-fit test to the hypothesis H that all *marked* individuals have the same probability p_i of being caught in the ith sample, as follows. Let p_{ij} be the probability that a member of R_i is caught in n_j, then H is the hypothesis that $p_{ij} = q_i \ldots q_{j-1}p_j$ $(j > i)$ and $p_{ii} = p_i$. This can be tested using (cf. Mitra [1958])

$$T=\sum_{i=1}^{s}\sum_{j=i}^{s+1}\frac{(m_{ij}-R_i\hat{p}_{ij})^2}{R_i\hat{p}_{ij}},$$

where $\hat{p}_{ij} = \hat{q}_i \ldots \hat{q}_{j-1}\hat{p}_j$, $\quad \hat{p}_{i,s+1} = \hat{q}_i\hat{q}_{i+1}\cdots\hat{q}_s$,

$$\hat{p}_i = m_i\Big/\left(\sum_{j=1}^{i}R_j-\sum_{j=1}^{i-1}m_j\right) \quad\text{and}\quad m_{i,s+1} = R_i - r_i .$$

When H is true and the R_i are large, T is approximately distributed as chi-squared with $\frac{1}{2}s(s+1) - s$ degrees of freedom.

CHAPTER 5

OPEN POPULATION: MARK RELEASES DURING SAMPLING PERIOD[*]

5.1 THE JOLLY–SEBER METHOD

5.1.1 The Schnabel census

In this chapter we shall apply the Schnabel method of **4.1.1** to an open population in which there is possibly death, recruitment, immigration and *permanent* emigration (i.e. animals enter and leave the population only once). We shall again consider a sequence of s samples of sizes $n_1, n_2, \ldots, n_s$ but with the added generalisation that, for each sample, only R_i of the n_i are marked and returned to the population. This more general model allows for accidental deaths due to marking and handling, and also includes the case when some of the R_i are zero, as in commercial exploitation where the sample is permanently removed from the population. The following discussion of this model is based mainly on Jolly [1965] and Seber [1965].

1 Assumptions and notation

We shall assume the following:

(a) Every animal in the population, whether marked or unmarked, has the same probability p_i $(= 1 - q_i)$ of being caught in the ith sample, given that it is alive and in the population when the sample is taken.
(b) Every marked animal has the same probability ϕ_i of surviving from the ith to the $(i+1)$th sample and of being in the population at the time of the $i+1$ sample, given that it is alive and in the population immediately after the ith release $(i = 1, 2, \ldots, s-1)$.
(c) Every animal caught in the ith sample has the same probability ν_i of being returned to the population: in many experiments $1 - \nu_i$ can be regarded as the probability of accidental death through handling, etc.
(d) Marked animals do not lose their marks and all marks are reported on recovery.
(e) All samples are instantaneous, i.e. sampling time is negligible, and each release is made immediately after the sample. Let

t_i = time when the ith sample is taken,
N_i = total number in the population just before time t_i,
M_i = total number of marked animals in the population just before time t_i,
U_i = $N_i - M_i$,
n_i = number caught in the ith sample,

---

[*]This chapter should be read in conjunction with **13.1**.

m_i = number of marked animals caught in the ith sample,

u_i = $n_i - m_i$,

m_{hi} = number caught in the ith sample last captured in the hth $(1 \leqslant h \leqslant i-1)$,

R_i = number of marked animals released after the ith sample,

r_i = number of marked animals from the release of R_i animals which are subsequently recaptured.

z_i = number of different animals caught before the ith sample which are not caught in the ith sample but are caught subsequently,

B_i = number of *new* animals joining the population in the interval from time t_i to time t_{i+1} which are still alive and in the population at time t_{i+1},

$N_i(h)$ = number in the population at time t_i which first joined the population between times t_h and t_{h+1}, that is, which are members of B_h $(1 \leqslant h \leqslant i-1)$,

ρ_i = M_i/N_i.

(The above notation is different from Jolly's in that his symbols N_{i0}, n_{i0}, n_{ji}, R_i, s_i, Z_i, α_i and η_i are replaced by U_i, u_i, m_{ij}, r_i, R_i, z_i, ρ_i and ν_i respectively.) Our symbols can be categorised as follows:

Unknown fixed parameters: p_i, ϕ_i, ν_i; for mathematical convenience we also define U_i to be a fixed parameter. We shall make use of the intermediary parameters

$$\alpha_i = \phi_i q_{i+1}, \quad \beta_i = \phi_i p_{i+1} \quad \text{and}$$

$$\begin{aligned}\chi_i &= (1-\phi_i) + \phi_i q_{i+1}(1-\phi_{i+1}) + \dots + \phi_i q_{i+1} \dots \phi_{s-1} q_s \\ &= 1 - \phi_i p_{i+1} - \phi_i q_{i+1} \phi_{i+1} p_{i+2} - \dots - \phi_i q_{i+1} \dots \phi_{s-2} q_{s-1} \phi_{s-1} p_s \\ &= 1 - \beta_i - \alpha_i \beta_{i+1} - \dots - \alpha_i \alpha_{i+1} \dots \alpha_{s-1} \beta_s, \end{aligned} \tag{5.1}$$

the last expression being the conditional probability that a marked animal released in the ith release of R_i animals is not caught again.

Known random variables: n_i, m_i, u_i $(u_1 = n_1)$, m_{hi}, R_i, r_i and z_i; we define $m_1 = r_s = z_1 = z_s = 0$. The random variables m_i, r_i and z_i are all functions of the m_{hi} and their calculation is discussed later.

Unknown random variables: N_i, M_i, B_i and ρ_i; we define $M_1 = 0$ and $B_0 = N_1$.

2 *Estimation*

PROBABILITY MODEL. To demonstrate how the probability model for the Schnabel census is set up, we shall consider first of all the simple case of $s = 3$. Using the notation of **4.1.1**, let w be a non-empty subset of the set of integers $\{1, 2, \dots, s\}$ and let a_w be the number of marked animals with capture history w. Suppose that on the last occasion when the group of a_w animals is captured, d_w are not returned to the population. Let $b_w = a_w - d_w$, then the conditional distribution of the random variables $\{b_w, d_w\}$ conditional on the $\{u_i\}$ is given by

$$f(\{b_w, d_w\} \mid \{u_i\})$$

$$= \left\{ \frac{u_1!}{\prod\limits_{w \supset 1} \{b_w!\, d_w!\}} (\nu_1 \chi_1)^{b_1} (\nu_1 \phi_1 p_2 \nu_2 \chi_2)^{b_{12}} (\nu_1 \phi_1 q_2 \phi_2 p_3 \nu_3)^{b_{13}} \right.$$

$$\times (\nu_1 \phi_1 p_2 \nu_2 \phi_2 p_3 \nu_3)^{b_{123}} (1-\nu_1)^{d_1} (\nu_1 \phi_1 p_2 (1-\nu_2))^{d_{12}}$$

$$\left. \times (\nu_1 \phi_1 q_2 \phi_2 p_3 (1-\nu_3))^{d_{13}} (\nu_1 \phi_1 p_2 \nu_2 \phi_2 p_3 (1-\nu_3))^{d_{123}} \right\}$$

$$\times \left\{ \frac{u_2!}{b_2! b_{23}! d_2! d_{23}!} (\nu_2 \chi_2)^{b_2} (\nu_2 \phi_2 p_3 \nu_3)^{b_{23}} (1-\nu_2)^{d_2} (\nu_2 \phi_2 p_3 (1-\nu_3))^{d_{23}} \right\}$$

$$\times \left\{ \frac{u_3!}{b_3! d_3!} \nu_3^{b_3} (1-\nu_3)^{d_3} \right\}. \tag{5.2}$$

For the general case of s samples it can be shown by induction (e.g. Seber [1965: 252]) that

$$f(\{b_w, d_w\} \mid \{u_i\}) = \frac{\prod\limits_{i=1}^{s} u_i!}{\prod\limits_{w} \{b_w!\, d_w!\}} \prod_{i=1}^{s-1} \left\{ \chi_i^{R_i - r_i}\ \alpha_i^{z_i+1}\ \beta_i^{m_i+1} \right\}$$

$$\times \prod_{i=1}^{s} \left\{ \nu_i^{R_i} (1-\nu_i)^{n_i - R_i} \right\} \tag{5.3}$$

$$= L_1 \times L_2, \text{ say.}$$

Since we have binomial sampling,

$$f(\{u_i\}) = \prod_{i=1}^{s} \left\{ \binom{U_i}{u_i} p_i^{u_i} q_i^{U_i - u_i} \right\} \tag{5.4}$$

and we finally have the unconditional distribution

$$f(\{b_w, d_w\}) = f(\{b_w, d_w\} \mid \{u_i\})\ f(\{u_i\}). \tag{5.5}$$

(This distribution was obtained indirectly by Jolly [1965: 234] using a different method.)

As mentioned above, m_i, r_i and z_i are functions of the m_{hi}, so that all the information about the fixed parameters is contained in the set of sufficient statistics $\{u_i, m_{hi}, R_i\}$. We also note that the marked population, represented by (5.3), supplies information on the $\{\phi_i, p_i\}$ via the parameters $\{\alpha_i, \beta_i\}$, and assuming that p_i is the same for marked and unmarked, the $\{u_i\}$ supply information on the parameters $\{U_i\}$.

Since (5.3) factorises into two components L_1 and L_2, where L_1 does not contain the parameters $\{\nu_i\}$, the maximum-likelihood estimates of the α_i and β_i will be unchanged if some of the ν_i are put equal to unity. On the other hand, if $\nu_j = 0$, so that there is a 100 per cent removal on the jth sample ($R_j = r_j = 0$), we find that not all the parameters can be estimated.

In particular, we find that α_{j-1}, α_j and β_j are no longer identifiable, only the products $\alpha_j' = \alpha_{j-1}\alpha_j$ and $\beta_j' = \alpha_{j-1}\beta_j$ can be estimated. To see this we note the following:

$$\begin{aligned}\chi_k &= 1 - \beta_k - \alpha_k\beta_{k+1} - \ldots - \alpha_k \ldots \alpha_{j-2}(\alpha_{j-1}\beta_j) - \\ &\qquad - \alpha_k \ldots \alpha_{j-2}(\alpha_{j-1}\alpha_j)\beta_{j+1} - \ldots\end{aligned}$$

for $k < j$, χ_j does not appear in (5.3), and using the relationship $z_{j+1} + m_{j+1} = z_j + r_j$ with $r_j = 0$ we have

$$\alpha_{j-1}^{z_j}\, \beta_{j-1}^{m_j}\, \alpha_j^{z_{j+1}}\, \beta_j^{m_{j+1}} = (\alpha_{j-1}\alpha_j)^{z_{j+1}}\, (\alpha_{j-1}\beta_j)^{m_{j+1}}\, \beta_{j-1}^{m_j}.$$

From the estimates of α_j' and β_j' we can then estimate p_{j+1} $(= \beta_j'/(\alpha_j' + \beta_j'))$ and $\phi_{j-1}q_j\phi_j$ $(= \alpha_j' + \beta_j')$, the probability of survival from immediately after time t_{j-1} to time t_{j+1}.

If two releases are zero, say $R_j = R_{j+1} = 0$, then it transpires that the parameters α_{j-1}, α_j, α_{j+1}, β_j and β_{j+1} are no longer identifiable, only the products β_j', $\alpha_{j+1}' = \alpha_{j-1}\alpha_j\alpha_{j+1}$ and $\beta_{j+1}' = \alpha_{j-1}\alpha_j\beta_{j+1}$ can be estimated. From the estimates of α_{j+1}' and β_{j+1}' we can then obtain estimates of p_{j+2} and $\phi_{j-1}q_j\phi_j q_{j+1}\phi_{j+1}$, the probability of survival from immediately after time t_{j-1} to time t_{j+2}. The question of zero releases is discussed further on p. 210.

Finally it should be stressed that (5.2)–(5.5) can only be applied to a population in which there is no emigration, or the emigration is "permanent" in the sense that if a marked animal moves outside the population it remains outside for the remainder of the experiment. This requirement is clearly seen by examining the probabilities associated with the b_w in (5.2). For example, the class of b_{13} individuals has a probability $\nu_1\phi_1 q_2\phi_2 p_3\nu_3\chi_3$ which only applies if the members of this class are actually in the population at time t_2 and are not temporarily outside the sampling area during the second sample.

MAXIMUM-LIKELIHOOD ESTIMATES. By differentiating the logarithm of (5.5), maximum-likelihood estimates of the unknown parameters can be derived in the usual fashion (cf. Seber [1965] for the special case of no losses on capture, i.e. $\nu_i = 1$). Also, simple moment type estimates of the unknown random variables N_i, M_i and B_i can then be obtained from the relations:

$$\begin{aligned}E[N_{i+1} \mid N_i, B_i] &= B_i + \phi_i(N_i - n_i + R_i), \\ E[n_i \mid N_i] &= N_i p_i,\end{aligned}$$

and

$$E[m_i \mid M_i] = M_i p_i.$$

However, the same set of estimates for the unknown parameters and random variables can also be obtained by an intuitive argument which we reproduce from Jolly [1965].

The number of marked animals M_i can be thought of as consisting of the number captured, m_i, plus the number $M_i - m_i$ not captured in the ith sample. Immediately after the ith sample there are two groups of marked animals, the

$M_i - m_i$, of which z_i are subsequently caught, and the R_i just released, of which r_i are subsequently caught. Since the chances of recapture are assumed to be the same for both groups, we would expect

$$\frac{z_i}{M_i - m_i} \approx \frac{r_i}{R_i}, \tag{5.6}$$

which leads to the estimate

$$\hat{M}_i = \frac{R_i z_i}{r_i} + m_i, \quad (i = 2, 3, \ldots, s-1). \tag{5.7}$$

Also the proportion of marked in the sample will represent the proportion of marked in the population, i.e.

$$\frac{m_i}{n_i} = \frac{\hat{M}_i}{\hat{N}_i}$$

or

$$\hat{N}_i = \hat{M}_i n_i / m_i, \quad (i = 2, 3, \ldots, s-1). \tag{5.8}$$

A natural estimate of ϕ_i would be the ratio of the marked animals alive at time t_{i+1} to the marked animals in the population just after the ith release, namely

$$\hat{\phi}_i = \frac{\hat{M}_{i+1}}{\hat{M}_i - m_i + R_i} \quad (i = 2, 3, \ldots, s-2) \tag{5.9}$$

and

$$\hat{\phi}_1 = \frac{\hat{M}_2}{R_1}. \tag{5.10}$$

Intuitive estimates of p_i, ν_i, χ_i, ρ_i, B_i and U_i are then given by

$$\hat{p}_i = \frac{n_i}{\hat{N}_i} = \frac{m_i}{\hat{M}_i}, \quad (i = 2, 3, \ldots, s-1), \tag{5.11}$$

$$\hat{\nu}_i = \frac{R_i}{n_i}, \quad (i = 1, 2, \ldots, s), \; \hat{\chi}_i = 1 - \frac{r_i}{R_i},$$

$$\hat{\rho}_i = \frac{\hat{M}_i}{\hat{N}_i} = \frac{m_i}{n_i}, \quad (i = 2, 3, \ldots, s),$$

$$\hat{B}_i = \hat{N}_{i+1} - \hat{\phi}_i(\hat{N}_i - n_i + R_i), \quad (i = 2, 3, \ldots, s-2), \tag{5.12}$$

and

$$\hat{U}_i = \hat{N}_i - \hat{M}_i.$$

It should be noted that $\hat{M}_i$ and $\hat{N}_i$ are not maximum-likelihood estimates but are simply used as intermediate steps in the calculation of the maximum-likelihood estimates $\hat{\phi}_i$, $\hat{p}_i$ and $\hat{U}_i$. Also the estimate of B_i is only valid if we replace assumption (b) by the stronger assumption that ϕ_i is the same for all individuals and not just for the marked.

To calculate the above estimates we require n_i, m_i, R_i, r_i and z_i. In particular, if

$$c_{ij} = \sum_{h=1}^{i} m_{hj} \quad (i < j),$$

the number in the jth sample last caught in the ith or before, then

$$m_i = c_{i-1,\,i},$$

$$r_i = \sum_{j=i+1}^{s} m_{ij},$$

and

$$z_i = \sum_{j=i+1}^{s} c_{i-1,\,j}$$

can be calculated from arrays of the m_{hi} and c_{hi} (cf. Example 5.1 below). The z_i can also be calculated iteratively from the relations

$$z_{i+1} = z_i + r_i - m_{i+1},$$

starting with $z_1 = 0$.

This method of grouping the recaptures according to when they were last caught (m_{hi}) is often known as Method B (Leslie and Chitty [1951]). In setting out the array of m_{hi} (cf. Table 5.1, p. 206), we shall use Leslie and Chitty's method of display (as it also applies to bird banding, p. 244) rather than that of Jolly who uses an array $\{m_{ji}\}$ i.e. the "transpose" of Table 5.1.

It is of interest to compare Leslie's estimates (cf. Leslie *et al.* [1953]) with those given above. For example, Leslie's estimate of M_i ($= \hat{\psi}_i$ in his notation) is obtained from the equation

$$\frac{c_{i-1,\,i+1}}{M_i - m_i} = \frac{m_{i,\,i+1}}{R_i},$$

i.e. by equating the proportions of the $M_i - m_i$ marked animals and the release R_i which are caught in just the $(i+1)$th sample (and not in all future samples as in equation (5.6)). Leslie's estimates of N_i and ϕ_i then follow from the above estimate of M_i in the same way that $\hat{N}_i$ and $\hat{\phi}_i$ follow from the $\hat{M}_i$.

One or two comments on the range of the suffix i in the above estimates are appropriate at this point. Since $M_1 = 0$ (i.e. no marked animals in the population before the first sample), N_1, p_1 and B_1 cannot be estimated from (5.8), (5.11) and (5.12). However, from the trend in $\hat{N}_2, \ldots, \hat{N}_{s-1}$ or the lack of trend (e.g. $N_1' = \hat{N}_2$) we can estimate p_1 and B_1 by

$$p_1' = n_1/N_1'$$

and

$$B_1' = \hat{N}_2 - \hat{\phi}_1(N_1' - n_1 + R_1).$$

Also z_s and r_s are both zero, so that M_s, and consequently N_s, p_s, ϕ_{s-1} and B_{s-1}, cannot be estimated. The product $\beta_{s-1} = \phi_{s-1}p_s$ can be estimated by[*]

$$\hat{\beta}_{s-1} = 1 - \hat{\chi}_{s-1} = r_{s-1}/R_{s-1},$$

[*] This unbiased estimate was pointed out to me by Stuart Crosbie.

and if an estimate of p_s is available (e.g. $p'_s = \hat{p}_i$, where the samples numbered s and i are taken under similar conditions and with the same effort), then ϕ_{s-1} can be estimated by

$$\phi'_{s-1} = \hat{\beta}_{s-1}/p'_s. \tag{5.13}$$

This leads to the estimates

$$N'_s = n_s/p'_s$$

and

$$B'_{s-1} = N'_s - \phi'_{s-1}(\hat{N}_{s-1} - n_{s-1} + R_{s-1}).$$

Alternatively, if ϕ_{s-1} can be estimated from the trend (or lack of trend) in $\hat{\phi}_1, \hat{\phi}_2, \ldots, \hat{\phi}_{s-2}$, then p_s can be estimated from (5.13). Finally we note that ϕ_s and B_s are undefined.

VARIANCES AND COVARIANCES. Every animal in the population belongs to just one of the mutually exclusive groups of B_i new animals (including $B_0 = N_1$). Also each of these groups can be split up into various multinomial classes which contribute to the random variables used in the above estimates. Therefore, treating the $\{B_i\}$ as fixed parameters, and utilising the fact that the multinomial distributions arising from different B_i are mutually independent, Jolly uses the delta method to derive asymptotic expressions for the variances and covariances of the estimates. His formulae reduce to the following when the expectations (conditional on $\{B_i\}$) of the random variables N_i, M_i, R_i, etc. are replaced by their "observed" values:

$$V[\hat{N}_i] = N_i(N_i - n_i)\left\{\frac{M_i - m_i + R_i}{M_i}\left(\frac{1}{r_i} - \frac{1}{R_i}\right) + \frac{1-\rho_i}{m_i}\right\} + \\ + N_i - \sum_{h=0}^{i-1}\frac{N_i^2(h)}{B_h}, \quad (i = 2, 3, \ldots, s-1), \tag{5.14}$$

$$\text{cov}[\hat{N}_i, \hat{N}_j] = \sum_{h=0}^{i-1}\left\{N_j(h) - \frac{N_i(h)N_j(h)}{B_h}\right\} \quad (i < j),$$

$$V[\hat{\phi}_i] = \phi_i^2\left\{\frac{(M_{i+1} - m_{i+1})(M_{i+1} - m_{i+1} + R_{i+1})}{M_{i+1}^2}\left(\frac{1}{r_{i+1}} - \frac{1}{R_{i+1}}\right) + \\ + \frac{M_i - m_i}{M_i - m_i + R_i}\left(\frac{1}{r_i} - \frac{1}{R_i}\right) + \frac{1-\phi_i}{M_{i+1}}\right\}, \quad (i = 1, 2, \ldots, s-2; \; M_1 = m_1 = 0), \tag{5.15}$$

$$\text{cov}[\hat{\phi}_i, \hat{\phi}_{i+1}] = -\frac{\phi_i\phi_{i+1}(M_{i+1} - m_{i+1})}{M_{i+1}}\left(\frac{1}{r_{i+1}} - \frac{1}{R_{i+1}}\right), \tag{5.16}$$

$$\text{cov}[\hat{\phi}_i, \hat{\phi}_j] = 0 \quad (j > i + 1), \tag{5.17}$$

$$V[\hat{B}_i] = \frac{B_i^2(M_{i+1}-m_{i+1})(M_{i+1}-m_{i+1}+R_{i+1})}{M_{i+1}^2}\left(\frac{1}{r_{i+1}}-\frac{1}{R_{i+1}}\right)+$$

$$+\frac{M_i-m_i}{M_i-m_i+R_i}\left\{\frac{\phi_i R_i(1-\rho_i)}{\rho_i}\right\}^2\left(\frac{1}{r_i}-\frac{1}{R_i}\right)+$$

$$+\frac{(N_i-n_i)(N_{i+1}-B_i)(1-\rho_i)(1-\phi_i)}{M_i-m_i+R_i}+N_{i+1}(N_{i+1}-n_{i+1})\left(\frac{1-\rho_{i+1}}{m_{i+1}}\right)+$$

$$+\phi_i^2 N_i(N_i-n_i)\left(\frac{1-\rho_i}{m_i}\right), \quad (i=2,3,\ldots,s-2),$$

$$\text{cov}[\hat{B}_i, \hat{B}_{i+1}] = -\phi_{i+1}(N_{i+1}-n_{i+1})(1-\rho_{i+1})$$

$$\times\left\{\frac{B_i R_{i+1}}{M_{i+1}}\left(\frac{1}{r_{i+1}}-\frac{1}{R_{i+1}}\right)+\frac{N_{i+1}}{m_{i+1}}\right\},$$

and

$$\text{cov}[\hat{B}_i, \hat{B}_j] = 0, \quad (j > i+1).$$

The above formulae are asymptotic and are only valid for large expectations of the random variables n_i, m_i, R_i, r_i and z_i; the small sample properties of the formulae need further investigation. Estimates of the variances and covariances are obtained by replacing each unknown by its estimate.

The first expression in (5.14) represents the error of estimation $V[\hat{N}_i \mid N_i]$, while the final terms represent an approximation for

$$V[N_i] = E[N_i] - \sum_{h=0}^{i-1}\frac{\{E[N_i(h)]\}^2}{B_h}, \qquad (5.18)$$

where the expectations are conditional on the $\{B_i\}$. Jolly shows that the terms $E[N_i(h)]$ in (5.18) are most readily obtained as successive products of $(E[N_{k+1}]-B_k)/E[N_k]$ using the relations

$$N_{h+1}(h) = B_h \qquad (5.19)$$

and

$$E[N_{i+1}(h)] = \frac{E[N_{i+1}]-B_i}{E[N_i]}\cdot E[N_i(h)], \quad (i > h). \qquad (5.20)$$

These relations give all the $E[N_i(h)]$ for $h > 1$, but we run into difficulty with $N_i(0)$ and $N_i(1)$ as neither B_0 nor B_1 are estimable. However, since N_2 and ϕ_1 can be estimated, B_0 $(= N_1)$ and B_1 $(\approx N_2 - \phi_1(N_1 - n_1 + R_1))$ can also be estimated if an estimate of N_1 is available from the trend (or lack in trend) in the $\hat{N}_i$. Estimates of the $N_i(h)$ then follow from (5.19) and (5.20) using $\hat{B}_i$ and $\hat{N}_i$ to estimate $\hat{B}_i$ and $E[N_i]$ respectively. However, in general, $V[N_i]$ will usually be much smaller than $V[\hat{N}_i \mid N_i]$ and can therefore be ignored in most cases (except possibly when p_i is large); this avoids the awkward computations just outlined.

In conclusion we note from $V[\hat{\phi}_i]$ and $V[\hat{N}_i]$ that the accuracies of

$\hat{\phi}_i$ and $\hat{N}_i$ depend both on the numbers of recaptures (r_i and m_i) and on the sampling intensities p_i. The dependence on p_i is reflected in the variance terms $N_i - n_i$ and $M_i - m_i$; as n_i approaches N_i, $V[\hat{N}_i \mid N_i]$ tends to zero. The above variance formulae have been studied for robustness by various authors (cf. **13.1.2**).

BIAS. In deriving $\hat{M}_i$ and $\hat{N}_i$ (equations (5.7) and (5.8)), we used arguments which are reminiscent of the Petersen method of **3.1.1**. For this reason it is natural to consider whether we can "patch up" our estimates to reduce their bias. To determine how this can be done we note that the Petersen estimate $\hat{N}$ for a two-sample experiment satisfies

$$\frac{u_2}{\hat{N} - n_1} = \frac{m_2}{n_1},$$

while the almost unbiased modification N^* satisfies

$$\frac{u_2}{N^* - n_1} = \frac{m_2 + 1}{n_1 + 1}. \tag{5.21}$$

Therefore, on comparison with (5.6) we are led to consider

$$M_i^* = \frac{R_i + 1}{r_i + 1} \cdot z_i + m_i \tag{5.22}$$

which, using Jolly's formula for the asymptotic bias of $\hat{M}_i$ (cf. Jolly [1965: 238]) is readily shown to have a bias of smaller order than that of $\hat{M}_i$. It can also be shown that the modified estimates

$$N_i^* = \frac{n_i + 1}{m_i + 1} \cdot M_i^*,$$

$$\phi_i^* = \frac{M_{i+1}^*}{\hat{M}_i - m_i + R_i}, \quad (i = 2, 3, \ldots, s-2),$$

$$\phi_1^* = \frac{M_2^*}{R_1},$$

and

$$B_i^* = N_{i+1}^* - \phi_i^*(N_i^* - n_i + R_i)$$

are all approximately unbiased. The corrections for bias will not affect the asymptotic variances, so that asymptotically $V[N_i^*] = V[\hat{N}_i]$, etc.

It was pointed out in **3.1.1** that the Petersen estimate and its modification may be unreliable for $m_2 < 10$. Therefore it is recommended that m_i and r_i should be greater than 10 for a satisfactory application of the above method.

ALTERNATIVE PROBABILITY MODELS. A number of deterministic formulations of the above capture–recapture model have been developed since the 1930's and these are reviewed in Cormack [1968] and compared in Parr [1965].

As these approaches have been largely superseded by the simpler and more general stochastic models of this chapter they will not be discussed. However, it should be mentioned that the most general stochastic model developed so far is a hypergeometric type model set up by Robson [1969] and Pollock [1975b]. This model, although complex, allows for survival and catchability to depend on the mark status of the animal, and therefore represents a major step forward in the field of capture–recapture analysis. Unfortunately the notation is difficult and the method requires large numbers of recaptures so that the sizes of the various capture classes are not too small. The model is discussed in some detail in **13.1.6** and estimates for various cases are derived intuitively.

The above Jolly–Seber method has not yet had much time to be used in practice. However, applications are given in Jolly [1965: capsids], Parr [1965: damselflies], Sadleir [1965: deer mice], Lidicker [1966: house mice], Delong [1966: house mice], Parker [1968: pink salmon; cf. Appendix, where the method is applied to a population of dead fish in the spawning area], Parr *et al.* [1968: grasshoppers, butterflies and damselflies], and White [1971: grasshoppers]. The method can also be applied to previous recapture experiments where the m_{hi} are tabulated, that is, where recaptures are grouped according to Leslie and Chitty's [1951] method B (e.g. Orians [1958: Table 3], Turner [1960a: Table 1], Sonleitner and Bateman [1963: Table 2], Kikkawa [1964: 296]). Further examples and recent developments are described in **13.1**.

Example 5.1 Black-kneed capsid (*Blepharidopterus angulatus*): Jolly [1965]

Thirteen successive samples were taken at alternatively 3- and 4-day intervals from an apple orchard population of female black-kneed capsid, and the values of m_{hi} are recorded in Table 5.1. The steps in calculating the estimates are as follows:

(i) Sum the rows in Table 5.1 to give the values r_h (the number from the release R_h subsequently recaptured), e.g. $r_7 = 108$.

(ii) Sum each column in Table 5.1 cumulatively from top to bottom, entering the accumulated totals in the columns of Table 5.2. For example, the fourth column of Table 5.2 is 5, 5 + 18, 5 + 18 + 33.

(iii) In Table 5.2 sum each of the rows, excluding the first entry, to obtain the z_i; the first entries are the m_i, e.g. $m_7 = 112$, $z_7 = 110$. The z_i's can be checked from the relation $z_{i+1} - z_i = r_i - m_{i+1}$.

(iv) Calculate $\hat{M}_i$, $\hat{\rho}_i$, $\hat{N}_i$, $\hat{\phi}_i$ and $\hat{B}_i$ as in Table 5.3. For example:

$$\hat{M}_7 = \frac{R_7 z_7}{r_7} + m_7 = \frac{243(110)}{108} + 112 = 359{\cdot}50,$$

$$\hat{\rho}_7 = \frac{m_7}{n_7} = \frac{112}{250} = 0{\cdot}4480,$$

TABLE 5.1

Tabulation of m_{hi}, the number caught in the ith sample last captured in the hth sample, for a black-kneed capsid population: data from Jolly [1965: Table 2].

	i 1	2	3	4	5	6	7	8	9	10	11	12	13	
	n_i 54	146	169	209	220	209	250	176	172	127	123	120	142	
	R_i 54	143	164	202	214	207	243	175	169	126	120	120	–	Total
h														r_h
1		10	3	5	2	2	1	0	0	0	1	0	0	24
2			34	18	8	4	6	4	2	0	2	1	1	80
3				33	13	8	5	0	4	1	3	3	0	70
4					30	20	10	3	2	2	1	1	2	71
5						43	34	14	11	3	0	1	3	109
6							56	19	12	5	4	2	3	101
7								46	28	17	8	7	2	108
8									51	22	12	4	10	99
9										34	16	11	9	70
10											30	16	12	58
11												26	18	44
12													35	35
Total	m_i 0	10	37	56	53	77	112	86	110	84	77	72	95	

TABLE 5.2

Tabulation of c_{hi}, the number caught in the ith sample last caught in or before the hth sample, for a black-kneed capsid population: from Jolly [1965: Table 3].

h \ i	1	2	3	4	5	6	7	8	9	10	11	12	13	Total	
1		10	3	5	2	2	1	0	0	0	1	0	0	14	z_2
2			37	23	10	6	7	4	2	0	3	1	1	57	z_3
3				56	23	14	12	4	6	1	6	4	1	71	z_4
4					53	34	22	7	8	3	7	5	3	89	z_5
5						77	56	21	19	6	7	6	6	121	z_6
6							112	40	31	11	11	8	9	110	z_7
7								86	59	28	19	15	11	132	z_8
8									110	50	31	19	21	121	z_9
9										84	47	30	30	107	z_{10}
10											77	46	42	88	z_{11}
11												72	60	60	z_{12}
12													95		

TABLE 5.3
Population estimates for a black-kneed capsid population: from Jolly [1965: Table 4].

i	$\hat{\rho}_i$	$\hat{M}_i$	$\hat{N}_i$	$\hat{\phi}_i$	$\hat{B}_i$	$\hat{\sigma}[\hat{N}_i]$	$\hat{\sigma}[\hat{\phi}_i]$	$\hat{\sigma}[\hat{B}_i]$	$\hat{\sigma}[\hat{N}_i \mid N_i]$
1	–	0	–	0·649	–	–	0·114	–	–
2	0·0685	35·02	511·2	1·015	263·2	151·2	0·110	179·2	150·8
3	0·2189	170·54	779·1	0·867	291·8	129·3	0·107	137·7	128·9
4	0·2679	258·00	963·0	0·564	406·4	140·9	0·064	120·2	140·3
5	0·2409	227·73	945·3	0·836	96·9	125·5	0·075	111·4	124·3
6	0·3684	324·99	882·2	0·790	107·0	96·1	0·070	74·8	94·4
7	0·4480	359·50	802·5	0·651	135·7	74·8	0·056	55·6	72·4
8	0·4886	319·33	653·6	0·985	– 13·8	61·7	0·093	52·5	58·9
9	0·6395	402·13	628·8	0·686	49·0	61·9	0·080	34·2	59·1
10	0·6614	316·45	478·5	0·884	84·1	51·8	0·120	40·2	48·9
11	0·6260	317·00	506·4	0·771	74·5	65·8	0·128	41·1	63·7
12	0·6000	277·71	462·8	–	–	70·2	–	–	68·4
13	0·6690	–	–	–	–	–	–	–	–

$$\hat{N}_7 = \hat{M}_7/\hat{\rho}_7 = 802{\cdot}5,$$

$$\hat{\phi}_7 = \frac{\hat{M}_8}{\hat{M}_7 - m_7 + R_7} = \frac{319{\cdot}33}{359{\cdot}50 - 112 + 243} = 0{\cdot}651,$$

and

$$\hat{B}_7 = \hat{N}_8 - \hat{\phi}_7(\hat{N}_7 - n_7 + R_7) = 135{\cdot}7.$$

(v) The calculation of the variances is straightforward, though tedious if not computerised. For example,

$$\hat{V}[\hat{N}_7 \mid N_7] = \hat{N}_7(\hat{N}_7 - n_7)\left\{\frac{\hat{M}_7 - m_7 + R_7}{\hat{M}_7}\left(\frac{1}{r_7} - \frac{1}{R_7}\right) + \frac{1 - \hat{\rho}_7}{m_7}\right\}$$
$$= 5241{\cdot}6$$

$$\hat{V}[N_7] = \hat{N}_7 - \sum_{h=0}^{6} \hat{N}_7^2(h)/\hat{B}_h.$$

As mentioned in the general theory, $V[N_7]$ can only be estimated if an estimate of N_1 is available. Jolly sets $\hat{N}_1 = 500$ ($= \hat{B}_0$), so that

$$\hat{B}_1 = \hat{N}_2 - \hat{\phi}_1(\hat{N}_1 - n_1 + R_1) = 186{\cdot}7,$$

and the $\hat{N}_7(h)$ can be obtained from the recurrence formulae (5.19) and (5.20). Thus,

$$\hat{V}[N_7] = 353{\cdot}28$$

and

$$\hat{V}[\hat{N}_7] = 5241{\cdot}76 + 353{\cdot}28 = 5595{\cdot}04.$$

Comparing $\hat{\sigma}[\hat{N}_i]$ with $\hat{\sigma}[\hat{N}_i \mid N_i]$ in Table 5.3, we see that it was not worth calculating the terms $V[\hat{N}_i]$.

In conclusion we note that, apart from $\hat{N}_2$, the coefficients of variation of the $\hat{N}_i$ vary from about 10 to 16 per cent, while those for the $\hat{\phi}_i$ are approximately 10 per cent. This accuracy was achieved with a sampling intensity ($100p_i$) of approximately 30 per cent. Cormack [1968] estimates that by the end of the experiment about 60 per cent of the living population was marked, despite the influx of an estimated 1500 new individuals.

Example 5.2 Damselfly (*Ischnura elegans* (Van der Linden)): Parr [1965]

A colony of the damselfly (*Ischnura elegans*) centred on a chain of four ponds at Maryborough Farm, Pembrokeshire, was studied from 24 June to 7 July, 1964. Butterfly nets having six-foot handles were used to catch the daily sample. The sampling was random to the extent that every individual that could be caught was taken, irrespective of age (teneral or adult) and sex; the whole area of the colony being covered as far as possible. Those individuals flying over the water or resting on low emergent vegetation were more difficult to capture unharmed than those over land. This is because dragonflies tend to fly very low over the water surface and, in attempting capture with a wet heavy net, there is danger of causing damage to the delicate insect. The wings and head suspension are most likely to be damaged in this way, particularly in teneral insects. However, the insects were successfully marked if handled carefully, and observations on freshly marked and released teneral insects failed to reveal wing damage or any degree of abnormal behaviour.

The insects were marked with small spots of quick-drying cellulose paint, and the method of handling and marking the insects is described in some detail by the author. In practice it was found most convenient to capture, mark, and release the damselflies from a specific area occupied by the colony before continuing the sampling in another part. The insects were thus held captive for only a relatively short time, and the number of individuals caught more than once on a particular day was small. Individuals incapable of normal flight after capture were not released, though this seldom happened. Marking, recording and releasing were mostly carried out as soon as a batch of approximately 30 insects had been captured: this would generally be done twice or three times during each day's collecting period.

Although the sampling was not strictly random, the marked insects dispersed rapidly, particularly in fine weather, so that there appeared to be a uniform mixing of marked and unmarked. Evidence for this dispersal came from the recapture of adult specimens within minutes of being marked at distances up to 150 yards away.

The population estimates for the male population only are set out in Table 5.4.

5.1.2 General applications

Jolly has pointed out the wide applicability of the general theory, and we shall now consider some variations in the above model. Much of the

TABLE 5.4
Estimates of population parameters for a male damselfly population: from Parr [1965: Table 15].

Day	$\hat{p}_i$	$\hat{M}_i$	$\hat{N}_i$	$\hat{\phi}_i$	$\hat{B}_i$	$\hat{\sigma}[\hat{N}_i \mid N_i]$
1	–	0	–	1·0710	–	–
2	0·0545	82·5	1514·0	0·4319	– 52·4	992·5
3	–	–	–	–	–	–
4	0·0953	57·2	600·7	1·1210	–181·3	257·3
5	0·2586	126·9	490·8	0·5296	+305·7	161·0
6	0·1600	90·5	565·6	1·1990	+ 36·5	208·4
7	0·2222	158·8	714·6	0·6417	– 8·8	263·0
8	0·2771	124·4	448·6	0·8481	+171·1	104·5
9	0·2836	156·4	551·4	1·1800	– 87·9	149·4
10	0·4287	241·2	562·8	0·4960	+166·5	182·2
11	0·3085	137·5	445·6	0·9184	+226·1	138·0
12	0·2927	186·0	635·4	–	–	484·2
13	0·1935	–	–	–	–	–

following discussion (except for the examples) is based on Jolly [1965: 239–41].

1 Release and future recapture operated independently

From (5.14) we see that, given M_i, m_i and R_i, $V[\hat{N}_i \mid N_i]$ will be a minimum when the number of recaptures r_i is as large as possible. To achieve this it could be advantageous (and cheaper) to have a separate organisation for recording future recaptures from that of marking and releasing animals. Since for each time t_i it is necessary to distinguish only two classes of marked animals in the future recaptures, namely those marked before t_i and not at t_i, and those released at t_i, a very simple code of marks might be used in specific situations, thus enabling untrained persons to recapture marked animals over a wide area. Such a recapture system could proceed continuously, since the time at which an animal is recaptured or even the number of times it is recaptured after time t_i is of no importance as far as time t_i is concerned; only the fact that it *is* recaptured after t_i is relevant. Releases, on the other hand, would only be made at the particular times for which the estimates $\hat{M}_i$ were required, the catching, marking, and releasing being done by more experienced staff.

Example 5.3

Suppose $s = 10$ with samples 1, 4, 7, giving rise to marked releases by experienced staff; individual marks or tags are used. In the remaining samples 2, 3, etc. the sample is taken from the population (or just observed) by untrained staff, the marked members are noted, and the sample returned *without* any additional marking. For this situation $R_i \leqslant m_i$ for $i \neq 1, 4, 7$ with equality if there are no casualties among the marked individuals caught. The two tables of m_{hi} and c_{hi}, respectively, can then be set up and all the parameters estimated as before.

In practice, the sampling between "official" releases may be an almost continuous process, so that the marked releases on these sampling occasions are small; in fact, R_i is zero if no marked individuals are caught. This means that for these samples the estimates of N_i, ϕ_i, etc. would either have large variances (because of the term $1/r_i$ in the variance formulae) or would not be applicable ($R_i = m_i = 0$). However, with such an experimental set-up one would normally be concerned with just the periods of time t_1 to t_4, t_4 to t_7 and t_7 to t_{10}, and the recaptures between official releases would simply be used to augment r_1, r_4, r_7, z_4 and z_7. Thus, if $\phi_{(k)}$, $B_{(k)}$ $(k = 1, 2, 3)$ refer to these three intermediate periods of recapture time, then the estimates of $\phi_{(1)}$, $\phi_{(2)}$ and $B_{(2)}$ are

$$\hat{\phi}_{(1)} = \hat{M}_{(2)}/R_1,$$

$$\hat{\phi}_{(2)} = \hat{M}_{(3)}/(\hat{M}_{(2)} - m_{(2)} + R_{(2)}),$$

and

$$\hat{B}_{(2)} = \hat{N}_{(3)} - \hat{\phi}_{(2)}(\hat{N}_{(2)} - n_{(2)} + R_{(2)}),$$

where $\hat{M}_{(2)} = \hat{M}_4$, $\hat{M}_{(3)} = \hat{M}_7$, $\hat{M}_{(2)} - m_{(2)} + R_{(2)} = \hat{M}_4 - m_4 + R_4$, etc. If an estimate $\hat{N}_1$ is available from the trend, or lack of trend, in $\hat{N}_4$, $\hat{N}_7$, then $B_{(1)}$ can be estimated by

$$\hat{B}_{(1)} = \hat{N}_4 - \hat{\phi}_{(1)}(\hat{N}_1 - n_1 + R_1).$$

The variances and covariances given on p. 202 still apply here by simply replacing i by (i) in the formulae and using the above notation.

The parameter $\phi_{(2)}$ represents the probability of a marked animal surviving from immediately after time t_4 to t_7, so that

$$\phi_{(2)} \leqslant \phi_4\phi_5\phi_6$$

with equality if there are no permanent removals through samples 5 and 6. In this case of no permanent removals we have $R_5 = m_5$, $R_6 = m_6$ and

$$\hat{\phi}_{(2)} = \frac{\hat{M}_7}{\hat{M}_4 - m_4 + R_4}$$

$$= \frac{\hat{M}_5}{\hat{M}_4 - m_4 + R_4} \cdot \frac{\hat{M}_6}{\hat{M}_5} \cdot \frac{\hat{M}_7}{\hat{M}_6}$$

$$= \hat{\phi}_4\hat{\phi}_5\hat{\phi}_6.$$

If the population is commercially exploited so that the removal is 100 per cent ($R_i = 0$ for $i \neq 1, 4, 7$), then a simple tagging code representing the three releases 1, 4, 7 is sufficient. We then have

$$\phi_{(2)} = \phi_4 q_5 \phi_5 q_6 \phi_6.$$

(This case was discussed on p. 199; see also p. 212). However, if the removal is not 100 per cent, individual tags should be used, so that an animal recaptured more than once is counted only once in the r_i and z_i.

2 Non-random sampling

All the information from the Schnabel census comes essentially from three types of samples; the first of size n_i, giving information on $\rho_i = M_i/N_i$, the second, of size $z_i + r_i$, concerned with future recaptures, and the third of size R_i released into the population. Although up till now we have assumed that all sampling is strictly random, there may be particular situations in which complete random sampling is not necessary or even possible. For example, as long as the probability of capturing a member of M_i is the same as that of capturing one of N_i, then $\hat{\rho}_i = m_i/n_i$ will be a satisfactory estimate of ρ_i. The same argument also applies to z_i/r_i as an estimate of $(M_i - m_i)/R_i$ when members of $M_i - m_i$ and R_i have the same probability of being caught after the ith sample. We shall now consider four types of non-randomness.

NON-RANDOM n_i. If the released animals mix freely and randomly with the population, and if the probability of selection is independent of "mark" status, then non-random samples will still be random with respect to mark status, and $\hat{\rho}_i$ will be a satisfactory estimate of ρ_i (cf. **3.1.2**). Therefore, if the n_i are likely to be non-random then the releases R_i should be arranged so that mixing can take place as much as possible.

POPULATION NOT ALL CATCHABLE. Suppose that the animals have been marked and released from random samples, and that recapturing is to be done during the breeding season when the animals are most easily observed. Supposing that a substantial proportion are either not breeding at all or are breeding in inaccessible places away from the main breeding areas, then a sample observed in the main breeding areas may differ in many respects from a random sample. The experimenter then faces the problem of whether the locality of an animal in the breeding season could in any way be associated with its mark status. For example, if the breeding members of the population have increased their numbers since the date of marking, either by immigration or by young members not born at the time of marking having reached breeding age, then the proportion of marked animals is likely to be higher among the non-breeders than the population as a whole. The estimate $\hat{\rho}_i$ would then be biased by the exclusion of non-breeders from the sample. If, on the other hand, the marking has been sufficiently recent for no further members to have reached breeding age and if immigration was considered to affect breeders and non-breeders alike, ρ_i could be validly estimated from the non-random sample. An example of this problem of accessibility is given in Phillips and Campbell [1970].

Whether or not additions to the breeding members had taken place since marking, the ratio z_i/r_i would not be biased by the exclusion of inaccessible breeders unless place of breeding was associated with age, the r_i being in general younger than the z_i (only r_i can include animals first marked as recently as time t_i). However, since non-breeders may well be older on the average than breeders, the exclusion of non-breeders would almost certainly bias z_i/r_i. Under these circumstances Jolly suggests that one way of removing bias would be to exclude from both R_i and r_i all individuals first marked at time t_i; larger numbers would have to be captured to compensate for the resulting loss in information. The R_i would then be a random selection of the *marked* animals in the population just after time t_i, and assuming (as has been done throughout) that capture does not affect behaviour, r_i would be a random sample from R_i, and z_i a random sample from $M_i - m_i$, even if both non-breeders and distant breeders were excluded. Thus r_i/R_i would estimate $z_i/(M_i - m_i)$ and $\hat{M}_i$ would be a valid estimate of M_i.

RELEASE INDEPENDENT OF SAMPLE. So far it has been assumed that the R_i released will consist of part or all of the n_i, and generally this will be the most convenient way of obtaining R_i. However, as Jolly points out, R_i could be *any* group of animals which is known to behave similarly to the remainder of the population. In fact, the R_i could be introduced from outside the population and have nothing to do with the n_i. For example, in commercial hunting and fishing the samples or catches, n_i, are completely removed from the population by the hunters, and the releases could be made independently by the scientists immediately after each commercial catch (except of course for the first release). Although losses on capture in this case are 100 per cent, this represents no loss of information, for once an animal released at time t_i is recaptured, it has yielded all its information about time t_i.

This particular situation in which release and capture are operated separately with a 100 per cent loss on capture has been considered by Seber [1962] under the title of the "multi-sample single-recapture" census (see also **13.1.4** for a more detailed exposition). If we count the sample after the first release as sample number 2, then, in terms of the above notation, Seber's model for the marked population is given by the distribution

$$f(\{m_{ij}\} \mid \{R_i\}) = \prod_{i=1}^{s-1} \left\{ \frac{R_i!}{[\prod_{j=i+1}^{s} m_{ij}!][(R_i - r_i)!]} \beta_i^{m_{i,i+1}} (\alpha_i\beta_{i+1})^{m_{i,i+2}} \cdots (\alpha_i\alpha_{i+1} \cdots \alpha_{s-2}\beta_{s-1})^{m_{is}} \chi_i^{R_i - r_i} \right\}. \quad (5.23)$$

Since animals can only be recaptured once, m_{ij} is now the number from the ith release caught in the jth sample.

It can be shown that the above probability model simplifies to

$$\frac{\prod_{i=1}^{s-1} R_i!}{\prod_{i=1}^{s-1} \{[\prod_{j=i+1}^{s} m_{ij}!][(R_i - r_i)!]\}} \prod_{i=1}^{s-1} \{\chi_i^{R_i - r_i} \alpha_i^{z_{i+1}} \beta_i^{m_{i+1}}\} \quad (5.24)$$

which, apart from a constant term, is the same as L_1 of (5.3). Therefore, since (5.24) contains the same information about $\{p_i, \phi_i\}$ as (5.5), the maximum-likelihood estimates of p_i and ϕ_i are unchanged and are still given by (5.9), (5.10) and (5.11); this was demonstrated by Jolly [1965: 243]. In addition, if we equate Seber's symbols $\{\phi_i, p_{i-1}, \alpha_i, \beta_i, \theta_i, x_{1i}, x_{2i}, x_{3i}, x_{4i}, x_{5i}\}$ with $\{\phi_i, p_i, \beta_i, \alpha_i, 1-\chi_i, z_{i+1}, r_{i+1}, m_{i+1}, r_i, z_i + r_i\}$, it can be shown that Seber's estimate of $V[\hat{\phi}_i]$ is equivalent to (5.15). Intuitively this is to be expected since in estimating variances the expected values of random variables are estimated by the observed values, so that, asymptotically, there is no difference between treating $\{R_i\}$ as fixed or random (cf. **1.3.4**). Therefore we would also expect the asymptotic variances of the other estimates $\hat{N}_i$ and $\hat{B}_i$ to have the same estimates as before when the R_i are fixed parameters.

The above model can be extended to the case when the sampling or exploitation is continuous and is carried on for some time after the last release.(*) Suppose that for s years a release of R_i individuals ($i = 1, 2, \ldots, s$) is made at the beginning of each year, and for t years ($t \geqslant s$) the population is continuously exploited. Let

α_i = probability of an individual surviving the ith year, given that it is alive at the beginning of the year;

β_i = probability of an individual being caught and removed from the population in the ith year, given that it is alive at the beginning of the year;

and m_{ij} = number from the ith release which are removed from the population in the jth year ($i = 1, 2, \ldots, s;\ j = i, i+1, \ldots, t$).

Then it transpires that, apart from a difference in the interpretation of the parameters α_i and β_i, the probability model for the above experimental situation is identical with one discussed by Seber [1970b] and analysed in **5.4.1.** Hence, from (5.36), the joint distribution of the $\{m_{ij}\}$ is proportional to

$$\left\{\prod_{i=1}^{s-1} \alpha_i^{z_{i+1}} \beta_i^{m_i}(1-\theta_i)^{R_i - r_i}\right\}\left\{\beta_s^{m_s}(1-\theta_s)^{R_s - r_s}\right\}\left\{\gamma_{s+1}^{m_{s+1}}\gamma_{s+2}^{m_{s+2}} \ldots \gamma_t^{m_t}\right\},$$

where $\gamma_j = \alpha_s\alpha_{s+1} \ldots \alpha_{j-1}\beta_j, \quad (j = s+1, s+2, \ldots, t),$

and $\theta_i = \beta_i + \alpha_i\beta_{i+1} + \ldots + \alpha_i\alpha_{i+1} \ldots \alpha_{t-1}\beta_t.$

Equating $\hat{\phi}_i$ in **5.4.1** with $\hat{\alpha}_i$, we find that the maximum-likelihood estimate α_i is

$$\hat{\alpha}_i = \frac{z_{i+1}}{r_i + z_i} \cdot \frac{r_i}{R_i} \cdot \frac{R_{i+1}}{r_{i+1}} \quad (i = 1, 2, \ldots, s-1)$$

and from (5.37),

$$V[\hat{\alpha}_i] \approx \alpha_i^2\left\{\frac{1}{E[r_i]} + \frac{1}{E[r_{i+1}]} + \frac{1}{E[z_{i+1}]} - \frac{1}{R_i} - \frac{1}{R_{i+1}} - \frac{1}{E[r_i + z_i]}\right\},$$

$$\operatorname{cov}[\hat{\alpha}_i, \hat{\alpha}_j] \approx 0, \quad j > i+1,$$

and

$$\operatorname{cov}[\hat{\alpha}_i, \hat{\alpha}_{i+1}] \approx -\alpha_i\alpha_{i+1}\left\{\frac{1}{E[r_{i+1}]} - \frac{1}{R_{i+1}}\right\}.$$

(*)See also **13.1.4** (4).

Here z_i is once again the number of tags recovered after the ith release from releases prior to the ith, r_i is the number of tags recovered from the ith release and m_j is the number of tags recovered in year j. Although r_i and z_i now have slightly different algebraic representations in terms of sums of the m_{ij} (e.g. $r_i = \sum_{j=i}^{t} m_{ij}$ instead of $r_i = \sum_{j=i+1}^{s} m_{ij}$, as the year's catch after the first release is treated as sample number 1 and not sample number 2 as in the general theory of **5.1.1**), the tabular methods described in Example 5.1 (p. 205) for calculating r_i and z_i still apply (cf. Example 5.11 on p. 248).

Robson [1963] has also considered the above model for the case $s = t$, and although he uses Poisson approximations to the multinomial distributions his model contains the same "information". Therefore, equating Robson's symbols $\{M_i, R_i, T_i\}$ with our symbols $\{R_i, r_i, z_i + r_i\}$, we are not surprised to find that Robson's estimate of survival, $\hat{S}$, is the same as $\hat{\alpha}_i$. Also, neglecting R_i^{-1} and R_{i+1}^{-1}, we find that $V[\hat{\alpha}_i]$ and cov $[\hat{\alpha}_i, \hat{\alpha}_j]$ reduce to Robson's formulae, as expected.

RELEASE IS A NON-RANDOM SUB-SAMPLE OF SAMPLE CAPTURED. In some situations (e.g. Cormack [1964]) it is not possible or practicable to obtain a random selection R_i from the sample n_i, and the question arises as to how much of the above general theory can still be used. Jolly points out that since M_i is estimated solely from counts of marked animals, we can imagine the existence of a subpopulation of which the R_i *are* a random sample and for which the $\hat{M}_i$ and $\hat{\phi}_i$ are satisfactory estimates. Whether $\hat{\phi}_i$ can now be applied to the parent population is a matter of conjecture and depends on further knowledge of the population. As far as further estimation is concerned, if members of the subpopulation have a different probability of being captured from those of the parent population, then we can go no further. However, if the probabilities are the same then we can estimate ρ_i and hence N_i, the size of the parent population – irrespective of whether or not the survival probabilities are the same for the two populations. Finally, Jolly notes that B_i cannot be estimated unless both populations have the same probability of capture and the same survival probability ϕ_i.

Example 5.4 Fulmar petrel (*Fulmarus glacialis*): Cormack [1964]

Cormack applied the above method to the study of a colony of over 100 breeding pairs of fulmars on a small island in Orkney. Each bird captured there was marked *individually* by a set of coloured leg bands which were clearly visible in flight. Sampling was carried out in successive years, and each sample, except for the first, consisted of two parts: the banded birds (m_i) which were simply observed and not recaptured, and the u_i unbanded birds which were caught on their nests, banded and released.

In carrying out an analysis of the data it was assumed that:

(i) All banded birds alive at the time of the ith sample have the same probability p_i of being "captured" (that is, seen) in the ith sample.
(ii) The capture and banding of a bird does not alter its expectation of life.

(iii) The probability ϕ_i of a banded bird surviving from the ith sample to the $(i+1)$th is independent of the age of the bird.

Since the banded members of the sample were simply observed in flight and the unbanded members captured in their nests, the usual assumption that p_i is the same for banded and unbanded did not hold. However, in support of assumption (i) there was a strong tendency for birds to use the same nest site each year, and even the inaccessibility of a nest did not affect the probability of a banded bird being sighted and identified in flight around the island. Assumption (ii) is not unreasonable as, in contrast to the usual multiple-recapture experiments, the birds were handled only once. Although a goodness-of-fit test seemed to indicate that the probability model based on the above assumptions was adequate, assumption (iii) was open to question as there were insufficient data to test this assumption.

Using the above assumptions, Cormack considered just the banded population and assumed that the numbers u_i of newly banded birds released into the population were not random variables but fixed parameters. Theoretically this amounts to dealing with just the conditional density function (5.3) which, since $R_i = n_i$, is given by L_1 (with $d_w = 0$). In fact, equating $\{r_i, R_i, r_i + z_i, m_i, R_i - m_i, R_i - r_i\}$ with Cormack's parameters $\{t_i, s_i, v_i, a_i, b_i, c_i\}$, we find that L_1 is proportional to Cormack's likelihood function [1964: 431] as expected. Now L_1 contains the same information about $\{p_i, \phi_i\}$ as the whole likelihood (5.5), so that the maximum-likelihood estimates of ϕ_i and p_i are unchanged. Hence

$$\hat{M}_i = \frac{R_i z_i}{r_i} + m_i, \quad (i = 2, 3, \ldots, s-1),$$

$$\hat{\phi}_i = \hat{M}_{i+1}/(\hat{M}_i - m_i + R_i), \quad (i = 1, 2, \ldots, s-2;\ m_1 = \hat{M}_1 = 0),$$

and $$\hat{p}_i = m_i/\hat{M}_i, \quad (i = 2, 3, \ldots, s-1).$$

But the variance formulae for the $\{\hat{\phi}_i\}$ will not be the same as those given by Jolly, since we want variances conditional on the $\{u_i\}$. Using the delta method, Cormack showed that asymptotically

$$V[\hat{\phi}_i \mid \{u_i\}]$$

$$= \phi_i^2 \left\{ \frac{\chi_{i+1}(1-p_{i+1})^2}{(1-\chi_{i+1})E[R_{i+1}]} + \frac{\chi_{i+1}^2 p_{i+1}^2 (1-p_{i+1})\phi_i}{(1-\chi_i)(1-\chi_{i+1})E[m_{i+1}]} + \frac{\chi_i}{(1-\chi_i)E[R_i]} \right\}, \tag{5.25}$$

where χ_i, defined in (5.1), satisfies the recurrence relation

$$(1-\chi_i) = \phi_i(1-\chi_{i+1} + \chi_{i+1}p_{i+1}). \tag{5.26}$$

However, using (5.26), replacing $E[R_i]$, $E[m_i]$ by R_i and m_i respectively, and replacing ϕ_i, p_i, χ_i by their estimates ($\hat{\chi}_i = (R_i - r_i)/R_i$), we find that the estimates of (5.25) and cov $[\hat{\phi}_i, \hat{\phi}_j \mid \{u_i\}]$ are the same as Jolly's estimates. This means that equations (5.15), (5.16) and (5.17) can also be used here. This is not surprising since we would expect variance estimates to be asymptotically the same, irrespective of whether we regard the u_i as random variables or fixed parameters.

Another parameter of interest is E_L, the expected life-span, which we now discuss in some detail. Let T_i be the time from the ith release to the $(i+1)$th sample and suppose that the mortality process in this interval is Poisson with parameter μ_i. Then, if there is no immigration, we have from (1.1)

$$\phi_i = e^{-\mu_i T_i},$$

and

$$\hat{\mu}_i = -\frac{1}{T_i} \log \hat{\phi}_i, \quad (i = 1, 2, \ldots, s-2),$$

is the maximum-likelihood estimate of μ_i. The average instantaneous mortality rate, μ, can be estimated by

$$\hat{\mu} = \sum_{i=1}^{s-2} \hat{\mu}_i/(s-2) = -\log \hat{\phi}, \quad \text{say},$$

where

$$\hat{\phi} = \{\prod_{i=1}^{s-2} \hat{\phi}_i^{1/T_i}\}^{1/(s-2)}$$

(which reduces to the geometric mean of the $\{\hat{\phi}_i\}$ when $T_i = 1$). If $\mu_i = \mu$ $(i = 1, 2, \ldots, s-2)$ then from (1.3)

$$E_L = 1/\mu,$$

which can be estimated by

$$\hat{E}_L = 1/\hat{\mu} \tag{5.27}$$

$$= -1/\log \hat{\phi}.$$

Using the delta method we find that

$$V[\hat{E}_L] \approx V[\hat{\phi}]\, \phi^{-2}(\log \phi)^{-4},$$

where

$$V[\hat{\phi}] = \frac{\phi^2}{(s-2)^2}\left\{\sum_{i=1}^{s-2} \frac{V[\hat{\phi}_i]}{\phi_i^2 T_i^2} + 2\sum_{i=1}^{s-3} \frac{\text{cov}[\hat{\phi}_i, \hat{\phi}_{i+1}]}{\phi_i \phi_{i+1} T_i T_{i+1}}\right\}.$$

An alternative estimate of $V[\hat{E}_L]$ is available if we use (5.27) and (1.10), namely

$$v[\hat{E}_L] = v[\hat{\mu}]/\hat{\mu}^4$$

where (defining $\hat{\mu}_{s-1} = \hat{\mu}_1$)

$$v[\hat{\mu}] = \frac{1}{(s-2)(s-5)}\{3\sum_{i=1}^{s-2} (\hat{\mu}_i - \hat{\mu})^2 - \sum_{i=1}^{s-2} (\hat{\mu}_{i+1} - \hat{\mu}_i)^2\}.$$

Table 5.5 was derived from Cormack's table, and we see that a coefficient of variation of about 4 per cent was obtained for the estimates $\hat{\phi}_i$. This accuracy was achieved by a sampling intensity (100 p_i) of about 60 per cent; this may be compared with 30 per cent in Jolly's study (Example 5.1). Also $\hat{\phi}_i$ appears to be fairly constant, so that it is appropriate to consider ϕ and E_L. Cormack shows that for a bird just starting to breed, $\hat{\phi}$ and $\hat{E}_L$, together with their standard deviations, are given by 0·9420 ± 0·01 and

TABLE 5.5
Recapture data and population estimates for a fulmar population: data from Cormack [1964: Table 1].

Year i	Captured u_i	Seen m_i	R_i	r_i	z_i	$\hat{M}_i$	$\hat{p}_i$	$\hat{\phi}_i$	$\hat{\sigma}[\hat{\phi}_i]$
1950	11	0	11	10	0	–	–	0·9697	0·097
1951	66	4	70	63	6	10·667	0·38	0·9287	0·040
1952	28	36	64	60	33	71·200	0·51	0·9735	0·039
1953	2	43	45	42	50	96·571	0·45	0·9619	0·041
1954	4	54	58	54	38	94·815	0·57	0·9593	0·036
1955	51	63	104	104	29	94·788	0·66	0·9664	0·035
1956	13	69	73	73	64	140·890	0·49	0·9419	0·040
1957	5	99	86	86	38	144·953	0·68	0·8546	0·040
1958	19	85	94	94	39	128·149	0·66	0·9444	0·038
1959	8	51	55	55	82	138·964	0·37	0·9662	0·036
1960	26	102	128	112	35	142·000	0·72	0·9028	0·032
1961	3	133	136	102	14	151·667	0·88	–	–
1962	18	116	134	–	–	–	–	–	–

16·7 ± 3·0 years respectively. He points out that the rather tentative start to the experiment with only 11 individuals marked in the first year caused $V[\hat{\phi}_1]$ to be much larger than any subsequent variance. Omitting all references to these 11 birds, the data were re-analysed, giving 0·9378 ± 0·0075 and 15·58 ± 1·93 years, a considerable increase in precision.

In conclusion, we raise the question of whether $\hat{\phi}_i$ applies to the whole population rather than to just the banded population. For example, some birds of the colony nest on sites totally inaccessible to the experimenter and therefore could not enter the banded population. Our estimate of $\hat{\phi}_i$ would then not apply to these birds if the choice of nesting site affected the survival of the adult bird. Although it is true that the choice of nesting site would probably affect the chance of successfully rearing young, it was only the adult breeding population that was considered in this investigation. Cormack also mentioned the fact that the pattern of attendance at and near the nest was different for the two sexes, so that the probability of capture and recapture was different for males and females. However, Cormack felt that this would not greatly affect the validity of the above analysis, provided the survival probabilities were the same for males and females. Unfortunately this assumption could not be tested as there were insufficient data.

5.1.3 Special cases

1 Enclosed populations

Suppose that the population area is bounded or enclosed, so that there is no migration. Then losses are due solely to deaths, and the influx of new individuals represents recruitment only. If, however, there is no recruitment, as for example in a non-breeding season, or if the new recruits can be distinguished from the others, for example by their size, then the unrecruited population can be analysed as one in which only mortality is operating (i.e. each $B_i = 0$, except $B_0 = N_1$). Assuming ϕ_i the same for both marked and unmarked, Jolly [1965] shows that the estimates are then given by

$$\hat{N}_i = \frac{R_i Z_i}{r_i} + n_i, \quad (i = 1, 2, \ldots, s-1),$$

$$\hat{\phi}_i = \frac{\hat{N}_{i+1}}{\hat{N}_i - n_i + R_i}, \quad (i = 1, 2, \ldots, s-2),$$

and

$$\hat{p}_i = \frac{n_i}{\hat{N}_i}, \quad (i = 1, 2, \ldots, s-1),$$

where $Z_i = z_i + \sum_{j=i+1}^{s} u_j$, the number of animals not caught in the ith sample but caught subsequently; $Z_i - z_i$ is the number of animals caught for the *first* time after the ith sample. These estimates were originally obtained by Darroch [1959] for the case of no losses on handling ($R_i = n_i$).

By analogy with (5.22), approximately unbiased estimates are given by

$$N_i^* = \frac{(R_i+1)}{(r_i+1)} Z_i + n_i$$

$$= \frac{(R_i+1)(Z_i+r_i+1)}{r_i+1} - 1 + (n_i - R_i)$$

and when $R_i = n_i$ this reduces to the unbiased estimate suggested by Darroch (called N'' in his notation): see also Robson [1979].

From Jolly we have (replacing expected values of random variables by "observed" values)

$$V[\hat{N}_i] = (N_i - n_i)(N_i - n_i + R_i)\left(\frac{1}{r_i} - \frac{1}{R_i}\right) + N_i - N_i^2/N_1,$$

$$\text{cov}[\hat{N}_i, \hat{N}_j] = N_j - \frac{N_i N_j}{N_1}, \quad (i < j),$$

$$V[\hat{\phi}_i] = \phi_i^2 \left\{ \frac{(N_{i+1} - n_{i+1})(N_{i+1} - n_{i+1} + R_{i+1})}{N_{i+1}^2}\left(\frac{1}{r_{i+1}} - \frac{1}{R_{i+1}}\right) + \right.$$

$$\left. + \frac{N_i - n_i}{N_i - n_i + R_i}\left(\frac{1}{r_i} - \frac{1}{R_i}\right) + \frac{1-\phi_i}{N_{i+1}} \right\},$$

$$\text{cov}[\hat{\phi}_i, \hat{\phi}_{i+1}] = -\phi_i \phi_{i+1} \frac{(N_{i+1} - n_{i+1})}{N_{i+1}}\left(\frac{1}{r_{i+1}} - \frac{1}{R_{i+1}}\right),$$

and

$$\text{cov}[\hat{\phi}_i, \hat{\phi}_j] = 0, \quad (j > i+1).$$

We note that it is now possible to estimate N_1 as $Z_1 \neq 0$.

2 No death or emigration

For completeness we shall consider the rather uncommon situation in which there is no death or emigration, so that the only changes in population size are due to recruitment and possibly immigration. In this case $\phi_i = 1$ $(i = 1, 2, \ldots, s-1)$ and from Jolly [1965: 242] we have

$$M_i = \sum_{h=1}^{i-1} (R_h - m_h), \quad \hat{\rho}_i = m_i/n_i,$$

leading to

$$\hat{N}_i = M_i/\hat{\rho}_i, \quad (i = 2, 3, \ldots, s)$$

and

$$\hat{B}_i = \hat{N}_{i+1} - (\hat{N}_i - n_i + R_i), \quad (i = 2, 3, \ldots, s-1).$$

We note that N_s and B_{s-1} are estimable but not N_1. Expressions for the asymptotic variances and covariances are given by Jolly, namely

$$V[\hat{N}_i] = N_i(N_i - n_i)\left(\frac{1-\rho_i}{m_i}\right) + N_i - \sum_{h=0}^{i-1} \frac{N_i^2(h)}{B_h},$$

$$\text{cov}[\hat{N}_i, \hat{N}_j] = \sum_{h=0}^{i-1}\left\{N_j(h) - \frac{N_i(h)N_j(h)}{B_h}\right\},$$

$$V[\hat{B}_i] = N_{i+1}(N_{i+1} - n_{i+1})\left(\frac{1-\rho_{i+1}}{m_{i+1}}\right) + N_i(N_i - n_i)\left(\frac{1-\rho_i}{m_i}\right),$$

$$\text{cov}[\hat{B}_i, \hat{B}_{i+1}] = -N_{i+1}(N_{i+1} - n_{i+1})\left(\frac{1-\rho_{i+1}}{m_{i+1}}\right),$$

$$\text{cov}[\hat{B}_i, \hat{B}_j] = 0, \quad (j > i+1)$$

and

$$V[\hat{N}_i \mid N_i] = N_i(N_i - n_i)\left(\frac{1-\rho_i}{m_i}\right).$$

The difference between $V[\hat{N}_i]$ and $V[\hat{N}_i \mid N_i]$, namely $V[N_i]$, arises entirely from stochastic death due to loss on capture, there being no death from other causes. Therefore, when $R_i = n_i$, $N_i(h) = B_h$ (for $h \leqslant i-1$) and $V[N_i] = \text{cov}[\hat{N}_i, \hat{N}_j] = 0$; this case was first considered by Darroch [1959] who also obtained $\hat{N}_i$, $\hat{B}_i$ and $V[\hat{N}_i]$.

3 Three-point census: triple-catch method

When $s = 3$ we have from the general theory the estimates

$$\hat{M}_2 = \frac{R_2 z_2}{r_2} + m_2, \tag{5.28}$$

$$\hat{N}_2 = n_2\hat{M}_2/m_2,$$

$$\hat{\phi}_1 = \hat{M}_2/R_1,$$

$$\hat{p}_2 = m_2/\hat{M}_2$$

and

$$\hat{\phi}_2\hat{p}_3 = r_2/R_2.$$

If an estimate $\hat{N}_1$ is available we can also add

$$\hat{B}_1 = \hat{N}_2 - \hat{\phi}_1(\hat{N}_1 - n_1 + R_1).$$

The asymptotic variances of $\hat{N}_2$ and $\hat{\phi}_1$ are given by

$$V[\hat{N}_2] = N_2(N_2 - n_2)\left\{\frac{M_2 - m_2 + R_2}{M_2}\left(\frac{1}{r_2} - \frac{1}{R_2}\right) + \frac{1 - \rho_2}{m_2}\right\} + $$
$$+ (N_2 - B_1)\{1 - (N_2 - B_1)/N_1\}$$

and

$$V[\hat{\phi}_1] = \phi_1^2\left\{\frac{(M_2 - m_2)(M_2 - m_2 + R_2)}{M_2^2}\left(\frac{1}{r_2} - \frac{1}{R_2}\right) + \frac{1 - \phi_1}{M_2}\right\}.$$

As pointed out in the general theory, the expression in $V[\hat{N}_2]$ involving $N_2 - B_1$ will usually be negligible.

If there is no emigration, and mortality is a Poisson process with parameter μ, then from (1.1)

$$\phi_i = e^{-\mu T_i}, \quad (i = 1, 2),$$

where T_i is the time from the ith release to the $(i+1)$th sample. We can now estimate the remaining parameters using the following chain of estimates:

$$\hat{\mu} = -\frac{1}{T_1} \log \hat{\phi}_1$$

$$\hat{\phi}_2 = e^{-\hat{\mu} T_2} = (\hat{\phi}_1)^{T_2/T_1},$$

$$\hat{p}_3 = r_2/(R_2\hat{\phi}_2)$$

$$\hat{M}_3 = m_3/\hat{p}_3, \quad \hat{N}_3 = n_3\hat{M}_3/m_3$$

and

$$\hat{B}_2 = \hat{N}_3 - \hat{\phi}_2(\hat{N}_2 - n_2 + R_2).$$

From the above discussion we see that if $T_1 = T_2$ then, on the assumption that $\phi_1 = \phi_2$, three samples is the minimum number needed to estimate all the unknowns (except B_1). Although such estimates may not be very accurate with such a short sample sequence, it may be all the samples that an experimenter can obtain.

Bailey [1951, 1952] was the first to introduce the so-called triple-catch method and considered the case of no losses on capture. His estimates were based on a deterministic model, and we shall briefly compare his estimate of M_2 with ours. Putting $R_2 = n_2$ and rearranging (5.28), we have

$$\frac{r_2 + z_2}{\hat{M}_2 + u_2} = \frac{r_2}{n_2},$$

where (denoting a_w as the number with capture history w)

$$\frac{r_2}{n_2} = \frac{a_{23} + a_{123}}{u_2 + m_2}.$$

In words, $\hat{M}_2$ is obtained by equating the proportions of the marked population $M_2 + u_2$ and the release n_2 which are caught in the third sample. It can be shown that Bailey's estimate M_2' of M_2 is given by

$$\frac{r_2 + z_2}{M_2' + u_2} = \frac{a_{23}}{u_2},$$

which ignores the recapture information a_{123} given by the release of m_2 previously marked individuals.

Example 5.5 Meadow grasshopper (*Chorthippus parallelus*): Parr *et al.* [1968]

A colony of the meadow grasshopper was studied, using the mark-recapture method, on 9, 10, 11 September, 1966. The small field supporting the colony contained meadow grasses cut short for hay and was bounded by a road on one side and thick scrubby vegetation on the remaining sides. As *C. parallelus* is unable to fly, it may be caught by sweeping on warm sunny days when it is very active; on dull days it may be more efficient to capture the insects by hand. An attempt to mark the grasshoppers with cellulose paint failed as in many cases it was seen to peel off the insect's cuticle in a matter of minutes. The use of black Indian ink proved more satisfactory, although this was liable to flake off partially, and was difficult to apply because of the greasy nature of the cuticle. (Oil paint would have been a better marking agent but was not available.) However, it was found that once the insect was marked, some particles of ink remained attached to the insect, and even if most of the ink flaked off, these particles could easily be seen using a hand-lens. The insects were marked in batches from limited parts of the colony area, and after marking were released as near to the point of capture as possible.

The recapture data for just the male population are summarised below:

$$(n_1, n_2, n_3) = (52, 41, 39), \quad (R_1, R_2) = (52, 40), \quad m_{12} = 5,$$

$$m_{13} = 7, \quad m_{23} = 8,$$

where m_{hi} is the number caught in the ith sample last captured in the hth. Then $r_1 = m_{12} + m_{13} = 12$, $r_2 = m_{23} = 8$, $z_2 = m_{13} = 7$, $m_2 = m_{12} = 5$, $m_3 = m_{13} + m_{23} = 15$,

$$\hat{M}_2 = \frac{40(7)}{8} + 5 = 40,$$

$$\hat{N}_2 = 40(41)/5 = 328,$$

$$\hat{\phi}_1 = 40/52 = 0{\cdot}769,$$

$$\sqrt{\{\hat{V}[\hat{N}_2 \mid N_2]\}} = 184{\cdot}0$$

and

$$\sqrt{\{\hat{V}[\hat{\phi}_1]\}} = 0{\cdot}317.$$

We note that both $\hat{N}_2$ and $\hat{\phi}_1$ have large coefficients of variation, and this is due to the small values of m_2 and r_2. It was recommended on p. 204 that these values should be at least as great as 10.

Since $T_1 = T_2 = 1$ day, then, assuming $\phi_1 = \phi_2$, we have

$$\begin{aligned}\hat{\phi}_2 &= \hat{\phi}_1,\\ \hat{M}_3 &= 75(40)/52 = 57{\cdot}7,\\ \hat{p}_3 &= 0{\cdot}26, \quad \hat{N}_3 = 150 \quad \text{and} \quad \hat{B}_2 = -96.\end{aligned}$$

From p. 204, approximately unbiased estimates are given by

$$M_2^* = \frac{41(7)}{9} + 5 = 36{\cdot}9,$$

$$N_2^* = 36{\cdot}9(42)/6 = 258{\cdot}3$$

and

$$\phi_1^* = 36{\cdot}9/52 = 0{\cdot}71.$$

4 Ricker's two-release method

Ricker [1958: 128] has suggested a useful method for estimating the probability of survival for a given period of time by making a release at the beginning and end of the period followed by a sample. For $i = 1, 2$ let R_i be the size of the ith release and let m_{i3} be the number from the ith release caught in the sample. Let α_1 be the probability of survival between releases, and assume that every marked individual alive just after the second release has the same probability β_2 of surviving to the time of the sample and being caught in the sample. Then, neglecting the complications of sampling without replacement, m_{13} and m_{23} will be independent binomial variables, so that

$$f(m_{13}, m_{23}) = \binom{R_1}{m_{13}}(\alpha_1\beta_2)^{m_{13}}(1-\alpha_1\beta_2)^{R_1-m_{13}}\binom{R_2}{m_{23}}\beta_2^{m_{23}}(1-\beta_2)^{R_2-m_{23}}.$$

It is readily shown that the maximum-likelihood estimates of α_1 and β_2 (which are also the moment estimates) are

$$\hat{\alpha}_1 = \left(\frac{m_{13}}{R_1}\right)\bigg/\left(\frac{m_{23}}{R_2}\right)$$

and

$$\hat{\beta}_2 = m_{23}/R_2.$$

However, a slight modification of $\hat{\alpha}_1$, namely

$$\tilde{\alpha}_1 = \frac{m_{13}(R_2+1)}{R_1(m_{23}+1)}$$

is almost unbiased, as

$$\begin{aligned}E[\tilde{\alpha}_1] &= E\left[\frac{m_{13}}{R_1}\right]E\left[\frac{R_2+1}{m_{23}+1}\right]\\ &= \alpha_1\beta_2 \cdot \frac{1}{\beta_2}(1-(1-\beta_2)^{R_2+1})\\ &\approx \alpha_1.\end{aligned}$$

Also

$$E\left[\frac{m_{13}(m_{13}-1)}{R_1(R_1-1)} \cdot \frac{(R_2+1)(R_2+2)}{(m_{23}+1)(m_{23}+2)}\right] = E\left[\frac{m_{13}(m_{13}-1)}{R_1(R_1-1)}\right] E\left[\frac{(R_2+1)(R_2+2)}{(m_{23}+1)(m_{23}+2)}\right]$$

$$= (\alpha_1\beta_2)^2 \cdot \frac{1}{\beta_2^2}[1-(1-\beta_2)^{R_2+2} - (R_2+2)\beta_2(1-\beta_2)^{R_2+1}]$$

$$\approx \alpha_1^2,$$

so that

$$v[\tilde{\alpha}_1] = \tilde{\alpha}_1^2 - \frac{m_{13}(m_{13}-1)(R_2+1)(R_2+2)}{R_1(R_1-1)(m_{23}+1)(m_{23}+2)}$$

is an almost unbiased estimate of $V[\tilde{\alpha}_1]$.

Ricker's scheme is very flexible as it can be used when exploitation is taking place continuously between releases and when the sample is not instantaneous but extends over a period of time. If the number m_{12} of tagged removed from the population between releases is recorded, then β_1, the probability of being caught between releases, is estimated by

$$\hat{\beta}_1 = m_{12}/R_1.$$

In this case the joint distribution of m_{12}, m_{13}, m_{23} is given by (5.23) with $s = 3$.

The case when the instantaneous natural mortality rate and the instantaneous exploitation rate are both constant is discussed in **6.4.2**(*2*); Ricker's method is also mentioned again in **9.2.2**(*1*).

Example 5.6 Bluegills (*Lepomis macrochirus*): Ricker [1958]

The above method was applied to a population of bluegills in Muskellunge Lake, Indiana. Of $R_1 = 230$ bluegills marked before the start of the 1942 fishing season, $m_{13} = 13$ were captured in 1943. Of $R_2 = 93$ marked before the start of the 1943 season, $m_{23} = 13$ were captured in 1943. Thus α_1, the probability of survival in 1942, is estimated by

$$\tilde{\alpha}_1 = \frac{13(94)}{230(14)} = 0{\cdot}379\ 50$$

and

$$v[\tilde{\alpha}_1] = (0{\cdot}3795)^2 - \frac{13(12)(94)(95)}{230(229)(14)(15)}$$

$$= 0{\cdot}144\ 02 - 0{\cdot}125\ 95 = 0{\cdot}018\ 07.$$

The approximate 95 per cent confidence interval, $\tilde{\alpha}_1 \pm 1{\cdot}96\sqrt{v}$, for α_1 is $(0{\cdot}10, 0{\cdot}66)$.

5.1.4 Underlying assumptions

1 Validity of underlying model

GOODNESS-OF-FIT TEST. Since (5.3) comes from a product of independent multinomial distributions (cf. 5.2), we can calculate a goodness-of-fit test for the adequacy of this model with regard to the *marked* population. For example, when $s = 3$, the test statistic is

$$\frac{(b_1 - u_1\hat{\nu}_1\hat{\chi}_1)^2}{u_1\hat{\nu}_1\hat{\chi}_1} + \frac{(b_{12} - u_1\hat{\nu}_1\hat{\phi}_1\hat{p}_2\hat{\nu}_2\hat{\chi}_2)^2}{u_1\hat{\nu}_1\hat{\phi}_1\hat{p}_2\hat{\nu}_2\hat{\chi}_2} + \ldots + \frac{(d_1 - u_1(1-\hat{\nu}_1))^2}{u_1(1-\hat{\nu}_1)} + \ldots +$$

$$+ \frac{(b_2 - u_2\hat{\nu}_2\hat{\chi}_2)^2}{u_2\hat{\nu}_2\hat{\chi}_2} + \ldots + \frac{(d_2 - u_2(1-\hat{\nu}_2))^2}{u_2(1-\hat{\nu}_2)} + \ldots +$$

$$+ \frac{(b_3 - u_3\hat{\nu}_3)^2}{u_3\hat{\nu}_3} + \frac{(d_3 - u_3(1-\hat{\nu}_3))^2}{u_3(1-\hat{\nu}_3)},$$

which is asymptotically distributed as chi-squared. For general s the goodness-of-fit statistic has $(f_1 - f_2)$ degrees of freedom, where f_1 is the number of squared terms (after any necessary pooling) and $f_2 = 4s - 3$ (since we have $3s - 3$ parameters $\nu_1, \ldots, \nu_s, \alpha_1, \ldots, \alpha_{s-2}, \beta_1, \ldots, \beta_{s-1}$ and s constraints – one for each multinomial distribution).

In calculating the goodness-of-fit statistic we note that

$$\hat{\chi}_i = \frac{R_i - r_i}{R_i}, \quad \hat{\nu}_i = \frac{R_i}{n_i}, \quad \hat{\beta}_{s-1} = 1 - \hat{\chi}_{s-1},$$

and

$$\hat{\alpha}_i = \hat{\phi}_i\hat{q}_{i+1} = (\hat{M}_{i+1} - m_{i+1})/(\hat{M}_i - m_i + R_i)$$

$$\hat{\beta}_i = \hat{\phi}_i\hat{p}_{i+1} = m_{i+1}/(\hat{M}_i - m_i + R_i) \quad (i = 1, 2, \ldots, s-2).$$

Since the d_w will generally be small it is recommended that they be pooled with the b_w. In fact in some situations many of the expected frequencies may be less than unity, so that the large-scale pooling required would lead to an approximate test only.

METHOD OF LESLIE, CHITTY AND CHITTY. Another method which indirectly tests for the validity of the underlying model with respect to the marked population has been proposed by Leslie *et al.* [1953]. This method, which we shall now discuss in detail, consists of comparing the increase in the marked population at time t_i due to the release of freshly marked individuals, with an estimate of this increase obtained from the marked population of animals which are caught at least twice.

Suppose we consider just the marked population; this is generated by the releases, v_i say, of individuals marked for the first time ($v_i \leqslant u_i$). We now regard these releases as constituting the "unmarked" population, and those captured from this population (i.e. those caught at least twice) constitute the "marked" population. To begin with, we have a population of R_1 "unmarked" individuals of which m_2 are caught in the second sample. These m_2 animals now represent the "marked" population, and v_2 "unmarked" animals are added to the population by "recruitment". In the third sample, animals previously caught twice are regarded as "recaptures", while those caught in only one of the previous samples are from the "unmarked" population and are regarded as having been caught for the "first" time: once again we have a "recruitment" of v_2 "unmarked". This new population of "marked" and "unmarked", consisting solely of marked animals, can be analysed using exactly the same method as demonstrated in Example 5.1 (p. 205). The only difference is that one excludes from the group m_{hi} all animals caught only once before. This new group we note by $m_{.hi}$, the number of animals caught in the ith sample which were last caught in the hth sample and at least once prior to the hth sample. From the array $\{m_{.hi}\}$ we can calculate the corresponding array $\{c_{.hi}\}$ and the quantities $r_{.i}$, $z_{.i}$. Using the "dot" notation to denote membership of the marked population, we find that $n_{.i} = m_i$; $m_{.i}$ is the number of marked caught at least twice before the ith sample and in the ith sample; $R_{.i} = m_i - d_i$, where d_i is the number of marked caught in the ith sample which are not returned to the population. The quantities $N_{.i}$, $\phi_{.i}$ and $\rho_{.i}$ now refer to the marked population while $M_{.i}$ refers to the "marked" population of those caught at least twice. We note that $N_{.1} = R_1$, and define $M_{.1} = M_{.2} = 0$, $m_{.1i} = 0$ for $i = 2, 3, \ldots, s$.

Since

$$E[N_{.i+1} \mid N_{.i}] = \phi_{.i}(N_{.i} + v_i - d_i),$$

we can "estimate" v_i by (cf. equations (5.6) to (5.12))

$$\begin{aligned}\hat{v}_i &= (\hat{N}_{.i+1}/\hat{\phi}_{.i}) - \hat{N}_{.i} + d_i \\ &= \frac{\hat{M}_{.i} - m_{.i} + R_{.i}}{\hat{\rho}_{.i+1}} - \frac{\hat{M}_{.i}}{\hat{\rho}_{.i}} + d_i, \quad (i = 3, 4, \ldots, s-1).\end{aligned}$$

To find the asymptotic variance of $\hat{v}_i$ we note from Jolly [1965: 237] that the component due to the errors of estimation in the covariance of any pair of $\hat{M}_{.i}$, $\hat{\rho}_{.i}$ and $\hat{\rho}_{.i+1}$, is zero. Therefore, using the delta method, treating d_i as a constant, and neglecting covariance terms, we have from Jolly

$$\begin{aligned}\sigma^2[\hat{v}_i] &= V[\hat{M}_{.i} \mid M_{.i}]\left(\frac{1}{\rho_{.i+1}} - \frac{1}{\rho_{.i}}\right)^2 + \\ &\quad + V[\hat{\rho}_{.i+1} \mid \rho_{.i+1}]\frac{(M_{.i} - m_{.i} + R_{.i})^2}{\rho_{.i+1}^4} + V[\hat{\rho}_{.i} \mid \rho_{.i}]\frac{M_{.i}^2}{\rho_{.i}^4} \\ &= (M_{.i} - m_{.i})(M_{.i} - m_{.i} + R_{.i})\left(\frac{1}{r_{.i}} - \frac{1}{R_{.i}}\right)\left(\frac{1}{\rho_{.i+1}} - \frac{1}{\rho_{.i}}\right)^2 + \\ &\quad + \rho_{.i+1}(1 - \rho_{.i+1})\left(\frac{1}{n_{.i+1}} - \frac{1}{N_{.i+1}}\right)\frac{(M_{.i} - m_{.i} + R_{.i})^2}{\rho_{.i+1}^4} + \\ &\quad + \rho_{.i}(1 - \rho_{.i})\left(\frac{1}{n_{.i}} - \frac{1}{N_{.i}}\right)\frac{M_{.i}^2}{\rho_{.i}^4}.\end{aligned}$$

It can be shown that this formula is equivalent to that given by Leslie *et al.* [1953: 168] if we ignore all terms involving $R_{.i}^{-1}$, $N_{.i}^{-1}$, $N_{.i+1}^{-1}$ (i.e. the finite population corrections) and use $c_{.i-1,i+1}/m_{.i,i+1}$ instead of $z_{.i}/r_{.i}$ in the estimation of $M_{.i} - m_{.i}$ (cf. p. 201).

Having calculated an estimate $\hat{v}_i$ we now see whether the observed value of v_i lies in the interval $\hat{v}_i \pm k\,\hat{\sigma}[\hat{v}_i]$, where k is to be chosen. We can also compare $\hat{N}_{\cdot i}$ and $\hat{M}_i$ as they are estimates of the marked population; $\hat{\phi}_{\cdot i}$ and $\hat{\phi}_i$ should be fairly similar as they are both estimates of ϕ_i.

Example 5.7 Six-spot Burnet moth (*Zygaena filipendula* L.): Manly and Parr [1968]

The above analysis was applied to mark-recapture data given by Manly and Parr. One sample was taken on each of five days, and sunny weather throughout the period of study ensured adequate mixing of the marked and unmarked between samples. The insects were marked with cellulose "dope" applied to the underside of the hindwings, and the following colour code was used:

Day 1: green (g)
Day 2: white (w)
Day 3: blue (b)
Day 4: orange (o)

There were no losses on capture (i.e. $R_i = n_i$, $R_{\cdot i} = n_{\cdot i} = m_i$, $v_i = u_i$) and using the above colour abbreviations the recapture data were recorded as follows:

Day 1 (19 July): 57 captured, marked and released.
Day 2 (20 July): 52 captured; 25g, 27 unmarked.
Day 3 (21 July): 52 captured; 8g, 9w, 11gw, 24 unmarked.
Day 4 (22 July): 31 captured; 2g, 3w, 4b, 5gb, 1wb, 2gwb, 14 unmarked.
Day 5 (24 July): 54 captured; 1g, 2w, 7b, 5o, 4gw, 2gb, 2go, 4wb, 1wo, 1bo, 5gbo, 1$gwbo$, 19 unmarked.

In working with such data it is helpful to convert the colours to numbers.

The arrays $\{m_{hi}\}$ and $\{m_{\cdot hi}\}$ are recorded in Table 5.6: for example $m_{34} = 4b + 5gb + 1wb + 2gwb = 12$ and $m_{\cdot 34} = 5gb + 1wb + 2gwb = 8$. From these data the arrays $\{c_{hi}\}$, $\{c_{\cdot hi}\}$ were obtained, and the estimates in Table 5.7 were then calculated. As there is a reasonable agreement between $\hat{N}_{\cdot i}$, $\hat{\phi}_{\cdot i}$ and $\hat{v}_i$, and $\hat{M}_i$, $\hat{\phi}_i$ and v_i respectively, we have no reason to reject the underlying model for the marked population.

In order to estimate $\sigma[\hat{v}_4]$ we must either ignore terms involving $1/N_{\cdot 5}$ or else estimate $N_{\cdot 5}$ from the trend, or lack of trend, in $\hat{N}_{\cdot 3}$ and $\hat{N}_{\cdot 4}$; for this experiment both methods led to almost the same value of $\hat{\sigma}[\hat{v}_4]$.

If a goodness-of-fit test is required for the above data, the first step is to find the b_w ($= a_w$ as $d_w = 0$). This can be done using, for example, a tree diagram for each release of u_i insects marked for the first time. An example of this is given in Fig. 5.1.

2. *Equal probability of capture for marked individuals*

METHOD OF LESLIE. Assumption (a) of p. 196 will be satisfied if all sampling is random or if there is random mixing of marked and unmarked and

TABLE 5.6
Tabulation of m_{hi} and $(m_{.hi})$ for a moth population: data from Manly and Parr [1968: 86].

	i	1	2	3	4	5	
	$n_i\ (n_{.i})$	57	52(25)	52(28)	31(17)	54(35)	Total
h							$r_h\ (r_{.h})$
1			25	8	2	1	36
2				20(11)	3(0)	6(4)	29(15)
3					12(8)	13(6)	25(14)
4						15(10)	15(10)
	u_i	57	27	24	14	19	

TABLE 5.7
Estimation of population parameters for a moth population from the data in Table 5.6.

i	$\hat{M}_i$	$\hat{\phi}_i$	$\hat{N}_{.i}$	$\hat{\phi}_{.i}$	v_i	$\hat{v}_i$	$\hat{\sigma}[\hat{v}_i]$
1	–	0·78	–	–	57	–	–
2	44·7	0·74	–	0·76	27	–	–
3	53·0	0·76	48·4	0·69	24	28	18
4	58·3	–	53·1	–	14	6	12
5	–	–	–	–	19	–	–

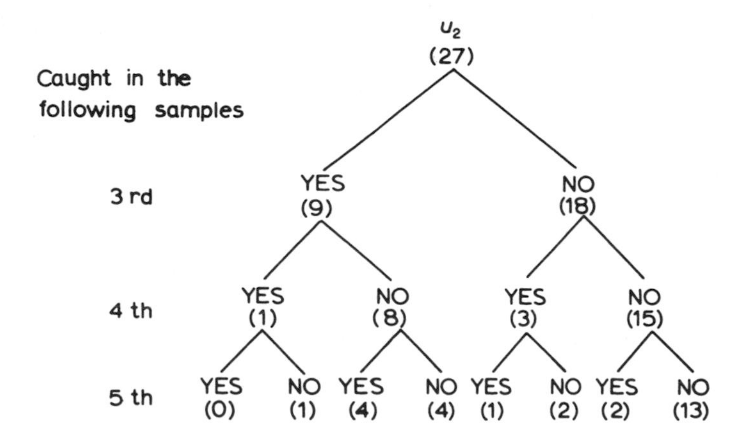

Fig 5.1 Tree diagram for calculating the b_w from the second release of newly marked individuals.

all individuals have the same catchability. Leslie's method described on p. 161 can be used for testing whether the sampling is random with respect to just the marked population by studying the capture frequencies of members of a group of animals known to be in the population over a given period of time. For example, consider the group of animals released after sample i

which had been marked for the first time, and suppose G of this group were captured for the last time in sample $i + t + 1$. Then members of G could have been recaptured up to t times during the intervening interval, and we can apply Leslie's chi-squared test to see if the sampling is random with respect to this group. By varying i and t subject to $i \geqslant 1$, $t \geqslant 3$ and $i + t + 1 \leqslant s$ we have a whole series of intervals each with its own chi-squared test. For example, if $s = 8$ we have the intervals defined by the pairs (1, 5), (1, 6), (1, 7), (1, 8); (2, 6), (2, 7), (2, 8); (3, 7), (3, 8); (4, 8). Since the groups of animals corresponding to these intervals are mutually exclusive, the various chi-squareds are independent and can be added. One way of doing this would be to add the chi-squareds which are based on the same release. However, if one uses the above intervals, the groups may be too small (say < 20 members) for a valid application of the chi-squared test. In this case it is preferable to pool the information from several samples. For example, if $s = 8$ we can use the pairs (1, 5), (1, 6) and (2, 6), where (1, 6) represents the group of individuals from the first release which are recaptured in at least one of the samples 6, 7 and 8.

Examples of the above method are given in Orians [1958], Turner [1960b], Nakamura *et al.* [1971], Murton *et al.* [1972: 860] and Sonleitner [1973].

Example 5.8

Using the data of Example 5.7, we have 5 samples and therefore just one interval defined by (1, 5). Here $G = 15$ (the number of animals caught in the 5th sample bearing a green mark) and, using the notation of **4.1.5** (*2*), Table 5.8 was derived. For example, g_i, for sample 3, is the number caught

TABLE 5.8

Leslie's test for random sampling using the data from Example 5.6.

Sample	g_i	g_i^2	x	f_x	$x f_x$	$x^2 f_x$
2	5	25	0	1	0	0
3	8	64	1	8	8	8
4	8	64	2	5	10	20
	21	153	3	1	3	9
				15	21	37

in all three samples 1, 3 and 5, namely $2gb + 5gbo + 1gwbo = 8$. We find that $\mu = 1{\cdot}4$, $\sigma^2 = 1{\cdot}4 - 0{\cdot}68 = 0{\cdot}72$ and

$$T = \sum_{x=0}^{3} f_x(x - \mu)^2/\sigma^2 = 10{\cdot}6,$$

which, at 14 degrees of freedom, indicates that the observed and expected variances are close. Although such a test should be viewed with caution as G is small, it does at least support the hypothesis of random sampling as far as just the first release in concerned.

UNIFORM MIXING. The question of random sampling is a crucial one when the animals are relatively immobile so that mixing of marked and unmarked is slow. The following example is a good illustration of this.

Example 5.9 Land snail (*Helix aspersa*): Parr *et al.* [1968]

In September 1966 an attempt was made to estimate the number in a colony of *H. aspersa* living on a wall. This species is known to have a low mortality rate and observations showed that there was very little migration. It was therefore expected that a mark-recapture study of this snail using the Jolly–Seber method could be used to demonstrate high survival probabilities ϕ_i, a series of similar daily estimates of population size, and a high overall proportion of the population caught by the end of the experiment. In fact, none of these expectations was fully realised, and although 935 different snails were marked in six days the average daily estimate of population size was only about 850. On the last day of observation, 102 out of a total of 230 snails had not been seen previously. This breakdown of the Jolly–Seber method was due to the fact that random mixing of marked and unmarked snails did not occur during the six-day sampling period owing to irregular spells of activity and aestivation (rest). Thus, if snails aestivated on the surface of the wall they were recaptured every day, but if they aestivated in holes in the wall they were inaccessible.

The experiment was then repeated on the same colony in September 1967, but for a period of ten days, in the hope that this might be long enough to allow mixing of marked and unmarked. A Petersen estimate (cf. **3.1.1**) of the population size based on the first two days of sampling gave a value of $\hat{N}$ = 743 which was far too low. Similar estimates on subsequent days based on the numbers of recaptures, y_i, of the individuals originally marked and released on the first day are shown in Table 5.9. These estimates are valid if there is negligible mortality or the mortality rates are the same for marked and unmarked (cf. **3.2.1** (*2*)). The authors felt that the slow upward drift in the estimates reflected the slow mixing of marked and unmarked snails. The later estimates would therefore be more reliable as the mixing became more uniform. The day-to-day fluctuations in the estimates were due to the relatively large standard deviations.

The population was again sampled on 28 September 1967 (16 days after the last previous sampling) when 357 snails were caught of which 254 were marked. Since there were 1077 + (318 − 206) = 1189 marked snails in the population the Petersen estimate is

$$\hat{N} = 1189(357)/254 = 1672$$

with a standard deviation of (cf. **3.1.1**) $\sqrt{v^*} = 49{\cdot}6$. A similar overall estimate was obtained from the last entry in Table 5.9 using the total population of marked, namely

$$\hat{N} = 318(1077)/206 = 1662{\cdot}5.$$

This value is very close to the above overall estimate and supports the

TABLE 5.9
Recapture data for a population of snails, September 3–12, 1967: from Parr *et al.* [1968: Table 10].

Day i	n_i	m_i	M_i	b_{1i}	$\hat{N} = \frac{n_1 n_i}{y_i}$
1	217	–	–	–	–
2	243	71	217	71	742·7
3	310	110	389	56	1201·3
4	326	198	590	71	996·4
5	371	248	717	69	1166·8
6	306	243	840	51	1302·0
7	274	233	903	40	1486·5
8	296	250	943	51	1259·5
9	258	170	989	32	1749·6
10	318	206	1077	48	1437·6

y_i = number captured on day i which were marked on day 1

authors' view that random mixing of the snails had occurred by the end of the period of regular sampling.

The authors finally mentioned that the weather was generally dry between 3 and 10 September and many snails may have aestivated throughout this period, but heavy rain caused considerable movement of snails on 10 September and provided the necessary mixing of marked and unmarked.

3 *Equal probability of survival for marked population*

Possible departures from assumption (b) of p. 196, that ϕ_i is the same for all marked individuals, are: (i) The method of catching (and tagging, if individual tags are not used) has a deleterious effect, so that animals caught many times have a higher mortality rate. (ii) If individual tags are used, so that only the untagged in each sample are given a tag, newly tagged animals may have a higher mortality rate than those tagged on a previous occasion. (iii) Different age-classes have different mortality rates.

We shall now describe contingency table methods for detecting the above three types of departures.

SURVIVAL INDEPENDENT OF MARK STATUS. Let v_{xi} be the number of animals released from the ith sample which have been caught x times in the first i samples ($x = 1, 2, \ldots, i$) and let $v_{xi,i+1}$ be the number from this group recaptured in the $(i+1)$th sample. Then, using an obvious notation,

$$E[v_{xi,i+1}/v_{xi}] = \phi_{xi}p_{xi}, \quad (x = 1, \ldots, i)$$

and we can test the hypothesis $H: \phi_{xi}p_{xi} = \phi_i p_i$ using the pairs $(v_{xi,i+1}, v_{xi} - v_{xi,i+1})$ in a 2-by-i contingency table; any groups which are too small can be pooled. If the sampling is random so that $p_{xi} = p_i$ $(x = 1, \ldots, i)$

then a test of H is a test that $\phi_{xi} = \phi_i$, that is, survival in the marked population is independent of mark status. Alternatively, if $\phi_{xi} = \phi_i$ then we are testing whether p_i is independent of mark status. A contingency table can be constructed for each $i = 2, 3, \ldots, s-1$, and since the chi-squareds are independent they can be added together to give a total chi-square. Further contingency tables can be constructed if we consider v_{xij}, the number of v_{xi} which are not caught again until the jth sample ($j > i$).

INITIAL TAGGING MORTALITY. In fishery investigations it is well known that the operation of attaching or inserting a tag places considerable stress on the fish. This stress may lead to an increase in mortality during the period immediately following release. To test for this initial mortality (called "type I losses" in the fisheries literature) Robson [1969] suggests using the 2-by-2 contingency table, Table 5.10; once again the tests are independent for $i = 2, 3, \ldots, s-1$. An essentially equivalent test has been proposed by Manly [1971b: 184].(*)

TABLE 5.10

Contingency table for detecting type I losses: from Robson [1969].

	Captured after ith sample	Not captured after ith sample
First captured in the ith sample and released	*	*
Recaptured in the ith sample and released	*	*

TABLE 5.11

Contingency table for detecting sustained type I losses: from Robson [1969].

	Captured in sample $i+1$ and	
	captured after sample $i+1$	not captured after sample $i+1$
First caught in sample i	*	*
Caught before sample i	*	*

TABLE 5.12

Contingency table for detecting sustained type I losses: from Robson [1969].

	Captured after sample * and	
	captured in sample $i+1$	not captured in sample $i+1$
First caught in sample i	*	*
Caught before sample i	*	*

(*)For further applications see Ericson [1977] and Van Noordwijk [1978].

If it is suspected that the mortality effect for those first tagged in the ith sample carries over after the $(i+1)$th release, Robson [1969] suggests two contingency tables, Table 5.11 and 5.12, for detecting this. These tests are independent within pairs as well as between pairs.

SURVIVAL INDEPENDENT OF AGE. If we suspect that survival depends on age (or sex), we can split up the release of R_i individuals into age–sex classes and record the subsequent recaptures from these classes in the $(i+1)$th sample or in all subsequent samples. Once again, contingency tables can be constructed for $i = 1, 2, \ldots, s-1$ and tests of homogeneity carried out. However, if survival is independent of mark status and probability of capture independent of age, the Jolly–Seber method will not be greatly affected by age-dependent mortality. This follows from the fact that the two groups of $M_i - m_i$ and R_i marked individuals will than have much the same age-distribution, so that the basic equation (cf. (5.6))

$$z_i/(M_i - m_i) \approx r_i/R_i$$

will still hold.

4 Special cases

COMMERCIALLY EXPLOITED POPULATIONS. When the multi-sample single-recapture census is used, the marked population has probability function (5.23). A goodness-of-fit test for this model is given by

$$T = \sum_{h=1}^{s} \sum_{i=h+1}^{s} \frac{(m_{hi} - R_h \hat{p}_{hi})^2}{R_h \hat{p}_{hi}},$$

where

$$\begin{aligned}\hat{p}_{hi} &= \hat{\phi}_h \hat{q}_{h+1} \hat{\phi}_{h+1} \hat{q}_{h+2} \cdots \hat{\phi}_{i-1} \hat{p}_i \\ &= \hat{\alpha}_h \hat{\alpha}_{h+1} \cdots \hat{\alpha}_{i-2} \hat{\beta}_{i-1}.\end{aligned}$$

T is asymptotically χ_k^2, where $k = \frac{1}{2}s(s-1) - (2s-3)$; $2s - 3$ degrees of freedom are subtracted as there are $2s - 3$ parameters estimated, namely $\alpha_1, \ldots, \alpha_{s-2}, \beta_1, \ldots, \beta_{s-1}$.

NO MIGRATION. In this situation each consecutive pair of samples may be regarded as a Petersen-type experiment and the discussion in **3.2.1** applies. For example, if there is no recruitment, and mortality is such that it removes a random portion of the population between samples, we can derive a chain of Petersen estimates. Procedures for eliminating recruits from samples are given in **3.2.1** (*4*), p. 72.

NO MORTALITY. When there is no mortality or emigration one can test for "dilution" by recruitment or immigration using Leslie's dilution test of p. 159.

5.2 THE MANLY–PARR METHOD

5.2.1 Population size

PETERSEN-TYPE ESTIMATE. Recently Manly and Parr [1968] suggested a simple method of estimating N_i without making any assumptions about the survival probabilities (cf. assumption (b) on p. 196). They point out that when all animals have the same probability of capture in the ith sample, we have

$$E[n_i \mid N_i] = N_i p_i.$$

Therefore, given an estimate $\widetilde{p}_i$ of p_i, we have the obvious estimate

$$\widetilde{N}_i = n_i/\widetilde{p}_i.$$

Here p_i can be estimated by picking out a class of C_i animals known to be in the population at the time of the ith sample, and then observing the number c_i of this class caught in the ith sample, i.e.

$$\widetilde{p}_i = c_i/C_i$$

and hence

$$\widetilde{N}_i = n_i C_i/c_i.$$

We note that $\widetilde{N}_i$ is basically a Petersen estimate in which the n_i individuals caught in the ith sample represent the "first" sample and the C_i animals represent the "second" sample. It seems that the above method was first used by Davis *et al.* [1964: 4–5] for the case of just three samples only (see also Meslow and Keith [1968: 814]).

MEAN AND VARIANCE. Manly [1969] has obtained asymptotic expressions for the mean and variance of $\widetilde{N}_i$ by considering the four categories set out in Table 5.13. Suppose that an animal has probability θ_i of being included in

TABLE 5.13

Animals alive at the time of the ith sample: from Manly [1969].

	Captured in sample i	Not captured in sample i	Total
In class C_i	c_i	$C_i - c_i$	C_i
Not in class C_i	$n_i - c_i$	$N_i - C_i - n_i + c_i$	$N_i - C_i$
Total	n_i	$N_i - n_i$	N_i

the class of C_i animals, then if the probability of capture in the ith sample is assumed to be independent of class status, the four random variables in Table 5.13 will have a joint multinomial distribution proportional to

$$\frac{N_i!}{(N_i - C_i - n_i + c_i)!}(p_i\theta_i)^{c_i}((1-p_i)\theta_i)^{C_i-c_i}(p_i(1-\theta_i))^{n_i-c_i}((1-p_i)(1-\theta_i))^{N_i-C_i-n_i+c_i}$$

$$= \frac{N_i!}{(N_i - C_i - n_i + c_i)!}\, p_i^{n_i}(1-p_i)^{N_i-n_i}\theta_i^{C_i}(1-\theta_i)^{N_i-C_i}. \qquad (5.29)$$

Manly shows that the maximum-likelihood estimates of p_i, N_i and θ_i are

$$\tilde{p}_i = c_i/C_i,$$
$$\tilde{N}_i = n_i/\tilde{p}_i$$

and

$$\tilde{\theta}_i = C_i/\tilde{N}_i.$$

Using the delta method he proves that asymptotically (i.e. for large $N_i\theta_i$ and $C_i p_i$)

$$E[\tilde{N}_i \mid N_i] = N_i + (1-p_i)(1-\theta_i)/(p_i\theta_i) \tag{5.30}$$

and

$$E[(\tilde{N}_i - N_i)^2 \mid N_i] = N_i(1-p_i)(1-\theta_i)/(p_i\theta_i). \tag{5.31}$$

An estimate of this last expression is given by

$$\begin{aligned} \tilde{V}_i &= \tilde{N}_i(\tilde{N}_i - n_i)(\tilde{N}_i - C_i)/(n_i C_i) \\ &= \tilde{N}_i(C_i - c_i)(n_i - c_i)/c_i^2. \end{aligned} \tag{5.32}$$

As $\tilde{N}_i$ is a Petersen-type estimate it is worth considering Chapman's modification (cf. **3.1.1**)

$$\begin{aligned} N_i^* &= \frac{(n_i+1)(C_i+1)}{(c_i+1)} - 1 \\ &\approx \frac{n_i C_i}{c_i}\left(1+\frac{1}{n_i}\right)\left(1+\frac{1}{C_i}\right)\left(1-\frac{1}{c_i}\right) - 1 \\ &\approx \frac{n_i C_i}{c_i} - \left(1-\frac{C_i}{c_i}\right)\left(1-\frac{n_i}{c_i}\right). \end{aligned}$$

Taking expectations and using the delta method we find that (cf. (5.30))

$$\begin{aligned} E[N_i^* \mid N_i] &\approx E[\tilde{N}_i \mid N_i] - \left(1-\frac{1}{p_i}\right)\left(1-\frac{1}{\theta_i}\right) \\ &\approx N_i, \end{aligned}$$

so that N_i^* is approximately unbiased. On the basis of the discussion in **3.1.1** it is recommended that c_i should be greater than 10 if $\tilde{N}_i$ or N_i^* is to give even the order of magnitude of the true population size. It is also noted that (5.29) and the approximate unbiasedness of N_i^* follow directly from Darroch [1958: 352–3 with $p_1 = p_i$, $p_2 = \theta_i$] : see also Robson [1979].

In conclusion we observe that since $\tilde{N}_i$ is a Petersen estimate, the general theory of **3.2.2** (*2*) can be applied here. This means that if p_i and θ_i are not the same for all members of the population but rather follow a bivariate distribution, then $\tilde{N}_i$ and N_i^* will still be asymptotically unbiased, and (5.32) will be a satisfactory estimate of the *true* variance, provided p_i and θ_i are independent. In this case the only change in (5.30) and (5.31) is to replace p_i and θ_i by $E[p_i]$ and $E[\theta_i]$ where the expectations are with respect to the bivariate distribution.

SPECIFYING THE CLASS C_i. We shall now turn our attention to the problem of specifying a suitable class of C_i individuals for a population in which all emigration is permanent (the case considered in the Jolly–Seber method) and there are no losses on sampling ($R_i = n_i$). Let $r_{\bullet i}$ be the number of individuals from the group of m_i marked animals caught in the ith sample which are later recaptured. Then defining $C_i = r_{\bullet i} + z_i$, the number of marked animals caught both before and after the ith sample, p_i can be estimated by

$$\widetilde{p}_i = r_{\bullet i}/(r_{\bullet i} + z_i).$$

This estimate can be compared with Jolly's estimate

$$\begin{aligned}\hat{p}_i &= \frac{m_i}{\hat{M}_i} \\ &= \frac{m_i}{\left(\dfrac{n_i z_i}{r_i} + m_i\right)} \\ &= \frac{\left(r_i \dfrac{m_i}{n_i}\right)}{\left(r_i \dfrac{m_i}{n_i} + z_i\right)}.\end{aligned}$$

Assuming random sampling, then, when assumption (b) of constant survival probability is true, $r_{\bullet i}/r_i$ will be approximately equal to m_i/n_i and the two estimates of p_i will give similar values. However, if assumption (b) is not true and survival depends on age, for example, $\hat{p}_i$ may be unsuitable if the age structure of the m_i marked individuals differs significantly from the age structure of the whole sample of n_i individuals. This could occur if sampling is not random with respect to age, so that mark status depends on age.

With this choice of C_i we find that θ_i, the probability of being captured both before and after the ith sample, is not the same for all the N_i animals. For example, new immigrants or recruits entering the population between the $(i-1)$th and ith samples will have zero probability of belonging to this class of C_i. However, as pointed out above, Manly's model can still be used if the events "caught in sample i" and "belonging to the class C_i" are independent for each individual. We note that if there are deaths through handling, then these two events are dependent, as a marked animal has a better chance of being recaptured after the ith sample, that is of belonging to C_i, if it is not caught in the ith sample.

Manly and Parr [1968] have given a method for setting out the recapture data and calculating $r_{\bullet i}$ and z_i. However, these quantities are more readily calculated using the methods of this chapter; z_i can be obtained from the table of m_{hi} (cf. Example 5.1 on p. 205) and $r_{\bullet i}$ is easily obtained from the table of $m_{\bullet hi}$ (cf. Example 5.7 on p. 226). For example, using the data of Example 5.7, we find that $(r_{\bullet 2}, r_{\bullet 3}, r_{\bullet 4}) = (15, 14, 10)$ and $(z_2, z_3, z_4) = (11, 12, 20)$. Hence the Manly–Parr estimates can be calculated as in Table

5.14, where they are compared with Jolly's estimates ($\tilde{\sigma}_i = \tilde{V}_i^{\frac{1}{2}}$). There is not a great deal of difference between the two sets of estimates.

TABLE 5.14
Comparison of estimates for a moth population: data from Manly and Parr [1968: 86].

	Manly–Parr method			Jolly–Seber method		
i	$\tilde{p}_i$	$\tilde{N}_i$	$\tilde{\sigma}_i$	$\hat{p}_i$	$\hat{N}_i$	$\hat{\sigma}[\hat{N}_i \mid N_i]$
1	–	–	–	–	–	–
2	0·5769	90·13	12·8	0·5590	93·03	13·1
3	0·5385	96·56	15·0	0·5287	98·35	14·2
4	0·3333	93·01	19·8	0·2914	106·37	23·6
5	–	–	–	–	–	–

5.2.2 Survival estimates

Suppose that $S_i n_i$ of the n_i animals released from the ith sample survive and remain in the population until the $(i+1)$th sample, and let $m_{i,i+1}$ be the number of animals caught in both the ith and $(i+1)$th samples. Then, if sampling is random,

$$E[m_{i,i+1} \mid S_i n_i] = S_i n_i p_{i+1},$$

and S_i can be estimated by

$$\tilde{S}_i = m_{i,i+1}/(n_i \tilde{p}_{i+1}).$$

This estimate may be compared with Jolly's estimate

$$\hat{\phi}_i = \frac{\hat{M}_{i+1}}{\hat{M}_i - m_i + n_i}$$

$$= \frac{m_{i+1}}{(\hat{M}_i - m_i + n_i)\hat{p}_{i+1}}$$

$$= \frac{m_{i+1}\left(\dfrac{r_i}{r_i + z_i}\right)}{n_i \hat{p}_{i+1}}.$$

If survival is independent of mark status then we would expect $m_{i,i+1}/r_i$ to to be approximately equal to $m_{i+1}/(r_i + z_i)$ and the two estimates will give similar results. For example, using the data of Example 5.7 (p. 226) we find that $(\tilde{S}_1, \tilde{S}_2, \tilde{S}_3) = (0{\cdot}76, 0{\cdot}71, 0{\cdot}69)$ which may be compared with $(\hat{\phi}_1, \hat{\phi}_2, \hat{\phi}_3) = (0{\cdot}78, 0{\cdot}74, 0{\cdot}76)$.

Manly and Parr also give an estimate of B_i, the number of new entries into the population between the ith release and $(i+1)$th sample, namely

$$\tilde{B}_i = \tilde{N}_{i+1} - \tilde{S}_i \tilde{N}_i.$$

Thus, for Example 5.7 we have $(\tilde{B}_2, \tilde{B}_3) = (32, 26)$ which are similar to $(\hat{B}_2, \hat{B}_3) = (30, 32)$.

Asymptotic variances for $\tilde{S}_i$ and $\tilde{B}_i$ have not yet been given.

5.3 REGRESSION METHODS

5.3.1 Instantaneous samples

In addition to the notation given in **5.1.1** (*1*) we shall define:

v_i = number of newly marked individuals in the ith release,

R_{ij} = number of animals from the group of v_i which are available for capture in the jth sample,

b_{ij} = number of animals from the group of v_i animals which are caught in the jth sample,

ϕ_{ij} = probability that an animal from the ith release is alive and in the population at the time of the jth sample,

t_{ij} = time from the ith release to the jth sample.

Then if sampling is random, we have, using Bailey's modification of the Petersen estimate (**3.1.2**),

$$E\left[\frac{R_{ij}(n_j+1)}{(b_{ij}+1)} \mid R_{ij}, n_j\right] \approx N_j,$$

and

$$E[R_{ij} \mid v_i] = v_i\phi_{ij}.$$

Therefore, combining these equations and taking logarithms, we are led to consider the regression model

$$E[y_{ij}] \approx \log N_j - \log \phi_{ij}. \tag{5.33}$$

where

$$y_{ij} = \log\left\{\frac{v_i(n_j+1)}{(b_{ij}+1)}\right\}.$$

Now, for given j and random n_j, the b_{ij} and hence the y_{ij} are independent as they come from different mark releases. Also, for given i, the y_{ij} are virtually independent, provided there are no losses on capture ($v_i = u_i$). Even if there are some losses on capture any dependence effect will be small in comparison with sampling errors, so that in general we may assume that the y_{ij} are all independently distributed. There is also some evidence that the logarithmic transformation will, in general, stabilise the variance. Therefore, defining the functional form of ϕ_{ij} and assuming the y_{ij} to be independently normally distributed, (5.33) can be analysed using multiple regression methods. Two examples are considered below.

PERMANENT EMIGRATION. If all emigration is permanent (as was assumed for the Jolly–Seber method) and the probability of surviving and remaining in the population for unit time is constant throughout the experiment (= ϕ say), then, for no losses on sampling,

$$\phi_{ij} = \phi^{t_{ij}}$$

and

$$E[y_{ij}] \approx \log N_j - (\log \phi)t_{ij}.$$

This multiple linear-regression model was proposed by Chapman [1954: 7] and least-squares estimates of log ϕ and $\{\log N_j\}$ can be obtained in the usual manner. As a check on the validity of the model we can plot y_{ij} against t_{ij} for each j. These plots should be approximately linear, approximately parallel, and have much the same scatter about their fitted lines if the assumption of constant variance is applicable.(*)

NO MIGRATION OR RECRUITMENT. When there is no migration or recruitment and N_0 is the population size at time zero, then

$$N_j = N_0 \phi^{t_j}$$

and

$$E[y_{ij}] \approx \log N_0 + (\log \phi)(t_j - t_{ij}).$$

5.3.2 Continuous sampling

We shall now consider the case of an exploited population in which the sampling (exploitation) is continuous. When there is no migration or recruitment and both the instantaneous natural mortality rate μ and the instantaneous exploitation rate μ_E are constant, Paulik [1963a] gives the following general regression model for estimating μ and μ_E (the underlying theory comes from pp. 272–4):

Since the sampling is continuous we shall divide up the total experimental period into equal time-intervals and consider the marked catch only in each interval. It is assumed that the releases are timed so that each release is made at the beginning of one of the time-intervals. Let

m_{ij} = number from the ith release caught in the jth time-interval, where j is now measured *relative* to the ith release ($i = 1, 2, \ldots, I$; $j = 1, 2, \ldots, J_i$), and

p_{ij} = probability that a marked individual from the ith release is caught in the jth time-interval after the release.

Then from (6.9),

$$\begin{aligned} p_{ij} &= \text{(probability of survival for } j-1 \text{ intervals)} \\ &\quad \times \text{(probability of capture in one interval)} \\ &= e^{-Z(j-1)} P_1, \end{aligned}$$

where $Z = \mu + \mu_E$ (the total instantaneous mortality rate) and, putting $J = 1$ in (6.6),

$$P_1 = \mu_E(1 - e^{-Z})/Z.$$

(*)A similar model based on times between successive captures, but assuming constant p is given by Van Noordwijk [1978].

Since $E[m_{ij}] = R_i p_{ij}$, we are led to consider the regression model

$$E[Y_{ij}] \approx -(Z+\log P_1) + Zj, \qquad (5.34,$$

where $Y_{ij} = \log(R_i/m_{ij})$. As P_1 and Z are functions of μ and μ_E, estimates of Z and $-(Z+\log P_1)$ will provide estimates of μ and μ_E.

Several procedures for estimation are available. For example, (5.34) can be analysed as a series of independent regression lines (one for each release), using the methods of **6.2.2**. The estimates of μ and μ_E can be averaged and sample estimates of the variances calculated, using the methods of **1.3.2.** Alternatively (5.34) can be analysed as a single multiple linear-regression model, using the model of **1.3.5** (*2*) with

$$\mathbf{Y}' = (Y_{11}, Y_{12}, \ldots, Y_{1J_1}, \ldots, Y_{I1}, Y_{I2}, \ldots, Y_{IJ_I})$$

$$\mathbf{X} = \begin{bmatrix} 1 & 1 \\ 1 & 2 \\ \multicolumn{2}{c}{\ldots} \\ 1 & J_1 \\ \hline \multicolumn{2}{c}{\ldots} \\ \hline 1 & 1 \\ 1 & 2 \\ \multicolumn{2}{c}{\ldots} \\ 1 & J_I \end{bmatrix}, \qquad \boldsymbol{\beta} = \begin{bmatrix} -(Z+\log P_1) \\ Z \end{bmatrix},$$

and $\mathbf{B}^{-1}$ given by (6.37) and (6.38) (except that j ranges from 1 to J_i and not 1 to J in $\mathbf{B}_i^{-1}$).

The case when the releases are made at different times *before* exploitation commences is considered in **6.4.**

5.4 TAG RETURNS FROM DEAD ANIMALS(*)

5.4.1 Survival independent of age: general model

We shall now discuss a variation of the multi-sample single-recapture census (p. 212) which is particularly useful for estimating survival probabilities for bird populations. Consider an experiment in which tagged animals are released at the beginning of each year for s consecutive years. The animals then die through natural mortality, predation, hunting, etc., and for t consecutive years ($t \geqslant s$) a record is kept of tags returned from dead animals.

(*)This section should be read in conjunction with **13.1.5.**

NOTATION AND ASSUMPTIONS. We shall assume that:

(i) Every tagged animal (irrespective of age or time of release) has the same probability ϕ_i of survival for year i, given that it is alive at the beginning of the year ($i = 1, 2, \ldots, t$).

(ii) Every tagged animal which dies in year i has the same probability λ_i of being found and its tag reported in year i; $100\lambda_i$ per cent is usually called the *recovery rate* for year i.

Assumption (i) implies that survival is time-dependent rather than age-dependent. This is not an unreasonable assumption for, say, adult bird populations where the survival rate is fairly constant except for fluctuations due to environmental changes in weather, hunting pressure, etc. The definition of $1 - \lambda_i$ includes two types of non-return: first, there are tagged individuals which die and are not found; and secondly, the tags may be found and not returned.

Let

R_i = number of tagged animals released at the beginning of year i ($i = 1, 2, \ldots, s$),

m_{ij} = number from the ith release which die in year j and are reported in the same year ($j = i, i+1, \ldots, t$),

m_j = number of tags recovered in year j,

D_j = number of tags returned from animals which die in their jth year after release, i.e. j is measured *relative* to the year of release,

$r_i = \sum_{j=i}^{t} m_{ij}$ (the number of tags recovered from the ith release),

$z_i = \sum_{j=i}^{t} m_{1j} + \sum_{j=i}^{t} m_{2j} + \ldots + \sum_{j=i}^{t} m_{i-1,j}$ (the total number of tags recovered after the ith release from releases prior to the ith; $i = 1, 2, \ldots, s$ and $z_1 = 0$),

$T_i = r_i + z_i$, (the total number of tags recovered after the ith release from the first i releases),

and define the "working" parameters

$$\alpha_i = \phi_i,$$
$$\beta_i = (1-\phi_i)\lambda_i,$$
$$\gamma_j = \alpha_s\alpha_{s+1} \ldots \alpha_{j-1}\beta_j, \quad (j = s+1, s+2, \ldots, t), \text{ and}$$
$$\theta_i = \beta_i + \alpha_i\beta_{i+1} + \ldots + \alpha_i\alpha_{i+1} \ldots \alpha_{t-1}\beta_t.$$

Here β_i is the probability that a tag is recovered from an animal in year i, given that the animal is alive at the beginning of the year; θ_i is the probability that a tag is recovered from an animal after the ith release, given that the animal is alive at the time of the ith release; and γ_j is the probability that a tag is recovered from an animal in year j, given that it is alive at the time of the sth release.

ESTIMATION. The joint probability function of the $\{m_{ij}\}$ is

$$f(\{m_{ij}\}|\{R_i\})$$

$$= \prod_{i=1}^{s} \left\{ \frac{R_i!}{[\prod_{j=i}^{t} m_{ij}!][(R_i - r_i)!]} ((1-\phi_i)\lambda_i)^{m_{ii}} (\phi_i(1-\phi_{i+1})\lambda_{i+1})^{m_{i,i+1}} \ldots \times (\phi_i\phi_{i+1} \ldots \phi_{t-1}(1-\phi_t)\lambda_t)^{m_{it}} (1-\theta_i)^{R_i - r_i} \right\}$$

$$\propto \prod_{i=1}^{s} \{\beta_i^{m_{ii}}(\alpha_i\beta_{i+1})^{m_{i,i+1}} \ldots (\alpha_i\alpha_{i+1} \ldots \alpha_{t-1}\beta_t)^{m_{it}}(1-\theta_i)^{R_i - r_i}\} \qquad (5.35)$$

$$= \{\prod_{i=1}^{s-1} \alpha_i^{z_{i+1}} \beta_i^{m_i} (1-\theta_i)^{R_i - r_i}\}\{\beta_s^{m_s} (1-\theta_s)^{R_s - r_s}\}\{\gamma_{s+1}^{m_{s+1}} \gamma_{s+2}^{m_{s+2}} \ldots \gamma_t^{m_t}\} \qquad (5.36)$$

and the maximum-likelihood estimates $\hat{\alpha}_i$, $\hat{\beta}_i$ and $\hat{\gamma}_i$ are obtained in the usual manner. However, Seber [1970b] has shown that these estimates can also be obtained by intuitive arguments. For example, the obvious estimate of θ_i is $\hat{\theta}_i = r_i/R_i$, so that from

$$\beta_i = \Pr[\text{recovered in year } i \mid \text{recovered after } i\text{th release}] \times \Pr[\text{recovered after } i\text{th release}]$$

we have the intuitive estimate

$$\hat{\beta}_i = \frac{m_i}{T_i} \cdot \hat{\theta}_i, \quad (i = 1, 2, \ldots, s).$$

As $\alpha_i = (\theta_i - \beta_i)/\theta_{i+1}$, α_i can be estimated by

$$\hat{\alpha}_i = \frac{\hat{\theta}_i - \hat{\beta}_i}{\hat{\theta}_{i+1}} = \frac{T_i - m_i}{T_i} \cdot \frac{\hat{\theta}_i}{\hat{\theta}_{i+1}},$$

and from

$$\frac{E[m_s]}{E[m_j]} = \frac{\beta_s}{\gamma_j}, \quad (j > s),$$

we have the estimate

$$\hat{\gamma}_j = \hat{\beta}_s m_j/m_s, \quad (j = s+1, \ldots, t).$$

Finally, as the transformation

$$\hat{\phi}_i = \hat{\alpha}_i \quad (i = 1, 2, \ldots, s-1)$$
$$\hat{\lambda}_i = \hat{\beta}_i/(1-\hat{\alpha}_i) \quad (i = 1, 2, \ldots, s-1)$$
$$\hat{\beta}_s = \hat{\beta}_s$$
$$\hat{\gamma}_j = \hat{\gamma}_j \quad (j = s+1, \ldots, t)$$

is one-to-one, $\hat{\phi}_i$ and $\hat{\lambda}_i$ are the maximum-likelihood estimates of ϕ_i and λ_i respectively. We note that ϕ_s and λ_s cannot be separately estimated through lack of identifiability; only the product $\beta_s = (1-\phi_s)\lambda_s$ is estimable. However, if $\hat{\phi}_1, \hat{\phi}_2, \ldots, \hat{\phi}_{s-1}$ are approximately the same, we could assume that ϕ_i is constant ($= \phi$ say), so that

$$\lambda_s = \beta_s/(1-\phi)$$

and

$$\lambda_{j+1} = \lambda_j \gamma_{j+1}/(\phi\gamma_j), \quad (j = s, \ldots, t-1: \gamma_s = \beta_s).$$

Estimating ϕ by ϕ^* (say the average of the $\hat{\phi}_i$), we have the estimates

$$\lambda_s^* = \hat{\beta}_s/(1-\phi^*)$$

and

$$\lambda_{j+1}^* = \lambda_j^* m_{j+1}/(\phi^* m_j), \quad (j = s, \ldots, t-1).$$

If we had assumed $\phi_i = \phi$ $(i = 1, 2, \ldots, t)$ right from the start, we would find that the maximum-likelihood equations cannot be solved explicitly, and iterative methods of solution are required.

CALCULATIONS. In performing the calculations we require r_i $(i = 1, 2, \ldots, s)$, m_j $(j = 1, 2, \ldots, t)$ and T_i $(i = 1, 2, \ldots, s)$; one method for obtaining these quantities is given by Seber (in his symbols they are called R_i, C_j and T_i respectively; also α_i and β_i are interchanged). However, it transpires that the tabular methods used for the Jolly–Seber method (Example 5.1, p. 206) can also be used here if we work with the random variables r_i, m_j and z_i $(z_1 = 0)$. We then have the sequence of estimates

$$\hat{\theta}_i = r_i/R_i,$$

$$\hat{\beta}_i = \hat{\theta}_i m_i/(r_i + z_i)$$

and, since $T_i - m_i = T_{i+1} - r_{i+1} = z_{i+1}$,

$$\hat{\phi}_i = \frac{z_{i+1}}{r_i + z_i} \cdot \frac{\hat{\theta}_i}{\hat{\theta}_{i+1}}, \quad (i = 1, 2, \ldots, s-1),$$

and

$$\hat{\lambda}_i = \hat{\beta}_i/(1-\hat{\phi}_i), \quad (i = 1, 2, \ldots, s-1).$$

It is noted that (5.35) is a slight extension of the model (5.23) with $\theta_i = 1 - \chi_i$. The main differences between the two models lie in the additional parameters γ_i and the ranges of the suffixes. In (5.23) the first sample after the first release (which is equivalent to the first recovery year in (5.35)) is called sample number 2, so that one has the term m_{i+1} in (5.24) but m_i in (5.36).

VARIANCES AND COVARIANCES. Using the delta method, Seber [1970b] shows that the asymptotic variance and covariance formulae for the estimates $\hat{\phi}_i$ and $\hat{\lambda}_i$ are

$$V[\hat{\phi}_i] = \phi_i^2\left\{\frac{1}{E[r_i]} + \frac{1}{E[r_{i+1}]} + \frac{1}{E[z_{i+1}]} - \frac{1}{R_i} - \frac{1}{R_{i+1}} - \frac{1}{E[r_i + z_i]}\right\}, \tag{5.37}$$

$$\text{cov}[\hat{\phi}_i, \hat{\phi}_j] = 0, \quad j > i+1,$$

$$\text{cov}[\hat{\phi}_i, \hat{\phi}_{i+1}] = -\phi_i\phi_{i+1}\left\{\frac{1}{E[r_{i+1}]} - \frac{1}{R_{i+1}}\right\},$$

and

$$V[\hat{\lambda}_i] = \frac{\lambda_i^2}{(1-\phi_i)^2}\left\{\frac{1}{E[r_i]} - \frac{1}{E[r_i+z_i]} - \frac{1}{R_i} + \frac{(1-\phi_i)^2}{E[m_i]} + \right.$$

$$\left. + \phi_i^2\left(\frac{1}{E[r_{i+1}]} + \frac{1}{E[z_{i+1}]} - \frac{1}{R_{i+1}}\right)\right\}. \qquad (5.38)$$

The above variances and covariances can be estimated by simply replacing the unknown parameters by their estimates and the expectations by the observed values of the random variables, e.g. r_i replaces $E[r_i]$, etc.

Since

$$\frac{1}{E[r_i]} - \frac{1}{E[r_i+z_i]} = \frac{1}{\theta_i}\left\{\frac{1}{R_i} - \frac{1}{\sum_{h=1}^{i} R_h\beta_h \ldots \beta_{i-1}}\right\},$$

and θ_i increases with t, $V[\hat{\phi}_i]$ and $V[\hat{\lambda}_i]$ both decrease as t increases. This means that by increasing t we can increase the number of recoveries and thereby increase the precision of our estimates.

It should be noted that the above formulae are asymptotic, and their usefulness will depend not only on the validity of the underlying model but also on the expected numbers of recaptures. For example, if the underlying model is appropriate, (5.37) is valid for large $E[r_i]$, $E[r_{i+1}]$, etc., while (5.38) generally holds for large $(1-\phi_i)E[r_i]$, $(1-\phi_i)E[r+z_i]$, etc. If $(1-\phi_i)$ is small (i.e. low mortality rates), there will be few dead animals available for recovery, so that $\hat{\lambda}_i$ will be based on small samples and $V[\hat{\lambda}_i]$ will be large. This sensitivity of the variance to $(1-\phi_i)$ is indicated by the presence of the term $(1-\phi_i)^2$ in the denominator of $V[\hat{\lambda}_i]$.

GOODNESS-OF-FIT TEST. Assuming the R_i to be large, a standard chi-squared goodness-of-fit test for the model can be carried out using the statistic (cf. Mitra [1958])

$$T = \sum_{i=1}^{s}\sum_{j=i}^{t}(m_{ij}-E_{ij})^2/E_{ij},$$

which is approximately distributed as chi-squared with $(s-1)(t-1) - \frac{1}{2}s(s-1)$ degrees of freedom (i.e. $st - \frac{1}{2}s(s-1)$ square terms minus $s+t-1$, the number of parameters estimated). Here,

$$E_{ii} = R_i\hat{\beta}_i \qquad (i = 1, 2, \ldots, s)$$

$$E_{ij} = \begin{cases} R_i\hat{\phi}_i\hat{\phi}_{i+1} \ldots \hat{\phi}_{j-1}\hat{\beta}_j & (j = i+1, \ldots, s) \\ R_i\hat{\phi}_i\hat{\phi}_{i+1} \ldots \hat{\phi}_{s-1}\hat{\gamma}_j & (j = s+1, \ldots, t), \end{cases}$$

and the terms involving $R_i - r_i$ are not included in T as their contribution is zero ($R_i - r_i - R_i(1-\hat{\theta}_i) = 0$).

Example 5.10 Pink-footed geese: Seber [1970b]

The above method was applied to banding data (Table 5.15) on adult

TABLE 5.15
Numbers of ringed, pink-footed geese m_{ij} recovered over a period of 4 years (from Boyd [1956: Table 1]); the expected recoveries E_{ij} are in brackets.

Release R_i	Recovery year (j) 1	2	3	4	Recoveries r_i
301	32 (32)	22 (23·79)	16 (15·35)	7 (5·85)	77
766		70 (68·21)	50 (44·01)	9 (16·78)	129
897			52 (58·64)	29 (22·36)	81
m_i	32	92	118	45	287

TABLE 5.16
Cumulated columns of Table 5.15 for the calculation of z_i.

Recovery year (j) 1	2	3	4	
32	22	16	7	$45 = z_2$
	92	66	16	$82 = z_3$
		118	45	$45 = z_4$

pink-footed geese from Boyd [1956: Table 1, adults only]. As a first step in the calculations, the r_i and z_i are obtained from Tables 5.15 and 5.16; the method for constructing Table 5.16 is described in Example 5.1 (p. 205). Then, for example,

$$\hat{\theta}_1 = 77/301, \quad \hat{\theta}_2 = 129/766,$$

$$\hat{\phi}_1 = \frac{45}{77+0} \cdot \frac{77}{301} \cdot \frac{766}{129} = 0{\cdot}887\ 74,$$

$$\hat{\beta}_1 = \frac{77}{301} \cdot \frac{32}{77+0} = 0{\cdot}106\ 31,$$

and

$$\hat{\lambda}_1 = 0{\cdot}106\ 31/(1-0{\cdot}887\ 74) = 0{\cdot}9470.$$

Estimates together with standard deviations are given in Table 5.17.

TABLE 5.17
Estimates of the survival and recovery probabilities for a population of pink-footed geese: from Seber [1970b: Table 5].

i	$\hat{\beta}_i$	$\hat{\phi}_i \pm \hat{\sigma}[\hat{\phi}_i]$	$\hat{\lambda}_i \pm \hat{\sigma}[\hat{\lambda}_i]$
1	0·106 31	0·8877 ± 0·141	0·9470 ± 1·18
2	0·089 04	0·8789 ± 0·137	0·7352 ± 0·83
3	0·065 37	–	–

Since $\hat{\phi}_1 \approx \hat{\phi}_2$, we can assume ϕ_i to be approximately constant and estimate λ_3 and λ_4 using the method given on p. 242. Let $\phi^* = (\hat{\phi}_1 + \hat{\phi}_2)/2 = 0{\cdot}8833$; then

$$\lambda_3^* = \hat{\beta}_3/(1-\phi^*) = 0{\cdot}5602$$

and

$$\lambda_4^* = \lambda_3^* m_4/(\phi^* m_3) = 0{\cdot}242.$$

To carry out the goodness-of-fit test we require

$$\hat{\gamma}_4 = m_4\hat{\beta}_3/m_3 = 0{\cdot}024\ 93.$$

The E_{ij} can then be calculated, e.g. $E_{13} = R_1\hat{\phi}_1\hat{\phi}_2\hat{\beta}_3$, $E_{14} = R_1\hat{\phi}_1\hat{\phi}_2\hat{\gamma}_4$, etc., and these are given in brackets in Table 5.15. It is found that $T = \Sigma\Sigma\ (m_{ij} - E_{ij})^2/E_{ij} = 7{\cdot}58$, which, for 3 degrees of freedom, is not significant at the 5 per cent level (though only just; $\chi_3^2[0{\cdot}05] = 7{\cdot}8$).

In conclusion we note from the values of $\hat{\sigma}[\hat{\lambda}_i]$ that the estimates $\hat{\lambda}_i$ are not reliable, in fact they seem far too high.

5.4.2. Survival independent of age: special cases(*)

1 Constant recovery probability

It was noted briefly in the previous section that for the case $\phi_i = \phi$ $(i = 1, 2, \ldots, t)$ the maximum-likelihood equations for ϕ and λ_i must be solved iteratively. The same is true when $\lambda_i = \lambda$ $(i = 1, 2, \ldots, t)$ and ϕ_i varies; both these models were used by Boyd [1956], though details of the calculations are not given. In the latter case, however, if t is much greater than s, so that the probability of a banded bird dying by the end of the tth year is virtually unity, i.e.

$$(1-\phi_i) + \phi_i(1-\phi_{i+1}) + \ldots + \phi_i\phi_{i+1} \ldots (1-\phi_t) \approx 1,$$

then

$$\beta_i = (1-\phi_i)\lambda,$$
$$\theta_i = \lambda,$$

and ϕ_i can be estimated by ϕ_i^*,

where $\quad 1 - \phi_i^* = \hat{\beta}_i/\hat{\theta}_i = m_i/T_i$.

This estimate has been used by several authors e.g. Hickey [1952: 14, Table 3, where T_i is calculated for $i = s, s+1, \ldots, t$ only] and Coulson [1960: 206]. For a further discussion of this method of estimation cf. Seber [1972].

2 Constant survival and recovery probabilities

If $\phi_i = \phi$ and $\lambda_i = \lambda$ $(i = 1, 2, \ldots, t)$, then from (5.36)

$$f(\{m_{ij}\}|\{R_i\}) \propto (1-\phi)^{m_{11}}(\phi(1-\phi))^{m_{12}} \ldots (\phi^{t-1}(1-\phi))^{m_{1t}} \lambda^{r_1}(1-\theta_1)^{R_1-r_1}$$
$$\times\ (1-\phi)^{m_{22}}(\phi(1-\phi))^{m_{23}} \ldots (\phi^{t-2}(1-\phi))^{m_{2t}}\lambda^{r_2}(1-\theta_2)^{R_2-r_2} \ldots,$$

where $\theta_i = \lambda(1-\phi^{t-i+1})$. Unfortunately the maximum-likelihood equations are again complicated by the presence of λ and must be solved iteratively. However, since the r_i are independent binomial variables with parameters

(*) See 13.1.5.

R_i and θ_i, it transpires that the conditional distribution $f(\{m_{ij}\}|\{r_i\})$ does not depend on λ. Thus

$$\begin{aligned} f(\{m_{ij}\}|\{r_i\}) &\propto (1-\phi)^{D_1}(\phi(1-\phi))^{D_2} \ldots (\phi^{t-1}(1-\phi))^{D_t} \\ &\times (1-\phi^t)^{-r_1}(1-\phi^{t-1})^{-r_2} \ldots (1-\phi^{t-s+1})^{-r_s} \\ &= (1-\phi)^{\sum_{j=1}^{t} D_j} \phi^{\sum_{j=1}^{t}(j-1)D_j} \prod_{i=1}^{s} \{(1-\phi^{t-i+1})^{-r_i}\}, \end{aligned} \quad (5.39)$$

where D_j is the number of tags returned from animals which die in their jth year after release (see Table 5.19 below).

The maximum-likelihood estimate $\hat{\phi}$ of ϕ for (5.39) is the solution of

$$\sum_{j=1}^{t} \frac{(j-1)D_j}{\phi} - \sum_{j=1}^{t} \frac{D_j}{1-\phi} + \sum_{i=1}^{s} \left\{\frac{r_i(t-i+1)\phi^{t-i}}{1-\phi^{t-i+1}}\right\} = 0, \quad (5.40)$$

which must be solved iteratively. For computational convenience (5.40) can be rewritten as

$$-\frac{\sum_{j=1}^{t} D_j}{1-\phi} + \sum_{i=1}^{s} \left\{\frac{r_i(t-i+1)}{1-\phi^{t-i+1}}\right\} = \sum_{i=1}^{s} r_i(t-i+1) - \sum_{j=1}^{t} jD_j \quad (5.41)$$

or

$$g(\phi) = c,$$

which can be solved using the methods of **1.3.8** (*1*).

If f represents (5.39) then from maximum-likelihood theory (cf. (1.4))

$$V[\hat{\phi}] \approx -1/E\left[\frac{\partial^2 \log f}{\partial \phi^2}\right],$$

and $V[\hat{\phi}]$ can be estimated by

$$\hat{V}[\hat{\phi}] = -1\bigg/\left[\frac{\partial^2 \log f}{\partial \phi^2}\right]_{\phi=\hat{\phi}}.$$

Therefore, carrying out the differentiation and using (5.40), we find that

$$\hat{V}[\hat{\phi}] = \left(\sum_{j=1}^{t} \frac{D_j}{\hat{\phi}(1-\hat{\phi})^2} - \sum_{i=1}^{s} \left\{\frac{r_i(t-i+1)^2\hat{\phi}^{t-i-1}}{(1-\hat{\phi}^{t-i+1})^2}\right\}\right)^{-1}.$$

When $t = s$, the case considered by Haldane [1955], we note that

$$\sum_{i=1}^{s} \frac{r_i(s-i+1)}{1-\phi^{s-i+1}} = \frac{r_s}{1-\phi} + \frac{2r_{s-1}}{1-\phi^2} + \ldots + \frac{sr_1}{1-\phi^s}$$

in (5.41).

A regression method for estimating both ϕ and λ by least squares is given in 5.4.3 (cf. equation (5.46)).

METHOD OF LACK. If t is much greater than s, so that $\phi^{t-s+1} \approx 0$, then from (5.39)

$$f(\{m_{ij}\}|\{r_i\}) \propto (1-\phi)^{\Sigma D_j} \phi^{\Sigma(j-1)D_j}.$$

Hence $\tilde{\phi}$, the maximum-likelihood estimate of ϕ, is now given by

$$\tilde{\phi} = 1 - (\sum_j D_j / \sum_j jD_j),$$

an estimate suggested by Lack [1943, 1951] and Farner [1945]. Using the same method as for $\hat{\phi}$, it is readily shown that

$$\tilde{V}[\tilde{\phi}] = (1-\tilde{\phi})^2 \tilde{\phi} / \sum_j D_j$$

is an estimate of the asymptotic variance of $\tilde{\phi}$.

We note that for the ith release, the probability that a tag is recovered in the jth year after release, given that it is recovered, is

$$f(j) = \frac{\phi^{j-1}(1-\phi)\lambda}{(1-\phi^{t-i+1})\lambda} \quad (j = 1, 2, \ldots, t-i+1) \tag{5.42}$$

$$\approx \phi^{j-1}(1-\phi) \tag{5.43}$$

for large t. Haldane [1953] also discussed the case of large t using (5.43) as the basis for his approach. He assumed that we have $\sum_j D_j$ ($= N$ say) observations from this geometric distribution and hence obtained $\tilde{\phi}$ and $\tilde{V}[\tilde{\phi}]$ above. He also suggested a goodness-of-fit test of the data by comparing the observed frequencies D_j with the expected frequencies

$$E_j = N\tilde{\phi}^{j-1}(1-\tilde{\phi}). \tag{5.44}$$

CHOOSING A FIRST APPROXIMATION. The main difficulty in solving (5.41) lies in the choice of a suitable first approximation for $\hat{\phi}$. One method for doing this consists of guessing the value of ϕ, say ϕ_1, then truncating the table of m_{ij} horizontally at release number s', say, so that $\phi_1^{t-s'+1} \approx 0$, and then estimating ϕ from this reduced table using Lack's estimate $\tilde{\phi}$.

APPLICATIONS TO BIRD POPULATIONS. The above methods of Haldane and Lack have been widely used for bird populations where the birds are banded as nestlings before they leave the nest. However, as mortality both in the nest and in the first few months of independent life is heavy, bands recovered from these birds should be ignored, as far as the above methods are concerned, for a suitable initial period. For example, as British birds are mostly hatched from April to July, Lack recommends that tables of band returns be based on recoveries during each calendar year (cf. Farner [1949: 68; 1955]); that is, birds whose corpses are discovered between the first and second New Year's Day of their lives are said to have died at the age of one year, and so on; thus D_j is the number of bands recovered from birds that die at age j. But a number of authors suggest earlier dates (Hickey [1952: 15–16]), while for the case of wood-pigeons, Murton [1966] recommends extending the initial period until March of the following year on the grounds that inexperienced birds tend to be more susceptible to shooting.

TABLE 5.18

Application of Lack's method to recoveries from lapwings ringed as nestlings from 1909 to 1939 inclusive: from Haldane [1955: Table 1].

Recovery age (j)	Number recovered (D_j)	jD_j	E_j	$\frac{(D_j-E_j)^2}{E_j}$
1	194	194	206·38	0·74
2	145	290	136·21	0·57
3	90	270	89·90	0·00
4	54	216	59·33	0·48
5	48	240	39·16	2·00
6	25	150	25·85	0·03
7	24	168	17·06	2·82
8	9	72	11·25	0·45
9	6	54	7·43	0·28
10	5	50	4·90	0·00
11	5 } 7	55	9·53	0·67
12	1	12		
13	0	0		
14	1	14		
Total	607	1785	607·00	8·04

Example 5.11 Lapwing (*Vanellus vanellus*): Haldane [1955]

The recoveries from lapwings ringed as nestlings from 1909 to 1939 inclusive are given in Table 5.18. Using Lack's method we have

$$\tilde{\phi} = 1 - (607/1785) = 0{\cdot}6599$$

and

$$\tilde{\sigma}[\tilde{\phi}] = (1-\tilde{\phi})[\tilde{\phi}/\sum_j D_j]^{\frac{1}{2}} = 0{\cdot}0112.$$

The expected frequencies E_j (cf. 5.43) were also calculated, and from Table 5.18 we have

$$\sum_j (D_j - E_j)^2/E_j = 8{\cdot}04,$$

which, for 9 degrees of freedom, indicates an excellent fit.

Table 5.19 gives the recovery data for lapwings ringed from 1940 to 1951; here $s = t = 12$ and Haldane's method of analysis can be used. As a first step we have to solve $g(\phi) = c$, where

$$c = \sum_{i=1}^{s} r_i(s-i+1) - \sum_{j=1}^{s} jD_j = 640 - 237 = 403$$

and, from (5.41),

$$g(\phi) = \frac{(-120+9)}{1-\phi} + \frac{36}{1-\phi^2} + \frac{45}{1-\phi^3} + \cdots + \frac{84}{1-\phi^{12}}.$$

As a first approximation we can use $\tilde{\phi} = 0{\cdot}66$; thus Haldane calculates

TABLE 5.19
Application of Haldane's method to recoveries from lapwings ringed as nestlings from 1940 to 1951: from Haldane [1955: Table 2].

Year ringed	i	Age on recovery (j) 1	2	3	4	5	6	7	8	9	10	11	12	(1) r_i	(2) $12-i+1$	(1) × (2)
1940	1	2	1	0	2	1	0	1	0	0	0	0	0	7	12	84
1941	2	1	0	1	0	0	0	1	0	0	0	0		3	11	33
1942	3	2	2	1	0	2	0	0	0	0	0			7	10	70
1943	4	2	1	0	0	0	0	0	0	0				3	9	27
1944	5	4	1	3	2	0	0	0	1					11	8	88
1945	6	7	3	0	0	0	1	0						11	7	77
1946	7	6	0	0	0	0	0							6	6	36
1947	8	4	3	5	2	1								15	5	75
1948	9	6	4	2	3									15	4	60
1949	10	10	2	3										15	3	45
1950	11	15	3											18	2	36
1951	12	9												9	1	9
Totals	D_j	68	20	15	9	4	1	2	1	0	0	0	0	120		640
	jD_j	68	40	45	36	20	6	14	8	0	0	0	0	237		

$g(0{\cdot}66) = 367{\cdot}7$, $g(0{\cdot}60) = 457{\cdot}9$ and $g(0{\cdot}63) = 398{\cdot}7$, which is a close enough approximation when one considers $\hat{\sigma}[\hat{\phi}]$. Thus

$$\hat{\phi} = 0{\cdot}63, \quad \hat{V}[\hat{\phi}] = 0{\cdot}001\ 567,$$

and hence

$$\hat{\sigma}[\hat{\phi}] = 0{\cdot}039\ 58.$$

Also the expectation of life at birth is estimated to be (cf. (1.3))

$$-1/\log \hat{\phi} = 2{\cdot}2 \text{ years}.$$

To compare $\hat{\phi}$ with $\tilde{\phi}$ we compute

$$z = \frac{|\hat{\phi} - \tilde{\phi}|}{\sqrt{\hat{V}[\hat{\phi}] + \tilde{V}[\tilde{\phi}]}} = \frac{0{\cdot}03}{\sqrt{0{\cdot}001\ 693}} = 0{\cdot}75$$

and enter this value in the tables of the unit normal distribution. As $z < 1{\cdot}96$ we do not reject the hypothesis that ϕ is different for the two periods 1909–1939 and 1940–1951. The two estimates may then be combined, using weights (cf. **1.3.2**) $w_1 = \{\hat{V}[\hat{\phi}]\}^{-1}$ and $w_2 = \{\tilde{V}[\tilde{\phi}]\}^{-1}$, thus

$$\phi' = (w_1\hat{\phi} + w_2\tilde{\phi})/(w_1 + w_2) = 0{\cdot}658,$$

$$V[\phi'] = (w_1 + w_2)^{-1} = 0{\cdot}000\ 112,$$

and

$$\sigma[\phi'] = 0{\cdot}0106.$$

A similar example is set out in some detail in Murton [1966: 185–191].

TABLE 5.20
Estimates of the average annual adult survival rates of starlings according to cause of death: from Coulson [1960: Table 1].

	Recoveries (D_j)										$\tilde{\phi} \pm \tilde{\sigma}[\tilde{\phi}]$
$j =$	1	2	3	4	5	6	7	8	9	10	(%)
"Found dead"	482	233	101	57	26	14	9	1	1	1	48·9 ± 1·2
Cat	57	26	19	8	2	2	1	1	0	0	50·6 ± 3·3
Shot	157	50	27	11	4	2	1	0	0	0	40·2 ± 2·4
Falling down chimney	22	9	5	4	1	0	0	0	0	0	46·0 ± 5·7
Other predators	12	4	1	2	0	1	0	0	0	0	46·0 ± 8·1
Other causes	7	15	4	3	0	0	0	1	0	0	
Total	737	337	157	85	33	19	11	3	1	1	47·2 ± 1·0

Example 5.12 Starling (*Sturnus vulgaris*): Coulson [1960]

In Great Britain there have been more starlings ringed and recovered than any other bird. Coulson has used the methods of Lack and Haldane to analyse some of these extensive data. With the large numbers it was possible to break down the data into various sub-categories, such as cause of death, regions where ringed and recovered, sex, and even country where recovered. For example, Table 5.20 uses Lack's method for starlings ringed before 1950 to relate survival estimates to cause of death. The estimates of ϕ are given in the final column, together with their standard deviations; there is a significant difference between the survival rates for those "found dead" and those shot.

ESTIMATING REPORTING RATES. Four methods have been suggested for estimating the band reporting rate (Tomlinson [1968], Henny [1967]):

(i) If the birds are hunted then the number of bands reported can be compared with the estimated number (from kill surveys) of banded birds bagged. This method was used by Geis and Atwood [1961], Martinson [1966] and Martinson and McCann [1966] to determine band-reporting rates for recovered waterfowl in the late 1950's and 1960's.

(ii) If some of the birds carry a band inscribed "reward", then the proportion of reward bands returned can be compared with the proportion of ordinary bands. Bellrose [1955] used this method to determine reporting rates for waterfowl, while Tomlinson [1968] applied it to mourning doves (cf. Table 5.21). However, Tomlinson points out that the reward band method is unsatisfactory in two respects: (a) If it becomes known that there are two types of bands, one that yields a reward and another that does not, there may be fewer non-reward bands reported with a consequent reduction in the overall band-reporting rate. (b) There is not necessarily a 100 per cent reporting of all reward bands recovered, so that the actual reporting rates may be less than the rates estimated by this method. For a recent study see Henny and Burnham [1976].

TABLE 5.21
The estimation of band reporting rates in U.S.A., using reward banding for states whose rates of recovery are probably not influenced by local game-management activities: from Tomlinson [1968: Table 2].

		Reward bands			Ordinary bands		
			Recoveries			Recoveries	
Location	Year	Bandings	No.	%	Bandings	No.	%
California	1965	35	2	5·71	71	3	4·23
Idaho	1966	43	2	4·65	147	2	1·36
Maryland	1965	50	9	18·00	159	7	4·40
Maryland	1966	50	10	20·00	254	13	5·12
Missouri	1966	50	3	6·00	99	4	4·04
South Carolina	1965	49	8	16·33	97	1	1·03
South Dakota	1966	50	3	6·00	150	1	0·67
Texas	1966	25	2	8·00	75	4	5·33
Virginia	1965	50	4	8·00	97	4	4·12
Washington	1965	50	6	12·00	100	3	3·00
Average				10·47			3·33

Band reporting rate = 100 (3·33)/10·47 = 31·81 per cent

(iii) A comparison of band serial numbers taken at check stations with band serial numbers reported will give an estimate of the reporting rate. Stair [1957] used this method for mourning doves in Arizona.

(iv) The reporting rate can be estimated by comparing two independent estimates of hunting mortality where one estimate is not affected by incomplete reporting. Details of one such method are given in Henny [1967].

BAND LOSSES. A serious problem in bird-banding studies is the loss of bands, loss of legibility of numbers and directions printed on the bands, etc. Hickey [1952: 18 ff.], Farner [1955: 422], Paynter [1966]). Apart, however, from the double-tagging methods of **6.3.3** for estimating tag loss, very little has been done in setting up suitable models for the estimation of such losses (c.f. Ludwig [1967] for a band wear model). A useful study of the effect of band loss is given by Nelson *et al.* [1980] (c.f. Additional References).

5.4.3 Age-dependent survival(*)

The case when survival depends on age rather than time has received little attention, from a mathematical point of view, in the literature so far. In the following discussion we shall consider a number of special cases only.

Suppose that the animals are banded and released at birth, and let ϕ_i now be defined as the probability that a live animal of exact age $i-1$ years survives for a further year. Then, using the same notation as that given in

(*)See also **13.1.5** (*3*).

5.4.1 (apart from ϕ_i), the joint distribution of the $\{m_{ij}\}$, the number of tags recovered in year j from the ith release, is proportional to

$$[(1-\phi_1)\lambda_1]^{m_{11}}[\phi_1(1-\phi_2)\lambda_2]^{m_{12}}\ldots[\phi_1\phi_2\ldots\phi_{t-1}(1-\phi_t)\lambda_t]^{m_{1t}}(1-\theta_1)^{R_1-r_1}$$

$$\times[(1-\phi_1)\lambda_2]^{m_{22}}[\phi_1(1-\phi_2)\lambda_3]^{m_{23}}\ldots[\phi_1\phi_2\ldots\phi_{t-2}(1-\phi_{t-1})\lambda_t]^{m_{2t}}(1-\theta_2)^{R_2-r_2}$$

$$\times[(1-\phi_1)\lambda_3]^{m_{33}}[\phi_1(1-\phi_2)\lambda_4]^{m_{34}}\ldots[\phi_1\phi_2\ldots\phi_{t-3}(1-\phi_{t-2})\lambda_t]^{m_{3t}}(1-\theta_3)^{R_3-r_3}$$

$$\cdots\cdots\cdots\cdots\cdots\cdots\cdots\cdots\cdots\cdots$$

$$\times[(1-\phi_1)\lambda_s]^{m_{ss}}[\phi_1(1-\phi_2)\lambda_{s+1}]^{m_{s,s+1}}\ldots$$

$$[\phi_1\phi_2\ldots\phi_{t-s}(1-\phi_{t-s+1})\lambda_t]^{m_{st}}(1-\theta_s)^{R_s-r_s}$$

$$=\{\prod_{j=1}^{t}(1-\phi_j)^{D_j}\lambda_j^{m_j}\}\{\prod_{j=1}^{t-1}\phi_j^{D_{j+1}+D_{j+2}+\ldots+D_t}\}\{\prod_{i=1}^{s}(1-\theta_i)^{R_i-r_i}\}$$

where $\theta_i = (1-\phi_1)\lambda_i + \phi_1(1-\phi_2)\lambda_{i+1} + \ldots + \phi_1\phi_2\ldots\phi_{t-i}(1-\phi_{t-i+1})\lambda_t$,

$D_j = m_{1j} + m_{2,j+1} + \ldots \quad (j = 1, 2, \ldots, t)$,

and

$$m_j = \sum_{i=1}^{\min(j,s)} m_{ij}.$$

Here θ_i is the probability that a tag is eventually recovered from the ith release, D_j is the number of tags recovered from animals which die in their jth year of life, and m_j is the number of tags recovered during the jth year. Unfortunately the maximum-likelihood equations for the above model do not appear to have explicit solutions and must be solved iteratively, using, for example, the Newton–Raphson method (cf. **1.3.8**(*3*)).(*)

CONSTANT RECOVERY RATE. A number of methods have been proposed for estimating the ϕ_i when the recovery rate is constant from year to year (i.e. $\lambda_i = \lambda$). For example, if $\phi_1\phi_2\ldots\phi_{t-s+1} \approx 0$ (i.e. virtually all tagged animals are dead by the end of the experiment), then $(D_j + D_{j+1} + \ldots + D_t)/\lambda$ is approximately the total number of tagged animals surviving for $j-1$ years, (D_j/λ) is approximately the number of deaths of tagged animals in their jth year of life, and

$$q_j = D_j/(D_j + D_{j+1} + \ldots + D_t), \quad (j = 1, 2, \ldots, t-s),$$

is an estimate of $1-\phi_j$, the probability of a tagged animal dying in its jth year of life. Although the above formula can be used for $j = t-s+1, \ldots, t$, the resulting estimates will usually be inaccurate as they are based on few (if any) recoveries; for example if $D_j = 0$ then $q_j = 0$.

A number of authors have used the estimates q_j (usually in the form of a composite life-table; cf. **10.1.2** (*3*)), even when $\phi_1\phi_2\ldots\phi_{t-s+1}$ is not approximately zero. In this case some tagged animals may still be alive after the tth year, so that D_{t+1}, $D_{t+2}\ldots$, etc. should be added to the denominator of q_j, i.e. $1-\phi_j$ will be overestimated. This can also be seen

(*) An explicit solution has been given by Seber [1971] for the case when $\lambda_i = \lambda$; see also Fordham [1970].

by noting from the relation

$$E[m_{ij}] = R_i\phi_1\phi_2\ldots\phi_{j-i}(1-\phi_{j-i+1})\lambda$$

that for $j \leqslant t-s+1$

$$\frac{E[D_j]}{E[D_j+\ldots+D_t]} = \frac{(R_1+R_2+\ldots+R_s)\phi_1\phi_2\ldots\phi_{j-1}(1-\phi_j)}{(R_1+R_2+\ldots+R_s)\phi_1\phi_2\ldots\phi_{j-1} - K_j}$$

$$< 1-\phi_j,$$

where $K_j = R_1\phi_1\phi_2\ldots\phi_t + R_2\phi_1\phi_2\ldots\phi_{t-1} + \ldots + R_s\phi_1\phi_2\ldots\phi_{t-s+1}$.

Another method for estimating $1-\phi_j$ has been proposed independently by Bellrose and Chase [1950], Paludan [1951] and Hickey [1952: 11, Table 2]. These authors use D'_j (usually expressed as a percentage) obtained by dividing D_j by the total number of animals which could contribute to D_j. For example, if $s = 3$, $t = 5$ we have

$$D'_1 = \frac{D_1}{R_1+R_2+R_3}, \quad D'_2 = \frac{D_2}{R_1+R_2+R_3}, \quad D'_3 = \frac{D_3}{R_1+R_2+R_3},$$

$$D'_4 = \frac{D_4}{R_1+R_2} \quad \text{and} \quad D'_5 = \frac{D_5}{R_1}.$$

Then $1-\phi_j$ is estimated by

$$q'_j = D'_j/(D'_j+D'_{j+1}+\ldots+D'_t).$$

To check on the validity of this method of estimation we note that

$$E[D'_j] = \phi_1\phi_2\ldots\phi_{j-1}(1-\phi_j)\lambda, \qquad (5.45)$$

$$E[D'_j+\ldots+D'_t] = \phi_1\phi_2\ldots\phi_{j-1}(1-\phi_j)\lambda + \ldots + \phi_1\phi_2\ldots\phi_{t-1}(1-\phi_t)\lambda$$

$$= (\phi_1\phi_2\ldots\phi_{j-1}-\phi_1\phi_2\ldots\phi_t)\lambda$$

and

$$\frac{E[D'_j]}{E[D'_j+\ldots+D'_t]} = \frac{1-\phi_j}{1-\phi_j\phi_{j+1}\ldots\phi_t}.$$

Therefore the above ratio is approximately equal to $1-\phi_j$ if and only if $\phi_j\phi_{j+1}\ldots\phi_t \approx 0$, so that as j increases, q'_j will tend to overestimate $1-\phi_j$; this is probably the case in Hickey's example set out in Table 5.22 (see also Imber and Williams [1968: 259, Table 2]).

Some authors compute the values $D'_j\ (R_1+\ldots+R_s)$, i.e. when $s = 3$, $t = 5$ we have

$$D_1,\ D_2,\ D_3,\ D_4\left(1+\frac{R_3}{R_1+R_2}\right), \quad D_5\left(1+\frac{R_2+R_3}{R_1}\right),$$

so that one only needs to calculate the corrections $D_4R_3/(R_1+R_2)$ and $D_5(R_2+R_3)/R_1$ (e.g. Westerskov [1963]). The above series can also be written as

TABLE 5.22
Mortality estimates for a hypothetical population: from Hickey [1952: 11].

Year banded	Number banded (R_i)	Number reported dead by age intervals 0–1	1–2	2–3	3–4
1940	1000	100	30	10	3
1941	1000	95	25	11	–
1942	1000	100	20	–	–
Total	$(D_j) =$	295	75	21	3
	$\Sigma R_i =$	3000	3000	2000	1000
	$100\ D_j' =$	9·83	2·50	1·05	0·30
	$100\ (D_j' + \ldots + D_t') =$	13·68	3·85	1·35	0·30
Ratio	$(q_j') =$	0·72	0·65	0·78	–
	$\log_{10}(100\ D_j') =$	0·993	0·398	0·021	–0·523
	$j-1 =$	0	1	2	3

$$D_1,\ D_2,\ D_3,\ D_4 \frac{(R_1+R_2+R_3)}{(R_1+R_2+R_3)-R_1}, \qquad D_5 \frac{R_1+R_2+R_3}{(R_1+R_2+R_3)-R_1-R_2},$$

which is the formulation used by Farner [1955: 402, equation (3)].

CONSTANT SURVIVAL RATE. A maximum-likelihood method for dealing with this special case of constant survival and reporting rates is given in **5.4.2** (*2*). However, from (5.45) with $\phi_j = \phi$ we have

$$E[100\ D_j'] = \phi^{j-1}(1-\phi)(100\ \lambda)$$

and we consider the linear regression model

$$Y_j = \log[100\ \lambda(1-\phi)] + (j-1)\log\phi + e_j, \qquad (5.46)$$

where $Y_j = \log(100\ D_j')$. Strictly speaking the e_j will be correlated, as each D_j' is based on random variables from the same multinomial distributions, though the degree of dependency will be small if the D_j' are small (which is usually the case). Therefore it is not unreasonable to assume that the e_i are approximately independently distributed with mean zero and constant variance, and to estimate ϕ and 100λ by least squares. We note that logarithms to the base 10 can be used in (5.46), and using Hickey's hypothetical data in Table 5.22 we find that the plot is approximately linear.

We should also mention briefly one other method of estimation sometimes used in the literature. From (5.44) we have

$$\frac{E[D_{j+1}']}{E[D_j']} = \frac{\phi_j(1-\phi_{j+1})}{(1-\phi_j)},$$

so that when $\phi_j = \phi$ $(j = 1, 2, \ldots, t)$, D_2'/D_1', D_3'/D_2', etc. are all estimates of ϕ. A slight variation of this method which does not utilize all the data is to work with appropriate pairs of D_j (modified so as to be based on the same tag releases). For example, from Table 5.22 one can use the ratios 75/295, 21/55 and 3/10; a further example is given in Jenkins *et al.* [1967: Table 6]. It is stressed that such ratios only apply when ϕ_j is constant.

For a more comprehensive review of the above two sections cf. Seber [1972].

CHAPTER 6

OPEN POPULATION: MARK RELEASES BEFORE SAMPLING PERIOD

6.1 SINGLE RELEASE: INSTANTANEOUS SAMPLES

The multi-release methods considered in the previous chapter, although providing maximum information about population changes, involve a considerable expenditure of effort. Also, such multiple releases may be impractical or uneconomic, particularly in the study of commercially exploited populations. It is natural, therefore, to consider what can be learnt from just a single tag–release followed by a series of "instantaneous" random samples *removed* from the population. For example, in fisheries the scientist could release a single batch of tagged fish, and the series of random samples would then be the hauls by the fishermen. However, since the information from a single tag–release is limited, we find that we must build into our statistical models extra assumptions about the population processes of death, migration, etc. if the effects of these processes are to be estimated. Obviously the ideal model will be the one with the minimum of assumptions and which is robust with regard to departures from those assumptions. We shall now consider a number of possible models.

6.1.1 Constant instantaneous mortality and constant recruitment

The first model we shall consider, due to Parker [1963], is for a population in which there is mortality and recruitment but no migration as, for example, in a bounded population area. This model is not unrealistic because of the existence of localised populations, even, for example, in fisheries which do not have any physical boundaries. Let

N_0 = size of the catchable population at time zero,
M_0 = number of marked or tagged individuals released into the population at time zero,
U_0 = $N_0 - M_0$ (size of unmarked population at time zero),
s = number of samples removed,
n_i = size of the ith sample ($i = 1, 2, \ldots, s$),
m_i = number of marked individuals in the ith sample,
u_i = $n_i - m_i$,
N_i = size of total population just before the ith sample,
M_i = size of marked population just before the ith sample,
U_i = $N_i - M_i$,

and t_i = time at which the ith sample is taken.

It should be noted that we are departing slightly from the notation of Chapter 5 where, for reasons of symmetry, it was convenient to define $n_1 = M_0$, i.e. the first sample provided the mark release.

We shall assume the following:

(1) Immigration and emigration are negligible.
(2) The instantaneous natural mortality rate μ is constant throughout the whole experiment and is the same for both marked and unmarked; thus from (1.1) the probability of an individual surviving for time t between samples is $\exp(-\mu t)$.
(3) The number recruited per unit time, r, is constant throughout the whole experiment.
(4) Every individual, whether marked or unmarked, has the same probability of being caught in the ith sample (given that it is alive at the time of the ith sample).
(5) Sampling is instantaneous, or at least takes a negligible period of time as far as population changes are concerned.
(6) Marked individuals do not lose their marks, and all marks are reported on recovery.

For $t < t_1$, let r_t be the number recruited and still surviving at time t. Then, if the recruits are subject to the same mortality rate as the original members of the population, we have the deterministic equation (cf. (1.2))

$$\frac{dr_t}{dt} = -\mu r_t + r$$

which, for the boundary condition $r_0 = 0$, has solution

$$r_t = r(1 - e^{-\mu t})/\mu .$$

Using this relationship, we can now build up the following sequence of deterministic equations:

$$\begin{aligned}
M_1 &= M_0 e^{-\mu t_1}, \\
M_2 &= (M_1 - m_1)e^{-\mu(t_2 - t_1)} \\
&= M_0 e^{-\mu t_2} - m_1 e^{-\mu(t_2 - t_1)}, \\
N_1 &= N_0 e^{-\mu t_1} + r(1 - e^{-\mu t_1})/\mu, \\
N_2 &= (N_1 - n_1)e^{-\mu(t_2 - t_1)} + r(1 - e^{-\mu(t_2 - t_1)})/\mu \\
&= N_0 e^{-\mu t_2} - n_1 e^{-\mu(t_2 - t_1)} + r(1 - e^{-\mu t_2})/\mu, \qquad \text{etc.}
\end{aligned}$$

Then, from assumption (4), if the n_i are fixed parameters and M_0/N_0 is small, m_i will be approximately distributed as Poisson with conditional mean

$$\begin{aligned}
a_i &= E[m_i \mid m_1, m_2, \ldots, m_{i-1}] \\
&= n_i M_i / N_i
\end{aligned}$$

TABLE 6.1
Recapture data for a population of bluegills, common sunfish and hybrids between the two in Flora Lake, Wisconsin, 1953–1955: from Parker [1963: Table 1].

	1953		1954		1955			1953		1954		1955	
t	n	m	n	m	n	m	t	n	m	n	m	n	m
1			260	69			30	250	20	280	99	580	46
2			439	109			31	443	47	240	59		
3	86	11	447	99			32					151	24
4	733	41	366	89	509	179	33	580	61	164	31		
5	618	51	634	141			34			249	36		
6	703	54	543	100	340	63	35			90	11		
7	673	93	840	98			37					290	15
8	948	123	540	125			38	628	43	141	16		
9	244	24					39	151	7	269	51	821	19
10			808	84			40	305	56	97	23		
11			241	26	221	38	41	36	7	543	66	470	17
12	225	26	836	89			43			592	57		
13	622	93			173	29	45	91	11	406	34		
14	540	78					46	1025	51	283	17	226	19
15	650	100					47	308	14	302	22		
16	907	103					48			363	41		
17	252	17			151	37	49	308	14			223	15
18	1019	113					50			475	76	131	2
19	549	48					52	51	5			221	3
20	567	78					54			582	85	280	8
21	438	45			105	24	55	78	5				
22	183	11			152	20	56	193	10			465	6
23	537	60	599	146			57			376	56		
24	98	7	320	62	215	35	58	290	39				
25	795	77	195	33			59					148	6
26			301	63	287	25	60			177	21		
27	387	15	166	26			83	54	4				
28	233	24					104	132	5				
29			112	33									

$$= \frac{n_i\left(M_0 - \sum_{j=1}^{i-1} m_j e^{\mu t_j}\right) e^{-\mu t_i}}{\left[N_0 - \sum_{j=1}^{i-1} n_j e^{\mu t_j} + r(e^{\mu t_i} - 1)/\mu\right] e^{-\mu t_i}}$$

and the joint probability function of $m_1, m_2, \ldots, m_s$ is

$$f(\{m_i\} \mid \{n_i\}) = \prod_{i=1}^{s} e^{-a_i} \frac{a_i^{m_i}}{m_i!} = L, \text{ say.}$$

When the periods of time between samples are the same, say $t_j = j$, then this model reduces to that given by Parker.

The maximum-likelihood estimates $\hat{N}_0$, $\hat{\mu}$ and $\hat{r}$ are solutions of the equations

$$\partial \log L/\partial N_0 = \partial \log L/\partial \mu = \log L/\partial r = 0,$$

which must be obtained by some iterative procedure such as the Newton–Raphson method (**1.3.8** (*3*)). This method also provides an estimate of the asymptotic variance–covariance matrix from the last iteration.

Example 6.1 Bluegills and common sunfish: Parker [1963]

The above model was applied to recapture data collected from a population of bluegills, sunfish, and hybrids between the two in Flora Lake, Wisconsin, during the summers of 1953 through 1955 (Table 6.1). The fish have a growing season which extends from mid-May to mid-October and the catchable population consisted of fish over 50 millimetres in total length. The marking of 3229, 2768 and 1111 individuals caught with fyke nets was concluded on 6 July, 25 June, and 1 July of 1953, 1954 and 1955, respectively, and during the three following summers, 16 943, 14 155 and 6159 individuals respectively were removed. A day was taken as the unit of time, and although instantaneous sampling was assumed in applying the above theory, each sample actually represented a day's catch.

Starting values for the three maximum-likelihood estimates were selected by choosing those which provided the minimum sum of squared deviations of the expected m_i from the observed m_i, utilising a grid of various combinations of N_0, μ and r which could reasonably be expected to bracket the true values. The computations were performed on an IBM 709 system and Parker's results are given in Table 6.2 in the form *estimate ± estimated standard deviation.*

TABLE 6.2

Maximum-likelihood estimates and their standard deviations, using Parker's method: from Parker [1963: Table 2].

Year	M_0	$\hat{N}_0$	$\hat{\mu}$	$\hat{r}$
1953	3 229	28 149 ± 1 114	0·0161 ± 0·0067	89·7 ± 43·4
1954	2 768	14 347 ± 432	0·0054 ± 0·0048	50·7 ± 18·1
1955	1 111	3 089 ± 267	0·0212 ± 0·0022	145·5 ± 18·3

From Table 6.2 we see that the estimated standard deviations for $\hat{\mu}$ and $\hat{r}$ are very large for 1953 and 1954, while for 1955 all estimated standard deviations are small. A value of $\hat{\mu} = 0{\cdot}02$ operating for 60 days means that only $100 \exp(-60\hat{\mu}) = 30$ per cent of the initial population survived the experiment in 1955, and Parker infers from this that μ must be small for much of the year for the population to maintain itself.

6.1.2 Variable mortality and constant instantaneous recruitment

Using the same notation as in **6.1.1**, we shall now consider a different model based on the following assumptions:

(1) Immigration and emigration are negligible.
(2) Marked and unmarked have the same instantaneous natural mortality rate μ_i during the interval $[t_{i-1}, t_i)$ for $i = 1, 2, \ldots, s$ $(t_0 = 0)$.
(3) The instantaneous recruitment rate λ is constant, i.e. if there was no natural mortality or exploitation the unmarked population would increase exponentially, so that the expected unmarked population size at time t would be $U_0 \exp(\lambda t)$ and the expected number of new recruits $U_0(\exp(\lambda t) - 1)$. This assumption of recruitment proportional to population size could reasonably apply, for example, to a population where the recruitment is from births (cf. Feller [1957: 405]).
(4) Sampling is instantaneous.
(5) Every individual, whether marked or unmarked, has the same probability p_i $(= 1 - q_i)$ of being caught in the ith sample (given that it is alive at the time of the ith sample (time t_i)).
(6) Marked individuals do not lose their marks or tags, and all marks are reported on recovery.

Given the above assumptions, we have the deterministic equations

$$U_i = U_0 e^{(\lambda-\mu_1)t_1} q_1 e^{(\lambda-\mu_2)(t_2-t_1)} \ldots q_{i-1} e^{(\lambda-\mu_i)(t_i-t_{i-1})},$$

$$M_i = M_0 e^{-\mu_1 t_1} q_1 e^{-\mu_2(t_2-t_1)} \ldots q_{i-1} e^{-\mu_i(t_i-t_{i-1})}$$

and hence

$$\frac{U_i}{M_i} = \frac{U_0 e^{\lambda t_i}}{M_0}.$$

Estimating U_i/M_i by u_i/m_i and taking logarithms, we are led to consider the regression model

$$E[y_i] \approx \log(U_0/M_0) + \lambda t_i \quad (i = 1, 2, \ldots, s)$$
$$= \beta_0 + \lambda t_i, \quad \text{say},$$

where $y_i = \log(u_i/m_i)$. Therefore, assuming the y_i to be approximately distributed as independent normal variables with constant variance, we can obtain a confidence interval for β_0 and hence for U_0.

If sampling is not instantaneous but takes a time Δ_i, say, for the ith sample, then u_i/m_i is now an estimate of $\bar{U}_i/\bar{M}_i$, the ratio of the average unmarked population size to the average marked population size during the ith period of sampling. Suppose that μ_{Ei} is the instantaneous rate of exploitation during the ith sample; then

$$q_i = e^{-\mu_{Ei}\Delta_i},$$

$$\bar{U}_i = \int_0^{\Delta_i} U_i \exp\{(\lambda-\mu_i-\mu_{Ei})t\}\,dt/\Delta_i$$

$$\approx U_i \exp\{\tfrac{1}{2}(\lambda-\mu_i-\mu_{Ei})\Delta_i\}$$

and similarly

$$\bar{M}_i \approx M_i \exp\{-\tfrac{1}{2}(\mu_i+\mu_{Ei})\Delta_i\}.$$

Hence

$$\frac{\bar{U}_i}{\bar{M}_i} \approx \frac{U_i}{M_i}\exp(\tfrac{1}{2}\lambda\Delta_i)$$

and our model becomes

$$E[y_i] \approx \log(U_0/M_0) + \lambda(t_i + \tfrac{1}{2}\Delta_i).$$

We note that if $\Delta_i = 1$, then rearranging,

$$E[y_i] \approx [\log(U_0/M_0) - \tfrac{1}{2}\lambda] + \lambda(t_i+1),$$

which, apart from a change of sign throughout, is Fischler's model [1965: 304].

Example 6.2 Blue crab (*Callinectes sapidus*): Fischler [1965]

The above regression method was applied to a population of blue crabs in the Neuse River, North Carolina, during 1958. Here recruitment referred to the process whereby a crab grew into the legal size-range, so that N_0 was the initial size of the *legally* exploitable population. A total of 361 tagged crabs were available at the beginning of the experiment (4 July) and sampling was continued through to 30 July. The unit of time was a day, and a sample consisted of a day's catch ($\Delta_i = 1$). It was found necessary, for practical reasons, to measure the recaptures m_i in numbers, but the total catches n_i in thousands of pounds, so that the estimate of U_0 is also in thousands of pounds. As the number of recaptures was very small we have effectively $n_i = u_i$ and $y_i = \log(n_i/m_i)$.

The data are set out in Table 6.3, and Fischler showed that a plot of y_i against $t_i + 1$ (Fig. 6.1) was approximately linear, thus supporting the above theoretical analysis. He obtained an estimate of 722 000 pounds for U_0 with a 95 per cent confidence interval of (516 000, 1 009 000) pounds. From the slope of the regression line λ was estimated to be 0·086.

For a full discussion of the assumptions underlying the regression model the reader is referred to the original article, but the following points should be noted: (a) Natural mortality appeared to be negligible during the period of the experiment. (b) Estimates of the daily recruitment rate (e^{λ}) from both the ratios of soft to hard crabs and the ratios of precommercial

TABLE 6.3
Data used in Fischler's method for estimating the size of the crab population in the Neuse River on 4 July 1958: from Fischler [1965: Table 15].

Date	m_i (number)	n_i (1000 lb)	y_i (log n_i/m_i)	t_i
7 July	13	21·8	0·546 96	3
8	10	16·5	0·500 77	4
9	7	16·0	0·826 68	5
10	7	27·5	1·368 28	6
11	6	15·0	0·916 29	7
14	3	32·0	2·367 13	10
15	1	22·0	3·091 04	11
16	6	22·3	1·312 83	12
17	2	23·4	2·459 59	13
18	3	18·0	1·791 76	14
21	3	31·4	2·348 20	17
22	1	16·6	2·809 40	18
23	3	18·6	1·824 55	19
24	1	22·7	3·112 36	20
25	2	17·8	2·186 05	21
28	3	34·6	2·445 24	24
29	3	31·7	2·357 71	25
30	1	18·4	2·912 35	26

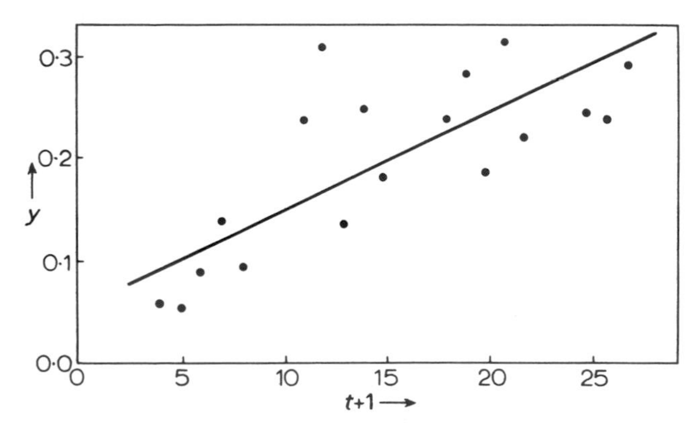

Fig. 6.1 Plot of y_i $(= \log (n_i/m_i))$ against $t_i + 1$ for a blue crab population: redrawn from Fischler [1965].

crabs (3·4 to 4·5 inches carapace width) to commercial size crabs (> 4·5 inches) did not seem to vary greatly, thus supporting the assumption of constant λ. However, it was felt that on the basis of these experiments

$\hat{\lambda} = 0{\cdot}086$ was too high, a more realistic value being something less than 0·05. This was put down to movement of the fishing fleet away from the release areas with the tagged crabs consequently becoming less catchable than the untagged. (c) From the length–frequency distribution of crabs in the catches there did not seem to be any size selectivity with the three major types of fishing gear used. (d) There was a lower percentage of tagged females recaptured than of tagged males. This was felt to be mainly due to the tendency of female crabs to move out of the river after mating into the more saline waters of Pamlico and Core Sounds. If this emigration was proportional to the population size then its effect would be to reduce λ. (e) The loss of carapace tags through moulting appeared to be negligible. (f) The tagging program was widely advertised and every effort was made to obtain a complete tag return. (g) The estimate of U_0 agreed favourably with two other estimates obtained from catch-effort data.

6.1.3 Variable mortality and recruitment

1 Parker's model

The model in **6.1.1** is effectively based on the simple equation

$$E\left[\frac{m_i}{n_i} \mid M_i, N_i\right] = \frac{M_i}{N_i}$$

and if recruitment is taking place, M_i/N_i will decrease as i increases. Parker [1955] suggests the simple expedient of plotting m_i/n_i against t_i and extending this curve to $t = 0$ to obtain an estimate of M_0/N_0, and hence of N_0. Provided there is a reasonably smooth trend in the m_i/n_i, and the above equation is valid, such a method will always give a graphical estimate of N_0 irrespective of the changes taking place in the population. In some circumstances, however, it may be possible to choose a suitable transformation of the m_i/n_i which will stabilise the variance, achieve some measure of normality and, hopefully, linearise the regression so that N_0 can be estimated by least squares. For example, in **6.1.2** it was found that a logarithmic transformation achieved linearity when the assumptions stated there were valid.

Example 6.3 Bluegills and common sunfish: Parker [1955]

Using a unit of time of one day, the above method was applied to marking data collected in 1953 from a population of bluegills, common sunfish, and hybrids between the two in Flora Lake, Wisconsin. The fish were caught in fyke nets between 2 and 6 July and marked by the removal of the left pelvic fin. Using the same net, subsequent samples were removed from the population between 9 July and 18 October, giving the data of Table 6.4.

Assuming the m_i to be approximately binomially distributed, the transformation $y_i = \arcsin\sqrt{(m_i/n_i)}$ is known to stabilise the variance and achieve some measure of normality (tables for the arcsin transformation are given in Hald [1952; $m/n = 0\ (0{\cdot}001)\ 1$, radians] and Snedecor [1946: $m/n - 0\ (0{\cdot}001)\ 1$,

TABLE 6.4

Data used in making a marked-fish regression analysis of a population of bluegills, common sunfish, and hybrids between the two: from Parker [1955: Table 1].

Date	t_i	Sample size n_i	Number marked m_i	Ratio $\left(\frac{m_i}{n_i}\right)$	arcsin $\sqrt{\text{ratio}}$ (degrees) y_i	Estimated nat. mort. of marked	No. marked remaining at day's end
July 6	0	–	–	–	–	9[a]	3220
9	3	86	11	0·1279	20·96	15	3194
10	4	733	41	0·0559	13·69	5	3148
11	5	618	51	0·0825	16·64	5	3092
12	6	703	54	0·0768	16·11	5	3033
13	7	673	93	0·1382	21·81	5	2935
14	8	948	123	0·1298	21·13	4	2808
15	9	244	24	0·0984	18·24	4	2780
18	12	225	26	0·1156	19·91	13	2741
19	13	622	93	0·1495	22·79	4	2644
20	14	540	78	0·1444	22·30	4	2562
21	15	650	100	0·1538	23·11	4	2458
22	16	907	113	0·1246	20·70	4	2341
23	17	252	17	0·0675	15·12	4	2320
24	18	1 019	113	0·1109	19·46	4	2203
25	19	549	48	0·0874	17·16	3	2152
26	20	567	78	0·1376	21·81	3	2071
27	21	438	45	0·1027	18·72	3	2023
28	22	183	11	0·0601	14·18	3	2009
29	23	537	60	0·1117	19·55	3	1946
30	24	98	7	0·0714	15·45	3	1936
31	25	795	77	0·0969	18·15	3	1856
Aug 2	27	387	15	0·0388	11·39	6	1835
3	28	233	24	0·1030	18·72	3	1808
5	30	250	20	0·0800	16·43	6	1782
6	31	443	47	0·1061	19·00	3	1732
8	33	580	61	0·1052	18·91	5	1666
13	38	628	43	0·0685	15·12	13	1610
14	39	151	7	0·0464	12·39	2	1601
15	40	305	56	0·1836	25·40	2	1543
16	41	36	7	0·1944	26·13	2	1534
20	45	91	11	0·1209	20·36	9	1514
21	46	1 025	51	0·0498	12·92	2	1461
22	47	308	14	0·0454	12·25	2	1445
24	49	308	14	0·0454	12·25	4	1427
27	52	51	5	0·0980	18·24	7	1415
30	55	78	5	0·0641	14·65	6	1404
31	56	193	10	0·0518	13·18	2	1392
Sept 2	58	290	39	0·1345	21·47	4	1349
27	83	54	4	0·0741	15·79	51	1294
Oct 18	104	132	5	0·0379	11·24	41	1248
Totals	1203	16 930	1701		712·83	280	

[a] During marking

degrees]. A plot of y_i against t_i (Fig. 6.2) was found to be approximately linear, so that a linear regression model

$$y_i = \beta_0 + \beta t_i + e_i, \quad (i = 1, 2, \ldots, s),$$

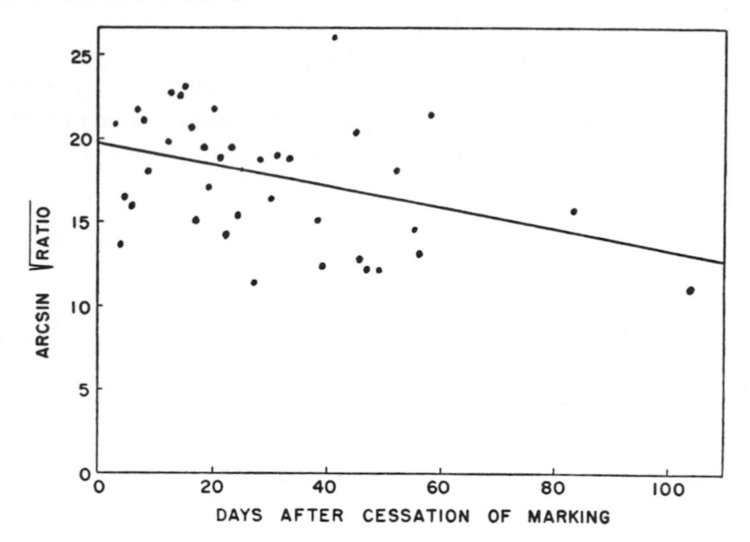

Fig. 6.2 Plot of y_i ($=$ arcsin $\sqrt{m_i/n_i}$) versus t_i, the time of the ith sample, for a population of bluegills and sunfish: from Parker [1955].

where the e_i are assumed to be independently and identically distributed as $\mathfrak{N}[0, \sigma^2]$, could be fitted. The usual least-squares estimates of β and β_0 are

$$\hat{\beta} = \frac{\Sigma\, y_i\,(t_i - \bar{t})}{\Sigma\,(t_i - \bar{t})^2},$$

$$\hat{\beta}_0 = \bar{y} - \hat{\beta}\bar{t},$$

and, since $\hat{\beta}_0$ is an estimate of arcsin $\sqrt{(M_0/N_0)}$, we have the estimate

$$\hat{N}_0 = M_0/(\sin \hat{\beta}_0)^2.$$

A $100(1-\alpha)$ per cent confidence interval for β_0 is given by

$$\hat{\beta}_0 \pm t_{s-1}[\alpha/2]\,[\hat{\sigma}^2 \Sigma\, t_i^2/n\, \Sigma\,(t_i-\bar{t})^2]^{1/2},$$

where $(s-2)\hat{\sigma}^2 = \Sigma\,(y_i - \bar{y})^2 - \hat{\beta}^2\, \Sigma\,(t_i - \bar{t})^2$.

By taking sines, squaring and inverting, this interval can be converted into a confidence interval for N_0.

Without going through the details, we have from Parker $M_0 = 3220$, $s = 40$, and the fitted regression line is

$$Y = 19{\cdot}7614^\circ - 0{\cdot}065\,269\, t.$$

Also

$$\hat{N}_0 = 3220/(\sin 19{\cdot}7614^\circ)^2 = 28\,159$$

and an approximate 95 per cent confidence interval for N_0 is (23 394, 34 681). If we ignore the effects of possible recruitment, we note that the conventional Petersen estimate given by Paloheimo in **3.7**, namely

$$M_0\, \Sigma\, n_i/\Sigma\, m_i = 3220\,(16\,930)/(1701) = 32\,049,$$

is about 14 per cent greater than $\hat{N}_0$.

An average annual survival rate of 0·573 was calculated from age data (cf. Chapter 10) and from this Parker was able to estimate the total recruitment for the whole experiment as follows. Assuming a constant instantaneous natural mortality rate μ throughout the year, a daily natural mortality rate of

$$1 - e^{-\mu} = 1 - (e^{-\mu t})^{1/t} = 1 - (0{\cdot}573)^{1/365} = 0{\cdot}001\,53$$

was used to estimate the number of marked fish dying between each sample. At the beginning of the experiment it was estimated that 9 of the 3229 tagged fish released died during the actual tagging period, so that M_0 was taken to be 3220. Putting $t = 104$ in the fitted regression line, we can estimate M_{40}/N_{40} which, using $M_{40} = 1248$, leads to $\hat{N}_{40} = 24\,468$ as the estimate of the total population size on October 18. Parker then estimated the total natural mortality throughout the experiment as 3457: we are not given the method of estimation, though it was no doubt similar to that used for estimating the mortality of marked fish (for example, an N_i sequence can be calculated from the M_i sequence using the fitted regression). Finally, assuming no migration, we have

$$\begin{aligned} \textit{recruitment} &= \textit{removal} + \textit{natural mortality} - (\hat{N}_0 - \hat{N}_{40}) \\ &= 16\,930 + 3457 - (28\,159 - 24\,468) \\ &= 16\,696. \end{aligned}$$

2 *Chapman's model*

If θ_i is the probability that a member of N_0 survives up to time t_i then, using the notation of **6.1.1**,

$$E[N_i] = N_0\theta_i + R_i$$

and

$$E[M_i] = M_0\theta_i,$$

where R_i is the net recruitment to time t_i (i.e. recruits that come into the population after time zero and are still alive at time t_i). Applying the Petersen method (**3.1.1**) to the ith sample, we have

$$E\left[\frac{n_i M_i}{m_i} \mid M_i, N_i\right] \approx N_i,$$

which, combined with the above two equations, gives us

$$E[y_i] \approx N_0 + R_i/\theta_i,$$

where $y_i = n_i M_0/m_i$. If θ_i is not too small, a small adjustment for bias can be made, such as (Chapman [1965: 536]) $y_i = (n_i+1)(M_0+1)/(m_i+1)$.

Since R_i and θ_i will both be functions of t_i, we are led to consider the regression model

$$E[y_i] = N_0 + g(t_i), \quad i = 1, 2, \ldots, s,$$

where g is a function of t_i. Chapman recommends approximating g by a polynomial function and estimating N_0 by least squares.

The basic assumptions underlying the above approach are as follows:

(1) All members of N_0 have the same probability of survival θ_i to time t_i. This means that the tagged and untagged members of N_0 have the same mortality rates and are removed by the sampling process at the same rate.
(2) The assumptions underlying the Petersen method must hold for each sample, i.e. samples are random or, if the samples are systematic, there is uniform mixing of tagged and untagged (including new recruits).
(3) Immigration and emigration are negligible. Alternatively these can be included if θ_i is defined to be the probability that a member of N_0 survives and is in the population at t_i, and R_i is defined to be the net influx of new life (which may be negative) rather than just net recruitment. In the two examples below, migration is negligible (lake population).
(4) Individuals do not lose their tags, and all tags found are reported.

Example 6.4 Bluegills and sunfish: Chapman [1965].

Using the 1953 data in Table 6.1 (p. 258), the values of y_i were calculated and entered in Table 6.5. A plot of y_i versus t_i (Fig. 6.3) revealed a

TABLE 6.5

Values of $y_i = (n_i+1)(M_0+1)/(m_i+1)$ calculated from Parker's 1953 data: from Chapman [1965: Table 2].

t_i (days)	y_i (thousands)	t_i (days)	y_i (thousands)	t_i (days)	y_i (thousands)
3	23·4	19	36·3	38	46·2
4	56·4	20	23·2	39	61·4
5	38·4	21	30·8	40	17·3
6	41·3	22	49·5	41	14·9
7	23·2	23	28·5	45	24·8
8	28·7	24	40·0	46	63·7
9	32·6	25	33·0	47	66·5
12	27·0	27	78·3	49	66·5
13	21·4	28	30·2	52	28·0
14	22·1	30	38·6	55	42·5
15	20·8	31	29·9	56	57·0
16	28·2	33	30·3	58	23·5
17	45·4				
18	28·9				

slight trend and, because of the wide variation in the y_i, the linear model

$$E[y_i] = N_0 + \beta t_i \tag{6.1}$$

seemed most appropriate. To test for linearity, Chapman suggested grouping the y_i into classes, so that the members of a class may be regarded as

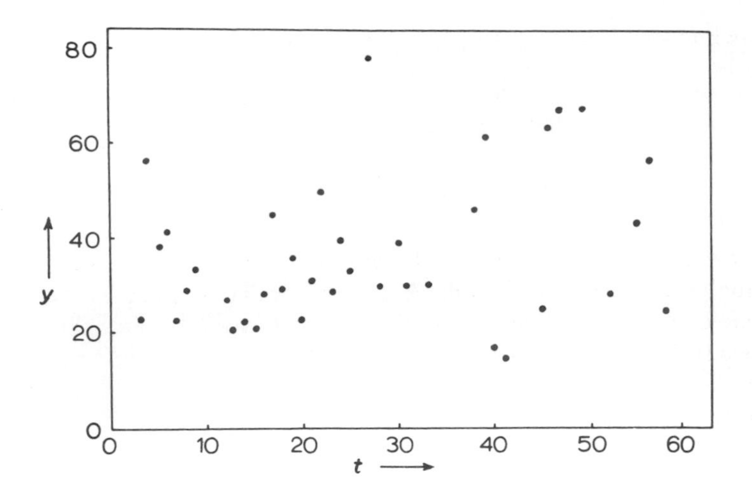

Fig. 6.3 Plot of y_i $(= (n_i + 1)(M_0 + 1)/(m_i + 1))$ versus t_i, the day of recapture, for Parker's 1953 data.

approximate replicates. A working unit of 7 days was used, and the class divisions are indicated by lines in the t_i column of Table 6.5. A straight line was then fitted to the class means, and from Table 6.6 the usual analysis of variance test for linearity was carried out (cf. Draper and Smith [1966: 26] or Keeping [1962: 342]). Chapman pointed out that since the

TABLE 6.6

Test for linearity of Parker's 1953 data: from Chapman [1965: Table 3].

Source	d.f.	Sum of squares	Mean square
Linear regression	1	843·88	843·88
Deviations from linear regression	5	1408·09	281·62
Error	31	6994·89	225·64

Test for linearity of regression: F = 1·25 with (5, 31) d.f.

replicates are not true replicates, there is a trend in each class, and the error sum of squares in Table 6.6 is inflated. An attempt to correct for this can be made by fitting a straight line within each class and then pooling the residual sums of squares to give a residual mean-square error of 183·06 with 24 degrees of freedom. If deviations from linear regression are tested against this, the F value is 1·54 which is still not significant.

From Fig. 6.3 the variance of y_i does not appear to depend on t_i, so that, assuming constant variance and making the usual normality assumptions, we have the least-squares estimates $\hat{\beta} = 0{\cdot}2954$ and $\hat{N}_0 = 28{\cdot}73$ *thousand* with a 95 per cent confidence interval for N_0 of 28 730 ± 9820. With stronger but unverifiable assumptions, Parker (Example 6.1) obtained the much narrower limits 28 149 ± 2183.

If we assume the linear regression model with $R(t)/\theta(t) = \beta t$ and define $r(t)$ to be the net recruitment on the tth day, then

$$\begin{aligned} r(t) &= R(t) - R(t-1) \, . \, \frac{\theta(t)}{\theta(t-1)} \\ &= \theta(t)[\beta t - \beta(t-1)] \\ &= \beta\theta(t) \, . \end{aligned}$$

Since $\theta(t)$ is the survival from both natural mortality and the sampling or catching process, it will be a decreasing function of t, and $r(t)$ will also be decreasing. If estimates of the natural mortality rate are available then it is possible to calculate $\theta(t)$, and hence $R(t)$, in a stepwise fashion for each value of t. Alternatively one could endeavour to represent $r(t)$ by some simple function such as

$$r(t) = r_0 e^{-r_1 t}.$$

We note in passing that if $\theta(t)$ is near unity for t near zero, then $\hat{r}(0) = \hat{\beta} = 0{\cdot}2954$ *thousand*, i.e. the initial recruitment is estimated at about 300 per day.

Finally Chapman points out that a test of $\beta = 0$ for (6.1) is not significant at the 5 per cent level (though it is at the 10 per cent level) and it is therefore possible to assert that the hypothesis $R(t) = 0$ is consistent with the data. However, he adds that, in spite of this, the regression model (6.1) seems to be the best simple model.

Example 6.5 Bluegills and sunfish: Chapman [1965]

Using the 1955 data of Table 6.1, Chapman constructed Table 6.7. In contrast to Example 6.4, we find that a plot of y_i versus t (Fig. 6.4) is now non-linear and the variance of y_i increases with t_i. We can therefore either try to fit, say, a quadratic using weighted least squares or else look for a

TABLE 6.7

Values of $y_i = (n_i + 1)(M_0 + 1)/(m_i + 1)$ calculated from Parker's 1955 data: from Chapman [1965: Table 2].

t_i (days)	y_i (thousands)	t_i (days)	y_i (thousands)	t_i (days)	y_i (thousands)
4	3·2	24	6·7	46	12·6
6	5·9	26	12·3	49	15·6
11	6·3	30	13·7	50	48·9
13	6·4	32	6·8	52	61·7
17	4·4	37	20·2	54	34·7
21	4·7	39	45·7	56	74·0
22	8·1	41	29·1	59	23·7

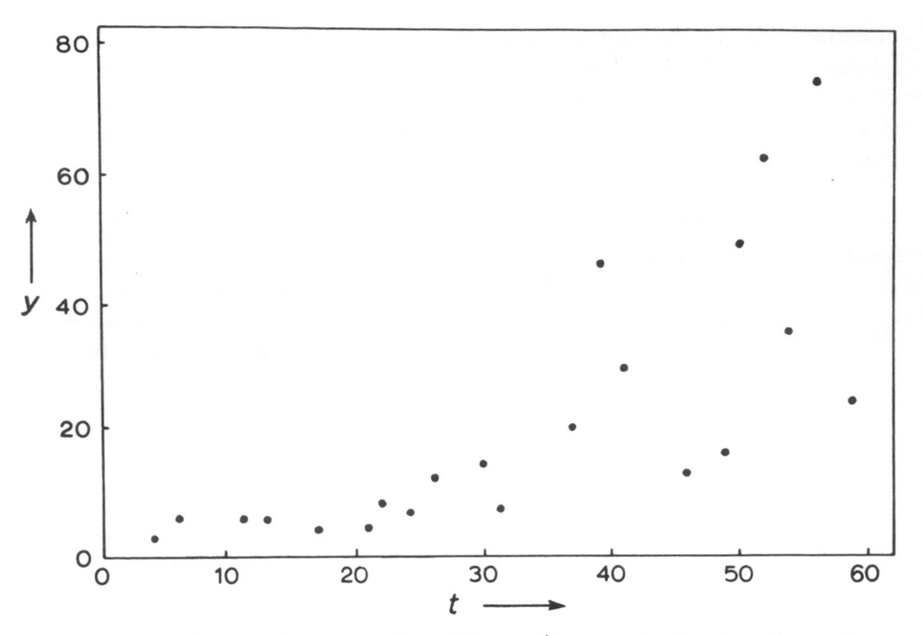

Fig. 6.4 Plot of y_i $(= (n_i + 1)(M_0 + 1)/(m_i + 1))$ versus t_i, the day of recapture, for Parker's 1955 data.

suitable transformation of the data which stabilises the variance and (hopefully) linearises the regression.

Chapman once again grouped the data into classes (the choice of class divisions in Table 6.7 being somewhat more arbitrary than for Table 6.5) and

TABLE 6.8

Preliminary analysis of Parker's 1955 data: from Chapman [1965: Table 4].

Group	Mean day of sampling	Mean of y_i	Standard deviation
1	8·5	5·45	1·52
2	21·5	7·24	3·21
3	30·5	23·10	15·08
4	49·0	34·70	24·39
5	56·3	44·13	26·44

TABLE 6.9

Test for linearity of 1955 data (log. transfomation): from Chapman [1965: Table 5].

Source	d.f.	Sum of squares	Mean square
Linear regression	1	2·0752	2·0752
Deviations from linear regression	3	2·0709	0·6903
Error	16	10·7728	0·6733

Test for linearity of regression: $F = 1{\cdot}02$ with (3, 16) d.f.

calculated Table 6.8. On the evidence of this table an analysis was first attempted with the standard deviation of y_i assumed proportional to t_i. However, this did not provide a satisfactory fit using either a linear or a quadratic regression, and a logarithmic transformation was subsequently tried. It was found that for the transformed data the variances within classes did not differ significantly from class to class, and the analysis of variance test for linearity (Table 6.9) was not significant. Therefore the model

$$\log y_i = \beta_0 + \beta t_i + e_i$$

was chosen, where the e_i are assumed to be independently and identically distributed as $\mathfrak{N}[0, \sigma^2]$. The least-squares estimates of β_0 and β are

$$\hat{\beta}_0 = 1{\cdot}2485, \quad \hat{\beta} = 0{\cdot}043\,96$$

and a 95 per cent confidence interval for β_0 is (0·673, 1·836). Now, from the above model we have the relationship

$$N_0 + g(t) = \exp(\beta_0 + \beta t),$$

so that putting $t = 0$ and $g(0) = 0$ (since $R(0) = 0$), we have $N_0 = \exp(\beta_0)$ and $\hat{N}_0 = \exp(\hat{\beta}_0) = 3{\cdot}485$ *thousand*. A 95 per cent confidence interval for N_0 is (exp (0·673), exp (1·836)) or (1·96, 6·27) *thousand*. Since

$$R(t)/\theta(t) = g(t) = e^{\beta_0}(e^{\beta t} - 1),$$

the daily net recruitment rate is given by

$$\begin{aligned} r(t) &= R(t) - R(t-1) \,.\, \theta(t)/\theta(t-1), \\ &= \theta(t)e^{\beta_0}(e^{\beta t} - e^{\beta(t-1)}) \\ &= e^{\beta_0}(1 - e^{-\beta})\,\theta(t)\,e^{\beta t} \end{aligned}$$

which is estimated by

$$\hat{r}(t) = (0{\cdot}1502)\,\theta(t)\,e^{0{\cdot}043\,96\,t}.$$

Knowledge of the natural mortality rate would then enable one to calculate $\theta(t)$ and hence $\hat{r}(t)$. For example, if

$$\theta(t) = e^{-\mu t - h(t)},$$

where μ is the instantaneous natural mortality rate and $h(t)$ represents the mortality due to the catching process, then $\hat{r}(t)$ is increasing or decreasing according as $(0{\cdot}043\,96 - \mu)t$ is greater or less than $h(t)$. Chapman estimates the daily instantaneous removal rate to be about 0·02 (i.e. $h(t) = 0{\cdot}02\,t$), so that for Parker's assumption of constant $r(t)$ to hold (cf. **6.1.1**), we must have $\mu \approx 0{\cdot}044 - 0{\cdot}02 = 0{\cdot}024$. This may be compared with Parker's estimate of 0·0212 (Table 6.2).

6.2 SINGLE RELEASE AND CONTINUOUS SAMPLING: MORTALITY ESTIMATES

6.2.1 Maximum-likelihood estimates

In **6.1** sampling was considered to be a discrete instantaneous process as, for example, in short intensive fisheries. We shall now consider a model discussed by Gulland [1955b], Chapman [1961] and Paulik [1963a] in which sampling is regarded as a continuous process and recaptures are comparatively so few that the recapture times of marked animals can be recorded. The assumptions we shall make are the following:

(1) Immigration and emigration are negligible.
(2) For tagged individuals the instantaneous natural mortality rate and the instantaneous exploitation rate are both constant and equal to μ and μ_E respectively (more commonly M and F in fishery research). If we can also assume that these rates are the same for tagged and untagged then the estimates obtained below can be applied to the whole population.
(3) Individuals do not lose their tags, and all tags found are reported.

Let $Z = \mu + \mu_E$ be the total instantaneous mortality rate and let M_0 be the size of the tagged population at time zero. Then the probability of a tagged individual remaining alive at time t is $\theta(t) = \exp(-Zt)$ and the probability of recapture in the interval $(t,\ t+dt)$ is $\mu_E\theta(t)dt$. Suppose that m tagged individuals are recaptured at times $T_1, T_2, \ldots. T_m$ respectively, and that these are the only recaptures up to time τ when exploitation ceases. Then the probability that a tagged individual is caught during the whole period is

$$\int_0^\tau \mu_E\,\theta(t)dt = \mu_E[1-\theta(\tau)]/Z = P_\tau, \quad \text{say.} \tag{6.2}$$

Here P_τ can be regarded as the proportion of the initial tagged population M_0 expected to be finally caught, and it is therefore called the exploitation rate for time τ. For example, if τ is one year, then P_1 is the annual exploitation rate.

Assuming that tagged individuals are independent of each other, m has a binomial distribution

$$f_1(m) = \binom{M_0}{m} P_\tau^m(1-P_\tau)^{M_0-m}.$$

Also, given that a tagged individual is caught during the experiment, then the conditional density function for its recovery time is

$$f_2(t) = \begin{cases} \mu_E\theta(t)/P_\tau & 0 \leqslant t \leqslant \tau, \\ 0 & \text{otherwise.} \end{cases} \tag{6.3}$$

Hence the joint density function of $T_1, T_2, \ldots, T_m$ and m is

$$\begin{aligned} f(T_1, T_2, \ldots, T_m, m) &= f(T_1, T_2, \ldots, T_m | m)f_1(m) \\ &= \left(\prod_{i=1}^m f_2(T_i)\right) f_1(m) \\ &= \binom{M_0}{m}(1-P_\tau)^{M_0-m}\prod_{i=1}^m \{\mu_E\theta(T_i)\}. \end{aligned} \tag{6.4}$$

For the case when τ is so large that $\theta(\tau)$ can be neglected (i.e. $P_\tau \approx P_\infty = \mu_E/Z$), Gulland [1955b] derives maximum-likelihood estimates of μ and μ_E

and shows that they are biased: his estimates are slightly modified (cf. Seber [1962: 346]) when a short period of time is allowed at the beginning for the dispersal of the tagged individuals. Chapman [1961: 156] notes that for this special case of τ large, no minimum variance unbiased estimates of μ and μ_E exist. However, he shows that for $m \geqslant 2$ (i.e. neglecting the possibility of no recaptures or just a single recapture) approximately unbiased estimates are given by(*)

$$\mu_E^* = m(m-1)/(M_0 T_.)$$

and

$$\mu^* = (M_0 - m)(m-1)/(M_0 T_.),$$

where $T_. = \sum_{i=1}^{m} T_i$. Approximate variances are given by

$$V[\mu_E^*] = \frac{\mu_E(\mu_E + 2\mu)}{M_0} + \frac{2(\mu_E + \mu)}{M_0^2} + O\left(\frac{1}{M_0^3}\right)$$

and

$$V[\mu^*] = \frac{\mu(\mu_E^2 + \mu_E \mu + \mu^2)}{M_0} + O\left(\frac{1}{M_0^2}\right).$$

If we simply wished to estimate $Z = \mu + \mu_E$ then (cf. (2.29), $r = 1$)

$$Z^* = (m-1)/T_.$$

is a minimum variance unbiased estimate of Z with variance

$$V[Z^*] = Z^2/(m-2).$$

As this estimate and its variance do not depend on knowing M_0 we can use any well-defined class in the population at the beginning of the experiment as the "tagged" portion, provided there is no further recruitment into this class during the sampling period τ. For example, the class of animals of a particular age or size, or recruitment group (Ssentongo and Larkin [1973]) can be used, and the times T_i would then refer to the capture times of animals from this class.

When $\theta(\tau)$ cannot be neglected, Paulik [1963a] shows that it is more convenient to work with P_τ and S $(= e^{-Z})$, the probability of survival for one unit of time, than with μ and μ_E. By writing μ_E in terms of P_τ and S (cf. (6.2)), we find from (6.4) that the maximum-likelihood estimates of P_τ and S are given by

$$\hat{P}_\tau = m/M_0$$

and

$$-\frac{1}{\tau \log \hat{S}} - \frac{\hat{S}^\tau}{1 - \hat{S}^\tau} = \frac{\bar{T}}{\tau}$$

where $\bar{T} = T_./m$ Writing $\hat{Z} = -\log \hat{S}$, this last equation becomes

$$\frac{1}{\tau \hat{Z}} - \frac{1}{\exp(\tau \hat{Z}) - 1} = \frac{\bar{T}}{\tau} \tag{6.5}$$

and when $0 \leqslant \bar{T} < \tau/2$, (6.5) can be solved using a table of $\bar{T}/\tau$ as a function of $\tau \hat{Z}$ given by Deemer and Votaw [1955: 50] and reproduced in **A5**; for

(*)The effect of truncation is considered by Nicholson and Pope [1977].

$\bar{T} \geqslant \mathcal{T}/2$ the maximum-likelihood estimate of Z is 0. For m large Paulik shows that

$$V[\hat{S}] \approx \frac{1}{M_0 P_{\mathcal{T}}} \left\{ \frac{1}{[S \log S]^2} - \frac{\mathcal{T}^2 S^{\mathcal{T}-2}}{[1-S^{\mathcal{T}}]^2} \right\}^{-1},$$

$$\to (S \log S)^2/(M_0 P_{\mathcal{T}}) \quad \text{as } \mathcal{T} \to \infty$$

and

$$V[\hat{P}_{\mathcal{T}}] = P_{\mathcal{T}}(1-P_{\mathcal{T}})/M_0.$$

GROUPED OBSERVATIONS. Very often the recovery times are not known exactly. For example, in a trap fishery the traps may be lifted only when a certain number of fish have been accumulated, so that a tagged fish could have been caught at any time since the last lift. Also in many fisheries the catch is not thoroughly examined for tags until the vessels are unloaded in port or until the fish are being processed. Suppose, then, that the recovery period $[0, \mathcal{T})$ is divided into J small intervals $[0, d_1)$, $[d_1, d_2), \ldots, [d_{J-1}, d_J)$, where $d_J = \mathcal{T}$, and suppose that m_j tagged individuals are caught in the jth interval ($j = 1, 2, \ldots, J$). Paulik suggests taking the middle point as representative of each interval and approximating $\bar{T}$ by

$$\sum_{j=1}^{J} m_j(d_j + d_{j-1})/2m$$

in (6.5). However, when the intervals are large, he gives a more efficient method as follows.

Suppose that the intervals are of equal length so that $d_j = j$ units of time ($j = 1, 2, \ldots, J$: $J = \mathcal{T}$) and, from (6.2),

$$P_J = \mu_E(1-e^{-ZJ})/Z \tag{6.6}$$

$$= -\mu_E(1-S^J)/\log S. \tag{6.7}$$

Then, neglecting the complications of sampling without replacement within each interval, the $\{m_j\}$ have a joint multinomial distribution

$$f(\{m_j\}) = \frac{M_0!}{\{\prod_j m_j!\}(M_0-m)!}(1-P_J)^{M_0-m} \prod_{j=1}^{J} \{p_j^{m_j}\}, \tag{6.8}$$

where p_j = (probability of surviving for $(j-1)$ intervals) $\times$ (probability of capture in the jth interval)

$$= e^{-Z(j-1)}P_1 \tag{6.9}$$

$$= S^{j-1}P_J \Big/ \sum_{j=0}^{J-1} S^j, \tag{6.10}$$

since, from (6.7),

$$P_1 = P_J(1-S)/(1-S^J).$$

Hence, substituting for p_j in (6.8) we find that $f(\{m_j\})$ is proportional to

$$S^x P_J^m (1-P_J)^{M_0-m} \left(\sum_{j=0}^{J-1} S^j\right)^{-m}$$

where $x = \sum_{j=1}^{J} (j-1)m_j$. The maximum-likelihood estimates of P_J and S are therefore given by

$$\tilde{P}_J = m/M_0$$

and

$$\left(\sum_{j=0}^{J-1} j\tilde{S}^j\right) \Big/ \left(\sum_{j=0}^{J-1} \tilde{S}^j\right) = x/m. \tag{6.11}$$

The solution of (6.11) is discussed in **A6** where a table is available for $3 \leqslant J \leqslant 9$. The asymptotic variances are given by

$$V[\tilde{P}_J] = P_J(1-P_J)/M_0$$

and

$$V[\tilde{S}] = \frac{S^2\left(\sum_{j=0}^{J-1} S^j\right)^2}{M_0 P_J\left[\left(\sum_{j=0}^{J-1} S^j\right)\left(\sum_{j=0}^{J-1} j^2 S^j\right) - \left(\sum_{j=0}^{J-1} j S^j\right)^2\right]}$$

$$= \frac{1}{M_0 P_J}\left[\frac{1}{S(1-S)^2} - \frac{J^2 S^{J-2}}{(1-S^J)^2}\right]^{-1}$$

If the number of recovery periods J is sufficiently large to make $S^J \to 0$, then

$$V[\tilde{S}] \to S(1-S)^2/(M_0 P_J)$$

When estimates of μ and μ_E are required, we can obtain maximum-likelihood estimates by solving for μ and μ_E in terms of S and P_J. For example, from (6.7) we have

$$\tilde{\mu}_E = -\tilde{P}_J \log \tilde{S}/(1-\tilde{S}^J)$$

and, using the delta method,

$$V[\tilde{\mu}_E] \approx \left\{\frac{\log S}{1-S^J}\right\}^2 V[\tilde{P}_J] + \left\{\frac{P_J}{S(1-S^J)}\right\}^2 \left\{1 + \frac{JS^J \log S}{1-S^J)}\right\}^2 V[\tilde{S}].$$

Paulik points out that the actual recovery data may be a mixture of "exact" times T_i and interval approximations as above. If only a few of the recovery times are known exactly, it does not seem worthwhile to treat them separately. However, if the exact times are known for a sizeable proportion of the recoveries he suggests that two estimates should be made: one for the exact time and one for the interval data. A suitable estimate of S would be a weighted mean of the two estimates where the weights are proportional to the estimated variances (cf. **1.3.2**) or more simply to the numbers of tagged individuals used in each estimate.

In concluding this section we note that $M_0 P_J$ can be estimated by m, so that $\hat{S}$, $\tilde{S}$ and estimates of their variances do not depend on knowledge of M_0.

Example 6.6 Plaice: Paulik [1962, 1963a]

Using a time-interval of three months and $J = 4$ recovery periods, the following recovery data on tagged plaice were obtained:

j	1	2	3	4	$M_0 = 1000$
m_j	139	91	52	40	$m = 322$

Then $x/m = [91 + 2(52) + 3(40)]/322 = 0{\cdot}9783$, and interpolating in Table **A6** with $K = J - 1 = 3$, (6.11) can be solved to give the *three-monthly* estimate $\tilde{S} = 0{\cdot}644$ with asymptotic variance 0·001 207; the *annual* survival rate is estimated to be $\tilde{S}^4 = 0{\cdot}172$. Also the *annual* exploitation rate is estimated by $\tilde{P}_4 = 322/1000 = 0{\cdot}322$ with variance estimate 0·000 218 3. Assuming approximate normality for $\tilde{S}$ and $\tilde{P}_4$, then approximate 95 per cent confidence intervals for S and P_4 are therefore (0·58, 0·71) and (0·292, 0·352) respectively. Finally we note that

$$\begin{aligned} \tilde{\mu}_E &= -\tilde{P}_4 \log \tilde{S}/(1-\tilde{S}^4) \\ &= (0{\cdot}322)(0{\cdot}440)/(1-0{\cdot}172) \\ &= 0{\cdot}171 \end{aligned}$$

is the three-monthly instantaneous exploitation rate (or an annual rate of 0·684). The large-sample variance of $\tilde{\mu}_E$ is estimated to be 0·000 882 5 and an approximate 95 per cent confidence interval for μ_E is (0·112, 0·230).

6.2.2 Regression estimates

Using grouped observations and J equal time-intervals once again, we have, from (6.8), $E[m_j] = M_0 p_j$. Therefore, taking logarithms and using (6.9), we are led to consider the linear regression model (Beverton [1954])

$$\log m_j = \log (M_0 P_1) + Z - Zj + e_j.$$

If we could assume the e_j to be independently and identically distributed as $\mathfrak{N}[0,\sigma^2]$ (Winsor and Clark [1940] give some support for this), then a straightforward confidence interval for the slope $-Z$ is available: in addition, since M_0 is known, P_1 (and hence μ_E) can be estimated. However, the m_j are approximately multinomially distributed (cf. (6.8)) and therefore correlated, so that Paulik [1963a] suggests the following analysis:

Consider the regression model (Paulik's model with the sign changed to make y_j positive)

$$E[y_j] \approx \beta_0 + \beta j, \tag{6.12}$$

where $y_j = \log (M_0/m_j)$, $\beta_0 = -(Z + \log P_1)$ and $\beta = Z$. Then, using the delta method, it can be shown that for large $M_0 p_i$ (greater than 10, say)

$$V[y_j] = V[\log m_j] \approx (1-p_j)/(M_0 p_j)$$

and

$$\text{cov}[y_j, y_k] \approx -1/M_0 \quad (j \neq k),$$

so that the variance–covariance matrix of the y_j is approximately $\sigma^2 \mathbf{B}$, where

$$\mathbf{B} = \frac{1}{M_0}\begin{bmatrix} p_1^{-1} - 1 & -1 & \dots & -1 \\ -1 & p_2^{-1} - 1 & \dots & -1 \\ \dots & \dots & \dots & \dots \\ -1 & -1 & \dots & p_J^{-1} - 1 \end{bmatrix}$$

and $\sigma^2 = 1$. Using the method of **1.3.5**(*2*) with

$$\mathbf{B}^{-1} = \frac{M_0}{1-\Sigma p_j}\begin{bmatrix} p_1(1-\Sigma p_j+p_1), & p_1p_2 & , \ldots, & p_1p_J \\ p_2p_1 & , p_2(1-\Sigma p_j+p_2), & \ldots, & p_2p_J \\ \ldots & , \ldots & , \ldots, & \ldots \\ p_Jp_1 & , p_Jp_2 & , \ldots, & p_J(1-\Sigma p_j+p_J) \end{bmatrix}$$

$$\mathbf{y} = \begin{bmatrix} y_1 \\ y_2 \\ \ldots \\ y_J \end{bmatrix}, \quad \mathbf{X} = \begin{bmatrix} 1 & 1 \\ 1 & 2 \\ & \ldots \\ 1 & J \end{bmatrix}, \quad \boldsymbol{\beta} = \begin{bmatrix} \beta_0 \\ \beta \end{bmatrix},$$

σ^2 unknown, and p_j estimated by m_j/M_0 in $\mathbf{B}^{-1}$, it transpires that the least-squares estimates of Z and β_0 are given by

$$\hat{Z} = \hat{\beta} = \frac{\Sigma\, jy_jm_j - (\Sigma\, y_jm_j)(\Sigma\, jm_j)/m}{\Sigma\, j^2m_j - (\Sigma\, jm_j)^2/m} = \frac{C}{D}, \quad \text{say},$$

and

$$\hat{\beta}_0 = (\hat{\beta}\,\Sigma\, jm_j - \Sigma\, y_jm_j)/m\,.$$

Variance estimates are:

$$\hat{V}[\hat{Z}] = \hat{\sigma}^2/D\,,$$

$$\hat{V}[\hat{\beta}_0] = \hat{\sigma}^2\left[\frac{1}{m} + \frac{(\Sigma\, jm_j)^2}{m^2D}\right],$$

and

$$\widehat{\text{cov}}\,[\hat{Z}, \hat{\beta}_0] = \hat{\sigma}^2\,\Sigma\, jm_j/(mD)\,,$$

where $\quad (J-2)\hat{\sigma}^2 = \Sigma\, y_j^2m_j - (\Sigma\, y_jm_j)^2/m - C^2/D\,.$

All the above summations are for $j = 1, 2, \ldots, J$; $\Sigma\, m_j = m$.

From a lemma of Chapman [1956] it is readily shown that for large M_0, $\mathbf{y}$ is approximately distributed as the multivariate normal. Hence an approximate $100(1-\alpha)$ per cent confidence interval for Z is

$$\hat{Z} \pm t_{J-2}[\alpha/2]\,\hat{\sigma}\,D^{-1/2}\,. \tag{6.13}$$

It transpires that this interval is functionally independent of M_0 and therefore does not depend on a correct knowledge of M_0 (though $\hat{\beta}_0$ does). This is what we would expect, as $\log M_0$ could have been included under β_0 in (6.12).

If estimates of S and P_1 are required we have

$$\hat{S} = e^{-\hat{Z}},$$

$$\hat{P}_1 = e^{-(\hat{Z}+\hat{\beta}_0)}$$

and, using the delta method,

$$V[\hat{S}] \approx S^2V[\hat{Z}]$$

and

$$V[\hat{P}_1] \approx P_1^2(V[\hat{Z}] + 2\,\mathrm{cov}\,[\hat{Z}, \hat{\beta}_0] + V[\hat{\beta}_0]),$$

which can be estimated from the formulae given above.

Estimates of μ and μ_E are given by (cf. (6.7))

$$\hat{\mu}_E = \frac{\hat{Z}\hat{P}_1}{1-\hat{S}} = \frac{\hat{Z}e^{-\hat{\beta}_0}}{e^{\hat{Z}} - 1}$$

and

$$\hat{\mu} = \hat{Z} - \hat{\mu}_E .$$

The asymptotic variances can be calculated in the usual manner using the delta method.

In comparing the above regression method with the maximum-likelihood approach of **6.2.1**, Paulik mentions that one objection to the regression method might be that for a small number of recovery periods J, the large t-value may lead to overwide confidence intervals. However, he feels that the wider regression interval is probably more realistic in practice, especially as the sampling may not be truly random (as required for (6.8)). When some of the m_j are small or equal to zero, regression estimates cannot be used, whereas maximum-likelihood estimates can.

Example 6.7 Plaice: Paulik [1963a]

Using the data of Example 6.6, Table 6.10 was calculated. A plot of y_j versus j is roughly linear, thus suggesting that the above regression

TABLE 6.10

Application of Paulik's regression method to tagged plaice.

j	m_j	jm_j	j^2m_j	y_j	y_jm_j	jy_jm_j
1	139	139	139	1·9733	274·29	274·29
2	91	182	364	2·3969	218·12	436·24
3	52	156	468	2·9565	153·74	461·22
4	40	160	640	3·2189	128·76	515·04
Total	322	637	1611		774·91	1686·79

method can be used. Hence

$$C = 1686{\cdot}79 - (774{\cdot}91)(637)/322 = 153{\cdot}82,$$

$$D = 1611 - (637)^2/322 = 350{\cdot}85,$$

$$\hat{Z} = 153{\cdot}82/350{\cdot}85 = 0{\cdot}4384,$$

$$\hat{\beta}_0 = [637(0{\cdot}4384) - 774{\cdot}91]/322 = -1{\cdot}5393,$$

$$\hat{\sigma}^2 = 0{\cdot}397\,86, \qquad \hat{V}[\hat{Z}] = 0{\cdot}001\,134,$$

$$\hat{V}[\hat{\beta}_0] = 0{\cdot}005\,676 \text{ and } \widehat{\mathrm{cov}}\,[\hat{Z}, \hat{\beta}_0] = 0{\cdot}002\,243.$$

The 95 per cent confidence interval for Z, the three-monthly total instantaneous mortality rate, is (0·2935, 0·5833). The survival and exploitation rates per three-month period are $\hat{S} = 0{\cdot}645$ and $\hat{P}_1 = 0{\cdot}1384$, and we note that these estimates are very close to the maximum-likelihood estimates obtained in Example 6.6. Finally, from the confidence interval for Z we obtain the approximate 95 per cent confidence interval (0·56, 0·75) for $S = e^{-Z}$.

DISCONTINUITIES IN SAMPLING EFFORT. In exploited populations the sampling may not be completely continuous, as for example in some fisheries which close down for the weekend. However, in such a case if the unit of time is taken as a *working* week and Δ represents the length of the closed period (say the weekend), then (6.9) becomes

$$p_j = P_1 \exp\{-[\mu(1+\Delta) + \mu_E](j-1)\} \tag{6.14}$$

and $\hat{\beta}$, the estimate of the slope of (6.12), will now be an estimate of $\mu(1+\Delta) + \mu_E$ instead of $Z = \mu + \mu_E$. Thus the calculation of $\hat{\beta}_0$, $\hat{\beta}$ and $\hat{P}_1$ $(= \exp(-\hat{\beta}_0 - \hat{\beta}))$ will be the same as before, but $\hat{S} = \exp(-\hat{\mu} - \hat{\mu}_E)$ $(\neq \exp(-\hat{\beta}))$ will be different. Least-squares estimates of μ and μ_E are now obtained by solving

$$\hat{\beta} = Z(1+\Delta) - \mu_E \Delta$$

and

$$\hat{P}_1 = \frac{\mu_E}{Z}(1 - e^{-Z})$$

for Z and μ_E; $\hat{S}$ can then be calculated.

If, however, the lengths of the closure periods are irregular but the actual sampling or working periods are still of equal length (say one time-unit) then

$$p_j = P_1 \exp\{-\mu_E(j-1) - \mu\gamma_{j-1}\},$$

where γ_{j-1} is the total number of time-units from time zero to the beginning of the jth sampling period. This leads to the regression model

$$E[y_j] = -\log P_1 + \mu_E(j-1) + \mu\gamma_{j-1}, \tag{6.15}$$

and least-squares estimates of $-\log P_1$, μ and μ_E can be obtained once again, using the method of **1.3.5** (*2*). Paulik points out that this model is only useful in practice if there is sufficient variability in the lengths of the closure, so that γ_{j-1} is not approximately proportional to $j-1$. If this is not the case the matrix to be inverted for the least-squares estimation (namely $\mathbf{X}'\mathbf{B}^{-1}\mathbf{X}$ in the notation of **1.3.5** (*2*)) will be ill-conditioned.

6.3 SINGLE RELEASE: UNDERLYING ASSUMPTIONS

6.3.1 Initial mortality of tagged

In many fishery experiments a large proportion, $1-\nu$ say, of the tagged fish die immediately after release through the traumatic experience of being handled and tagged (cf. Paulik [1963b]). For the models in **6.1**, using M_0 instead of νM_0 leads to the overestimation of N_0, though in **6.1.2** the estimation of the instantaneous recruitment rate (λ) is unaffected. However, in the continuous sampling models of **6.2** the estimates of Z and S do not depend on knowing M_0, so that the above type of error will not affect these estimates or their variance estimates.

The parameter ν can sometimes be estimated by retaining some of the tagged fish in tanks and noting the proportion $\hat{\nu}$ which survive an initial period. Alternatively, all the tagged fish can be retained for a short period after tagging to allow for any tagging mortality (North Atlantic Fish Marking Symposium [1963: 8]).

6.3.2 Non-reporting of tags

Another source of error which also has the effect of reducing M_0 is the non-reporting of tags by the fishermen. Suppose that ρ is the proportion of tags recaptured which are subsequently reported to the research agency by the fishermen, and assume that this proportion remains constant during the entire sampling period. Let $c = \nu\rho$ $(= r$ in Paulik's notation), the effective proportion of M_0 utilised. Then, as already mentioned above, the estimates of Z and S for the models in **6.2** do not depend on knowing M_0 and are therefore not affected by a *constant* reporting rate of less than unity. However, as far as the other parameters are concerned, Paulik [1963a] shows that the procedures of **6.2** now estimate cP_J (or cP_1 in the regression models) instead of P_J and, since

$$\mu_E = P_J Z/(1-e^{-ZJ}),$$

$c\mu_E$ instead of μ_E. For example, if $\nu = 0{\cdot}7$ and $\rho = 0{\cdot}7$ then $c \approx 0{\cdot}5$ and the true μ_E would be about twice the estimated μ_E. Unfortunately c cannot be estimated separately, except for the model (6.15) which now becomes

$$E[y_j] = -(\log c + \log P_1) + \mu_E(j-1) + \mu\gamma_{j-1}.$$

Since P_1 is a function of μ and μ_E only, we can now estimate the three parameters μ, μ_E and c.

Paulik suggests that for seasonal fisheries, an obvious way to separate c and μ_E is to take advantage of the change in the total mortality rate that occurs when the fishery season begins. This may be done by making one or more additional releases of tagged fish before fishing begins and using the methods of **6.4**.

By letting n_2 denote the catch from a particular sample or sampling period and n_1 and N denote the size of the marked and total population, respectively, just before the sample is taken, then ρ can be estimated for each sample using the theory of **3.2.4**. The method described there consists of dividing the catch from each sample into two parts; one part is inspected by trained observers ($\rho = 1$) and the other is inspected by the fishermen ($\rho \leqslant 1$). ρ is then estimated by the ratio of the marked fraction of the catch as reported by the fishermen to the marked fraction reported by the observers; a test for $\rho = 1$ can also be carried out. If the same proportion of the catch is inspected by trained observers on each occasion, then, since ρ is related to a binomial parameter p ($p = (1+\rho\gamma)^{-1}$ in the notation of **3.2.4**), a contingency table can be set up to test for constant p (and hence for constant ρ).

6.3.3. Tag losses[(*)]

Tags eventually become detached through wear and tear and the rate of tag loss can be estimated using, for example, double tagging (cf. **3.2.3**). Suppose that the two types of tag are denoted by A and B and let

$\pi_k(t)$ = probability that a tag of type k comes off by time t after the initial tag release ($k = A, B$),

$\pi_{AB}(t)$ = probability that both tags come off by time t,

$m_k(t)$ = number caught at time t bearing tag k only,

$m_{AB}(t)$ = number caught at time t bearing both tags,

$m_0(t)$ = number caught at time t which have lost both tags,

and

$m(t) = m_A(t) + m_B(t) + m_{AB}(t) + m_0(t)$ (the number belonging to the release M_0 that are caught at time t).

Then assuming that the tags are independent of each other (i.e. $\pi_{AB}(t) = \pi_A(t)\pi_B(t)$), we can estimate $\pi_A(t)$, $\pi_B(t)$ and $m(t)$ using the moment estimates (cf. **3.2.3**(2)):

$$\hat{\pi}_A(t) = m_B(t)/[m_B(t) + m_{AB}(t)],$$

$$\hat{\pi}_B(t) = m_A(t)/[m_A(t) + m_{AB}(t)],$$

and

$$\hat{m}(t) = [m_A(t) + m_{AB}(t)][m_B(t) + m_{AB}(t)]/m_{AB}(t)$$

$$= c_t[m_A(t) + m_B(t) + m_{AB}(t)], \text{ say.}$$

This means that the observed recaptures given in the above square bracket must be corrected by a factor of

$$c_t = \left[1 - \frac{m_A(t)m_B(t)}{[m_A(t) + m_{AB}(t)][m_B(t) + m_{AB}(t)]}\right]^{-1} = \frac{1}{1 - \hat{\pi}_{AB}(t)}$$

to give an estimate of the actual number of recaptures $m(t)$. The estimates $\hat{\pi}_A(t)$, $\hat{\pi}_B(t)$ and $\hat{m}(t)$ (or suitable transformations) can also be plotted against t; in the case of continuous sampling one would normally take t as the midpoint of the sampling interval.

If the loss of tags follows a Poisson process with parameter L_k, say, for type k tag, then (cf. (1.1))

$$1 - \pi_k(t) = e^{-L_k t} \tag{6.16}$$

and the plot of $\log(1-\hat{\pi}_k(t))$ against t should be approximately linear. This type of model has been used, for example, by Beverton and Holt [1957: 204], Gulland [1963] and Backiel [1964].

If single tagging is used (say tag A) and (6.16) holds, then

$$\textit{observed recaptures} = m(t)\, e^{-L_A t}.$$

Hence, for this case, if y_j is calculated from observed recaptures then the only change in the regression model (6.12) is that $\beta = Z + L_A$.

(*)See also **13.1.4** (5).

TAGS INDISTINGUISHABLE. In some situations the only information recorded is the number of tags for each tagged individual, so that just the numbers $m_{AB}(t)$ and $m_C(t)$ $(= m_A(t) + m_B(t))$ are available. For this case we can still estimate $m(t)$ if we can assume that the tags are independent and that $\pi_A(t) = \pi_B(t)$ $(= \pi(t)$ say). Then from **3.2.3** (2), $m(t)$ and $\pi(t)$ are estimated by

$$\tilde{m}(t) = [m_C(t) + 2m_{AB}(t)]^2/4m_{AB}(t)$$

and

$$\tilde{\pi}(t) = m_C(t)/[m_C(t) + 2m_{AB}(t)],$$

CONSTANT LOSS RATE. If one of the tags or marks is permanent, then Robson and Regier [1966] give the following method for estimating rate of tag retention for the other tag.

Suppose that tag A is permanent, so that $\pi_A(t) = \pi_{AB}(t) = 0$, $m_B(t) = m_0(t) = 0$ and $m(t) = m_A(t) + m_{AB}(t)$. Assuming that r, the probability of retaining a type B tag for unit time, is constant, then

$$1 - \pi_B(t) = r^t$$

and $m_{AB}(t)$ is a binomial variable with parameters $m(t)$ and r^t. Hence, if a series of instantaneous samples is taken at times t_j $(j = 1, 2, \ldots, J)$ and we define $m_j = m(t_j)$ and $b_j = m_{AB}(t_j)$, then the joint distribution of the $\{b_j\}$ given the $\{m_j\}$ is

$$f(\{b_j\} \mid \{m_j\}) = \prod_{j=1}^{J} \binom{m_j}{b_j} (r^{t_j})^{b_j} (1 - r^{t_j})^{m_j - b_j}. \qquad (6.17)$$

The maximum-likelihood estimate $\hat{r}$ of r is readily shown to be the solution of

$$\sum_{j=1}^{J} \left\{ \frac{(b_j - m_j r^{t_j}) t_j}{(1 - r^{t_j})} \right\} = 0 \qquad (6.18)$$

which must be solved iteratively. If $\hat{r}$ is close to unity, Robson and Regier suggest that an approximate solution r_0 is given by

$$\log(1 - r_0) = \frac{\sum m_j t_j \log\left(\dfrac{m_j - b_j}{m_j t_j}\right)}{\sum m_j t_j}, \qquad (6.19)$$

which may be used as a trial value in the iterative process. The successive approximations r_i are then given by

$$r_{i+1} = r_i \left\{ 1 + \frac{\sum_j (b_j - m_j r_i^{t_j})\, t_j/(1 - r_i^{t_j})}{\sum_j m_j t_j^2 r_i^{t_j}/(1 - r_i^{t_j})} \right\}. \qquad (6.20)$$

From standard maximum-likelihood theory, $\hat{r}$ is asymptotically normal with mean r and variance

$$V[\hat{r}] = r^2 / \sum_j \left(\frac{m_j t_j^2 r^{t_j}}{1 - r^{t_j}} \right). \qquad (6.21)$$

When r is close to unity and t_j is small, as when t is measured in years, then

$$V[\hat{r}] \approx \frac{r(1-r)}{\sum m_j t_j}.$$

Robson and Regier infer from this that the estimator $\hat{r}$ behaves approximately as the proportion of tag retentions occurring in a single sample of size $\sum m_j t_j$ taken one year after release. That is, the sample of size m_j taken t_j years after release contributes as much information as a sample of size $m_j t_j$ taken one year after release.

Since the b_j are binomial variables, a goodness-of-fit statistic for the validity of the model (6.17) is given by

$$T = \sum_{j=1}^{J} \left\{ \frac{(b_j - m_j \hat{r}^{t_j})^2}{m_j \hat{r}^{t_j}(1-\hat{r}^{t_j})} \right\}, \tag{6.22}$$

where T is approximately distributed as χ^2_{J-1}.

For the special but not atypical case of only two samples taken $t_1 = 1$ and $t_2 = 2$ years after the tag–release, an explicit solution to equation (6.18) is available in the form

$$\hat{r} = \frac{b_1 - m_1}{2(m_1 + 2m_2)} + \sqrt{\left\{ \left[\frac{b_1 - m_1}{2(m_1 + 2m_2)} \right]^2 + \frac{b_1 + 2b_2}{m_1 + 2m_2} \right\}}.$$

In conclusion we note that if (6.16) holds, then $r = \exp(-L_B)$. Cucin and Regier [1965: 255] used this relationship to correct recaptures for tag losses in an application of the Petersen method.

Example 6.8 Lake whitefish (*Coregonus clupeaformis*): Robson and Regier [1966]

The above method of estimation was applied to data (Table 6.11) from a study of lake whitefish reported by Cucin and Regier (unpublished). Type *B*

TABLE 6.11

Recoveries of fish that were both tagged and fin-clipped at time of release: from Robson and Regier [1966: Table 1].

	Year of capture					
	1962		1963		1964	
Year of release	Tag retained	Tag lost	Tag retained	Tag lost	Tag retained	Tag lost
1961	79	4	20	0	3	1
1962	–	–	66	1	5	1

tag consisted of a plastic streamer type attached through the anterior base of the dorsal fin by means of monofilament nylon. Fish tagged in 1961 had their left pectoral fins removed by clipping, while those tagged in 1962 had their left pelvics removed. A biologist carefully examined over 10 000 whitefish taken in gill nets and pound nets from 1961 to 1964 and noted the number of clipped fish that had lost tags. A number of lines of evidence, all indirect, indicated that tagged fish were not appreciably more vulnerable to the gear than untagged fish. From previous study it was known that very few, if any, of the fish in this population were missing fins due to natural causes.

The fishing season was 7 months long each year, and fish were examined throughout the season. For purposes of this example the authors treated the data as though all sampling within a year was considered to have been instantaneous and on anniversaries of the mean tagging date. For greater accuracy and with sufficiently large samples, the recapture data could be summarised by months.

According to the binomial model used for the distribution of b_j (cf. (6.17)), the fractions $79/(79+4)$ and $66/(66+1)$ are both estimates of r, the annual tag retention rate; the fractions $20/(20+0)$ and $5/(5+1)$ are both estimates of r^2; and $3/(3+1)$ is an estimate of r^3. A test of the model (6.17) for the two releases can be carried out in two stages: first, testing the homogeneity of the two estimates of r and the two estimates of r^2 in separate 2-by-2 contingency tables, and, second, testing the goodness of fit of the pooled estimates of 1-, 2-, and 3-year retention rates to the predicted values of $\hat{r}$, $\hat{r}^2$ and $\hat{r}^3$ respectively, using T (6.22).

The homogeneity of the two estimates of r was tested in Table 6.12, giving a chi-square value of $\chi_1^2 = 1{\cdot}27$ ($P > 0{\cdot}20$). Homogeneity of the two

TABLE 6.12

A test of homogeneity for the two estimates of r: from Robson and Regier [1966].

	1961–62	1962–63	Total
Tag retained	79	66	145
Tag lost	4	1	5
Total	83	67	150

TABLE 6.13

A test of homogeneity for the two estimates of r^2: from Robson and Regier [1966].

	1961–63	1962–64	Total
Tag retained	20	5	25
Tag lost	0	1	1
Total	20	6	26

estimates of r^2 was tested in Table 6.13. Here the numbers are too small for the use of the chi-square approximation, but the exact probability is readily computed in this case. Given that one tag loss occurred among the 26 fish, the probability that it occurred among the 6 fish of the 1962–64 sample rather than among the 20 fish of the 1961–63 sample is simply 6/26 or 0·23; hence once again $P > 0{\cdot}20$. For later reference we note from the chi-square table that $P = 0{\cdot}23$ corresponds to a χ_1^2 value of approximately 1·47.

Since homogeneity was not rejected at this stage of the test, $\hat{r}$ can be calculated using the pooled data in the form:

	$t_1 = 1$	$t_2 = 2$	$t_3 = 3$
m_j	150	26	4
b_j	145	25	3

Clearly $\hat{r}$ must be near to unity, so that a first approximation r_0 can be calculated from (6.19). Thus

$$\log(1-r_0) = \frac{150(1)\log\left(\dfrac{150-145}{150(1)}\right)}{150(1)+26(2)+4(3)} + \frac{26(2)\log\left(\dfrac{26-25}{26(2)}\right) + 4(3)\log\left(\dfrac{4-3}{4(3)}\right)}{150(1)+26(2)+4(3)}$$

$$= -1{\cdot}512,$$

or $r_0 = 0{\cdot}969$. Substituting this value into (6.18) we get

$$\frac{[145-150(0{\cdot}969)](1)}{1-0{\cdot}969} + \frac{[25-26(0{\cdot}969)^2](2)}{1-(0{\cdot}969)^2} + \frac{[3-4(0{\cdot}969)^3](3)}{1-(0{\cdot}969)^3}$$

$$= -11{\cdot}2903(1) + 9{\cdot}6170(2) - 7{\cdot}0930(3),$$

which adds to $-13{\cdot}335$ instead of 0. The value 0·969 is therefore slightly larger than $\hat{r}$, and applying the iteration formula (6.20) we divide $-13{\cdot}335$ by

$$\sum_j \frac{m_j t_j^2 r_0}{(1-r_0^{t_j})} = \frac{150(1)^2(0{\cdot}969)}{1-0{\cdot}969} + \frac{26(2)^2(0{\cdot}969)^2}{1-(0{\cdot}969)^2} + \frac{4(3)^2(0{\cdot}969)^3}{1-(0{\cdot}969)^3}$$

$$= 6651{\cdot}86$$

to give

$$r_1 = 0{\cdot}969\left[1 - \frac{13{\cdot}335}{6651{\cdot}86}\right] = 0{\cdot}967.$$

Substituting this value into (6.18) we get $+2{\cdot}6374$ instead of 0, indicating that 0·967 is slightly smaller than $\hat{r}$. Going through the iteration once again using r_1 as the trial value leads to

$$r_2 = 0{\cdot}967\left[1 + \frac{2{\cdot}6374}{6233{\cdot}54}\right] = 0{\cdot}9674,$$

so that to three decimal places, $\hat{r} = 0{\cdot}967$.

The estimated standard deviation of $\hat{r}$ calculated from (6.21) is

$$\hat{\sigma}[\hat{r}] = 0{\cdot}967/\sqrt{6233{\cdot}54} = 0{\cdot}0122,$$

so that an approximate 95 per cent confidence interval for r is $\hat{r} \pm$ 1·96(0·0122) or (0·943, 0·991).

From (6.22)

$$T = \frac{[145-150(0{\cdot}967)]^2}{150(0{\cdot}967)(1-0{\cdot}967)} + \frac{[25-26(0{\cdot}967)^2]^2}{26(0{\cdot}967)^2[1-(0{\cdot}967)^2]} + \frac{[3-4(0{\cdot}967)^3]^2}{4(0{\cdot}967)^3[1-(0{\cdot}967)^3]}$$
$$= 1{\cdot}399,$$

which, for two degrees of freedom, is non-significant ($P > 0{\cdot}70$). We can now pool all the tests of significance by summing the three independent chi-square values which we have computed, namely

$$1{\cdot}27 + 1{\cdot}47 + 1{\cdot}40 = 4{\cdot}14$$

with $1 + 1 + 2 = 4$ degrees of freedom. This overall test ($P \approx 0{\cdot}40$) is also non-significant, so that we have found no grounds for rejecting the underlying model.

6.3.4 Exponential distribution of recapture times

From (6.3) we find that the density function for the recovery time of a tagged individual, given that it is caught in the time-interval $[0, \tau]$, is

$$f_2(t) = \frac{Z\,e^{-Zt}}{1-e^{-Z\tau}}, \quad 0 \leqslant t \leqslant \tau.$$

Therefore, given a random sample of m recapture times from this truncated exponential distribution, one can carry out a standard chi-squared goodness-of-fit test based on comparing observed and expected frequencies. For example, if the interval $[0, \tau]$ is split up into J intervals $[d_0, d_1)$, $[d_1, d_2)$, $\ldots$, $[d_{J-1}, d_J)$ (where $d_0 = 0$, $d_J = \tau$) and m_j tagged are caught in the jth interval, then the goodness-of-fit statistic is

$$\sum_{j=1}^{J} (m_j - E_j)^2/E_j,$$

where $$E_j = m\left\{\frac{\exp(-\hat{Z}d_{j-1}) - \exp(-\hat{Z}d_j)}{1-\exp(-\hat{Z}\tau)}\right\}$$

and $\hat{Z}$ is a suitable estimate of Z. When $\hat{Z}$ is the solution of (6.5) then, for large m, the above goodness-of-fit statistic is approximately distributed as χ^2_{J-1} (one degree of freedom is not subtracted for the estimation of Z as $\hat{Z}$ is calculated from the ungrouped data: Kendall and Stuart [1973: 447]). We note that if τ is large, so that the effect of truncation on $f_2(t)$ above is negligible, then a wide variety of tests for the negative exponential distribution are available (see p. 46).

When μ is zero, or negligible compared to μ_E, the recovery process of tagged individuals is Poisson with parameter μ_E. This means that the distribution of recapture times is negative exponential and, as Paulik [1963a] has pointed out, the theory of life-testing can be applied here. For example, we start with M_0 tagged individuals, and an individual is said to have "failed" in time $[0, \tau]$ if it is recaptured. Hence all the techniques associated with the exponential distribution and life-testing given in Epstein and Sobel [1953, 1954] and Epstein [1954, 1960a, b, c, d] (see also Buckland [1964] for a general bibliography) can be used here.

6.4 MULTI-RELEASE MODELS: CONTINUOUS SAMPLING

6.4.1 Notation and assumptions

In **6.3.2** it was mentioned that one way of separating μ_E and c is to take advantage of the change in the total mortality rate that occurs when the season begins. This can be accomplished by making one or more tag releases before the fishing begins. Let

I = number of pre-season releases,

R_i = number of tagged fish released τ_i time-units before the opening date ($i = 1, 2, \ldots, I$),

m_{ij} = number of tagged fish from the ith release of R_i individuals caught in the jth sampling period ($j = 1, 2, \ldots, J$),

$m_{i.} = \sum_i m_{ij}$,

$m_{..} = \sum_i \sum_j m_{ij}$,

and

$x_. = \sum_i \sum_j (j-1) m_{ij} = \sum_j (j-1) m_{.j}$.

We shall assume that

(1) The instantaneous natural mortality rate and the instantaneous exploitation rate are constant and equal to μ and μ_E, respectively; we define $Z = \mu + \mu_E$.
(2) The pre-season instantaneous natural mortality rate is the same as the during-season rate μ.
(3) Immigration and emigration are negligible.
(4) All tagged fish have the same probability of being caught in a particular sampling period (given that they are alive at the beginning of the period), irrespective of which tag release they are from.
(5) Sampling periods are each of one time-unit duration as in **6.2.**
(6) Fish do not lose their tags.
(7) A constant proportion ν of each tag release survives the tagging process. It is assumed further that the tagging mortality is complete within a short time after tagging.
(8) The probability ρ that a recaptured tag is reported remains constant during the entire season for all types of recapture gear used.

In the following models it is convenient to use the parameters

$$c = \nu\rho, \quad S = e^{-Z}, \quad \phi = e^{-\mu}, \quad \text{and} \quad \beta = cP_J,$$

where P_J is given by (6.7); β is the probability that a tagged individual survives any initial tagging mortality, is recaptured, and is reported on capture.

6.4.2 Maximum-likelihood estimates

1 General theory

Working with the parameters S, ϕ and β instead of μ, μ_E and c, Paulik [1963a] showed that the joint distribution of the $\{m_{ij}\}$ is proportional to (cf. (6.8))

$$\prod_{i=1}^{I}\left\{\left[\frac{\beta\phi^{\tau_i}}{\sum_{j=0}^{J-1} S^j}\right]^{m_{i1}}\left[\frac{\beta\phi^{\tau_i}S}{\sum_{j=0}^{J-1} S^j}\right]^{m_{i2}}\cdots\left[\frac{\beta\phi^{\tau_i}S^{J-1}}{\sum_{j=0}^{J-1} S^j}\right]^{m_{iJ}}(1-\beta\phi^{\tau_i})^{R_i-m_{i.}}\right\}$$

and the maximum-likelihood estimates are the solutions of

$$\left(\sum_{j=0}^{J-1} j\hat{S}^j\right)\Big/\left(\sum_{j=0}^{J-1}\hat{S}^j\right) = x_{.}/m_{..}, \tag{6.23}$$

$$\sum_{i=1}^{I}\left\{\frac{\tau_i(m_{i.}-R_i\hat{A}_i)}{1-\hat{A}_i}\right\} = 0, \tag{6.24}$$

and

$$\sum_{i=1}^{I}\left\{\frac{(m_{i.}-R_i\hat{A}_i)}{1-\hat{A}_i}\right\} = 0, \tag{6.25}$$

where $\hat{A}_i = \hat{\beta}\hat{\phi}^{\tau_i}$. Equation (6.23) can be solved for $\hat{S}$ using the method described in **A6**, while (6.24) and (6.25) must be solved iteratively for $\hat{\phi}$ and $\hat{\beta}$, using, for example, the methods of **1.3.8.**

The estimate $\hat{S}$, which is uncorrelated with $\hat{\phi}$ and $\hat{\beta}$, has asymptotic variance

$$V[\hat{S}] = \left\{\left[\sum_i R_iA_i\right]\left[\frac{1}{S(1-S)^2} - \frac{J^2S^{J-2}}{(1-S^J)^2}\right]\right\}^{-1}, \tag{6.26}$$

while the asymptotic variance–covariance matrix for $\hat{\phi}$ and $\hat{\beta}$ is $\mathbf{V}^{-1}$, where (cf. (1.7))

$$\mathbf{V} = \begin{bmatrix} \sum_i\left\{\dfrac{\tau_i^2R_iA_i}{\phi^2(1-A_i)}\right\}, & -\sum_i\left\{\dfrac{\tau_iR_i\phi^{\tau_i}}{\phi(1-A_i)}\right\} \\ \cdots, & \sum_i\left\{\dfrac{R_i\phi^{\tau_i}}{\beta(1-A_i)}\right\}\end{bmatrix}.$$

(The (1, 1) element in $\mathbf{V}^{-1}$ is $V[\hat{\phi}]$).

The maximum-likelihood estimates of μ, μ_E and c are given by

$$\hat{\mu} = -\log\hat{\phi},$$

$$\hat{\mu}_E = -\log\hat{S} - \hat{\mu}$$

and, writing

$$\hat{P}_J = -\hat{\mu}_E(1-\hat{S}^J)/\log\hat{S},$$

$$\hat{c} = \hat{\beta}/\hat{P}_J.$$

Approximate variances of these estimates can be calculated using the delta method and $\mathbf{V}^{-1}$. Thus

$$V[\hat{\mu}] \approx \frac{1}{\phi^2} V[\hat{\phi}]$$

and

$$V[\hat{\mu}_E] \approx \frac{1}{S^2} V[\hat{S}] + \frac{1}{\phi^2} V[\hat{\phi}].$$

DISCONTINUITIES IN SAMPLING EFFORT. The above theory can be readily modified, as on p. 279, to deal with the case when the sampling is not completely continuous: for example, some fisheries close down for the weekend. Suppose that the unit of time is now taken as the *working* week and let Δ be the length of the closed period (assumed to be constant from week to week). Then $\hat{S}$, $\hat{\phi}$, $\hat{\beta}$, $\hat{\mu} = -\log\hat{\phi}$, and their variances, are calculated as above, but now S and β are different functions of μ, μ_E and c. For example, S, the probability of surviving for one calendar week, is given by

$$S = \exp(-\mu_E - (1+\Delta)\mu).$$

Now P_1, the probability of capture in a *working* week, is still given by (cf. (6.6))

$$P_1 = \mu_E(1-e^{-Z})/Z$$

and the probability that a tagged individual is recaptured in the jth week of the season *given* that it is alive at the beginning of the season is

$$p_j = P_1 S^{j-1}.$$

Therefore P_J, the probability that a tagged individual is recaptured during the season given that it is alive at the beginning of the season, is given by

$$P_J = \sum_{j=1}^{J} p_j = P_1(1-S^J)/(1-S).$$

Hence from $\hat{S}$, $\hat{\mu}$ and $\hat{\beta}$ we obtain

$$\hat{\mu}_E = -\log\hat{S} - (1+\Delta)\hat{\mu}, \tag{6.27}$$

$$\hat{Z} = \hat{\mu} + \hat{\mu}_E,$$

$$\hat{P}_J = \frac{\hat{\mu}_E}{\hat{Z}} \cdot (1-e^{-\hat{Z}}) \cdot \frac{(1-\hat{S}^J)}{(1-\hat{S})} \tag{6.28}$$

and finally

$$\hat{c} = \hat{\beta}/\hat{P}_J. \tag{6.29}$$

Also, from the delta method,

$$V[\hat{\mu}_E] \approx \frac{1}{S^2} V[S] + \left(\frac{1+\Delta}{\phi}\right)^2 V[\hat{\phi}]. \tag{6.30}$$

2 *One release before the season*

ONE RELEASE AT START OF SEASON. For the simple case of just two releases, when the first release is made some time (τ_1) before the season begins and the second immediately preceding the season ($\tau_2 = 0$), equations (6.24) and (6.25) have solutions (Paulik [1962])

$$\hat{\phi}^{\tau_1} = \left(\frac{m_{1.}}{R_1}\right) \bigg/ \left(\frac{m_{2.}}{R_2}\right) \quad (= \hat{\alpha} \text{ say})$$

and

$$\hat{\beta} = m_{2.}/R_{2.}$$

Defining $\alpha = \phi^{\tau_1} = e^{-\mu\tau_1}$, the probability of survival up to the beginning of the season, it is readily seen that $m_{1.}$ and $m_{2.}$ have independent binomial distributions with parameters $(R_1, \alpha\beta)$ and (R_2, β) respectively. Hence, using the delta method

$$V[\hat{\alpha}] \approx \alpha^2 \left[\frac{1-\alpha\beta}{R_1\alpha\beta} + \frac{1-\beta}{R_2\beta}\right], \tag{6.31}$$

$$V[\hat{\beta}] = \beta(1-\beta)/R_2, \tag{6.32}$$

$$V[\hat{\mu}] \approx \frac{1}{(\tau_1\alpha)^2} V[\hat{\alpha}], \tag{6.33}$$

and

$$V[\hat{\mu}_E] \approx \frac{1}{S^2} V[\hat{S}] + (1+\Delta)^2 V[\hat{\mu}]. \tag{6.34}$$

By setting $\alpha = \alpha_1$ and $\beta = \beta_2$ we see that the above model is a special case of Ricker's two-release method in **5.1.3** (*4*). Hence an approximately unbiased estimate of α is

$$\tilde{\alpha} = \frac{m_{1.}(R_2+1)}{R_1(m_{2.}+1)},$$

and

$$v[\tilde{\alpha}] = \tilde{\alpha}^2 - \frac{m_{1.}(m_{1.}-1)(R_2+1)(R_2+2)}{R_1(R_1-1)(m_{2.}+1)(m_{2.}+2)}$$

is an approximately unbiased estimate of $V[\tilde{\alpha}]$.

Example 6.9 Pink salmon: Paulik [1962]

Paulik applied the above two-release method to some data reported by Elling and Macy [1955]; the m_{ij} for the three sampling periods ($J = 3$) are given in Table 6.14. Here the basic unit of time is the number of days (5.5)

TABLE 6.14

Tag recoveries, m_{ij}, from two releases where the second release is made at the beginning of the season: from Paulik [1962: Table 1].

	Recoveries in jth period			
	1	2	3	$m_{i.}$
R_1 = 1195	304	63	12	379
R_2 = 1083	566	86	26	678
$m_{.j}$	870	149	38	1057

fished during 1 week, and as the first tag release was made 6 days before the fishing began, $\tau_1 = 6/5{\cdot}5 = 1{\cdot}0909$. From Table 6.14,

$$x_{.}/m_{..} = [149 + 2(38)]/1057 = 0{\cdot}2129$$

and, from **A6** with $K = 2$, (6.23) has solution $\hat{S} = 0{\cdot}1825$. We also have

$$\hat{\beta} = 678/1083 = 0{\cdot}6260$$

$$\hat{\alpha} = \frac{379}{1195} \cdot \frac{1083}{678} = 0{\cdot}5066$$

and

$$\hat{\mu} = -(1/\tau_1)\log(\hat{\phi}^{\tau_1}) = -(1/\tau_1)\log\hat{\alpha} = 0{\cdot}6234.$$

The period of closure is $\Delta = 1{\cdot}5/5{\cdot}5 = 0{\cdot}2727$, so that from (6.27), (6.28) and (6.29),

$$\hat{\mu}_E = 1{\cdot}7010 - 0{\cdot}7934 = 0{\cdot}9076,$$

$$\hat{P}_J = \frac{0{\cdot}9076}{1{\cdot}5310}(1 - e^{-1{\cdot}5310})\frac{(1 - 0{\cdot}1825^3)}{(1 - 0{\cdot}1825)}$$

$$= 0{\cdot}5648,$$

and

$$\hat{c} = 0{\cdot}6260/0{\cdot}5648 = 1{\cdot}11.$$

To find the variance estimates we note first of all that

$$\sum_i R_i \hat{A}_i = R_1 \hat{\alpha}\hat{\beta} + R_2\hat{\beta} = m_{1.} + m_{2.} = m_{..}\,. \qquad (6.35)$$

Hence, replacing parameters by estimates, we have from (6.26), (6.32) and (6.31),

$$\hat{V}[\hat{S}] = 0{\cdot}000\ 144\ 7,$$

$$\hat{V}[\hat{\beta}] = \hat{\beta}(1-\hat{\beta})/R_2 = 0{\cdot}000\ 216\ 2,$$

and

$$\hat{V}[\hat{\alpha}] = \hat{\alpha}^2\left[\frac{1}{m_{1.}} + \frac{1}{m_{2.}} - \frac{1}{R_1} - \frac{1}{R_2}\right]$$

$$= 0{\cdot}000\ 604\ 0$$

Therefore, from (6.33),

$$\hat{V}[\hat{\mu}] = \frac{1}{(\tau_1\hat{\alpha})^2}\hat{V}[\hat{\alpha}] = 0{\cdot}001\ 978$$

and, from (6.34),

$$\hat{V}[\hat{\mu}_E] = \frac{1}{\hat{S}^2}\hat{V}[\hat{S}] + (1+\Delta)^2\,\hat{V}[\hat{\mu}]$$

$$= 0{\cdot}007\ 549.$$

The approximate 95 per cent confidence intervals, *estimate* $\pm 1{\cdot}96$ (*standard deviation*), for S, μ and μ_E are (0·159, 0·206), (0·536, 0·711) and (0·737, 1·078), respectively. (In the original example Paulik neglected the effect of the closed period Δ in estimating μ_E and c.)

ONE RELEASE DURING SEASON. Paulik [1962] points out that the above method can still be used if the second release is made some time during the season instead of at the beginning of the season. In this case the recaptures from the second release are recorded according to the sampling period numbered from the time of this release (the coded period) rather than from the beginning of the season, and the data from the first release are truncated, so that the number of recording periods is the same for each release. For example, suppose that the second release in the previous example had been made at the beginning of the second week of fishing. Then the number of recovery periods is now 2 and the recoveries m_{ij} are arranged as in Table 6.15. The analysis is the same as above, the only difference being that $J = 2$ instead of 3.

TABLE 6.15

Tag recoveries, m_{ij}, from two releases where the second release is made during the season: from Paulik [1962].

	Recoveries in jth coded period		
	1	2	$m_{i.}$
$R_1 = 1195$	304 (first week)	63 (second week)	367
$R_2 = 1083$	566 (second week)	86 (third week)	652
$m_{.j}$	870	149	1019

SEVERAL RELEASES DURING SEASON. If several releases are made during the season, then the recoveries from these releases can be grouped according to the coded recovery period. Paulik [1962] points out that if the same number J of coded recovery periods are used for each release then the grouped data can be treated as a single release, and the problem reduces to the above case of just one release during the season.

6.4.3 Regression estimates

Returning to the problem of **6.4.1**, where all releases are made before the fishing season begins, we see that p_{ij}, the probability that a tagged fish from the ith release is recaptured and reported in the jth sampling interval, is given by

$$p_{ij} = cP_1 \exp\{-Z(j-1)-\mu\tau_i\}.$$

Now $E[m_{ij}] = R_i p_{ij}$, and putting $y_{ij} = \log(R_i/m_{ij})$ we are led to consider the regression model

$$E[y_{ij}] \approx -(\log c + \log P_1 + Z) + \mu\tau_i + Zj. \quad (6.36)$$

This model is an extension of (6.12) and can be analysed the same way.

Thus, by writing

$$\mathbf{Y}' = (y_{11}, y_{12}, \ldots, y_{1J}, \ldots, y_{I1}, y_{I2}, \ldots, y_{IJ})$$

$$\mathbf{X} = \left[\begin{array}{ccc} 1 & \tau_1 & 1 \\ 1 & \tau_1 & 2 \\ \cdots & \cdots & \cdots \\ 1 & \tau_1 & J \\ \hline 1 & \tau_2 & 1 \\ 1 & \tau_2 & 2 \\ \cdots & \cdots & \cdots \\ 1 & \tau_2 & J \\ \hline \cdots & \cdots & \cdots \\ \hline 1 & \tau_I & 1 \\ 1 & \tau_I & 2 \\ \cdots & \cdots & \cdots \\ 1 & \tau_I & J \end{array}\right], \qquad \boldsymbol{\beta} = \left[\begin{array}{c} -(\log c + \log P_1 + Z) \\ \mu \\ Z \end{array}\right]$$

and given $\sigma^2 \mathbf{B}$, the variance–covariance matrix of $\mathbf{Y}$, the least-squares estimate of $\boldsymbol{\beta}$ can be found using the method of **1.3.5**(*2*). Paulik [1963a] shows that

$$\mathbf{B}^{-1} = \left[\begin{array}{cccc} \mathbf{B}_1^{-1} & \mathbf{0} & \cdots & \mathbf{0} \\ \mathbf{0} & \mathbf{B}_2^{-1} & \cdots & \mathbf{0} \\ \cdots & \cdots & \cdots & \cdots \\ \mathbf{0} & \cdots & \cdots & \mathbf{B}_I^{-1} \end{array}\right], \tag{6.37}$$

where $\mathbf{B}_i^{-1} =$

$$\frac{R_i}{1-\sum_j p_{ij}} \left[\begin{array}{cccc} p_{i1}(1-\sum_j p_{ij}+p_{i1}), & -1, & \ldots, & -1 \\ -1, & p_{i2}(1-\sum_j p_{ij}+p_{i2}), & \ldots, & -1 \\ \cdots, & \cdots, & \ldots, & \cdots \\ -1, & -1, & \ldots, & p_{iJ}(1-\sum_j p_{ij}+p_{iJ}) \end{array}\right] \tag{6.38}$$

and p_{ij} is estimated by m_{ij}/R_i. A qualitative check on the linearity of the regression model is given by plotting y_{ij} against j for each i.

Explicit least-squares estimates of $\log(cP_1)$, μ and μ_E, together with the estimate of their variance–covariance matrix, are given by Paulik in the appendix to his paper. However, it should be noted that he uses the model

$$E[\log(m_{ij}/R_i)] = \log c + \log P_1 + Z - \mu\tau_i - Zj$$

or, in terms of his notation,

$$E[\log(n_{ij}/N_i)] = \log r + \log\left\{\frac{F}{F+X}(1-e^{-(F+X)})\right\} + F + X - X\rho_i - (F+X)j.$$

Example 6.10 Pink salmon: Paulik [1963a, b]

In 1950 Elling and Macy [1955] conducted a tagging programme in the northern part of south-eastern Alaska, when daily releases of tagged pink salmon were made for $I = 12$ consecutive days immediately preceding the opening of the commercial fishery. During the fishery, recaptures m_{ij} from each of the releases were recorded for three consecutive weeks ($J = 3$), and the data are set out in Table 6.16. In this experiment the fishery was closed

TABLE 6.16

Weekly recoveries of tagged pink salmon in the commercial fishery for twelve pre-season releases: from Paulik [1963a: Table 1].

i	$5{\cdot}5\tau_i$	R_i	m_{i1}	m_{i2}	m_{i3}	i	$5{\cdot}5\tau_i$	R_i	m_{i1}	m_{i2}	m_{i3}
1	12	784	96	5	2	7	6	351	93	19	5
2	11	574	88	15	5	8	5	1 509	475	96	24
3	10	862	137	29	6	9	4	1 003	352	81	20
4	9	1 097	219	33	6	10	3	1 938	705	155	30
5	8	1 146	305	51	17	11	2	1 661	783	117	34
6	7	1 195	304	63	12	12	1	1 083	566	86	26
		$m_{..} = 5\ 060$				Total		13 203	4 123	750	187

for $1\frac{1}{2}$ days in each week, so that a working week of 5·5 days was taken as the unit of time. However, apart from scaling the values of τ_i, the effect of the closures was otherwise ignored in the application of the maximum-likelihood and regression methods (though this point is raised in Paulik [1963b: 235]).

Using the regression method, Paulik shows that the least-squares estimate of

$$\begin{bmatrix} \log(cP_1) \\ \mu \\ \mu_E \end{bmatrix} \quad \text{is} \quad \begin{bmatrix} -0{\cdot}6868 \\ 0{\cdot}6410 \\ 0{\cdot}9660 \end{bmatrix}$$

with variance–covariance matrix estimated by

$$\begin{bmatrix} 0{\cdot}000\ 233\ 0 & 0{\cdot}000\ 481\ 5 & -0{\cdot}000\ 094\ 4 \\ \cdots & 0{\cdot}000\ 927\ 2 & -0{\cdot}000\ 933\ 3 \\ \cdots & \cdots & 0{\cdot}002\ 706\ 5 \end{bmatrix}.$$

Assuming asymptotic normality, approximate 95 per cent confidence intervals for μ and μ_E are (0·5814, 0·7007) and (0·8641, 1·0681) respectively.

Estimates of S, β and c are obtained from the relations

$$S = \exp(-\mu - \mu_E),$$

$$\beta = cP_J,$$
$$= (cP_1)(1-S^J)/(1-S),$$

and

$$c = \beta/P_J.$$

Thus $\hat{S} = 0{\cdot}2005$, $\hat{\beta} = 0{\cdot}6243$ and $\hat{c} = 1{\cdot}046$. Using the delta method, an approximate 95 per cent confidence interval for c is (0·957, 1·137).

The above estimates may be compared with the corresponding maximum-likelihood estimates (cf. **6.4.2**)

$$\hat{S} = 0{\cdot}1970, \quad \hat{\beta} = 0{\cdot}6230 \quad \text{and} \quad \hat{c} = 1{\cdot}0515.$$

The closeness of the two sets of estimates provides quantitative support for the validity of the assumptions underlying the two models. Graphical evidence for the validity of the regression model is given in Paulik [1963b].

In conclusion we note that the hypothesis $c = 1$ is not rejected by the above data as the confidence interval for c contains unity.

DISCONTINUITIES IN SAMPLING EFFORT. If the unit of time is the working week and the length of the closed period is constant (Δ say) from week to week, then

$$p_{ij} = c\,e^{-\mu\tau_i}\, S^{j-1}P_1,$$

where $S = e^{-(\mu_E + (1+\Delta)\mu)}$.

Hence the regression model now becomes

$$E[y_{ij}] = -\log(cP_1) + \mu\tau_i + (-\log S)(j-1),$$

where $P_1 = \mu_E(1-e^{-Z})/Z$.

If the closed period is variable but the working period is constant (of duration one time-unit), then the appropriate regression model is

$$E[y_{ij}] = -\log(cP_1) + \mu(\tau_i + \gamma_{j-1}) + \mu_E(j-1),$$

where γ_{j-1} is the total number of time-units from the beginning of the season to the beginning of the jth recovery period. When the natural mortality rate is not constant but changes from μ' to μ when the season begins, then

$$E[y_{ij}] = -\log(cP_1) + \mu'\tau_i + \mu\gamma_{j-1} + \mu_E(j-1)$$

and we can use this model to test the hypothesis that $\mu' = \mu$.

RELEASES DURING THE SEASON. Paulik [1963a] also indicates briefly how the above regression methods can be simply extended to deal with releases actually made during the fishing season. For such releases we put $\tau_i = 0$ and $j = 1, 2, \ldots, J_i$ where, for a given i, j always refers to the recovery period *relative* to the ith release (i.e. the coded recovery period); this model is also mentioned in **5.3.2.**, and from equation (13.22) in **13.1.4**(*4*).

CHAPTER 7

CATCH-EFFORT METHODS: CLOSED POPULATION

7.1 VARIABLE SAMPLING EFFORT

7.1.1 Introduction

Catch-effort methods are based on the general assumption that the size of a sample caught from a population is proportional to the effort put into taking the sample. More specifically, this means that one unit of sampling effort is assumed to catch a fixed proportion of the population, so that if samples are permanently removed, the decline in population size will produce a decline in catch per unit effort. Such techniques, first used in 1914 for bears in Norway (Hjort and Ottestad [1933]), are now widely used in the study of fish and small-mammal populations, where effort is usually measured in such units as line or trap per unit time. Let

N = initial population size,

n_i = size of ith sample removed from the population ($i = 1, 2, \ldots, s$),

$x_i = \sum_{j=1}^{i-1} n_j \; (i = 2, 3, \ldots, s+1; \; x_1 = 0)$,

f_i = units of effort expended on the ith sample,

and

$$F_i = \sum_{j=1}^{i-1} f_j \; (i = 2, 3, \ldots, s+1; \; F_1 = 0).$$

We shall assume that:

(1) The population is closed, except for the removals.

(2) Sampling is a Poisson process with regard to effort. Mathematically this means that the probability of a given individual being caught when the population is subjected to δf units of sampling effort is $k\delta f + o(\delta f)$. Here k, usually called the (Poisson) *catchability coefficient* (or q in fishery research), is assumed to be constant throughout the whole experiment and the same for each individual. Also the units of effort are assumed to be independent, i.e. traps do not compete with each other.

(3) All individuals have the same probability p_i $(= 1-q_i)$ of being caught in the ith sample. It can be shown from assumption (2) that (cf. (1.1)) $q_i = \exp(-kf_i)$.

Using the above assumptions we shall now consider a number of possible models.

7.1.2 Maximum-likelihood estimation

The joint distribution of the $\{n_i\}$ is given by

$$f(\{n_i\}) = \prod_{i=1}^{s} \binom{N-x_i}{n_i} p_i^{n_i} q_i^{N-x_{i+1}} \tag{7.1}$$

or, rearranging, the multinomial distribution

$$\frac{N!}{\left(\prod_{i=1}^{s} n_i!\right)(N-x_{s+1})!} p_1^{n_1}(q_1p_2)^{n_2}\ldots(q_1q_2\ldots q_{s-1}p_s)^{n_s}(q_1q_2\ldots q_s)^{N-x_{s+1}} \tag{7.2}$$

It is readily shown that the maximum-likelihood estimates $\hat{N}$ and $\hat{k}$ are solutions of

$$kF_{s+1} = -\log(1-(x_{s+1}/N)) \tag{7.3}$$

and

$$NF_{s+1} = \sum_{i=1}^{s} f_i(x_i + n_i p_i^{-1}). \tag{7.4}$$

These two equations can be solved iteratively, using, for example, the regression estimates of **7.1.3** as first approximations. Usually kf_i is small so that $p_i = 1 - \exp(-kf_i) \approx kf_i(1-\frac{1}{2}kf_i)$ and (7.4) reduces to

$$NF_{s+1} \approx \sum_{i=1}^{s} f_i(x_i + \tfrac{1}{2}n_i) + (x_{s+1}/k).$$

Using standard maximum-likelihood theory (cf. (1.7)) and Stirling's approximation for large factorials, it can be shown that the asymptotic variance–covariance matrix of $\hat{N}$ and $\hat{k}$ is $\mathbf{V}^{-1}$, where

$$\mathbf{V} = \begin{bmatrix} E\left[\dfrac{x_{s+1}}{N(N-x_{s+1})} + \dfrac{1}{2N^2} - \dfrac{1}{2(N-x_{s+1})^2}\right], & F_{s+1} \\ F_{s+1}, & E\left[\sum_{i=1}^{s} n_i f_i^2 q_i/p_i^2\right] \end{bmatrix}$$

The (1, 1) element of $\mathbf{V}^{-1}$ is the asymptotic variance of $\hat{N}$. $\mathbf{V}$ can be estimated by replacing the expectations by random variables, and the unknown parameters by their estimates.

A suitable statistic for testing the adequacy of the model (7.2) is (cf. **1.3.6**(*3*))

$$\sum_{i=1}^{s}\left\{\frac{(n_i - \hat{N}\hat{q}_1\hat{q}_2\ldots\hat{q}_{i-1}\hat{p}_i)^2}{\hat{N}\hat{q}_1\hat{q}_2\ldots\hat{q}_{i-1}\hat{p}_i}\right\}$$

which is asymptotically distributed as chi-squared with $s-2$ degrees of freedom.

7.1.3 Regression estimates

1 Leslie's method

From (7.1) we have

$$\begin{aligned} E[n_i \mid x_i] &= p_i(N-x_i) \\ &\approx kf_i(N-x_i), \end{aligned}$$

when kf_i is small. The relationship $p_i \approx kf_i$ can be made exact if we redefine k to be the average probability that an individual is captured with one unit of effort and assume units of effort to be independent and additive (cf. **3.2.2**(*3*)): we shall call this new coefficient, K, the "binomial catchability coefficient". Therefore, defining $Y_i = n_i/f_i$, the catch per unit effort, we have

$$E[Y_i \mid x_i] = K(N - x_i), \quad (i = 1, 2, \ldots, s). \tag{7.5}$$

This linear regression model was first given by Leslie and Davis [1939] and De Lury [1947], and then developed more fully by De Lury [1951] and Chapman [1954: 10].

Now, conditional on $\{x_i\}$, the $\{Y_i\}$ are independently distributed and

$$\begin{aligned} V[Y_i \mid x_i] &= (N - x_i) p_i q_i / f_i^2 \\ &= (N - x_i) K q_i / f_i \\ &\approx (N - x_i) K / f_i . \end{aligned}$$

Therefore, setting $\theta = N$ and $\gamma = K$, we can carry out a weighted least-squares analysis using the method of **1.3.5**(*3*) with, for example, $w_i = f_i/(N - x_i)$. However, the weights contain N and must be estimated iteratively. To start the cycle we can set $w_i = 1$, estimate N, and then substitute this estimate in w_i, etc. However, as Ricker [1958: 147] points out, factors other than sample size (e.g. day-to-day variations in the catchability) can also play a large part in producing the scatter of points about the regression line. It is therefore recommended that one carries out an unweighted least-squares analysis, i.e. the variance of Y_i is assumed to be constant ($= \sigma^2$, say). Then, setting $w_i = 1$ $(i = 1, 2, \ldots, s)$ in **1.3.5**(*3*), the least-squares estimates of K and N are

$$\tilde{K} = -\sum_{i=1}^{s} Y_i (x_i - \bar{x}) \Big/ \sum_{i=1}^{s} (x_i - \bar{x})^2$$

and

$$\tilde{N} = \bar{x} + (\bar{Y}/\tilde{K}).$$

The estimate $\tilde{N}$ is approximately unbiased, and from (1.18)

$$V[\tilde{N}] \approx \frac{\sigma^2}{K^2}\left[\frac{1}{s} + \frac{(N - \bar{x})^2}{\Sigma (x_i - \bar{x})^2}\right].$$

A plot of Y_i versus x_i will provide a rough visual check on the adequacy of the regression model, including the assumption of constant variance. However, such graphical evidence should not be taken as final, for a straight-line fit is still possible in some situations, even when the assumptions do not hold. For example, a linear model is still possible even with natural mortality (cf. p. 307) or migration (Ketchen [1953]) taking place.

We note that $N = x$ when $Y = 0$, so that a simple graphical estimate of N is obtained by extending the fitted regression line to cut the x-axis. Also the intercept on the Y-axis will give an estimate of KN.

Assuming the Y_i to be approximately normally distributed, we can use the general method of **1.3.5**(*3*) to obtain an approximate $100(1-\alpha)$ per cent confidence interval for N, namely $(\bar{x}+d_1,\ \bar{x}+d_2)$. Here d_1 and d_2 are the roots of

$$d^2\left(\widetilde{K}^2 - \frac{\widetilde{\sigma}^2 t_{s-2}^2[\alpha/2]}{\Sigma(x_i-\bar{x})^2}\right) - 2d\bar{Y}\widetilde{K} + \left(\bar{Y}^2 - \frac{\widetilde{\sigma}^2 t_{s-2}^2[\alpha/2]}{s}\right) = 0, \qquad (7.6)$$

where
$$(s-2)\widetilde{\sigma}^2 = \Sigma[Y_i - \widetilde{K}(\widetilde{N}-x_i)]^2$$
$$= \Sigma(Y_i-\bar{Y})^2 - [\Sigma Y_i(x_i-\bar{x})]^2/\Sigma(x_i-\bar{x})^2.$$

Also a 100 $(1-\alpha)$ per cent confidence interval for the slope K is

$$\widetilde{K} \pm t_{s-2}[\alpha/2]\,(\widetilde{\sigma}^2/\Sigma(x_i-\bar{x})^2)^{\frac{1}{2}}. \qquad (7.7)$$

Apart from one or two isolated examples (e.g. Eberhardt *et al.* [1963], Van Etten *et al.* [1965], Lewis and Farrar [1968]), the Leslie regression method has been used mainly in fishery research. For example, a number of interesting graphs for fish populations are given in Omand [1951].

Example 7.1 Lobsters: De Lury [1947]

Table 7.1 gives a day-by-day record for 17 days of the lobster catch in a particular area. The unit of effort is taken as a thousand traps fished for one day, and the catch n_i is measured in units of 1000 lb. De Lury mentions

TABLE 7.1

Leslie's regression method applied to data from a lobster population: from De Lury [1947: Table 1].

Date	n_i (1000 lb)	f_i (1000 traps)	y_i $(= n_i/f_i)$	x_i (1000 lb)
May 23	6·995	8·470	0·8259	0
24	5·851	7·770	0·7530	6·995
25	3·221	3·430	0·9391	12·846
26	6·345	7·970	0·7961	16·067
27	3·035	4·740	0·6403	22·412
29	6·271	8·144	0·7700	25·447
30	5·567	7·965	0·6989	31·718
31	3·017	5·198	0·5804	37·285
June 1	4·559	7·115	0·6408	40·302
2	4·721	8·585	0·5499	44·861
5	3·613	6·935	0·5210	49·582
6	0·473	1·060	0·4462	53·195
7	0·928	2·070	0·4483	53·668
8	2·784	5·725	0·4863	54·596
9	2·375	5·235	0·4537	57·380
10	2·640	5·480	0·4818	59·755
12	3·569	8·300	0·4300	62·395

that the distribution of lobster size remained fairly constant during the 17 days, so that identifying "pounds" with "number" of individuals in the above theory is not a serious misrepresentation. This is equivalent to redefining K as the average probability that 1000 pounds of lobster are caught by 1000 traps in one day.

Assumption (1) that the population is closed is a reasonable one, as migration and natural mortality were negligible and the sampling period excluded moulting times when there would be recruitment through growth to legal size. A plot of Y_i versus x_i (Fig. 7.1) is approximately linear, thus suggesting that the assumptions underlying (7.5) are approximately satisfied.

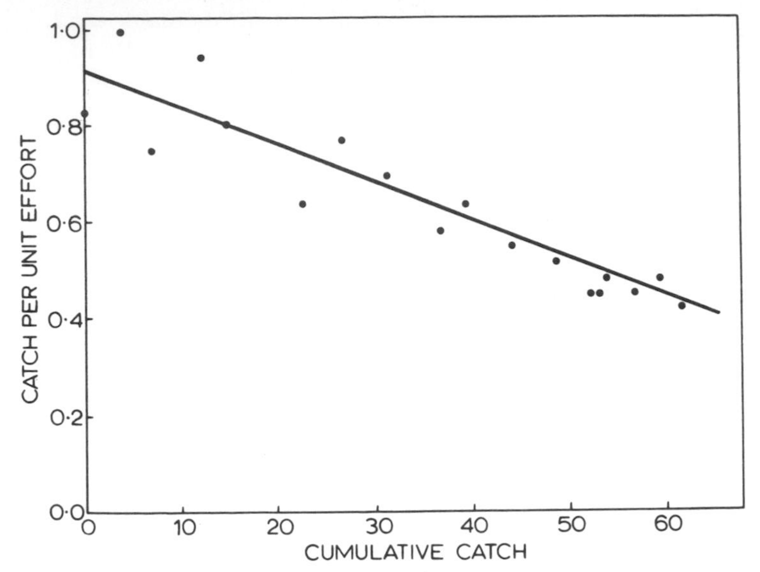

Fig. 7.1 Plot of catch per unit effort (n_i/f_i) versus the cumulative catch (x_i) for a lobster population: redrawn from De Lury [1947].

From Table 7.1 it was found that $\tilde{N} = 120{\cdot}5$ *thousand* pounds with 95 per cent confidence interval (77, 327), and $\tilde{K} = 0{\cdot}0074$ per 1000 traps with 95 per cent confidence interval (0·0058, 0·0090).

Example 7.2 Blue crab (*Callinectes sapidus*): Fischler [1965]

The total catches per week, for 12 weeks, from a commercial-size male crab population are given in Table 7.2. Three different kinds of gear were used, namely trot-lines, crab trawls and shrimp trawls. From the ratios of total catch over total effort for each gear, the relative "fishing powers" of the three gears were found to be 1·00, 1·42 and 1·06 respectively; for example, one crab trawl per day was equivalent to 1·42 trot-lines operating for one day. These powers were then used to convert the weekly fishing efforts of the crab and shrimp trawls into standard units of trot-lines per day.

The assumption of a closed population seems reasonable because of the following: (i) Natural mortality was negligible. (ii) The catches n_i in

TABLE 7.2

Catch-effort data for a population of commercial-size male crabs: from Fischler [1965: Table 11].

i (week)	n_i (lb)	f_i (effort)	y_i $(= n_i/f_i)$	x_i (lb)
1	33 541	194	172·9	0
2	47 326	248	190·8	33 541
3	36 460	243	150·0	80 867
4	33 157	301	110·2	117 327
5	29 207	357	81·8	150 484
6	33 125	352	94·1	179 691
7	14 191	269	52·8	212 816
8	9 503	244	38·9	227 007
9	13 115	256	51·2	236 510
10	13 663	248	55·1	249 625
11	10 865	234	46·4	263 288
12	9 887	227	43·6	274 153

Table 7.2 have been corrected for recruitment from precommercial to commercial size, so that the population under study was the population of commercial-size male crabs at the beginning of the experiment. (iii) Only the male population is considered, as tagging studies indicated that commercial-size females were emigrating out of the area while the fishery was operating.

The approximate linearity of Y_i versus x_i in Fig. 7.2 gives qualitative support for the adequacy of the regression model, and in particular for the

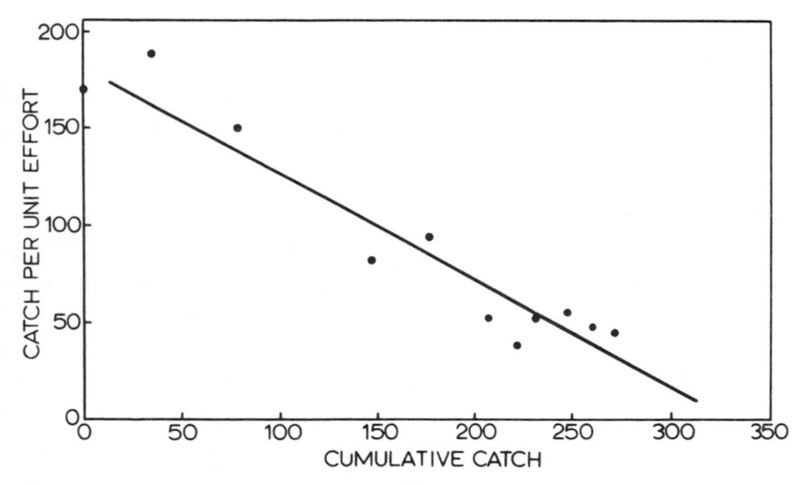

Fig. 7.2 Plot of catch per unit effort (n_i/f_i) versus the cumulative catch (x_i) for a population of male crabs: redrawn from Fischler [1965].

assumption of constant catchability K. Obviously K could be affected by weather conditions, as adverse weather would reduce successful fishing time each day and consequently reduce the daily catch per unit of effort. Fischler felt that this was the case at the beginning and end of the 12 weeks and also for weeks 7 and 8, when there were north-east winds of from 5 to 11 knots. In order to study some of the weather effects the trot-line was singled out as the gear most likely to be affected. It was found that the multiple regression of trot-line catch per unit effort with respect to average wind speed and average wind direction each day during the season was not significant. This suggests that weather generally had little effect on catchability, except possibly at the three periods mentioned above.

Fischler points out that if any increase or decrease in the number of units of effort is followed by an immediate decrease or increase in catch per unit effort, then competition between fishing units may be indicated. As there was no evidence of this for all the catch-effort data examined, it was assumed that gear competition did not affect catchability.

From Table 7.2 it was found that $\widetilde{N} = 330\ 300$ with (cf. (7.6)) 95 per cent confidence interval (299 600, 373 600), and $\widetilde{K} = 0{\cdot}000\ 56$ with 95 per cent confidence interval (0·000 45, 0·000 67).

2 *Ricker's method*

Treating the whole experiment as F_{s+1} samples, each with one unit of effort, then

$$E[N-x_i] = N(1-K)^{F_i}. \tag{7.8}$$

Therefore, combining this equation with (7.5), and taking logarithms, leads to Ricker's regression model [1958: 149].

$$E[y_i] \approx \log(KN) + [\log(1-K)]F_i, \tag{7.9}$$

where $y_i = \log Y_i = \log(n_i/f_i)$. For this model the least-squares estimates K' and N' are given by

$$\log(1-K') = \sum_{i=1}^{s} y_i(F_i-\bar{F})/\sum_{i=1}^{s}(F_i-\bar{F})^2$$

and

$$\log N' = \bar{y} - \bar{F}\log(1-K') - \log K',$$

where $\bar{F} = \sum_{i=1}^{s} F_i/s$. If the y_i are assumed to be independently distributed with constant variance σ_y^2 then, using the delta method,

$$\begin{aligned} V[K'] &\approx (1-K)^2\, V[\log(1-K')] \\ &= (1-K)^2\, \sigma_y^2/\Sigma(F_i-\bar{F})^2 \end{aligned}$$

and

$$V[N'] \approx N^2\left\{\frac{\sigma_y^2}{s} + \left(\frac{\bar{F}}{1-K} - \frac{1}{K}\right)^2 \cdot V[K']\right\},$$

where σ_y^2 can be estimated by

$$\sigma_y'^2 = \frac{1}{s-2}\{\Sigma(y_i-\bar{y})^2 - [\Sigma y_i(F_i-\bar{F})]^2/\Sigma(F_i-\bar{F})^2\}.$$

Assuming normality, a confidence interval for K can be obtained from the corresponding interval for $\log(1-K)$, the slope of the regression line.

Ricker also suggests the moment-type estimate (cf. (7.8) with $i = s+1$)

$$N^* = \frac{x_{s+1}}{1-(1-K')^{F_{s+1}}}.$$

This estimate is used, for example, by Libosvárský [1962] for estimating the weight of fish stock in small streams.

3 De Lury's method

From (7.2),

$$E[n_i] = Nq_1q_2\ldots q_{i-1}p_i$$
$$= Np_i \exp(-kF_i). \qquad (7.10)$$

If kf_i is small, so that $p_i \approx kf_i$, then, taking logarithms, we are led to consider De Lury's [1947] regression model

$$E[y_i] \approx \log(kN) - kF_i. \qquad (7.11)$$

We note that when K is small, $\log(1-K) \approx -K$, $K \approx k$ and (7.9) reduces to (7.11).

The least-squares estimates k'' and N'' for this model are given by

$$k'' = -\sum_{i=1}^{s} y_i(F_i-\bar{F})/\sum_{i=1}^{s}(F_i-\bar{F})^2$$

and

$$\log(k''N'') = \bar{y} + k''\bar{F}.$$

Assuming the y_i to be independently distributed with constant variance σ_y^2, then, using the delta method,

$$V[N''] \approx \sigma_y^2N^2\left[\frac{1}{s} + \left(\frac{k\bar{F}-1}{k}\right)^2 \cdot \frac{1}{\Sigma(F_i-\bar{F})^2}\right],$$

where σ_y^2 is estimated by $\sigma_y'^2$ in section *2*, above.

We note that the $\{n_i\}$ have the multinomial distribution (7.2), so that the $\{y_i\}$ are not strictly independent and the method of **6.2.2** should be used. However, as $N \to \infty$, the joint distribution of the $\{y_i\}$ tends to the multivariate normal, and for small $\{p_i\}$ the covariances are small compared with the variances; the combination of these two implies approximate independence. Also, as in the Leslie method, the day-to-day fluctuations in catchability, etc. would generally rule out a weighted least-squares analysis, so that the above assumptions of independence and constant variance are perhaps not unreasonable.

The model (7.11) can be modified using an approximation suggested by

Paloheimo [1961] and Chapman [1961]. From (7.10),

$$E[Y_i] = Nf_i^{-1}p_i\, e^{-kF_i}$$
$$= kN\, e^{-kF_i}\,(1-e^{-kf_i})/(kf_i),$$

and using the approximation

$$\log[(1-e^{-w})/w] \approx -w/2$$

for small w, we have

$$E[Y_i] \approx kN\exp\{-k(F_i+\tfrac{1}{2}f_i)\}.$$

Therefore, taking logarithms, our regression model is now

$$E[y_i] \approx \log(kN) - k(F_i+\tfrac{1}{2}f_i).$$

Braaten [1969] shows that this model is more robust with respect to changes in k than (7.11).

4 Comparison of methods

Ricker recommends the use of (7.11) when k and K are small – say less than 0·02 – and (7.9) when K is larger. The reason for this is that when K is of moderate size, so that one unit of effort catches a reasonable proportion of the population (say, greater than 2 per cent), then the effects of sampling without replacement *during* each sample can no longer be ignored.

Although Leslie's method is superior to De Lury's in that it is less dependent on the underlying assumptions and provides a graphical estimate of N, both lines could be plotted as a check on the underlying assumptions. De Lury points out that any departures from the underlying assumptions are likely to be reflected in different ways in the two models, and a substantial agreement between the two would give qualitative support to the assumptions.

In conclusion we make the obvious remark that the above methods will only work if sufficient individuals are removed from the population, so that there is a significant decline in the catch per unit effort. For example, Omand [1951] found no such decline in several fish populations.

7.1.4 Estimating total catch

In this section we mention briefly a slightly different problem from that discussed above. Suppose that the total catch for a given period of time is unknown (e.g. as in a sports fishery), then, if the catch per unit effort is constant during the period and can be estimated by means of a suitable sampling procedure, and the total effort can be estimated, the product of these two estimates will provide an estimate of the total catch (e.g. Havey [1960]). A brief review and examples of this method as applied to a sports fishery are discussed in Rose and Hassler [1969]; a general review covering game-kill and creel census procedures has been prepared by Schultz [1959]. Two detailed models for carrying out a creel census of fishermen are given by Robson [1960, 1961].

7.1.5 Departures from underlying assumptions

We shall now consider the main types of departure from the underlying assumptions of **7.1.1**; much of the following discussion is based on Ricker [1958: Chapters 1 and 6], which should be consulted for further details.

1 Population not closed

Any recruitment, natural mortality, or migration in the population could introduce serious errors into the above methods unless opposing tendencies happened to balance out. Obviously these effects will be minimised if the whole experiment is concentrated in one area for as short a period of time as possible. Such considerations should be taken into account in determining s, the number of samples. A model allowing for a constant instantaneous natural mortality rate μ is discussed in section 7 below.

2 Variation in catchability coefficient

In many populations the catchability may vary with size (or species, sex, time, locality etc.; cf. **3.2.2**(*1*)), so that K is effectively the average coefficient over all sizes appearing in the catch. This average will be constant only if the distribution of size in the population remains constant, a condition that may not be met. One could overcome this difficulty by stratifying the sample according to size-classes and estimating each class separately (see Hamley [1975] for a survey of gillnet selectivity).

De Lury [1951: 296] points out that the constancy of k or K does not refer to day-to-day variation, which may be treated as error, but to the absence of trends during the whole experiment. However, environmental conditions can produce trends in the catchability coefficient, and Paloheimo [1963: 73], for example, considers a model for a lobster population where the catchability is assumed to vary linearly with water temperature. Thus, if K_i is the catchability coefficient for the ith sample, he assumes on the basis of experimental studies (McLeese and Wilder [1958]) that $K_i = K(T_i - T)$, where T_i is the temperature at the time of the ith sample and T is the "threshold" temperature at which the lobsters become inactive. Leslie's regression model now becomes

$$E[Y_i \mid x_i] = K(T_i - T)(N - x_i)$$

and K, T and N can be estimated by non-linear least-squares (cf. Draper and Smith [1966]) or by regressing Y_i on x_i, T_i and $T_i x_i$. However, the T_i are subject to error and Paloheimo suggests an alternative method of estimation. For a certain range of temperatures lobster activity is linearly related to temperature, so that an alternative model is $K_i = K(a_i - a)$ where a_i is an index of acitivity (say walking rate).[*]

3 Variation in trap efficiency

In fisheries many kinds of gear decrease in efficiency the longer they are left before being lifted and reset, so that the catch per unit time begins

[*]For a similar example see Mercer [1975].

to fall off. For example, as more fish are caught on a set line, the less vacant hooks there are, and eventually a point of saturation may be reached. On the other hand, with net fishing, the presence of some fish already in the net may tend to scare others away, so that saturation may be reached long before the net is full.

In some situations the catches can be corrected for gear saturation. For example, Murphy [1960] gives a comprehensive deterministic model for doing this for the case of the baited long line. His model is developed in terms of instantaneous rates of bait loss, of hooking fish, and of losing hooked fish; the reader is referred to his article for details.

4 *Units of effort not independent*

This can happen when traps are so close together that "physical" competition exists between them which is independent of population size. For example, too many anglers at a pool may frighten the fish; or setting a new gill net near one already in operation may scare the fish away from the latter.

5 *Non-random sampling*

In fisheries, for example, not all the population may be subjected to the catching process; some parts of the fishing grounds may be inaccessible or too sparsely populated to warrant fishing there.

6 *Incomplete record of effort*

In commercially exploited populations information on effort may be missing or unavailable for some parts of the catch. To overcome this, Ricker suggests computing the catch per unit effort for as much data as possible and then dividing this value into the residual catch to give an estimate of effort for this residual. Adding this estimate to the known effort will then give the total effort f_i.

Sometimes effort records are complete and catch records incomplete, so that the above procedure can be used in reverse.

7 *Different catching methods*

When different fishing gears (e.g. trawl and drift-net) are used for the same population, the problem arises of converting all the different units of effort into a standard unit. This is a major problem, and for a discussion of some of the difficulties involved see Gulland [1955a], Beverton and Holt [1957: 172–7], Parrish [1962] and Pope [1975]. In some cases standardising is impossible because the gears are so unlike, e.g. they may select different sizes of fish or may operate at different times of the year.

To examine the effect of two types of gear operating in the same area at the same time, let f_i', n_i', x_i' and Y_i' be the effort, catch, cumulated catch, and catch per unit effort, respectively, for a second type of gear. Then, using Leslie's model (7.5) we have

$$E[Y_i \mid x_i, x_i'] = K(N - x_i - x_i'), \qquad (7.12)$$

$$E[Y_i' \mid x_i, x_i'] = K'(N - x_i - x_i')$$

and

$$E[n_i'] = \frac{f_i'K'}{f_iK} E[n_i].$$

If $(f_i'K')/(f_iK)$ remains approximately constant throughout the whole experiment as indicated by the constancy of the sample ratios n_i'/n_i then, following De Lury [1951: 296],

$$E[Y_i \mid x_i] \approx KN - K\left(1 + \frac{K'f_i'}{Kf_i}\right)x_i \qquad (7.13)$$

$$= KN - K_0 x_i \text{ say.} \qquad (7.14)$$

Let $\hat{K}_0$ be the least-squares estimate of the slope K_0 of (7.14), then estimating $(K'f_i')/(Kf_i)$ by x_{s+1}'/x_{s+1} $(= \Sigma n_i'/\Sigma n_i)$, we have the estimate (Dickie [1955: 810])

$$\hat{K} = \hat{K}_0/[1 + (x_{s+1}'/x_{s+1})].$$

N can then be estimated from the least-squares estimate of KN.

Similar equations can also be developed for De Lury's model as follows. Let p_i, p_i' be the probabilities of capture in the ith sample by the two gears, and assume that the two gears are operating independently. Then

$$E[N - x_i - x_i'] = Nq_1q_1'q_2q_2' \ldots q_{i-1}q_{i-1}'$$

$$= N\exp(-kF_i - k'F_i'),$$

and when k is small, so that $k \approx K$ (cf. (7.10) and (7.11)), the above equation may be combined with (7.12) to give De Lury's model

$$E[\log Y_i] \approx \log(kN) - kF_i - k'F_i'. \qquad (7.15)$$

If f_i'/f_i is constant (and therefore equal to F_i'/F_i), then (7.15) becomes

$$E[\log Y_i] \approx \log(kN) - k_0F_i, \qquad (7.16)$$

where $k_0 = k[1 + (k'F_i'/kF_i)]$.

The above theory can be applied to two important situations (Dickie [1955]). The first is when several gears are used but one is dominant and the rest are lumped together under f_i'; if x_i and x_i' are known for each sample, then the appropriate model for estimating N is (7.12). The second is when data are available for only part of the effort, or when good records are kept for part of the catch only; in this case (7.14) and (7.16) are appropriate.

When sampling is continuous, so that n_i is the number caught in the ith sample interval, then n_i' may also be interpreted as the number dying from natural mortality in this interval. If the instantaneous natural mortality rate is constant and equal to μ say, then (cf. (1.1))

$$E[n_i' \mid x_i, x_i'] = (1 - e^{-\mu t_i})(N - x_i - x_i')$$

$$\approx \mu t_i(N - x_i - x_i'),$$

where t_i is the length of the ith sampling interval. Therefore, if t_i/f_i remains constant ($= \Sigma t_i / \Sigma f_i = T_{s+1}/F_{s+1}$ say), K_0 in (7.14) is given by

$$K_0 = K\left(1 + \frac{\mu T_{s+1}}{KF_{s+1}}\right) = K + \frac{\mu T_{s+1}}{F_{s+1}},$$

and, given an estimate of μ, K can be estimated.

When there are two types of gear as well as natural mortality, then

$$K_0 = K\left(1 + \frac{K' f_i'}{K f_i} + \frac{\mu t_i}{K f_i}\right).$$

In this case we can also use the model (Dickie [1955])

$$\begin{aligned} E[Y_i \mid x_i, x_i'] &= KN - K\left(1 + \frac{\mu t_i}{K f_i + K' f_i'}\right)(x_i + x_i') \\ &= KN - K_1(x_i + x_i'), \end{aligned} \tag{7.17}$$

where $K_1 = K\left\{1 + \frac{\mu t_i}{K f_i} \cdot \frac{1}{[1 + (K' f_i')/(K f_i)]}\right\}$.

Therefore, if the ratios $f_i' : f_i : t_i$ are constant, then, given an estimate $\hat{\mu}$ of μ, K can be estimated by

$$\hat{K} = \hat{K}_1 - \frac{\mu T_{s+1}}{F_{s+1}} \cdot \frac{1}{[1 + (x_{s+1}'/x_{s+1})]},$$

where $\hat{K}_1$ is the least-squares estimate of the slope of (7.17). Some examples using the above model are given in Dickie [1955: 816–19].

7.1.6 Single mark release

1 Regression methods

Suppose that M individuals are captured, marked and released at the beginning of the experiment, so that $U = N - M$ is the initial unmarked population size. We shall let the suffixes m and u denote membership of the marked and unmarked populations respectively. Then, treating these populations separately, two Leslie regression lines can be considered:

$$E[m_i/f_i] = K_m(M - x_{mi}) \tag{7.18}$$

and

$$E[u_i/f_i] = K_u(U - x_{ui}), \tag{7.19}$$

where m_i and u_i ($= n_i - m_i$) are the numbers of marked and unmarked in the ith sample respectively. With the usual normality assumptions we can now derive a t-statistic (e.g. Wetherill [1967: 236]) for testing the hypothesis that the two lines have the same slope, i.e. that marked and unmarked are equicatchable. We note that apart from actual variations in catchability, any migration or recruitment will affect the two models differently. For example, recruitment and immigration will have no effect on the marked population; under certain conditions immigration and emigration rates can be estimated from the two slopes (Ketchen [1953]).

Since M is known, the values $m_i/[f_i(M - x_{mi})]$ can be used to provide a check on the constancy of K_m. Sometimes, however, there is an initial mortality of d marked individuals through the effects of handling and marking, so that the number of marked alive before the first sample is actually $M - d$. Fitting the line (7.18) will then provide an estimate of $M - d$, and hence of d.

In conclusion we note that De Lury's method can also be generalised in the same way to give two regression lines.

2 *Maximum-likelihood estimation*

When $K_m = K_u$ ($= K$ say) and the assumptions underlying the Petersen method (cf. **3.1.1**) hold for each sample, then

$$E[n_i \mid x_i] = Kf_i(N - x_i)$$

and

$$E[m_i \mid n_i] = n_i(M - x_{mi})/(N - x_i) = n_iM_i/N_i \text{ say.}$$

If we can assume that n_i has a Poisson distribution and that m_i, conditional on n_i, is also Poisson, then we have Chapman's model [1954: 12]

$$\prod_{i=1}^{s}\left\{e^{-Kf_iN_i}\frac{(Kf_iN_i)^{n_i}}{n_i!}\, e^{-n_iM_i/N_i}\frac{(n_iM_i/N_i)^{m_i}}{m_i!}\right\}.$$

For this model the maximum-likelihood estimates of N and K are solutions of

$$KF_{s+1} = \sum_{i=1}^{s}\left\{\left(\frac{n_i - m_i}{N - x_i}\right) + \frac{n_iM_i}{(N - x_i)^2}\right\}$$

and

$$NF_{s+1} = \sum_{i=1}^{s} x_if_i + (x_{s+1}/K),$$

with asymptotic variance–covariance matrix (expressed in terms of the N_i and M_i) the inverse of

$$\begin{bmatrix} \sum_{i=1}^{s}\left\{\frac{Kf_i}{N_i}\left(1 + \frac{M_i}{N_i}\right)\right\}, & F_{s+1} \\ F_{s+1}, & \sum_{i=1}^{s} f_iN_i/K \end{bmatrix}.$$

7.2 CONSTANT SAMPLING EFFORT: REMOVAL METHOD

7.2.1 Maximum-likelihood estimation(*)

In **7.1** we assumed that p_i, the probability of capture in the ith sample, varied from sample to sample. However, in some carefully controlled experiments in which the same effort is used for each sample under almost identical conditions it is reasonable to assume that p_i is constant and equal to p, say. We therefore consider a model for which the following assumptions are true:

(*)See also section **12.9**.

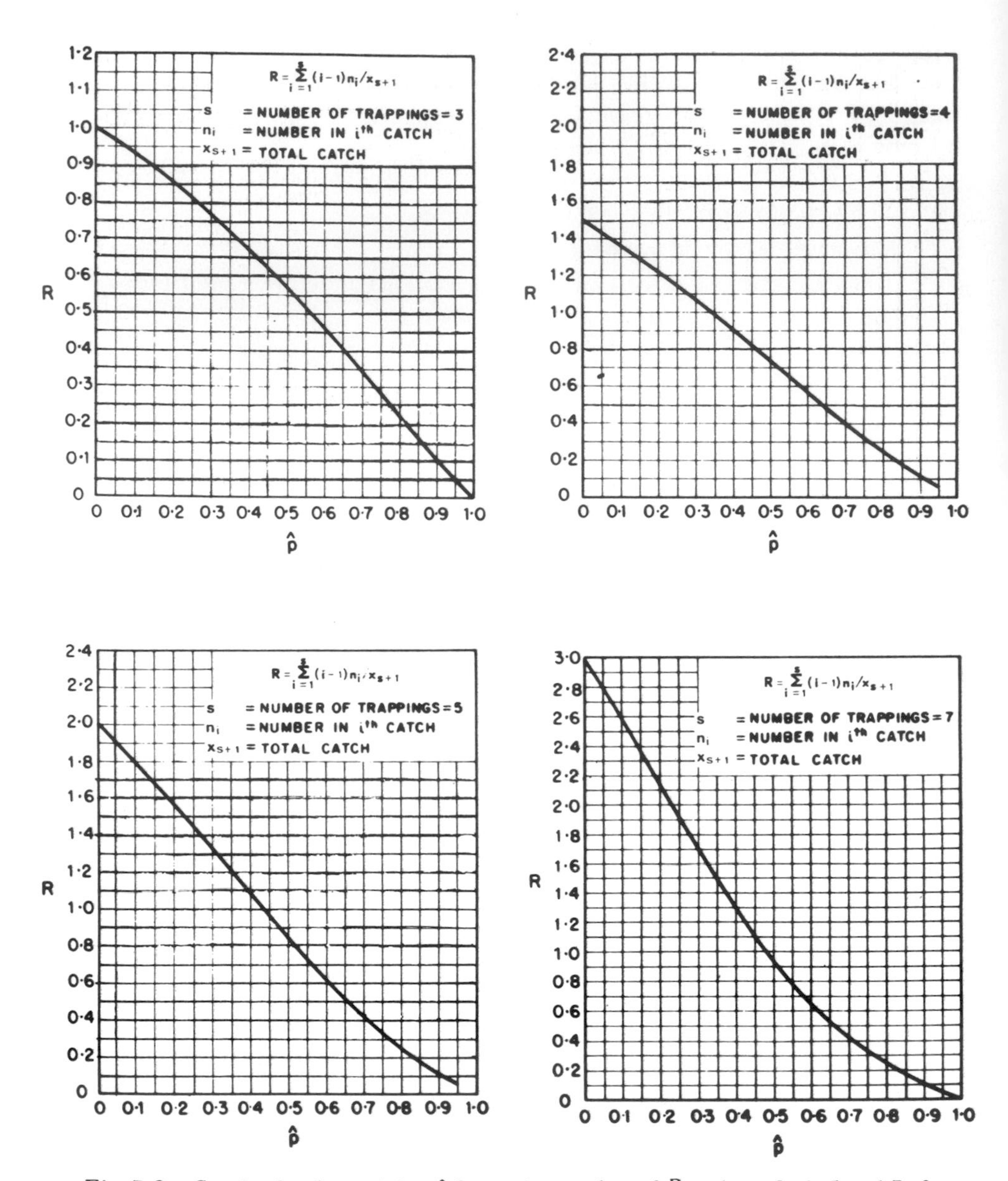

Fig. 7.3 Graphs for determining $\hat{p}$ for a given value of R and $s = 3, 4, 5$ and 7: from Zippin [1956].

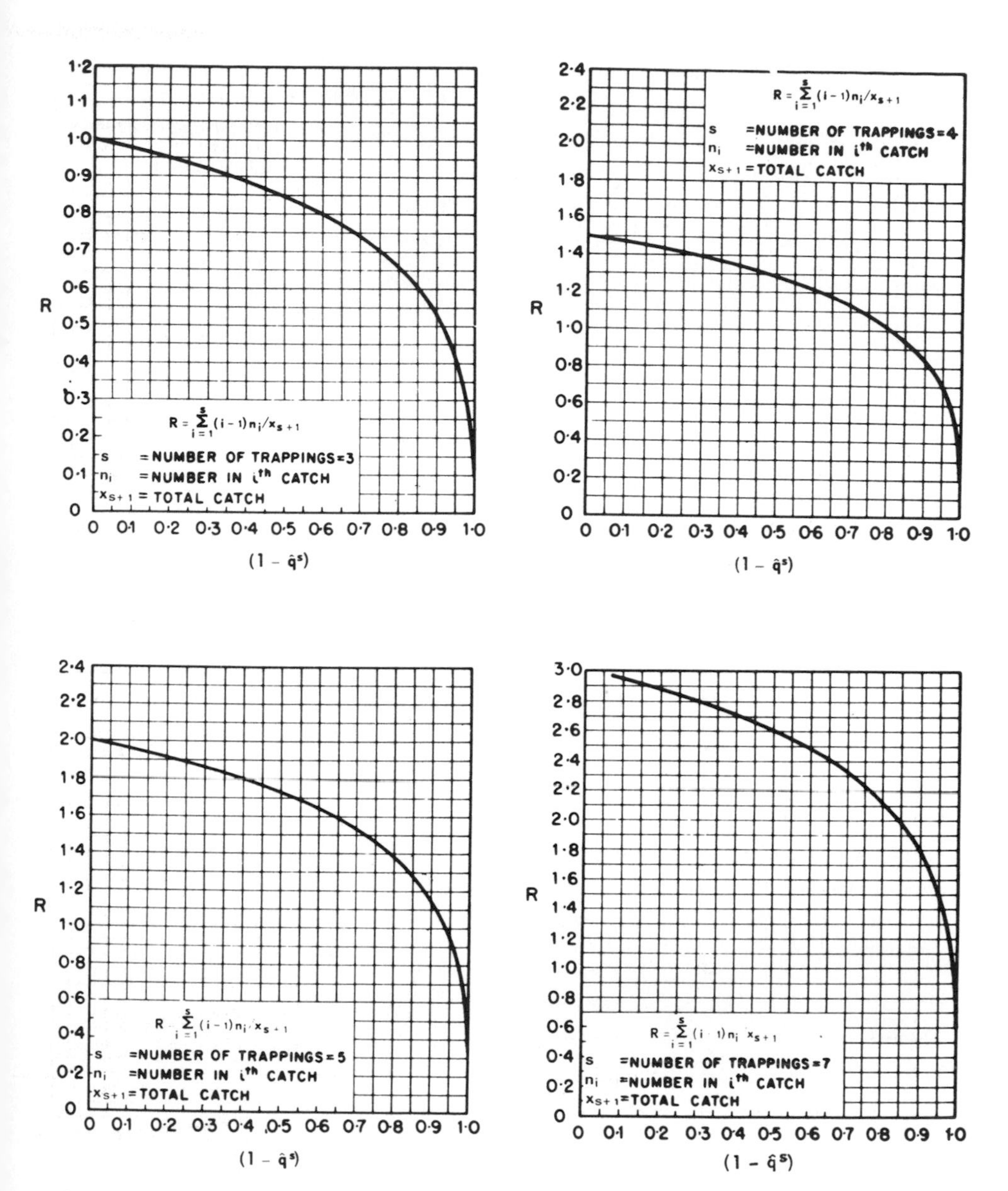

Fig. 7.4 Graphs for determining $(1-\hat{q}^s)$ for a given value of R and s = 3, 4, 5 and 7: from Zippin [1956].

(1) The population is closed.
(2) The probability of capture in the ith sample is the same for each individual exposed to capture.
(3) The probability of capture p remains constant from sample to sample.

Given the above assumptions, Moran [1951] and Zippin [1956, 1958] obtained maximum-likelihood estimates of N and p; the following discussion is from Zippin [1956]. However see **12.9.2** for further comments.

Setting $p_i = p$ in (7.2), the joint probability function of the $\{n_i\}$ becomes

$$\frac{N!}{(N-x_{s+1})!\prod_{i=1}^{s} n_i!}\, p^{x_{s+1}}\, q^{sN - \sum_{i=1}^{s+1} x_i}. \tag{7.20}$$

The maximum-likelihood estimates $\hat{N}$ and $\hat{p}$ are given by the equations

$$\hat{N} = x_{s+1}/(1-\hat{q}^s) \tag{7.21}$$

and

$$\frac{\hat{q}}{\hat{p}} - \frac{s\hat{q}^s}{(1-\hat{q}^s)} = \frac{\sum_{i=1}^{s}(i-1)n_i}{x_{s+1}} \quad (= R \text{ say}), \tag{7.22}$$

and Zippin gives graphs (reproduced as Fig. 7.3 and 7.4) for facilitating their solution. It can be shown (cf. Seber and Whale [1970]) that (7.22) has a unique solution $\hat{q}$ in the range $[0, 1]$ for $0 \leqslant R \leqslant (s-1)/2$; when $R > (s-1)/2$ the above method is not applicable and the experiment "fails". This condition for failure can be rewritten as

$$\sum_{i=1}^{s} (s+1-2i)n_i < 0,$$

and the probability of this happening will generally be small for large N and will decrease as s increases.

For N large, $\hat{N}$ and $\hat{p}$ are asymptotically unbiased with asymptotic variances

$$V[\hat{N}] = \frac{N(1-q^s)q^s}{(1-q^s)^2 - (ps)^2 q^{s-1}} \tag{7.23}$$

and

$$V[\hat{p}] = \frac{(qp)^2(1-q^s)}{N[q(1-q^s)^2 - (ps)^2 q^s]}.$$

Assuming asymptotic normality, confidence limits for N and p can be calculated in the usual manner.

In designing an experiment for a particular population the biologist will be interested in knowing what proportion of the total population he must catch in order to estimate N with a given precision. Also, given that a certain proportion is to be caught, the question arises as to whether it is better to set a large number of traps and take a few samples or set a few traps and take a large number of samples. In studying this latter question first, we shall assume that the home ranges of the animals overlap sufficiently for the number of animals exposed to capture during the first trapping to be the same, regardless of the number of traps set.

Suppose that t traps are set on each sampling occasion, and let p_0 $(= 1 - q_0)$ represent the probability that an animal will be captured by a single trap during a single sampling period. Assuming that p_0 is the same for each trap and that the traps operate independently, the probability that an animal is not caught in a particular sample is $q = q_0^t$. Thus fixing $1 - q^s$ $(= 1 - q_0^{st})$, the total proportion of the population expected to be caught, is equivalent to fixing st, the total number of traps set throughout the whole experiment. Zippin shows that as t and s vary subject to fixed st, $V[\hat{N}]$ decreases only slightly as s increases beyond 3. This means that for practical purposes $V[\hat{N}]$ depends almost entirely on N, the trap efficiency p_0, and the total trapping effort st. Hence, for a given accuracy or precision the biologist would normally choose t and s to minimise the cost. However, if migration and natural death are possible factors, a short intensive programme would be the most appropriate.

Zippin gives a table (reproduced as Table 7.3) showing the total proportion of the population which must be captured in order that $\hat{N}$ may have

TABLE 7.3

Proportion of total population required to be trapped for a specified coefficient of variation of $\hat{N}$: from Zippin [1956: Table 1].

N	Coefficient of variation			
	30%	20%	10%	5%
	Proportion (to nearest 0·05) of population to be captured (in 100 or fewer trappings)			
200	0·55	0·60	0·75	0·90
300	0·50	0·60	0·75	0·85
500	0·45	0·55	0·70	0·80
1 000	0·40	0·45	0·60	0·75
10 000	0·20	0·25	0·35	0·50
100 000	0·10	0·15	0·20	0·30

a specified coefficient of variation. A cursory glance at the table reveals that a relatively large proportion of the population must be captured (particularly when $N < 200$) in order to obtain reasonably accurate estimates. This is a serious limitation on the above so-called "removal method", except in the situation where it is desirable to reduce a population sharply, as in the case of a crop-damaging rodent. Alternatively the problem of depletion can be avoided if the animals are live-trapped and "removed" by tagging before releasing them. In this case N can also be estimated from the recapture data using the methods of Chapter 5. However, on the debit side, the trapping effort used in catching untagged animals may fall off significantly as tagged animals take up more and more of the traps.

If the population is physically depleted, other problems may arise. For example, p may not remain constant as the more catchable animals are caught first (Libosvárský [1962: 517]). Also, as numbers decrease, territorial competition on the boundary of the population may lead to immigration into the depopulated sampling area; Chew and Butterworth [1964] give an example where home range appears to vary inversely with population density.

To test the validity of the model (7.20) we can use a standard goodness-of-fit statistic (cf. **1.3.6** (*3*))

$$T_1 = \sum_{i=1}^{s} (n_i - E_i)^2/E_i,$$

where $E_i = \hat{N}\hat{p}\hat{q}^{i-1}$ $(i = 1, 2, \ldots, s)$. When (7.20) is valid, T_1 is asymptotically distributed as χ^2_{s-2}. Zippin also suggests the test statistic (cf. (7.1))

$$T_2 = \sum_{i=1}^{s} [n_i - (\hat{N} - x_i)\hat{p}]^2/[(\hat{N} - x_i)\hat{p}\hat{q}]$$

and demonstrates that it is asymptotically equivalent to T_1.

Example 7.3 Zippin [1956, 1958]

In a three-night trapping programme the following catches were made: $n_1 = 165$, $n_2 = 101$, $n_3 = 54$. Then

$$R = \sum_{i=1}^{s} \frac{(i-1)n_i}{x_{s+1}} = \frac{n_2 + 2n_3}{n_1 + n_2 + n_3} = \frac{20}{320} = 0{\cdot}65,$$

and entering the graphs, Fig. 7.3 and 7.4, with $s = 3$ and $R = 0{\cdot}65$ gives $\hat{p} = 0{\cdot}42$, $(1-\hat{q}^3) = 0{\cdot}80$ and

$$\hat{N} = x_{s+1}/(1-\hat{q}^3) = 320/0{\cdot}80 = 400.$$

Estimating N and q by $\hat{N}$ and $\hat{q}$ in (7.23) leads to

$$\hat{V}[\hat{N}] = \frac{400(0{\cdot}80)(0{\cdot}20)}{(0{\cdot}80)^2 - (1{\cdot}26)^2(0{\cdot}20/0{\cdot}58)} = 691{\cdot}1$$

and

$$\hat{\sigma}[\hat{N}] = 26{\cdot}3.$$

Hence an approximate 95% confidence interval for N is $400 \pm 1{\cdot}96\,(26{\cdot}3)$ or (348, 452).

To calculate T_1 we compare

$$(n_1, n_2, n_3) = (165, 101, 54)$$

and

$$(E_1, E_2, E_3) = (\hat{N}\hat{p}, N\hat{p}\hat{q}, \hat{N}\hat{p}\hat{q}^2)$$
$$= (168, 97{\cdot}44, 56{\cdot}52).$$

Hence

$$T_1 = \Sigma(n_i - E_i)^2/E_i = 0{\cdot}29$$

which, at one degree of freedom, indicates an excellent fit.

Further examples of the above method are given in Davis [1957: mice], Menhinick [1963: insects], Kikkawa [1964: 262, small rodents] and Johnson [1965: freshwater fish]. The removal method has also been widely used by Polish ecologists under the name of "Standard Minimum" method (Grodzinski *et al.* [1966]) using a standard method of laying traps, and prebaiting.

7.2.2 Maximum-likelihood estimation: special cases

As the above method is widely used in studying small mammal populations and (with the advent of efficient electric fishing methods – cf. Moore [1954]) fish populations in streams, we shall now consider the three cases $s = 3, 2, 1$ in more detail.

1 Three-sample method

It was mentioned above that under certain conditions $V[\hat{N}]$ decreases only slightly as s increases beyond 3. Since standard trap-lines operating for three nights have been widely used, and three samples is the minimum number of samples for which the preceding goodness-of-fit statistics T_1 and T_2 can be applied, it is of practical interest to consider this special case more closely.

When $s = 3$, (7.21) and (7.22) now have explicit solutions (Junge and Libosvárský [1965])

$$\hat{N} = \frac{6X^2 - 3XY - Y^2 + Y(Y^2 + 6XY - 3X^2)^{\frac{1}{2}}}{18(X-Y)} \tag{7.24}$$

and

$$\hat{p} = \frac{3X - Y - (Y^2 + 6XY - 3X^2)^{\frac{1}{2}}}{2X}$$

where $X = 2n_1 + n_2$ and $Y = n_1 + n_2 + n_3$; we note that the experiment "fails"(*) if $n_3 > n_1$, for then $\hat{N}$ is negative. From sampling experiments using binomial tables and a table of random numbers, Zippin [1956] came to the conclusion that for $p = 0{\cdot}4$ and $N > 200$ the large-sample normal theory gave reasonable confidence intervals for N. For $50 < N < 200$ he found that $V[\hat{N}]$ of (7.23) was too small, and the distribution of N was skewed, so that $\hat{N} \pm 1{\cdot}96\, \hat{\sigma}[\hat{N}]$ was more like a 90 per cent than a 95 per cent confidence interval. However, the adequacy of this confidence interval needs further investigation for other values of p, particularly when p is small.

Another aspect worthy of investigation is the effect of varying catchability in the population on the validity of $\hat{N}$ and $V[\hat{N}]$. Junge and Libosvárský [1965] indicated by means of an example that $\hat{N}$ will not be greatly affected by any variation in catchability if the range of p over the population is less than $0{\cdot}05$ and $p > \frac{1}{2}$. Their conclusion is borne out by the following theoretical analysis given by Seber and Whale [1970].

Let x_j $(j = 1, 2, \ldots, N)$ be the probability that the jth member of the population is caught in a sample, given that it is in the population, and suppose that this probability remains constant from sample to sample. Then, assuming that the population represents a random selection of such

(*) See also Carle and Strub [1978: Fig. 1].

probabilities with regard to the species as a whole and the particular trapping method used, $x_1, x_2, \ldots, x_N$ may be regarded as a random sample from a probability density function $f(x)$. Hence the probability of obtaining a given outcome of the experiment is

$$P_u = \left\{\prod_{i=1}^{n_1} x_i\right\}\left\{\prod_{j=n_1+1}^{n_1+n_2}(1-x_j)x_j\right\}\left\{\prod_{k=n_1+n_2+1}^{n_1+n_2+n_3}(1-x_k)^2x_k\right\}\left\{\prod_{l=n_1+n_2+n_3+1}^{N}(1-x_l)^3\right\}$$

and the conditional probability function of n_1, n_2 and n_3 given the particular sample of probabilities $\{x_j\}$ is

$$f(n_1, n_2, n_3 \mid \{x_j\}) = \sum_u P_u,$$

where $\sum_u$ denotes summation over all possible groupings of the N animals such that n_1, n_2 and n_3 are the numbers in each of the three categories, i.e. u represents a permutation of N objects such that n_i fall in the ith category ($i = 1, 2, 3$). Since the $\{x_j\}$ are independent they can be integrated out and the unconditional probability function is given by:

$$\begin{aligned} f(n_1, n_2, n_3) &= E[f(n_1, n_2, n_3 \mid \{x_j\})] \\ &= E[\sum_u P_u] \\ &= \sum_u E[P_u] \\ &= \sum_u a_1^{n_1}(a_1-a_2)^{n_2}(a_1-2a_2+a_3)^{n_3}(1-3a_1+3a_2-a_3)^{N-Y} \\ &= \frac{N!}{\left(\prod_{i=1}^{3} n_i!\right)(N-Y)!}\, a_1^{n_1}(a_1-a_2)^{n_2}(a_1-2a_2+a_3)^{n_3} \\ &\qquad \times (1-3a_1+3a_2-a_3)^{N-Y}, \end{aligned} \tag{7.25}$$

where $a_k = E[x^k]$. This distribution can be used to examine the asymptotic mean of $\hat{N}$ for different density functions $f(x)$. For example, if $f(x)$ is uniform on $[c, d]$ $(0 \leqslant c \leqslant d \leqslant 1)$ and $w = c/d$, then $a_1 = \frac{1}{2}d(1+w)$, $a_2 = \frac{1}{3}d^2(1+w+w^2)$ and $a_3 = \frac{1}{4}d^3(1+w)(1+w^2)$. The asymptotic mean of $\hat{N}$, obtained by replacing X and Y in (7.24) by their expected values with respect to (7.25), is given by BN, where B is tabulated for different values of w and d in Table 7.4. When $w = 1$, then $B = 1$; this is to be expected since a maximum-likelihood estimate is asymptotically unbiased. From the table we find that for $w < 1$, $B < 1$, and B is sensitive to the value of w; when $w \geqslant 0{\cdot}5$ the asymptotic bias is negligible, and when $c > 0{\cdot}01$, $B > 0{\cdot}8$. If $f(x)$ has a unimodal (e.g. beta type) distribution, defined on $[c, d]$, then the corresponding values of B will be larger. Therefore we can conclude that, asymptotically, $\hat{N}$ is fairly insensitive to variations in catchability.

If there is a considerable variation in catchability, the more catchable individuals will be caught first, so that the average probability of capture will decrease from one trapping to the next and $\hat{N}$ will underestimate N.

TABLE 7.4

Asymptotic value of $B = E[\hat{N}]/N$ when the catchability distribution $f(x)$ is uniform on $[c, d]$: from Seber and Whale [1970: Table 2].

w (= c/d)	d	c	B
0·5	0·25	0·125	0·968
	0·50	0·250	0·972
	0·75	0·375	0·976
	1·00	0·500	0·983
0·2	0·25	0·050	0·879
	0·50	0·100	0·888
	0·75	0·150	0·899
	1·00	0·200	0·913
0·01	0·25	0·0025	0·766
	0·50	0·005	0·776
	0·75	0·0075	0·788
	1·00	0·010	0·803
0·0	0·25	0	0·758
	0·50	0	0·768
	0·75	0	0·780
	1·00	0	0·795

TABLE 7.5

Expected percentage underestimation of N by $\hat{N}$ when the probability of capture decreases by a constant amount on three successive trappings: from Zippin [1958: Table 2].

p	Absolute decrease in p on successive trappings				
(First trapping)	0·01	0·02	0·03	0·05	0·10
0·1	47·4	65·7	75·3	85·2	–
0·2	15·7	27·6	37·3	51·4	71·6
0·3	6·1	12·0	17·4	26·9	45·7
0·4	2·8	5·6	8·4	13·9	26·6
0·5	1·4	2·8	4·2	7·2	14·8
0·6	0·7	1·4	2·2	3·7	7·9

For example,

$$\begin{aligned}\frac{E[n_2]}{E[n_1]} &= \frac{a_1 - a_2}{a_1} \\ &= (1-a_1)\left(\frac{a_1 - a_2}{a_1 - a_1^2}\right) \\ &\leqslant (1-a_1),\end{aligned}$$

since $a_2 = E[x^2] \geqslant (E[x])^2 = a_1^2$, with equality if and only if x is constant (i.e. $a_k = p^k$). Such a decrease in average catchability can also occur when climate or other factors decrease the activity of the animals during the course of the experiment (e.g. Gentry and Odum [1957]). Zippin [1958] gives a table (Table 7.5) showing the expected percentage underestimation of N when p decreases by a constant amount.

2 *Two-sample method*

ESTIMATION. In some situations it may be unwise or impracticable to take more than just two samples. For example, if climatic conditions have considerable effect on p then a short intensive survey may be needed if p is to remain constant. Again, if p is large, three samples may severely deplete the population unless tagging can be used to represent "removal". On the other hand, if one wishes to reduce the population as much as possible, then for p large it may be a waste of time or uneconomic to take a third sample. For example, Le Cren points out (Seber and Le Cren [1967]) that the removal method can also be applied to quite different situations such as the sorting of animals out of samples of mud, soil or grain. Here a third sorting may be a waste of effort when p, the probability of being found in a given searching, is large.

When $s = 2$ and $n_1 > n_2$, the maximum-likelihood equations (7.21) and (7.22) have simple solutions

$$\hat{N} = n_1^2/(n_1 - n_2)$$

and

$$\hat{q} = n_2/n_1,$$

which are also the moment estimates. Using the delta method, we have (Seber and Le Cren [1967])

$$E[\hat{N}] \approx N + \frac{q(1+q)}{p^3} = N + b[N], \text{ say,} \tag{7.26}$$

$$V[\hat{N}] \approx \frac{Nq^2(1+q)}{p^3}, \tag{7.27}$$

$$E[\hat{p}] \approx p - \frac{q}{Np} \tag{7.28}$$

and

$$V[\hat{p}] \approx \frac{q(1+q)}{Np}. \tag{7.29}$$

Replacing N and p by their estimates leads to the variance estimates

$$\hat{V}[\hat{N}] = \frac{n_1^2 n_2^2 (n_1 + n_2)}{(n_1 - n_2)^4} \tag{7.30}$$

and

$$\hat{V}[\hat{p}] = \frac{n_2(n_1 + n_2)}{n_1^3}. \tag{7.31}$$

We note that $\hat{p}$ may be adjusted for bias; thus

$$p^* = 1 - [n_2/(n_1+1)]$$

is almost unbiased as

$$\begin{aligned} E[p^*] &= \underset{n_1}{E}\, E[p^* \mid n_1] \\ &= \underset{n_1}{E}\left[1 - \frac{(N-n_1)p}{(n_1+1)}\right] \\ &= \underset{n_1}{E}\left[1 - \frac{(N+1)p}{(n_1+1)} + p\right] \\ &= p + 1 - (1-q^{N+1}) \\ &= p + q^{N+1} \\ &\approx p\,. \end{aligned}$$

Robson and Regier [1968] have considered a slight modification of $\hat{N}$, namely

$$N^* = (n_1^2 - n_2)/(n_1 - n_2)\,.$$

This estimate, based on the relationship

$$E[n_1^2 - n_2] = (E[n_1])^2,$$

is also biased, though the bias will probably be less than $b[N]$ of (7.26) for small samples.

Example 7.4 Trout: Seber and Le Cren [1967]

Two successive electric fishings in a section of a small stream yielded 79 and 28 young trout respectively. Thus,

$$\hat{N} = 79^2/(79-28) = 122\,,$$

$$\hat{q} = 28/79 = 0{\cdot}35\,,$$

$$\hat{\sigma}[\hat{N}] = 79(28)(\sqrt{107})/51^2 = 8{\cdot}8, \quad \hat{b}[\hat{N}] = 2\,,$$

and

$$\hat{N} \pm 1{\cdot}96\,\hat{\sigma}[\hat{N}] = 122 \pm 18 \quad \text{or } 104 < N < 140\,.$$

VALIDITY OF VARIANCE FORMULAE. The bias terms are included in (7.26) and (7.28) to indicate the accuracy of the delta method. Thus by considering the proportional biases ($b[N]/N$, etc.) we find that (7.26) and (7.27) only hold for large values of Np^3, while (7.28) and (7.29) are valid for large Np^2. This means that for quite reasonable values of N and p, the formula for $V[\hat{N}]$ may be unsatisfactory. For example, if $n_1 = 27$ and $n_2 = 20$, then $\hat{N} = 104$, $\hat{p} = 0{\cdot}25$, $\hat{b} = 74$ and $\hat{\sigma}[\hat{N}] = 76$, the last two estimates indicating that (7.26) and (7.27) may not be valid.

When Np^3 is small, not only is the formula for $V[\hat{N}]$ unreliable but the estimates $\hat{N}$ and $\hat{\sigma}[\hat{N}]$ may be totally misleading. For example, Robson and Regier [1968: 147] in a simulation experiment with $N = 500$ and $p = 1/10$ obtained "catches" of $n_1 = 60$ and $n_2 = 35$, thus yielding $\hat{N} = 144$ and

$\hat{\sigma}[\hat{N}] = 35{\cdot}7$. Here N is drastically underestimated as the first catch happened to exceed substantially its expected value of $Np = 50$, and the second catch was smaller than expected. Also $\sigma[\hat{N}] = 877$ and $b[\hat{N}] = 1710$, so that the asymptotic formulae are open to question and the value of $\hat{\sigma}[\hat{N}]$ is completely misleading.

By considering a coefficient of variation of 25 per cent it is suggested, as a rough guide, that the $100(1-\alpha)$ per cent confidence interval

$$\hat{N} \pm z_{\alpha/2}\,\hat{\sigma}[\hat{N}] \tag{7.32}$$

will be satisfactory if

$$Np^3 \geqslant 16q^2(1+q). \tag{7.33}$$

Also computer studies indicate that on account of the skewness of the distribution of $\hat{N}$, the bias correction $\hat{b}[\hat{N}]$ is best ignored when calculating this confidence interval. However, when (7.33) is not satisfied the following method can be used:

Let

$$u = \frac{n_1 - n_2 + 2}{(n_1+1)(n_1+2)},$$

then from Seber and Whale [1970]

$$E[u] = \frac{1}{N+1} + \frac{Nq^{N+1}}{N+1},$$

$$\approx \frac{1}{N+1},$$

and

$$E[u^2] \approx (1+p)^2 b_2 - [2Np(1+p) + 2 + 9p + 5p^2]\,b_3 + {} + (N+2)p[Np+9p+9]\,b_4 - 6(N+2)^2p^2 b_5,$$

where $$b_k = \frac{1}{N^k}\cdot\frac{(N-2)(N-3)\ldots(N-k-1)}{[(N-1)p-1][(N-1)p-2]\ldots[(N-1)p-k]}.$$

Hence

$$V[u] \approx E[u^2] - (N+1)^{-2},$$

and estimating Np, p, N and $(N+1)^{-1}$ by n , p^*, $u^{-1}-1$ and u respectively leads to an estimate $\hat{V}[\hat{u}]$. Assuming u to be approximately normal, an approximate $100(1-\alpha)$ per cent confidence interval for $(N+1)^{-1}$ is given by

$$u \pm z_{\alpha/2}\,\hat{\sigma}[\hat{u}]\,. \tag{7.34}$$

To compare this interval with (7.32) the following probabilities were computed for different values of N and p and for $z_{\alpha/2} = 1{\cdot}64,\ 1{\cdot}96$ (i.e. 90 per cent and 95 per cent confidence intervals respectively):

$$A_1 = \Pr\left[\frac{|u-(N+1)^{-1}|}{\hat{\sigma}[\hat{u}]} < z_{\alpha/2} \quad\text{and}\quad n_1 > n_2\right],$$

$$A_2 = \Pr\left[\frac{|\hat{N}-N|}{\hat{\sigma}[\hat{N}]} < z_{\alpha/2} \quad\text{and}\quad n_1 > n_2\right]$$

and

$$C = \Pr[n_1 > n_2],$$

where C is the probability that the experiment is "successful". Since the confidence intervals are only calculated when the experiment is successful, the true "worth" of an interval should be measured by the conditional probabilities A_i/C; these are given in Table 7.6. From this table we see that

TABLE 7.6

A comparison of the exact probabilities (A_i) for two large-sample confidence intervals based on the two-sample removal method for different population sizes (N) and probabilities of capture (p): C is the probability that the first sample is greater than the second ($n_1 > n_2$).

					90% interval			95% interval		
N	p	Np^3	$16q^2(1+q)$	C	A_1	A_2	A_1/C	A_1	A_2	A_1/C
20	0·50	2·5	6	0·88	0·82	0·67	0·93	0·84	0·68	0·95
	0·75	8·5	1·3	0·98	0·98	0·83	1·00	0·98	0·85	1·00
40	0·25	0·6	15·8	0·68	0·57	0·44	0·84	0·58	0·45	0·85
	0·50	5	6	0·96	0·88	0·79	0·91	0·93	0·80	0·96
	0·75	17	1·3	1·00	1·00	0·84	1·00	1·00	0·89	1·00
80	0·05	0·0	28·2	0·44	0·43	0·06	1·00	0·43	0·06	1·00
	0·10	0·1	24·6	0·53	0·49	0·21	0·93	0·49	0·21	0·93
	0·25	1·3	15·8	0·78	0·60	0·56	0·77	0·64	0·59	0·83
	0·50	10	6	1·00	0·88	0·84	0·88	0·89	0·88	0·90
	0·75	34	1·3	1·00	0·96	0·87	0·96	0·99	0·91	0·99
100	0·05	0·0	28·2	0·46	0·46	0·09	0·99	0·46	0·10	0·99
	0·10	0·1	24·6	0·55	0·45	0·23	0·82	0·45	0·25	0·82
	0·25	1·6	15·8	0·81	0·64	0·60	0·79	0·68	0·63	0·84
	0·50	12·5	6	1·00	0·85	0·86	0·85	0·90	0·90	0·90
	0·75	42	1·3	1·00	0·92	0·89	0·92	0·97	0·91	0·97
200	0·05	0·0	28·2	0·50	0·40	0·13	0·79	0·40	0·16	0·80
	0·10	0·2	24·6	0·60	0·40	0·31	0·66	0·44	0·33	0·74
	0·25	3	15·8	0·90	0·74	0·72	0·82	0·78	0·75	0·86
	0·50	25	6	1·00	0·79	0·89	0·79	0·85	0·92	0·85
	0·75	85	1·3	1·00	0·71	0·90	0·71	0·80	0·93	0·80
400	0·05	0·1	28·2	0·53	0·33	0·20	0·62	0·38	0·21	0·71
	0·10	0·4	24·6	0·66	0·50	0·40	0·76	0·54	0·42	0·82
	0·25	6	15·8	0·97	0·82	0·82	0·85	0·86	0·85	0·89
	0·50	50	6	1·00	0·74	0·90	0·74	0·81	0·93	0·81
	0·75	169	1·3	1·00	0·36	0·90	0·36	0·43	0·94	0·43

the values of A_1 fluctuate (particularly as N increases), while those for A_2 steadily increase up to 0·90 and 0·94. This shows that the maximum-likelihood method is satisfactory for $N > 200$ and p not too small. In particular, the simple rule (7.33) seems reasonable, though the coefficient 16 may be too small. We also note from Table 7.6 that (7.34), although requiring considerable calculation without a computer, can be used for small values of p, and for $N < 200$. However, such cases should be avoided in practice, as when p is small, C is also small, and the resulting confidence interval may be so wide as to be almost useless.

VARIABLE PROBABILITY OF CAPTURE. The effect of variation in p from animal to animal on the maximum-likelihood estimate $\hat{N}$ can be investigated using the method of p. 316. From (7.25) the joint distribution of n_1 and n_2 is

$$\frac{N!}{n_1!\,n_2!\,(N-n_1-n_2)!}\,a_1^{n_1}(a_1-a_2)^{n_2}(1-2a_1+a_2)^{N-n_1-n_2}. \tag{7.35}$$

Using the delta method, we find that, asymptotically,

$$E_B[\hat{N}] = \frac{(E[n_1])^2}{E[n_1-n_2]} = \frac{Na_1^2}{a_2} = NB, \text{ say,} \tag{7.36}$$

$$E_B[\hat{p}] = a_2/a_1 \tag{7.37}$$

and

$$V_B[\hat{N}] = Na_1^3(4a_2^2-5a_1a_2-a_1a_2^2+2a_1^2)/a_2^4,$$

where the suffix B denotes expectation with respect to (7.35). We note that when x_j is constant ($= p$, say), $a_k = p^k$, and $V_B[\hat{N}]$ reduces to $V[\hat{N}]$ as expected.

From (7.27) our estimate of the variance of $\hat{N}$ is

$$\hat{V}[\hat{N}] = \hat{N}\hat{q}^2(1+\hat{q})/\hat{p}^3$$

and replacing each estimate by its asymptotic expected value (cf. (7.36) and (7.37)), we find that, asymptotically,

$$E[\hat{V}[\hat{N}]] = Na_1^3(4a_2^2-5a_1a_2-a_1a_2^2B^{-1}+2a_1^2)/a_2^4.$$

Since

$$1-B = \frac{a_2-a_1^2}{a_2} = \frac{V[x]}{a_2} > 0,$$

B is less than unity and the net effect of variable catchability is that $\hat{N}$ and $\hat{V}[\hat{N}]$ will (asymptotically) underestimate N and $V_B[\hat{N}]$ respectively, while $\hat{p}$ will overestimate a_1 ($= E[x]$).

The effect of the shape of $f(x)$ on B has already been discussed in **3.2.2**(*2*). For example, when $f(x)$ is uniform on $[c, d]$ ($0 \leqslant c < d \leqslant 1$) and $w = c/d$, then $B = 3(1+w)^2/[4(1+w+w^2)]$: for $w = 1, \frac{1}{2}, \frac{1}{5}$ and 0, the respective values of B are 1, $\frac{27}{28}$ $\frac{27}{31}$ and $\frac{3}{4}$. Unfortunately, $\hat{V}[\hat{N}]$ is also sensitive to the value of w; for the "worst" case $c = 0$, $d = 1$ we have asymptotically

$$G = E[\hat{V}[\hat{N}]]/V_B[\hat{N}] = \tfrac{2}{3}.$$

Sometimes there are definite subgroups in the population due to age, sex, habitat preference, etc., so that $f(x)$ may be regarded as a probability function, namely

$$f(x_j) = \theta_j \quad (j = 1, 2, \ldots, J;\ 0 \leqslant x_j \leqslant 1;\ \sum_j \theta_j = 1),$$

where θ_j is the proportion of the population with probability of capture x_j. Then

$$\begin{aligned} 1 - B &= 1 - (\Sigma\, x_j\theta_j)^2/(\Sigma\, x_j^2\theta_j) \\ &= (\sum_{j<k}\sum \theta_j\theta_k(x_j - x_k)^2)/(\Sigma\, x_j^2\theta_j), \end{aligned}$$

which will be small if the x_j are not too different. For example, when $J = 3$, $\theta_j = \frac{1}{3}$ $(j = 1, 2, 3)$ and $x_1 = \frac{1}{4}$, $x_2 = \frac{1}{2}$, $x_3 = \frac{3}{4}$, then $B = \frac{6}{7}$ and $G = 0{\cdot}90$.

REMOVAL BY MARKING. We shall now briefly consider the use of marking as a means of "removal". Suppose that n_1 animals caught in the first sample are marked and released back into the population and that m_2 are recaptured in the second sample of n_2 animals. Let p be the constant probability of capture of an unmarked animal and let p_0 be the probability that a marked animal is caught in the second sample. Then using (7.20) with $s = 2$, and u_2 $(= n_2 - m_2)$ instead of n_2, the joint probability function of n_1, u_2 and m_2 is

$$\begin{aligned} f(n_1, u_2, m_2) &= f(n_1, u_2) f(m_2 \mid n_1, u_2) \\ &= \frac{N!}{n_1!\, u_2!\, (N - n_1 - u_2)!}\, p^{n_1 + u_2} q^{2N - 2n_1 - u_2} \binom{n_1}{m_2} p_0^{m_2} q_0^{n_1 - m_2} \end{aligned}$$

which can be rearranged as the multinomial distribution

$$\frac{N!}{u_1!\, u_2!\, m_2!\, (N - n_1 - u_2)!} (pq_0)^{u_1} (qp)^{u_2} (pp_0)^{m_2} (q^2)^{N - n_1 - u_2}, \qquad (7.38)$$

where $u_1 = n_1 - m_2$. Estimates of N and p are the same as before, with u_2 replacing n_2; p_0 is estimated by m_2/n_1. However, the above model can be used for testing the hypothesis H that $p = p_0$. Setting $p = p_0$ in (7.38), the maximum-likelihood estimates of N and p are

$$\tilde{N} = (n_1 + n_2)^2/4m_2$$

and

$$\tilde{p} = 2m_2/(n_1 + n_2),$$

and from **1.3.6** (*3*) the multinomial goodness-of-fit statistic for testing H is

$$\begin{aligned} T &= \frac{(u_1 - \tilde{N}\tilde{p}\tilde{q})^2}{\tilde{N}\tilde{p}\tilde{q}} + \frac{(u_2 - \tilde{N}\tilde{q}\tilde{p})^2}{\tilde{N}\tilde{p}\tilde{q}} + 0 + 0 \\ &= \frac{(u_1 - u_2)^2}{(u_1 + u_2)}, \end{aligned}$$

which is approximately χ_1^2 when H is true.

When $p = p_0$ we can compare the efficiencies of the removal estimate $\hat{N} = n_1^2/(n_1 - n_2)$ and the Petersen estimate $\hat{N}_p = n_1 n_2/m_2$. Thus, from **4.1.6** (*1*) (with $s = 2$) and (7.27), we have

$$\frac{V[\hat{N}_p]}{V[\hat{N}]} \approx \frac{Nq^2/p^2}{Nq^2(1+q)/p^3} = \frac{p}{1+q} < 1,$$

and when p is small the Petersen estimate is much more efficient than the removal estimate (see also **14.1.2**(*5*)). However, the cost of marking the first sample increases with p as more individuals are caught in the first sample, so that for large p the removal estimate may give more value for money. If the removal is not permanent, as is sometimes the case for large p, then the cost of marking is offset to some extent by the cost of storing the first sample until the second sample is taken.

Example 7.5 Hypothetical data: Seber and Le Cren [1967]

Electric fishing is carried out in an enclosed section of a small brook on two occasions under identical conditions. On the first fishing 49 trout are caught, marked and released alive. On the second fishing 50 trout are caught of which 24 are found to be marked. Thus $u_1 = 25$, $u_2 = 26$ and $m_2 = 24$ and the two-sample removal estimate of N is

$$\hat{N} = n_1^2/(n_1 - u_2) = 49^2/23 = 104$$

with

$$\hat{\sigma}[\hat{N}] = \frac{n_1 u_2 \sqrt{(n_1 + u_2)}}{(n_1 - u_2)^2} = \frac{49(26)\sqrt{75}}{23^2} = 20.$$

From the recapture data we can also calculate the modified Petersen estimate (cf. **3.1.1**)

$$N^* = \frac{(n_1+1)(n_2+1)}{m_2+1} - 1 = \frac{50(51)}{25} - 1 = 101$$

with standard deviation estimated by

$$\sqrt{v^*} = \left\{\frac{(n_1+1)(n_2+1)u_1 u_2}{(m_2+1)^2(m+2)}\right\}^{\frac{1}{2}} = 10.$$

We note that N^* has a smaller standard deviation than $\hat{N}$, as expected.

To test $p = p_0$ we calculate

$$T = (25 - 26)^2/(25 + 26) = 0{\cdot}02$$

which, for one degree of freedom, indicates an excellent fit.

3 One-sample method

Suppose we take a single catch of size n from a population of size N. If an estimate $\hat{p}$ of p ($= 1/\theta$), the probability of capture, is available from other sources, then, assuming n to be binomially distributed, an estimate of N is

$$\hat{N} = n/\hat{p} = n\hat{\theta}, \text{ say.}$$

Assuming $\hat{\theta}$ to be statistically independent of n, we have

$$E[\hat{N}] = NpE[\hat{\theta}] = N + Np\, E[\hat{\theta}-\theta]$$

and

$$V[\hat{N}] = (N^2p^2 + Npq)\, E[(\hat{\theta}-\theta)^2] + Nq(2E[\hat{\theta}-\theta]+\theta) - N^2p^2(E[\hat{\theta}-\theta])^2.$$

Usually $\hat{\theta}$ is obtained from another population with approximately the same value of p or from different areas of the same population. However, as pointed out by Seber and Le Cren [1967: 636], unless fairly precise estimates of θ are available, $\hat{N}$ will have a large coefficient of variation. The reader is referred to that article for a number of different methods of estimating θ; in each case $E[(\hat{\theta}-\theta)^2]$ is given (asymptotically) and $E[\hat{\theta}-\theta]$ is assumed to be negligible. For further examples see Mann [1971].

7.2.3 Regression estimates

When the probability of capture, p, is constant we have from (7.1) that

$$E[n_i \mid x_i] = p(N - x_i), \tag{7.39}$$

a model considered by Hayne [1949b]. If a plot of n_i versus x_i, the accumulated catch, is approximately linear, then we have some qualitative evidence as to the stability of p. Once again the methods of **1.3.5** (*3*) provide point and interval estimates of p and N; the intercept of the fitted line on the x-axis will provide a graphical estimate of N.

Since

$$E[n_i] = Npq^{i-1},$$

then taking logarithms we have an alternative regression model

$$E[y_i] \approx \log(pN) + (i-1)\log q, \tag{7.40}$$

where $y_i = \log n_i$. This model is of the same form as (7.9), so that the approximate formulae given in that section apply here. We also note that logarithms to the base 10 can be used in (7.40).

Example 7.6 Whitefish (*Coregonus clupeaformis*): Ricker [1958: 150]

A small lake on an island in Lake Nipigon, Ontario, was fished by gill nets in an identical manner for 7 successive weeks; the same sizes of nets, positions and lengths of sets were repeated each week. For whitefish of fork length 13–14 inches (33–35 cm) the weekly catches and their logarithms are given in Table 7.7.

A plot of n_i against x_i (Fig. 7.5) suggests that the model (7.39) is not unreasonable. Therefore least-squares estimates of p and N are

$$\tilde{p} = -\Sigma\, n_i(x_i - \bar{x})/\Sigma\,(x_i - \bar{x})^2 = 0{\cdot}1895$$

and

$$\tilde{N} = \bar{x} + (\bar{n}/\tilde{p}) = 142.$$

TABLE 7.7

Weekly catches (n_i) from a whitefish population: data from Ricker [1958: 150].

Week (i)	1	2	3	4	5	6	7
Catch (n_i)	25	26	15	13	12	13	5
Cumulative catch (x_i)	0	25	51	66	79	91	104
$\log_{10} n_i$	1·40	1·42	1·18	1·11	1·08	1·11	0·70
$i - 1$	0	1	2	3	4	5	6

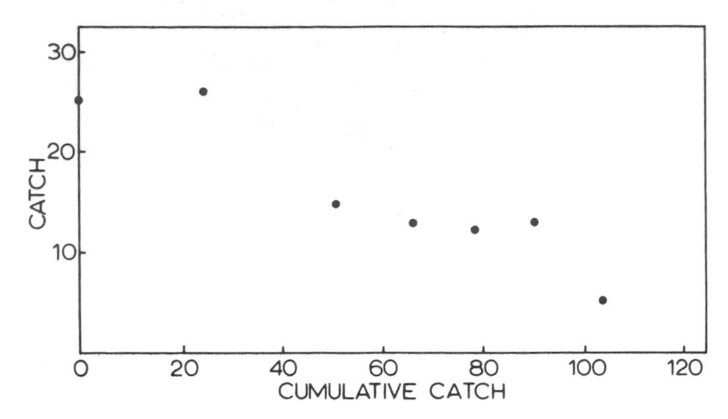

Fig. 7.5 Plot of catch (n_i) versus cumulative catch (x_i) for a population of whitefish: data from Ricker [1958].

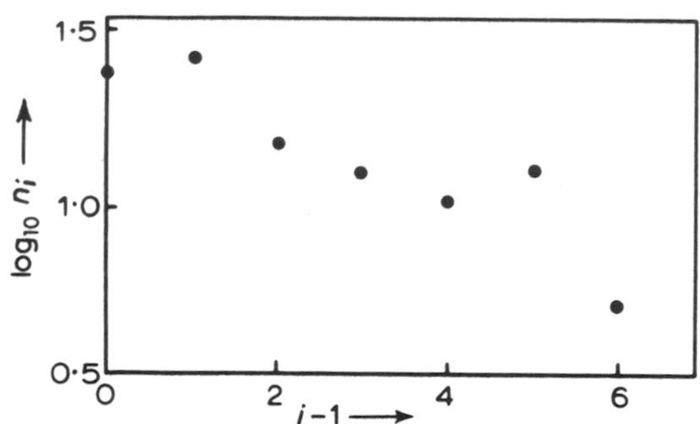

Fig. 7.6 Plot of $\log_{10} n_i$ versus $i - 1$ for a population of whitefish: data from Ricker [1958].

Using (7.6) and (7.7) with n_i and p instead of Y_i and K, approximate 95 per cent confidence intervals for N and p are (114, 207) and (0·108, 0·271), respectively.

The plot of $y_i = \log_{10} n_i$ against $(i-1)$ is approximately linear (Fig. 7.6), so that (7.40) can also be used. Least-squares estimates of $\log_{10}(1-p)$ and $\log_{10}(pN)$ can be obtained in the usual manner, leading to the estimates

$$p' = 0·2070 \quad \text{and} \quad N' = 134.$$

Assuming normality, and taking the antilogarithm of the confidence interval for the slope of (7.40), gives a confidence interval for $1 - p$; thus the 95 per cent confidence interval for p is found to be (0·101, 0·300). Also from p. 302,

$$V[N'] \approx N^2\sigma_y^2\left[\frac{1}{s}+\left(\frac{\frac{1}{2}(s+1)p-1}{p}\right)^2\cdot\frac{12}{s(s-1)(s+1)}\right],$$

where σ_y^2 is estimated from the usual residual sum of squares about the regression line. Assuming N' to be approximately normally distributed, an approximate 95 per cent confidence interval for N is given by $N' \pm 1{\cdot}96\sqrt{v'}$, where v' is the estimate of $V[N']$. This interval (which is of doubtful validity) is found to be (122, 146).

Finally, applying Zippin's maximum-likelihood method (cf. Example 7.3 on p. 314),

$$R = \frac{n_2 + 2n_3 + \ldots + 6n_7}{n_1 + n_2 + \ldots + n_7} = \frac{238}{109} = 2{\cdot}18,$$

and entering this value in Fig. 7.3 and 7.4 with $s = 7$, we find that $\hat{p} = 0{\cdot}19$ and $(1-\hat{q}^7) = 0{\cdot}77$. Hence

$$\hat{N} = 109/0{\cdot}77 = 142$$

and an approximate 95 per cent confidence interval for N is (109, 175).

Further examples of the above regression methods, together with graphs, are given in Hayne [1949b], Barbehenn [1958], Menhinick [1963], Webb [1965], Golley *et al.* [1965], Southwood [1966: 182], Andrzejewski and Jezierski [1966], Buchalczyk and Pucek [1968], and Pucek [1969].

UTILISING TAG DATA. If a tag release is made at the beginning of the experiment then the respective regression lines from the tagged and untagged populations can be compared as in **7.1.6**. For example, when the two graphs are approximately linear we can test the hypothesis that the two regression lines are parallel, i.e. that p is the same for tagged and untagged.

CHAPTER 8

CATCH-EFFORT METHODS: OPEN POPULATION

8.1 CONTINUOUS SAMPLING

8.1.1 Mortality only[*]

1 Catch equations

Consider a population which is closed, apart from depletion through continuous exploitation and natural mortality. Let

$$N(t) = \text{total population size at time } t,$$
$$n(t) = \text{number caught in the interval } [0, t],$$

and

$$\mathring{f}(t) = \text{effort } per\ unit \text{ time operating at time } t.$$

We shall assume that the natural mortality and mortality due to exploitation are both Poisson processes with parameters μ and $k\mathring{f}(t)$ respectively. We note in passing that in **7.1.1** we defined the sampling (exploitation) process as one in which the probability of an individual being caught when subject to δf units of effort is $k\delta f + o(\delta f)$, where k is the Poisson catchability coefficient (cf. **7.1.1**). When the effort depends on the time spent, we have $\delta f = \mathring{f}(t)\delta t$.

From the above definitions and assumptions Chapman [1961] shows that (cf. (1.1))

$$E[N(t)] = N(0) \exp(-\mu t - k\mathring{F}(t)), \tag{8.1}$$

where $\mathring{F}(t) = \int_0^t \mathring{f}(x)\, dx$ is the accumulated effort up to time t, and

$$E[n(t)] = \int_0^t k\mathring{f}(x) N(x)\, dx. \tag{8.2}$$

Suppose that the whole sampling period can be divided up into s periods $[0, t_1), [t_1, t_2), \ldots, [t_{s-1}, t_s)$ during which the effort per unit time is constant and equal to $\mathring{f}_1, \mathring{f}_2, \ldots, \mathring{f}_s$ respectively; thus $\mathring{f}(t) = \mathring{f}_i$ for $t_{i-1} \leqslant t < t_i$ $(i = 1, 2, \ldots, s;\ t_0 = 0)$. Let

N_i = size of population at the *beginning* of the ith period,
n_i = size of catch in the ith period,
Δ_i = $t_i - t_{i-1}$,
f_i = $\Delta_i \mathring{f}_i$ (total effort in ith period),
μ_i = instantaneous natural mortality rate in the ith period,

[*]See also **13.2.1**.

$$\mu_{Ei} = k\overset{o}{f}_i \text{ (instantaneous exploitation rate in the } i\text{th period)},$$
$$Z_i = \mu + \mu_{Ei} \text{ (total instantaneous mortality rate)},$$

and

$$F_i = \sum_{j=1}^{i} f_j \text{ (accumulated effort up to } \textit{and including} \text{ the } i\text{th period, i.e. accumulated effort up to time } t_i).$$

(We note that $N_1 = N(0)$, and the definition of F_i differs from that in **7.1.1** where the ith period is not included. A number of authors (e.g. Chapman [1961]) use the symbol f_i instead of $\overset{o}{f}_i$ to represent the effort per unit time, and sometimes the distinction is not clearly made). Then, from (8.1) and (8.2) it is readily shown that

$$E[N_i] = N_1 \exp(-\mu t_{i-1} - kF_{i-1}), \tag{8.3}$$

$$E[N_{i+1} | N_i] = N_i \exp(-Z_i \Delta_i), \tag{8.4}$$

and

$$E[n_i | N_i] = k\overset{o}{f}_i \int_0^{\Delta_i} N_i \exp(-Z_i x)\, dx$$

$$= k\overset{o}{f}_i N_i [1 - \exp(-Z_i\Delta_i)]/Z_i. \tag{8.5}$$

We note that (8.5) can also be written in the form

$$E[n_i | N_i] = kf_i \bar{N}_i, \tag{8.6}$$

where
$$\bar{N}_i = \int_0^{\Delta_i} N_i \exp(-Z_i x) dx \Big/ \int_0^{\Delta_i} dx$$
$$= N_i [1 - \exp(-Z_i\Delta_i)]/(Z_i\Delta_i)$$

is the average population size during the ith period. The above equations, often referred to as the "catch" equations, have been widely used in fishery work and go back to Baranov [1918] and Ricker [1940, 1944].

2 *Estimation*

MORTALITY AND CATCHABILITY. We shall now consider a number of methods for estimating μ and k given the $\{n_i\}$ and $\{\overset{o}{f}_i\}$.

From (8.5), with $i + 1$ instead of i, and (8.4), we have

$$E[n_{i+1} | N_i] = \frac{\Delta_{i+1} k\overset{o}{f}_{i+1}}{\Delta_{i+1} Z_{i+1}} \cdot N_i \exp(-Z_i\Delta_i)\{1 - \exp(-Z_{i+1}\Delta_{i+1})\}.$$

Dividing by (8.5) and taking logarithms, we are led to consider the model

$$E[(y_i/\Delta_i)] \approx \frac{1}{\Delta_i} \log\left[\frac{Z_{i+1}\Delta_{i+1}\{1 - \exp(-Z_i\Delta_i)\}}{Z_i\Delta_i\{1 - \exp(-Z_{i+1}\Delta_{i+1})\}}\right] + \mu + k\overset{o}{f}_i, \tag{8.7}$$

where $y_i = \log(n_i/f_i) - \log(n_{i+1}/f_{i+1})$. Widrig [1954] and, independently, Beverton and Holt [1957: 192] first developed this model and suggested regressing y_i/Δ_i on $\overset{o}{f}_i$ by least squares; Beverton and Holt also gave an iterative procedure to take into account the non-linear term. However, if $W = Z\Delta$ is small for each period then

$$\log\left[1-e^{-W})/W\right] \approx -\tfrac{1}{2}W, \qquad (8.8)$$

and, to a first order of approximation, the non-linear term of (8.7) is given by

$$\tfrac{1}{2}(Z_{i+1}\Delta_{i+1}-Z_i\Delta_i)/\Delta_i = \tfrac{1}{2}\left[(\Delta_{i+1}-\Delta_i)\mu + (\overset{o}{f}_{i+1}-\overset{o}{f}_i)k\right]/\Delta_i.$$

Paloheimo [1961] shows that (8.8) is a satisfactory approximation for the likely range $0{\cdot}10 \leqslant W \leqslant 1{\cdot}00$. Using the above approximation, (8.7) becomes

$$E[y_i] \approx \mu(\Delta_i+\Delta_{i+1})/2 + k(\overset{o}{f}_i+\overset{o}{f}_{i+1})/2, \quad i = 1, 2, \ldots, s-1, \qquad (8.9)$$

which, for the common situation $\Delta_i = 1$ (e.g. one year), reduces to Paloheimo's [1961] linear model:(*)

$$E[y_i] \approx \mu + k(\overset{o}{f}_i+\overset{o}{f}_{i+1})/2. \qquad (8.10)$$

We note that (8.9) can be analysed as a multiple linear regression, provided the $\overset{o}{f}_i$ vary. When $\overset{o}{f}_i = \overset{o}{f}$, say, then (8.9) reduces to

$$E[y_i] \approx (\mu+k\overset{o}{f})(\Delta_i+\Delta_{i+1})/2$$

and μ and k cannot be estimated separately; only $Z = \mu + k\overset{o}{f}$ is estimable. In this case, when $\Delta_i = 1$, the instantaneous natural mortality and exploitation rates are both constant and the method of **6.2.2** can also be used here for estimating Z (with m_i and M_0 replaced by n_i and N_1; (6.13) does not depend on knowing M_0).

In analysing (8.9) and (8.10) we have to take into account the fact that the y_i will be correlated even if the n_i are assumed to be independent. Several approaches are possible; if n_i is not small compared with N_1 we can assume the n_i to be multinomially distributed and use the delta method to derive the variance–covariance matrix (cf. **6.2.2**); if n_i is small compared with N_i we can ignore the dependence between the n_i and, following Chapman [1961: 160], assume the $\log n_i$ to be independently normally distributed with common variance σ^2. Then, since

$$V[y_i] = V[\log n_i - \log n_{i+1}] = 2\sigma^2, \text{ etc.},$$

we find that the variance–covariance matrix of $\mathbf{y}' = (y_1, y_2, \ldots, y_{s-1})$ is $\sigma^2\mathbf{B}$, where

$$\mathbf{B} = \begin{bmatrix} 2 & -1 & 0 & 0 & \ldots & 0 & 0 \\ -1 & 2 & -1 & 0 & \ldots & 0 & 0 \\ 0 & -1 & 2 & -1 & \ldots & 0 & 0 \\ \ldots & \ldots & \ldots & \ldots & \ldots & \ldots & \ldots \\ 0 & 0 & 0 & 0 & \ldots & -1 & 2 \end{bmatrix}$$

and

(*)The robustness of this model is considered by Tanaka [1975].

$$\mathbf{B}^{-1} = \frac{1}{s}\begin{bmatrix} s-1, & s-2, & s-3 & \ldots & 3, & 2, & 1 \\ s-2, & 2(s-2), & 2(s-3) & \ldots & 6, & 4, & 2 \\ s-3, & 2(s-3), & 3(s-3) & \ldots & 9, & 6, & 3 \\ \ldots & \ldots & \ldots & \ldots & \ldots & \ldots & \ldots \\ 3, & 6, & 9 & \ldots & (s-3)3, & (s-3)2, & s-3 \\ 2, & 4, & 6 & \ldots & (s-3)2, & (s-2)2, & s-2 \\ 1, & 2, & 3 & \ldots & (s-3), & s-2, & s-1 \end{bmatrix}.$$

Thus, in vector notation (8.10) is $\mathbf{y} = \mathbf{X\beta} + \mathbf{e}$, where

$$\mathbf{X} = \begin{bmatrix} 1, & (f_1+f_2)/2 \\ 1, & (f_2+f_3)/2 \\ \ldots, & \ldots \\ 1, & (f_{s-1}+f_s)/2 \end{bmatrix}, \qquad \boldsymbol{\beta} = \begin{bmatrix} \mu \\ k \end{bmatrix},$$

and $\mathbf{e}$ has a multivariate normal distribution $\mathfrak{N}[\mathbf{0}, \sigma^2\mathbf{B}]$. This model can now be analysed using the general method of **1.3.5**(*2*).

INITIAL POPULATION SIZE. We now turn our attention to the problem of estimating N_1, the initial population size. From (8.5) and (8.3) we have

$$E[n_i] = \Delta_i\, k\overset{o}{f}_i N_1 \exp(-\mu t_{i-1} - kF_{i-1})[1-\exp(-Z_i\Delta_i)]/Z_i\Delta_i. \tag{8.11}$$

Taking logarithms and using the approximation (8.8), we arrive at the multiple linear regression

$$\begin{aligned} E[\log(n_i/f_i)] &\approx \log(N_1 k) - \mu(t_{i-1} + \tfrac{1}{2}\Delta_i) - k(F_{i-1} + \tfrac{1}{2}f_i) \\ &= \log(N_1 k) - \mu(t_{i-1}+t_i)/2 - k(F_{i-1}+F_i)/2, \end{aligned} \tag{8.12}$$

for which least-squares estimates of k, μ, $\log(N_1 k)$ and hence N_1 can be obtained in the usual way. Estimates of the N_i can then be calculated iteratively from (8.4). The model (8.12) was first derived by Chapman [1961: 161], though there is a small misprint; in terms of our notation the left-hand side of his equation is $E[\log(n_i/\overset{o}{f}_i)]$.

3 Seasonal fishery

Chapman [1961: 155] and Muir and White [1963] have discussed the problem of a seasonal fishery where exploitation takes place only for a fraction τ_i of each period at the beginning of the period. Equations (8.4) and (8.5) become

$$\begin{aligned} E[N_{i+1} \mid N_i] &= N_i \exp(-Z_i\tau_i\Delta_i) \exp\{-\mu(1-\tau_i)\Delta_i\} \\ &= N_i \exp\{-(\mu + k\,\overset{o}{f}_i\tau_i)\Delta_i\} \end{aligned}$$

and

$$E[n_i \mid N_i] = k \overset{o}{f}_i N_i \{1 - \exp(-Z_i \tau_i \Delta_i)\}/Z_i.$$

Using the same method as above and redefining the total effort in the ith period, namely $f_i = \Delta_i \tau_i \overset{o}{f}_i$, (8.9) becomes

$$E[y_i] \approx \mu[\Delta_i + \tfrac{1}{2}(\Delta_{i+1}\tau_{i+1} - \Delta_i \tau_i)] + k(f_i + f_{i+1})/2.$$

When $\Delta_i = 1$ and $\tau_i = \tau$ this equation reduces to (8.10) which, as pointed out by Muir and White [1963], can therefore be used for a seasonal fishery, provided the efforts f_i are correctly calculated.

4 *Age-composition data*

In using (8.10) the different y_i values can be based on different segments of the population. For example, let N_1 be the size of the age-class of age a or older (designated a+) and let N_2 be the size of the same age-class a year later (i.e. of age $(a+1)+$). If n_1 and n_2 are the respective catches from this class in the first two years, then from (8.10) we have

$$E[\log(n_1 f_2/n_2 f_1)] \approx \mu + k(f_1 + f_2)/2.$$

For the second and third years we now redefine N_2 to be the size of the age-class a+ in the second year, and N_3 the size of this class one year later. If n'_2 and n'_3 are the respective catches from this class in the second and third years, then

$$E[\log(n'_3 f_2/n'_2 f_3)] \approx \mu + k(f_2 + f_3)/2.$$

Thus by recording the number caught in the age-classes a+ and $(a+1)+$ for each year, a regression model of the form (8.10) can be used.

If there is a large number of individuals growing into the age-class a each year, the statistical dependence between the ratios $n_1 f_2/n_2 f_1$, $n'_2 f_3/n'_3 f_2$, etc. will not be as strong as between $n_1 f_2/n_2 f_1$, $n_2 f_3/n_3 f_2$, etc. For this situation an unweighted least-squares estimation of k and μ may be more appropriate than Chapman's method using the variance—covariance matrix $\sigma^2 \mathbf{B}$. However, although equations (8.4) and (8.5) apply in this situation, (8.3) (and hence (8.12)) no longer apply as N_i refers to a different population for each pair of consecutive years.

The essential feature of the above approach is that the same cohort is considered for each pair of consecutive catches. Also the same efforts are used throughout, though strictly speaking f_i is the effort applied to the whole population and not just the cohort. This means that if trapping is used to catch the individuals, the effort with respect to the cohort will begin to fall off as the traps fill up with individuals not belonging to the cohort. It would seem desirable, therefore, to choose the age a so as to include as much of the total catch data as possible in the analysis, or else modify the effort data f_i in some reasonable manner. For example, one possibility is to estimate the "effective" number of traps as the average of the total number of traps and the number of traps at the end of a sampling period not filled by individuals from outside the cohort.

If some of the age-classes are not fully recruited (i.e. catchable) then a should be chosen as the youngest fully recruited age-group.

When sample sizes are large, as in commercial fishing, it requires considerable time and expense to age each individual in the sample. However, if age and some easily measured quantity such as length are correlated, then a simpler method is to measure the length of each individual and then either choose a random subsample from the whole sample for age determinations, or else stratify the sample according to length-classes and choose a subsample from each class (Tanaka [1953]). Let

X_i = number in the sample belonging to the ith length-class $(i = 1, 2, \ldots, I)$,

X_{ai} = number of age a in the ith length-class,

X_a = $\sum_i X_{ai}$ (number of age a in the sample),

x_i = number in the subsample belonging to the ith length-class,

and x_{ai} = number from the group of x_i individuals of age a.

We shall introduce a random variable y_a which is defined to be equal to unity if an individual chosen at random is of age a and zero otherwise. Let y_{aij} be the value of y_a for the jth member of the ith class in the subsample; then, irrespective of which of the above two subsampling methods is used, an unbiased estimate of X_a is given by

$$\hat{X}_a = \sum_{i=1}^{I} \frac{X_i}{x_i} \sum_{j=1}^{x_i} y_{aij}.$$

5 Variable catchability coefficient

If the catchability k is not constant, we can work with the parameters μ and μ_{Ei} instead. By the same method as was used in deriving (8.5) it is readily shown that

$$E[n_i \mid N_i] = \mu_{Ei} N_i [1 - \exp(-Z_i \Delta_i)]/Z_i. \qquad (8.13)$$

From the ratios n_i/n_{i+1} Murphy [1965] gives an iterative method for estimating all the μ_{Ei}, given that one has estimates of μ and just one of the μ_{Ei}. If we define $\mu_{Ei} = k_i \overset{o}{f_i}$ then k_i can also be estimated.

AGE DATA. If the catchability varies with age (e.g. Muir [1963]), Muir [1964]) shows how to modify the method in section *4* above to allow for this. Let k_{a+} and $k_{(a+1)+}$ be the (constant) catchabilities for the age-classes a+ and $(a+1)+$ respectively; then, using the general catch equations we find that (8.10) now becomes

$$E[y_i] \approx \mu - \log\left(\frac{k_{(a+1)+}}{k_{a+}}\right) + \tfrac{1}{2} k_{a+} \left(f_i + \frac{k_{(a+1)+}}{k_{a+}} f_{i+1}\right). \qquad (8.14)$$

If an independent estimate of $k_{(a+1)+}/k_{a+}$ is available, then the above linear regression can be analysed as in section *4*.

Paloheimo and Kohler [1968: 564, 568] suggest setting up a linear regression for each pair of age-classes a and $a+1$ and give a combined non-linear model which allows the natural mortality to be age-dependent as well as the catchability.

Example 8.1 Maskinonge (*Esox masquinongy*): Muir [1964]

Working with the age-group a = IV and an estimate $k_{V+}/k_{IV+} = 1\cdot2$, Muir used (8.14) to analyse the data in Table 8.1. Here

$$y_i = \log\left(\frac{n_{i,\text{IV}+}/f_i}{n_{i+1,\text{V}+}/f_{i+1}}\right) \quad \text{and} \quad x_i = \tfrac{1}{2}(f_i + 1\cdot2\, f_{i+1});$$

for example, $y_1 = \log(1\cdot42/0\cdot82) = 0\cdot5481$ and $x_1 = \frac{1}{2}(90 + 1\cdot2(102)) = 106$.

TABLE 8.1

Catch and effort data for age-groups IV+ and V+ for a population of maskinonge: from Muir [1964: Table 4].

Year i	Age-class	Catch n_i	Effort f_i	$\frac{n_i}{f_i}$	y_i	x_i
1952	IV+	128	90	1·42	0·5481	106
	V+	89		0·99		
1953	IV+	103	102	1·01	0·5710	114
	V+	84		0·82		
1954	IV+	104	105	0·99	−0·0726	144
	V+	60		0·57		
1955	IV+	215	152	1·41	0·4187	168
	V+	164		1·07		
1956	IV+	217	153	1·42	1·0438	352
	V+	143		0·93		
1957	IV+	369	458	0·81	1·3987	475
	V+	231		0·50		
1958	IV+	138	410	0·34	0·8198	385
	V+	81		0·20		
1959	IV+	84	299	0·28	0·2390	292
	V+	44		0·15		
1960	IV+	129	237	0·54		
	V+	53		0·22		

Writing (8.14) in the form

$$E[y_i] = \beta_0 + \beta x_i,$$

where $\beta_0 = \mu - \log 1\cdot2$ and $\beta = k_{a+}$, we find that the least-squares estimates are

$$\hat{\mu} = \hat{\beta}_0 + \log 1{\cdot}2 = 0{\cdot}0008 + 0{\cdot}1823 = 0{\cdot}1831$$

and

$$\hat{k}_{a+} = 0{\cdot}0024.$$

Assuming the y_i to be independently normally distributed with constant variance, the usual 95 per cent confidence interval for the slope of the regression line yields $0{\cdot}0024 \pm 0{\cdot}0022$. This wide interval reflects the wide dispersion about the fitted line (Fig. 8.1); from the graph the dispersion appears to be less when the fishing effort is greater (the trend in the graph is accentuated by the compression of the x-axis).

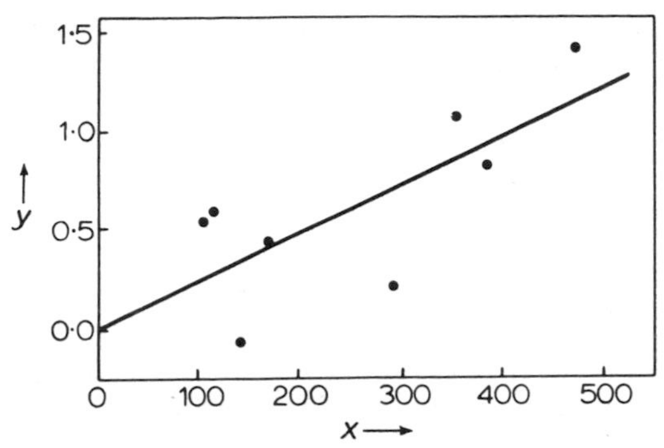

Fig. 8.1 Application of Muir's regression model (8.14) to a population of maskinonge: redrawn from Muir [1964].

8.1.2. Recruitment only(*)

Consider a population in which recruitment and removal through exploitation are the only processes affecting the population size. Then from Fischler [1965] we have the following deterministic modification of Leslie's regression model (**7.1.3** (*1*)) to allow for recruitment:

Using the notation of **8.1.1**, let r_i be the number recruited in the ith interval. Then, for the catchable (fully recruited) population.

$$N_{i+1} = N_i - n_i + r_i \tag{8.15}$$

and from (8.6) we have the deterministic relationship

$$n_i = k\, f_i \bar{N}_i \tag{8.16}$$

where $\bar{N}_i$ is the average population size during the ith interval. Assuming recruitment to be proportional to population size, i.e. $r_i = a_i \bar{N}_i$, and approximating $\bar{N}_i = \frac{1}{2}(N_i + N_{i+1})$, we have from (8.15) the recurrence relationship

$$N_{i+1} = \left(\frac{2 + a_i}{2 - a_i}\right) N_i - \frac{2n_i}{2 - a_i}. \tag{8.17}$$

Substituting for N_{i+1} in (8.16) leads to

$$\frac{n_i}{f_i} = k\left[\frac{2N_i}{2 - a_i} - \frac{n_i}{2 - a_i}\right] \tag{8.18}$$

(*) For models of mortality and recruitment see 13.2.2.

and applying (8.17) recursively gives us:

$$\frac{n_1(2-a_1)}{f_1\,2} = k\left(N_1 - \frac{n_1}{2}\right),$$

$$\frac{n_2}{f_2}\,\frac{(2-a_2)}{2}\,\frac{(2-a_1)}{(2+a_1)} = k\left(N_1 - \frac{2n_1}{2+a_1} - \frac{(2-a_1)n_2}{(2+a_1)2}\right),$$

and, for $i > 2$,

$$\frac{n_i}{f_i}\cdot\frac{(2-a_i)(2-a_{i-1})\dots(2-a_1)}{2(2+a_{i-1})\dots(2+a_1)} = k\left(N_1 - \frac{2n_1}{2+a_1} - \frac{2(2-a_1)}{(2+a_2)(2+a_1)}\cdot n_2\right.$$

$$- \dots - \frac{2(2-a_{i-2})(2-a_{i-3})\dots(2-a_1)}{(2+a_{i-1})(2+a_{i-2})(2+a_{i-3})\dots(2+a_1)}\cdot n_{i-1}$$

$$\left. - \frac{(2-a_{i-1})(2-a_{i-2})\dots(2-a_1)}{(2+a_{i-1})(2+a_{i-2})\dots(2+a_1)}\cdot\frac{n_i}{2}\right).$$

If independent estimates of the a_i are available, such as from age data, then substituting these estimates in the above equations leads to a regression model of the form

$$y_i = k(N_1 - x_i) + e_i, \tag{8.19}$$

which can be analysed as in **1.3.5**(*3*). Once least-squares estimates of N_1 and k are calculated, estimates of the N_i can be obtained iteratively from (8.17).

Example 8.2 Blue crab (*Callinectes sapidus*): Fischler [1965]

Using a time-interval of one week, estimates $\hat{a}_i$ of the weekly recruitment rate were calculated from such information as the ratio of pre-commercial to commercial size crabs in catches and the moulting rate. From these estimates the values of y_i and x_i above were calculated and set out in Table 8.2. A plot of y_i against x_i is fairly linear (Fig. 8.2), thus supporting the use of (8.19). Therefore, using the method of **1.3.5**(*3*) with $w_i = 1$, we find that the least-squares estimates of k and N_1 are $\hat{k} = 0{\cdot}000\ 427$ and $\hat{N}_1 = 454\ 413$ with 95 per cent confidence intervals (0·000 375, 0·000 479) and (432 749, 480 583) respectively.

8.1.3 Variable catchability: recruitment estimates

If the catchability k is not constant or effort data are not available, then we must work with the instantaneous natural mortality and exploitation rates μ and μ_{Ei} respectively. Allen [1966, 1968] gives the following deterministic method for estimating the level of recruitment each year: by recruitment we shall mean the process whereby an individual becomes catchable.

Suppose that exploitation is carried out continuously for s consecutive periods of time, which we shall call years for convenience, and suppose that

TABLE 8.2

Weekly crab catch in pounds weight per standard unit of effort, and estimates of weekly recruitment rate: from Fischler [1965: Table 13].

i (week)	n_i/f_i (pounds)	$\hat{a}_i$ (recruitment rate)	y_i	x_i
1	186·0	0·147	172·33	2 167·00
2	244·5	0·147	195·51	72 032·12
3	226·7	0·161	155·27	124 512·05
4	203·9	0·245	113·42	170 864·93
5	196·4	0·245	85·40	212 897·56
6	292·5	0·245	99·43	256 123·54
7	210·0	0·245	55·80	287 210·05
8	194·5	0·175	42·02	302 997·27
9	293·3	0·161	53·57	318 657·09
10	370·0	0·126	58·60	336 895·08
11	350·6	0·112	49·32	353 735·97
12	373·7	0·112	46·99	368 371·92
13	266·1	0·112	29·91	380 038·30
14	274·2	0·105	27·66	389 870·93
15	283·7	0·084	26·04	399 069·71
16	258·8	0·084	21·84	405 681·19
17	238·4	0·175	17·62	410 620·66
18	246·6	0·175	15·29	415 131·36
19	202·3	0·091	11·01	448 464·97
20	170·2	0·077	8·52	420 692·93

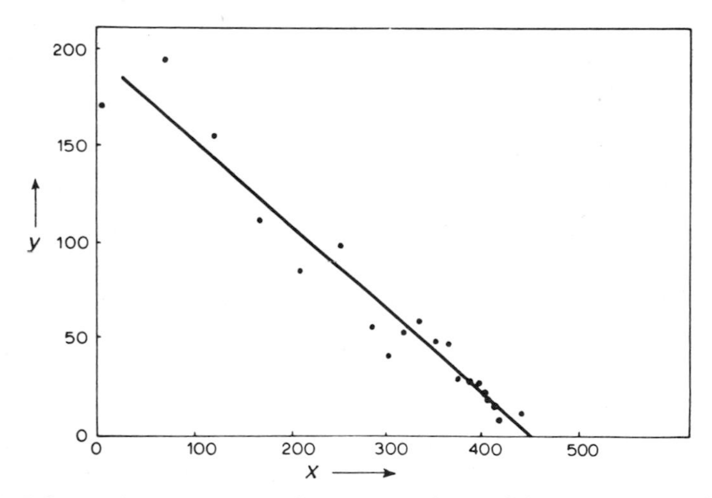

Fig.8.2 Application of Fischler's regression model (8.19) to a population of blue crabs: redrawn from Fischler [1965].

all animals caught can be assigned to one of $J+1$ age-groups. We shall assign an individual to the age-class a if its age lies in the interval $[a, a+1)$. Although recruitment is sometimes a continuous process, we shall assume that all recruiting takes place at the beginning of each year.

For $i = 1, 2, \ldots, s$ and $a = 0, 1, \ldots, J$ let:

N_i = size of the total population, including both catchable and non-catchable (i.e. non-recruited) members, at the beginning of year i,

$N_{i,a}$ = size of the age-class a at the beginning of year i,

n_i = total catch in year i,

$n_{i,a}$ = number caught in year i which belonged to age-class a at the beginning of the year,

$P_{i,a}$ = $n_{i,a}/n_i$,

$\rho_{i,a}$ = proportion of the group of size $N_{i,a}$ which is catchable at the beginning of year i,

$R_{i,a}$ = proportion of the catchable group of $\rho_{i,a} N_{i,a}$ individuals newly recruited at the beginning of year i,

$\bar{r}_{i,a}$ = average number of catchable individuals alive during year i from the group $\rho_{i,a} N_{i,a}$ individuals,

p_i = probability that a catchable member of the N_i individuals is caught in the ith year,

Z_i = $\mu_i + \mu_{Ei}$ = total instantaneous mortality rate for the catchable part of the partially recruited age-classes in year i,

Z_i' = $\mu_i' + \mu_{Ei}'$ = total instantaneous mortality rate for the fully recruited age-classes in year i,

and μ_i'' = instantaneous natural mortality rate for the unrecruited part of the population in year i.

Assuming that the age-classes $a = A, A+1, \ldots, J$ are fully recruited (i.e. $\rho_{i,a} = 1$), we also define

$$N_{i(A)} = \sum_{a=A}^{J} N_{i,a};$$

$n_{i(A)}$, $P_{i(A)}$, $\rho_{i(A)}$ and $\bar{r}_{i(A)}$ are all similarly defined. The aim of the following analysis is to provide estimates of the $R_{i,a}$.

For $a < A$ we have the deterministic equation

$$N_{i+1,a+1} = \rho_{i,a} N_{i,a} e^{-Z_i} + (1-\rho_{i,a}) N_{i,a} e^{-\mu_i''},$$

while for the fully recruited age-classes,

$$N_{i+1(A+1)} = N_{i(A)} e^{-Z_i'}.$$

Also

$$\bar{r}_{i,a} = \left(\int_0^1 \rho_{i,a} N_{i,a} e^{-Z_i t} dt\right) \Big/ \left(\int_0^1 dt\right)$$

$$= \rho_{i,a} N_{i,a} (1 - e^{-Z_i})/Z_i$$

and, by the same argument,

$$\bar{N}_{i(A)} = N_{i(A)}(1-e^{-Z_i'})/Z_i',$$

where $\bar{N}_{i(A)}$ is the average number alive, during year i, from the group represented by $N_{i(A)}$. Therefore

$$E[n_{i,a}] = p_i \bar{r}_{i,a}, \qquad E[n_{i(A)}] = p_i \bar{N}_{i(A)}, \tag{8.20}$$

and equating random variables to the expectations, we have the deterministic equations

$$\frac{P_{i,a}}{P_{i(A)}} = \frac{n_{i,a}}{n_{i(A)}} = \frac{\bar{r}_{i,a}}{\bar{N}_{i,a}}.$$

For the next year we have a similar expression

$$\frac{P_{i+1,a+1}}{P_{i+1(A+1)}} = \frac{\bar{r}_{i+1,a+1}}{\bar{N}_{i+1(A+1)}}.$$

Using the proportions $P_{i,a}$ rather than the actual numbers $n_{i,a}$ means that these proportions can be estimated from a subsample of the n_i without having to collect age data for the whole sample. From the above equations we have (introducing $B_{i,a}$ as a convenient intermediary statistic)

$$\begin{aligned} B_{i,a} &= \frac{P_{i(A)} P_{i+1,a+1}}{P_{i,a} P_{i+1(A+1)}} \\ &= \frac{\rho_{i+1,a+1}[\rho_{i,a} e^{-Z_i} + (1-\rho_{i,a}) e^{-\mu_i''}] \left[\frac{1-e^{-Z_{i+1}}}{Z_{i+1}}\right] \left[\frac{1-e^{-Z_i'}}{Z_i'}\right]}{\rho_{i,a} e^{-Z_i'} \left[\frac{1-e^{-Z_i}}{Z_i}\right] \left[\frac{1-e^{-Z_{i+1}'}}{Z_{i+1}'}\right]}. \end{aligned} \tag{8.21}$$

Allen points out that although Z_i may vary considerably from year to year, it is unlikely that the relation between the mortality rates for the partially recruited age-classes and the fully recruited age-classes will change appreciably from one year to the next. In particular, since $[1-\exp(-Z_i')]/Z_i'$ is the average annual probability of survival for the fully recruited age-classes, we would expect

$$\left(\frac{1-e^{-Z_i'}}{Z_i'}\right) \Big/ \left(\frac{1-e^{-Z_i}}{Z_i}\right) \approx \left(\frac{1-e^{-Z_{i+1}'}}{Z_{i+1}'}\right) \Big/ \left(\frac{1-e^{-Z_{i+1}}}{Z_{i+1}}\right)$$

and these terms can be cancelled out in (8.21). Also it is not unreasonable to assume that $\mu_i'' = \mu_i$, so that

$$\begin{aligned} B_{i,a} &= \frac{\rho_{i+1,a+1}[\rho_{i,a} e^{-Z_i} + (1-\rho_{i,a}) e^{-\mu_i}]}{\rho_{i,a} e^{-Z_i'}} \\ &= \frac{\rho_{i+1,a+1}}{\rho_{i,a}} T_i [\rho_{i,a} + (1-\rho_{i,a}) e^{\mu E_i}], \end{aligned} \tag{8.22}$$

where $T_i = e^{Z_i' - Z_i}$. From the definition of $R_{i+1,a+1}$, we have

$$1 - R_{i+1,a+1} = \frac{\rho_{i,a} N_{i,a} e^{-Z_i}}{\rho_{i+1,a+1} N_{i+1,a+1}}$$

$$= \frac{\rho_{i,a}}{\rho_{i+1,a+1}} \cdot \frac{1}{\rho_{i,a} + (1-\rho_{i,a}) e^{\mu E_i}} \tag{8.23}$$

and substituting from (8.22) leads to

$$1 - R_{i+1,a+1} = T_i / B_{i,a}.$$

For most practical purposes it is probably satisfactory to take $T_i = 1{\cdot}0$, though an approximate estimate is available if certain assumptions can be made. For example, assuming that the total instantaneous mortality rate of the youngest fully recruited age-class A is the same as the total instantaneous mortality rate of the recruited part of the partially recruited age-classes, we have, using (8.20),

$$\frac{P_{i+1,A+1}}{P_{i,A}} = \frac{n_{i+1,A+1}}{n_{i+1}} \cdot \frac{n_i}{n_{i,A}}$$

$$= \frac{n_i}{n_{i+1}} \cdot \frac{p_{i+1} \bar{N}_{i+1,A+1}}{p_i \bar{N}_{i,A}}$$

$$= \frac{n_i p_{i+1}}{n_{i+1} p_i} \cdot \frac{(1 - e^{-Z_{i+1}'}) Z_i'}{(1 - e^{-Z_i'}) Z_{i+1}'} \cdot \frac{N_{i+1,A+1}}{N_{i,A}}$$

$$= c\, e^{-Z_i}, \text{ say,}$$

since we have assumed that

$$N_{i+1,A+1} = N_{i,A}\, e^{-Z_i}.$$

Using a similar argument, we find that

$$\frac{P_{i+1(A+1)}}{P_{i(A)}} = c\, e^{-Z_i'}$$

so that T_i can be estimated by

$$\hat{T}_i = \frac{P_{i+1,A+1} P_{i(A)}}{P_{i,A} P_{i+1(A+1)}}.$$

This leads to the estimate(*)

$$(B_{i,a} - \hat{T}_i)/B_{i,a} = 1 - P_{i+1,A+1} P_{i,a} / P_{i,A} P_{i+1,a+1} \qquad \text{for } R_{i+1,a+1}.$$

Allen [1966] also gives a number of methods for estimating k and N_i, and the reader is referred to **13.2.2**(*1*) for further details.

8.2 INSTANTANEOUS SAMPLES

8.2.1 Mortality only: regression method

We shall now consider a model in which samples may be regarded as

(*)This can be obtained directly (Chapman [1974a], with different notation for B).

"instantaneous" or point samples removed at times $t_1, t_2, \ldots, t_s$. Let

N = size of the population just before the first sample,
p_i = probability that an individual in the population at time t_i is caught in the ith sample,
n_i = size of ith sample removed from the population, and
ϕ_i = probability of survival of an individual from time t_i to time t_{i+1} given that it is alive just after the ith sample ($i = 1, 2, \ldots, s-1$).

Assuming that sampling and natural mortality are the only processes affecting the population size, then the joint probability function of the n_i is given by (cf. (7.2))

$$\frac{N!}{(\prod_{i=1}^{s} n_i!)(N-n)!} p_1^{n_1}(q_1\phi_1 p_2)^{n_2} \ldots (q_1\phi_1 q_2\phi_2 \ldots q_{s-1}\phi_{s-1}p_s)^{n_s}\, \theta^{N-n}$$

where $n = \sum_{i=1}^{s} n_i$ and

$$\theta = 1 - p_1 - q_1\phi_1 p_2 \ldots - q_1\phi_1 q_2\phi_2 \ldots q_{s-1}\phi_{s-1}p_s.$$

However, we have more parameters than random variables, so that additional assumptions must be made if all the parameters are to be estimated. Possible assumptions are:

(i) $q_i = \exp(-k f_i)$ where f_i is the effort expended on the ith sample.
(ii) Natural mortality is a Poisson process with parameter μ, so that $\phi_i = \exp[-\mu(t_{i+1} - t_i)]$.

With these assumptions it transpires that the maximum-likelihood equations for N, μ and k do not have explicit solutions and must be solved iteratively. However,

$$\begin{aligned} E[n_i] &= N\, q_1\phi_1 q_2\phi_2 \ldots q_{i-1}\phi_{i-1}p_i \\ &= N\, q_1 q_2 \ldots q_{i-1}p_i\phi_1\phi_2 \ldots \phi_{i-1} \\ &= N \exp(-kF_{i-1})\{1 - \exp(-kf_i)\} \exp\{-\mu(t_i - t_1)\}, \end{aligned}$$

where $F_{i-1} = \sum_{j=1}^{i-1} f_j$, so that using the approximation (8.8) and taking logarithms, we obtain the linear regression model

$$E[\log(n_i/f_i)] \approx \log(Nk) - k(F_{i-1} + F_i)/2 - \mu(t_i - t_1),$$

which is of the same form as (8.12).

If assumption (i) is replaced by (i)$'$: $p_i = p$ ($i = 1, 2, \ldots, s$), then

$$E[n_i] = N q^{i-1} p \exp\{-\mu(t_i - t_1)\}$$

and a possible regression model is

$$E[\log n_i] \approx \log(Np) + (i-1)\log q - \mu(t_i - t_1).$$

This can be analysed by least squares in the usual manner, provided t_i is not proportional to i.

8.2.2 Mortality only: pairwise removal method

Another method for estimating survival probabilities is to take a series of pairs of removals. For example, Seber and Le Cren [1967] discuss the following special case of just two pairs which may be useful for studying small populations.

1 First two catches returned

Consider two two-sample experiments in which the time taken by each experiment is sufficiently short for natural mortality to be negligible during the experiment. Let N_1, N_2 be the respective sizes of the population before each experiment, let n_i $(i = 1, 2, 3, 4)$ be the respective catches, let p_1 be the constant probability of capture for the first two samples, and let p_2 be the constant probability of capture for the second two samples. After the first experiment the total catch $n_1 + n_2 - d$ is returned to the population, d being the number which are deliberately removed or die accidentally through trapping. Then the joint probability function of the n_i is (cf. (7.20) with $s = 2$)

$$\frac{N_1!}{n_1!\, n_2!\, (N_1 - n_1 - n_2)!} p_1^{n_1+n_2} q_1^{2N_1 - 2n_1 - n_2}$$

$$\times \frac{N_2!}{n_3!\, n_4!\, (N_2 - n_3 - n_4)!} p_2^{n_3+n_4} q_2^{2N_2 - 2n_3 - n_4}$$

and the maximum-likelihood estimates of the unknown parameters are

$\hat{q}_1 = n_2/n_1$, $\hat{q}_2 = n_4/n_3$, $\hat{N}_1 = n_1^2/(n_1 - n_2)$ and $\hat{N}_2 = n_3^2/(n_3 - n_4)$.

The probability of survival ϕ between the two experiments is estimated by

$$\hat{\phi} = \hat{N}_2/(\hat{N}_1 - d).$$

Using the delta method it is readily shown that

$$V[\hat{\phi}] \approx \frac{N_1 q_1^2(1+q_1)\phi^2}{p_1^3(N_1 - d)^2} + \frac{N_2 q_2^2(1+q_2)}{p_2^3(N_1 - d)^2}.$$

Example 8.3 Trout: Seber and Le Cren [1967]

Suppose the following catches are observed: 91, 54, 53 and 12 ($d = 0$); then

$$\hat{q}_1 = 54/91, \quad \hat{q}_2 = 12/53, \quad \hat{N}_1 = 91^2/37, \quad \hat{N}_2 = 53^2/41$$

and

$$\hat{\phi} = 53^2(37)/91^2(41) = 0{\cdot}31.$$

Replacing unknown parameters by their estimates, $\hat{V}[\hat{\phi}] = 0{\cdot}0037$ and, assuming $\hat{\phi}$ to be approximately normal, an approximate 95 per cent confidence interval for ϕ is $0{\cdot}31 \pm 0{\cdot}12$.

If $p_1 = p_2 = p$, say, the maximum-likelihood estimates are now given by (Seber and Le Cren [1967: 639])

$$\tilde{q} = (n_2+n_4)/(n_1+n_3),$$
$$\tilde{N}_1 = (n_1+n_2)/(1-\tilde{q}^2),$$
$$\tilde{N}_2 = (n_3+n_4)/(1-\tilde{q}^2),$$

and

$$\tilde{\phi} = \tilde{N}_2/(\tilde{N}_1-d).$$

When d is negligible we have from the delta method

$$V[\tilde{\phi}] = V[(n_3+n_4)/(n_1+n_2)] \approx \frac{q^2N_2(N_1+N_2)}{N_1^3(1-q^2)}.$$

If we can assume that $p_1 = p_2$ we would expect intuitively that the estimate of ϕ for this model would have a smaller variance than that obtained for the model with $p_1 \neq p_2$, since the information from the catches is shared among fewer parameters, i.e. we would expect $V[\tilde{\phi}] < V[\hat{\phi}]$.

One can test the hypothesis $p_1 = p_2$ by assuming that $\hat{p}_i$ is asymptotically normal. Since $\hat{p}_1$ and $\hat{p}_2$ are statistically independent, $\hat{p}_2-\hat{p}_1$ is then asymptotically normally distributed with mean $p_2 - p_1$ and variance estimated by (cf. (7.31))

$$[n_2(n_1+n_2)/n_1^3] + [n_4(n_3+n_4)/n_3^3].$$

Example 8.4 Using the data of the previous example,

$$\hat{p}_2-\hat{p}_1 = 0{\cdot}3670, \quad \hat{V}[\hat{p}_2-\hat{p}_1] = 0{\cdot}0156$$

and

$$z = \frac{0{\cdot}3670}{\sqrt{(0{\cdot}0156)}} = 2{\cdot}93.$$

Since $|z| > 2{\cdot}58$ we reject the hypothesis that $p_1 = p_2$ at the 1 per cent level of significance.

2 *First two catches not returned*

It was assumed above that after the first experiment the total catch was returned to the population. If, however, this catch is not returned, the joint probability function of the n_i now becomes (with $n = \Sigma n_i$)

$$\frac{N_1!}{(\prod_{i=1}^{4} n_i!)(N_1-n)!} p_1^{n_1}(q_1p_1)^{n_2}(q_1^2\phi p_2)^{n_3}(q_1^2\phi q_2p_2)^{n_4}[q_1^2(1-\phi(1-q_2^2))]^{N_1-n}, \tag{8.24}$$

and the maximum-likelihood estimates (also the moment estimates) are

$$\hat{q}_1 = n_2/n_1, \quad \hat{q}_2 = n_4/n_3, \quad \hat{N}_1 = n_1^2/(n_2-n_1)$$

and

$$\hat{\phi} = \frac{n_3^2(n_1 - n_2)}{n_2^2(n_3 - n_4)}.$$

Using the delta method we find that

$$V[\hat{\phi}] \approx \phi^2\left[\frac{4}{N_1 q_1^2 \phi p_2} + \frac{5}{N_1 p_1^2} + \frac{4}{N_1 p_1 q_1} - \frac{3}{N_1 q_1^2 \phi p_2^2}\right].$$

If $p_1 = p_2 = p$ the maximum-likelihood estimates are now

$$\tilde{q} = (n_2 + n_4)/(n_1 + n_3),$$

$$\tilde{N}_1 = (n_1 + n_2)/(1 - \tilde{q}^2),$$

and

$$\tilde{\phi} = (n_3 + n_4)/[\hat{q}^2(n_1 + n_2)];$$

also

$$V[\tilde{\phi}] \approx \frac{\phi^2}{N_1 p q}\left[\frac{1 - \phi q^2}{\phi q(1+q)} + \frac{4(1+q)}{(1+\phi q^2)}\right].$$

Once again we would expect $V[\tilde{\phi}] < V[\hat{\phi}]$.

Standard goodness-of-fit statistics for the multinomial distribution (8.24), and for (8.24) with $p_1 = p_2 = p$, can be calculated using the method of **1.3.6** (*3*). If both statistics are calculated, then the latter minus the former (with one degree of freedom) will give a test of the hypothesis $p_1 = p_2$, *given* that (8.24) is valid.

8.3 SINGLE TAG RELEASE

8.3.1 Continuous sampling

Consider an open population in which continuous exploitation, natural mortality, and possibly recruitment, are the only processes affecting the population size. If a single tag release is made at the beginning of the sampling, then by studying the tagged population only, the effects of recruitment can be ignored. When Z, the total instantaneous mortality rate, is constant throughout the whole experiment, then the methods of **6.2** are applicable here. If Z is not constant, then effort data can be used along with tag returns to estimate k and μ in $Z_i = \mu + k\overset{o}{f_i}$ as follows.

Adopting the same notation and assumptions as in **8.1.1** with the additions

M_i = size of tagged population at the beginning of the ith period,

and

m_i = number of tagged individuals caught in the ith period,

we have from (8.12) Chapman's [1961: 157] regression model

$$E[\log(m_i/f_i)] \approx \log(M_1 k) - \mu(t_{i-1} + t_i)/2 - k(F_{i-1} + F_i)/2.$$

If the proportion of the release recaptured is small, then the dependence between the m_i can be neglected. Therefore, assuming the $\log(m_i/f_i)$ to be independently distributed with approximately constant variance, least-squares estimates of k and μ can be obtained by a straightforward multiple linear regression. Chapman points out that we do not have to use the information on M_1, the size of the tag release, which is an advantage if there is (i) initial tag loss, (ii) initial mortality caused by the tagging procedure, or (iii) a time-lapse before tagged individuals are fully "vulnerable" to the catching process.

Chapman also gives an alternative method when the actual recapture times of tagged individuals are recorded as in **6.2.1.** For each period of length Δ_i, the distribution of the recapture times T_{ij} ($j = 1, 2, \ldots, m_i$) measured from the start of the period is the truncated exponential (cf. p. 272)

$$f(T) = \frac{Z_i \exp(-Z_i T)}{1 - \exp(-Z_i \Delta_i)}, \quad 0 \leqslant T \leqslant \Delta_i.$$

If $\bar{T}_i = \sum_{j=1}^{m_i} T_{ij}/m_i$, then, from Deemer and Votaw [1955], the maximum-likelihood estimate $\hat{Z}_i$ of Z_i is the solution of

$$\frac{1}{Z_i\Delta_i} - \frac{1}{\{\exp(Z_i\Delta_i)\} - 1} = \frac{\bar{T}_i}{\Delta_i}$$

when $0 \leqslant \bar{T}_i < \Delta_i/2$ and 0 otherwise. The authors also provide a table (reproduced as Table **A5**) to facilitate the solution of this equation. For large m_i we have approximately

$$E[\hat{Z}_i] = Z_i = \mu + k\overset{o}{f}_i \tag{8.25}$$

and

$$V[\hat{Z}_i] = \frac{1}{m_i}\left[\frac{1}{Z_i^2} - \frac{\Delta_i^2}{\exp(Z_i\Delta_i)\{1 - \exp(-Z_i\Delta_i)\}^2}\right]^{-1}. \tag{8.26}$$

Therefore, assuming the $\hat{Z}_i$ to be independent, $\hat{Z}_i$ can be plotted against $\overset{o}{f}_i$, and μ and k can be estimated by weighted least squares (cf. **1.3.5**(*1*)), the appropriate weights being estimated from (8.26).

We note that for both the above methods, separate estimation of μ and k is possible only if the effort per unit time $\overset{o}{f}_i$ varies from period to period; in general, the greater the variation, the smaller the variances of the estimates. It was also mentioned above that M_1 can be estimated, so that any well-defined class in the population at the beginning of the sampling can be used as a "tagged" population. In particular, if ageing can be done accurately, a fully recruited age-class can be used. This ensures that M_1 is a reasonable fraction of the population, which is a difficult goal to achieve with tagging experiments in large populations. It also means that the group M_1 is well mixed with the rest of the population, and the assumption that individuals are not affected by tagging as far as catchability is concerned is avoided. However, methods of ageing are time-consuming, and some other correlated variable such as weight or length is often used for age classification (cf. p. 333). This will mean that in applying the second method above, the variances given by (8.26) will be underestimates due to errors in the ageing process and the loss of information through the inevitable grouping of the catch-times T_{ij} which goes with large m_i. Also there is the problem of obtaining f_i, the effort associated with just the age-class only; one such method is suggested in **8.1.1**(*4*).

NON-REPORTING OF TAGS. If the total effort f_i can be classified into two categories, one which has a known proportional tag return of unity or nearly unity, and the other with an unknown reported ratio ρ_i, then a technique due to Paulik [1961] can be used to test whether ρ_i is significantly less than unity. In fisheries the former category would usually be the fraction of the effort inspected by special observers, or it may be the units of gear operated by fishermen who, by virtue of special training, long experience, or a keen interest in the experiment, could be relied upon to turn in all the tags that appear in their catches.

Adopting the notation of **8.1.1**, let $f_i = f_{ia} + f_{ib}$ $(i = 1, 2, \ldots, s)$, where the suffixes a and b denote the above two categories "inspected" and "uninspected" respectively; m_i the number of tags recaptured and r_i the number of tags reported in the ith interval are split up in the same way $(r_{ia} = m_{ia})$. From equations (8.3) and (8.5), with M_i and m_i instead of N_i and n_i, the probabilities that a particular tag is reported from the f_{ia} and f_{ib} units of gear are $kf_{ia}A_i/(Z_i\Delta_i)$ and $k\rho_i f_{ib}A_i/(Z_i\Delta_i)$ respectively, where

$$A_i = \exp(-\mu t_{i-1} - k F_{i-1})[1 - \exp(-Z_i \Delta_i)].$$

Therefore, given that a tag has been reported in the ith interval, the probability that it comes from the f_{ia} units of gear is

$$p_i = f_{ia}/(f_{ia} + \rho_i f_{ib}), \tag{8.27}$$

and the conditional probability function of r_{ia} given r_i is

$$f(r_{ia} \mid r_i) = \binom{r_i}{r_{ia}} p_i^{r_{ia}} q_i^{r_{ib}}.$$

This model is of the same form as (3.28), and the discussion in **3.2.4** concerning the estimation of ρ_i and testing $\rho_i = 1$ is applicable here with p_0 now defined to be f_{ia}/f_i.

If $f_{ib}/f_{ia} = \gamma$ $(i = 1, 2, \ldots, s)$ then $p_i = (1 + \rho_i\gamma)^{-1}$ and we can test the hypothesis $H : \rho_i = \rho$ using the statistic

$$\sum_{i=1}^{s} \frac{(r_{ia} - r_i\hat{p})^2}{r_i\hat{p}\hat{q}},$$

where $\hat{p} = (\Sigma\, r_{ia})/\Sigma\, r_i$ $(= r_a/r$ say). When H is true, the above statistic is approximately χ^2_{s-1} and the recoveries for the whole experiment can be pooled. In this case

$$f(r_a \mid r) = \binom{r}{r_a} p^{r_a} q^{r_b},$$

where $p = (1+\rho\gamma)^{-1}$, and we can test the hypothesis that $\rho = 1$ using the same approach as in **3.2.4** (with $p_0 = (1+\gamma)^{-1}$).

As p_i in (8.27) is unchanged if k is replaced by k_i, the above theory is still valid when the catchability varies from interval to interval.

Example 8.5 Trout: adapted from Paulik [1961]

A biologist wishes to estimate the percentage of tags detected and reported by sportsmen and would like to be 90 per cent sure of discovering a reported ratio of 80 per cent or lower. 2000 marked trout are released in a large lake shortly before the opening of the fishing season, and an unknown proportion of them are expected to survive until opening day. It is assumed that (i) the population is closed, apart from natural mortality and removals by the fishermen, (ii) the catchability of the marked trout is constant within a week but may vary from week to week, and (iii) ρ is constant throughout the season. If at least 1000 tags are expected to be returned, what percentage of the fishermen should be sampled each day? From **A2** with $\alpha = 0{\cdot}05$, $1-\beta = 0{\cdot}90$, $\rho = 0{\cdot}8$ and $r = 1000$, we have $p_0 = 0{\cdot}20$ $(= f_{ia}/f_i)$, so that 20 per cent of the fishermen should be interviewed.

If it can also be assumed that marked and unmarked are equally vulnerable to capturing and have the same natural mortality rate, so that the proportion of tagged in the population remains constant throughout the season, then the method of **3.2.4** is directly applicable here. The biologist must then decide whether to sample the catch or the effort, and this can be done by computing appropriate "cost" curves for the two methods as in Paulik [1961: 826].

8.3.2 Instantaneous samples

Suppose now that "instantaneous" or point samples are removed from the population at times $t_1, t_2, \ldots, t_s$. For a population in which there is just mortality and recruitment (i.e. no migration) Chapman [1965] gives the following method, based on a single release of M_0 marked individuals at time zero, for estimating the various unknown parameters. Let

M_i = size of marked population just before the ith sample,
N_i = size of total population just before the ith sample,
U_i = $N_i - M_i$,
n_i = size of the ith sample,
m_i = number found marked in the ith sample,
u_i = $n_i - m_i$,
K_m = binomial catchability of marked individuals (cf. **7.1.3**(*1*)),
K_u = binomial catchability of unmarked individuals,

f_i = effort applied to the ith sample,
r_i = number of recruits added to the population in the interval $[t_{i-1}, t_i)$ which are alive at time t_i ($t_0 = 0$),
$\Delta_i = t_i - t_{i-1}$

and

U_0 = size of unmarked population at time zero.

We shall assume that:

(1) $K_m = K_u$ (= K say), i.e. all animals are equally catchable, whether marked or unmarked, old or new recruits. Chapman points out that this assumption would not hold if the animals acquire some skill in avoiding being caught, for then new recruits would be more easily caught than "old" members of the population. We also note that for this assumption to hold, either the marked animals or the sampling effort must be distributed randomly.
(2) All individuals, whether marked or unmarked, have the same instantaneous natural mortality rate μ throughout the whole experiment.
(3) Units of effort are additive and are distributed randomly over the population, so that

$$E[u_i \mid U_i] = K f_i U_i$$

and

$$E[m_i \mid M_i] = K f_i M_i.$$

(4) Conditional on fixed M_i, the random variable m_i has a Poisson distribution ($i = 1, 2, \ldots, s$).
(5) There is no loss of marks, and all marks are reported on recovery.

From the above assumptions we have the following deterministic relationships:

$$U_i = (U_{i-1} - u_{i-1})\, e^{-\mu\Delta_i} + r_i \tag{8.28}$$

and

$$M_i = (M_{i-1} - m_{i-1})\, e^{-\mu\Delta_i}, \tag{8.29}$$

for $i = 1, 2, \ldots, s$ ($u_0 = m_0 = 0$). The basic parameters to be estimated are k, U_0, μ and $r_1, r_2, \ldots, r_s$ via the intermediary unknowns $\{M_i\}$ and $\{U_i\}$. We note that although μ and U_1 $(= U_0 \exp(-\mu\Delta_1) + r_1)$ are estimable, U_0 and r_1 cannot be estimated separately without further assumptions or additional information. For example, Chapman suggests that it may be possible to estimate r_1 by extrapolation from $r_2, r_3, \ldots, r_s$ using a low-degree polynomial.

Since the m_i involve only the two parameters k and μ, Chapman derives maximum-likelihood equations from the joint probability function of the m_i:

$$\prod_{i=1}^{s} \left\{ e^{-K f_i M_i} \frac{(K f_i M_i)^{m_i}}{m_i!} \right\},$$

where the M_i satisfy (8.29). It is readily shown that $\hat{K}$ and $\hat{\mu}$, the maximum-likelihood estimates, are solutions of

$$K = \left(\sum_{i=1}^{s} m_i\right) / \left(\sum_{i=1}^{s} f_i M_i\right) \tag{8.30}$$

and

$$\left(\sum_{i=1}^{s} m_i\right)\left(\sum_{i=1}^{s} f_i \frac{dM_i}{d\mu}\right) \Bigg/ \left(\sum_{i=1}^{s} f_i M_i\right) = \sum_{i=1}^{s} \frac{m_i}{M_i} \frac{dM_i}{d\mu}, \tag{8.31}$$

where $\dfrac{dM_1}{d\mu} = -\Delta_1 M_1$

and $\dfrac{-dM_i}{d\mu} = e^{-\mu\Delta_i}\left[\Delta_i(M_{i-1} - m_{i-1}) \dfrac{dM_{i-1}}{d\mu}\right], \quad (i = 2, 3, \ldots, s).$

Equation (8.31) may be solved iteratively as it involves just the single unknown μ and the known parameter M_0. The asymptotic variance–covariance matrix of $\hat{K}$ and $\hat{\mu}$ is the inverse of (cf. (1.7))

$$\begin{bmatrix} \sum_{i=1}^{s} \dfrac{f_i M_i}{K}, & \sum_{i=1}^{s} f_i \dfrac{dM_i}{d\mu} \\ \sum_{i=1}^{s} f_i \dfrac{dM_i}{d\mu}, & \sum_{i=1}^{s} K \dfrac{f_i}{M_i} \cdot \left(\dfrac{dM_i}{d\mu}\right)^2 \end{bmatrix} \tag{8.32}$$

Since the observations u_i alone involve the unknown r_i they must provide all the information necessary to estimate the r_i. This suggests equating each u_i to its expected value, so that, from assumption (3),

$$\hat{U}_i = u_i/(\hat{K} f_i), \quad (i = 1, 2, \ldots, s).$$

and from (8.28)

$$\begin{aligned} \hat{r}_i &= \hat{U}_i - (\hat{U}_{i-1} - u_{i-1}) \exp(-\hat{\mu}\Delta_i) \\ &= \frac{u_i}{\hat{K} f_i} - u_{i-1}\left(\frac{1}{\hat{K} f_{i-1}} - 1\right) \exp(-\hat{\mu}\Delta_i), \quad (i = 2, 3, \ldots, s). \end{aligned} \tag{8.33}$$

Assuming u_i and m_i to be statistically independent, we have, using the delta method (there is a small misprint in Chapman's equation: one f_i should be f_{i-1}),

$$\begin{aligned} V[\hat{r}_i] \approx\ & \frac{1}{K^2 f_i^2} V[u_i] + \frac{1}{K^4}\left(\frac{E[u_i]}{f_i} - \frac{E[u_{i-1}]}{f_{i-1}} e^{-\mu\Delta_i}\right)^2 V[\hat{K}] + \\ & + \left(\frac{1}{K f_{i-1}} - 1\right)^2 e^{-2\mu\Delta_i} V[u_{i-1}] + (E[u_{i-1}])^2 \Delta_i^2 \left(\frac{1}{K f_{i-1}} - 1\right)^2 e^{-2\mu\Delta_i} V[\hat{\mu}] - \\ & - \left[\frac{E[u_i]E[u_{i-1}]\Delta_i}{K^2 f_i}\left(\frac{1}{K f_{i-1}} - 1\right) e^{-\mu\Delta_i} - \frac{(E[u_{i-1}])^2 \Delta_i}{K^2 f_{i-1}}\left(\frac{1}{K f_{i-1}} - 1\right) e^{-2\mu\Delta_i}\right] \operatorname{cov}[\hat{K}, \hat{\mu}]. \end{aligned}$$

If the u_i are assumed to have a Poisson distribution then

$$V[u_i] = E[u_i] = K f_i U_i .$$

The N_i can be estimated either iteratively from

$$\begin{aligned}\hat{N}_i &= \hat{U}_i + \hat{M}_i \\ &= (\hat{N}_{i-1} - n_i) \exp(-\hat{\mu}\Delta_i) + \hat{r}_i,\end{aligned}$$

or more simply by the moment estimate

$$N_i^* = n_i/(\hat{K} f_i).$$

If u_i is very much greater than m_i, so that n_i ($\approx u_i$) is approximately independent of the m_i (and hence of $\hat{K}$), then, using the delta method,

$$V[N_i^*] \approx N_i^2 \left(\frac{V[n_i]}{K^2 f_i^2 N_i^2} + \frac{V[\hat{K}]}{K^2} \right).$$

We also note that the above theory can be applied to unbounded populations if μ represents natural mortality plus permanent emigration and r_i represents recruitment plus immigration of unmarked individuals.

Finally Chapman discusses the problem of testing for what Ricker [1958: 122] calls type C error, in which the marked individuals do not have the same catchability as the unmarked immediately after marking but eventually settle down to have the same catchability at a later stage. Ricker mentions that this error may be due to (i) abnormal behaviour of the marked during the season of their marking, or (ii) initial non-random distribution of the marked combined with a (possibly only temporary) non-random distribution of sampling effort. If the catchability (K_1, say) of marked individuals is different for the first sample, then the above model is modified to the extent of just the single equation

$$E[m_1 | M_1] = K_1 f_1 M_1 .$$

The maximum-likelihood estimates $\tilde{K}$ and $\tilde{\mu}$ for this modified model, together with asymptotic variances and covariances, are still obtained from (8.30), (8.31) and (8.32) but with all summations now running from 2 to s instead of 1 to s. Estimates $\tilde{M}_i$, $\tilde{U}_i$ and $\tilde{r}_i$ are the same functions of $\tilde{\mu}$ and $\tilde{K}$ as before, except for $i = 1$. For this case $\tilde{K}$ is replaced by $\tilde{K}_1 = m_1/(f_1 \tilde{M}_1)$, where $\tilde{M}_1 = M_0 \exp(-\tilde{\mu}\Delta_1)$, so that $\tilde{U}_1 = u_1/(\tilde{K}_1 f_1)$ in the equation (8.33) for $\tilde{r}_2$: $V[\tilde{r}_2]$ has to be modified accordingly.

Chapman gives an approximate test of the hypothesis $K_1 = K$ based on the statistic

$$x = m_1 - \tilde{E}[m_1] = m_1 - \tilde{K}_1 f_1 M_0 \exp(-\tilde{\mu}\Delta_1).$$

When the hypothesis is true, then for large samples x will be approximately normally distributed with mean zero and approximate variance

$$V[x] = V[m_1] + (f_1 M_0)^2 e^{-2\mu\Delta_1} (V[\tilde{K}] - 2K\Delta_1 \operatorname{cov}[\tilde{K}, \tilde{\mu}] + K^2 \Delta_1^2 V[\tilde{\mu}]).$$

Assuming that m_1 has a Poisson distribution then K, μ and $V[m_1]$ can be estimated by $\tilde{K}$, $\tilde{\mu}$ and $\tilde{E}[m_1]$ respectively in $V[x]$.

Example 8.6 Halibut: Chapman [1965]

The data in Table 8.3 are taken from a report of the International Pacific Halibut Commission [1960: Appendix 2]. $M_0 = 1051$ marked fish were released in 1951 and the catches n_i were taken in the successive seasons 1952 to 1957. The catches may be treated as point samples since the duration of the fishing was less than 10 per cent of the year for the first four

TABLE 8.3

Catch and effort data for a halibut fishery (Upper Hecate Strait, 1951–1957): from Chapman [1965: Table 1].

t_i (years)	n_i (million lb)	m_i	f_i (thousand skates)	$\tilde{r}_i$ (million lb)	$\tilde{N}_i$ (million lb)	$\tilde{M}_i$
0	(30·6)	–	(320·8)	–	(58·7)	–
1	30·8	214	251·8	(55·5)	75·3	739·9
2	33·0	139	228·6	57·5	88·8	370·2
3	36·7	65	244·2	53·2	92·5	162·8
4	28·7	22	219·9	41·0	80·3	68·9
5	35·4	14	263·2	46·5	82·8	33·0
6	30·6	7	283·6	33·0	66·4	13·4

$M_0 = 1051$, $\Delta_i = 1$ (year)

years and about 15 per cent of the year for the last two years, and $\Delta_i = 1$ year. The effort data f_i are given in thousands of "skates", while n_i, $\tilde{r}_i$ and $\tilde{N}_i$ in Table 8.3 are recorded in millions of pounds. Since n_i and m_i are measured in different units, namely millions of pounds and numbers respectively, the u_i cannot be calculated. However, Chapman points out that since the m_i by weight amount to no more than a few thousand pounds out of a catch of at least 30 million pounds, n_i is a good approximation to u_i which he used in the following calculations.

To test whether there was less than full availability of tagged fish in 1952 (the first season after tagging), K and μ were estimated on the basis of the 1953 to 1957 recoveries. Thus

$$\tilde{K} = 1{\cdot}6248 \times 10^{-3}, \quad e^{-\tilde{\mu}} = 0{\cdot}704 \quad \text{or } \tilde{\mu} = 0{\cdot}35,$$

and the asymptotic variance–covariance matrix is

$$10^{-8} \times \begin{bmatrix} 1{\cdot}654\,34 & 72{\cdot}385\,33 \\ 72{\cdot}385\,33 & 8948{\cdot}254\,7 \end{bmatrix}.$$

Also

$$\tilde{E}[m_1] = 1{\cdot}6248 \times 10^{-3}\,(251{\cdot}8)(1051)(0{\cdot}704) = 302{\cdot}7,$$

and $V[x]$ is estimated by

$$\widetilde{V}[x] = 302{\cdot}7 + 501{\cdot}7 = 804{\cdot}4.$$

Therefore, to test $K_1 = K$ against the alternative $K_1 < K$ we calculate

$$z = (214 - 302{\cdot}7)/\sqrt{804{\cdot}4} = -3{\cdot}12$$

and, since $z < -2{\cdot}58$, we reject the hypothesis at the 1 per cent level of significance. By comparison, $\widetilde{K}_1 = 0{\cdot}001\ 15$, which is only about 70 per cent of $\widetilde{K}$.

The estimates $\widetilde{M}_i$ and $\widetilde{r}_i$ are given in Table 8.3 and the N_i are estimated by

$$\widetilde{N}_i = n_i/(\widetilde{K} f_i) \quad (i = 1, 2, \ldots, 6).$$

In this experiment catch and effort data were available for time zero (before the mark release), so that N_0 and r_1 could be estimated by

$$\widetilde{N}_0 = n_0/(\widetilde{K} f_0) = 58{\cdot}7$$

and

$$\begin{aligned}\widetilde{r}_1 &= \widetilde{U}_1 - (\widetilde{U}_0 - u_0)\, e^{-\tilde{\mu}} \\ &\approx \widetilde{N}_1 - (\widetilde{N}_0 - n_0)\, e^{-\tilde{\mu}} \\ &= 75{\cdot}3 - (58{\cdot}7 - 30{\cdot}6)(0{\cdot}704) \\ &= 55{\cdot}5.\end{aligned}$$

Finally we note that if the data for the first year are included, thus ignoring the differences in catchability, then the estimates of K and μ are

$$\hat{K} = 1{\cdot}257 \times 10^{-3} \quad \text{and} \quad \hat{\mu} = 0{\cdot}323.$$

CHAPTER 9

CHANGE IN RATIO METHODS

9.1 CLOSED POPULATION

9.1.1 Introduction

Changes in observed sex ratios, age ratios and marked-to-unmarked ratios (the Petersen method) have been widely used to estimate population abundance, productivity and survival probabilities (Hanson [1963], Kelker and Hanson [1964]). For example, the idea that population numbers could be estimated from a knowledge of the sex ratios before and after a differential kill of the sexes was first noted by Kelker [1940, 1944]. A number of variations of this method were then used by Allen [1942], Rasmussen and Doman [1943], Riordan [1948], Petrides [1949], Lauckhart [1950] and Dasmann [1952]; a summary of this early work is given in Hanson [1963]. However, these early methods were based on intuitive notions, so that no estimates of variance were given. It seems that the first stochastic models were developed by Chapman [1954, 1955] for the closed population, and this work was followed up by Lander [1962] and Chapman and Murphy [1965] who considered the problem of estimating mortality rates.

Recently Paulik and Robson [1969] gave an excellent survey of the whole field, pointing out the wide applicability of the change in ratio (CIR) method (variously known as the "dichotomy", "change in composition", "survey–removal" method) and demonstrating how a number of different methods are all variations on the same theme. They suggest that one factor which has inhibited the proper statistical treatment of CIR methods, particularly in the derivation of variances and confidence intervals, is the terminology associated with applications in fish and wildlife management. For example, there is the widespread practice of using one component as a base and expressing the other component or components in terms of this base, which is sometimes arbitrarily standardised; e.g. number of males per 100 females or number of subadults per adult. As also pointed out by Smirnov [1967: 216], this is in direct contrast to the statistical practice of formulating such problems in terms of fractions, probabilities or proportions, for which sampling errors of estimation are more readily calculated.

NOTATION. Consider a closed population consisting of two types of animals. These are designated as x-type and y-type animals; for example, x-type may be males and y-type females, or x-type may be sport fish and y-type coarse fish. Suppose there is a differential change in the ratio of

x-type to y-type animals between time t_1 and time t_2. Let

$$\begin{aligned}
X_i &= \text{number of } x\text{-type animals in the population at time } t_i,\\
Y_i &= \text{number of } y\text{-type animals in the population at time } t_i,\\
N_i &= X_i + Y_i \text{ (total population size at time } t_i),\\
P_i &= X_i/N_i \ (= 1-Q_i),\\
R_x &= X_1 - X_2,\\
R_y &= Y_1 - Y_2,\\
R &= R_x + R_y \ (= N_1 - N_2)
\end{aligned}$$

and $f = R_x/R$ (f is only defined when R_x and R_y have the same sign).

The quantities R_x and R_y, which are assumed to be known, are called the "removals", though, as pointed out by Rupp [1966], the theory below still holds when the removals are negative, that is additions. (Paulik and Robson [1969] define $R_x = X_2 - X_1$, etc., though they use the above notation when discussing the planning of CIR experiments.)

ESTIMATING POPULATION SIZE. Now

$$P_2 = \frac{X_1 - R_x}{N_1 - R} = \frac{P_1 N_1 - R_x}{N_1 - R},$$

or, solving for N_1,

$$N_1 = \frac{R_x - RP_2}{P_1 - P_2}. \tag{9.1}$$

Therefore, if $\hat{P}_1$ and $\hat{P}_2$ are estimates of P_1 and P_2, N_1, X_1 and N_2 can be estimated by

$$\hat{N}_1 = \frac{R_x - R\hat{P}_2}{\hat{P}_1 - \hat{P}_2}, \tag{9.2}$$

$$\hat{X}_1 = \hat{P}_1 \hat{N}_1,$$

and

$$\hat{N}_2 = \hat{N}_1 - R = \frac{R_x - R\hat{P}_1}{\hat{P}_1 - \hat{P}_2}.$$

Example 9.1 Mule deer: Rasmussen and Doman [1943]

In their words: "... on a crowded range near Logan, Utah, a severe loss occurred during the winter of 1938–39. Age counts made before the loss showed 83 fawns per 100 adults and after the loss 53 fawns per 100 adults. In a complete coverage of the area 248 dead fawns and 60 dead adults were counted". Let X and Y represent the fawn and adult populations respectively. Then,

$$\hat{P}_1 = 83/183 = 0{\cdot}4536, \quad \hat{P}_2 = 53/153 = 0{\cdot}3464, \quad R_x = 248, \quad R_y = 60,$$

so that

$$\hat{N}_1 = \frac{248 - 308(0{\cdot}3464)}{0{\cdot}4536 - 0{\cdot}3464} = 1318$$

and

$$\hat{X}_1 = \hat{P}_1 \hat{N}_1 = 598 \text{ fawns.}$$

Example 9.2 Trout: Rupp [1966]

Suppose that 500 sub-catchable trout are stocked in a small pond the same day that anglers remove 160 catchable trout. The proportions of catchable-size trout before and after this day were estimated to be $\hat{P}_1 = 0{\cdot}40$ and $\hat{P}_2 = 0{\cdot}10$ respectively. If X = size of catchable population, then $R_x = 160$, $R_y = -500$, so that

$$\hat{N}_1 = \frac{160 - (160-500)(0{\cdot}10)}{0{\cdot}40 - 0{\cdot}10} = 647$$

and

$$\hat{X}_1 = 0{\cdot}40(647) = 259.$$

VARIANCES. Using the delta method it is readily shown that for independent estimates of P_1 and P_2,

$$V[\hat{N}_1] \approx (P_1 - P_2)^{-2} \{N_1^2 V[\hat{P}_1] + N_2^2 V[\hat{P}_2]\} \tag{9.3}$$

and

$$V[\hat{X}_1] \approx (P_1 - P_2)^{-2} \{N_1^2 P_2^2 V[\hat{P}_1] + N_2^2 P_1^2 V[\hat{P}_2]\}. \tag{9.4}$$

We shall now consider various methods of obtaining the estimates $\hat{P}_i$.

9.1.2 Binomial sampling(*)

1 Chapman's model

Suppose that a random sample of n_i animals is taken with replacement at time t_i, and x_i animals are found to be of x-type. Then $\hat{P}_i = x_i/n_i$ is an unbiased estimate of P_i with variance

$$V[\hat{P}_i] = P_i Q_i/n_i$$

and (9.3) reduces to

$$V[\hat{N}_1] \approx (P_1 - P_2)^{-2} \left\{\frac{X_1 Y_1}{n_1} + \frac{X_2 Y_2}{n_2}\right\}. \tag{9.5}$$

Equation (9.2) can also be written as

$$\hat{N}_1 = \frac{n_1(R_x y_2 - R_y x_2)}{x_1 y_2 - x_2 y_1},$$

where $y_i = n_i - x_i$, which avoids the round-off errors which may arise in calculating the $\hat{P}_i$. However, the formula (9.2) is simple to use and is quite satisfactory, provided that sufficient decimal places are carried over; for simplicity we shall quote estimates to 4 decimal places throughout this chapter, though in practice 5 or 6 decimal places are often more appropriate.

For the case of binomial sampling, $\hat{N}_1$ and $\hat{X}_1$ were first obtained by Chapman [1954] who showed that they are maximum-likelihood estimates for

(*) For an extension see Otis [1980], Additional References.

the binomial model

$$f(x_1, x_2 \mid \{X_i, Y_i, n_i\}) = \prod_{i=1}^{2} \binom{n_i}{x_i} \left(\frac{X_i}{N_i}\right)^{x_i} \left(\frac{Y_i}{N_i}\right)^{y_i}. \tag{9.6}$$

If $\hat{Y}_i = \hat{N}_i - \hat{X}_i$, the estimates can also be derived from the intuitive relations

$$\frac{x_1}{\hat{X}_1} = \frac{y_1}{\hat{Y}_1} \left(= \frac{n_1}{\hat{N}_1}\right) \tag{9.7}$$

and

$$\frac{x_2}{\hat{X}_1 - R_x} = \frac{y_2}{\hat{Y}_1 - R_y} \left(= \frac{n_2}{\hat{N}_1 - R}\right). \tag{9.8}$$

The binomial model is applicable when, for example, animals are just observed, so that a particular animal may be seen more than once. It can also be used as a reasonable approximation to the more realistic situation of sampling without replacement, provided that less than about 10 per cent of the population is sampled. When sampling is without replacement the probability function (9.6) must be replaced by

$$\prod_{i=1}^{2} \binom{X_i}{x_i}\binom{N_i - X_i}{n_i - x_i} \Big/ \binom{N_i}{n_i}.$$

However, it transpires that the maximum-likelihood estimates of N_1 and X_1 for this model are still $\hat{N}_1$ and $\hat{X}_1$. The asymptotic variances are now given by (9.3) and (9.4), with

$$V[\hat{P}_i] = \frac{P_i Q_i}{n_i} \cdot \frac{(N_i - n_i)}{(N_i - 1)},$$

which can be estimated by

$$\hat{V}[\hat{P}_i] = \frac{\hat{P}_i \hat{Q}_i}{n_i - 1} \left(1 - \frac{n_i}{\hat{N}_i}\right).$$

So far it has been assumed that the sample sizes n_1 and n_2 are chosen in advance, and this will be the case if the experiment is planned using the methods of section 5 below. However, if the size of the sample is determined by the amount of effort available for sampling, then n_1 and n_2 are strictly random variables. In this case Chapman and Murphy [1965] assume that x_1, x_2, y_1 and y_2 are all independent Poisson variables, and we find that under this assumption the conditional distribution of x_i given $n_i = x_i + y_i$ is binomial (cf. **1.3.7** (*1*)). But when n_1 and n_2 are random variables it can be shown, using arguments similar to those given in **3.1.3**, that the variance estimates of $\hat{N}_1$ and $\hat{X}_1$ are virtually unchanged.

Example 9.3 Pheasants: Paulik and Robson [1969]

For a hypothetical pheasant (*Phasianus colchicus*) population in a pre-season survey, 600 of 1400 mature birds seen are cocks. In a post-season survey of 2000 mature birds only 200 are cocks. From a complete check it was found that during the hunting season 8000 cocks and 500 hens were killed. Therefore if X and Y, respectively, represent the cocks and hens, we have

$$\hat{P}_1 = 600/1400 = 0{\cdot}4286, \quad \hat{P}_2 = 200/2000 = 0{\cdot}1000, \quad R_x = 8000, \quad R_y = 500.$$

Hence
$$\hat{N}_1 = \frac{800 - 8500(0{\cdot}1)}{0{\cdot}4286 - 0{\cdot}1000} = 21\,761,$$
$$\hat{X}_1 = (0{\cdot}4286)\,21\,761 = 9326,$$
and
$$\hat{N}_2 = 21\,761 - 8500 = 13\,261.$$

Assuming binomial sampling:
$$\hat{V}[\hat{P}_1] = (0{\cdot}4286)(0{\cdot}5714)/1400 = 0{\cdot}000\,174\,9,$$
and
$$\hat{V}[\hat{P}_2] = (0{\cdot}1000)(0{\cdot}9000)/2000 = 0{\cdot}000\,045\,0.$$

so that, substituting in (9.3), $V[\hat{N}_1]$ is estimated by

$$\hat{V}[\hat{N}_1] = (0{\cdot}4286 - 0{\cdot}1000)^{-2}[(21\,761)^2(0{\cdot}000\,174\,9) + (13\,261)^2(0{\cdot}000\,045\,0)]$$
$$= 840\,571$$

and hence
$$\hat{\sigma}[\hat{N}_1] = \sqrt{\hat{V}[\hat{N}_1]} = 917.$$

2 *Underlying assumptions*

The basic assumptions underlying (9.6) are:

(i) The population is closed.

(ii) All animals have the same probability of being caught in the ith sample ($i = 1, 2$).

(iii) The removals R_x and R_y are known exactly.

ASSUMPTION (i). With regard to the first assumption, (9.6) will still hold when there is mortality both between time t_1 and the time of removal, and between the time of removal and time t_2, provided that in each period the survival rates are the same for both types of animal (i.e. P_1 and P_2 remain constant). However, N_1 now refers to the population size just prior to the removal, as in the following example.

Example 9.4 Silver salmon: Paulik and Robson [1969]

Suppose a hatchery marks and releases 250 000 silver salmon smolts that are about 18 months old. Silver salmon mature at age -2 and age -3, and the fish maturing at age -2 are predominantly males. These precocious males or "jacks" are usually too small to be taken by the fishery, so that the number returning to the hatchery represents the total number of males "removed" from the ocean population of marked age -2 silver salmon. Suppose that 5000 marked age -2 jacks and no age -2 females return to the hatchery, and the following year both the catch and the escapement to the hatchery are sampled to determine the fraction of marked age -3 silvers that are males; four hundred of 1000 marked age -3 silvers sampled are found to be males. Assume that the sex ratio in the smolts is known to be 1:1 and assume further that there is no difference between the ocean survival of male and female marked silver salmon during both their first 6 months (i.e. P_1 constant) and their last 12 months (i.e. P_2 constant) in the ocean. Then $P_1 = \frac{1}{2}$, $\hat{P}_2 = 400/1000 = 0{\cdot}40$, $R_x = 5000$, $R_y = 0$, and the estimated number of marked age -2 silver salmon in the ocean just before the jacks left the population is

$$\hat{N}_1 = \frac{5000(1-0{\cdot}40)}{0{\cdot}50 - 0{\cdot}40} = 30\ 000.$$

Assuming binomial sampling and setting $V[\hat{P}_1] = 0$ (as P_1 is known), we have from (9.3)

$$\hat{V}[\hat{N}_1] = (0{\cdot}10)^{-2}\{25\ 000^2(0{\cdot}4)(0{\cdot}6)/1000\} = 15\times 10^6,$$

and $$\hat{\sigma}[\hat{N}_1] = 3873.$$

An estimate of the probability of survival during early ocean life is

$$\hat{N}_1/(250\ 000) = 0{\cdot}12.$$

(The above example is based on Murphy [1952].)

ASSUMPTION (ii). This assumption implies that (a) the probability of capture is the same for all animals of a given type, and (b) the two types of animal have the same probability of capture. (a) may be realistic if care is exercised in the sampling procedure, though (b) may not be true in some situations. For example, cock pheasants are more easily seen than hen pheasants; if the two types of animal refer to different species they may be sampled at different rates, using possibly different trapping methods. Suppose then that λ is the ratio of the probability of sampling y-type animals to the probability of sampling x-type animals, and let $d = 1 - \lambda$. Then x_i now has a binomial distribution with parameters n_i and $P_{i\lambda} = X_i/(X_i + \lambda Y_i) = X_i/(N_i - dY_i)$. Therefore, replacing random variables by their expected values, we find that, for large n_i, $\hat{N}_1$, $\hat{X}_1$ and $\hat{N}_2$ are asymptotically unbiased estimates of (cf. Chapman [1955: 281])

$$N_1' = \frac{R_x - RP_{2\lambda}}{P_{1\lambda} - P_{2\lambda}} = (X_1 + \lambda Y_1)\left(\frac{N_1R_x - X_1R - dY_1R_x + dR_xR_y}{N_1R_x - X_1R - dY_1R_x + dX_1R_y}\right),$$

$$X_1' = P_{1\lambda}N_1' = X_1\left(\frac{N_1R_x - X_1R - dY_1R_x + dR_xR_y}{N_1R_x - X_1R - dY_1R_x + dX_1R_y}\right),$$

and

$$N_2' = \frac{R_x - RP_{1\lambda}}{P_{1\lambda} - P_{2\lambda}} = (X_2 + \lambda Y_2)\left(\frac{N_1R_x - X_1R - dY_1R_x}{N_1R_x - X_1R - dY_1R_x + dX_1R_y}\right).$$

In general, $X'_1 \neq X_1$, except in one important case, i.e. when $R_y = 0$. Unfortunately, for this case $N'_i = X_i + \lambda Y_i$, so that N'_i may be widely different from N_i when $\lambda \neq 1$. However, when $R_y = 0$, $\hat{X}_1$ is not only asymptotically unbiased but the usual estimate of variance, namely (cf. (9.4))

$$\hat{V}[\hat{X}_1] = (\hat{P}_1 - \hat{P}_2)^{-2}\{\hat{N}_1^2 \hat{P}_2^2 \hat{V}[\hat{P}_1] + \hat{N}_2^2 \hat{P}_1^2 \hat{V}[\hat{P}_2]\},$$

is also an asymptotically unbiased estimate of

$$(P_{1\lambda} - P_{2\lambda})^{-2}\{(X_1 + \lambda Y_1)^2 P_{2\lambda}^2 V[\hat{P}_1] + (X_2 + \lambda Y_2)^2 P_{1\lambda}^2 V[\hat{P}_2]\}$$

which, by the delta method, is the true asymptotic variance of $\hat{X}_1$.

Example 9.5 Brook trout and cisco: Paulik and Robson [1969]

Paulik and Robson consider the following hypothetical data for a lake containing brook trout (*Salvelinus fontinalis*) and cisco (*Coregonus* sp.). Gill nets set before the fishing season caught 150 brook trout and 50 cisco. During the first three weeks of the season 700 brook trout were taken out of the lake. Gill nets set at the end of the three weeks captured 30 cisco and 30 brook trout. As the brook trout are about three times as susceptible to the gill nets as the cisco, N_1 cannot be estimated. However, $\hat{P}_1 = 150/200 = 0{\cdot}75$, $\hat{P}_2 = 30/60 = 0{\cdot}50$, $R_x = 700$, $R_y = 0$ and the number of brook trout before the season is estimated to be

$$\hat{X}_1 = \frac{0{\cdot}75(700)(1 - 0{\cdot}50)}{0{\cdot}75 - 0{\cdot}50} = 1050.$$

Assuming binomial sampling, we have $\hat{\sigma}[\hat{X}_1] = 93$.

Example 9.6 Mule deer (*Odocoileus hemonius*): Riordan [1948]

The adult sex ratios were estimated before and after the hunting season using aerial counts. Letting X and Y denote the buck and doe populations respectively, the following data were obtained for a particular area: $n_1 = 657$, $x_1 = 220$; $n_2 = 1011$, $x_2 = 129$; $R_x = 5500$ and $R_y = 0$. Then

$$\hat{N}_1 = 23\,150, \quad \hat{X}_1 = 7752$$

$$\hat{\sigma}[\hat{N}_1] = 2240 \quad \text{and} \quad \hat{\sigma}[\hat{X}_1] = 398.$$

Using aerial sighting, the assumption of "equicatchability" for the two sexes becomes the assumption that bucks and does are equally distinguishable. Possible departures from this assumption can occur through the difficulty of distinguishing between yearling bucks with spike antlers and does, and between does and fawns when the deer are bunched together in a large group. However, as shown above, since $R_y = 0$, $\hat{X}_1$ and $\hat{\sigma}[\hat{X}_1]$ are both robust with regard to such departures.

For a number of areas, ground surveys were also carried out and compared with the aerial survey data. Although in some areas there were significant differences in the estimated proportions of bucks, there seemed to be little difference in overall accuracy between the two methods. However, the aerial census has the following advantages over the ground census:

(1) All areas are equally accessible by air, thus allowing for more uniform sampling; this is not true to the same extent for ground observers.

(2) The sampling is more likely to be random by air since ground observers tend to drive the animals ahead, thus increasing the opportunity for the more wary ones to get away without being counted.

(3) The aerial method provides a larger number of animals for classification in a given time or at a given cost.

(4) In classifying animals on the ground, using binoculars at long distances, there is the possible error introduced by the fact that antlers make the bucks distinguishable at a greater distance than does and fawns can be distinguished from one another. Since only animals positively sex-identified are tallied, this would lead to an underestimate of the does and fawns in comparison with the number of bucks. However, with aerial counts one can simply cruise over the area at altitudes between 30 and 200 feet.

(5) If sex-ratio studies are made in the rutting season, the bucks, and especially the older ones, appear to be less cautious and are much easier to approach on foot or by automobile than at other times of the year. This would lead to an overestimate of the proportion of bucks.

ASSUMPTION (iii). This assumption may be false as R_x and R_y could be underevaluated because of unknown natural mortality, unreported kills, and possible "crippling losses". The main method of determining the total kill in a hunting season is by questionnaire. However, this can introduce all kinds of bias which must be allowed for if one is to have any confidence in the final values of R_x and R_y. To begin with, there is the non-response bias due to some hunters not returning their questionnaires (cf. Sen [1971a, b]); this may be minimised using repeated mailings (MacDonald and Dillman [1968]). Secondly, there may be certain biases present in the completed questionnaires. For example, Atwood [1956] detected three sorts of bias: (1) Prestige bias, which arose from pride and caused hunters to claim a higher number of daily-bag limits than they had actually obtained. The effect of this bias was greatest where the daily bag limit was low. (2) Type I-Memory bias, which was due to failure of memory acting together with overstatement to cause hunters to report their bag as a number ending in zero or five, and somewhat above the true figure. (3) Type II-Memory bias, which was due to the fact that hunters recalled small numbers more accurately than large ones, and further, when the total seasonal bag was large, it was reported as larger still. The above types of response bias can be partly allowed for by, for example, disregarding the replies from hunters who reported their kill as multiples of five or the daily-bag limit. MacDonald and Dillman [1968] suggest evaluating the questionnaires in the light of those returned by hunters whose performance is known. Reward banding has also been used as a means of checking reporting rates (Tomlinson [1968]).

The hunting kill can also be determined from bag checks in the field or by using checking stations. Bag checks are useful in determining the kill from a particular population or area and are commonly used for small game with wardens or conservation officers making the check in the course of their patrols. Possible sources of error are: inadequacy of the selection of hunters to be questioned; inadequacy of the time of questioning with regard to the whole hunting season; and interviewer bias. Checking stations are essentially roadblocks where it may or may not be compulsory for the hunter to stop. This method is particularly useful if a large area is accessible by only a few roads. However, check-station operation is expensive in manpower if the season is long; also, since hunters tend to move *en masse*, time-consuming lineups may form at the station in the evenings. Where it is not compulsory for a hunter to stop at a checking station, the tendency will be for the successful hunters to stop, and for those who are unsuccessful, or have inferior animals or over-bag animals, to pass on.

In addition to the animals which are killed and taken home, there are those which are killed, but lost or abandoned. Some crippled animals are subsequently found or shot by other hunters and so appear in the bag, while other crippled animals subsequently recover. The "crippling loss", defined to be those animals which are shot, do not recover, and do not appear in the bag, can be estimated from hunter reports or by searching sample plots or transects for carcases (e.g. Whitlock and Eberhardt [1956], Robinette *et al.* [1954, 1956]). The Petersen method using tagged carcases randomly placed has also been used in conjunction with other sampling methods. Small game found dead after the hunting season can be X-rayed for the presence of lead shot; X-ray methods have been used for estimating illegal kill (e.g. Chesness and Nelson [1964]).

The above discussion is based on Geis and Taber [1963] and the reader is referred to their chapter and MacDonald and Dillman [1968] for further references and details. To evaluate the effect of bias on our estimates $\hat{N}_1$ and $\hat{X}_1$, suppose that the actual removals are $R_x + \Delta_x$ and $R_y + \Delta_y$. Then (from equation (9.1)),

$$\begin{aligned}
N_1 &= \frac{R_x + \Delta_x - P_2(R + \Delta_x + \Delta_y)}{P_1 - P_2} \\
&= \frac{R_x - P_2 R}{P_1 - P_2} + \frac{\Delta_x - P_2(\Delta_x + \Delta_y)}{P_1 - P_2} \\
&= \frac{R_x - P_2 R}{P_1 - P_2}\left\{1 + \frac{\Delta_x - P_2(\Delta_x + \Delta_y)}{R_x - P_2 R}\right\} \\
&= \frac{R_x - P_2 R}{P_1 - P_2}\{1 + b_1\}, \text{ say.}
\end{aligned}$$

Therefore, asymptotically (i.e. large n_i),

$$E[\hat{N}_1] = \frac{R_x - P_2 R}{P_1 - P_2} = \frac{N_1}{(1 + b_1)},$$

and

$$E[\hat{X}_1] = P_1 E[\hat{N}_1] = \frac{X_1}{(1+b_1)}.$$

Also, using a similar argument, it can be shown that asymptotically

$$E[\hat{N}_2] = N_2/(1+b_2),$$

where

$$b_2 = \frac{\Delta_x - P_1(\Delta_x + \Delta_y)}{R_x - P_1 R}.$$

Chapman [1955] has evaluated b_1 for two special cases:

Suppose that the percentage of unreported kills is the same for both classes so that

$$\frac{\Delta_x}{R_x} = \frac{\Delta_y}{R_y} = \frac{(\Delta_x + \Delta_y)}{R} = k, \text{ say.}$$

Then $b_1 = b_2 = k$, and if k is small and positive, N_1, X_1 and N_2 are all slightly underestimated.

Alternatively, if $R_y = 0$, $\Delta_x/R_x = k$ and the percentage mortality or unknown removal is the same for both classes (i.e. $\Delta_x/X_1 = \Delta_y/Y_1$), then it can be shown that

$$(1+b_1)^{-1} = 1 - (\Delta_x/X_1) = 1 - (kR_x/X_1),$$

(which is of smaller order than $(1+k)^{-1} \approx 1-k$) and $b_2 = 0$. Therefore if k is small, any unreported removals will have little effect on $\hat{N}_1$, $\hat{X}_1$ and their variance estimates (which require both $\hat{N}_1$ and $\hat{N}_2$; cf. (9.3) and (9.4)).

Example 9.7 Pheasants: Paulik and Robson [1969]

In Example 9.3 (p. 356), $\hat{N}_1 = 21\,761$. However, if the game biologist thinks that the estimated number of hens shot is about 15 per cent too low because of some unreported illegal kills and that the estimated number of cocks is 15 per cent low because of unobserved deaths of wounded birds, then $b_1 = k = 0{\cdot}15$ and, correcting $\hat{N}_1$ for bias, we have

$$\hat{N}_1(1+b_1) = 21\,761\,(1{\cdot}15) = 25\,069.$$

If the illegal kill of hens is considered to be much greater, so that $k_x = 0{\cdot}15$ and $k_y = 1{\cdot}00$, say, then

$$\Delta_x = k_x R_x = (0{\cdot}15)\,8000 = 1200,$$

$$\Delta_y = k_y R_y = (1{\cdot}00)\,500 = 500,$$

and recalling that $\hat{P}_2 = 0{\cdot}10$, b_1 is estimated by

$$\hat{b}_1 = \frac{1200 - (0{\cdot}10)(1200+500)}{8000 - (0{\cdot}10)(8000+500)} = 0{\cdot}144.$$

Hence

$$\hat{N}_1(1+\hat{b}_1) = 21\,761(1{\cdot}144) = 24\,895.$$

3 Optimum allocation

If N_1 is to be estimated with minimum variance, subject to $n_1 + n_2$ fixed, then, using (9.5), Chapman [1955] shows that the optimal sample allocation is

$$\frac{n_2}{n_1} = \left(\frac{X_2 Y_2}{X_1 Y_1}\right)^{\frac{1}{2}} = \left(1 - \frac{R_x}{X_1}\right)^{\frac{1}{2}} \left(1 - \frac{R_y}{Y_1}\right)^{\frac{1}{2}},$$

and n_2 should be chosen smaller than n_1. For the special case $R_y = 0$,

$$\frac{n_2}{n_1} = \left(1 - \frac{R_x}{X_1}\right)^{\frac{1}{2}}$$

and Chapman's table of values

R_x/X_1	0·5	0·4	0·3	0·2	0·1	0·05	0·02	0·01
n_2/n_1	0·71	0·77	0·84	0·89	0·95	0·975	0·990	0·995

suggests that n_2 should then be only slightly less than n_1.

However, if X_1 is to be estimated, $V[\hat{X}_1]$ is minimised, subject to $n_1 + n_2$ fixed, if

$$\frac{n_2}{n_1} = \frac{N_2}{N_1}\left(\frac{X_1 Y_2}{X_2 Y_1}\right)^{\frac{1}{2}}.$$

For the case $R_y = 0$, $Y_1 = Y_2$ and

$$\frac{n_2}{n_1} = \left(1 - \frac{R_x}{N_1}\right)\left(1 - \frac{R_x}{X_1}\right)^{-\frac{1}{2}}.$$

Table 9.1 (reproduced from Chapman) gives the ratio n_2/n_1 for different values of R_x/N_1 and $P_1 = X_1/N_1$, and we see that when $P_1 < 0{\cdot}5$, the second sample should be slightly larger than the first. Therefore if $R_y = 0$ and both $\hat{N}_1$ and $\hat{X}_1$ are required, it would seem that choosing $n_1 = n_2$ represents a near optimum allocation of effort.

4 Confidence intervals

Paulik and Robson [1969] present three methods of constructing confidence intervals for N_1 which we now discuss.

The simplest confidence interval for N_1 is obtained by assuming $\hat{N}_1$ to be asymptotically normal, so that an approximate $100(1-\alpha)$ per cent confidence interval is given by

$$\hat{N}_1 \pm z_{\alpha/2}\,\hat{\sigma}[\hat{N}_1]. \tag{9.9}$$

However, when the samples are too small for the delta method to apply in the derivation of $V[\hat{N}_1]$, it is still possible to obtain confidence intervals by employing modified versions of the delta method. One such modification is to find a monotonic function of $\hat{N}_1$ that approaches normality faster than $\hat{N}_1$;

TABLE 9.1

Optimum sample ratio n_2/n_1 for the estimate of X_1 when there is no removal of y-type animals: from Chapman [1955: Table 2].

	R/N_1						
P_1	0·25	0·20	0·15	0·10	0·05	0·02	0·01
0·1	–	–	–	–	1·344	1·096	1·043
0·2	–	–	1·700	1·273	1·097	1·033	1·015
0·3	1·838	1·386	1·202	1·102	1·041	1·014	1·007
0·4	1·225	1·132	1·075	1·039	1·016	1·005	1·003
0·5	1·061	1·032	1·016	1·007	1·001	1·000	1·000
0·6	0·982	0·980	0·982	0·986	0·993	0·997	0·998
0·7	0·935	0·947	0·959	0·972	0·985	0·994	0·997
0·8	0·905	0·924	0·943	0·963	0·981	0·993	0·996
0·9	0·882	0·907	0·931	0·954	0·977	0·991	0·996

the monotonicity property is then employed to convert the limits for the same function of N_1 into limits for N_1. (This procedure has already been used a number of times, e.g. **3.1.4** and equation (4.8).) In the present situation the distribution of $1/\hat{N}_1$ approaches normality faster than the distribution of $\hat{N}_1$; as Paulik and Robson point out, it is generally advisable to avoid using estimates with high variability in the denominator of a fraction (in this case in $\hat{P}_1 - \hat{P}_2$). Therefore, using the delta method,

$$V[1/\hat{N}_1] \approx \frac{(R_x - P_1 R)^2}{(R_x - P_2 R)^4} V[\hat{P}_2] + \frac{1}{(R_x - P_2 R)^2} V[\hat{P}_1]$$

and the approximate $100(1-\alpha)$ per cent confidence interval for $1/N_1$ is

$$(1/\hat{N}_1) \pm z_{\alpha/2}\, \hat{\sigma}[1/\hat{N}_1]. \tag{9.10}$$

This can be inverted and reversed to give an interval for N_1.

A minor objection to the above interval is that when it is inverted it may be quite skewed about the point estimate $\hat{N}_1$. One way of circumventing this difficulty is to use asymmetrical limits for $(1/N_1)$, so that when inverted, $\hat{N}_1$ will be in the centre of the interval obtained. To do this we must choose α_1 so that

$$\hat{N}_1 - \{(1/\hat{N}_1) + z_{\alpha-\alpha_1} \hat{\sigma}[1/\hat{N}_1]\}^{-1} = \{(1/\hat{N}_1) - z_{\alpha_1} \hat{\sigma}[1/\hat{N}_1]\}^{-1} - \hat{N}_1.$$

(Our definition of z_{α_1}, given in **1.2.1**, differs from that of Paulik and Robson.) Rearranging the above equation, we find that α_1 is the solution of

$$h(\alpha_1) = 2\hat{N}_1 \hat{\sigma}(1/\hat{N}_1) \quad (= d, \text{ say}) \tag{9.11}$$

where $$h(\alpha_1) = \frac{z_{\alpha-\alpha_1} - z_{\alpha_1}}{z_{\alpha_1} z_{\alpha-\alpha_1}}.$$

Paulik and Robson point out that since the denominator of $h(\alpha_1)$ changes slowly in comparison with the numerator, (9.11) can usually be solved by linear interpolation between $h(\alpha_{11})$ and $h(\alpha_{12})$ using two initial guesses α_{11} and α_{12}, such that $h(\alpha_{11}) < d$ and $h(\alpha_{12}) > d$. Table 1 of the Kelley Statistical Tables (Kelley [1948: 37]) tabulates tail probabilities such as α_1 by 0·0001 increments and this provides a useful aid in the solution of (9.11). We note that the symmetrical interval will always be wider than the unsymmetrical interval based on (9.10).

When R_x and R_y have the same sign, a more exact confidence interval can be derived. Let $f = R_x/R$; then, from (9.1),

$$N_1 = \frac{R(f-P_2)}{P_1 - P_2}.$$

Since

$$w = R(f-\hat{P}_2) - N_1(\hat{P}_1 - \hat{P}_2)$$

is a linear function of binomial random variables, it is approximately normally distributed with mean zero and variance

$$V[w] = N_1^2 V[\hat{P}_1] + (N_1 - R)^2 V[\hat{P}_2].$$

Therefore, from the probability statement

$$\Pr[w^2 \leqslant z_{\alpha/2}^2 V[w]] \approx 1 - \alpha, \tag{9.12}$$

the upper and lower limits for N_1 can be found as the roots of a quadratic equation, namely

$$N_L = (a-b)/c, \quad N_U = (a+b)/c,$$

where $$a = \hat{N}_1 - R\left(\frac{z_{\alpha/2}\,\hat{\sigma}[\hat{P}_2]}{\hat{P}_1 - \hat{P}_2}\right)^2,$$

$$b = \left[\left(\frac{z_{\alpha/2}\,\hat{\sigma}[\hat{P}_2]}{\hat{P}_1 - \hat{P}_2}\right)^2 \left((\hat{N}_1 - R)^2 + \frac{\hat{V}[\hat{P}_1]}{\hat{V}[\hat{P}_2]}\left\{\hat{N}_1^2 - \left(\frac{z_{\alpha/2}\,R\,\hat{\sigma}[\hat{P}_2]}{\hat{P}_1 - \hat{P}_2}\right)^2\right\}\right)\right]^{\frac{1}{2}},$$

and

$$c = 1 - \left(\frac{z_{\alpha/2}}{\hat{P}_2 - \hat{P}_1}\right)^2 (\hat{V}[\hat{P}_1] + \hat{V}[\hat{P}_2]).$$

Example 9.8

Paulik and Robson demonstrate the calculation of the above three confidence intervals for $\alpha = 0{\cdot}05$ using the data of Example 9.3. We recall that $R_x = 8000$, $R_y = 500$, $f = 0{\cdot}9412$, $\hat{P}_1 = 0{\cdot}4286$, $\hat{P}_2 = 0{\cdot}1$, $\hat{V}[\hat{P}_1] = 0{\cdot}000\,174\,9$, $\hat{V}[\hat{P}_2] = 0{\cdot}000\,045\,0$, $\hat{N}_1 = 21\,761$ and $\hat{\sigma}[\hat{N}_1] = 917$.

The interval (9.9) is $21\,761 \pm 1{\cdot}96(917)$, i.e. $(19\,964 < N_1 < 23\,558)$.

The interval (9.10) is $(0{\cdot}4216 \times 10^{-4} < 1/N_1 < 0{\cdot}4975 \times 10^{-4})$ which inverted leads to $(20\,101 < N_1 < 23\,720)$. To calculate the symmetrical 95 per cent confidence interval for N_1, we calculate $2\hat{N}_1\,\hat{\sigma}[1/\hat{N}_1] = 0{\cdot}084\,25$, and interpolating between $h(0{\cdot}030) = 0{\cdot}0448$ and $h(0{\cdot}35) = 0{\cdot}0911$ obtain

$\alpha_1 = 0{\cdot}0343$ ($h(0{\cdot}0343) = 0{\cdot}0845$). Table I of Kelley [1948: 131, 134] gives $z_{0\cdot 0343} = 1{\cdot}8210$ and $z_{0\cdot 0157} = 2{\cdot}1520$, so that the 95 per cent confidence interval

$$(1/\hat{N}_1) - z_{0\cdot 0343}\,\hat{\sigma}[1/\hat{N}_1] < 1/N_1 < (1/\hat{N}_1) + z_{0\cdot 0157}\,\hat{\sigma}[1/\hat{N}_1]$$

is found to be $(0{\cdot}4242 \times 10^{-4} < 1/N_1 < 0{\cdot}5013 \times 10^{-4})$ which, inverted, yields $(19\,948 < N_1 < 23\,574)$.

Using the third interval with $z_{\alpha/2} = 1{\cdot}96$, we find that $a = 21\,747$, $b = 1797$ and $c = 0{\cdot}9922$; $N_L = 20\,108$ and $N_U = 23\,730$. Thus the quadratic method leads to the confidence interval $(20\,108 < N_1 < 23\,730)$.

COMPARISON OF INTERVALS. Of the three methods of interval estimation, the interval derived from (9.12) is more accurate than the intervals obtained from (9.9) and (9.10). The quadratic method is more accurate in the sense that if all the underlying assumptions are satisfied, particularly the assumption of binomial sampling, the probability that the interval (N_L, N_U) covers the true population size will be closer to the nominal probability of $1 - \alpha$ than either of the other two intervals. For most situations the interval found from using the inverse is only slightly less accurate than that derived from the quadratic. When sample sizes are large, the three methods will yield practically identical confidence intervals, while for small samples the accuracy of (9.9) is significantly poorer than either of the other two intervals.

Paulik and Robson give a helpful numerical comparison of the three intervals by using the estimates given in Example 9.8, but allowing n_1 and n_2 to vary. Suppose that L1 is the length of the interval (9.9), L2 is the length of the interval obtained by inverting (9.10) and L3 is the quadratic interval. Then from Paulik and Robson we have Table 9.2.

TABLE 9.2

Comparison of the lengths of three confidence intervals: from Paulik and Robson [1969: 22].

$n_1 = n_2$	L1	L2	L3
100	13 696	15 202	15 516
200	9 685	10 189	10 287
500	6 125	6 249	6 272
1000	4 331	4 374	4 382

5 Planning CIR experiments

We shall now discuss the problem of determining the number of animals that must be sampled in order to obtain an estimate of N_1 with a given accuracy when R_x and R_y are both positive and the x-category is chosen so that $f = R_x/R$ is greater than P_1 (i.e. $f > P_1 > P_2$). The following discussion is based on Paulik and Robson [1969: 22–27].

Let $(1-\alpha)$ be the probability that $\hat{N}_1$ will not differ from N_1 by more than 100 ϵ per cent; that is

$$\Pr\left[-\epsilon < \frac{\hat{N}_1 - N_1}{N_1} < \epsilon\right] \geqslant 1 - \alpha, \tag{9.13}$$

where α and ϵ are to be chosen by the experimenter (cf. **3.1.5**(*1*) for suggested levels: there $\epsilon = A$). Then substituting for $\hat{N}_1$, (9.13) becomes

$$\Pr\left[-\epsilon < \frac{u(f-\hat{P}_2)}{(\hat{P}_1 - \hat{P}_2)} - 1 < \epsilon\right] \geqslant 1 - \alpha, \tag{9.14}$$

where $u = R/N_1$, the rate of exploitation. To evaluate the left-hand side of (9.14) we define a random variable

$$w = uf + \epsilon(P_1 - P_2) + \hat{P}_2(1+\epsilon-u) - \hat{P}_1(1+\epsilon),$$

which has mean zero and variance

$$V[w] = (1+\epsilon-u)^2 V[\hat{P}_2] + (1+\epsilon)^2 V[\hat{P}_1].$$

Then

$$\Pr\left[\frac{\hat{N}_1 - N_1}{N_1} < \epsilon\right] = \Pr[w < \epsilon(P_1 - P_2)]$$

and a similar random variable can be defined to deal with the negative ϵ in (9.14). Therefore, if the sampling is binomial, w is asymptotically normal, and for the case $n_1 = n_2 = n$ Paulik and Robson show that

$$\Pr\left[\left|\frac{\hat{N}_1 - N_1}{N}\right| < \epsilon\right] \approx \phi\left(\frac{\epsilon\sqrt{n}(P_1 - P_2)}{\sqrt{\{(1+\epsilon-u)^2 P_2 Q_2 + (1+\epsilon)^2 P_1 Q_1\}}}\right) - \phi\left(\frac{-\epsilon\sqrt{n}(P_1 - P_2)}{\sqrt{\{(1-\epsilon-u)^2 P_2 Q_2 + (1-\epsilon)^2 P_1 Q_1\}}}\right) = 1 - \alpha, \tag{9.15}$$

where ϕ is the cumulative unit normal distribution. This approximation is reasonably good even for small sample sizes. The choice of $n_1 = n_2$ is prompted by the discussion on p. 363.

Equation (9.15) cannot be solved explicitly, and Paulik and Robson programmed an iterative routine to obtain approximate solutions which were then used to construct the charts in Fig. 9.1–9.3. In using these charts it is convenient to think in terms of $\Delta P = P_1 - P_2$; thus, $f - P_2 > \Delta P$ and $u = \Delta P/(f-P_2)$. Also, writing $P_2 = P_1 - \Delta P$, then

$$u = 1/[1 + (f-P_1)/\Delta P] \tag{9.16}$$

and, since $f \leqslant 1$,

$$u \geqslant 1/[1 + (1-P_1)/\Delta P] = u\ (\min), \text{ say.} \tag{9.17}$$

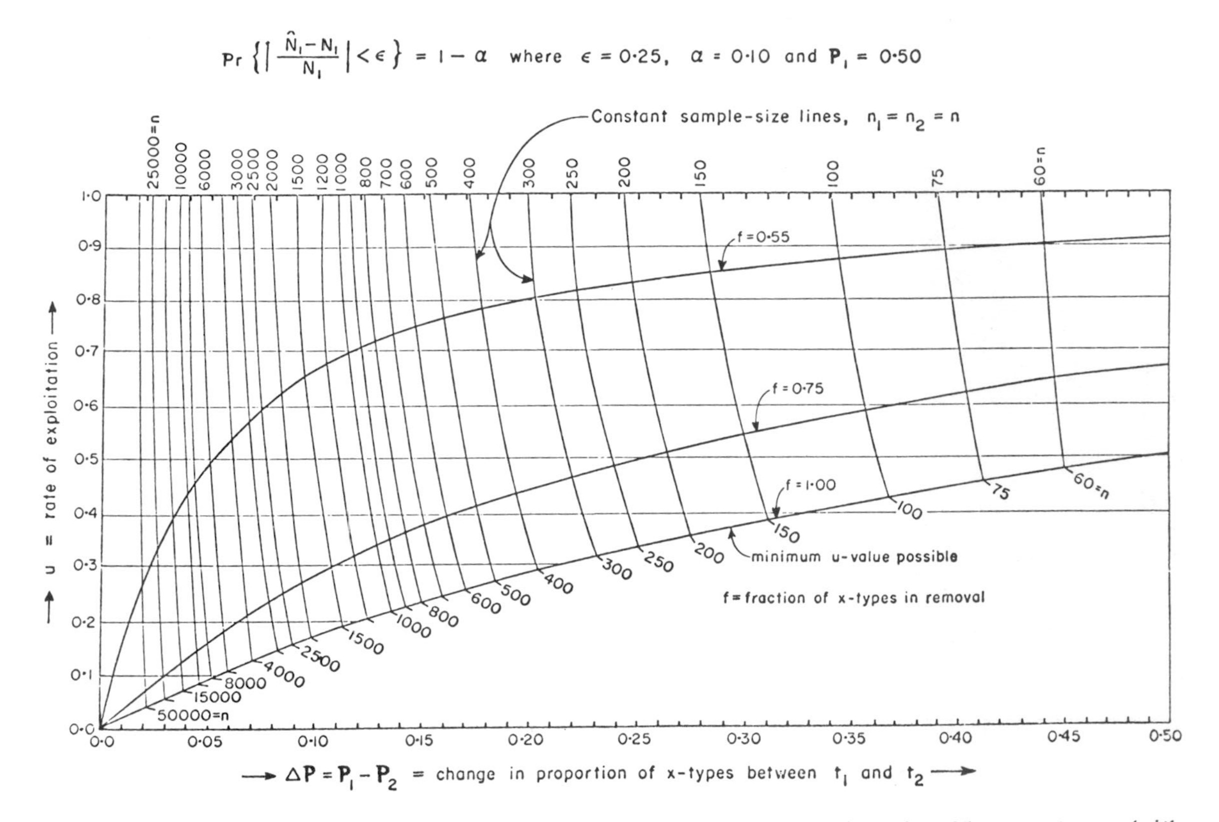

Fig. 9.1 Sample sizes required for CIR estimates of population size with less than 25 per cent error (with probability 0·90) when the initial proportion P_1 of x-types is 0·50: from Paulik and Robson [1969].

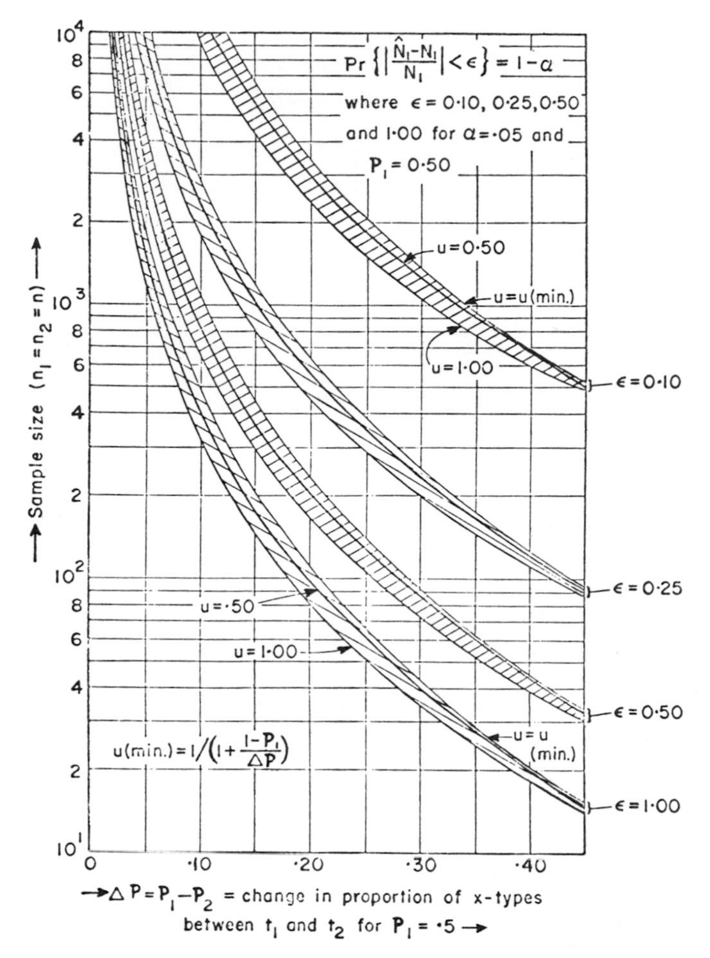

Fig. 9.2 Sample sizes required for CIR estimates of population size with less than 100 ϵ per cent error (with probability 0·95) for ϵ = 0·10, 0·25, 0·50 and 1·00, when the initial proportion P_1 of x-types is 0·50: from Paulik and Robson[1969].

For P_1 = 0·50, α = 0·10, ϵ = 0·25 and given ΔP, Fig. 9.1 gives combinations of n and u satisfying (9.15). If the investigator also knows the value of u, a unique n can be determined by interpolating between two sample size lines. A more conservative approach which avoids interpolation is to read n from the first line to the left of the $(u, \Delta P)$-point; a sample size so determined will be slightly larger than the minimum required. Alternatively, if f is known, the $(u, \Delta P)$-point can be found by interpolation using (9.16) and the u-lines for f values of 1·0, 0·75 and 0·55. For example, if the entire removal is of x-types (f = 1) and the expected ΔP is 0·25, u is approximately 0·33 and the sample size needed to estimate the population size within 25 per cent with 0·90 probability is 250 animals in each sample.

In using Fig. 9.1 it should be noted that n is much more sensitive to

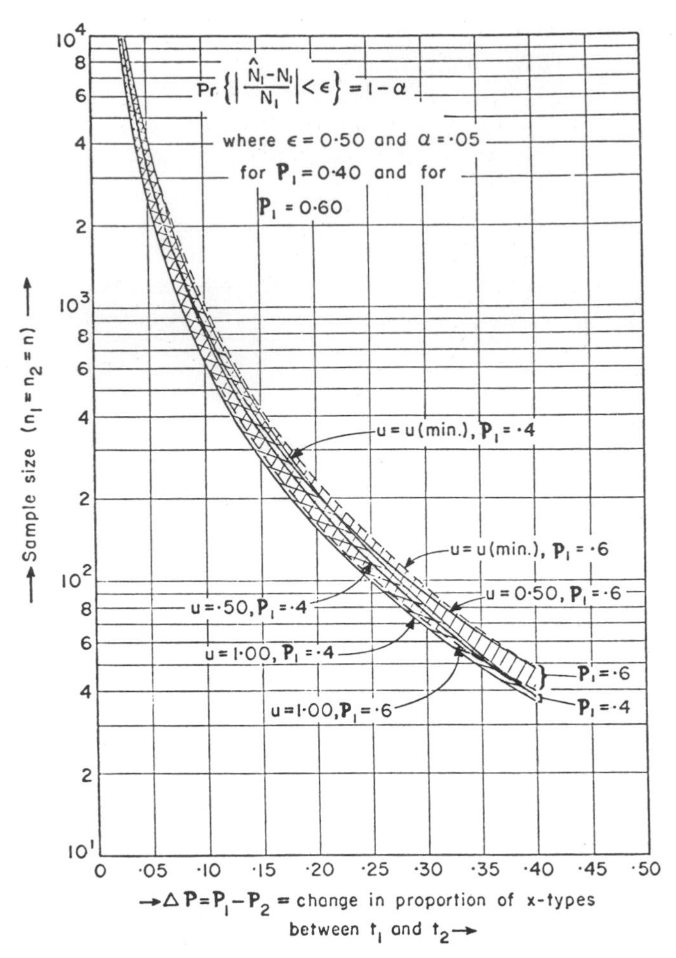

Fig. 9.3 Sample sizes required for CIR estimates of population size with less than 50 per cent error (with probability 0·95) for initial proportions of x-types $P_1 = 0{\cdot}40$ and $P_1 = 0{\cdot}60$: from Paulik and Robson [1969].

ΔP than to u. For example, when ΔP is in the vicinity of 0·015 to 0·020, sample sizes of around 50 000 are needed. This rapid increase in sample size as ΔP decreases below 0·05 indicates that the CIR method is so unreliable for ΔP's in this range that it is almost useless as a means of estimating N_1. For ΔP's between 0·05 and 0·10 use of this method is questionable, and the assumptions upon which the model is based become critical.

The four curves in Fig. 9.2 enable a planner to estimate sample sizes for $\epsilon = 0{\cdot}10, 0{\cdot}25, 0{\cdot}50, 1{\cdot}00$, $P_1 = 0{\cdot}50$ and $\alpha = 0{\cdot}05$. In each band $u = 1{\cdot}00$ on the lower boundary and $u = u$ (min) (cf. (9.17)) on the upper boundary. As ΔP approaches its maximum value of $P_1 = 0{\cdot}5$, the minimum value of u, u (min), approaches 0·50.

For the special case $\epsilon = 0{\cdot}50$, Fig. 9.3 gives the same bands as Fig. 9.2 except for $P_1 = 0{\cdot}4$ and $P_1 = 0{\cdot}6$. We see that for ΔP's between 0·05 and 0·30, the range of most practical importance, these bands differ only slightly. This means that for a fixed ΔP, the sample size is insensitive to the value of P_1 for P_1 in the vicinity of 0·50.

On p. 363 formulae are given for the optimum value of n_2/n_1. It was shown there that although a value of unity, which is used in Fig. 9.1–9.3, is not optimal it will be satisfactory in most cases. However. if u is large, the nominal accuracy of the estimate can be increased by partitioning the total sample size of $2n$ obtained from one of the three figures according to the optimum ratio.

Example 9.9

Consider the problem of designing an experiment to estimate the size of the pheasant population described in Example 9.3 (p. 356). How many pheasants should be sexed in pre- and post-season samples to ensure that, with probability 0·95, $\hat{N}_1$ will not be in error by more than 25 per cent? Suppose we know from previous years that the kill of cocks can be expected to change the proportion of cocks by about 0·30. Then assuming $P_1 = \frac{1}{2}$, u will be slightly greater than $u(\text{min}) = 1/[1 + (0{\cdot}5/0{\cdot}3)] = 0{\cdot}38$ as most of the removals will be males ($f \approx 1$). Therefore, entering Fig. 9.2 with $\epsilon = 0{\cdot}25$, $\Delta P = 0{\cdot}30$ and $u = 0{\cdot}40$, we obtain $n \approx 250$, so that a total of 500 pheasants must be sexed in the two samples.

Alternatively, if $P_1 = 0{\cdot}50$, $\Delta P = 0{\cdot}2$, $f = 0{\cdot}70$, then from (9.16) $u = 0{\cdot}5$. Therefore for $\epsilon = 0{\cdot}10$, $n \approx 3000$.

6 *Removals estimated*

In many situations the removals R_x, R_y are not known exactly and must be estimated. Therefore, if $\hat{R}_x$ and $\hat{R}_y$ are independent unbiased estimates of R_x and R_y, and $\hat{R} = \hat{R}_x + \hat{R}_y$, then N_1 can be estimated by

$$\hat{N}_1 = \frac{\hat{R}_x - \hat{P}_2\hat{R}}{\hat{P}_1 - \hat{P}_2}.$$

Using the delta method,

$$V[\hat{N}_1] \approx (P_1 - P_2)^{-2}\{N_1^2 V[\hat{P}_1] + N_2^2 V[\hat{P}_2] + (1 - P_2)^2 V[\hat{R}_x] + P_2^2 V[\hat{R}_y]\}.$$

Alternatively we may have independent estimates of R and $f = R_x/R$, so that

$$\hat{N}_1 = \frac{\hat{R}(\hat{f} - \hat{P}_2)}{\hat{P}_1 - \hat{P}_2}$$

and

$$V[\hat{N}_1] \approx (P_1 - P_2)^{-2}\{N_1^2 V[\hat{P}_1] + N_2^2 V[\hat{P}_2] + R^2 V[\hat{f}] + (f - P_2)^2 V[\hat{R}]\}. \quad (9.18)$$

A possible situation is when R is known but f has to be estimated by means of a sample from the R removals. In this case $V[\hat{R}] = 0$ and $\hat{f}$ will usually be a binomial proportion.

7 The Petersen estimate

It was first pointed out by Rupp [1966] that the Petersen method can be regarded as a special case of the CIR method. To see this, let X and Y denote the marked and unmarked populations respectively ($X_1 = P_1 = 0$) and suppose that M animals are caught for marking and then released. Then $R_y = M$, $R_x = -M$, and

$$\begin{aligned}\hat{N}_1 &= \frac{R_x - R\hat{P}_2}{\hat{P}_1 - \hat{P}_2} \\ &= \frac{-M - (M-M)\hat{P}_2}{0 - \hat{P}_2} \\ &= \frac{M}{\hat{P}_2},\end{aligned}$$

the Petersen estimate (cf. **3.1.1**).

8 Several removals

Chapman [1955] has considered the less common situation in which there are several selective removals, each followed by a random sample. For $i = 1, 2, \ldots, s$ let R_{xi}, R_{yi} be the total (cumulative) removals from the X and Y classes respectively prior to the ith sample, and let

$$\begin{aligned}R_i &= R_{xi} + R_{yi}, \quad (R_1 = 0), \\ X_i &= X_1 - R_{xi}, \\ N_i &= N_1 - R_i,\end{aligned}$$

and

$$P_i = X_i/N_i.$$

If the sampling is binomial then the maximum-likelihood estimates $\hat{X}_1$ and $\hat{N}_1$ are solutions of

$$\sum_{i=1}^{s} (x_i/X_i) = \sum_{i=1}^{s} (n_i/N_i)$$

and

$$\sum_{i=1}^{s} (y_i/Y_i) = \sum_{i=1}^{s} (n_i/N_i)$$

which can be obtained iteratively (cf. **1.3.8**). The asymptotic variances for these solutions are

$$V[\hat{X}_1] = \sum_{i=1}^{s} \left\{\frac{n_i P_i^2}{X_i Y_i d}\right\}$$

and

$$V[\hat{N}_1] = \sum_{i=1}^{s} \left\{\frac{n_i}{X_i Y_i d}\right\},$$

where $$d = \left(\sum_{i=1}^{s} \frac{n_i}{X_i Y_i}\right)\left(\sum_{i=1}^{s} \frac{n_i P_i^2}{X_i Y_i}\right) - \left(\sum_{i=1}^{s} \frac{n_i}{N_i Y_i}\right)^2$$

$$= \sum_{i<j}\sum \left\{\frac{n_i n_j (P_i - P_j)^2}{X_i Y_i X_j Y_j}\right\}.$$

A pair of first approximations for the iterative process is given, for instance, by the first and last samples, i.e.

$$\tilde{N}_1 = \frac{R_s - R_{xs}\hat{P}_s}{\hat{P}_1 - \hat{P}_s}, \quad \tilde{X}_1 = \hat{P}_1 \tilde{N}_1.$$

If, however, R_{xi}, R_{yi} are small relative to X_1, Y_1, Chapman shows that approximations to the maximum-likelihood estimates are given by solving

$$(1 - \hat{P}_1)\,\Sigma\, x_i R_{xi} + \hat{P}_1\,\Sigma\, y_i R_{yi} - \hat{P}_1(1-\hat{P}_1)\,\Sigma\, n_i R_i = 0 \tag{9.19}$$

and

$$\frac{\Sigma\, x_i}{\hat{P}_1} + \frac{\Sigma\, x_i R_{xi}}{\hat{N}_1 \hat{P}_1^2} - \frac{\Sigma\, y_i}{1-\hat{P}_1} - \frac{\Sigma\, y_i R_{yi}}{\hat{N}_1(1-\hat{P}_1)^2} = 0, \tag{9.20}$$

where the summation in each case is $i = 1, 2, \ldots, s$. $\hat{P}_1$ $(= \hat{X}_1/\hat{N}_1)$ may be determined from the quadratic (9.19), and this substituted in (9.20) gives $\hat{N}_1$, and finally $\hat{X}_1$.

An interesting application of the above method is when the sampling process is combined with the removal process. For instance, the sampler might only return some of the animals examined (e.g. just y-types). Alternatively he can effectively "remove" animals by tagging them before they are released back into the population. Chapman suggests the possibility of setting up a sequential procedure for sampling and removing x-types from the sample for s steps, where s is determined by the actual observations.

9.1.3 Other sampling procedures

1 Subsampling

By regarding the x-type animals as "marked", the procedures given in **3.4** can be used for estimating P_i and $V[\hat{P}_i]$. For example, suppose that the total population area is divided up into K subareas of which k are selected at random. Let the suffix j ($j = 1, 2, \ldots, k$) denote membership of the jth subarea sampled; thus x_{ij} is defined to be the number of x-types in the ith sample from the jth subarea, and P_{ij} is the proportion of x-types in the jth subarea at time t_i. Then two estimates of P_i are available, namely the pooled estimate

$$\hat{P}_i = \left(\sum_{j=1}^{k} x_{ij}\right)\bigg/\left(\sum_{j=1}^{k} n_{ij}\right) = x_i/n_i$$

and the average

$$\bar{P}_{i\cdot} = \frac{1}{k}\sum_{j=1}^{k}\left(\frac{x_{ij}}{n_{ij}}\right) = \frac{1}{k}\sum_{j=1}^{k}\hat{P}_{ij}, \quad \text{say.}$$

If $P_{ij} = P_i$ $(j = 1, 2, \ldots, k)$, as indicated by a chi-squared test of homogeneity using

$$T = \sum_{j=1}^{k} \left\{ \frac{(x_{ij} - n_{ij}\hat{P}_i)^2}{n_{ij}\hat{P}_i(1-\hat{P}_i)} \right\}$$

with $k - 1$ degrees of freedom, then one would use $\hat{P}_i$ with variance estimate $\hat{P}_i(1-\hat{P}_i)/n_i$, and we are back to the case considered in **9.1.2**. However, if the hypothesis of homogeneity is strongly rejected, then $\bar{P}_{i.}$ should be used along with an empirical estimate of variance. For example, if k/K is small then an approximately unbiased estimate of $V[\bar{P}_{i.}]$ is (cf. **3.4.2**)

$$\hat{V}[\bar{P}_{i.}] = \frac{1}{k(k-1)} \sum_{j=1}^{k} (\hat{P}_{ij} - \bar{P}_{i.})^2.$$

We note that when $n_{ij} = n$, say, for all i, j, the two estimates of P_i are identical.

The above subsampling procedure can be generalised by allowing k_1 sampling locations in the pre-season survey and k_2, possibly different, locations in the post-season survey. In this case the above discussion still holds, except that for given i, $j = 1, 2, \ldots, k_i$. Paulik and Robson [1969: 15] point out that if the variability of the $\hat{P}_{ij}$ about $\bar{P}_{i.}$ is the same for each survey then the two sets of data can be combined to give an overall estimate of the variance of $\hat{P}_{ij}$, namely

$$v = \sum_{i=1}^{2} \sum_{j=1}^{k_i} (\hat{P}_{ij} - \bar{P}_{i.})^2/(k_1 + k_2 - 2).$$

Then $V[\bar{P}_{i.}]$ is estimated by v/k_i.

2 *Using a single sample*

Suppose that the removal is due to a hunting season and that the two classes are males and females. With some populations it may not be possible to take a pre-season sample, and P_1 must be estimated by some sort of extrapolation from a post-season sample of the previous year. One such method is as follows:

Let P = last year's post-season proportion of males,
b = proportion of births since the last season which are males,
and L = average number of young per female.

Assuming that the mortality in the period from the end of last season to the beginning of this season is negligible (or proportionally the same for males and females), then it is readily shown that the proportion of males before this season is

$$P_1 = \frac{P + QbL}{1 + QL},$$

where $Q = 1 - P$. Therefore, if the above parameters are all estimated independently by unbiased estimates $\hat{P}$, $\hat{b}$, $\hat{L}$ respectively, and $\hat{P}_1$ is the resulting estimate of P_1, then, using the delta method,

$$V[\hat{P}_1] \approx \frac{(1+L-bL)^2}{(1+QL)^4} V[\hat{P}] + \frac{Q^2L^2}{(1+QL)^2} V[\hat{b}] + \frac{Q^2(Q+b-1)^2}{(1+QL)^4} V[\hat{L}].$$

However, in general, such estimates of P_1 will not be very accurate.

3 Utilising age ratios

Occasionally the P_i can be estimated from age ratios using the methods of Kimball [1948: 303] and Severinghaus and Maguire [1955] (cf. Hanson [1963: 42]) as follows. Let X, Y, X_s, Y_s be the number of male and female adults, and male and female subadults in the population (cf. **9.2.1** for a definition of subadult); then if $r = X/Y$ we have the identity

$$r = \frac{X_s}{Y_s} \cdot \frac{(Y_s/Y)}{(X_s/X)}.$$

Therefore if X_s/Y_s, the subadult sex ratio, is known (say equal to 1) and the ratios of subadults to adults for both males and females are estimated from a trapped sample, say, then r, and hence

$$P = \frac{X}{X+Y} = \frac{r}{r+1},$$

can be estimated. This method could be particularly useful if sample estimates of adult sex ratios are strongly biased through a differential catchability between sexes or through the difficulty of obtaining a proper random sample with regard to sexes (particularly if the sexes tend to segregate). The main assumption underlying the method is that for each sex, the subadults and adults have the same probability of being caught, so that although sexes may have different catchabilities, unbiased estimates of Y_s/Y and X_s/X are obtainable.

Suppose that a sample from the population yielded numbers x, y, x_s, y_s in each of the four categories respectively. Then P is estimated by

$$\hat{P} = \hat{r}/(\hat{r}+1),$$

where $$\hat{r} = \left(\frac{X_s}{Y_s}\right) \cdot \frac{(y_s/y)}{(x_s/x)}.$$

Assuming the random variables x, y, x_s and y_s to be independent Poisson variables, and using the delta method, an estimate of $V[\hat{P}]$ is

$$\hat{V}[\hat{P}] = \frac{r^2}{(\hat{r}+1)^4}\left\{\frac{1}{x}+\frac{1}{y}+\frac{1}{x_s}+\frac{1}{y_s}\right\}.$$

4 Known sampling effort

Let

$$E[x_i] = a_i X_i \quad \text{and} \quad E[y_i] = b_i Y_i, \tag{9.21}$$

then in **9.1.2**, with binomial sampling, we effectively assumed that

$a_i = b_i = n_i/N_i$. However, if the sample sizes are determined by sampling effort, so that the n_i are now random variables, we can make the alternative assumption (Chapman and Murphy [1965]) that $a_i = b_i = Kf_i$, where K is the binomial catchability (cf. **7.1.3**(*1*)) and the sampling efforts f_1 and f_2 are known. Then replacing $E[n_i]$ by n_i we have the equations

$$\frac{n_1/f_1}{N_1} = \frac{n_2/f_2}{N_1 - R} = K \left(= \frac{(n_1/f_1) - (n_2/f_2)}{R} \right),$$

leading to the estimates

$$\hat{N}_1 = \frac{Rc_1}{c_1 - c_2} \quad \text{and} \quad \hat{N}_2 = \frac{Rc_2}{c_1 - c_2}, \tag{9.22}$$

where $c_i = n_i/f_i$. The estimate $\hat{N}_i$ was first given by Petrides [1949] and is particularly useful when it is not possible to distinguish between the two classes, e.g. deer tracks in the snow (Davis [1963]). We also note that

$$\hat{N}_1 = R \bigg/ \left(1 - \frac{c_2}{c_1}\right),$$

so that animal signs can be used to measure the number of animals, provided the index signs-per-animal is the same for both samples.

Using the delta method, we have

$$V[\hat{N}_1] \approx (E[c_1] - E[c_2])^{-2} \{N_1^2 V[c_2] + N_2^2 V[c_1]\}.$$

If n_1 and n_2 are assumed to be independent Poisson random variables then $E[c_i]$ and $V[c_i]$ can be estimated by c_i and c_i/f_i, respectively.

When the two classes are distinguishable, X_1 is estimated by

$$\hat{X}_1 = \frac{x_1}{n_1} \hat{N}_1.$$

Example 9.10 Pheasants (adapted from Petrides [1949])

A pre-season roadside count over 50 miles yields 150 pheasants, while a post-season count over 30 miles yields 45 pheasants. The season's kill is known to be 5600 birds. Then $f_1 = 50$, $f_2 = 30$, $c_1 = 150/50 = 3$ and $c_2 = 45/30 = 1{\cdot}5$, so that

$$\hat{N}_1 = \frac{5600\ (3)}{3 - 1{\cdot}5} = 11\ 200$$

and

$$\hat{N}_2 = 11\ 200 - 5600 = 5600.$$

Assuming Poisson sampling,

$$\hat{V}[\hat{N}_1] = (3 - 1{\cdot}5)^{-2} \left\{ 11\ 200^2 \left(\frac{1{\cdot}5}{30}\right) + 5600^2 \left(\frac{3}{50}\right) \right\} = 3\ 623\ 800$$

and

$$\hat{\sigma}[\hat{N}_1] = 1903.$$

Since c_i is the sample mean of so many birds per mile, $V[c_i]$ can be estimated from replicated data using (1.9) if the number of birds seen each mile is recorded. This should be done where possible.

5 *Constant class probability of capture*

Chapman and Murphy consider the experimental situation in which $a_1 = a_2 = a$ and $b_1 = b_2 = b$ in (9.21), i.e. the classes X and Y are not equally catchable, but the chances of catching an x-type animal remain the same for the two surveys, and similarly for y-type animals. Then our estimating equations become (cf. (9.7) and (9.8) on p. 356)

$$\frac{x_1}{\hat{X}_1} = \frac{x_2}{\hat{X}_1 - R_x} \quad \text{and} \quad \frac{y_1}{\hat{Y}_1} = \frac{y_2}{\hat{Y}_1 - R_y},$$

which have solutions

$$\hat{X}_1 = \frac{R_x x_1}{(x_1 - x_2)} \quad \text{and} \quad \hat{Y}_1 = \frac{R_y y_1}{(y_1 - y_2)}.$$

Using the delta method,

$$V[\hat{X}_1] \approx (E[x_1 - x_2])^{-2}\{X_1^2 V[x_2] + X_2^2 V[x_1]\}$$

and

$$V[\hat{Y}_1] \approx (E[y_1 - y_2])^{-2}\{Y_1^2 V[y_2] + Y_2^2 V[y_1]\}.$$

Here $E[x_1 - x_2]$ is estimated by $x_1 - x_2$, and if x_1, x_2 are independent Poisson variables then $V[x_i]$ is estimated by x_i.

When the classes are equally catchable ($a = b$) then the data can be pooled to give

$$\frac{n_1}{\hat{N}_1} = \frac{n_2}{\hat{N}_1 - R}$$

or

$$\hat{N}_1 = \frac{Rn_1}{n_1 - n_2}.$$

This estimate is the same as $\hat{N}_1$ of (9.22) when $f_1 = f_2$.

9.1.4 Utilising tag information

Chapman [1955: 284] showed that unless tagging is very much more expensive than the cost of classification, the Petersen recapture method will yield more information, that is, provide estimates with smaller variances, for the same amount of effort (cf. **14.1.2**(*6*)). The recapture method in this situation would consist of tagging the n_1 animals from the pre-season survey and then observing the number of tagged animals, m ($= m_x + m_y$ say), in the removal R. The tagging estimate of N_1 would then be (cf. **3.1.1**)

$$N_{1t}^* = \frac{(n_1 + 1)(R + 1)}{m + 1} - 1$$

with $V[N^*_{1t}]$ estimated by

$$v[N^*_{1t}] = \frac{(n_1+1)(R+1)(n_1-m)(R-m)}{(m+1)^2(m+2)}.$$

If R has to be estimated independently by $\hat{R}$, say, then we simply add $N_1^2 V[\hat{R}]/R^2$ to $V[N^*_{1t}]$.

When the animals in the first sample and in the m recaptures are also classified according to whether they are x-type or y-type, then several estimates of X_1 are available (cf. **3.2.5**(*1*)). For example, if the probabilities of capture in the first sample are the same for both types then the appropriate estimate of X_1 is

$$\tilde{X}_{1t} = \frac{x_1}{n_1} \cdot N^*_{1t} = \hat{P}_1 N^*_{1t}$$

with variance (cf. (1.11))

$$V[\tilde{X}_{1t}] \approx P_1^2 V[N^*_{1t}] + N_1^2 V[\hat{P}_1] + V[N^*_{1t}]\, V[\hat{P}_1].$$

But if the probabilities of capture in the first sample are different then X_1 should be estimated by

$$X^*_{1t} = \frac{(x_1+1)(R_x+1)}{(m_x+1)} - 1$$

with variance estimate

$$v[X^*_{1t}] = \frac{(x_1+1)(R_x+1)(x_1-m_x)(R_x-m_x)}{(m_x+1)^2(m_x+2)}.$$

If both R and $f = R_x/R$ are estimated independently by $\hat{R}$ and $\hat{f}$, then

$$X^*_{1t} = \frac{(x_1+1)(\hat{f}\hat{R}+1)}{(m_x+1)} - 1$$

and the variance is increased by approximately

$$X_1^2 \left\{ \frac{V[\hat{f}]}{f^2} + \frac{V[\hat{R}]}{R^2} \right\}.$$

When tagging is used along with classification, the first problem of interest is to determine whether there is a reasonable agreement between the two estimates $\hat{N}_1$ and N^*_{1t}. Since the two sources of information are essentially independent (for a given first sample size n_1), a large-sample test for agreement is to calculate

$$\frac{|\hat{N}_1 - N^*_{1t}|}{\{\hat{V}[\hat{N}_1] + \hat{V}[N^*_{1t}]\}^{\frac{1}{2}}} \qquad (9.23)$$

and compare its value with the appropriate significance levels of the unit normal distribution. If the estimates are not compatible, it may be due to failure of the assumptions in one or both models. For example, as Chapman points out, the assumptions may be correct in regard to expectations, but the sampling may not be random, so that the variances are larger than those given above and the denominator of (9.23) is underestimated. Chapman also notes that the most important sources of error in $\hat{N}_1$, such as the under-estimation of R_x and R_y, will tend to bias $\hat{N}_1$ downwards, while factors such as trap-shyness and tag-mortality will tend to bias N^*_{1t} upwards. This means that if the estimates are compatible, there is a suggestion, but not necessarily proof, that the assumptions of the two models are fulfilled, and in this case we could obtain a new estimate, $\hat{N}_{1C}$ say, based on the combined tagging and removal information. The most efficient way of combining the two estimates is to take the weighted average (cf. **1.3.2**)

$$\hat{N}_{1C} = (w_1\hat{N}_1 + w_2 N^*_{1t})/(w_1 + w_2),$$

where $w_1 = \{V[\hat{N}_1]\}^{-1}$ and $w_2 = \{V[N^*_{1t}]\}^{-1}$ have to be estimated. We then have

$$V[\hat{N}_{1C}] = (w_1 + w_2)^{-1} = \frac{V[\hat{N}_1]V[N^*_{1t}]}{V[\hat{N}_1] + V[N^*_{1t}]}$$

and $\hat{N}_{1C}$ has smaller variance than either $\hat{N}_1$ or N^*_{1t}. Unfortunately the same method cannot be used to obtain $\hat{X}_{1C}$ since $\hat{X}_1$ and X^*_{1t} are not independent (both contain x_1).

If binomial sampling is used, an alternative approach suggested by Chapman is to use the combined binomial model (cf. (9.6) and (3.3))

$$\left\{\prod_{i=1}^{2}\binom{n_i}{x_i}\left(\frac{X_i}{N_i}\right)^{x_i}\left(1-\frac{X_i}{N_i}\right)^{y_i}\right\}\binom{n_1}{m}\left(\frac{R}{N_1}\right)^{m}\left(1-\frac{R}{N_1}\right)^{n_1-m}$$

for which the maximum-likelihood estimates N'_{1C} and X'_{1C} are solutions of

$$\frac{x_1}{X_1} + \frac{x_2}{X_1 - R_x} = \frac{y_1}{Y_1} + \frac{y_2}{Y_1 - R_y} \tag{9.24}$$

and

$$\frac{x_1}{X_1} + \frac{x_2}{X_1 - R_x} = \frac{2n_1}{N_1} + \frac{n_2 - n_1 + m}{N_1 - R}. \tag{9.25}$$

These equations do not have simple solutions and must be solved iteratively (the solutions given by Chapman are incorrect). One method would be the following: start with a trial value of X_1 (say $\hat{X}_1$, X^*_{1t} or the average of the two), solve equations (9.24) and (9.25) as quadratics in Y_1 and N_1 respectively, choosing the larger root in each case, evaluate the new trial value $N_1 - Y_1$, and repeat the procedure. By inverting the information matrix, Chapman shows that asymptotically

$$V[N'_{1C}] = \left\{\frac{(P_1 - P_2)^2}{\dfrac{X_1Y_1}{n_1} + \dfrac{X_2Y_2}{n_2}} + \frac{n_1R}{N_1^2N_2}\right\}^{-1};$$

also

$$V[X'_{1C}] = \frac{\dfrac{X_1Y_1}{n_1}P_2^2 + \dfrac{X_2Y_2}{n_2}P_1^2 + \dfrac{n_1R}{N_1^2N_2}\left[\dfrac{X_1Y_1}{n_1}\cdot\dfrac{X_2Y_2}{n_2}\right]}{(P_1-P_2)^2 + \dfrac{n_1R}{N_1^2N_2}\left[\dfrac{X_1Y_1}{n_1}+\dfrac{X_2Y_2}{n_2}\right]}.$$

9.1.5 Exploitation rate

The rate of exploitation u is defined by Ricker [1958] as the fraction of the initial population removed by man during a specified time-interval. In terms of our notation, Paulik and Robson [1969: 16] show that

$$\begin{aligned} u &= \frac{R}{N_1} \\ &= \frac{R(P_1-P_2)}{R_x - P_2R} \\ &= \frac{P_1-P_2}{f-P_2}, \end{aligned}$$

where $f = R_x/R$. Similarly the rates of exploitation for x-type and y-type animals are respectively

$$u_x = \frac{R_x}{X_1} = \frac{fR}{P_1N_1} = \frac{fu}{P_1}$$

and

$$u_y = \frac{(1-f)u}{(1-P_1)}.$$

To estimate u we do not need to know R_x; an estimate of f is sufficient. Thus

$$\hat{u} = \frac{\hat{P}_1-\hat{P}_2}{\hat{f}-\hat{P}_2}, \quad \hat{u}_x = \frac{\hat{f}\hat{u}}{\hat{P}_1}$$

and, using the delta method,

$$V[\hat{u}] \approx (f-P_2)^{-4}\{(f-P_2)^2V[\hat{P}_1] + (f-P_1)^2V[\hat{P}_2] + (P_1-P_2)^2V[\hat{f}]\}. \quad (9.26)$$

We note that the above estimates of u and u_x are essentially the same as those derived by Robinette [1949], Petrides [1954], Selleck and Hart [1957] and discussed by Hanson [1963]. These earlier papers, however, do not give any variance formulae.

Example 9.11 Pronghorn antelope (*Antilocapra americana*): Paulik and Robson [1969]

To estimate the rate of hunting mortality in a hypothetical pronghorn antelope population, a pre-season field survey yielded 550 mature females and 167 female fawns. In a sample of the kill, 63 mature females and 16 female fawns were found. The post-season survey yielded 370 mature females and 130 female fawns. Thus

$\hat{P}_1 = 550/717 = 0{\cdot}7671$, $\hat{P}_2 = 370/500 = 0{\cdot}7400$, $\hat{f} = 63/79 = 0{\cdot}7975$,

so that

$$\hat{u} = \frac{0{\cdot}7671 - 0{\cdot}7400}{0{\cdot}7975 - 0{\cdot}7400} = 0{\cdot}4713.$$

This result would commonly be expressed as either a hunting mortality of 47 per cent or an exploitation rate of 0·47.

Assuming binomial sampling in all three cases, we have

$$\hat{V}[\hat{P}_1] = (0{\cdot}7671)(0{\cdot}2329)/717 = 0{\cdot}000\,249\,2,$$
$$\hat{V}[\hat{f}] = (0{\cdot}7975)(0{\cdot}2025)/79 = 0{\cdot}002\,044, \text{ etc.},$$

so that, substituting in (9.26),

$$\hat{V}[\hat{u}] = 0{\cdot}2455 \quad \text{and} \quad \hat{\sigma}[\hat{u}] = 0{\cdot}4955.$$

The estimate of $\sigma[\hat{u}]$ is very large, thus indicating, as pointed out by the authors, that $\hat{u}$ is almost useless as an estimate of u. The fault lies in the small observed value $\hat{P}_1 - \hat{P}_2 = 0{\cdot}0271$; for this difference it would be necessary to obtain impracticably large samples (cf. p. 370) to give an estimate of u with a reasonable coefficient of variation.

9.2 OPEN POPULATION

9.2.1 Estimating productivity

Hanson [1963] considers three age-classes in a population: juveniles, subadults, and adults. Juveniles are animals which are less than fully grown; subadults are essentially fully grown, but the majority of their cohort have not completed their first breeding season; adults are fully grown, and the majority of their cohort have completed one or more breeding seasons. Using these age-classes, Hanson defines productivity as the ratio of subadult to adult animals. Since the CIR method holds for additions to the population (that is, R_x and R_y negative), $\hat{u}$ defined in the previous section can be used as an estimate of productivity; the only difference being that $\hat{u}$ is now negative.

Example 9.12 Quail (*Lophortyx gambelii*): Hanson [1963]

A careful survey of Gambel's quail on a study area in early August (time t_1) found 325 adult males and 250 adult females, as well as a number of juvenile birds. A second survey in early October (time t_2), after all the juveniles had grown sufficiently to be considered subadults and to be part of the total population of mature birds, found 1230 mature males and 1145 mature females. It was assumed that the survival rates of adult males and females from August to October were the same, and that there was an equal sex ratio ($f = \frac{1}{2}$) in the juveniles recruited to the population of mature birds as subadults between August and October. Thus $\hat{P}_1 = 325/575 = 0{\cdot}565\,22$, $\hat{P}_2 = 1230/2375 = 0{\cdot}517\,89$ and

$$\hat{u} = \frac{0{\cdot}565\ 22 - 0{\cdot}517\ 89}{0{\cdot}5000 - 0{\cdot}517\ 89} = -2{\cdot}65.$$

In the calculation of $\hat{V}[\hat{u}]$ (cf. 9.26) we note that $V[\hat{f}] = 0$, as f is assumed known. Then, assuming binomial sampling,

$$\hat{V}[\hat{u}] = 5.70 \quad \text{and} \quad \hat{\sigma}[\hat{u}] = 2{\cdot}4.$$

The high value of $\hat{\sigma}[\hat{u}]$ reflects the general inaccuracy of productivity estimates when $f - P_2$ is small. In fact, as f is often near 0·5, the CIR method for estimating u is of limited usefulness; it will only apply when the sex ratio is widely different from unity.

9.2.2 Natural mortality

1 Survival ratios

Consider a population in which there are removals due to exploitation and possibly natural mortality. Let s_x and s_y be the fractions of x-type and y-type animals respectively surviving from time t_1 to time t_2. Then, using the notation of **9.1.1**, Paulik and Robson [1969] give the following equations:

$$X_2 = X_1 s_x, \quad Y_2 = Y_1 s_y,$$

and hence

$$\theta = \frac{s_x}{s_y} = \left(\frac{Y_1}{X_1}\right) \Big/ \left(\frac{Y_2}{X_2}\right) = \left(\frac{1 - P_1}{P_1}\right)\left(\frac{P_2}{1 - P_2}\right).$$

Therefore, if $\hat{P}_i$ is an estimate of P_i, θ is estimated by

$$\hat{\theta} = \left(\frac{1 - \hat{P}_1}{\hat{P}_1}\right)\left(\frac{\hat{P}_2}{1 - \hat{P}_2}\right), \tag{9.27}$$

and, using the delta method,

$$V[\hat{\theta}] \approx [P_1(1 - P_2)]^{-4}\{(1 - P_2)^2 P_2^2 V[\hat{P}_1] + (1 - P_1)^2 P_1^2 V[\hat{P}_2]\}. \tag{9.28}$$

If x_i and y_i are the numbers of x-type and y-type animals in the sample taken at time t_i, then (9.27) reduces to

$$\hat{\theta} = (y_1 x_2)/(y_2 x_1). \tag{9.29}$$

Paulik and Robson give several variations of the above method when additional information is available. For example, if it is known that the ratio of the two kinds of animals at time t_1 is 1 : 1, then

$$\hat{\theta} = \hat{P}_2/(1 - \hat{P}_2), \tag{9.30}$$

and, setting $P_1 = \frac{1}{2}$, $V[\hat{P}_1] = 0$, in (9.28),

$$V[\hat{\theta}] \approx (1 - P_2)^{-4} V[\hat{P}_2]. \tag{9.31}$$

In some situations it is possible to ensure that $s_y = 1$, so that $\hat{\theta}$ is an estimate of s_x. One way of doing this is to introduce Y_1 animals into the population just before time t_2, so that no removals of y-types can occur before the survey for estimating P_2, i.e. $Y_2 = Y_1$. In practice this scheme is nearly always used under circumstances in which the values of X_1 and Y_1 are known (e.g. two tag releases) but either P_2 or the ratio (Y_2/X_2) must be estimated by sampling (Ricker [1958: 128–9], Geis and Taber [1963: 286], see also **5.1.3**(*4*) and **6.4.2**(*2*)). When X_1 and Y_1 are known,

$$\hat{\theta} = \frac{Y_1}{X_1}\frac{\hat{P}_2}{(1-\hat{P}_2)} = \left(\frac{x_2}{X_1}\right)\bigg/\left(\frac{y_2}{Y_1}\right) \tag{9.32}$$

and setting $V[\hat{P}_1] = 0$ in (9.28),

$$V[\hat{\theta}] \approx (1-P_2)^{-4}\left(\frac{Y_1}{X_1}\right)^2 V[\hat{P}_2]. \tag{9.33}$$

Minor variations of this method are used by Eberhardt *et al.* [1963: 15] and Parker [1965: 1541]; further applications of the method are given in Meslow and Keith [1968: 822], Barkalow *et al.* [1970] and Phinney [1974: 135].

Example 9.13 Pheasants: Paulik and Robson [1969]

On a hypothetical study area, field surveys of mature pheasants in the fall before hunting began found 830 cocks and 1000 hens. Similar surveys after the hunting season found 230 cocks and 1000 hens. Then, from (9.29) an estimate of the survival rate of cocks relative to hens is

$$\hat{\theta} = \frac{1000\ (230)}{830\ (1000)} = 0{\cdot}277.$$

Assuming binomial sampling with $\hat{V}[\hat{P}_i] = \hat{P}_i(1-\hat{P}_i)/n_i$, we find, using equation (9.28), that

$$\hat{V}[\hat{\theta}] = 0{\cdot}000\,580\,0 \quad\text{and}\quad \hat{\sigma}[\hat{\theta}] = 0{\cdot}024.$$

If there had been no pre-hunting season survey and it was known from past experience that the sex ratio in the pre-season mature pheasant population was 1 : 1, then from (9.30) and (9.31)

$$\hat{\theta} = x_2/y_2 = 230/1000 = 0{\cdot}230,$$
$$\hat{V}[\hat{\theta}] = 0{\cdot}000\,283\,3 \quad\text{and}\quad \hat{\sigma}[\hat{\theta}] = 0{\cdot}017.$$

(The above example was adapted from Hanson [1963].)

Example 9.14 Striped bass (*Roccus saxatilis*): Paulik and Robson [1969]

The application of equation (9.32) is demonstrated using some mark–recapture data on striped bass. In the spring of 1958 (time t_1) $X_1 = 3891$ bass were tagged with disk-dangler tags and released in the Sacramento–San Joaquin delta area of San Francisco Bay. Approximately one year later (time t_2) a second group of $Y_1 = 2965$ bass were tagged using the same equipment and procedures. During the next 8 years from the spring of 1959 to the spring of 1967, $x_2 = 430$ recoveries from the first release and $y_2 = 1026$ recoveries from the second release were made. Assuming that both groups of tagged fish have the same probability of recovery and the same survival rate over the 8 years, so that P_2 remains constant, then y_2/x_2 will estimate Y_2/X_2, the ratio of the tagged population sizes at time t_2. Therefore, since $Y_2 = Y_1$ (i.e. $s_y = 1$), an estimate of s_x, the proportion of striped bass surviving for a year beginning in the spring of 1958 and ending in the spring of 1959 is

$$\hat{s}_x = \left(\frac{430}{3891}\right) \Big/ \left(\frac{1026}{2965}\right) = 0{\cdot}3194.$$

Assuming binomial sampling with

$$\hat{P}_2 = 430/(430+1026) = 0{\cdot}2953,$$
$$\hat{V}[\hat{P}_2] = (0{\cdot}2953)(0{\cdot}7047)/1456 = 0{\cdot}000\,142\,9.$$

Hence, from (9.33),

$$\hat{V}[\hat{s}_x] = 0{\cdot}000\,336\,6 \quad \text{and} \quad \hat{\sigma}[\hat{s}_x] = 0{\cdot}0184.$$

EXPLOITATION AND NATURAL MORTALITY. In the above discussion no distinction was made between mortality due to exploitation and mortality due to natural causes. The next problem, then, is to separate the total mortality into these two components. To do this, Chapman and Murphy [1965] suggest that it is useful to consider two cases: "instantaneous" removal, when the removal takes place in a short period of time, and "continuous" uniform removal, when the removal takes place uniformly over an extended period. We shall now consider these two cases separately.

2 *Instantaneous removals*

Suppose that the removals R_x, R_y take place instantly at time zero. We shall now have a slight change in notation and assume that the two samples are taken time t_1 before the removal and time t_2 after the removal. Let:

ϕ_{1x} = fraction of x-types surviving from time $(-t_1)$ to time zero, and
ϕ_{2x} = fraction of x-types surviving from time zero to time t_2.

ϕ_{1y} and ϕ_{2y} are similarly defined. Then

$$X_2 = (X_1\phi_{1x} - R_x)\phi_{2x} \tag{9.34}$$

and

$$Y_2 = (Y_1\phi_{1y} - R_y)\phi_{2y}. \tag{9.35}$$

However, we now have too many parameters to estimate, and as a first step in simplifying the model we shall assume $\phi_{ix} = \phi_{iy} = \phi_i$ $(i = 1, 2)$. Then from (9.34) and (9.35) we have

$$P_2 = \frac{X_2}{X_2 + Y_2} = \frac{P_1N_1\phi_1 - R_x}{N_1\phi_1 - R}.$$

and rearranging,

$$N_1\phi_1 = \frac{R_x - RP_2}{P_1 - P_2}.$$

This means that $\hat{N}_1$ defined in (9.2) is now an estimate of $N_1\phi_1$, the population size just prior to the removal, though the exploitation rate $u = R/(\phi_1 N_1)$ is still given by

$$u = \frac{f - RP_2}{P_1 - P_2},$$

a fact noted by Chapman and Murphy [1965]. However, these authors suggest a different model, using knowledge of sampling effort, which leads to the estimation of N_1, ϕ_1 and ϕ_2 as follows.

We shall assume that the instantaneous natural mortality rate is constant throughout the period under investigation and equal to μ; hence

$$\phi_i = e^{-\mu t_i}.$$

In addition we make the same assumptions as those given in **9.1.3**(*4*) with regard to sampling effort, namely,

$$E[x_i] = Kf_i X_i \quad \text{and} \quad E[y_i] = Kf_i Y_i \quad (i = 1, 2), \tag{9.36}$$

where K is the binomial catchability and f_i, the sampling effort, is assumed known. We now have four unknown parameters X_1, Y_1, μ and K which can be estimated from the four observations x_1, y_1, x_2 and y_2. As a first step we substitute (9.36) in (9.34) and (9.35), and defining $\alpha = \phi_1/(Kf_1)$, $\beta = 1/(\phi_2 Kf_2)$, we have

$$\alpha E[x_1] - R_x = \beta E[x_2]$$

and

$$\alpha E[y_1] - R_y = \beta E[y_2].$$

Replacing the expectations by observed values, we have two equations in α and β which have solutions

$$\hat{\alpha} = \frac{R_x y_2 - R_y x_2}{x_1 y_2 - x_2 y_1}$$

and

$$\hat{\beta} = \frac{R_x y_1 - R_y x_1}{x_1 y_2 - x_2 y_1}.$$

Hence

$$\exp\{(-\hat{\mu}(t_1 + t_2)\} = \hat{\phi}_1 \hat{\phi}_2 = (f_1 \hat{\alpha})/(f_2 \hat{\beta}),$$

leading to the estimate

$$\hat{\mu} = -\frac{1}{(t_1 + t_2)} \log\left\{\frac{f_1}{f_2} \cdot \frac{R_x y_2 - R_y x_2}{R_x y_1 - R_y x_1}\right\}.$$

Also

$$1/\hat{K} = \{(f_1 \hat{\alpha})^{t_2} (f_2 \hat{\beta})^{t_1}\}^{1/(t_1 + t_2)},$$

so that from (9.36),

$$\hat{X}_i = x_i/(\hat{K} f_i), \quad \hat{Y}_i = y_i/(\hat{K} f_i)$$

and

$$\hat{N}_i = \hat{X}_i + \hat{Y}_i.$$

Assuming x_1, x_2, y_1 and y_2 to be independent Poisson variables, we have, using the delta method and estimating both $E[x_i]$ and $V[x_i]$ by x_i, the following variance estimates:

$$\hat{V}[\hat{\mu}] = \frac{1}{(t_1+t_2)^2}\left\{\frac{R_x^2 y_2 + R_y^2 x_2}{(R_x y_2 - R_y x_2)^2} + \frac{R_x^2 y_1 + R_y^2 x_1}{(R_x y_1 - R_y x_1)^2}\right\}$$

and

$$\frac{\hat{V}[\hat{N}_1]}{\hat{N}_1^2} = x_1\left\{\frac{(x_2+y_2)y_1}{(x_1+y_1)(x_1y_2-x_2y_1)} + \frac{R_y t_1}{(t_1+t_2)(R_x y_1 - R_y x_1)}\right\}^2 +$$

$$+ y_1\left\{\frac{(x_2+y_2)x_1}{(x_1+y_1)(x_1y_2-x_2y_1)} + \frac{R_x t_1}{(t_1+t_2)(R_x y_1 - R_y x_1)}\right\}^2 +$$

$$+ x_2\left\{\frac{y_1}{(x_1y_2-x_2y_1)} - \frac{R_y t_2}{(t_1+t_2)(R_x y_2 - R_y x_2)}\right\}^2 +$$

$$+ y_2\left\{\frac{x_1}{(x_1y_2-x_2y_1)} - \frac{R_x t_2}{(t_1+t_2)(R_x y_2 - R_y x_2)}\right\}^2 .$$

In conclusion we note that in the above theory n_i $(= x_i + y_i)$ is no longer a fixed parameter but a random variable; one cannot fix both effort and sample size in advance.

3 Continuous removals

We shall now assume that the samples are taken at the beginning and end of a continuous uniform removal carried out over a period of time T. Let μ_x, μ_y be the instantaneous natural mortality rates and let μ_{Ex}, μ_{Ey} be the instantaneous removal (exploitation) rates for x-type and y-type animals respectively. We can now write the following equations:

$$E[x_i\,|X_i] = a_i X_i, \quad E[y_i|Y_i] = b_i Y_i, \tag{9.37}$$

$$E[X_2] = X_1 \exp\{-(\mu_x+\mu_{Ex})T\}, \tag{9.38}$$

$$E[Y_2] = Y_1 \exp\{-(\mu_y+\mu_{Ey})T\}, \tag{9.39}$$

and from (8.13) we have further

$$E[R_x] = \frac{X_1\mu_{Ex}}{(\mu_x+\mu_{Ex})}(1-\exp\{-(\mu_x+\mu_{Ex})T\}) \tag{9.40}$$

and

$$E[R_y] = \frac{Y_1\mu_{Ey}}{(\mu_y+\mu_{Ey})}(1-\exp\{-(\mu_y+\mu_{Ey})T\}). \tag{9.41}$$

A particular example of the above is the constant level fishery (a C.C.U. fishery in the terminology of Paulik [1963a]), where μ_{Ex} and μ_{Ey} are the instantaneous fishing rates. Typically the sampling might be done annually; each sample can be considered either as the second sample as far as the previous year is concerned or as the first sample for the next year.

Once again we have too many parameters to estimate, and as a first simplification Chapman and Murphy [1965] make the following assumptions:

(i) $\mu_x = \mu_y = \mu$, and
(ii) $a_i = b_i \quad (i = 1, 2)$.

Although we still cannot estimate all the parameters individually it is now possible to obtain approximate estimates of μ_{Ex} and μ_{Ey}. Replacing expected values by observed values and defining the unit of time so that $T = 1$, we have from equations (9.37) to (9.41):

$$\left(\frac{R_x}{x_1}\right)\Big/\left(\frac{R_y}{y_1}\right) = \frac{[1 - \exp\{-(\mu+\mu_{Ex})\}]\,\mu_{Ex}/(\mu+\mu_{Ex})}{[1 - \exp\{-(\mu+\mu_{Ey})\}]\,\mu_{Ey}/(\mu+\mu_{Ey})} \qquad (9.42)$$

$$(= r_1, \quad \text{say}),$$

and

$$\left(\frac{x_2}{x_1}\right)\Big/\left(\frac{y_2}{y_1}\right) = \exp\{-(\mu_{Ex}-\mu_{Ey})\} \quad (= r_2, \quad \text{say}). \qquad (9.43)$$

We note that (9.43) is equivalent to (9.29). If μ was known (or zero as in Lander [1962]) it would be possible to solve the above equations for μ_{Ex} and μ_{Ey}. Fortunately the dependence of r_1 on μ is slight, as seen from Table 9.3; for μ varying from 0·0 to 0·4 the variation in r_1 is about 1 per cent.

TABLE 9.3

r_1 evaluated for various μ_{Ex}, μ_{Ey} and μ: from Chapman and Murphy [1965: Table 1].

μ_{Ex}	μ_{Ey}	μ	r_1	μ_{Ex}	μ_{Ey}	μ	r_1
0·35	0·05	0·0	6·05	0·50	0·20	0·0	2·17
0·35	0·05	0·2	6·09	0·50	0·20	0·2	2·18
0·35	0·05	0·4	6·11	0·50	0·20	0·4	2·19
0·50	0·10	0·0	4·13	0·60	0·20	0·0	2·49
0·50	0·10	0·2	4·16	0·60	0·20	0·2	2·51
0·50	0·10	0·4	4·19	0·60	0·20	0·4	2·52

In fact, approximating each exponential in (9.42) by the first three terms of its power-series expansion and using the further approximation $(1-w)^{-1} \approx 1 + w$, we find that μ cancels out of (9.42). This approximation for (9.42), together with (9.43), are readily solved to give the estimates

$$\hat{\mu}_{Ex} = \frac{r_1 \log r_2}{1 + \frac{1}{2}(\log r_2) - r_1} \qquad (9.44)$$

and

$$\hat{\mu}_{Ey} = \hat{\mu}_{Ex} + \log r_2. \qquad (9.45)$$

To find asymptotic variances of these estimates we assume once again that x_1, x_2, y_1 and y_2 are independent Poisson variables. In addition the catches or removals R_x and R_y are now random variables, and we assume them to be independent Poisson variables also. Therefore, using the delta method, Chapman and Murphy show that

$$\hat{V}[\hat{\mu}_{Ex}] = \left\{\frac{(2\log r_2)(2+\log r_2)}{(2+\log r_2 - 2r_1)^2}\right\}^2 \hat{V}[r_1] + \\ + \left\{\frac{(8\log r_2)(2+\log r_2)r_1(1-r_1)}{r_2(2+\log r_2-2r_1)^4}\right\}\widehat{\text{cov}}[r_1, r_2] + \\ + \left\{\frac{4r_1(1-r_1)}{r_2(2+\log r_2-2r_1)^2}\right\}^2 \hat{V}[r_2]$$

and

$$\hat{V}[\hat{\mu}_{Ey}] = \left\{\frac{(2\log r_2)(2+\log r_2)}{(2+\log r_2-2r_1)^2}\right\}\hat{V}[r_1] + \\ + \frac{4}{r_2}\left\{\frac{(\log r_2)(2+\log r_2)}{(2+\log r_2-2r_1)}\right\}\left\{1+\frac{4r_1(1-r_1)}{(2+\log r_2-2r_1)^2}\right\}\widehat{\text{cov}}[r_1, r_2] + \\ + \frac{1}{r_2^2}\left\{1+\frac{4r_1(1-r_1)}{(2+\log r_2-2r_1)^2}\right\}^2 \hat{V}[r_2],$$

where $\hat{V}[r_1] = r_1^2\left\{\frac{1}{R_x}+\frac{1}{x_1}+\frac{1}{R_y}+\frac{1}{y_1}\right\}$,

$$\hat{V}[r_2] = r_2^2\left\{\frac{1}{x_1}+\frac{1}{x_2}+\frac{1}{y_1}+\frac{1}{y_2}\right\}$$

and

$$\widehat{\text{cov}}[r_1, r_2] = r_1 r_2\left\{\frac{1}{x_1}+\frac{1}{y_1}\right\}.$$

As a check on the approximate solutions (9.44) and (9.45), and to provide more accurate estimates, Table 9.4 (reproduced from Chapman and Murphy [1965]) provides a tabulation of the function

$$r_1 = \frac{\mu_{Ex}(\mu+\mu_{Ex}+\log r_2)[1-\exp(-\mu-\mu_{Ex})]}{(\mu_{Ex}+\log r_2)(\mu+\mu_{Ex})[1-\exp(-\mu-\mu_{Ex}-\log r_2)]}$$

for $\mu_{Ex} = 0{\cdot}05\ (0{\cdot}05)\ 1$, $\mu = 0{\cdot}0, 0{\cdot}2, 0{\cdot}4$, and for admissible values of r_2. To use this table we choose the x-class as the more heavily exploited class, i.e. $\mu_{Ex} > \mu_{Ey}$ and $r_2 < 1$, then we choose μ, find r_1 in the appropriate r_2 column and read off μ_{Ex}, the required estimate; the estimate of μ_{Ey} is then given by (9.45). Although the most appropriate value of μ should be chosen, it is clear that the choice of μ makes little difference (this was first noted by Lander [1962]). In fact, if the assumption (i) of equal natural mortality was not true, we would expect the estimate of μ_{Ex} to be little affected, provided

TABLE 9.4

Tabulation of r_1 as a function of μ, μ_{Ex} and r_2: from Chapman and Murphy [1965: Table 2].

						μ = 0·00								
r_2 =	0·40	0·45	0·50	0·55	0·60	0·65	0·70	0·75	0·80	0·85	0·90	0·95	1·00	μ_{Ex}
													1·0000	0·05
												2·0018	1·0000	0·10
											3·1905	1·4820	1·0000	0·15
										4·9275	2·0074	1·3119	1·0000	0·20
									8·3474	2·6408	1·6426	1·2275	1·0000	0·25
								21·1709	3·5035	2·0178	1·4654	1·1771	1·0000	0·30
								4·8880	2·4787	1·7274	1·3608	1·1436	1·0000	0·35
							7·7755	3·1032	2·0338	1·5596	1·2918	1·1198	1·0000	0·40
						19·0385	4·0669	2·4186	1·7854	1·4504	1·2430	1·1021	1·0000	0·45
						5·8836	2·9467	2·0569	1·6270	1·3737	1·2067	1·0883	1·0000	0·50
					11·0121	3·7643	2·4066	1·8335	1·5173	1·3169	1·1786	1·0773	1·0000	0·55
				208·8195	5·2886	2·8983	2·0890	1·6820	1·4370	1·2733	1·1563	1·0684	1·0000	0·60
				9·4038	3·6787	2·4280	1·8801	1·5725	1·3756	1·2388	1·1381	1·0610	1·0000	0·65
			73·7130	5·1836	2·9208	2·1329	1·7324	1·4899	1·3273	1·2108	1·1231	1·0548	1·0000	0·70
			9·5470	3·7381	2·4804	1·9307	1·6225	1·4254	1·2884	1·1876	1·1105	1·0494	1·0000	0·75
		369·2827	5·4338	3·0085	2·1929	1·7837	1·5378	1·3736	1·2563	1·1682	1·0997	1·0449	1·0000	0·80
		11·4086	3·9442	2·5690	1·9906	1·6721	1·4704	1·3312	1·2294	1·1517	1·0904	1·0409	1·0000	0·85
		6·1488	3·1758	2·2756	1·8408	1·5846	1·4157	1·2960	1·2067	1·1375	1·0824	1·0374	1·0000	0·90
	18·5009	4·3625	2·7073	2·0660	1·7254	1·5142	1·3704	1·2662	1·1872	1·1252	1·0754	1·0343	1·0000	0·95
	7·8719	3·4639	2·3922	1·9090	1·6339	1·4564	1·3323	1·2407	1·1703	1·1145	1·0691	1·0316	1·0000	1·00

TABLE 9.4 (continued)

Tabulation of r_1 as a function of μ, μ_{Ex} and r_2: from Chapman and Murphy [1965: Table 2].

μ = 0·20

r_2 = 0·40	0·45	0·50	0·55	0·60	0·65	0·70	0·75	0·80	0·85	0·90	0·95	1·00	μ_{Ex}
												1·0000	0·05
											2·0035	1·0000	0·10
										3·1961	1·4832	1·0000	0·15
									4·9408	2·0109	1·3130	1·0000	0·20
								8·3785	2·6479	1·6455	1·2285	1·0000	0·25
							21·2722	3·5165	2·0232	1·4680	1·1781	1·0000	0·30
							4·9114	2·4879	1·7321	1·3632	1·1446	1·0000	0·35
						7·8216	3·1180	2·0413	1·5638	1·2941	1·1208	1·0000	0·40
					19·1748	4·0910	2·4301	1·7920	1·4543	1·2452	1·1030	1·0000	0·45
					5·9256	2·9641	2·0667	1·6330	1·3774	1·2088	1·0892	1·0000	0·50
				11·1054	3·7911	2·4208	1·8422	1·5229	1·3205	1·1806	1·0782	1·0000	0·55
			210·8905	5·3333	2·9189	2·1013	1·6899	1·4422	1·2767	1·1583	1·0693	1·0000	0·60
			9·4968	3·7098	2·4452	1·8911	1·5800	1·3807	1·2421	1·1401	1·0619	1·0000	0·65
		74·5589	5·2347	2·9453	2·1480	1·7425	1·4969	1·3322	1·2140	1·1250	1·0556	1·0000	0·70
		9·6563	3·7749	2·5012	1·9444	1·6320	1·4320	1·2930	1·1908	1·1124	1·0503	1·0000	0·75
	374·1552	5·4958	3·0380	2·2112	1·7962	1·5467	1·3800	1·2608	1·1713	1·1016	1·0457	1·0000	0·80
	11·5587	3·9891	2·5942	2·0072	1·6838	1·4789	1·3374	1·2339	1·1547	1·0923	1·0417	1·0000	0·85
	6·2295	3·2118	2·2978	1·8560	1·5956	1·4238	1·3020	1·2110	1·1405	1·0842	1·0383	1·0000	0·90
18·7797	4·4196	2·7380	2·0861	1·7397	1·5247	1·3782	1·2720	1·1914	1·1281	1·0771	1·0352	1·0000	0·95
7·9901	3·5090	2·4192	1·9275	1·6474	1·4665	1·3399	1·2464	1·1744	1·1173	1·0709	1·0324	1·0000	1·00

TABLE 9.4 (concluded)

Tabulation of r_1 as a function of μ, μ_{Ex} and r_2: from Chapman and Murphy [1965: Table 2].

$\mu = 0{\cdot}40$

$r_2 =$ 0·40	0·45	0·50	0·55	0·60	0·65	0·70	0·75	0·80	0·85	0·90	0·95	1·00	μ_{Ex}
												1·0000	0·05
											2·0052	1·0000	0·10
										3·2017	1·4845	1·0000	0·15
									4·9541	2·0144	1·3141	1·0000	0·20
								8·4094	2·6550	1·6483	1·2295	1·0000	0·25
							21·3734	3·5295	2·0286	1·4705	1·1791	1·0000	0·30
							4·9346	2·4970	1·7367	1·3655	1·1456	1·0000	0·35
						7·8675	3·1327	2·0488	1·5679	1·2963	1·1217	1·0000	0·40
					19·3109	4·1149	2·4416	1·7985	1·4581	1·2473	1·1039	1·0000	0·45
					5·9676	2·9815	2·0764	1·6389	1·3810	1·2108	1·0901	1·0000	0·50
				11·1986	3·8179	2·4349	1·8508	1·5284	1·3239	1·1826	1·0791	1·0000	0·55
			212·9613	5·3779	2·9394	2·1135	1·6978	1·4474	1·2801	1·1602	1·0702	1·0000	0·60
			9·5898	3·7407	2·4623	1·9020	1·5873	1·3856	1·2453	1·1420	1·0627	1·0000	0·65
		75·4052	5·2858	2·9698	2·1630	1·7525	1·5038	1·3369	1·2171	1·1269	1·0565	1·0000	0·70
		9·7656	3·8116	2·5219	1·9579	1·6414	1·4386	1·2976	1·1938	1·1142	1·0512	1·0000	0·75
	379·0338	5·5578	3·0675	2·2294	1·8087	1·5555	1·3864	1·2653	1·1743	1·1034	1·0466	1·0000	0·80
	11·7089	4·0339	2·6192	2·0237	1·6954	1·4873	1·3435	1·2382	1·1577	1·0941	1·0426	1·0000	0·85
	6·3101	3·2478	2·3198	1·8712	1·6066	1·4319	1·3079	1·2153	1·1434	1·0860	1·0391	1·0000	0·90
19·0589	4·4765	2·7685	2·1060	1·7538	1·5351	1·3860	1·2778	1·1956	1·1310	1·0789	1·0360	1·0000	0·95
8·1084	3·5541	2·4460	1·9458	1·6607	1·4764	1·3474	1·2520	1·1785	1·1201	1·0727	1·0332	1·0000	1·00

that μ_x and μ_y were not too different and were small relative to μ_{Ex}, which for example, is a common fishery situation.

If we replace assumption (ii) by the stronger assumption

(ii′) $a_i = b_i = Kf_i \quad (i = 1, 2), \quad f_1, f_2$ known,

as in the instantaneous removal model, we can estimate μ from the relations (cf. (9.38) and (9.39))

$$\exp\{-(\mu+\mu_{Ex})\} = \frac{X_2}{X_1} = \frac{x_2 f_1}{x_1 f_2},$$

that is,

$$\hat{\mu} = \log\left(\frac{x_1 f_2}{x_2 f_1}\right) - \hat{\mu}_{Ex}, \tag{9.46}$$

and finally estimate X_1 and Y_1 from (9.40) and (9.41) (with $T = 1$). The procedure is iterative; μ is guessed, $\hat{\mu}_{Ex}$ determined from (9.44) or Table 9.4, and then $\hat{\mu}$ (from 9.46) may be used to re-enter the table. One cycle will usually be sufficient as the estimates of μ_{Ex} and μ_{Ey} are essentially independent of $\hat{\mu}$.

CHAPTER 10

MORTALITY AND SURVIVAL ESTIMATES FROM AGE DATA

10.1 LIFE TABLES

10.1.1 Introduction

A life table is basically a table used for determining the mortality rate and the life expectancy of an individual at a given age. Detailed techniques for constructing and analysing such tables (cf. Keyfitz [1968], Goodman [1969]) have been developed for human populations, where they are largely used for determining insurance premiums. The success of such methods is due to the fact that human populations are large and stable, and information on various categories such as age, occupation, sex, etc. is readily obtainable by census, sample survey, records, etc. In recent years life tables have also been used for investigating the dynamics of animal populations and for determining policies in population management (Quick [1963]). However, animal populations tend to fluctuate with changes in environment, and the data for the life table are usually based on samples which may not be as representative as one would hope. Therefore, although animal life tables may help to give an overall picture of the population, they are often of limited accuracy and, where possible, should be backed up by other methods of estimation (e.g. the methods of **5.4**).

Life table methods are discussed, for example, by Deevey [1947], Allee *et al.* [1949: 294–301], Hickey [1952: birds], Eberhardt [1969a: wildlife management], Southwood [1966: insects], Varley and Gradwell [1970: insects] and Caughley [1966: mammals]. General methods for ageing animals are described in Taber [1969], Tesch [1968: fish], Southwood [1966: insects].

NOTATION. An animal life table usually contains the following columns:

(i) age, x: measured in some convenient unit of time such as day, month, year, etc.

(ii) l_x: the number surviving to age x from a "cohort" of l_0 animals, i.e. from l_0 animals born at the same time. The usual convention of scaling l_x so that $l_0 = 1000$ or 1, although useful in standardising survivorship curves (see below), is not recommended when samples are small, as it can give a false impression of the accuracy of the table. If scaling is used then the raw data should also be quoted.

(iii) $d_x = l_x - l_{x+1}$: the number of deaths in the age-class $[x, x+1)$. If w is the last age in the table, then

$$l_x = (l_x - l_{x+1}) + (l_{x+1} - l_{x+2}) + \ldots + (l_w - 0)$$
$$= d_x + d_{x+1} + \ldots + d_w.$$

(iv) $q_x = d_x/l_x$: the observed mortality rate at age x. Here

$$q_x = d_x/(d_x + d_{x+1} + \ldots + d_w). \quad (10.1)$$

(v) e_x: the observed mean expectation of life remaining for animals of age x.

Two further columns can be included to facilitate the calculation of e_x:

(vi) L_x: the average number of animals alive during the interval $[x, x+1)$. Thus

$$L_x = \int_x^{x+1} l_t \, dt$$

which is usually approximated by $\frac{1}{2}(l_x + l_{x+1})$, the average of the numbers alive at the beginning and end of this interval (this approximation is used in Table 10.1).

(vii) T_x, where

$$\begin{aligned} T_x &= \int_x^w l_t \, dt \\ &= L_x + L_{x+1} + \ldots + L_w \\ &\approx \tfrac{1}{2}(l_x + l_{x+1}) + \tfrac{1}{2}(l_{x+1} + l_{x+2}) + \ldots + \tfrac{1}{2}(l_w + 0) \\ &= \tfrac{1}{2} l_x + \sum_{y=x+1}^{w} l_y . \end{aligned} \quad (10.2)$$

This column is usually obtained by summing the L_x column cumulatively from the bottom up. As T_x is the total number of time-intervals lived by the group of l_x animals, we have from (10.2)

$$e_x = T_x/l_x \approx \tfrac{1}{2} + \sum_{y=x+1}^{w} (l_y/l_x). \quad (10.3)$$

An example of the calculations is given in Table 10.1.

TABLE 10.1

Artificial life table illustrating the calculation of q_x and e_x.

x	l_x	d_x	$100\ q_x$	L_x	T_x	e_x
0–1	1000	750	75	625	804	0·80
1–2	250	210	84	145	179	0·72
2–3	40	30	75	25	34	0·85
3–4	10	7	70	6·5	9	0·90
4–5	3	2	67	2	2·5	0·83
5–6	1	1	100	0·5	0·5	0·50
6–7	0	–	–	–	–	–

A number of other parameters are sometimes estimated from life tables. For example, if Z_x is the observed total instantaneous mortality rate during the interval $[x, x+1)$, then, from (1.1),

$$l_{x+1} = l_x \, e^{-Z_x}. \tag{10.4}$$

Hence the proportion surviving, S_x $(= 1 - q_x)$, is given by

$$S_x = l_{x+1}/l_x = e^{-Z_x},$$

so that

$$Z_x = -\log S_x.$$

If q_x is small, then $-\log S_x \approx q_x$ and hence $q_x \approx Z_x$.

A plot of l_x versus x is called a "survival" or "survivorship curve", and the plot of d_x versus x is called a "mortality curve", or sometimes "kill curve". Deevey [1947] gives three basic types of survival curves (Fig. 10.1):

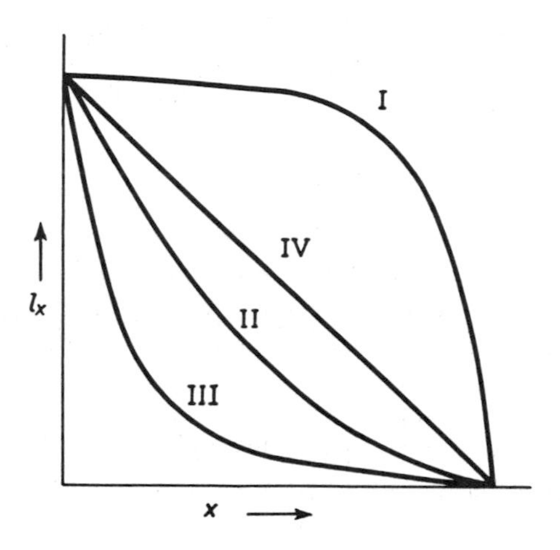

Fig. 10.1 Types of survival curves: after Slobodkin [1962].

Type I: mortality acts most heavily on the old individuals.

Type II: a constant fraction of the cohort dies at each age. This means that the survival rate is constant, so that $l_x = S\, l_{x-1} = S^x l_0$ and the plot of $\log l_x$ against x is a straight line. Such a survival curve is typical of the adult stages of many birds and fishes, though it may not hold strictly for all ages of a species' life. Indeed, as pointed out by Hickey [1952], many of the species that Deevey [1947] grouped under this category may exhibit a type I curve when their complete span of life is finally investigated.

Type III: mortality acts most heavily on the early stages of life. This curve reflects the common situation where a single female produces an enormous number of young, but there is a very high initial mortality through predation, etc. Although this type of curve is no doubt characteristic of many fishes and invertebrates it is very difficult to obtain accurate information on survival in the early stages of life.

The various types of curves are also discussed by Slobodkin [1962: Chapter 4] who adds a fourth curve, type IV, say, representing a population in which a constant number die per unit time, regardless of population size. This type of curve appears to be common in animals receiving some human protection, as for example in a laboratory or a zoo (Comfort [1957: 361]).

It should be noted that the above notation (which follows actuarial practice) differs from that commonly used in the construction of animal life tables. Usually x denotes the age-class $[x-1, x)$, and l_x is defined to be the number surviving at the beginning of this interval, i.e. of age $x - 1$: a survivorship curve is then the plot of l_x against $x - 1$. One reason for the popularity of the latter notation is that in applying life-table methods to tagging experiments it is convenient to think of d_x as the number of tagged animals found dead in their xth year of life (called D_x in **5.4.3**). Since both notations are used, they can in practice be combined by the simple expedient of labelling the age-intervals 0–1, 1–2, etc. instead of 0, 1, 2, etc. or 1, 2, 3, etc. For example, the first entry in the d_x column would refer to the number dying in the interval $[0, 1)$, irrespective of whether we call it d_0 or d_1.

LIFE AND FERTILITY TABLE. This table consists of the columns x and l_x (usually with $l_0 = 1$), as for the life table, but l_x now refers to the female population only. A new column is then added on the basis of observations: this is the m_x or *age-specific fertility* column that records the number of *living* females born per female in each age-interval; sometimes the sex at birth is not readily distinguishable, and m_x must be estimated from total births per female of age x and, say, a 50 : 50 sex ratio. If one is interested in numbers of births (and not just live births), e.g. numbers of eggs, then the fertility column is often called the *fecundity* column.

Columns l_x and m_x are then multiplied together to give the total number of females born in each age category; this is the $l_x m_x$ column of Table 10.2.

TABLE 10.2

Life and fertility table for the beetle, *Phyllopertha horticola*: **modified from Laughlin [1965].**

x (weeks)	l_x	m_x	$l_x m_x$	$x l_x m_x$
0	1·00	–	Immature stages	–
49	0·46	–		–
50	0·45	–		–
51	0·42	1·0	0·42	21·42
52	0·31	6·9	2·13	110·76
53	0·05	7·5	0·38	20·14
54	0·01	0·9	0·01	0·54
Total			2·94	152·86

From this column one can calculate, for example, the *net reproductive rate*

$$R_0 = \sum_x l_x m_x / l_0 = 2{\cdot}94 ,$$

or estimate the *generation length* (i.e. mean lapse of time between a female's date of birth and the mean date of birth of her offspring) by

$$\frac{\sum_x x l_x m_x}{\sum_x l_x m_x} = \frac{152{\cdot}86}{2{\cdot}94} = 52{\cdot}0 . \qquad (10.5)$$

Further examples of such tables are given in Turner *et al.* [1970]. Other parameters which may be of interest are discussed in Southwood [1966: 287–91] and Mertz [1971].

MEAN MORTALITY RATE. If the observed mortality rate is fairly constant, then we can combine the estimates q_x in several ways to give a mean mortality rate. For example, we can use the average

$$q_{\text{ave}} = \frac{1}{w} \sum_{x=0}^{w-1} q_x = \frac{1}{w} \sum_{x=0}^{w-1} \frac{d_x}{l_x} \qquad (10.6)$$

(the last age w is not included as $d_w = l_w$ and $q_w = 1$), or the more popular pooled version

$$q_{\text{pool}} = \sum_{x=0}^{w-1} d_x \Big/ \sum_{x=0}^{w-1} l_x . \qquad (10.7)$$

Farner [1955: 405] discusses both these mortality rates and points out that unweighted averages such as q_{ave} give equal weight to age-classes of different sizes. To avoid this disproportionate emphasis one can take a weighted average of the q_x with weights proportional to the numbers l_x in the various classes; this leads to q_{pool}.

If Z_x is constant ($= Z$ say), then from (10.4) the mean expectation of life at birth is

$$\left(\int_0^\infty l_t \, dt\right)/l_0 = \int_0^\infty e^{-Zt} \, dt = 1/Z = -1/\log(1-q), \text{ say,}$$

which is commonly estimated by

$$e_0' = -1/\log(1-q_{\text{pool}}) .$$

A number of approximations have been widely used, particularly in ornithological work, e.g. when q is small $-\log(1-q) \approx q$ and $-1/\log(1-q) \approx 1/(q+\frac{1}{2}q^2) \approx (1-\frac{1}{2}q)/q$; the latter approximation leads to Burkitt's [1926] estimate $(2-q)/2q$.

LIFE EQUATIONS. For some populations it is helpful to classify individuals according to stage of development rather than according to age. In this case, if a cohort of individuals (and possibly their offspring) are followed through the various stages, and the numbers in each stage recorded, one can set up for the cohort a so-called *life equation* (or commonly called "budget" in entomological studies; Southwood [1966: 277 ff.]); two examples of life equations are given in Tables 10.3 and 10.4. Quick [1963] describes

TABLE 10.3

Hypothetical life equation for a stationary turkey population: adapted from Mosby [1967: Table 6.12].

Stage	Factor causing change in population number		Population
(1)	Spring breeding population; 50 : 50 sex-ratio assumed.		1000
(2)	Number of successful nests (number females) × (% nesting success)/100 500 × 0·351	= 175·5	
(3)	Number poults hatched (successful nests) × (av. clutch) × (hatchability %)/100 175·5 × 12·3 × 0·93	= 2008	
(4)	Number of poults in fall (poults hatched) × (% poult survival)/100 2008 × 0·755	= 1516	
(5)	Fall adult population (adults in spring) × (% adult survival)/100 1000 × 0·993	= 993	
(6)	Total pre-hunting season population (5) + (4)	=	2509
(7)	Hunting season harvest (pre-hunting season pop.) × (% harvest)/100 2509 × 0·333	= 836	
(8)	Post-hunting season population (6) – (7)	=	1673
(9)	Overwinter loss (8) – (10)	= 673	
(10)	Spring breeding population (assuming stationary population) (1)	=	1000

the life equation as a "book-keeping system that involves intensive observation of living animals" and points out that the method requires a trained observer to spend a great deal of time in the field keeping a careful record of the development of the cohort.

As it is not always possible to count all the individuals in a given cohort, sampling methods may be needed for keeping track on the numbers entering each stage. In this case a difficulty can arise as the age-ranges for the different stages may overlap, so that different members of the cohort may be in different stages at the same time (e.g. Beaver [1966], Berryman

TABLE 10.4
Model life equation for spruce budworm, 1952 and 1953 (numbers per 10 sq.ft of branch surface): abridged from Morris and Miller [1954].

Stage	l_x	Mortality factor	d_x	100 q_x
Eggs	174	Parasites	3	2
		Predators	15	9
		Other	1	1
		Total egg loss	19	12
Instar I	155	Dispersion loss	74·4	48
Hibernacula	80·6	Winter loss	13·7	17
Instar II	66·9	Dispersion loss	42·2	63
Instars III and IV	24·7	Parasites	8·92	36
		Disease	0·54	2
		Birds	3·39	14
		Other	10·57	43
		Total larvae loss	23·42	95
Pupae	1·28	Parasites	0·10	8
		Predators	0·13	10
		Other	0·23	18
		Total pupae loss	0·46	36
Moths	0·82		0	0
Generation:		$\Sigma\, d_x = 173{\cdot}18$	173·18/174 = 99·53	

[1968], Hughes and Gilbert [1968], Ashford *et al.* [1970]). This means that a single sample taken at a time of overlap will not, in itself, provide useful information about any of the overlapping stages. Only by examining a series of samples can the numbers in each stage be estimated. A number of methods for estimating the various parameters, including the stage specific mortality rates, are available and these are described in **13.3.5**.

A life equation can also be described graphically in terms of a "survival curve"; examples of such curves are given in Quick [1963].

10.1.2 Types of life tables

1 Age-specific (horizontal) life tables

The age-specific life table (also known as the dynamic life table) is based on the fate of a given cohort of l_0 individuals. Data for such a table can be collected in two ways: (i) by recording the ages at death of animals in the cohort, thus giving a d_x series, or (ii) by recording the number of animals from the cohort still alive at various times, thus giving an l_x series.

The age-specific method is most readily applied to laboratory populations of relatively short-lived species such as voles (Leslie and Ranson [1940], Leslie *et al.* [1955]), or to zoo populations (Comfort [1957]). However, for natural as opposed to laboratory populations the initial cohort can be identified by tagging, and either of the above methods for collecting the data can be used; obviously the population must be enclosed, otherwise emigration is confounded with mortality (Brooks [1967]). If method (i) is used, then a problem can arise with regard to predation, as some of the tagged individuals may disappear without a trace or may be difficult to find; radioactive tagging can be useful here (Schnell [1968]). If live-trapping is used and the probability of capture is constant throughout the period of investigation, then the number caught on each trapping occasion will be a constant proportion of the live population, and the sequence of captures will form an l_x series, except for l_0 (e.g. Marsden and Baskett [1958: though emigration is confounded with mortality]).

2 *Time-specific (vertical) life tables*

If N_x, the number of animals of age x in a population, remains constant with respect to time for each x, then the population is said to be "stationary" and the age-structure of the population at any instant of time t will reflect the fate of a cohort of N_0 animals born at time t. Thus the cohort of N_0 individuals will reduce to N_1 individuals at time $t + 1$, to N_2 at time $t + 2$, etc., and the sequence $\{N_x\}$ forms an l_x series. This series can then be used to construct a life table as in the age-specific case above. Such a life table, determined from the age-structure of a stationary population at a given point in time, is called a "time-specific" life table. In practice the age structure must be determined by a sample, so that time-specific tables involve sampling errors. Therefore although a time-specific table is usually easier to obtain (particularly for long-lived animals), it is less accurate than an age-specific table and can only be used when the population is stationary. Some of the problems associated with obtaining a suitable sample are considered by Caughley [1966], and the following discussion is based on his article.

The concept of a stationary population has been developed from demographic research on man and is useful for species which, like man, have no seasonally restricted period of births. However, very few species breed at the same rate throughout the year, and the stationary age distribution must be redefined if it is to include seasonal breeders. For species with one restricted breeding season each year a stationary population can be defined as one in which the numbers in each age-group do not vary at intervals of one year. The stationary age distribution can then be defined for such populations as the distribution of ages at a given time of the year, and there will be an infinite number of different age distributions, depending on the time chosen. But the distribution of ages will only form an l_x series with integral values of x if all the births for the year occur instantaneously and the sample is taken at that instant. This situation does not occur in practice, though it is approximated when the population has a restricted season of births, the age structure is sampled over this period, and the ratio of live births to the numbers in the other age-groups can be determined, for example from the number of females either pregnant or suckling young, or from the size of egg-clutch. Birth rates can also be determined from dead animals using egg follicle counts (e.g. ducks, pheasants), corpora lutea counts (e.g. deer, elk, moose), placental scar counts (e.g. muskrat, beaver), and embryo counts (all mammals).

If the age distribution is sampled at time t $(0 < t < 1)$ after the breeding season then we obtain a series, l_t, l_{t+1}, ... In this case the series q_{x+t} can sometimes be fitted by a regression curve, and the values of q_x for integral values of x obtained by interpolation. For example, Hickey [1960] constructed a life table for the domestic sheep by recording the ages at death of 83 113 females on selected farms in the North Island of New Zealand. Using the age series 1½, 2½, 3½ years, etc. in preference to integral ages, he found that the q_x series conformed closely to the regression

$$\log q_x = 0{\cdot}156x + 0{\cdot}24\,;$$

q_0 was calculated from a knowledge of the number of lambs dying before 1 year of age out of 85 309 born alive.

SAMPLING METHOD. Caughley [1966] points out that because a sample consists of dead animals its age frequencies do not necessarily form a d_x series. Such a series is obtained only when the sample represents the frequencies of ages at death in a stationary population. Many published samples treated as if they formed a d_x series are not actually appropriate to this form of analysis. For instance, if the animals are obtained by shooting which is unselective with respect to age, the sample gives the age structure of the living population at that time and leads to an l_x series; that the animals were killed to get the data is irrelevant. Similarly, groups of animals killed by avalanches, fires or floods – catastrophic events that preserve a sample of the age-frequencies of animals during life – can sometimes be used for an l_x series.

A sample may include both l_x and d_x components. For example, it could consist of a number of dead animals, some of which have been unselectively shot, whereas the deaths of others are attributable to "natural" mortality. Alternatively it could be formed by a herd of animals killed by an avalanche in an area where carcasses of animals that died naturally were also present. In both these cases the l_x and d_x data are confounded and these heterogeneous samples of ages at death cannot be treated as either an l_x or a d_x series.

Summarising, there are three main methods for obtaining data for a time-specific life table:

(i) Recording ages at death by ageing a random sample of carcasses (d_x series).

(ii) Recording ages at death of a random sample killed by some catastrophic event; the frequencies form an l_x series (provided there are no carcasses from natural causes).

(iii) Recording the ages of a random sample obtained alive or dead by live-trapping or unselective hunting (l_x series). Common sources of such information with regard to hunting are (Quick [1963]): (a) Game inspection stations; particularly adaptable to procuring data on big game. These stations also provide opportunity to classify the source of data with respect to area and to the time-period in which the kills were made. (b) Game-bird wing and tail samples obtained by the co-operation of the hunter. This method is usable with upland game birds in the U.S.A., such as ruffed grouse, when the game department provides a stamped addressed and blood-proof envelope in which the hunter can send samples. (c) Hunter-bag checks. This procedure requires that the warden or game biologist personally examines the hunter's bag; it has been widely used to get data on water-fowl and to some extent on pheasants. (d) Fur buyer collections; examination of pelts and carcases taken by trappers or in the possession of fur buyers.

BIAS. Unfortunately both methods (i) and (iii) are subject to bias with regard to the frequency of the first-year class. For example, dead immature animals, especially those dying soon after birth, tend to decay faster than the adults, so that they are under-represented in the count of carcases. Also the ratio of juveniles to adults in a shot sample is usually biased because the two age-classes have different susceptibilities to hunting. However, if a d_x series is used, we have from (10.1)

$$q_x = d_x \Big/ \left(\sum_{y=x}^{w} d_y \right),$$

and the value of q at any age x, say, is independent of the frequencies of the younger age-classes. Alternatively, if an l_x series is used,

$$q_x = 1 - (l_{x+1}/l_x),$$

so that even if the l_x series is calculated from age-frequencies in which the initial frequency is inaccurate, the ratio l_{x+1}/l_x for the older ages will not be affected. Thus for methods (i) and (iii) the biases mentioned above will only affect the first one or two values of q_x.

When using an l_x series another possible source of error lies in the ageing technique. For example, l_x strictly refers to the number in the sample of *exact* age x years, whereas it may only be possible to age an animal to within one or two months. In this case l_x refers to the number in the age range $[x - \delta_1, x + \delta_2]$, where $\delta_i \geqslant 0$ $(i = 1, 2)$. However, this error of age classification will have little effect on the ratios l_{x+1}/l_x, provided the mortality rate does not change too abruptly from one age-class to the next, so that

$$l_{x+1+t}/l_{x+t} \approx l_{x+1}/l_x$$

for all t in the interval $[-\delta_1, \delta_2]$. Even when the animals are aged exactly, the same problem can still arise because a sample is rarely instantaneous, often taking several months to collect.

TEST FOR STATIONARITY. Caughley [1966] mentions five methods used for determining whether the age structure of a sample is consistent with its having been drawn from a stationary age distribution:

(a) Comparison of the "mean mortality rate", calculated from the age distribution of the sample, with the proportion represented by the first age-class (Kurtén [1953: 51]).
(b) Comparison of the annual female fecundity of a female sample with the sample number multiplied by the life expectancy at birth, the latter statistic being estimated from the age structure (Quick [1963: 210]).
(c) Calculation of instantaneous birth rates and death rates, respectively, from a sample of the population's age distribution and a sample of ages at death (Hughes [1965]).
(d) Comparison of the age distribution with a prejudged notion of what a stationary age distribution should be like (Breakey [1963]).
(e) Examination of the l_x and d_x series, calculated from the sampled age distribution, for evidence of a common trend (Quick [1963: 204]).

Caughley points out that the above methods are invalid as they are based on circular arguments, e.g. methods (a) to (c) are tautological because they assume that the sampled age distribution is either an l_x or a d_x series. In fact, given no information other than a single age distribution, it is impossible to test whether the distribution is from a stationary population. Stationarity can only be tested by examining a chronological sequence of age distributions.

STABLE AGE DISTRIBUTION. We shall see in 13.3.4 that when the birth and mortality rates are constant over time for each age-class, then the age distribution eventually assumes a stable form and the population is said to be stable. If, in addition, the population remains constant in size, the size of each age-class remains constant and the population is said to be stationary.

Example 10.1 Himalayan thar (*Hemitragus jemlahicus*): Caughley [1966]

The Himalayan thar, a hollow-horned ungulate introduced into New Zealand in 1904, now occupies 2000 square miles of mountainous country in the South Island. Thar were liberated at Mount Cook and have since spread mostly north and south along the Southern Alps. They are still spreading at a rate of about 1·1 miles a year, so that the populations farthest from the point of liberation have been established only recently and have not yet had time to increase greatly in numbers. Closer to the site of liberation the density is higher, and around the point of liberation itself there is evidence that the population has decreased.

The thar can be accurately aged from the growth rings on its horns which are laid down in each winter of life other than the first. An l_x series was obtained from a sample of 623 females, older than 1 year, shot in the Godley and Macaulay valleys between November 1963 and February 1964. Preliminary work on behaviour indicated that there was very little dispersal of females into or out of this region, both because the females have distinct home ranges and because there are few ice-free passes linking the valley heads.

The first question Caughley considers is whether or not the population is stationary. Obviously it is impossible to determine the stationary nature of a population by examining the age structure of a single sample only, even when birth rates are known. However, in some circumstances, a series of age structures will give the required information. Caughley utilises this fact in investigating the stability of the population sampled as follows.

The sample was taken about halfway between the point of liberation and the edge of the range, that is, in the region between increasing and decreasing populations where one would expect to find a stationary population. The animals came into the Godley valley from the southwest and presumably colonised this side of the valley before crossing the two miles of river bed to the northeast side. Having colonised the northeast side, the thar would then cross the Sibald Range to enter the Macaulay valley, which is a further six miles northeast. The sample can therefore be divided into three subsamples corresponding to the different periods of time that the animals have been present in the three areas. A 10×3 contingency test for differences between the three age distributions of females 1 year of age or older gave no indication that the three subpopulations differed in age structure ($\chi^2_{18} = 22{\cdot}34$; $P = 0{\cdot}2$). This information can be interpreted in two ways: either the three subpopulations were constant in size and were hence likely to have stationary age distributions, or the subpopulations were increasing at the same rate, so that they could have identical stable age distributions. The second alternative implies that the subpopulations would have different densities because they had been increasing for differing periods of time. However, an analysis of the three densities gave no indication that they were different and Caughley rejected the second alternative. Independent subjective evidence based on observation also suggests that the populations were not increasing or decreasing.

The second question Caughley considers is whether the age structure of the sample is an unbiased estimate of the age structure of the population. The most obvious source of bias is behavioural or range differences between males and females. For instance, should males tend to occupy terrain which is more difficult to hunt over than that used by females, they would be under-represented in a sample obtained by hunting. During the summer, thar range in three main kinds of groups: one consisting of females, juveniles and kids, a second consisting of young males, and the third of mature males. The task of sampling these three groupings in the same proportion as they occur throughout the area is complicated by their preferences for terrain that differs in slope, altitude, and exposure. Consequently the attempt to take an unbiased sample of both males and females was abandoned, and the hunting was directed towards sampling only the nanny–kid herds in an attempt to take a representative sample of females.

To determine whether some age-classes were more susceptible than others to shooting, several chi-squared tests were carried out. Females other than kids were divided into two groups: those from herds in which some members were aware of the presence of the shooter before he fired, and those from herds which were undisturbed before shooting commenced. If any age-group is particularly wary, its members should occur more often in the "disturbed" category. However, a chi-squared test ($\chi_9^2 = 7{\cdot}28$; $P = 0{\cdot}6$) revealed no significant differences between the age structures of the two categories. The sample was also divided into those females shot at ranges less than 200 yards and those shot outside this range. If animals in a given age-class are more easily stalked than others, they will tend to be shot at closer ranges. Alternatively, animals which present small targets may be under-represented in the sample of those shot at ranges over 200 yards. This is certainly true of kids which are difficult to see, let alone to shoot, at ranges in excess of 200 yards. The kids, therefore, were not included in the analysis because their under-representation was an acknowledged fact. However, for the older females, a chi-squared test ($\chi_9^2 = 9{\cdot}68$; $P = 0{\cdot}4$) did not indicate any significant difference between the age structures of the two groups. This did not imply that no biases existed – the yearling class could well be under-represented beyond 200 yards – but that no bias could be detected from the sample of the size used.

The shooting yielded 623 females 1 year old or older, and the numbers (f_x) at each age are shown in Table 10.5. Here l_x refers to the (scaled) number of animals in the age interval x years $-\frac{1}{2}$ month to x years $+ 2\frac{1}{2}$ months rather than the number with exact age x. However, as pointed out on p. 402, this will not have much effect on the q_x series.

Up to the age of 12 years (beyond this age the frequencies dropped below 5) it was found that the age frequencies were closely fitted by a quadratic regression (Fig. 10.2)

$$E[\log f_x] = \beta_0 + \beta_1 x + \beta_2 x^2, \quad (x = 1, 2, \ldots, 12),$$

the cubic term being non-significant. The fitted regression is

$$\log Y_x = 1{\cdot}9673 + 0{\cdot}0246x - 0{\cdot}01036x^2,$$

and Caughley used it to obtain "smoothed" age frequencies Y_x. He felt that by using this curve he would greatly reduce the "noise" resulting from

TABLE 10.5

Life table and fecundity table for the thar *Hemitragus jemlahicus* (females only): from Caughley [1966: Table 2].

Age (years)	Frequency in sample f_x	Adjusted frequency Y_x	Female live births per female (m_x)	l_x	d_x	100 q_x
0–1	–	205	0·000	1000	533	53·3
1–2	94	95·83	0·005	467	6	0·3
2–3	97	94·43	0·135	461	28	6·1
3–4	107	88·69	0·440	433	46	10·6
4–5	68	79·41	0·420	387	56	14·5
5–6	70	67·81	0·465	331	62	18·7
6–7	47	55·20	0·425	269	60	22·3
7–8	37	42·85	0·460	209	54	25·8
8–9	35	31·71	0·485	155	46	29·7
9–10	24	22·37	0·500	109	36	33·0
10–11	16	15·04	0·500	73	26	35·6
11–12	11	9·64	} 0·470	47	18	38·2
12–13	6	5·90		29		
13–14	3		} 0·350			
14–15	4					
15–16	3					
16–17	0					
17–18	1					

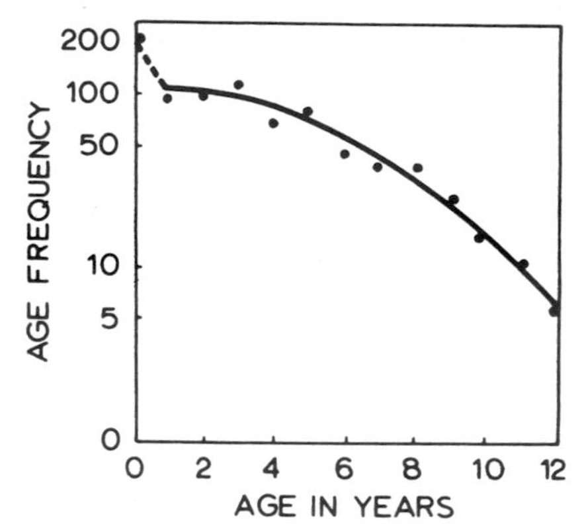

Fig. 10.2 Age frequency (on a logarithmic scale) plotted against age for a sample of female thar: from Caughley [1966].

sampling variation, the differential effect on mortality of different seasons, and the minor heterogeneities which, although not detectable, were almost certain to be present.

The frequencies of births were estimated from the observed mean numbers of female kids per female at each age (m_x in Table 10.5). They were calculated as the number of females at each age in the shot sample either carrying a foetus or lactating, divided by the number of females of that age. These values were then halved because the sex ratio of late foetuses and kids did not differ significantly from 1: 1 (93 males : 97 females). The method is open to a number of objections: it assumes that all kids were born alive, that all females neither pregnant nor lactating were barren for that season, and that twinning did not occur. The first assumption, if false, would give rise to a positive bias, and the second and third to a negative bias. However, the ratio of females older than 2 years that were neither pregnant nor lactating did not differ significantly between the periods November to December and January to February ($\chi_1^2 = 0{\cdot}79, P = 0{\cdot}4$), suggesting that still-births and mortality immediately after birth were not common enough to bias the calculations seriously. Errors were unlikely to be introduced by temporarily barren females suckling yearlings, because no female shot in November that was either barren (as judged by the state of the uterus), or pregnant, was lactating. Errors resulting from the production of twins would be very small: Caughley found no evidence of twinning in the area.

The potential number of female kids produced by the females in the sample was estimated by

$$Y_0 = \sum_{x=1}^{12} Y_x m_x = 205,$$

and the series Y_x gave an l_x series. This series was standardised to give $l_0 = 1000$ by multiplying by 4·878; d_x and q_x were then calculated as in Table 10.1.

For a further study of the same population see Caughley [1970].

3 *Composite life tables*

These life tables are based on mortality data obtained by recording the ages of animals at death, irrespective of their year of birth. As the animals are born at different times they do not come from the same cohort, so that the numbers d_x dying in each age-class are not strictly amenable to an age-specific analysis. Also the d_x series does not refer to the numbers of deaths in each age-class at a given instant of time, so that a time-specific analysis is not appropriate. However, if we treat all the animals as though they were born at the same time, an age-specific life table can be constructed for this hypothetical population. Such a method was first used in 1902 by Karl Pearson who estimated life expectancies for ancient Egyptians using the age of deaths recorded on 141 mummy cases. This approach has also been widely used in analysing returns from bird-banding operations, and further details have been given in **5.4.3**.

10.2 AGE-SPECIFIC DATA: STOCHASTIC MODELS

10.2.1 General model(*)

In **10.1** the approach to life-table analysis is basically deterministic in that one talks about observed rates rather than estimates of probabilities. However, this deterministic approach, although providing an intuitive basis for estimation (as we shall see below), does not provide variances; these are essential if mortality rates are to be compared either within or between populations. Chiang [1960a, b] has set up a stochastic model for the age-specific situation, and the following theory is based on his articles.

We shall assume that all individuals of age x from a given cohort have the same probability P_x $(= 1 - Q_x)$ of surviving to age $x + 1$. Then, given l_x, l_{x+1} is a binomial variable with probability function

$$f(l_{x+1} \mid l_x) = \binom{l_x}{l_{x+1}} P_x^{l_{x+1}} Q_x^{l_x - l_{x+1}}, \tag{10.8}$$

so that the joint probability function of $l_1, l_2, \ldots, l_w$, given l_0 and w, is

$$f(l_1, l_2, \ldots, l_w \mid l_0, w) = \prod_{x=0}^{w-1} f(l_{x+1} \mid l_x).$$

Since $d_x = l_x - l_{x+1}$ and $l_x = d_x + d_{x+1} + \ldots + d_w$ (for $x = 0, 1, \ldots, w-1$; $d_w = l_w$), we can readily obtain the joint probability function of the d_x, given l_0, from (10.8), namely

$$f(d_0, d_1, \ldots, d_w \mid l_0, w)$$

$$= \frac{l_0!}{\prod_{x=0}^{w} d_x!} Q_0^{d_0} (P_0 Q_1)^{d_1} \ldots (P_0 P_1 \ldots P_{w-2} Q_{w-1})^{d_{w-1}} (P_0 P_1 \ldots P_{w-1})^{d_w}. \tag{10.9}$$

The maximum-likelihood estimates $\hat{P}_x$ of the P_x for the multinomial distribution (10.9) are also the moment estimates, i.e. from (10.8)

$$\hat{P}_x = l_{x+1}/l_x.$$

This estimate is unbiased as

$$\begin{aligned} E[\hat{P}_x] &= \underset{l_x}{E}\, E[\hat{P}_x \mid l_x] \\ &= \underset{l_x}{E}\, [P_x] \\ &= P_x, \end{aligned}$$

and, using (1.12) and the delta method,

$$\begin{aligned} V[\hat{P}_x] &= \underset{l_x}{E}\, \{V[\hat{P}_x \mid l_x]\} + \underset{l_x}{V}\, \{E[\hat{P}_x \mid l_x]\} \\ &= \underset{l_x}{E}\, [P_x Q_x / l_x] + 0 \\ &= P_x Q_x E[1/l_x] \end{aligned}$$

(*)See also **13.3.2**.

$$\approx \frac{P_x Q_x}{E[l_x]}\left\{1 + \frac{V[l_x]}{(E[l_x])^2}\right\}$$

$$= \frac{P_x Q_x}{E[l_x]}\left\{1 + \frac{1}{E[l_x]} - \frac{1}{l_0}\right\}. \quad (10.10)$$

The last step follows from the fact that l_x is (unconditionally) binomially distributed with parameters l_0 and $P_0 P_1 \ldots P_{x-1}$.

Turning our attention to Q_x, we note that

$$\hat{Q}_x = 1 - \hat{P}_x = d_x/l_x$$

is the same as q_x of **10.1.1**. Since

$$E[\hat{Q}_x] = 1 - E[\hat{P}_x] = Q_x,$$

and

$$V[\hat{Q}_x] = V[1-\hat{P}_x] = V[\hat{P}_x],$$

we see that $\hat{Q}_x$ is unbiased and $V[\hat{Q}_x]$ can be estimated by

$$\hat{V}[\hat{Q}_x] = \frac{(1-\hat{Q}_x)\hat{Q}_x}{l_x}\left\{1 + \frac{1}{l_x} - \frac{1}{l_0}\right\}, \quad (x = 0, 1, \ldots, w-1). \quad (10.11)$$

It is also readily shown that

$$\text{cov}[\hat{Q}_x, \hat{Q}_y] = 0 \quad (x \neq y) \quad (10.12)$$

so that the estimates $\hat{Q}_x$ are uncorrelated.

In conclusion we note that the above theory can still be applied when the sequence l_x is given for an arbitrary sequence of x values $x_0, x_1, \ldots, x_w$ (not necessarily integral) by simply replacing P_x by P_i, say, the probability of an individual of age x_i surviving to age x_{i+1}. For example, x may be integral, but the sequence of integers 0, 1, 2, . . . may be incomplete as l_x may not be available for some values of x. Also, x could refer to the *coded* age, so that the above theory can be applied to any group of l_0 individuals of the same age.

Example 10.2 Patagonian cavy (*Dolichotis patagonica*): Comfort [1957]

The above theory was applied to a zoo population of $l_0 = 56$ animals (sexes combined), and the data and mortality estimates are set out in Table 10.6: here x is measured in units of 100 days. We note that the standard deviations of the estimates are comparatively large as l_0 is small.

INSTANTANEOUS MORTALITY RATE. If Z_x is the total instantaneous mortality rate for individuals of age x, then

$$P_x = \exp[-\int_0^1 Z_{x+t}\, dt]$$

$$\approx \exp[-Z_{x+\frac{1}{2}}].$$

Hence $Z_{x+\frac{1}{2}}$ can be estimated by

$$\hat{Z}_{x+\frac{1}{2}} = -\log \hat{P}_x = \log(l_x/l_{x+1})$$

TABLE 10.6

Age-specific life table for a zoo population of 56 Patagonian cavies: data from Comfort [1957: Table 5].

Age (days)	d_x	l_x	$\hat{Q}_x$	$\hat{\sigma}[\hat{Q}_x]$
0–99	13	56	0·2321	0·056
100–	4	43	0·0930	0·044
200–	3	39	0·0769	0·043
300–	1	36	0·0278	0·028
400–	4	35	0·1143	0·054
500–	3	31	0·0968	0·054
600–	1	28	0·0357	0·035
700–	1	27	0·0370	0·037
800–	4	26	0·1538	0·072
900–	3	22	0·1364	0·074
1000–	1	19	0·0526	0·052
1100–	2	18	0·1111	0·075
1200–	5	16	0·3125	0·12
1300–	1	11	0·0909	0·09
1400–	1	10	0·1000	0·10
1500–	4	9	0·4444	0·17
1600–	1	5	0·2000	0·19
1700–	0	4	0·0000	–
1800–	0	4	0·0000	–
1900–	3	4	0·7500	0·24
2000–	1	1	1·0000	–

and, using the delta method,

$$V[\hat{Z}_{x+\frac{1}{2}}] \approx V[\hat{P}_x]/P_x^2 .$$

Kimball [1960] has compared this estimate with the classical actuarial estimate

$$Z'_{x+\frac{1}{2}} = d_x/[\tfrac{1}{2}(l_x + l_{x+1})] .$$

He shows that both estimates are biased and that for small samples there is not a great deal to choose between them. However, if l_0 is large and $Z_{x+t} = Z_x$ for $0 \leqslant t < 1$ then $-\log \hat{P}_x$ is the maximum-likelihood estimate of Z_x and should be used instead of $Z'_{x+\frac{1}{2}}$.

For an arbitrary sequence of x values $x_0, x_1, \ldots, x_w$, and defining $l_i = l_{x_i}$ and $t_i = x_{i+1} - x_i$, we have

$$P_i = \exp[-\int_0^{t_i} Z_{i+t}\, dt]$$

$$\approx \exp[t_i Z_{i+\frac{1}{2}t_i}].$$

The corresponding estimates of $Z_{i+\frac{1}{2}t_i}$ are now

$$\hat{Z}_{i+\frac{1}{2}ti} = -(1/t_i) \log \hat{P}_i = (1/t_i) \log (l_i/l_{i+1})$$

and

$$Z'_{i+\frac{1}{2}t_i} = \frac{d_i}{\frac{1}{2}t_i(l_i + l_{i+1})},$$

where $d_i = l_i - l_{i+1}$.

When the exact age at death of each individual is known, as for example in some laboratory or zoo populations, and the sequence $x_0, x_1, \ldots, x_w$ is not chosen in advance but rather determined by the number dying in each interval $[x_i, x_{i+1})$ (i.e. the d_i are predetermined and the x_i are now random variables), Kimball gives two methods for estimating the instantaneous mortality rates: one method leads to a maximum-likelihood estimate of $\mu_{i+\frac{1}{2}t_i}$ while the other (due to Seal [1954]) leads to an unbiased estimate of μ_i when $\mu_{i+t} = \mu_i$ for $0 \leqslant t < t_i$.

EXPECTATION OF LIFE. If we choose an individual at random from a cohort of size l_0, then, given that the individual lives to age x, the *further* length of life lived by the individual will be a random variable with expected value E_x, say. From (10.3) the usual estimate of E_x is

$$\begin{aligned} e_x &= \tfrac{1}{2} + \sum_{y=x+1}^{w} (l_y/l_x) \\ &= \tfrac{1}{2} + \sum_{y=x+1}^{w} (d_y + d_{y+1} + \ldots + d_w)/l_x \\ &= (\tfrac{1}{2}d_x + 1\tfrac{1}{2}\, d_{x+1} + \ldots + (w - x + \tfrac{1}{2})\, d_w)/l_x \end{aligned} \qquad (10.13)$$

for $x = 0, 1, \ldots, w$ ($e_w = \frac{1}{2}$). Now if the distribution of deaths in each interval $[x, x+1)$ is uniform, so that, as far as expected values are concerned, the d_x individuals may be regarded as having lived for half a year in the interval, then Y_x, the further length of life lived by an individual of age x, may be treated as a random variable taking values $\frac{1}{2}, 1\frac{1}{2}, \ldots, w - x - \frac{1}{2}$, $w - x + \frac{1}{2}$, with respective probabilities $Q_x, P_x Q_{x+1}, \ldots, P_x P_{x+1} \ldots P_{w-2} Q_{w-1}, P_x P_{x+1} \ldots P_{w-1}$. Hence

$$\begin{aligned} E_x &= E[Y_x] \\ &= \tfrac{1}{2}Q_x + 1\tfrac{1}{2}P_x Q_{x+1} + \ldots + (w - x + \tfrac{1}{2})P_x P_{x+1} \ldots P_{w-1}. \end{aligned}$$

Therefore, since

$$E[d_y \mid l_x] = l_x P_x P_{x+1} \ldots P_{y-1} Q_y, \quad (y = x, x+1, \ldots, w-1)$$

and

$$E[d_w \mid l_x] = l_x P_x P_{x+1} \ldots P_{w-1},$$

we have from (10.13) that

$$\begin{aligned} E[e_x] &= \mathop{E}_{l_x} E[e_x \mid l_x] \\ &= \mathop{E}_{l_x}[\tfrac{1}{2}Q_x + \ldots + (w - x + \tfrac{1}{2})P_x P_{x+1} \ldots P_{w-1}] \\ &= E_x, \end{aligned}$$

and e_x is an unbiased estimate of E_x. Also from (10.13) we see that e_x is a sample mean based on the frequencies d_y for l_x $(= d_x + d_{x+1} + \ldots + d_w)$ individuals. Hence, by the central limit theorem, e_x is asymptotically normal for large l_x and, given an estimate of $V[e_x]$, a large-sample confidence interval for E_x can be calculated in the usual manner (cf. (1.5)).

Chiang [1960a] proves that

$$V[e_x] = E[1/l_x] \sum_{y=x}^{w-1} \{\phi_{xy}[E_{y+1} + \tfrac{1}{2}]^2 P_y Q_y\}, \quad x = 0, 1, \ldots, w-1,$$

where ϕ_{xy} is the probability that an individual of age x survives to age y (i.e. $\phi_{xx} = 1$ and $\phi_{xy} = P_x P_{x+1} \ldots P_{y-1}$ for $y > x$). Hence estimating $E[1/l_x]$ by $1/l_x$, and ϕ_{xy} by l_y/l_x, we see that $V[e_x]$ can be estimated by

$$\hat{V}[e_x] = (1/l_x^2) \sum_{y=x}^{w-1} \{l_y[e_{y+1} + \tfrac{1}{2}]^2 \hat{P}_y \hat{Q}_y\}.$$

However, since e_x is a sample mean, an unbiased estimate of $V[e_x]$ is given by (cf. (1.9))

$$\begin{aligned} v[e_x] &= \frac{(\tfrac{1}{2} - e_x)^2 d_x + (1\tfrac{1}{2} - e_x)^2 d_{x+1} + \ldots + (w - x + \tfrac{1}{2} - e_x)^2 d_w}{l_x(l_x - 1)} \\ &= \left\{ \sum_{y=x}^{w} [(y + \tfrac{1}{2} - x) - e_x]^2 d_y \right\} \Big/ \{l_x(l_x - 1)\} \\ &= \left\{ \sum_{y=x}^{w} [(y + \tfrac{1}{2})^2 d_y] - l_x(e_x + x)^2 \right\} \Big/ \{l_x(l_x - 1)\}. \end{aligned}$$

If the term $l_x - 1$ in the above expression is replaced by l_x, we find from Chiang [1960b: 225] that $v[e_x] = \hat{V}[e_x]$.

Chiang's formulation of the above problem is slightly more general than the one given here. He considers the more general situation when the sequence $\{l_x\}$ is defined for the arbitrary sequence $x_0, x_1, \ldots, x_w$. For this this case let

$$\begin{aligned} l_i &= \text{number surviving to age } x_i, \\ d_i &= l_i - l_{i+1} \text{ (the number dying in the age interval } [x_i, x_{i+1})), \\ t_i &= x_{i+1} - x_i, \\ P_i &= \text{probability that an individual of age } x_i \text{ survives to age } x_{i+1}, \\ \phi_{ij} &= P_i P_{i+1} \ldots P_{j-1} \quad (j > i;\ \phi_{ii} = 1) \end{aligned}$$

and

$$E_i = \text{expected value of the further expectation of life for an individual of age } x_i.$$

Suppose that the distribution of deaths in each interval is such that, on the average, each of the d_i individuals lives $a_i t_i$ years in the interval $[x_i, x_{i+1})$, and define

$$c_j = (1 - a_{j-1}) t_{j-1} + a_j t_j.$$

Then Chiang shows that

$$e_i = \sum_{j=i}^{w} (x_j - x_i + a_j t_j) d_j / l_i$$

$$= a_i t_i + \left(\sum_{j=i+1}^{w} c_j l_j\right) / l_i$$

is an unbiased estimate of E_j with variance

$$V[e_i] = E[1/l_i] \sum_{j=i}^{w-1} \{\phi_{ij}[E_{j+1} + (1-a_j) t_j]^2 P_j Q_j\}.$$

This variance can be estimated by

$$\hat{V}[e_i] = (1/l_i^2) \sum_{j=i}^{w-1} \{l_j [e_{j+1} + (1-a_j) t_j]^2 \hat{P}_j \hat{Q}_j\}$$

$$= (1/l_i^2)[\sum_{j=i}^{w} (x_j + a_j t_j)^2 d_j - l_i (e_i + x_i)^2],$$

$$(i = 0, 1, \ldots, w-1).$$

10.2.2 Constant probability of survival

When P_x, the probability of surviving from age x to age $x+1$, is constant ($= P$, say), then using (10.8) we find that (10.9) reduces to

$$\frac{l_0!}{\prod_{x=0}^{w} d_x!} P^{l_1 + l_2 + \ldots + l_w} Q^{l_0 - l_w}.$$

Hence the maximum-likelihood estimate of Q $(= 1-P)$ is readily shown to be

$$\hat{Q} = (l_0 - l_w)/(l_0 + l_1 + \ldots + l_{w-1})$$

$$= \left(\sum_{x=0}^{w-1} d_x\right) \Big/ \left(\sum_{x=0}^{w-1} l_x\right),$$

which is the same as q_{pool} of (10.7). Using standard maximum-likelihood theory, we find that the asymptotic variance of $\hat{Q}$ is estimated by (cf. (1.27))

$$\hat{V}[\hat{Q}] = \left(\sum_{x=1}^{w} l_x\right)(l_0 - l_w) \Big/ \left(\sum_{x=0}^{w-1} l_x\right)^3.$$

Alternatively one can use the average estimate of (10.6), namely

$$q_{\text{ave}} = \sum_{x=0}^{w-1} \hat{Q}_x / w.$$

Then, from (10.12) and (1.9), an unbiased estimate of $V[q_{\text{ave}}]$ is given by

$$v[q_{\text{ave}}] = \frac{1}{w(w-1)} \sum_{x=0}^{w-1} (\hat{Q}_x - q_{\text{ave}})^2.$$

REGRESSION MODELS. Since

$$E[d_x] = l_0 P^x Q, \quad (x = 0, 1, \ldots, w-1),$$

we can take logarithms and consider the linear model

$$\log_{10}(d_x) = \log_{10}(l_0 Q) + x \log_{10} P + \epsilon_x. \qquad (10.14)$$

Here the ϵ_x are correlated, though the degree of correlation will usually be small. Therefore, assuming the ϵ_x to be independently normally distributed with zero mean and constant variance, least-squares estimates of $\log_{10} P$

and $\log_{10}(l_0 Q)$ can be obtained in the usual manner. If l_0 is unknown, as is sometimes the case, then it may be estimated from this model.

Another regression model which is often used can be obtained from the equation

$$E[l_x] = l_0 P^x$$

i.e., taking logarithms,

$$\log_{10} l_x = \log_{10} l_0 + x \log_{10} P + \epsilon_x .$$

However, the degree of correlation between the l_x variables is much greater than that between the d_x variables, so that, from a least-squares point of view, the model (10.14) is preferred.

10.3 TIME-SPECIFIC DATA: CONSTANT SURVIVAL RATE

10.3.1 Geometric model

Life-table methods for calculating mortality rates (and the complementary survival rates) for time-specific data are discussed in some detail in **10.1.2**(*2*). But when the mortality rate is constant over all age-classes we can either combine the rates to give q_{ave} or q_{pool}, or else use the following techniques of Chapman and Robson [1960] and Robson and Chapman [1961].

1 Chapman–Robson survival estimate

We shall assume that:

(1) S, the proportion of a given age-group surviving for one year, is the same for each age-group and remains constant from year to year.(*)
(2) N_x, the number of animals of age x in the population at the time of the year when reproduction occurs (called time "zero") is constant from year to year (N_0 = annual number of births).
(3) Sampling from the population is random with respect to age.

Then, from assumption (2), the number of animals N_x growing out of age x will be balanced by the number SN_{x-1} growing into this age-group. Hence

$$N_x = SN_{x-1} = S^x N_0 ,$$

and since

$$N = \sum_{x=0}^{\infty} N_x = N_0(1+S+S^2+\ldots) = N_0/(1-S),$$

we have

$$N_x = (1-S)S^x N, \quad (x = 0, 1, 2, \ldots).$$

If an animal is chosen at random at time zero, then the probability that it is of age x is N_x/N, i.e. x is a random variable with probability function

$$f(x) = (1-S)S^x \quad (x = 0, 1, 2, \ldots). \qquad (10.15)$$

In fact, even if the animal is chosen at random at some other time of the year, so that it is found to be of age $x+$, x will still have probability function (10.15), provided that the survival rate is the same for all age-groups

(*) We use S rather than ϕ to distinguish between proportion and probability.

throughout the year (though this rate may vary with the time of the year), i.e. provided that N_x/N remains constant. Therefore, given a random sample of size n $(=\sum_x n_x)$ in which n_x are observed to be of age x $(x = 0, 1, 2, \ldots, r)$, then, ignoring the complications of sampling without replacement in each age-group, the likelihood function for the sample is

$$(1-S)^n S^{\Sigma x n_x} = (1-S)^n S^X,$$

where $X = \sum_{x=0}^{r} x n_x$ is a sufficient statistic for S. Since X is a complete statistic, a uniformly minimum-variance unbiased estimate $\hat{S}$ of S exists, and this estimate is a unique function of X, say $h(X)$. From the identity $E[h(X)] = S$ in S, Chapman and Robson show that

$$\hat{S} = X/(n+X-1).$$

Although the exact variance of $\hat{S}$ is not expressible in closed form,

$$v[\hat{S}] = \hat{S}\left(\hat{S} - \frac{X-1}{n+X-2}\right)$$

is the minimum-variance unbiased estimate of this exact variance. But when n is large, $\hat{S}$ is approximately equal to $X/(n+X)$, the maximum-likelihood estimate, which has asymptotic variance $[S(1-S)^2]/n$; hence, for large n,

$$V[\hat{S}] \approx S(1-S)^2/n. \qquad (10.16)$$

Although the equations leading up to (10.15) are deterministic, the above theory can be justified on other grounds using a stochastic model (cf. Chapman and Robson [1960: 357]). We note that n is assumed fixed.

ALTERNATIVE ESTIMATES.(*) Various alternative estimates have been have been used in practice. For example, Jackson's [1939] estimate

$$\begin{aligned} S' &= (n_1+n_2+\ldots+n_r)/(n_0+n_1+\ldots+n_{r-1}) \\ &= (n-n_0)/(n-n_r) \end{aligned}$$

is widely used although it is biased, can be greater than unity, and depends on the behaviour of the two extreme age-classes n_0 and n_r in the sample. A more useful modification (Heincke [1913]), which differs only slightly from S' when n_r is small compared with n, is

$$\hat{S}_1 = (n-n_0)/n.$$

Since n_0 is binomially distributed with parameters n and $(1-S)$, $\hat{S}_1$ is unbiased with variance

$$V[\hat{S}_1] = S(1-S)/n.$$

We note from (10.16) that $V[\hat{S}] \approx (1-S)V[\hat{S}_1]$, so that $V[\hat{S}]$ will be much smaller than $V[\hat{S}_1]$ when $1-S$ is small.

Chapman and Robson [1960] have considered a more general class of estimates, namely

$$\hat{S}_p = [(n-n_0-n_1\ldots-n_{p-1})/n]^{1/p}, \quad p = 1, 2, \ldots,$$

(*)See also **13.3.3.**

where, to terms of order $1/n$,

$$E[\hat{S}_p] = S \mp \frac{(1-p)(1-S^p)}{np^2 S^{p-1}}$$

and

$$V[\hat{S}_p] = \frac{(1-S^p)}{np^2 S^{p-2}}.$$

This variance, considered as a function of p, is minimised for the value of p, say $\hat{p}$, satisfying

$$1 - S^p = \tfrac{1}{2}\,|\log S|\,p.$$

If this were to be used as a guide in practice, assuming some prior information on the general magnitude of S, p would be chosen as one of the integers near $\hat{p}$. Since $\hat{p}$ is an increasing function of S, $\hat{S}_1$ is preferred among this class when S is small, while p should be chosen fairly large when S is large. For example, when $S = 0{\cdot}5$, $\hat{p}$ lies between 2 and 3. Choosing $p = 2$ we note that the bias of $\hat{S}_2$ is approximately $-(1-S^2)/4Sn$, which is negligible for S near 1, but may not be otherwise. Also $V[\hat{S}_2] < V[\hat{S}_1]$ $(= S(1-S)/n)$ when $S > \frac{1}{3}$. However it should be stressed that when the geometric probability model (10.15) is valid, $\hat{S}$ is the most efficient estimate.

THRESHOLD AGE. In many populations there is a "threshold" age above which S is independent of age. For example, in fisheries there are many types of fishing gear which are selective against smaller sizes and consequently younger ages. In this case the exploitation rate for the younger fish will be smaller, and this in turn will affect the probability of survival of the younger classes. For this situation it is not unreasonable to assume that there is some age A, such that for all ages $x \geqslant A$, $N_{x+1} = SN_x$ and $N_x = S^{x-A} N_A$. Thus by simply relabelling the ages so that $A = 0$ (i.e. using a "coded" age) the above theory can still be applied. However, there may be no way of determining A in practice, so that the data in the youngest age-class may be suspect. Chapman and Robson [1960] give a useful test of the validity of the geometric probability model which gives particular emphasis to this zero class. They show that for large n (say greater than 100)

$$z = (\hat{S}_1 - \hat{S}) \Big/ \left\{ \frac{X(X-1)(n-1)}{n(n+X-1)^2(n+X-2)} \right\}^{\frac{1}{2}} \tag{10.17}$$

is asymptotically distributed as the unit normal when (10.15) is valid. Since $\hat{S}$ and $\hat{S}_1$ are both unbiased estimates of S under the above model, any significant difference in these estimates will indicate that, compared with the older age-classes, the $x = 0$ class is not properly represented. If a two-tailed test is required then we can use z^2 as χ_1^2. The authors also give a small-sample test based on the hypergeometric distribution (or its binomial approximation).

Example 10.3 Rock bass: Robson and Chapman [1961]

For a sample of rock bass trap-netted from Cayuga Lake, New York, during a single summer season, the threshold age A was arbitrarily chosen as VI and Table 10.7 was obtained. From this table we have

$$\hat{S} = \frac{X}{n+X-1} = \frac{196}{243+196-1} = 0{\cdot}4475$$

or, expressed as a percentage survival rate, 45 per cent. Also,

$$v[\hat{S}] = \frac{X}{n+X-1}\left(\frac{X}{n+X-1} - \frac{X-1}{n+X-2}\right)$$
$$= \frac{196}{438}\left(\frac{196}{438} - \frac{195}{437}\right) = 5{\cdot}66 \times 10^{-4}$$

and hence

$$\sqrt{v[\hat{S}]} = 0{\cdot}0238.$$

TABLE 10.7

Age frequencies, n_x, from a sample of rock bass: from Robson and Chapman [1961: 182].

Age	Coded age (x)	n_x	$x n_x$
VI+	0	118	0
VII+	1	73	73
VIII+	2	36	72
IX+	3	14	42
X+	4	1	4
XI+	5	1	5
Total		$n = 243$	$X = 196$

An approximate 95 per cent confidence interval for S is given by $0{\cdot}4475 \pm 1{\cdot}96\ (0{\cdot}0238)$ or $(0{\cdot}40,\ 0{\cdot}50)$.

Heincke's estimate

$$\hat{S}_1 = \frac{n-n_0}{n} = \frac{243-118}{243} = 0{\cdot}5144$$

is greater than $\hat{S}$, which suggests that there is a deficit in the zero age-group relative to the older age-group. To test whether the difference is significant, we calculate the chi-squared statistic

$$z^2 = (0{\cdot}4475 - 0{\cdot}5144)^2 \Big/ \left\{\frac{196\,(195)\,(242)}{243\,(438^2)\,(437)}\right\}$$
$$= 9{\cdot}858,$$

which is highly significant ($P < 0{\cdot}002$). This leads us to suspect the validity of the geometric model. Unfortunately the chi-squared test cannot tell us which of the three basic assumptions on p. 414 is at fault, as it merely establishes whether or not there is a reasonable agreement between the observed frequency in the age-group 0 and the frequency expected on the basis of the data in the older age-groups. If we suspect that age-group 0 is at fault we can eliminate it and recode the remaining age-groups as in Table 10.8. For this recoded set of data

$$\hat{S} = 71/(125+71-1) = 0{\cdot}3641,$$

$$\hat{S}_1 = (125-13)/125 = 0{\cdot}4160,$$

and

$$z^2 = (0{\cdot}3641-0{\cdot}4160)^2 \bigg/ \left\{\frac{71(70)(124)}{125(195^2)194}\right\} = 4{\cdot}032$$

TABLE 10.8

Recoded age frequencies, n_x, from Table 10.7.

Age	Recoded age (x)	n_x	$x\,n_x$
VII+	0	73	0
VIII+	1	36	36
IX+	2	14	28
X+	3	1	3
XI+	4	1	4
Total		$n = 125$	$X = 71$

which is still significant at the 5 per cent level. Therefore, eliminating age-class VII+ and recoding, we finally get a non-significant test.

For a further application of the above theory see Johnson [1968: male fur seal population].

INSTANTANEOUS MORTALITY RATE. Suppose that mortality is a Poisson process with parameter Z, then using 1 year, say, as the time unit, $S = \exp(-Z)$ or

$$Z = -\log S.$$

A reasonable estimate of Z is $\hat{Z} = -\log \hat{S}$ and, using the delta method, we have asymptotically

$$E[\hat{Z}] = S + \tfrac{1}{2}(1-S)^2/(nS)$$

and

$$V[\hat{Z}] = (1-S)^2/(nS).$$

Chapman and Robson [1960] suggest a modified estimate

$$Z^* = -\log \hat{S} - \frac{(n-1)(n-2)}{n(n+X-1)(X+1)}$$

which has negligible bias.

We note that in fisheries the instantaneous mortality rate Z is generally the instantaneous natural mortality rate plus the instantaneous exploitation rate.

2 Older age-groups pooled

In addition to the random sampling errors of the n_x there is usually an error in the process of measuring age. For example, in fishery work with commonly used procedures such as scale or otolith reading, the age determination becomes more time-consuming and more subject to error with increasing age. To alleviate both these problems Robson and Chapman [1961] suggest that exact ageing should be attempted for just the younger age-groups and that the remaining age-groups be combined. Although the pooling of older age-groups represents a loss in information, the time saved in the ageing process can be utilised for larger samples.

Suppose that animals J years old or less are aged exactly, and all animals $J+1$ years old or more are pooled. Let

$$n_{(J)} = \sum_{x=J+1}^{r} n_x$$

and

$$X = \sum_{x=0}^{J} x n_x + (J+1) n_{(J)},$$

then the maximum-likelihood estimate of S is now

$$\hat{S}_{\text{pool}} = \frac{X}{n - n_{(J)} + X}$$

with asymptotic variance

$$V[\hat{S}_{\text{pool}}] = S(1-S)^2/[n(1-S^{J+1})].$$

In this case no minimum-variance unbiased estimate of S exists. When $J=0$, the above estimate reduces to simply Heincke's estimate $\hat{S}_1$.

We note that, for large n (cf. (10.16)),

$$\frac{V[\hat{S}]}{V[\hat{S}_{\text{pool}}]} \approx 1 - S^{J+1}$$

is the efficiency of this method relative to ageing all the fish exactly. Thus, for every 100 fish aged exactly, the present pooling method would require

$$n = 100/(1-S^{J+1}) \tag{10.18}$$

fish to obtain the same accuracy of estimation. Robson and Chapman present a table of this relationship (Table 10.9) and conclude that when S is in the neighbourhood of 0·50, little is gained by attempting ageing on more than the first two or three age-groups if the above method of estimation is to be used.

Example 10.4 Using the data of Table 10.7 with $J=3$ we have Table 10.10. Thus

$$\hat{S}_{\text{pool}} = 193/(243-16+193) = 0{\cdot}46,$$

and

$$\hat{V}[\hat{S}_{\text{pool}}] = 0{\cdot}000\,612$$

which is only slightly larger than the value of 0·000 566 for the case when all the fish are aged exactly.

TABLE 10.9
The sample size n given by equation (10.18) for different values of S and J: from Robson and Chapman [1961: Table 1].

J \ S	0·25	0·50	0·75
0	133	200	400
1	107	133	228
2	101	114	173
3	100	107	146
4	100	103	132
5	100	101	122
...	...	...	...
∞	100	100	100

TABLE 10.10
Age data from a rock bass sample with age IX+ and older pooled: from Robson and Chapman [1961: 184].

Age	Coded age (x)	n_x	$x n_x$
VI+	0	118	0
VII+	1	73	73
VIII+	2	36	72
IX+ and older	$\geqslant 3$	16	48
Total		$n = 243$	$X = 193$

3 Age range truncated at both ends

In some experimental circumstances it may be necessary to truncate the age data on the right as well as on the left. For example, in fisheries the sampling gear may be effective only for a limited range of fish size, and therefore age, so that different gear is used for different age ranges. In addition the exploitation rate may vary for different size-classes, so that the probability of survival may vary with age. Such circumstances therefore lead to a partitioning of the age range into segments and a separate analysis for each segment, as follows.

Suppose that the age in a particular segment runs from 0 to K on the *coded* scale and assume that the assumptions on p. 414 are valid for this segment. Then the age distribution for the segment is given by the truncated probability function (cf. (10.15))

$$f(x) = S^x \Big/ \sum_{j=0}^{K} S^j, \quad x = 0, 1, 2, \ldots, K.$$

Chapman and Robson [1960: 361] mention that an unbiased estimate of S does not exist for this model, and the asymptotically unbiased maximum-likelihood estimate, $\hat{S}_{seg}$, say, is the solution of

$$X/n = \sum_{k=0}^{K} kS^k \Big/ \sum_{k=0}^{K} S^k$$
$$= [S/(1-S)] - (K+1)S^{K+1}/(1-S^{K+1}), \qquad (10.19)$$

where $X = \sum_{x=0}^{K} xn_x$ and $n = \sum_{x=0}^{K} n_x$. Methods for solving the above equation, using a table reproduced from Robson and Chapman [1961], are given in **A6**.

The asymptotic variance of $\hat{S}_{seg}$ is given by

$$V[\hat{S}_{seg}] = \frac{1}{n}\left[\frac{1}{S(1-S)^2} - \frac{(K+1)^2 S^{K-1}}{(1-S^{K+1})^2}\right]^{-1}$$

which decreases to the variance of the "non-truncated" estimate $S(1-S)^2/n$ as $K \to \infty$. For small K there can be a considerable loss in efficiency through an unnecessary truncation of the data.

Heincke's estimate $\hat{S}_1$ can also be used, though with a different mean and variance; n_0 is now binomial with parameters n and $(1-S)/(1-S^{K-1})$, so that

$$E[\hat{S}_1] = S - S^{K+1}(1-S)/(1-S^{K+1})$$
$$= S - b(\hat{S}_1), \text{ say,}$$

and

$$V[\hat{S}_1] = S(1-S)(1-S^K)/[n(1-S^{K+1})^2].$$

Although $\hat{S}_1$ is now biased the bias will usually be small. Unfortunately a simple test statistic similar to (10.17) is not available, though a comparison of the confidence interval based on $\hat{S}_1 + \hat{b}(\hat{S}_1)$ with that based on $\hat{S}_{seg}$ will give some idea as to the reliability of the zero class.

Example 10.5 Using the data of Table 10.7 with $K = 3$, we have Table 10.11. Here,

$$X/n = 187/241 = 0{\cdot}775\,93,$$

and entering Table **A6** with $K = 3$ and using linear interpolation leads to

$$\hat{S}_{seg} = 0{\cdot}5247,$$

TABLE 10.11

Age data for a segment of the age-range only: from Robson and Chapman [1961: 188].

Age	Coded age (x)	n_x	xn_x
VI+	0	118	0
VII+	1	73	73
VIII+	2	36	72
IX+	3	14	42
Total		$n = 241$	$X = 187$

with variance estimated by

$$\frac{1}{241}\left[\frac{1}{(0{\cdot}525)(1-0{\cdot}525)^2} - \frac{16(0{\cdot}525)^2}{[1-(0{\cdot}525)^4]^2}\right]^{-1} = 0{\cdot}001\ 266.$$

Without truncation on the right, the variance estimate obtained in Example 10·3 is 0·000 566, which is less than half the above estimate, thus indicating a considerable loss in precision through truncation.

An approximate 95 per cent confidence interval for S is given by $0{\cdot}525 \pm 1{\cdot}96\ (0{\cdot}001\ 266)^{\frac{1}{2}}$ or (0·45, 0·60).

Heincke's estimate is

$$\hat{S}_1 = (241-118)/241 = 0{\cdot}510,$$

and since

$$\hat{S}_1 + \hat{b}(\hat{S}_1) = 0{\cdot}510 + 0{\cdot}035 = 0{\cdot}545$$

lies within the above confidence interval, there is no reason for suspecting the zero class.

4 *Use of age–length data*

When n, the sample size, is large, as in commercial fishing, or when age determinations are time-consuming, it may be impractical or inefficient to age the whole sample. If age and length are correlated, a simpler method would be to stratify the sample by length, and then age just a subsample from each length-class (Ketchen [1950]). The sum of the coded ages for the subsample could then be used to provide an estimate X' say of X, the sum for the whole sample. In this case Robson and Chapman [1961] show that an almost unbiased estimate of S is

$$\hat{S}_L = \frac{X'}{n+X'-1} + \frac{(n-1)V[X']}{(n+X'-1)(n+X'-2)(n+X'-3)}$$

with asymptotic variance

$$V[\hat{S}_L] = \frac{S(1-S)^2}{n} + \frac{(1-S)^4}{(n-1)^2}V[X']. \tag{10.20}$$

The estimate X' and its variance can be obtained by one of the following two methods, depending on whether individual lengths in the sample are known or not.

GROUPED DATA. Consider the simpler case where the lengths are not determined accurately but are merely assigned to one of I length-classes. For $i = 1, 2, \ldots, I$ let

L_i = number in the sample belonging to the ith length-class,
l_i = number in the ith class subsampled for age determination,
X_i = (unknown) total coded age for the L_i individuals in the ith class,
$\bar{X}_i = X_i/L_i$,

x_{ij} = jth coded age determination in the subsample from the ith class ($j = 1, 2, \ldots, l_i$),

$x_i = \sum_j x_{ij}$ (total coded age for ith subsample),

and

$\bar{x}_i = x_i/l_i$.

Then

$$X = \sum_{i=1}^{I} X_i = \sum_{i=1}^{I} L_i \bar{X}_i, \qquad (10.21)$$

and from Cochran [1963: Chapter 5] we have the stratified-sample estimate of X, namely

$$X'_{\text{strat}} = \sum_{i=1}^{I} L_i \bar{x}_i.$$

The variance of X'_{strat} is estimated by

$$v[X'_{\text{strat}}] = \sum_{i=1}^{I} L_i(L_i - l_i)\, \hat{\sigma}_i^2/l_i, \qquad (10.22)$$

where $\hat{\sigma}_i^2$ is the usual unbiased estimate of the variance σ_i^2 of the age of individuals in the ith class, namely

$$\hat{\sigma}_i^2 = \sum_{j=1}^{l_i} (x_{ij} - \bar{x}_i)^2/(l_i - 1).$$

Example 10.6 Lemon soles: Ricker [1958]

To demonstrate the calculations, the first few rows of a table from Ricker [1958] (cf. Table 10.15) are reproduced in Table 10.12. Thus, for

TABLE 10.12

Extract from a table of age-length data given by Ricker [1958: 79].

Length-class (cm)	L_i	l_i	Coded age 0	1	2	3	4
27	6	6	5	1	0	0	0
28	9	9	3	4	2	0	0
29	30	10	4	4	1	1	0
30	51	10	1	5	4	0	0

example,

$$L_2\bar{x}_2 = x_2 = 3(0) + 4(1) + 2(2) = 8$$

and

$$L_3\bar{x}_3 = 3x_3 = 3[4(0) + 4(1) + 1(2) + 1(3)] = 27.$$

The contribution of the first two length-classes to $v[X'_{\text{strat}}]$ is zero as $L_i = l_i$ ($i = 1, 2$). The contribution of the third class is $30(30-10)\hat{\sigma}_3^2/10$ or

$60\hat{\sigma}_3^2$, where

$$\begin{aligned}\hat{\sigma}_3^2 &= \tfrac{1}{9}\{\Sigma\, x_{3j}^2 - x_3^2/l_3\}\\ &= \tfrac{1}{9}\{4(0)+4(1^2)+1(2^2)+1(3^2)-81/10\}\\ &= 0{\cdot}99.\end{aligned}$$

Example 10.7 Halibut: Robson and Chapman [1961]

A sample from a halibut population was stratified by length and subsampled for age determinations; the data are set out in Table 10.13. Here,

$$n = 1612, \quad X'_{\text{strat}} = 5660{\cdot}95$$

TABLE 10.13

Sample from a halibut population stratified by length and subsampled for age determinations: adapted from Robson and Chapman [1961: Table 2].

Length-class (i)	L_i	l_i	x_i	$\hat{\sigma}_i^2$	$L_i\bar{x}_i$	$\hat{\sigma}_i^2 L_i(L_i - l_i)/l_i$
1	25	13	28	3·31	53·84	76·38
2	37	18	42	6·59	86·33	257·38
3	79	40	57	3·79	112·57	291·92
4	146	75	90	3·51	175·20	485·13
5	204	103	191	5·95	378·29	1 190·23
6	259	132	296	7·15	580·78	1 781·70
7	224	103	291	7·89	632·85	2 076·22
8	177	81	389	4·74	850·03	994·35
9	163	74	340	8·76	748·91	1 717·32
10	116	61	363	12·78	690·29	1 336·66
11	87	41	266	16·01	564·43	1 562·73
12	37	20	168	20·46	310·80	643·47
13	26	16	107	21·83	173·87	354·74
14	32	13	123	20·02	302·76	936·32
Total	1 612	790			5 660·95	13 704·55

$$v[X'_{\text{strat}}] = 13\,704{\cdot}55,$$

$$\begin{aligned}\hat{S}_L &= \frac{5660{\cdot}95}{1612 + 5660{\cdot}95 - 1} + \frac{1611(13\,704{\cdot}55)}{7270{\cdot}67(7269{\cdot}67)(7268{\cdot}67)}\\ &= 0{\cdot}7785\end{aligned}$$

and

$$\begin{aligned}\hat{V}[\hat{S}_L] &= \frac{(0{\cdot}7785)(1-0{\cdot}7785)^2}{1612} + \frac{(1-0{\cdot}7785)^4(13\,704{\cdot}55)}{1611^2}\\ &= 0{\cdot}000\,036.\end{aligned}$$

An approximate 95 per cent confidence interval for S is

$$0{\cdot}779 \pm 1{\cdot}96\,(0{\cdot}006) \quad \text{or} \quad (0{\cdot}767,\, 0{\cdot}791).$$

INDIVIDUAL MEASUREMENTS. Suppose now that the lengths of all the members of the sample are carefully measured and let $\bar{Z}_i$ be the average of all the lengths in the ith length-class. Let z_{ij} be the length of the individual with coded age x_{ij}, and let

$$\bar{z}_i = \sum_{j=1}^{l_i} z_{ij}/l_i .$$

Then if the regression of x_{ij} on z_{ij} is approximately linear within the ith class, a simple regression estimate of X_i is available, namely (Cochran [1963: Chapter 7])

$$\bar{X}_i'' = \bar{x}_i + b_i(\bar{Z}_i - \bar{z}_i),$$

where $b_i = \sum_{j=1}^{l_i} (x_{ij}-\bar{x}_i)(z_{ij}-\bar{z}_i) \Big/ \sum_{j=1}^{l_i} (z_{ij}-\bar{z}_i)^2$

is the usual estimate of slope. Thus X can now be estimated by (cf. (10.21))

$$X'_{\text{reg}} = \sum_{i=1}^{I} L_i \bar{X}_i''$$

and the variance of this estimate is again estimated by (10.22) but with

$$\hat{\sigma}_i^2 = \frac{1}{l_i - 2} \left\{ \sum_j (x_{ij}-\bar{x}_i)^2 - \frac{[\sum_j (x_{ij}-\bar{x}_i)(z_{ij}-\bar{z}_i)]^2}{\sum_j (z_{ij}-\bar{z}_i)^2} \right\}.$$

OPTIMUM ALLOCATION OF EFFORT. Let c_A and c_L be the costs of making an age determination and a length reading, respectively, and suppose that c is the total finance available for the experiment, i.e.

$$c = nc_L + n'c_A ,$$

where $n' = \sum_{i=1}^{I} l_i$. If proportional subsampling is used, so that $l_i/L_i = f$, then

$$c = n(c_L + fc_A) \tag{10.23}$$

and one would choose n and f subject to (10.23), which minimise $V[\hat{S}_L]$ of (10.20). To do this, Robson and Chapman put the two components of variance in (10.20) on a per-unit basis, so that

$$V[\hat{S}_L] = \frac{V_1}{n} + \frac{V_2}{n'}(1-f) = \frac{1}{n}\left(V_1 - V_2 + \frac{V_2}{f}\right), \tag{10.24}$$

where $V_1 = S(1-S)^2$
and

$$V_2 = n' \frac{(1-S)^4}{(n-1)^2} \sum_{i=1}^{I} \frac{L_i^2 \sigma_i^2}{l_i},$$

which is approximately constant as $n'/(n-1) \approx l_i/L_i$ and $L_i/(n-1)$ $(\approx L_i/\Sigma L_i)$ is approximately equal to the population proportion in the ith length-class when the sampling is random. Thus, minimising (10.24) subject to (10.23) leads to

$$f_{\text{opt}} = \left(\frac{V_2}{V_1 - V_2} \cdot \frac{c_L}{c_A}\right)^{\frac{1}{2}};$$

if $V_2 \geqslant V_1$ then 100 per cent subsampling is required and $f_{opt} = 1$. Here σ_i^2 (the ith class variance), V_1 and V_2 can be estimated from a pilot sample or from previous experiments, using either of the two methods given above (provided that proportional subsampling is used). For example, using the data of Example 10.7, we have $f \approx \frac{1}{2}$,

$$\hat{V}_1 = (0{\cdot}7785)(1-0{\cdot}7785)^2 = 0{\cdot}038\ 159,$$

$$\hat{V}_2 = \frac{790(1-0{\cdot}7785)^4}{1611^2} \sum_i \hat{\sigma}_i^2 \frac{L_i^2}{l_i} = 0{\cdot}019\ 244,$$

and hence

$$f_{opt} = 1{\cdot}0077\,(c_L/c_A)^{\frac{1}{2}}.$$

Thus for $c_A/c_L = 10$, the optimum sampling ratio for future experiments is one age-reading to every three length-readings.

10.3.2 Regression model (catch curve)

If the probability p of catching an individual is the same for each individual irrespective of age, then

$$E[n_x \mid N_x] = pN_x. \tag{10.25}$$

Also, if S is now interpreted as the probability of survival, we have that

$$E[N_x] = N_0 S^x. \tag{10.26}$$

Combining (10.25) and (10.26), and taking logarithms, leads to the linear regression model

$$E[y_x] \approx \log(pN_0) + x \log S, \tag{10.27}$$

where $y_x = \log n_x$ (logarithms to the base 10 can also be used). In fishery research the plot of y_x versus x is called a "catch curve", and for a full discussion on the interpretation of such curves the reader is referred to Ricker [1958: Chapter 2]; further examples of catch curves are given in Kennedy [1954], Tester [1955] and Healey [1975: table 1].

Assuming the y_x to be independent with constant variance, the usual least-squares estimate of $\log S$ (and hence of S) can be obtained. In the case of fisheries there is some empirical evidence to suggest that for haul data, the log transformation stabilizes the variance (Jones [1956], Winsor and Clark [1940]).

If the n_x are assumed to be Poisson random variables, Chapman and Robson [1960: 366] suggest a modification of (10.27) to allow for bias, namely $y_x = \log n_x - [1/(n_x+1)]$ and recommend that the sample be truncated on the right for values of n_x less than 5. They also show that under certain conditions the regression method can still be used when S and N_0 are *independent* random variables, as follows.

Let $S_1, S_2, \ldots, S_x$ be the x realisations of the random variable S associated with N_x, then

$$E[N_x \mid N_0, S_1, \ldots, S_x] = N_0 S_1 S_2 \ldots S_x. \tag{10.28}$$

Combining (10.28) and (10.25), and assuming the variances of the n_x, N_0 and S to be small compared with their means, it transpires that (10.27) now becomes

$$E[\log n_x] \approx \log(pE[N_0]) + x\log(E[S]),$$

so that $E[S]$ can be estimated by least squares.

The above regression methods are particularly useful when the age distribution in the population is geometric but either the sampling is not random – when for example there is age segregation – or length data are used to obtain n_x (cf. Example 10.9). However, in any circumstances a graph is always useful as a check on the underlying assumptions; in particular the threshold age A above which S is approximately constant can usually be determined empirically from the graph.

Example 10.8 Herring: Tester [1955]

The age data in Table 10.14 are from Tester [1955: Table 5]. Here the graph of $\log n_x$ against x (Fig. 10.3) indicates that the threshold age is IV, so that S appears to be approximately constant for fish of age IV or older.

TABLE 10.14

Age-composition data for a herring population: from Tester [1955: Table 5].

Age (x)	I	II	III	IV	V	VI	VII	VIII	IX
n_x	34	608	6141	2607	497	91	17	4	1
$\log_{10} n_x$	1·53	2·78	3·79	3·42	2·70	1·96	1·23	0·60	0·00
Coded age				0	1	2	3	4	5

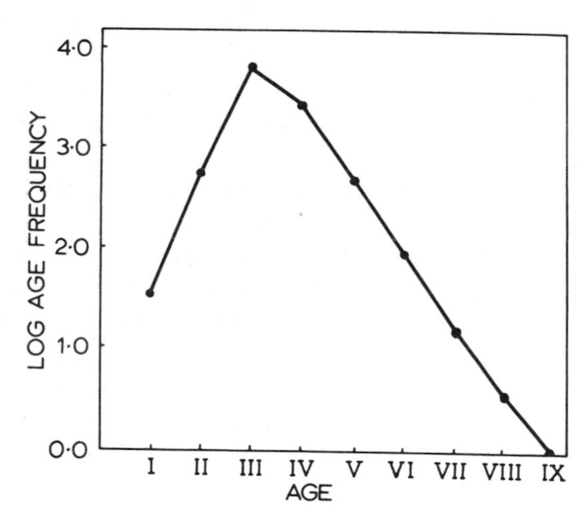

Fig. 10.3 Catch curve for a sample of herring: redrawn from Tester [1955].

TABLE 10.15
Age–length data for a population of lemon soles: from Ricker [1958: 79].

Length-class (i)	L_i	l_i	Age-class IV	V	VI	VII	VIII	IX
1	6	6	5 (5·0)	1 (1·0)	–	–	–	–
2	9	9	3 (3·0)	4 (4·0)	2 (2·0)	–	–	–
3	30	10	4 (12·0)	4 (12·0)	1 (3·0)	1 (3·0)	–	–
4	51	10	1 (5·1)	5 (25·5)	4 (20·4)	–	–	–
5	54	10	–	8 (43·2)	2 (10·8)	–	–	–
6	48	10	1 (4·8)	7 (33·6)	1 (4·8)	1 (4·8)	–	–
7	41	10	1 (4·1)	3 (12·3)	3 (12·3)	2 (8·2)	1 (4·1)	
8	27	10	–	2 (5·4)	6 (16·2)	1 (2·7)	1 (2·7)	–
9	13	10	–	1 (1·3)	4 (5·2)	3 (3·9)	–	2 (2·6)
10	6	6	–	–	1 (1·0)	3 (3·0)	2 (2·0)	–
11	3	3	–	–	1 (1·0)	1 (1·0)	1 (1·0)	–
12	1	1	–	–	–	–	1 (1·0)	–
Total ($\hat{n}_x$) =			(34·0)	(138·3)	(76·7)	(26·6)	(10·8)	(2·6)
$\log_{10} \hat{n}_x$ =			1·53	2·14	1·88	1·42	1·03	0·41
x =			0	1	2	3	4	5

Example 10.9 Lemon sole: Ricker [1958]

Table 10.15 is taken from Ricker [1958: 79] and the numbers in brackets for the ith length-class, say, are simply the observed age frequencies in the subsample from the ith class multiplied by L_i/l_i. Summing over each length-class leads to the estimates $\hat{n}_x$, and a plot of log $\hat{n}_x$ versus x is given in Fig. 10.4. From this graph we see that the regression method can be used for the four age-classes V–VIII inclusive; the last age-class is not used as $\hat{n}_5 < 5$.

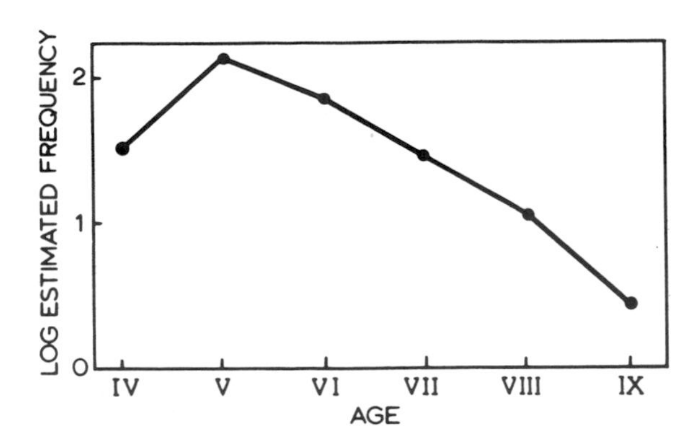

Fig. 10.4 Catch curve for a sample of lemon soles using subsampling for age determinations: data from Ricker [1958].

10.4 TIME-SPECIFIC DATA: AGE-DEPENDENT SURVIVAL[*]

Suppose that S is no longer constant with respect to age, and let S_x be the proportion of the group of age x which survives to age $x+1$ ($x = 1, 2, \ldots, K-1$). Assuming stationarity, N_x will remain constant from year to year, so that

$$N_{x+1} = S_x N_x = S_0 S_1 \ldots S_x N_0 \quad (x = 0, 1, \ldots, K-1)$$

and

$$\sum_{x=0}^{K} N_x = (1 + S_0 + S_0 S_1 + \ldots + S_0 S_1 S_2 \ldots S_{K-1}) N_0 .$$

If K is chosen in advance, then, by considering $N_x \Big/ \sum_{x=0}^{K} N_x$, the probability function for x, the age of an individual selected at random (conditional on $0 \leqslant x \leqslant K$), is (Chapman and Robson [1960])

$$f(x) = \frac{S_0 S_1 \ldots S_{x-1}}{(1 + S_0 + S_0 S_1 + \ldots + S_0 S_1 S_2 \ldots S_{K-1})} \quad (x = 0, 1, \ldots, K),$$
$$= \theta_x, \text{ say.}$$

Therefore if a random sample is taken and n_x are found to be of age x, then the joint probability function of $n_0, n_1, \ldots, n_K$, given $n \left(= \sum_{x=0}^{K} n_x\right)$, is

$$f(\{n_x\} \mid n) = \frac{n!}{\prod_{x=0}^{K} n_x!} \prod_{x=0}^{K} \theta_x^{n_x}.$$

It is readily shown that the maximum-likelihood estimate of S_x is the moment estimate

$$\hat{S}_x = n_{x+1}/n_x, \quad (x = 0, 1, \ldots, K-1; n_x \neq 0),$$

and using the delta method we find that asymptotically

$$V[\hat{S}_x] = \frac{S_x^2}{n}\left[\frac{1}{\theta_{x+1}} + \frac{1}{\theta_x}\right],$$

$$\operatorname{cov}[\hat{S}_x, \hat{S}_{x+1}] = -S_x S_{x+1}/(n\theta_{x+1}),$$

and

$$\operatorname{cov}[\hat{S}_x, \hat{S}_y] = 0, \quad (y > x+1).$$

Chapman and Robson suggest a slight modification of the maximum-likelihood estimate, namely

$$S_x^* = n_{x+1}/(n_x + 1),$$

which is almost unbiased since

$$E[S_x^*] = S_x[1 - (1 - \theta_x)^n].$$

Thus an almost unbiased estimate of the average survival rate

$$S_{\text{ave}} = (S_0 + S_1 + \ldots + S_{K-1})/K$$

is

$$S_{\text{ave}}^* = (S_0^* + S_1^* + \ldots + S_{K-1}^*)/K$$

---

[*]See also **13.3.3**.

with asymptotic variance

$$V[S^*_{\text{ave}}] = \frac{1}{K^2}\left\{\sum_{x=0}^{K-1} V[\hat{S}_x] + 2\sum_{x=0}^{K-2} \text{cov}[\hat{S}_x, \hat{S}_{x+1}]\right\}.$$

To estimate this variance we replace each S_x by S^*_x and each θ_x by n_x/n. Alternatively we can use the estimate given by (1.10).

One other estimate of interest is

$$\hat{S}_{\text{geo}} = (\hat{S}_0 \hat{S}_1 \ldots \hat{S}_{K-1})^{1/K} = (n_k/n_0)^{1/K},$$

an estimate of the geometric mean. If Z_x is the annual instantaneous mortality rate, so that $S_x = \exp(-Z_x)$, then $\sum_x Z_x/K$, the average instantaneous mortality rate, can be estimated by $-\log \hat{S}_{\text{geo}}$. However, $\hat{S}_{\text{geo}}$ depends only on the two extreme age-classes, so that it will be subject to greater fluctuations than S^*_{ave}.

To test the hypothesis $H: S_x = S$ $(x = 0, 1, \ldots, K-1)$ we can calculate a standard multinomial goodness-of-fit statistic

$$T = \sum_{x=0}^{K} (n_x - n\hat{\theta}_x)^2/(n\hat{\theta}_x)$$

which is asymptotically χ^2_{K-1} when H is true. Here

$$\hat{\theta}_x = \hat{S}^x_{\text{seg}} \Big/ \sum_{x=0}^{K} \hat{S}^x_{\text{seg}}$$

$$= \hat{S}^x_{\text{seg}}(1-\hat{S}_{\text{seg}})/(1-\hat{S}^{K+1}_{\text{seg}})$$

$$= \frac{1}{K+1}\left[1 - \frac{X(1-\hat{S}_{\text{seg}})}{n\hat{S}_{\text{seg}}}\right],$$

where $\hat{S}_{\text{seg}}$ is given by equation (10.19). An example of this is given by Kimball and Wolfe [1974].

Finally we note that there are matrix methods for studying stationary populations (cf. 13.3.4).

CHAPTER 11

POPULATIONS STRATIFIED GEOGRAPHICALLY

11.1 CLOSED POPULATION

11.1.1 The Petersen method

Very often a population is stratified geographically, so that the total population may be regarded as consisting of separate strata living in different areas. The experimenter may then be interested not only in the sizes of the separate strata but also in the degree of mixing between strata. One method of estimating the various population parameters is to use a generalisation of the Petersen method in which marked animals are released in each stratum, using a different mark for each stratum. After allowing for the dispersal of the marked, a sample is then taken from each stratum and the recaptures recorded. This model has been investigated by Schaefer [1951: population stratified temporally rather than spatially], Chapman and Junge [1956: moment estimation], Beverton and Holt [1957: deterministic approach with a mark release in just one stratum], Darroch [1961: extensive maximum-likelihood theory], and Overton and Davis [1969: 437–41]. The theory in this chapter is based on Darroch's article.

Let

s = number of strata,
A = total number of marked animals released,
a_i = number of marked released in stratum i,
U = total number in the unmarked population,
$\overset{*}{U}_i$ = number of unmarked in stratum i at the time of the mark release,
U_j = number of unmarked in stratum j at the time of the sampling,
N = $U + A$ (total population size),
n_j = size of sample from stratum j,
m_{ij} = marked members of n_j which were released in stratum i,
u_j = number of unmarked in n_j,
θ_{ij} = probability that a member of a_i moves to stratum j, and
ψ_{ij} = probability that a member of a_i is caught in stratum j.

The random variables are m_{ij}, u_{ij} and $n_{ij} = \sum_{i=1}^{s} m_{ij} + u_{ij}$: the remaining symbols represent fixed parameters with s, A and a_i known.

We shall assume that:

(i) The population is closed, i.e. there is no mortality, so that

$$\sum_j \theta_{ij} = 1 \quad (i = 1, 2, \ldots, s).$$

(ii) All individuals in the jth stratum, whether marked or unmarked, have the same probability p_j of being caught in the sample.

(iii) Marked individuals behave independently of one another in regard to moving between strata and being caught.

(iv) $\psi_{ij} = \theta_{ij} p_j$.

(v) The matrix $\Theta = [(\theta_{ij})]$ is non-singular: this implies that the transformation in (iv) subject to (i) is one-to-one, so that maximum-likelihood estimates of the θ_{ij} and p_j can be obtained from estimates of the ψ_{ij}.

MAXIMUM-LIKELIHOOD ESTIMATES. Given the above assumptions, the joint probability function of the random variables $\{m_{ij}\}$, $\{u_j\}$ is the product of

$$f(\{m_{ij}\}|\{a_i\}) = \prod_{i=1}^{s} \left\{ \frac{a_i!}{(a_i - m_{i.})! \prod_{j=1}^{s} m_{ij}!} (1 - \sum_j \theta_{ij}p_j)^{a_i - m_{i.}} \prod_{j=1}^{s} (\theta_{ij}p_j)^{m_{ij}} \right\} \tag{11.1}$$

and

$$f(\{u_j\}) = \prod_{i=1}^{s} \left\{ \binom{U_j}{u_j} p_j^{u_j}(1 - p_j)^{U_j - u_j} \right\}, \tag{11.2}$$

where $m_{i.} = \sum_i m_{ij}$. As the $\{m_{ij}\}$ and $\{u_j\}$ are independent, maximum-likelihood estimation for the joint distribution amounts to obtaining maximum-likelihood estimates $\hat{\theta}_{ij}$, $\hat{p}_j$, from (11.1) and then using (11.2) to write down the moment equations $\hat{U}_j\hat{p}_j = u_j$. Therefore, if Θ is non-singular, we find that

$$\hat{\theta}_{ij}\hat{p}_j = \hat{\psi}_{ij} = m_{ij}/a_i, \quad (i, j = 1, 2, \ldots, s), \tag{11.3}$$

$$\sum_{j=1}^{s} \hat{\theta}_{ij} = 1, \quad (i = 1, 2, \ldots, s), \tag{11.4}$$

and

$$\hat{U}_j = u_j/\hat{p}_j, \quad (j = 1, 2, \ldots, s). \tag{11.5}$$

Cross-multiplying in (11.3), these equations can be written in matrix form, namely

$$\mathbf{D}_a\hat{\Theta}\hat{\mathbf{D}}_p = [(m_{ij})] = \mathbf{M},$$

$$\Theta\mathbf{1} = \mathbf{1},$$

and

$$\hat{\mathbf{U}} = \mathbf{D}_u\hat{\mathbf{p}}. \tag{11.6}$$

Here $\mathbf{1}$ is a column vector of ones; $\mathbf{U} = [(U_j)]$, $\mathbf{a} = [(a_i)]$, $\mathbf{p} = [(p_j^{-1})]$ are column vectors; $\mathbf{D}_a = [(\delta_{ij}a_j)]$, $\mathbf{D}_u = [(\delta_{ij}u_j)]$ and $\mathbf{D}_p = [(\delta_{ij}\hat{p}_j)]$ are diagonal matrices. If Θ is non-singular and the sample sizes n_j large, a singular matrix $\mathbf{M}$ will be extremely unlikely. Therefore, assuming $\mathbf{M}$ to be non-singular,

$$\hat{\mathbf{D}}_\rho = \hat{\mathbf{D}}_p^{-1} = \mathbf{M}^{-1}\mathbf{D}_a\hat{\Theta}$$

and

$$\begin{aligned}\hat{\mathbf{p}} &= \hat{\mathbf{D}}_p^{-1}\mathbf{1} \\ &= \mathbf{M}^{-1}\mathbf{D}_a\hat{\Theta}\mathbf{1} \\ &= \mathbf{M}^{-1}\mathbf{D}_a\mathbf{1} \\ &= \mathbf{M}^{-1}\mathbf{a}.\end{aligned} \tag{11.7}$$

This means that the $\hat{p}_i$ are given by

$$\begin{bmatrix} \hat{p}_1^{-1} \\ \hat{p}_2^{-1} \\ \cdots \\ \hat{p}_s^{-1} \end{bmatrix} = \begin{bmatrix} m_{11} & m_{12} & \cdots & m_{1s} \\ m_{21} & m_{22} & \cdots & m_{2s} \\ \cdots & \cdots & \cdots & \cdots \\ m_{s1} & m_{s2} & \cdots & m_{ss} \end{bmatrix}^{-1} \begin{bmatrix} a_1 \\ a_2 \\ \cdots \\ a_s \end{bmatrix}$$

and the $\hat{\theta}_{ij}$, $\hat{U}_j$ then follow from equations (11.3) and (11.5), namely

$$\hat{\theta}_{ij} = a_i^{-1} m_{ij} \hat{p}_j^{-1}, \quad \hat{U}_j = u_j \hat{p}_j^{-1}$$

or, in matrix notation,

$$\hat{\boldsymbol{\Theta}} = \mathbf{D}_a^{-1}\mathbf{M}\hat{\mathbf{D}}_\rho \tag{11.8}$$

and

$$\hat{\mathbf{U}} = \mathbf{D}_u \hat{\boldsymbol{\rho}} = \mathbf{D}_u \mathbf{M}^{-1}\mathbf{a}. \tag{11.9}$$

Estimates of U and N are also available; thus

$$\hat{U} = \mathbf{1}'\hat{\mathbf{U}} = \mathbf{1}'\mathbf{D}_u\mathbf{M}^{-1}\mathbf{a} = \mathbf{u}'\mathbf{M}^{-1}\mathbf{a} \quad (= \mathbf{u}'\hat{\boldsymbol{\rho}}) \tag{11.10}$$

and

$$\begin{aligned} \hat{N} &= \hat{U} + A \\ &= \mathbf{u}'\mathbf{M}^{-1}\mathbf{a} + \mathbf{1}'\mathbf{a} \\ &= (\mathbf{u}'\mathbf{M}^{-1} + \mathbf{1}'\mathbf{M}\mathbf{M}^{-1})\mathbf{a} \\ &= \mathbf{n}'\mathbf{M}^{-1}\mathbf{a}, \end{aligned}$$

where $\mathbf{n} = [(n_j)]$. We note that the estimate $\hat{N}$ was given by Chapman and Junge [1956: their $\hat{N}_3$] and has been applied, for example, to the Alaska fur-seal herd (Kenyon *et al.* [1954]).

VARIANCES AND COVARIANCES. Darroch proves that the above estimates are consistent (for $\boldsymbol{\Theta}$ non-singular) and approximately unbiased (for large a_i). Also $\hat{\boldsymbol{\rho}}$ has negligible bias so that its variance–covariance matrix is approximately

$$E[(\hat{\boldsymbol{\rho}} - \boldsymbol{\rho})(\hat{\boldsymbol{\rho}} - \boldsymbol{\rho})'] = \boldsymbol{\Sigma}, \quad \text{say.}$$

Using the delta method, Darroch shows that

$$\boldsymbol{\Sigma} \approx \mathbf{D}_\rho \boldsymbol{\Theta}^{-1}\mathbf{D}_\mu \mathbf{D}_a^{-1}\boldsymbol{\Theta}'^{-1}\mathbf{D}_\rho$$

and

$$E[(\hat{\mathbf{U}} - \mathbf{U})(\hat{\mathbf{U}} - \mathbf{U})'] \approx \mathbf{D}_U \boldsymbol{\Theta}^{-1}\mathbf{D}_\mu \mathbf{D}_a^{-1}\boldsymbol{\Theta}'^{-1}\mathbf{D}_U + \mathbf{D}_U(\mathbf{D}_\rho - \mathbf{I}),$$

where $\mathbf{I}$ is the unit matrix, and the diagonal elements of $\mathbf{D}_\mu$ are

$$\mu_i = (\sum_j \theta_{ij}/p_j) - 1.$$

Hence

$$\begin{aligned} E[(\hat{N} - N)^2] &= E[(\hat{U} - U)^2] \\ &= \mathbf{1}'E[(\hat{\mathbf{U}} - \mathbf{U})(\hat{\mathbf{U}} - \mathbf{U})']\mathbf{1} \\ &\approx \mathbf{U}'\boldsymbol{\Theta}^{-1}\mathbf{D}_\mu \mathbf{D}_a^{-1}\boldsymbol{\Theta}'^{-1}\mathbf{U} + \mathbf{U}'(\boldsymbol{\rho} - \mathbf{1}) \\ &= \sum_i \eta_i^2 \mu_i / a_i + \sum_j U_j (p_j^{-1} - 1), \end{aligned}$$

where

$$\boldsymbol{\eta}' = [(\eta_i)]' = \mathbf{U}'\boldsymbol{\Theta}^{-1}. \tag{11.11}$$

The vector $\boldsymbol{\eta}$ has a simple interpretation when the movement pattern for marked and unmarked is the same. Assuming that θ_{ij} applies also to the unmarked individuals, we have the approximate deterministic equations

$$U_j \approx \sum_i \overset{*}{U}_i \theta_{ij}. \tag{11.12}$$

Writing these in matrix form and post-multiplying by $\boldsymbol{\Theta}^{-1}$, we have

$$\overset{*}{\mathbf{U}}' \approx \mathbf{U}'\boldsymbol{\Theta}^{-1}$$

which, together with (11.11), implies that $\eta_i \approx \overset{*}{U}_i$, the number of unmarked in stratum i at the time of the mark release.

11.1.2 Validity of assumptions

Although assumption (i) that there is no natural mortality cannot be tested for the above model, the theory is readily generalised in **11.2**, below, to deal with the case $\sum_j \theta_{ij} < 1$. Assumption (ii) of constant p_j will be satisfied if marking and point of release do not affect catchability and if the stratum area is sufficiently small, so that either the sampling effort is uniform over it and every member has an equal chance of capture, or there is uniform mixing of marked and unmarked, regardless of sampling uniformity (cf. **3.1.2**).

In the definition of ψ_{ij} and the derivation of (11.1) it was assumed that the a_i individuals released in the ith stratum move and are caught (or not caught) (1) independently of those released in any other stratum, and (2) independently of each other. (1) is very reasonable, but (2) is less likely to hold true in practice. For example, if the marked individuals are released close together at some random point in the stratum, they may tend to move to the same stratum (e.g. schooling in fish). On the other hand, if they are released over a carefully spaced grid of points so that they start with maximum possible distances between them, they may tend to move in different strata.

As well as "contagious" movement there is the possibility of "contagious" catching. For, example, if sampling in the jth stratum is not uniform but is concentrated in one or more subareas and if, having a common stratum of origin, the marked individuals from the ith stratum are not uniformly distributed in the jth stratum, there will be a certain amount of positive dependence on their being caught or not caught. However, Darroch [1961: 250–2] shows that the estimates are still consistent when there is both contagious movement and catching and that contagious movement has little effect on the variance–covariance matrix of $\hat{\boldsymbol{\rho}}$. Positive dependence in catchability will tend to increase the variances, though this effect will also be small if every effort is made to ensure that both the release and the sampling procedures are as nearly random as possible.

If $\boldsymbol{\Theta}$ is singular, the above theory is no longer valid. Instead of working through the parameters $\{\psi_{ij}\}$ we must now work with the $\{\theta_{ij}\}$ and $\{p_j\}$ directly. Unfortunately this direct approach of maximising the likelihood subject to the constraints $\boldsymbol{\Theta}\mathbf{1} = \mathbf{1}$ does not yield simple solutions of the resulting maximum-likelihood equations, so that a singular $\boldsymbol{\Theta}$ should be avoided. In fact, the further $\boldsymbol{\Theta}$ is from singularity the more reliable are the estimates. This is reflected in the variance formulae which depend on $\boldsymbol{\Theta}^{-1}$; the elements of $\boldsymbol{\Theta}^{-1}$ will be large if the determinant of $\boldsymbol{\Theta}$ is near zero. However, when (11.12) is true, $E[(\hat{U} - U)^2]$ will be insensitive to near singularity, for then η_i, although defined in terms of $\boldsymbol{\Theta}^{-1}$, is approximately equal to $\overset{*}{U}_i$ and is therefore virtually independent of $\boldsymbol{\Theta}^{-1}$. Even if (11.12) is not true, the η_i will tend to be positive and therefore not too large, as $\sum \eta_i = U$. (This last equation follows from the sequence $\sum \eta_i = \boldsymbol{\eta}'\mathbf{1} = \boldsymbol{\eta}'\boldsymbol{\Theta}\mathbf{1} = \mathbf{U}'\mathbf{1} = U$.)

Although it is impossible to predetermine the strata so that $\boldsymbol{\Theta}$ is guaranteed to be non-singular, one or two trivial cases may be foreseeable. For example, if it is suspected that $\theta_{hj} = \theta_{ij}$ for all j, then the hth and ith strata can be combined. However, any near singularity in $\boldsymbol{\Theta}$ will generally be reflected in the near singularity of $\mathbf{M}$, and $\mathbf{M}$ can be used to decide which strata to pool.

Finally it is noted that the estimate $\hat{U}$ does not depend on the assumption $\psi_{ij} = \theta_{ij}p_j$ when ψ_{ij} is the same for both marked and unmarked, i.e. when marked and unmarked have the same behaviour patterns. To see this we consider the moment equations

$$\sum_i \overset{*}{U}_i \psi_{ij} = u_j \quad \text{and} \quad a_i\psi_{ij} = m_{ij},$$

which lead to

$$\sum_i \overset{*}{U}_i m_{ij}/a_i = u_j \tag{11.13}$$

or

$$\overset{*}{\mathbf{U}}{}'\mathbf{D}_a^{-1}\mathbf{M} = \mathbf{u}'. \tag{11.14}$$

Thus a moment estimate of U is

$$\begin{aligned} \overset{*}{\mathbf{U}}{}'\mathbf{1} &= \mathbf{u}'\mathbf{M}^{-1}\mathbf{D}_a\mathbf{1} \\ &= \mathbf{u}'\mathbf{M}^{-1}\mathbf{a} \end{aligned}$$

which, from (11.10), is simply $\hat{U}$ once again. It is also readily shown from Darroch that $E[(\hat{U} - U)^2]$ can be written in terms of the ψ_{ij} and p_j only.

11.1.3 Variable number of strata

In some situations the number of strata may change between the times of the mark release and of sampling. Suppose that the marked individuals are released into s strata, and sampling is carried out in t strata. Then (11.1) and (11.2) are still valid, provided we define $i = 1, 2, \ldots, s$ and $j = 1, 2, \ldots, t$.

When $s < t$, the parameters $\{\theta_{ij}\}$, $\{p_j\}$ are no longer identifiable and the U_j cannot be estimated. But an estimate of U is still available if ψ_{ij} is the same for marked and unmarked. In this case, with $j = 1, 2, \ldots, t$ we have, from (11.13), t equations for the s unknowns $\{\overset{*}{U}_i\}$. We can then replace these t equations by s linear combinations of them, giving, from (11.14),

$$\overset{*}{\mathbf{U}}'\mathbf{D}_a^{-1}\mathbf{MQ} = \mathbf{u}'\mathbf{Q}$$

where $\mathbf{Q}$ is a t-by-s matrix of rank s such that $\mathbf{MQ}$ is non-singular. Thus our estimate of U, namely

$$\begin{aligned}\overset{*}{\mathbf{U}}'\mathbf{1} &= (\mathbf{Q}'\mathbf{u})'[\mathbf{MQ}]^{-1}\mathbf{D}_a\mathbf{1} \\ &= (\mathbf{Q}'\mathbf{u})'[\mathbf{MQ}]^{-1}\mathbf{a}\end{aligned}$$

takes the same form as $\hat{U}$ (cf. (11.10)). Its asymptotic variance can be found, using Darroch's technique [1961: 83], by writing $\hat{\mathbf{p}} = [\mathbf{MQ}]^{-1}\mathbf{a}$.

When $s > t$ the maximum-likelihood estimates of the $\{\theta_{ij}\}$, $\{p_j\}$ are not readily found. However, we can once again obtain a moment-type estimate of U using a method similar to one given by Chapman and Junge [1956]. Consider the moment equations

$$U_j p_j = u_j \quad \text{and} \quad m_{ij}/p_j = a_{ij},$$

where a_{ij} is the number from the group of a_i released in the ith stratum which move to stratum j. Then, substituting for p_j in the second equation and summing on j gives us s equations

$$\sum_j m_{ij}U_j/u_j = a_i, \quad (i = 1, 2, \ldots, s),$$

or

$$\mathbf{MD}_u^{-1}\mathbf{U} = \mathbf{a} \tag{11.15}$$

in the t unknowns $\{U_j\}$. Once again when $s = t$, (11.15) leads to $\hat{U}$, while for $s > t$ we look for a t-by-s matrix $\mathbf{R}$ such that $\mathbf{RM}$ is non-singular. Then $\mathbf{U}$ is estimated by $\mathbf{D}_u(\mathbf{RM})^{-1}\,\mathbf{Ra}$ and U by $\mathbf{u}'(\mathbf{RM})^{-1}\mathbf{Ra}$; expressions for asymptotic variances and covariances can be obtained using the methods of Darroch. In conclusion we mention that more general methods for dealing with the case $s > t$ (which also allow for mortality) are given in **11.2** below.

11.1.4 The Petersen estimate using pooled data

CONSISTENCY. If the experimenter does not know how the population is stratified, or he is unable to use different marks for different strata, he will pool the data and use the Petersen estimate of N, namely

$$\hat{N}_P = An/m,$$

where $u = \sum_{j=1}^{t} u_j$, $m = \sum_{i=1}^{s}\sum_{j=1}^{t} m_{ij}$ and $n = u + m$. In discussing the properties of this estimate it is convenient to consider

$$\hat{U}_P = \hat{N}_P - A = Au/m,$$

the Petersen estimate of the unmarked population size. The first property we consider is that of consistency. Letting $a_i \to \infty$, $U_j \to \infty$ in such a way that a_i/A, U_j/U and A/U are all constant, we find that $\hat{U}_P$ is a consistent estimate of

$$U_P = AE[u]/E[m]$$
$$= AB/C, \quad \text{say},$$
$$= \frac{A \sum U_j p_j}{U \sum\sum a_i \theta_{ij} p_j} \cdot U. \qquad (11.16)$$

Although the equation $U_P = U$ can be satisfied in an infinite number of ways, only the following four special cases are listed because of their simple physical interpretations:

(1) $p_j = p$ for $j = 1, 2, \ldots, t$.

(2) $a_i/\overset{*}{U}_i = A/U \quad (i = 1, 2, \ldots, s)$, (11.17)

and the movement pattern for marked and unmarked is the same, i.e.

$$U_j = \sum_i \overset{*}{U}_i \theta_{ij}. \qquad (11.18)$$

(3) Equation (11.18) holds and $\theta_{ij} = \theta_j$ for all i, j.

(4) $\sum_i a_i \theta_{ij} = kU_j, \quad (j = 1, 2, \ldots, t)$.

Condition (1) implies that a constant proportion of each stratum is sampled, while (11.17) implies that a constant proportion of each stratum is marked. (11.17) is the marking analogue of (1) and corresponds to a constant probability of capture over the whole population when the animals are initially caught for marking. If (3) is true, we have a complete mixing of the whole population and, when $s = t$, $\boldsymbol{\Theta}$ is singular. Condition (4) is equivalent to saying that the expected number of marked in the jth stratum is proportional to the number of unmarked; when $\sum_j \theta_{ij} = 1$, $k = A/U$. We note that conditions (1)–(3) were first considered by Chapman and Junge [1956].

EFFICIENCY. Using the delta method it can be shown that

$$E[(\hat{U}_P - U_P)^2] \approx (A^2B^2/C^4)V[m] + (A^2/C^2)V[u].$$

Now from the inequality $(\sum x_i y_i)^2 \leqslant (\sum x_i^2)(\sum y_i^2)$ with $x_i = \sqrt{a_i}$, $y_i = \sqrt{a_i} \sum_j \theta_{ij} p_j$ we find that (cf. (11.1))

$$V[m] = \sum_i V[m_{i\cdot}]$$
$$= \sum_i a_i (\sum_j \theta_{ij} p_j)(1 - \sum_j \theta_{ij} p_j)$$
$$\leqslant (\sum_i \sum_j a_i \theta_{ij} p_j)(1 - \sum_i \sum_j a_i \theta_{ij} p_j / A)$$
$$= A(C/A)(1 - C/A)$$
$$= \text{var}[m], \quad \text{say}.$$

Similarly,

$$V[u] = \sum_j V[u_j]$$
$$= \sum_j U_j p_j (1 - p_j)$$

$$\leqslant (\sum_j U_j p_j)(1-\sum_j U_j p_j/U)$$
$$= U(B/U)(1-B/U)$$
$$= \text{var}[u], \quad \text{say.}$$

Therefore, if the experimenter uses the Petersen estimate (or its approximately unbiased version $(A+1)u/(m+1)$, cf. (5.21)) and treats u and m as binomial variables, his expression for the mean-square error, namely

$$(A^2B^2/C^4)\,\text{var}\,[m] + (A^2/C^2)\,\text{var}\,[u]$$

will be conservative in that it will slightly overestimate the true mean-square error $E[(\hat{U}_P - U_P)^2]$.

When $s = t$, and $\hat{U}_P$ and the maximum-likelihood estimate $\hat{U}$ are both consistent, Darroch shows that asymptotically

$$E[(\hat{U}_P - U)^2] = Y \leqslant Z = E[(\hat{U} - U)^2], \tag{11.19}$$

with equality when $p_i = p$ and $\eta_i/a_i = U/A$ for $i = 1, 2, \ldots, s$ (see (11.11) for definition of η_i). These two conditions for equality are satisfied if, for example, (1) and (2) above are true.

TESTS FOR CONSISTENCY. A goodness-of-fit test for the hypothesis $H_1 : \theta_{ij} = \theta_j$ is equivalent to a test of homogeneity for the rows of the s-by-$(t+1)$ contingency table:

m_{11}	m_{12}	...	m_{1t}	$a_1 - m_{1.}$	a_1
m_{21}	m_{22}	...	m_{2t}	$a_2 - m_{2.}$	a_2
...	...	...	...	...	...
m_{s1}	m_{s2}	...	m_{st}	$a_s - m_{s.}$	a_s
$m_{.1}$	$m_{.2}$	...	$m_{.t}$	$A - m$	A

A "pooled" version of this test given by both Chapman and Junge, and Darroch, is a test of homogeneity on the columns of the 2-by-s contingency table:

$m_{1.}$	$m_{2.}$	...	$m_{s.}$	m
$a_1 - m_{1.}$	$a_2 - m_{2.}$	...	$a_s - m_{s.}$	$A - m$
a_1	a_2	...	a_s	A

using

$$T = \frac{\sum_i (m_{i.} - a_i m/A)^2}{a_i \dfrac{m}{A}\left(1-\dfrac{m}{A}\right)}.$$

Here T is asymptotically distributed as chi-squared with $s-1$ degrees of freedom when the hypothesis $H_2 : \sum_j \theta_{ij}p_j = d$ is true; in this case $\hat{U}_P$ is consistent when (11.18) is also true.

Two special cases of H_2 are H_1 and $H_3 : p_j = p$ (all j). When T is not significant, one can test H_1 against H_2 using the s-by-t contingency table obtained from the s-by-$(t+1)$ table above by deleting the last column. Unfortunately no straightforward test of H_3 against H_2 is available.

$H_4 : \sum_i a_i\theta_{ij} = kU_j$ can be tested by a test of homogeneity on the columns of the 2-by-t contingency table:

$m_{\cdot 1}$	$m_{\cdot 2}$	...	$m_{\cdot t}$	m
u_1	u_2	...	u_t	u
n_1	n_2	...	n_t	n

In conclusion we mention briefly another estimate of N due to Schaefer [1951], namely

$$\hat{N}_S = \sum_{i=1}^{s}\sum_{j=1}^{t} \frac{n_j a_i m_{ij}}{m_{\cdot i} m_{\cdot j}}.$$

We find, as first pointed out by Chapman and Junge, that $\hat{N}_S$ is consistent when either of the above conditions (1) or (2) is satisfied. The asymptotic variance of $\hat{N}_S$, which would require considerable computation, does not seem to be given in the literature.

11.2 OPEN POPULATION: NATURAL MORTALITY(*)

CASE I: $s = t$. If natural mortality is taking place between the times of marking and sampling, $\sum_j \theta_{ij}$ ($= \phi_i$ say) will be the probability of survival for members of a_i. In this case (11.1) is still valid but $\{\theta_{ij}\}$, $\{p_j\}$ are non-identifiable, as $\{\lambda\theta_{ij}\}$, $\{\lambda^{-1} p_j\}$ also lead to the same probability function (11.1). To tie down this non-identifiability, Darroch assumes that $\phi_i = \phi$ ($i = 1, 2, \dots, s$) and defines $\beta_{ij} = \theta_{ij}/\phi$ and $P_j = \phi p_j$. Here β_{ij} is the probability that a member of a_i is in stratum j at the time of sampling, given that it is alive at that time, and P_j is the probability of an animal surviving and, if it is in the jth stratum, of being caught there.

This re-parametrisation leads to the same distribution (11.1) but with θ_{ij} and p_j replaced by β_{ij} and P_j. Therefore, when $\mathbf{B} = [(\beta_{ij})]$ is non-singular, the maximum-likelihood estimates of β_{ij} and P_j will satisfy (cf. (11.3) and (11.4))

$$\hat{\beta}_{ij}\hat{P}_j = m_{ij}/a_i \quad (= \hat{\psi}_{ij})$$

and

$$\sum_j \hat{\beta}_{ij} = 1, \quad (i = 1, 2, \dots, s).$$

From (11.7) and (11.8) these equations have solutions

$$\hat{\mathbf{p}} = \mathbf{M}^{-1}\mathbf{a}$$

(*)See also **13.4**.

and

$$\hat{\mathbf{B}} = \mathbf{D}_a^{-1}\mathbf{M}\hat{\mathbf{D}}_\rho,$$

where $\hat{\rho}_i = 1/\hat{P}_i$. However, although (11.1) is re-parametrised, (11.2) remains unchanged, and we now have the problem of relating P_j to p_j so that U_j can be estimated. Fortunately we can get round this problem if we proceed by analogy with (11.5) and consider

$$\hat{U}_j \hat{p}_j = u_j$$

or, writing $W_j = U_j/\phi$,

$$\hat{W}_j \hat{P}_j = u_j.$$

This means that we can estimate W_j and $W = \Sigma W_j$ using the same methods as those used in **11.1.1**; for example, from (11.9) we have

$$\hat{\mathbf{W}} = \mathbf{D}_u\mathbf{M}^{-1}\mathbf{a}$$

and

$$\hat{W} = \mathbf{u}'\mathbf{M}^{-1}\mathbf{a}.$$

We note that $\hat{W}$ is the same as $\hat{U}$, though $\hat{W}$ is now an estimate of W (which is approximately the total unmarked population size at the time of the mark release) rather than of U, the final unmarked population size. This fact could have been deduced directly from (11.14) and the following equations where it was shown that, irrespective of any assumptions about the ψ_{ij}, $\hat{U}$ is an estimate of $\overset{*}{\mathbf{U}}'\mathbf{1}$, the initial size of the unmarked population.

As far as variance estimation is concerned, the only changes in the variance–covariance formulae are the addition of non-estimable second-order correction terms in $E[(\hat{\mathbf{W}}-\mathbf{W})(\hat{\mathbf{W}}-\mathbf{W})']$ and $E[(\hat{W}-W)^2]$ involving $1-\phi^{-1}$. The appropriate formulae can be obtained from equations (11.20) to (11.23) below by dropping out $\mathbf{X}$, and setting $\bar{\phi} = \phi$ and $\gamma_i = 1$. However, when ϕ is not too small these non-estimable corrections will be negligible.

We note that although the estimates $\hat{\psi}_{ij}$ $(= m_{ij}/a_i)$ are less than unity, in transforming from $\hat{\psi}_{ij}$ to $\hat{\beta}_{ij}$ it is possible that $\hat{P}_j$ may be greater than unity. If this anomaly takes place, Darroch points out that as the transformation is made by imposing the constraints $\sum_j \beta_{ij} = 1$ it is these constraints which are the root of the trouble. He suggests two diagnoses: (i) the ϕ_i are not equal, or (ii) the ϕ_i are equal but have large sampling errors. Here (i) can be allowed for by reducing the number of sample strata and applying the theory of case II below when $s > t$ (see Example 11.1 below). (ii) means that the $\hat{P}_j$ and $\hat{W}_j$ are virtually useless as estimates of P_j and W_j, and this would be confirmed by their having large variances. But this does not mean that the variance of $\hat{W}$ is also large, as the covariances of $\hat{W}_j$ may be large and negative.

Replacing θ_{ij}, p_j and U_j by β_{ij}, P_j and W_j, we find that the general theory of **11.1.4** still holds, though the Petersen estimate $\hat{U}_P$ is now an estimate of W rather than of U. Also the comments in **11.1.2** still apply here.

CASE II: $s > t$. When $s > t$ an alternative approach to that given in

11.1.2 is available. Allowing the ϕ_i to differ and defining $\bar{\phi} = \Sigma\, \phi_i/s$, we now redefine the parameters $\beta_{ij} = \theta_{ij}/\bar{\phi}$, $P_j = \bar{\phi}\, p_j$, $W_j = U_j/\bar{\phi}$ and introduce $\gamma_i = \phi_i/\bar{\phi}$. This means that we have $st - 1 + t$ independent parameters $\{\beta_{ij}\}$, $\{P_j\}$, the subtraction of 1 being due to the single constraint $\Sigma\Sigma\, \beta_{ij} = s$. However, instead of imposing $s-1$ constraints $\gamma_1 = \gamma_2 = \ldots = \gamma_s$, as in the case $s = t$ above, Darroch suggests imposing just $t-1$ constraints, thus reducing the number of independent parameters to st, the number of $\{\psi_{ij}\}$. The $t-1$ constraints will usually take the form $\gamma_i - \gamma_j = 0$ or possibly $\gamma_i + \gamma_j - 2\gamma_k = 0$, and they can be written in the general form

$$\sum_{k=1}^{s} x_{jk}\gamma_k = 0, \quad (j = 1, 2, \ldots, t-1).$$

To this set we add $\sum_k \gamma_k = s$ in the form

$$\sum_{k=1}^{s} x_{tk}\gamma_k = 1, \quad (x_{tk} = 1/s).$$

For a suitable choice of constraints the matrix $\mathbf{X} = [(x_{jk})]$ will be a t-by-s matrix of rank t, and

$$\mathbf{XB1} = \mathbf{X\gamma} = \mathbf{v},$$

where $\mathbf{v}'$ is the row vector $(0, 0, \ldots, 0, 1)$. If the s-by-t matrices $\mathbf{B}$ and $\mathbf{M}$ are also of rank t, Darroch shows that the maximum-likelihood estimates are now

$$\hat{\mathbf{p}} = [\mathbf{XD}_a^{-1}\mathbf{M}]_a^{-1}\mathbf{v}$$

$$\hat{\mathbf{B}} = \mathbf{D}_a^{-1}\mathbf{M}\hat{\mathbf{D}}_\rho, \quad \hat{\gamma} = \hat{\mathbf{B}}\mathbf{1}$$

$$\hat{\mathbf{W}} = \mathbf{D}_u[\mathbf{XD}_a^{-1}\mathbf{M}]^{-1}\mathbf{v}$$

and

$$\hat{W} = \mathbf{u}'[\mathbf{XD}_a^{-1}\mathbf{M}]^{-1}\mathbf{v}$$

where

$$\hat{\mathbf{\rho}} = [(\hat{P}_j^{-1})].$$

Also if

$$E[(\hat{\mathbf{\rho}} - \mathbf{\rho})(\hat{\mathbf{\rho}} - \mathbf{\rho})'] = \mathbf{\Sigma},$$

Darroch proves that

$$\mathbf{\Sigma} \approx \mathbf{D}_\rho(\mathbf{XB})^{-1}\mathbf{XD}_\mu\mathbf{D}_a^{-1}\mathbf{X}'(\mathbf{XB})'^{-1}\mathbf{D}_\rho,$$

$$E[(\hat{\mathbf{W}} - \mathbf{W})(\hat{\mathbf{W}} - \mathbf{W})'] \approx \mathbf{D}_W(\mathbf{XB})^{-1}\mathbf{XD}_\mu\mathbf{D}_a^{-1}\mathbf{X}'(\mathbf{XB})'^{-1}\mathbf{D}_W + \mathbf{D}_W(\mathbf{D}_\rho - \mathbf{I}\bar{\phi}^{-1}), \quad (11.20)$$

and

$$E[(\hat{W} - W)^2] \approx \sum_i \eta_i^2\mu_i/a_i + \sum_j W_j(\rho_j - \bar{\phi}^{-1}) \quad (11.21)$$

where $\mu_i = \sum_j \beta_{ij}P_j^{-1} - \gamma_i^2$ (11.22)

and

$$\mathbf{\eta}' = \mathbf{W}'(\mathbf{XB})^{-1}\mathbf{X}. \quad (11.23)$$

As far as the Petersen estimate $\hat{U}_P$ is concerned, the general theory of **11.1.4** still applies if we replace θ_{ij}, p_j, U_j and U by β_{ij}, P_j, W_j and W. For example, U_P can now be written in the form

$$U_P = \frac{A\, \Sigma\, W_j P_j}{W\, \Sigma\Sigma\, a_i\beta_{ij}P_j} \cdot W.$$

However, Darroch shows that of the four conditions given in **11.1.4** (suitably interpreted in terms of our new parameters) only (3) implies that $U_P = W$. For the other three conditions $U_P \neq W$, though if the ϕ_i are not too different, the difference $U_P - W$ will be small. This means that the goodness-of-fit tests given in **11.1.4** will now test for approximate consistency. Once again, when $\hat{U}_P$ and $\hat{W}$ are both consistent, the Petersen estimate will generally have a smaller asymptotic variance; the asymptotic variances will be equal if $P_j = P$ and $\eta_i/a_i = W/A_i$, where η_i is now defined by (11.23).

Example 11.1 Sockeye salmon: Darroch [1961]

Darroch considers an example given by Schaefer [1951] in which both stratifications are with respect to time instead of place. The population consisted of all adult sockeye salmon passing a certain point of a river during a period of $s = 8$ weeks on their way upstream to their spawning grounds. The fish were sampled and tagged according to the week in which they passed this point. Provided they succeeded in reaching the spawning grounds, most adult salmon died after spawning. In this case, the deaths took place over a period of $t = 9$ weeks and during each of these weeks, a number of dead fish were recorded, presumably very soon after death. The data from the experiment are set out in Table 11.1. Here m_{ij} is the number of fish tagged in the

TABLE 11.1
Schaefer's data on sockeye salmon: m_{ij}, a_i and u_j for $s = 8$, $t = 9$ (from Darroch [1961: Table 1]).

Week of tagging (i)	Week of recovery (j) 1	2	3	4	5	6	7	8	9	Total	a_i
1	1	–	2	–	–	–	–	–	–	3	15
2	1	3	7	–	–	–	–	–	–	11	59
3	1	11	33	24	5	1	–	1	–	76	410
4	–	5	29	79	52	3	2	7	3	180	695
5	–	–	11	67	77	2	16	7	3	183	773
6	–	–	–	14	25	3	10	6	2	60	335
7	–	–	–	–	–	–	1	5	–	6	59
8	–	–	–	–	–	–	1	–	–	1	5
Total	3	19	82	184	159	9	30	26	8	520	2351
u_j	16	113	718	2664	3317	635	1217	904	368	9952	

ith sampling week and recovered dead on the spawning ground in the jth recovery week; p_j is the probability of being recovered on the spawning ground in the jth recovery week; θ_{ij} is the probability that a fish tagged in the ith week dies on the spawning ground in the jth recovery week, and ϕ_i is the probability that a fish tagged in the ith week dies on the spawning ground during the 9-week period. $(1-\phi_i)$ represents the probability of dying before reaching the grounds or of surviving until after this period. Evidently a small percentage of salmon do manage to reach the sea alive and return to spawn again (Jones [1959]).

Since the m_{ij} in some of the outer weeks are too small to be used in what is essentially a large-sample theory, Darroch reduces s and t to four by grouping the first three and last three weeks of tagging into single strata and the first three and last four weeks of recovery into single strata. The new values of $\{m_{ij}\}$, $\{a_i\}$ and $\{u_j\}$ are given in Table 11.2, and assuming that $\phi_i = \phi$ we proceed as in case I above.

TABLE 11.2

Schaefer's data on sockeye salmon: reduced to $s = 4$, $t = 4$ (from Darroch [1961: Table 2]).

	m_{i1}	m_{i2}	m_{i3}	m_{i4}	$m_{i.}$	$a_i - m_{i.}$	a_i	$m_{i.}/a_i$
m_{1j}	59	24	5	2	90	394	484	0·186
m_{2j}	34	79	52	15	180	515	695	0·259
m_{3j}	11	67	77	28	183	590	773	0·237
m_{4j}	0	14	25	28	67	332	399	0·168
$m_{.j}$	104	184	159	73				
u_j	847	2664	3317	3124				

As a first step in the calculations, we investigate the consistency of the Petersen estimate using the tests outlined in **11.1.4**. We find that T = 16·91, which is significant at 0·1 per cent level of significance, and the vectors $[(m_{.j})]$ and $[(u_j)]$ are so obviously not proportional that there is no need to apply a chi-squared test. Since H_1 and H_4 are both rejected, it would appear that the Petersen estimate is unsatisfactory.

Evaluating $\hat{\boldsymbol{\rho}} = \mathbf{M}^{-1}\mathbf{a}$, and recalling that $\hat{P}_i = \hat{\rho}_i^{-1}$, we find that

$$\hat{P}_1 = 0{\cdot}1318, \quad \hat{P}_2 = 1{\cdot}9461, \quad \hat{P}_3 = 0{\cdot}1947, \quad \hat{P}_4 = 0{\cdot}1063$$

and

$$\hat{W} = \mathbf{u}'\hat{\boldsymbol{\rho}} = 54\,200.$$

The unsatisfactory value of $\hat{P}_2$ may be just a symptom of the general inaccuracy of the estimates, or it may indicate that the model is incorrect in assuming the ϕ_i equal. As Darroch points out, both of these explanations are probably correct, and although nothing can be done about the first we can act on the second. The $m_{i.}/a_i$ indicate where the possible differences in the ϕ_i lie, the middle two being appreciably larger than the outer two (cf. Table 11.2). Darroch therefore suggests estimating subject to just the two constraints: $\phi_1 = \phi_4$, $\phi_2 = \phi_3$. In order to apply the general theory of case II above, we must reduce t from four to three by pooling two of the recovery periods. It is permissible to group the jth and kth periods if (i) $(\beta_{ij}P_j + \beta_{ik}P_k)/(\beta_{ij}+\beta_{ik})$ is independent of i, in particular if (ii) $P_j = P_k$ or (iii) β_{ij}/β_{ik} is independent of i. (ii) cannot be tested, but (iii) can as it implies proportionality of the jth and kth columns of $\mathbf{M}$. In this case, the columns which are nearest to being proportional are the third and fourth, which we now pool. (Although the hypothesis of proportionality is rejected at the 0·1 per cent level, (i) will still be approximately true if P_3 is not too different from P_4.) Proceeding as in case II above, we have

$$\mathbf{X} = \begin{bmatrix} 1 & 0 & 0 & -1 \\ 0 & 1 & -1 & 0 \\ \frac{1}{4} & \frac{1}{4} & \frac{1}{4} & \frac{1}{4} \end{bmatrix}, \qquad \mathbf{M} = \begin{bmatrix} 59 & 24 & 7 \\ 34 & 79 & 67 \\ 11 & 14 & 105 \\ 0 & 14 & 53 \end{bmatrix},$$

$$\mathbf{a}' = (484, 695, 773, 399), \quad \mathbf{u}' = (847, 2664, 6441),$$

and evaluating $\hat{\boldsymbol{\rho}} = [\mathbf{X}\mathbf{D}_a^{-1}\mathbf{M}]^{-1}\mathbf{v}$, we obtain

$$\hat{\boldsymbol{\rho}}' = (6{\cdot}021, 1{\cdot}607, 6{\cdot}397) \quad \text{or} \quad \hat{\mathbf{P}}' = (0{\cdot}1661, 0{\cdot}6223, 0{\cdot}1563).$$

Although $\hat{P}_2$ now lies in [0, 1] it is still rather high. The other estimates are

$$\hat{\mathbf{B}} = \mathbf{D}_a^{-1}\mathbf{M}\hat{\mathbf{D}}_\rho = \begin{bmatrix} 0{\cdot}7339 & 0{\cdot}0797 & 0{\cdot}0925 \\ 0{\cdot}2945 & 0{\cdot}1827 & 0{\cdot}6167 \\ 0{\cdot}0857 & 0{\cdot}1393 & 0{\cdot}8689 \\ 0{\cdot}0000 & 0{\cdot}0564 & 0{\cdot}8497 \end{bmatrix}$$

and, summing the rows,

$$\hat{\gamma}_1 = \hat{\gamma}_4 = 0{\cdot}9061, \quad \hat{\gamma}_2 = \hat{\gamma}_3 = 1{\cdot}0939.$$

Also,

$$\hat{\mathbf{W}}' = \hat{\boldsymbol{\rho}}'\mathbf{D}_u = (5099, 4282, 41\,204)$$

and

$$\hat{W} = 50\,585.$$

The estimated variance–covariance matrix of $\hat{\boldsymbol{\rho}}$ is

$$\hat{\boldsymbol{\Sigma}} = \begin{bmatrix} 9{\cdot}96 & -14{\cdot}84 & 6{\cdot}31 \\ -14{\cdot}85 & 23{\cdot}58 & -10{\cdot}32 \\ 6{\cdot}31 & -10{\cdot}32 & 4{\cdot}78 \end{bmatrix}$$

and we note that the variance of $\hat{\rho}_2$ is large. It is unlikely that there was any catch dependence in this experiment, so that we need not make any mental reservations about $\hat{\boldsymbol{\Sigma}}$ underestimating the true variance–covariance matrix (see **11.1.2**). Finally the variance–covariance matrix of $\hat{W}$ is estimated by

$$\hat{E}[(\hat{\mathbf{W}} - \mathbf{W})(\hat{\mathbf{W}} - \mathbf{W})'] = 10^6 \begin{bmatrix} 7{\cdot}168 & -33{\cdot}474 & 34{\cdot}441 \\ -33{\cdot}474 & 167{\cdot}347 & -177{\cdot}113 \\ 34{\cdot}441 & -177{\cdot}113 & 198{\cdot}694 \end{bmatrix},$$

and hence

$$\hat{E}[(\hat{W}-W)^2] = 20{\cdot}916 \times 10^6.$$

It is interesting to note that the estimates of $V[\hat{W}_2]$ and $V[\hat{W}_3]$ are both much larger than the estimate of $V[\hat{W}]$.

Although the Petersen estimate is invalid, it is of interest to evaluate it and its variance. Using the approximately unbiased version,

$$\begin{aligned}\hat{U}_P &= (A+1)u/(m+1)\\ &= 2352(9952)/521\\ &= 44\ 927\end{aligned}$$

and

$$\hat{E}[(\hat{U}_P - U_P)^2] = 3{\cdot}181 \times 10^6.$$

The latter is a good deal smaller than $\hat{E}[(\hat{W}-W)^2]$, which might be expected from (11.19) since the $\hat{P}_j$ differ and the $\hat{\eta}_i/a_i$ vary considerably; here

$$\hat{\boldsymbol{\eta}}' = \mathbf{u}'[\mathbf{X}\mathbf{D}_a^{-1}\mathbf{M}]^{-1}\mathbf{X} = (9896, -20\ 728, 46\ 020, 15\ 397).$$

CHAPTER 12

RECENT DEVELOPMENTS: CLOSED POPULATION

12.1 ESTIMATING ABSOLUTE DENSITY

12.1.1 Converting totals to densities

A major problem in population studies which has received insufficient attention in the literature is the conversion of total counts, or counts on sample plots, to density estimates. If grid trapping is used then simply dividing the total count (N) on the grid by the area enclosed (A) by the grid will generally lead to a severe overestimation of the population density. This problem is due to what is commonly known as "edge effect", that is traps on the boundary of the trapping area tend to catch more animals than inner traps. This edge effect is due to immigrants, and animals living outside the trapping area which have home ranges overlapping the trapping region.

Following Dice [1938] a common technique for calculating the effective trapping area is to add to A a strip of width W, where $2W(= R)$ is some linear measure such as the average diameter of the home range of an animal (cf. p. 51).

1 Home range estimation

CIRCLES OR ELLIPSES. Although numerous methods of estimating the shape and area of the home range from recapture data have been suggested, particularly for small mammals, they are generally unsatisfactory as they are basically ad hoc and the results tend to vary with trap spacing (Faust *et al.* [1971]), number of captures (see below), species, season (Briese and Smith [1974]) and size of study area (Wierzbowska [1975: 17]). The concept of the centre of activity has also been criticized on the grounds that it may not have any biological significance (Siniff and Jessen [1969], Wierzbowska [1972], Smith *et al.* [1973], Koeppl *et al.* [1975: 86]):it is simply an average of points of contact (Hayne [1949c]). In the past circular home ranges were widely used, though ellipses now appear to be more popular (cf. Jennrich and Turner [1969], Tanaka [1972] and, for further references, Mazurkiewicz [1971] and Wierzbowska [1975]). However, Metzgar [1973a], using a smoked-paper tracking and Kolmogorov–Smirnov test, judged that 10 of his 22 home ranges were circular. Also Maza *et al.* [1973] "failed to find evidence to indicate that the home range is other than circular". A number of authors effectively assume a circular home range by their use of the term "recapture radius" (Burge and Jorgensen [1973]).

BIVARIATE DISTRIBUTIONS. As the probability of capture tends to decrease with increasing distance from the "centre" of the home range, various bivariate distributions have been fitted to the distribution of probability: this concept has recently been extended to three dimensional home ranges (Koeppl *et al.* [1977], Meserve [1977]). Although the bivariate normal provides a reasonable model in many situations (e.g. circles–Calhoun and Casby [1958], Maza *et al.* [1973]; circles and ellipses– Van Winkle [1975; ellipses– Mazurkiewicz [1971], Koeppl *et al.* [1975], Dunn and Gipson [1977], Randolph [1977], Hawes [1977), there are cases when it is not appropriate (e.g. Metzgar [1972, 1973a]) and a more general bivariate model is required. For example, for a non-homogeneous habitat such as an ecotone, Van Winkle *et al.* [1973] present a non-normal model with independent marginal distributions. However several authors (Tanaka [1974: 126] feel that the above bivariate models are inappropriate for mammal populations, particularly small rodents, and recommend methods which do not require the concept of centre of activity, or any assumptions about the shape of the home range; for example the method of Wierzbowska [1972]. In comparing various methods Wierzbowska [1975] concludes that her own 1972 method and a method of Tanaka [1972], who proposes using $W = \sqrt{(ab)}$ where $2a$ and $2b$ are the observed range length and width (ORL and ORW respectively), are the most versatile methods. Wierzbowska uses the concept of a random walk which also forms the basis of further, but laborious, methods proposed by Morisita (cf. Tanaka [1974]). However the assumption of random walking over a home range may be unrealistic (Ambrose [1969], Siniff and Jessen [1969]). A further problem is that there may be parts of the grid area which are seldom or never entered by animals (Wallin [1971]).

TRAP-REVEALED RANGE. One of the difficulties of the trapping method is that the home range of an animal may change during the course of an experiment. This could lead to an animal visiting more different traps with a consequent overestimate of the home range area (Andrzejewski and Wierzbowska [1970]). Even if the home range remains unchanged there will be a tendency for the trap-revealed range to grow as the number of captures increases, but it will level off after a certain number of captures. Some authors put this figure at 10; other at 6, 5 or even 2 if the population is large enough (Wierzbowska [1975:56]). It is clear that stochastic methods are needed for assessing trapping data rather than using such simple expedients as joining up outermost points (and possibly adding a strip of width equal to half the trap spacing) to obtain a map of the home range. Another problem in home range studies is to distinguish between the permanent resident and the immigrant. At present there seems to be a lack of objective methods for distinguishing the occasional sally from the normal home range movements.

We conclude that it is not easy to get an accurate picture of the shape and size of the home range using trapping as trapping can affect animal behaviour. For this reason the home ranges of small mammals have also been studied using radio telemetry (e.g. Doebel and McGinnes [1974], Trent and Rongstad [1974], Banks *et al.* [1975] and, for a mathematical analysis, Dunn and Gipson [1977]); remote censusing (Marten [1972b, 1973]); radioactive tracers (Ambrose [1969, 1973], Gentry, Smith and Beyers [1971]); footprints on smoked paper

Metzgar [1973a]) or in the sand (Sarrazin and Bider [1973]); and bait with fluorescent pigment (Franz [1972]) or coloured wool (Ryszkowski [1971]). An interesting comparison of trap- and track-revealed home ranges in *Peromyscus* is given by Metzgar (1973b): see also Metzgar [1973c]. An index of home range size based on an exponential model is given by Metzgar and Sheldon [1974].

2 Radio telemetry

In recent years telemetry has been used extensively for studying the movements and behaviour of animals. At present I have a bibliography of over 80 articles published in the 1970's describing the application of radio tracking to many species of birds, mammals and fish (for a review of the latter see Stasko and Pincock [1977]). Although the bibliography is incomplete it does at least indicate the considerable possiblities of such a method. Some of the papers refer to the design and monitoring aspects of a radio transmitter (e.g. design–Kolz *et al.* [1972, 1973], Corner and Pearson [1972], Luke *et al.* [1973], Bray and Corner [1972], Pedersen [1977]; monitoring–Gilmer *et al.* [1971, 1973], Lund [1974], Hutton *et al.* [1976], Hoskinson [1976], Whitehouse and Steven [1977], Wolcott [1977]), and transmitters are sometimes used to relay other physiological information such as body temperature and heart rate (Lonsdale *et al.* [1971], Skutt *et al.* [1973], Kolz *et al.* [1973], Langman [1973]). However, transmitters should be properly designed for the particular species being investigated as they can modify the behaviour of an animal. This has been demonstrated clearly in the case of birds (cf. Gilmer *et al.* [1974]) where a package which appears satisfactory for one sex or one avian species may not be tolerated by the other sex or another even closely related species. Unfortunately there do not appear to be many articles on the effects of radio packages and comparing them with other forms of tagging: some of these studies are Boag [1972], Boag *et al.* [1973], Gilmer *et al.* [1974], Brand *et al.* [1975], McCleave and Stred [1975], Lance and Watson [1977], and Facey *et al.* [1977].

3 Assessment lines

Even if the home range is known accurately there is still the problem of choosing W, some appropriate linear measure of the home range. Clearly the simplest approach is to treat W as an unknown parameter and devise a model for the joint estimation of N and W. One such approach involves the use of *assessment lines* which are generally lines of equally spaced traps cutting the sides of the trapping grid (usually at right angles) and extending from within the grid to out beyond the effective trapping area. An assessment line can hopefully be used to "calibrate" the trapping rate across the boundary of the grid, and the points of discontinuity of this rate along the line provide an estimate of W (Smith *et al.* [1971], O'Farrell *et al.* [1977]; see also section **12.9.4**). However this method is somewhat subjective and an alternative method of estimating the effective trapping area is given by Swift and Steinhorst [1976]. They make use of the plausible assumption that the marked proportion of animals caught in a segment of an assessment line falls off to zero as the edge of the effective trapping area is approached.

A different assessment method is given by Smith *et al.* [1972] who surround the grid by a dense band of traps, which they call a *dense line*. These traps provide information on other parameters such as mortality and migration numbers as well as population size. An extension of this method which is supposed to apply even if the band is omitted is given by Jorgensen *et al.* [1975]. These various methods are also surveyed in Smith *et al.* [1975]. However the above methods of determining the effective trapping area are rather subjective and depend very much on trial and error. Several authors (Hagen *et al.* [1973], Tanaka [1974], Barbehenn [1974] and Tanaka and Murakami [1977: 118]) do not support the use of assessment lines on the ground of unrealistic assumptions and the labour involved in the field. The use of assessment lines is discussed further in section **12.9.4** in relation to removal trapping.

4 Method of selected grids

An alternative approach to the problem is to estimate W directly from a series of selected grids. The key assumption is that W is independent of the grid size so that for the kth grid ($k = 1, 2, \dots, K$)

$$N_k = D(A_k + P_k W + \pi W^2), \qquad (12.1)$$

where D is the population density, and N_k, A_k and P_k are the "effective" population size, the area and perimeter respectively of the kth grid. MacLulich [1951] suggested using two such grids and solving the pair of equations for D and W, with each N_k replaced by its estimate $\hat{N}_k$ (p. 51).

Hansson [1969] suggests using a single grid and assumes that there is an inner subgrid for which the edge effect is negligible. However, Tanaka [1972] gives evidence that the edge effect, although decreasing as you move into the centre of the trapping grid, does not vanish for some central area.

Another method which avoids much of the subjectivity associated with the previous method is described by Otis *et al.* [1978]. They suggest using a single large grid and selecting K subgrids nested within each other inside the grid. If $y_k = \hat{N}_k / A_k$ then we have a non-linear regression model of the form

$$y_k = D(1 + a_k W + b_k W^2) + e_k, \quad k = 1, 2, \dots, K) \qquad (12.2)$$

where the a_k and b_k are known constants (cf. (12.1)), and $E[e_k] = 0$ if $\hat{N}_k$ is unbiased. As the grids are nested, the y_k, and therefore the "errors" e_k, will be correlated. The authors suggest putting the correlation between y_j and y_k equal to the proportion of over-lapping area between grids j and k (including their boundary strip), and carrying out a generalised non-linear least squares to estimate D and W. The estimates $\hat{N}_k$ can be based on either removal trapping or capture-recapture data. Useful practical details about the design of such an experiment with regard to choosing the number of traps, trap-spacing, etc. are given in Otis *et al.* [1978].

12.1.2 Sample plots

1 The importance of random sampling

Suppose there are N animals in an area A and let P_j be the probability of choosing the jth animal ($j = 1, 2, \ldots, N$) by some sampling scheme. Let $y_j = 1$ with probability P_j and $y_j = 0$ with probability $1 - P_j$. If n animals are chosen as a result of the sample scheme then

$$n = \sum_{j=1}^{N} y_j \text{ so that}$$

$$E[n] = \sum_j E[y_j] = \sum_j P_j \qquad (12.3)$$

and

$$V[n] = \sum_j V[y_j] + 2 \sum_{j<k}\sum \text{cov } [y_j, y_k]. \qquad (12.4)$$

where

$$\sum_j V[y_j] = \sum_j P_j(1 - P_j).$$

For example, suppose we choose at random a sample region of area pA and let n be the number of animals counted in the region. Then, since the region is randomly selected, the probability that it contains the jth animal is p: thus $P_j = p$ for all j and $E[n] = Np$. If the animals tend to be clustered, then cov $[y_j, y_k] > 0$ and $V[n] > Np(1 - p)$. On the other hand, if the objects are randomly distributed then the y_j are mutually independent, cov $[y_j, y_k] = 0$, and $V[n] = Np(1 - p)$. In this case n has a binomial distribution and N can be estimated by $\hat{N} = n/p$ (cf. p. 22). However, since random sampling is used, $E[n] = Np$ holds true in general, so that $\hat{N}$ is unbiased *irrespective* of whether the population is randomly distributed or not. This lesson applies to many of the procedures in this chapter. Provided random sampling is used, suitable estimates of population size or density can be obtained from (12.3): variance estimates can then be calculated using replication (as demonstrated on p. 23).

If the population is randomly distributed and s sample plots of area a are selected from S population plots, then the joint distribution of the plot counts $x_i (i = 1, 2, \ldots, s)$ is the multinomial distribution

$$f(x_1, x_2, \ldots, x_s) = \frac{N!}{x_1!x_2!\ldots x_s!(N - \Sigma x_i)!} \pi_1^{x_1} \pi_2^{x_2} \ldots \pi_s^{x_s} (1 - \Sigma\pi_i)^{N - \Sigma x_i} \qquad (12.5)$$

where $\pi_i = a/A (i = 1, 2, \ldots, s)$. Thus the x_i are correlated, though for large N and small a/A the x_i are approximately independently distributed as Poisson with mean $Na/A (= Da)$. We note that $n = \Sigma x_i$ is binomial with parameters N and $\Sigma\pi_i (= sa/A = p)$, as before.

2 Negative binomial model

On p. 25 we introduced the negative binomial distribution as a useful alternative to the usual binomial model when the population is not randomly distributed. Thus if x is the count on a sample plot (or sample unit such as plant or leaf, etc.) then x is assumed to have a negative binomial distribution with parameters k and P, where

$$kP = D, \quad (12.6)$$

the density per plot. Also if $\pi_1 = \Pr[x > 0]$, then

$$1 - \pi_1 = \Pr[x = 0] = (1 + P)^{-k}. \quad (12.7)$$

Now beginning with Anscombe [1949] some authors have stated or demonstrated (e.g. Bowden *et al.* [1969], Wilson and Gerrard [1971], Stormer *et al.* [1977]) that k is stable for many insect populations and does not depend on plot size. Thus if a pooled estimate of this constant value of k is available from previous experiments on the same or similar populations, then we can obtain a quick estimate of D from (12.6) and (12.7) namely

$$\widetilde{D} = k[(1 - \widetilde{\pi}_1)^{-1/k} - 1],$$

where $\widetilde{\pi}_1$, the proportion of nonempty plots or units, is an estimate of π_1. An approximation for $V[\widetilde{D}]$ is given by Wilson and Gerrard [1971]. However Shiyomi *et al.* [1976] give a theoretical argument and a practical example in which P is constant and k varies. If a prior estimate of P is available, then eliminating k from (12.6) and (12.7) leads to the estimate

$$\widetilde{D} = -P \log(1 - \widetilde{\pi}_1)/\log(1 + P).$$

Abrahamsen and Strand [1970] give a useful discussion on the negative binomial and the use of transformations in relationship to sampling soil animals. Bliss [1971] also considers the problem of fitting the negative binomial or its zero-truncated version. It should be noted that the appropriate method of analysing plot data is given on p. 23 (cf. equation (2.3)), and this straightforward application of sample survey theory does not require any assumptions about the underlying distribution; only that the plots are randomly selected. However the negative binomial model is helpful at the design stage when prior estimates of k, and N or D, are available (p. 25): for an application see Suzuki [1973]. An interesting example relating to insect density in grain is given by Hunter and Griffiths [1978]. Using a Bayesian approach, they obtain the negative binomial distribution from a uniform prior.

12.2 RELATIVE DENSITY

12.2.1 Experimental design

Eberhardt [1976] reminds us of the need for careful experimental design and adequate modelling in the use of population indices and measures of relative density. The sampling design should not be ignored just because an estimate is an index and not an absolute density. For example, a new method of stratifying mourning dove call-count routes reduced the error variance by about 30 percent (Blankenship *et al.* [1971]). Again, stratification in the aerial survey of a nesting population of bald eagles (Grier [1977]) reduced the variance by about 22 percent. Also more attention needs to be directed to such questions as sample size and the stability of the index with respect to weather and environmental conditions (Sauder *et al.* [1971: roadside counts for breeding waterfowl]). As mentioned on p. 53, analysis of variance models can be useful for studying indices. For example a suitable model for studying roadside counts taken over several years in different areas might be $E[Y_{ij}] = \mu + \alpha_i + \beta_j$, where i refers to year and j to area. Here the interactions are assumed to be negligible, and generally Y_{ij} will be some transformation of the actual counts such as the logarithm. We note that indices can also be used for estimating survival rates (Gross *et al.* [1974: 46, line transect indices for jack-rabbits] or studying population trends (Grieb [1970: 31, aerial census of Canada geese].

Eberhardt [1978b] gives a helpful discussion and survey of the variability generally expected from indices. He states that "regardless of any theoretical justification, the coefficient of variation of many kinds of index data seems sufficiently constant in practice to supply an approximate guide for planning purposes". In using his summary (reproduced as Table 12.1) it should be noted that he refers to the population coefficient of variation, and not the sample coefficient based on the sample mean and its standard deviation

TABLE 12.1
Summary of coefficients of variation for index methods: from Eberhardt [1978b: Table 2].

Group of organisms	Coefficient of variation
Aquatic biota	
Plankton	0·70
Benthos	
Surber sampler, counts	0·60
Surber sampler, biomass, volume	0·80
Grabs, corers	0·40
Shellfish	0·40
Finfish	0·50–2·00
Terrestrial biota	
Roadside counts	0·80
Call counts	0·70
Transects (on foot)	0·50–2·00
Pellet counts	1·00

Regression techniques can be useful for converting a relative index to an absolute density, or for making comparisons between indices (Bergerud [1972: ptarmigan], Wagner and Stoddart [1972: jackrabbits], Calef [1973: tadpoles], McCaffery [1973: road kills of deer], Fischer and Keith [1974: 596, ruffed grouse], Gross *et al.* [1974: jackrabbits], Schwartz [1974: quail], Brown and Smith [1976: doves], Dzieciolowski [1976a: ungulate track counts] and McCaffery [1976: deer trails]. A possible model, for example, is $E[Y_i] = \beta E[X_i]$, where X_i and Y_i are approximately unbiased estimates of the relative and absolute population density in the ith sampling area, and β is the correction factor. This equation is commonly referred to as a functional relationship and various methods are available for estimating β (cf. Moran [1971] and Seber [1977: 210]). However, it is not always easy to obtain a reliable correction factor, for example to convert counts on haul-out areas and in breeding colonies of marine mammals (Eberhardt *et al.* [1979]). For this reason pellet groups do not seem to have been very successful as an index method, though they can be useful in comparing habitat usage (Dzieciolowski [1976b]).

The roadside count index of so many birds heard calling per route has been widely used in the U.S.A. for studying population trends, for example mourning doves (Gates *et al.* [1975], Ruos [1974, 1977]) and woodcocks (Artmann [1975, 1977]). Surveys based on this technique have slowly adopted the use of random rather than systematically chosen routes. Some of the sampling problems associated with roadside counts of mourning dove calls are discussed by Gates and Smith [1972] who give two Poisson-based models for the call rate: one of the models allows for the rate to be time-dependent. Gates *et al.* [1974] found that a variety of contagious models can be fitted to their frequency distributions of call counts, thus indicating a non-random distribution. Francis [1973] gives a helpful discussion on Hewitt's roadside count method (p. 53): Albers [1976] found that this method gave estimates which were about half those of area counts on sample plots.

Other indices recently used are road-killed deer (McCaffery [1973], Allen and McCullough [1976]) and "scent stations" (Linhart and Knowlton [1975: coyotes], Lindzey *et al.* [1977: bears]). Small quadrats have been suggested as a method of rapidly assessing the abundance of small mammals and may provide a useful index (Myllymäki *et al.* [1971], Hansson [1972, 1975]).

Sen [1971a] gives an interesting application of sample survey techniques to the estimation of Red Spider mite incidence in the field. A multi-stage sampling plan is described which provides a quick and reliable estimate of incidence in a region by combining eye estimates of incidence from a large area with that based on counts from a subsample of bushes from the area. Knowledge of the spatial distribution may also be incorporated in the design of a multi-stage sampling experiment (Kuno [1976]).

12.2.2 Echo counts for fisheries

Echo surveys have been widely used for estimating the relative abundance of fish. If the population is sparse, echo counting based on "point" signals can be used, and techniques for analysing such data are given by MacNeill [1971a, b: 1972], Shibata [1971] and Lord [1973]. For high-density populations such as schools, a cumulative method using an echo integrator is more appropriate (cf. Thorne [1971], Thorne *et al.* [1971], Moose and Ehrenberg [1971] and Ford [1974]). If other sampling methods such as netting are used for calibrating the counts, then the estimates can be converted to absolute estimates of abundance. Recent studies using echo-counting are described by Spigarelli *et al.* [1973: fish density in relation to thermal plumes from a nuclear power station], Kelso *et al.* [1974: juvenile sockeye salmon], Thorne and Dawson [1974: escapement of sockeye salmon], Bazigos [1975, 1976: fish populations in Lake Tanganyika, Africa], Chapman [1976: estimation with an inexpensive echo sounder], Thorne [1977: herring], Mathisen *et al.* [1974: distribution and abundance of fish in relation to environmental factors] and Mathisen *et al.* [1977: juvenile sockeye salmon]. The last two papers give several analysis of variance models taking into account various fixed and random effects due to location, depth, day, etc. In addition to sampling errors we have the following problems (Sen and Southward [1977]): coverage of the cone of the signal, minimum depth at which the signal produces quantifiable fish echos, avoidance behaviour of fish, measurement errors of the equipment and its operation, errors due to the pitching and rolling of the vessel, and the orientation of individual fish towards the electrical impulse.

12.3 AERIAL SURVEY

12.3.1 Introduction

Aerial survey is the only practicable means of estimating the number of large animals inhabiting an extensive area on land or in the sea. Although the estimate is usually inaccurate (biased) and often imprecise (i.e. has large variance), it can be used to answer a broad range of ecological and management questions to an acceptable level of approximation. The precision of an estimate can be controlled by careful experimental design, and one of the first papers to give adequate attention to the design aspect is that by Siniff and Skoog [1964]. Subsequently the important papers by Jolly [1969a, b], which apply the sample theory of Cochran [1977] to aerial censusing, have encouraged a rigorous application of sound sample survey principles. Jolly [1969a] selected three designs as being particularly suited to aerial survey: (i) simple random sampling with equal-sized units, (ii) simple random sampling with unequal-sized units using the ratio method, and (iii) equal- or unequal-sized units selected with probability proportional to size (pps). Formulae for sampling with or without replacement are also given by Jolly. A helpful, more accessible, discussion of these methods is given by Caughley [1977a] who also considers such questions as systematic versus random sampling, and quadrat versus transect sampling units.

It appears that apart from Siniff and Skoog [1964], the use of sampling methods in aerial censusing began in the mid 1960's in East Africa in the Serengeti National Park, Tanzania (Jolly and Watson [1979]). Previously in East Africa several attempts were made at total counts from the air of certain species, and photography was sometimes used instead of visual counts. Watson evidently first used sample survey techniques in the Serengeti to estimate the zebra population in 1966. He used stratified random sampling with five ecologically defined strata, and within each stratum parallel strips extending the full width of the stratum were chosen at random. Since then, various workers in East Africa have adopted stratified strip (transect) sampling as a standard procedure for monitoring wildlife populations (cf. *East African Wildlife Journal* from about 1969 and the special 1969 issue of the *East African Agricultural and Forestry*

Journal: some references are Watson *et al.* [1969a, b], Sinclair [1972], Bell *et al.* [1973], Norton-Griffiths [1973, 1975b], Caughley and Goddard [1975] and Eberhardt *et al.* [1979: table 1, marine censuses]. Random strip sampling is also being used in Australia (cf. CSIRO publications, for example Frith [1964], Bailey [1971]).

The theory of stratified random strip sampling is given by Jolly [1969a] for the two cases (i) equal probability of selection within a stratum which leads to a ratio-type estimator, and (ii) pps sampling or probability of selection proportional to size, that is proportional to the length of the sample strip. In comparing these two methods Jolly and Watson [1979] point out that there is little difference in the precision of the two methods when the sampling fractions are small. However, case (ii) leads to an unbiased estimate, while the ratio estimate of (i) leads to an estimate with a bias of order s^{-1}, where s is the total number of strips sampled. For this and other reasons the proportional-to-size method is the one recommended by Jolly and Watson [1979]. Thus if N_j and A_j are the population size and area of the jth stratum ($j = 1, 2, \ldots, J$), x_{ij} is the count on the ith strip of area $a_{ij}(i = 1, 2, \ldots, s_j)$ in the jth stratum, and $d_{ij} = x_{ij}/a_{ij}$, then an unbiased estimate of the total population size, $N = \sum_j N_j$ is given by

$$\tilde{N} = \sum_j \hat{N}_j = \sum_j A_j \bar{d}_j \tag{12.8}$$

where $\bar{d}_j = \sum_i d_{ij}/s_j$. An unbiased estimate of the variance of $\tilde{N}$ is then given by

$$\tilde{\sigma}^2_N = \sum_j \hat{\sigma}^2_{N_j} = \sum_j \left\{ \frac{A_j^2}{s_j(s_j - 1)} \sum_{i=1}^{s_j} (d_{ij} - \bar{d}_j)^2 \right\}. \tag{12.9}$$

To determine the population of strips in a stratum the simplest method is to draw a baseline, usually along the "ecological axis" or direction of least ecological change in the stratum, and then draw parallel lines one strip width apart at right angles to this baseline. Selection with probability proportional to the length of the strip can be accomplished by choosing a point at random in a rectangle on the map enclosing the stratum. Those strips in which the sample points fall then constitute the sample from that stratum. In order to achieve pps sampling, the sampling must be with replacement. If the same strip is chosen say k times, then the strip is flown just the once but the count is included k times in the above formula for $\hat{N}_j$. In practice the strip is identified from the plane by flying the plane at a given altitude and direction, and counting all animals seen between two markers on a wing strut. This raises the question of calibration and the problem of variations in flight path: the reader is referred to Jolly and Watson [1979] for a useful discussion of these points. Further practical comments on some of the field problems associated with aerial censusing are given by Larsen [1972: polar bears], Norton-Griffiths [1975a], Caughley [1977a, b] and Eberhardt *et al.* [1979: marine mammals].

In practice, transects seem to have a clear advantage over quadrats (or irregular shaped blocks determined by natural boundaries) in terms of flying costs, ease of navigation, boundary effects, observer fatigue and sample error (Norton-Griffiths [1975a: 26–28], Caughley [1977a]). Quadrat or block counts are more appropriate in rough country, where transect flying is difficult; or where vegetation is very thick and or patchy; or where animals occur in very large and conspicuous herds. Some examples of quadrat counts are: moose (Evans *et al.* [1966], Le Resche and Rausch [1974], Peek *et al.* [1976]), coyotes (Nellis and Keith [1976]), caribou (Parker [1972]) and a nesting population of eagles (Grier [1977]). Admittedly animals are more readily seen using quadrat counts (Laws *et al.* [1975: 340]), but it is of little advantage since a significant proportion of animals still tend to be missed, no matter what sampling unit is used. Although there seems to be some reluctance on the part of investigators to admit the fact, there is ample evidence (e.g. Graham and Bell [1969], Hornocker [1970: 20–21], Bergerud [1971: 10], Stott and Olson [1972], Le Resche and Rausch [1974] and, in particular, Caughley [1974a: 922; 1977b: 35]) that even experienced observers can overlook as much as 20 percent or more of the animals, so that all estimates, whether based on quadrats or transects are underestimates. A number of methods for correcting these underestimates have been proposed. These are described below.

12.3.2 Method of correction factors

If accurate ground counts can be made over some of the sample strips then a correction factor can be calculated (Le Resche and Rausch [1974], Stott and Olson [1972], Hopper *et al.* [1975: 15], Caughley *et al.* [1976], Eberhardt *et al.* [1979]). For example, if x_i is an accurate count on sample strip i, y_i is the aerial count on the same strip, and P, the probability of an animal being seen from the plane, is constant, then we have the linear regression model $E[y_i|x_i] = Px_i$ and P can be estimated by least squares or by the ratio estimate $\bar{y}/\bar{x}$ (Jolly [1969b], Jolly and Watson [1979]). This estimate can then be used to correct the total aerial count. Unfortunately ground counts usually suffer from the same visibility bias, though to a lesser extent (Henny *et al.* [1972: 4–5]).

Correction factors can also be provided by (i) photography (Watson [1969], Sinclair [1969, 1972, 1973], Norton-Griffiths [1973] and Kerbes [1975]; (ii) infrared scanning (Graves *et al.* [1972], but see Caughley [1974a: 930]); (iii) ultraviolet photography (white coats appear black, e.g. Lavigne and Oritsland [1974a, b: polar bears], Lavigne *et al.* [1975: harp seals]; (iv) simulation experiments (Watson *et al.* [1969c], Caughley *et al.* [1976]) and (v) conspicuous tags, where P is estimated from the proportion of tagged animals seen from the air (Nellis and Keith [1976], Rice and Harder [1977]). Further examples of the use of correction factors are given by Goddard [1967, 1969], Watson *et al.* [1969a, b] and Pennycuick and Western [1972]. If counts are obtained solely by studying photographs under a low-powered microscope, a correction factor can be obtained by observing some of the photographs under a higher magnification.

Caughley [1977a: 612] states, from personal experience, that "quadrats are a nightmare when it comes to estimating correction factors. The counts are sensitive to variations in piloting skill, the time spent over the quadrat, the state of the observer's stomach, and the rapport between pilot and observer. The counts from transects are more robust to these influences".

12.3.3 Repeat data

The true count, n say, and the probability P of sighting an animal in a given transect can be estimated by flying over the transect several times. Several methods of estimation are then possible. For example if $\bar{y}$ and v are the sample mean and variance of k repeated counts, then assuming a binomial distribution (i.e. random distribution of animals) we have $E[\bar{y}] = nP$ and $E[v] = nP(1 - P)$. Thus moment estimates of n and P are (Caughley and Goddard [1972])

$$\hat{n} = \bar{y}^2/(\bar{y} - v) \tag{12.10}$$

and

$$\hat{P} = 1 - (v/\bar{y}). \tag{12.11}$$

Estimates of P can then be averaged over several such transects to obtain an overall correction factor for the total count. Since

$$E[v] = \frac{nk}{nk - 1}E[\bar{y}] - \frac{k}{nk - 1}E[\bar{y}^2],$$

$v = a\bar{y} + b\bar{y}^2 + e$, where $E[e] = 0$: Caughley and Goddard [1972] propose a similar quadratic model for handling non-randomly distributed populations. Another method of estimating n is by the bounded counts method (p. 58): $\bar{y}/n$ will then give an estimate of P.

12.3.4 Multiple regression method

An interesting regression method for estimating the true density from the "apparent" or observed density is given by Caughley (cf. Caughley [1974a], Caughley *et al.* [1976, 1977]). There is clear evidence that sightability goes down as the speed, altitude and strip width are increased (Pennycuick and Western [1972]). Therefore Caughley suggests regressing apparent density $\hat{D}$ on these factors and then extrapolating to zero. Thus if the model is of the form $E[\hat{D}] = \beta_0 + \beta_1 x_1 + \ldots + \beta_k x_k$, then β_0 represents the true density. This method shows some promise though the validity of the extrapolation needs further investigation. There are also other factors that need to be considered as, for example, fatigue and time of day can affect counts by observers (Larsen [1972], Norton-Griffiths [1976]): clearly observers need to be properly trained (Sinclair [1973]).

12.3.5 Combining data from independent observers

Suppose there are two independent observers and it is possible to map the locations of the animals or their signs so that the numbers seen by observer 1 and not by observer 2, seen by observer 2 and not by observer 1, and seen by both observers are available. Then, by regarding the animals seen by observer i as being "caught" in sample i, a Petersen estimate of the true count can then be made, provided observers are independent. This technique of using two incomplete lists to estimate effectively the number missing from both lists has a long history and is discussed further, for two or more lists, in section **12.8.3**. The first application of this method to ecology seems to be that of Magnusson *et al.* [1978:

crocodile nests]. If there are several observers then the bounded counts method (p. 58) could be used: Bergerud [1971], for example, uses the maximum count of three observers.

12.3.6 Use of group sizes

Several models have been proposed by Cook and Martin [1974] and Jolly and Watson [1979][(*)] for calculating correction factors and density estimates for populations which tend to cluster in large groups. The main assumption underlying these models is that, conditional on observing at least one member of a group of animals, the entire group is observed with certainty. This assumption could be achieved using a "two level" sampling procedure: on encountering a group of one or more animals the observer counts or photographs in the usual way, and then the pilot descends to a very low level so that the observer can re-count or re-photograph until he is satisfied that all the animals in the group have been accounted for. A problem with groups of animals is that a group may overlap more than one strip transect. In this case the group count is divided by the number of population strips that contain part of the group (cf. Jolly [1979a] for details).

12.3.7 Eberhardt's method

Eberhardt [1978a, c] introduced a very promising method for estimating P from counts on substrips $(0, \Delta]$ and $(\Delta, 2\Delta]$. This method, described in section **12.4.3**(*2*) for ground counts, can also be used for aerial counts. However there is a practical difference: if there is an observer counting on each side of the plane there will be a strip directly under the plane that will be ignored because of lack of visibility. This means that in the theory of **12.4.3**, $y = 0$ corresponds to the inner edge of the strip that an observer locates through the marks on the wing struts. As an observer may not see all the animals on the inner boundary we have the possibility that $g(0) < 1$, and we will need to estimate $g(0)$, which may be as difficult as estimating P itself.

12.3.8 Population indices

Another approach to the problem of underestimation is to simply recognise that the estimates are biased and treat them as relative rather than absolute measures of abundance. If the bias can be held constant by rigorously standardising the methods (e.g. fixed speed and altitude, fixed strip width) then the indices obtained can be used for monitoring changes in the population size and distribution, and determining preferences for different habitats (Boeker and Bolen [1972], Bailey [1971], Sinclair [1972], Murphy and Whitten [1976], Kimball and Wolfe [1974]). If an index is all that is required then clearly transects are superior to quadrats or blocks. However, the wildlife manager frequently has to know absolute densities so that he can translate a permissible harvest rate into hunting quotas.

Aerial censusing has been used not only for game animals but also for nesting sea-birds (Kadlec and Drury [1968]), gregarious waterfowl, and animal

(*)See also Cook and Johnson [1979], Additional References.

signs such as rabbit warrens (Parker and Myers [1974], Martin and Zickefoose [1976]), fox dens (Sargeant *et al.* [1975]), eagles' nests (Hickman [1972]) and crocodile nests (Magnusson *et al.* [1978]). Coastal censuses are also being used, though there are sampling problems if the coastline is very broken (King *et al.* [1972]).

Colour and false colour photography, infra-red sensing and multi-spectral imagery can be useful tools in fauna surveys and management, particularly relating to questions about habitats, habitat edges, land systems, breeding grazing areas and migration routes (Howard [1970], Schuerholz [1974], Cowardin and Myers [1974], Myers and Parker [1975], Tucker *et al.* [1975]). A recent innovation is the use of data from satellites such as LANDSAT (Gilmer *et al.* [1975], Kerbes and Moore [1975], Reeves and Marmelstein [1975]) and TIROS (Reeves *et al.* [1976]). LANDSAT imagery has two special attributes: firstly it is multispectral, that is, relative differences in the reflectance of the earth's surface are recorded in four spectral regions, and these can be manipulated to maximise contrast between and within areas of interest; and secondly, it is multitemporal so that contrasts can be compared over time.

12.3.9 Marine mammals

Aerial survey methods have also been used for counting marine animals such as sirenians, sea otters, polar bears, seals, dolphins, porpoises, walruses and whales (cf. the bibliography of Gilbert *et al.* [1976]; also Holdgate [1970], Schevill [1974] and, in particular, Eberhardt *et al.* [1979]). However, with both shipboard and aerial counts of marine animals there is an added complication: some of the individuals may be submerged at the time of the count and a correction factor is required for this. McLaren [1961] proposed a model for the shipboard count of ringed seals which assumed that all surfaced seals could be seen out to a fixed distance from the vessel. He gave a formula for P, the probability of seeing a seal on the surface given that it is in the strip being searched, in terms of the parameters (assumed constant) of diving time, time on the surface, and speed of the ship. This model and a modification are derived by Eberhardt [1978a:18]. Further models allowing for a decrease in visibility with distance were proposed by McLaren [1966] and Doi [1974] (cf. Eberhardt *et al.* [1979]). The main problem is that of analysing diving behaviour, and further research is needed, using, for example, radio telemetry (Evans [1974: whales]) or sequential photographs (Tayler and Saayman [1974]). Diving and surface times could perhaps be studied using a combination of direct visual observation and radio telemetry. However, diving behaviour can vary with the time of the day (Evans [1974]). Because different species have rather distinct behaviour patterns, habitat preferences, etc., Eberhardt *et al.* [1979] do not recommend multispecies counts for the aerial survey of marine mammals.

In recent years helicopters have been used for aerial surveys (e.g. Nellis and Keith [1976], Siniff *et al.* [1977: table 6]). Although they have the advantage of low flying, clear visibility and ease of operation from ships as, for example, in the Antarctic (Holdgate [1970]), they tend to be noisy and disturb the population.

12.4 TRANSECT METHODS

A helpful review of transect methods in general, along with a useful tree diagram for classifying these methods and their applications, is given by Eberhardt [1978a]. There are basically three methods: the line intercept method, the strip transect, and the line transect.

12.4.1 Line intercept method

The line intercept method consists of choosing a line of length L at random in a population area and measuring, for example, the length of line intersected by each member of the population. Using this and related measurements, we can obtain estimates of population density and the proportion of the ground covered by members of the population (Lucas and Seber [1977]). Clearly the probability that an object is intersected will depend on the two-dimensional size of the object, and examples are plant populations, shrub canopies, objects on the seashore or seabed, food fragments under a microscope (Seber and Pemberton [1979]), animal "signs" such as den sites (Eberhardt [1978a]) and geographical features such as lakes, fields and forests (Jolly [1979a]). If the population area is irregular in shape, and the transect line runs right across the population area, then the length of a randomly selected transect will be a random variable: this case is considered by Seber [1979].

Line intercept sampling is extensively used in forestry (Hazard and Pickford [1979], Warren [1979]) and the reader is referred to De Vries [1979: see also 1973, 1974]) for an excellent review. The main technique used there consists of replacing each object by a unique "needle", and the sample then consists of all objects whose needles intersect the transcent line. We note that in line intercept sampling the probability that an object is sampled is proportional to the size of the object. This kind of sampling leads to the theory of size-biased sampling and weighted distributions (Cox [1969], Patil and Ord [1976], Patil and Rao [1978]).

12.4.2 Strip transect method

The strip transect method, which simply amounts to counting animals or their signs on a strip of prescribed width $2W$, is appropriate when the population is fairly numerous and readily visible. If a simple random sample of such strips in the population area (or stratum) is taken, then the theory of **2.1.2** (cf. p. 28) can be used. When the strips are of different lengths, more efficient methods of sampling are available and some of these are discussed under aerial censusing (cf. **12.3**).

If measurements such as the right-angle distance y, the radial distance r, and the angle of sighting with respect to the centre line of the strip are recorded for each animal, then the counts can be corrected for visibility if not all the animals in the strip are sighted. This problem of visibility is particularly acute with marine mammal surveys when a certain fraction of the population is submerged during counting (cf. Eberhardt [1978a]). Methods of calculating a correction factor are described below in **12.4.3**.

In using a strip transect, we note from section **12.1.2** that we do not require the assumption that the objects are randomly and independently distributed over the population area for $E[n] = NP$ to be true, provided the transect is randomly placed. However, if we require a theoretical variance for the estimate $n/\hat{P}$ of N, then we need to postulate a model for the distribution of the objects. Typically it is assumed that the population is random, and this leads to the binomial distribution or its Poisson approximation. In this case, if an object is in the strip, its right-angled distance y has the uniform distribution on $[0, W]$ and the y's will be independent. However if the objects are not randomly distributed but the transect is randomly placed, y still has the same uniform distribution but the different y values will now be correlated. For example, we can envisage the extreme case where all the animals lie on a line parallel to the transect: y is now the same for all the animals.

For the case when animals are seen through flushing, it is usually assumed that the animals flush independently. However, this assumption is not required: all we need is that the (unconditional) probability P of flushing is the same for each animal.

12.4.3 Line transect

In contrast to the strip transect, the line transect, which can be regarded conceptually as a strip of infinite width, is perhaps more appropriate when the counts are likely to be low. The observer would then wish to record all the animals seen and not just those out to a distance W. The assumptions underlying the theory of line transects are listed on p. 29. However, as already mentioned, we do not require assumption (1) that the animals are randomly and independently distributed over the population area for $E[n] = NP$ to be true, provided the transect is randomly placed.

Following p. 29, let $g(y)$ be the probability that an object is seen, given that it is a right-angle distance y from the transect. It is now convenient to generalise the theory of **2.1.3** and assume that distances out to W only are recorded. Then arguing as on p. 29 (cf. Burnham and Anderson [1976]), the probability density function of y, *given* that the object is seen, is (denoting the uniform random variable y by z when it is conditional on being observed)

$$f(z) = g(z)/\mu_W, \quad 0 \leqslant z \leqslant W, \tag{12.12}$$

where

$$\mu_W = \int_0^W g(z)\, dz. \tag{12.13}$$

If the transect is randomly placed, the probability that an object is on the strip is $P_L = 2LW/A$, and the conditional probability of seeing the object, given that it is on the strip, is

$$\begin{aligned} P_W &= E_y\,[\Pr\{\text{object seen} \mid \text{on the strip and at distance } y\}] \\ &= E_y[g(y)] \\ &= \frac{1}{W}\int_0^W g(y)\,dy = \mu_W/W. \end{aligned} \tag{12.14}$$

Thus if n objects are actually seen on the strip then, from (12.3),

$$E[n] = NP = NP_L P_W = 2DL\mu_W, \qquad (12.15)$$

and the density $D(= N/A)$ can be estimated by $n/(2L\hat{\mu}_W)$, where $\hat{\mu}_W$ is an estimate of μ_W. It transpires that equations (12.12), (12.13) and (12.15) still hold if we formally let $W \to \infty$ (see p. 29 with $c = \mu_\infty$)

1 Parametric models

The parametric approach, described on pages 29–35, is to postulate a model for g, and hence for f. If the y's are independently and identically distributed with the uniform distribution on $[0, W]$, which will be the case for a random distribution of animals, we can obtain the maximum likelihood estimate of μ_W based on the n independent observed values of y (called z above). In addition to the exponential, power law and uniform distributions, the following models for f in (12.12) have also been considered: the truncated linear model (Järvinen and Väisänen [1975]), which is a special case of the power law; the incomplete gamma distribution (Sen *et al.* [1974, 1978a] see also their corrections in Sen *et al.* [1978b] and the comments by Burnham and Anderson [1976: 330]); the logistic model (Eberhardt [1978a: Appendix D]); the half-normal distribution (Hemingway [1971], Järvinen and Väisänen [1975]) and the family of exponential power series distributions (Pollock [1978] G.P. Patil *et al.* [1979], Quinn and Gallucci [1979]). The exponential model was also considered by Anderson, Burnham *et al.* [1979] for the case of grouped data: Gates and Smith [1980] (Additional References) give a polynomial model. A number of other estimation procedures have been investigated experimentally by Robinette *et al.* [1974].

If we redefine g as kg $(0 < k < 1)$, then $f(z)$ is still given by (12.12) as k cancels out. This means that there is an indeterminancy in models which are developed solely in terms of f. This indeterminancy is usually removed by postulating $g(0) = 1$ (assumption (7) on p. 29), that is, all animals directly in front of the observer are seen(*). In this case we have, from (12.12),

$$1/\mu_W = f(z)/g(z) = f(0)/g(0) = f(0),$$

and D is estimated by (cf. 12.15))

$$\hat{D} = n\hat{f}/(2L), \qquad (12.16)$$

where $\hat{f}$ is an estimate of $f(0)$. For example Pollock [1978] considers the model

$$f(z) = \frac{\exp\,[-(z/\lambda)^p]}{\lambda\Gamma[(1/p)+1]} \qquad (z > 0,\ p > 0,\ \lambda > 0)$$

for the case $W = \infty$ and gives an iterative procedure for finding the maximum likelihood estimate of $f(0)$ and its asymptotic variance. This flexible family of distributions contains several of the distributions mentioned above as special cases, for example the exponential $(p = 1)$, half-normal $(p = 2)$ and uniform distributions $(p = \infty)$. G.P. Patil *et al.* [1979] give a modification to allow for the

(*)See Ramsey [1979] (Additional References) for another approach.

effect of using photography for counting. We note that Pollock gives an approximate regression method for estimating $f(0)$ when the data are grouped: this method also applies to strip transects (finite W).

A general variance formula for $\hat{D}$ of (12.16) can be found as follows. Assuming that, for large n, $E[\hat{f}|n] \approx f(0)$, then, from (12.15),

$$E[\hat{D}] = \underset{n}{E}E[\hat{D}|n] \approx \underset{n}{E}[nf(0)/2L] = D,$$

and

$$\text{cov}[n, \hat{f}] \approx \underset{n}{E}E[(n-E[n])(\hat{f}-f(0))|n] = 0.$$

Hence, by the delta method,

$$V[\hat{D}] \approx D^2 \left\{ \frac{V[n]}{(E[n])^2} + \frac{V[\hat{f}]}{[f(0)]^2} \right\} \tag{12.17}$$

$$= D^2[C_n^2 + C_{\hat{f}}^2], \tag{12.18}$$

or

$$C^2 \approx C_n^2 + C_{\hat{f}}^2, \tag{12.19}$$

where C, C_n and $C_{\hat{f}}$ are the coefficients of variation of $\hat{D}$, n and $\hat{f}$ respectively. Equation (12.19) shows that, for large n, n and $\hat{f}$ behave as though they are independent. When the objects are randomly distributed and P is small, n is approximately Poisson with mean $NP = 2DL/f(0)$, and $C_n^2 \approx 1/E[n]$. Also, if $V[\hat{f}|n] \approx \hat{\sigma}^2/n$ for some constant σ^2 (cf. Burnham and Anderson [1976: 329]) then

$$V[\hat{D}] \approx \frac{Df(0)}{2L}\left\{1 + \frac{\sigma^2}{[f(0)]^2}\right\}, \tag{12.20}$$

and $V[\hat{D}]$ is inversely proportional to L. Thus if we have r sample transects of each of length L_i ($i = 1, 2, \ldots, r$) then the appropriate weighted mean is (cf. p. 6 with $w_i = L_i$)

$$\bar{D} = \sum_{i=1}^{r} L_i\hat{D}_i / \sum_{i=1}^{r} L_i$$

with approximately unbiased variance estimate

$$v[\bar{D}] = \frac{\Sigma L_i(\hat{D}_i - \bar{D})^2}{(r-1)\,\Sigma L_i}. \tag{12.21}$$

If $g(0) < 1$, as may be the case in aerial censuses, the estimate of D is $n\hat{f}/(2L\hat{g})$ where $\hat{g}$ is an estimate of $g(0)$.

2 Non-parametric models

In addition to the parametric models listed above, a number of non-parametric methods have been developed recently (cf. Anderson and Pospahala [1970], Emlen [1971], Burnham and Anderson [1976], Anderson *et al.* [1978], Eberhardt [1978a, c], Crain *et al.* [1978] and G.P. Patil *et al.* [1979].

EBERHARDT'S METHOD. Following Cox [1969], Edberhardt [1978a: Appendix C] suggests estimating $f(0)$ by

$$\hat{f} = \frac{1}{2\Delta} \frac{3n_1 - n_2}{n}, \tag{12.22}$$

and D by

$$\hat{D} = \frac{3n_1 - n_2}{4L\Delta}, \tag{12.23}$$

where n_1 and n_2 are the numbers of animals on either side of the line transect, at distances which fall within the intervals $(0, \Delta]$ and $(\Delta, 2\Delta]$ respectively, and n (which is not strictly required for (12.23)) is the total number of animals seen from the line transect ($W = \infty$). The choice of Δ depends on a compromise between the bias and the precision of $\hat{f}$: the larger Δ, the larger the bias but the greater the precision as a greater proportion of the observations ($= (n_1 + n_2)/n$) is utilised. In fact, $E[\hat{f}] \approx f(0) - \frac{1}{3}\Delta^2 f''(0) - \frac{1}{4}\Delta^3 f'''(0) \ldots$ so that the flatter the function f is at $z = 0$, the larger the value of Δ that can be tolerated. Thus for a given bias we would use a smaller proportion of the observations for the "spiked" exponential distribution than for the flatter half-normal and logistic distributions. Some useful graphs for determining the appropriate proportion are given by Eberhardt [1978c]: as a rough working rule the bias is less than 10 percent if the proportion ranges from about 75 percent for the exponential up to about 90 percent for flatter distributions. An application of this method to a case where counts are obtained from photographs is given by G.P. Patil *et al.* [1979].

If $\hat{f}$ is approximately unbiased then, making the reasonable assumption that the probability of an observed animal being in $(0, \Delta]$ is at least $\frac{1}{2}$, that is $\Delta/\mu_W = \Delta f(0) \geq \frac{1}{2}$, Eberhardt [1978a] shows that $C_{\hat{f}}^2 \approx 3\mu_W/(2\Delta E[n]) \leq 3/E[n]$. Thus if the animals are randomly distributed we have, from (12.19),

$$C^2 \leq 4/E[n], \tag{12.24}$$

which may be regarded as a rough bound: C_n^2 will generally be greater than $1/E[n]$ for a non-random distribution.

RANKED DATA. Another method of estimating $f(0)$, for the case $W = \infty$, is provided by the theory of S.A. Patil *et al.* [1979] (cf. **12.5.2**). If $z_{(1)} < z_{(2)} < \ldots < z_{(n)}$ are ranked right-angle distances, then $f(0)$ can be estimated by

$$\tilde{f} = k(n)/(nz_{(k)}), \tag{12.25}$$

where k is the integral part of $k(n)$, an appropriate sequence of real numbers satisfying $k(n) \to \infty$ and $k(n)/n \to 0$ as $n \to \infty$. The authors suggest setting $k(n) = \sqrt{n}$, though other sequences may lead to more efficient estimates and further research is in progress (Burnham, personal communication). They show that, asymptotically, $E[\tilde{f}|n] \approx f(0)$ and $V[\tilde{f}|n] \approx f^2(0)/k(n)$.

If all the n animals seen from the transect are noted but only distances out to W are recorded, then from S.A. Patil *et al.* [1979] (cf. **12.5.2**) we have the estimate

$$\tilde{f}_W = \frac{n_1}{n} \tilde{f}_W(0), \tag{12.26}$$

where n_1 is the number seen in the strip of width $2W$, n_1/n is an estimate of $\pi_W = \mu_W/\mu_\infty$, and $\tilde{f}_W(0)$ is of the same form as (12.25), except that it is based on the ranked z_i for just the animals seen in the strip. Thus

$$\tilde{f}_W(0) = k(n_1)/(n_1 z_{(k_1)}),$$

where k_1 is the integral part of $k(n_1)$. Also

$$V[\tilde{f}_W|n] \approx \frac{f^2(0)}{k(n_1)} + \frac{f^2(0)}{n}\left[\frac{1-\pi_W}{\pi_W}\right]\left[1+\frac{1}{k(n_1)}\right].$$

LOG-LINEAR MODEL. Anderson and Pospahala [1970] presented data on waterfowl nests which is well described by taking

$$g(x) = \exp(b_0 + b_1 x + b_1 x^2), \quad 0 \leqslant x \leqslant W,$$

with $b_0 = 0$ as we have assumed $g(0) = 1$. Estimation of N using this model is described by Anderson *et al.* [1978] and the details are summarised below.

Suppose that the interval $[0, W]$ is partitioned into k subintervals $I_1, I_2, \ldots, I_k$ each of width W/k, and let n_j be the number of individuals seen in the strip with perpendicular distances lying in I_j. Now π_j, the probability of an individual lying in I_j given that the individual is in the strip, is (cf. (12.14))

$$\pi_j = \frac{1}{W}\int_{I_j} g(y)dy \quad (= \bar{g}_j/k \text{ in the authors' notation})$$

$$\approx g(u_j)/k,$$

where u_j is the midpoint of I_j. In particular $\pi_1 \approx 1/k$ as $g(0) = 1$. Since $E[n_j|N_T] = N_T\pi_j$, where N_T is the total number in the transect, then

$$\frac{E[n_j|N_T]}{E[n_1|N_T]} = \frac{\pi_j}{\pi_1} \approx g(u_j) = \exp(b_1 u_j + b_2 u_j^2).$$

Thus we have the polynomial regression model

$$\log(n_j/n_1) = b_1 u_j + b_2 u_j^2 + e_j, \quad (j = 1, 2, \ldots, k),$$

where e_j is the "error" and $E[e_j] \approx 0$. However, the n_j, having a multinomial distribution, are correlated so that a generalised least squares estimate, $\hat{\mathbf{b}} = (\hat{b}_1, \hat{b}_2)'$, say, using an estimated variance–covariance matrix, is required (cf. Seber [1977: 61]). It transpires that

$$\hat{\mathbf{b}} = (\mathbf{X}'\hat{\mathbf{V}}^{-1}\mathbf{X})^{-1}\mathbf{X}'\hat{\mathbf{V}}^{-1}\mathbf{Y}$$

where $\mathbf{Y}$ is the vector of the $\log(n_j/n_1)$,

$$\mathbf{X} = \begin{bmatrix} u_2 & u_2^2 \\ u_3 & u_3^2 \\ \cdots & \cdots \\ u_k & u_k^2 \end{bmatrix},$$

$\hat{\mathbf{V}}^{-1}$ is an estimate of

$$\mathbf{V}^{-1} = \begin{bmatrix} \pi_2(\pi - \pi_2), & -\pi_2\pi_3, & \dots, & -\pi_2\pi_k \\ -\pi_3\pi_2, & \pi_3(\pi - \pi_3), & \dots, & -\pi_3\pi_k \\ \dots, & \dots & \dots, & \dots, \\ -\pi_k\pi_2, & -\pi_k\pi_3, & \dots, & \pi_k(\pi - \pi_k) \end{bmatrix}$$

and $\pi = \sum_{j=1}^{k} \pi_j$. The variance–covariance matrix of $\mathbf{Y}$ is approximately $\sigma^2\mathbf{V}$, where $\sigma^2 (= N_T/\pi)$ is unknown, and $\mathbf{V}$ can be estimated using $\hat{\pi}_1 = 1/k$, $\hat{\pi}_j = n_j/(n_1 k)$ and $\hat{\pi} = \sum_{j=1}^{k} \hat{\pi}_j$. The authors also suggest several methods of smoothing the estimate of $\mathbf{V}$. Once $\hat{b}_1$ and $\hat{b}_2$ are calculated we have the estimate

$$\hat{\mu}_W = \int_0^W \hat{g}(y)\, dy$$

$$= \int_0^W \exp(\hat{b}_1 y + \hat{b}_2 y^2)\, dy,$$

which can be expressed in terms of the standard normal distribution function. Finally

$$\hat{D} = n/(2L\hat{\mu}_W).$$

A variance estimate is best obtained by replication.

FOURIER SERIES METHOD. Another very promising method for a fixed width transect (finite W) has been proposed by Crain *et al.* [1978] who assume a Fourier cosine series for $f(z)$, namely

$$f(z) = \frac{1}{W} + \sum_{k=1}^{\infty} a_k \cos\left(\frac{k\pi z}{W}\right), \quad 0 \leqslant z \leqslant W,$$

where

$$a_k = \frac{2}{W}\int_0^W f(z) \cos\left(\frac{k\pi z}{W}\right) dz = \frac{2}{W} E\left[\cos\left(\frac{k\pi z}{W}\right)\right].$$

Then

$$f(0) \approx \frac{1}{W} + \sum_{k=1}^{m} a_k,$$

where the truncation point m is to be selected (usually $m \leqslant 6$ is satisfactory). Unbiased estimates of the a_k are $\hat{a}_k = 2\bar{u}_k/W$, where

$$\bar{u}_k = \frac{1}{n}\sum_{j=1}^{n} \cos\left(\frac{k\pi z_j}{W}\right),$$

and an approximately unbiased estimate of $f(0)$ is given by

$$\hat{f}_m = \frac{1}{W} + \sum_{k=1}^{m} \hat{a}_k.$$

The problem of grouped data is considered briefly by the authors and by G.P. Patil *et al.* [1979].

It may be necessary to choose W after collecting the data because (1) all observations are recorded, as in the usual line transect experiment; or (2) W may have been fixed at too large a value; or (3) there may be a few very large values of z which might best be regarded as outliers and ignored. Crain *et al.* [1978] suggest choosing W so that only 2–3 percent (or at most 10 percent) of the data are discarded. They also give an approximate expression for the variance of $\hat{D}_m = n\hat{f}_m/2L$ in the form of (12.20) with $\sigma^2 = \mathbf{1}'\mathbf{\Sigma}\mathbf{1}$, where $\mathbf{\Sigma}/n$ is the variance–covariance matrix of the $\hat{a}_k$'s and $\mathbf{1}$ is a vector of ones.

CHOICE OF METHOD. Whether one should use a parametric or non-parametric method is still an open question, and the answer will depend to a large extent on the nature of the population. As a general principle the non-parametric methods are much more robust with regard to the underlying assumptions but are less efficient, that is, they give a higher coefficient of variation. For example, if $\hat{C}$ is an estimate of the coefficient of variation of $\hat{D}$, $\hat{C}^2 \approx 2/n$ for Gates' exponential model (p. 31), while $\hat{C}^2 \approx 4/n$ for Eberhardt's nonparametric method. I believe that these values of $\hat{C}^2$ reflect the general range for randomly distributed populations.

A useful comparative study of the various methods has been made recently by Burnham *et al.* [1980] and it supports the Fourier series method as a leading contender. The problem of schooling is considered by Quinn [1979] (Additional References) who show that the nonparametric methods tend to fare better when schooling is present. Gates [1979] gives a further survey of the line transect literature, and Anderson, Laake *et al.* [1979] suggest a number of practical guidelines for setting up a transect study.

It should be noted that a key assumption underlying the above transect methods is that the observed perpendicular sighting distances (z_i) are assumed to be independent with the same density function $f(z)$ (irrespective of where the transect is placed). This is equivalent to assuming that the animals are independently sighted and are randomly distributed over the population area – strong assumptions, not always recognised by researchers. I believe that any mild dependence in the z_i will not seriously bias any of the parametric or nonparametric estimates of density, though theoretical variance formulae may be severely affected. This bias insensitivity should follow from the fact that the basic equation $E[n] = NP$ only requires that P is the same for each animal, irrespective of the distribution of the animals.

We note one other problem relating to the placing of line transects. It is often convenient, in practice, to have the transect running the full length of the population area so that the length L of the transect is a random variable if the transect is randomly located and the population area is irregular in shape. This problem applies to all transect techniques and appropriate methods of estimation are discussed in Seber [1979].

In conclusion we mention two other studies which require modifications of the above theory, namely circular plot surveys at points along the transect (Ramsey and Scott [1979]), and movement of the population (Smith 1979]).

3 Radial distances

CIRCULAR FLUSHING REGION. Methods of estimation have also been developed which use radial rather than right-angle distances. One such model, which we might call Hayne's circular flushing-region model, is considered by Eberhardt [1978a]. It is assumed that, for a given survey, an animal is flushed out into the open as soon as the observer crosses the boundary of a circle radius r centred on the animal. From the line intercept theory (cf. Lucas and Seber [1977] with $w_i = 2r_i$), it follows that Hayne's [1949a] estimator

$$\hat{D}_2 = \frac{n}{2L}\left(\frac{1}{n}\sum_{i=1}^{n}\frac{1}{r_i}\right)$$

$$= \frac{1}{2L}\sum_{i=1}^{n}\frac{1}{r_i} \tag{12.27}$$

is unbiased, that is $E[\hat{D}_2 \mid \{r_i\}] = D$. Since $\hat{D}_2$ is also unconditionally unbiased, it is unbiased irrespective of whether we regard the r_i as fixed or random, provided the transect is randomly placed. For this reason I prefer not to use Eberhardt's dichotomy of fixed- and variable-distance models.

Since the transect is randomly placed, the conditional density function of z given r for the above model is

$$f(z \mid r) = \frac{1}{r}, \quad 0 \leqslant z \leqslant r, \tag{12.28}$$

so that, given r, $\sin\theta = z/r$ has a uniform distribution on [0, 1]. Thus θ has density function

$$f_2(\theta) = \cos\theta, \quad 0 \leqslant \theta \leqslant \pi/2 \tag{12.29}$$

and, for the above circular model, (2.20) is true in general, and not just for the exponential model. However the flushing region need not be circular and detection may depend on the observer, so that other models have been suggested for the distribution of θ. For example, Robson has suggested that θ is perhaps uniform on $[0, \pi/2]$ when the searching is for inanimate objects. In the case of (12.29), $E[\theta] = 32{\cdot}7^\circ$ (p. 38) while for Robson's model $E[\theta] = 45^\circ$. Clearly $\bar{\theta}$, the average flushing angle will shed some light on the appropriateness of a given model. For example, Robinette *et al.* [1974] carried out a number of simulated experiments to compare ten methods of estimation and found that for 11 studies the Hayne estimator had an average bias of about 41 percent. Burnham and Anderson [1976] point out that this is not surprising as $\bar{\theta}$ ranged from $36{\cdot}0^\circ$ to $52{\cdot}4^\circ$ with an overall average of $43{\cdot}6^\circ$: this is well removed from $32{\cdot}7^\circ$. Gross *et al.* [1974] also tested out (12.28) by comparing their data on jackrabbits with the cosine curve given by (12.29). The data from one observer seemed to follow the curve fairly well, but the data from the second observer was a poor fit. In addition to an observer difference they found seasonal and yearly differences, and demonstrated a positive relationship between flushing distance and population density.

As mentioned on p. 38, (12.20) is best examined by comparing the $\sin\theta_i$ with the uniform distribution. For example, a goodness of fit test for lizard data is given by Eberhardt [1978a: table 3]. However, a word of caution. Although the θ_i will each have the same marginal density function they may not be independent if the population is not randomly distributed.

Up till now we have made no assumptions about the r_i ($i = 1, 2, \ldots, N$) other than they are constants. Suppose, however, that the r_i are a random sample from a distribution with density function $f_1(r)$. Then, arguing as on p. 29, it is readily shown that for the line transect ($W = \infty$), the probability of an animal being seen is

$$P = \frac{2L}{A}\int_0^\infty rf_1(r)\,dr \tag{12.30}$$

$$= \frac{2Ld}{A}, \text{ say,} \tag{12.31}$$

and the density function of r for animals *actually seen* is

$$f_2(r) = rf_1(r)/d. \tag{12.32}$$

Thus, conditional on n, the *observed* r_i are independently distributed with density function $f_2(r)$ and

$$E[r^{-1}\,|\,\text{animal seen}] = \int_0^\infty r^{-1}f_2(r)\,dr = 1/d.$$

Hence, from (12.27) and (12.31),

$$\begin{aligned} E[\hat{D}_2] &= \underset{n}{E}\, E[\hat{D}_2\,|\,n] \\ &= \underset{n}{E}[n/2Ld] \\ &= NP/(2Ld) \\ &= D, \end{aligned}$$

so that Hayne's estimator is unbiased; a fact already noted. Also from the argument leading to (12.29) we see that the conditional distribution of θ given r does not depend on r, that is r and θ are independently distributed.

The case $W = \infty$ was considered above, for two reasons. Firstly there is not much point in making W finite as there will be values of r that are greater than W. Secondly the distributional theory for finite W is much more difficult.

Finally we note that there is an "edge effect" to be considered. As the observer approaches the beginning of the transect a number of animals will be flushed whose flushing regions overlap the beginning of the transect, but these will not be counted. However, these will be compensated for at the other end of the transect, as animals will be counted there which are not in the region perpendicular to the transect, that is which do not lie on some perpendicular from the transect and therefore do not, strictly, have an observed y. This situation is identical to the problem of partial chords at the ends of the transect in line intersect sampling: chords are completed at one end and ignored at the other.

ELLIPTICAL FLUSHING REGION. Equation (12.28) follows from a circular flushing region and a random placing of the transect. However, Burnham and Anderson [1976] suggest an alternative class of flushing models based on the assumption that

$$f(z\,|\,r) = r^{-1}c(z/r),$$

where $c(z/r)$ is a function of z/r which does not depend on r. In this case $f(0|r) = c(0)/r$ and an unbiased estimate of $f(0)$ is now given by $c(0)(\Sigma r_i^{-1})/n$. The corresponding density estimate is then

$$\hat{D}_c = \frac{1}{2L}\hat{c}\sum_{i=1}^{n}\frac{1}{r_i}, \tag{12.33}$$

which is Hayne's estimate multiplied by $\hat{c}$, an estimate of $c(0)$. Since $y = r \sin\theta = rs$, say, then $f(s\,|\,r) = c(s)$, so that $c(s)$ is the density function of $\sin\theta$, and θ is statistically independent of r. Thus, if a form of $c(s)$ is postulated then $c(0)$ can be estimated from the observed θ_i. For example, Burnham [1979] introduces a model with an elliptical flushing region which leads to the density function

$$c(s) = \alpha[1 + (\alpha^2 - 1)s^2]^{-3/2},$$

where $\alpha = c(0)$, the parameter to be estimated. He gives a scheme for finding the maximum likelihood estimate of α and an estimate of $V[\hat{D}_c]$.(*)

STOCHASTIC FLUSHING MODEL. Up till now we have not needed any assumptions about the r_i ($i = 1, 2, \ldots, N$) other than they are constants. An alternative approach, which we shall call the stochastic flushing model, is to assume that the probability of an animal being flushed, given that its distance from the observer lies in the interval $[r, r + dr]$, is $h_1(r)dr$. The density function h_1 is conceptually different from f_1 of (12.30) as we have made no reference to the notion of a flushing region. Following (12.32), Burnham and Anderson [1976] develop a joint distribution for z and r, and show that if $f(z|r)$ is the conditional density function of z given r then

$$\hat{f}_r = \frac{1}{n}\sum_{i=1}^{n} f(0|r_i) \tag{12.34}$$

is an unbiased estimate of $f(0)$. The estimate of D based on $\hat{f}_r$, namely $n\hat{f}_r/(2L)$, reduces to the Hayne estimate if (12.28) is true, that is if $f(0|r) = 1/r$. Gates [1969] postulated the exponential models $f(z) = \lambda \exp(-\lambda z)$ and $f(r|z) = \lambda \exp[-\lambda(r - z)]$ which lead to (12.28) once again, so that Hayne's estimate is unbiased for this case.

(*) For a similar model see De Vries [1979], Additional References.

Following (12.32) we can define $h_2(r) = rh_1(r)/d$, the density function for an *observed* r. Gates' models then lead to the gamma distribution

$$h_2(r) = \lambda^2 r \exp(-\lambda r),$$

which provides the (modified) maximum likelifood estimate (cf. p. 37)

$$\hat{\lambda} = (2n-1)/\Sigma r_i. \qquad (12.35)$$

Since $f(0) = \lambda$, $\hat{\lambda}$ is an unbiased estimate of $f(0)$. Using Gates' model, Kovner and Patil [1974] compared (12.35), the minimum variance unbiased estimate of λ, with several other well-known estimators.

Eberhardt [1978a: Appendix C] also considers the stochastic flushing model and points out the following relationships (see also Burnham and Anderson [1976: 330]):

$$g(y) = 1 - \int_0^y h_1(r)dr$$

$$= 1 - H_1(y),$$

and, from (12.13) and using integration by parts,

$$\mu_\infty = \int_0^\infty g(y)dy$$

$$= \int_0^\infty rh_1(r)dr.$$

However, he then assumes that once the observer passes the animal the flushing probability drops to zero: this is what happens in the circular flushing region model.

In practice it is not possible to distinguish between $f_1(r)$ and $h_1(r)$ from the data, with the exception that f_1 is violated when the animals flush behind the observer. A basic assumption in both approaches is that the sighting (flushing) of one animal is independent of the sighting (flushing) of any other. However, as pointed out by Eberhardt, "group flushes" only pose a problem if there is a correlation between flushing distance and group size. Otherwise we can assume that the groups flush independently and estimate the number of groups using the above methods; we can then multiply this estimate by the average group size. Clearly further research is needed on this problem of group flushing and any model used should distinguish between any non-randomness of the population and dependence of flushing distances. For example, it is not clear which is the case in the models suggested by Sen *et al.* [1974: 338]. As already mentioned, non-randomness need not bias the density estimate if the transect is random, though the theoretical variance will be affected.

12.5 DISTANCE METHODS

12.5.1. Robustness of procedures for Poisson model

In **2.1.4** we considered a number of earlier methods for estimating population density from point–object (closest–individual) and object–object (nearest-neighbour) distances. However, in the 1970's there was an upsurge of interest in these distance methods, so that established methods have been examined for robustness and new robust methods have been developed. The following is a survey of these developments, and much of the discussion is based on the excellent review of Diggle [1979a] (see also Diggle [1979b]).

1 Hopkins' test

If the N objects are randomly distributed over the population region of area A then, ignoring boundary problems, we saw that the distribution of the distance $X_{(r)}$ from a randomly chosen point to the rth nearest neighbour has density function (2.36) on p. 47. Letting $N \to \infty$ and $A \to \infty$ with $D = N/A$ remaining constant, then (2.36) tends to the Poisson based model (2.27) for which $2D\pi X_{(r)}^2$ is χ_{2r}^2. Also, for (2.27), $2\pi D X_{(1)}^2$, $2\pi D(X_{(2)}^2 - X_{(1)}^2)$, ... , $2\pi D(X_{(k)}^2 - X_{(k-1)}^2)$ are independently and identically distributed as χ_2^2 (Pollard [1971]). In this case the underlying distribution of the objects is Poisson, the so-called Poisson "forest", and (2.27) can be deduced directly from the axioms of a two-dimensional homogeneous Poisson process. As pointed out on p. 41, the Poisson forest strictly applies to a population of density D spread over an infinite area, though it will be a reasonable approximation for the "real life" finite case provided the population area is large (cf. Holgate [1965b] for a discussion of this problem). However, this model is clearly not relevant to the case of a small population area where every object is important, and it is not appropriate to construct a buffer strip (p. 43) to get round the edge effect. It would appear that the method of Craig [1953a] on p. 47 and the paper by Brown and Rothery [1978] are the only approaches which take into account the effect of the boundary.

We noted on p. 50 that we obtain the same density function (2.27) if the distance, $X'_{(r)}$ say, is measured from a randomly chosen object, rather than a random point, to the rth nearest neighbour. This stochastic equivalence is unique to the Poisson forest and has been used as a basis for constructing tests of the Poisson model. For example, if we define $X = X_{(1)}$ and $X' = X'_{(1)}$ then Hopkins [1954] and Moore [1954] independently proposed the test statistic

$$F = \sum_{i=1}^{s} X_i^2 / \sum_{i=1}^{s} (X'_i)^2, \qquad (12.36)$$

where the $\{X_i\}$ are the point–object distances for a sample of s randomly located points and the $\{X'_i\}$ are object–object distances for an *independent* sample of s randomly located objects. As $2D\pi\Sigma X_i^2$ and $2D\pi\Sigma(X'_i)^2$ are independently distributed as χ_{2s}^2 under H, the hypothesis of a Poisson forest, F of (12.36) has an $F_{2s,2s}$ distribution and

$$B = \Sigma X_i^2 / [\Sigma X_i^2 + \Sigma(X'_i)^2]$$
$$= F/(F + 1)$$

has a beta distribution, $B(s, s)$ say, when H is true. This approach has intuitive appeal as X_i^2 tends to be increased by aggregation but decreased by regularity in the pattern; the reverse is true with $(X_i')^2$. Hence significantly large values of B (or F) suggest aggregation or clustering, while significantly small values of B suggest regularity in the underlying pattern.

We note, for completeness, a paper by Paloheimo [1972] which discusses the application of a spatial bivariate Poisson distribution to nearest neighbour distances when there are two species.

2 Maximum likelihood estimation

Given the density function (2.26) for $X_{(1)}$, the maximum likelihood estimate of D, corrected for bias by replacing s by $s-1$, is (cf. (2.29) and (2.30) with $r = 1$):

$$\hat{D} = (s-1)/\sum_i \pi X_i^2$$

with variance

$$V[\hat{D}] = D^2/(s-2).$$

Although the maximum likelihood estimate is most efficient for the Poisson model, it is not robust and is sensitive to variations in D (Pollard [1971: 1000–1]). Also, in the light of previous comments, $\hat{D}$ will overestimate D for a regular pattern and underestimate D for a clustered pattern. Persson [1964, 1965, 1971] and Holgate [1972] have also considered $\hat{D}$ and various less efficient estimators and show that they may be severely biased, particularly if there is regularity in the population pattern. Clearly $\hat{D}$ will be a poor estimator for most practical (non-random) situations.

The non-robustness of $\hat{D}$ has also been demonstrated by Batcheler (Batcheler and Bell [1970], Batcheler [1971, 1973]) who gives complicated empirical formulae for correcting for the bias (cf. Batcheler [1975a] for the formulae, and Batcheler [1975b] for an application to deer pellet groups). Batcheler and Hodder [1975] applied these formulae successfully to pine stands and demonstrated that for the same efficiency of estimation, distance methods are less time-consuming than quadrat sampling. A comparison of several distance methods is given by Laycock and Batcheler [1975]. However Batcheler's method is empirical and has no theoretical basis.

If X_i is only measured if $X_i < R$ so that for the s sample points only s_1 measurements are actually made, then the maximum likelihood estimate of D for this truncated model is given by Dr Darwin in Batcheler [1971], namely

$$\hat{D}_R = s_1 \Bigg/ \left\{ \sum_{i=1}^{s_1} \pi X_i^2 + (s - s_1)R^2 \right\}.$$

Computer calculations indicate that the bias of $\hat{D}_R$ is evidently small for a Poisson forest. From simulation studies Batcheler suggests that R should be chosen so that $s_1 \approx \frac{1}{2}s$. The estimator $\hat{D}_R$ suffers from the same lack of robustness as $\hat{D}$.

Another approach is to use the maximum likelihood estimate of $\theta(= 1/D)$, namely

$$\hat{\theta} = \pi[\Sigma X_i^2 + \Sigma(X_i')^2]/2s \qquad (= \tfrac{1}{2}(\hat{\theta}_1 + \hat{\theta}_2), \text{ say}), \tag{12.37}$$

which is unbiased with variance $\theta^2/(2s)$. As pointed out by Diggle [1975], $\hat{\theta}$ is essentially the average of two estimates of θ whose biases will tend to be in opposite directions when H is false. Diggle [1975] considers a second estimate

$$\theta^* = \sqrt{(\hat{\theta}_1 \hat{\theta}_2)} = (\pi/s)[\Sigma X_i^2 \Sigma(X_i')^2]^{1/2}, \tag{12.38}$$

and from moment studies for particular aggregated and regular type patterns concludes that θ^* is more robust than $\hat{\theta}$. Also, as expected, $\hat{\theta}$ is more robust than either $\hat{\theta}_1$ or $\hat{\theta}_2$, without loss of efficiency.

3 Holgate's tests

Unfortunately Hopkins' test of (12.36) and θ^* have a limited usefulness as the X_i' object–object measurements require the random selection of an object which can only be made if all the objects in the population can be counted (cf. p. 50). However, they have led to other tests and estimates based on replacing X_i' by suitable practical alternatives but with some loss of power or efficiency. For instance Holgate [1965c] suggested a test based on the pairs of distances $X(= X_{(1)})$ and $W(= X_{(2)})$ from each of s randomly located points to the nearest and second nearest objects respectively. It was stated above in section 1 that $2\pi D X_{(1)}^2$ and $2\pi D(X_{(2)}^2 - X_{(1)}^2)$ are independently distributed as χ_2^2 when H is true. Thus $W^2 - X^2$ plays the same role as $(X')^2$ in the Hopkins test, so that we have the test statistics

$$h_F = \Sigma X_i^2 / \Sigma(W_i^2 - X_i^2)$$

or

$$h_B = \Sigma X_i^2 / \Sigma W_i^2,$$

which are distributed as $F_{2s,2s}$ and $B(s, s)$, respectively, when H is true. Because the pair (X_i, W_i) come from the same sample point, an alternative test is given by Holgate, namely

$$h_N = s^{-1} \Sigma(X_i^2/W_i^2),$$

which, for $s \geqslant 10$, is approximately distributed as $\mathfrak{N}(\tfrac{1}{2}, 1/(12s))$ when H is true. Unfortunately Holgate's two tests, h_B (or h_F) and h_N, are particularly suspect against the alternatives of regularity (Diggle *et al.* [1976]). For example, in the extreme case when an object is located at each node of a regular square lattice, X_i/W_i may be arbitrarily close to 1, which incorrectly suggests aggregation, whereas for Hopkins' approach $X_i^2/[X_i^2 + (X_i')^2] \leqslant \tfrac{1}{3}$ (Diggle [1979a], which correctly suggests regularity.

4 Combining closest-individual and nearest-neighbour distances

An alternative strategy is to measure X, the distance from a random point P to the nearest object, at Q say, and then measure Y the distance from Q to its nearest neighbour. However, we find that X and Y are not independent and Y has a different distribution from that of X; in fact, $E[Y] \approx 1{\cdot}19E[X]$. (Diggle [1979a] points out that the final expression for $E[Y]$ in Kendall and Moran [1963: ch. 2] is incorrect by a factor of 2). The joint distribution of X and Y

has been further investigated by Cox and Lewis [1976], Cox [1976] and Cormack [1977]. They find it convenient to denote the pair (X, Y) by (X_1, Y_1) when $Y \leqslant 2X$ and by (X_2, Y_2) when $Y > 2X$: similarly a sample pair (X_i, Y_i) is designated (X_{1i}, Y_{1i}) or (X_{2i}, Y_{2i}). Thus for any experiment the data pairs $(X_1, Y_1), (X_2, Y_2), \ldots, (X_s, Y_s)$ from s random points P can be split into two sets

$$A = \{(X_{1i}, Y_{1i}), \quad i = 1, 2, \ldots, s_1\} \quad \text{and} \quad B = \{(X_{2j}, Y_{2j}), \quad j = 1, 2, \ldots, s_2\},$$

where $s_1 + s_2 = s$. Cox and Lewis [1976] prove that $4X_2^2/Y_2^2$ has a uniform distribution on [0, 1] for the Poisson model and several regular patterns; Cormack [1977] shows that this result is true for any pattern, provided P is chosen at random, thus suggesting that information about the pattern is essentially contained in the set A. Cox [1976] defines the random viariable (using Z_1 instead of T_1)

$$T_1 = X_1 [2\pi + \sin B_1 - (\pi + B_1) \cos B_1]^{1/2}/\sqrt{\pi},$$

where $\sin (B_1/2) = Y_1/(2X_1)$, and shows that for a Poisson forest T_1 and Y_2 are independently and identically distributed according to (2.27) on p. 42 with $k = 2$. He then deduces that the maximum likelihood estimate of θ (= $1/D$) is

$$\hat{\theta} = \pi\left(\sum_{i=1}^{s_1} T_{1i}^2 + \sum_{j=1}^{s_2} Y_{2j}^2\right) \bigg/ 2s. \tag{12.39}$$

This estimate is unbiased with minimum variance $\theta^2/(2s)$ and it is readily seen that $4sD\hat{\theta}$ is distributed as $\chi^2_{4(s_1+s_2)}$, that is as χ^2_{4s}, when H is true. Cox discusses modifications of $\hat{\theta}$ which are approximately unbiased for certain non-Poisson models. He also considers the estimate

$$\tilde{D} = 4(\pi s)^{-2} \sum_{j=1}^{s_2} (Y'_{2j})^{-2}, \tag{12.40}$$

where $Y'_{2j} = Y_{2j}$, given that Y_{2j} is greater than some small fixed value ϵ. Although this estimator uses only s_2/s of the data points (less than $\frac{1}{4}$ for the Poisson or aggregated patterns, though about 0·9, 0·8 and 0·6 for triangular, square and hexagonal lattices respectively), it seems to be approximately unbiased for a wide range of spatial patterns. Generally it appears to work particularly well for patterns of a more regular type.

5 *T-square sampling method*

Besag and Gleaves [1973] suggest a "T-square" sampling procedure which gives rise to a simpler distribution theory. Here X is defined as above, but Y is now replaced by $Z/\sqrt{2}$, where Z is the distance from Q to the nearest object within the half-plane delimited by the perpendicular to PQ passing through Q, and which excludes P. It then transpires that X and $Z/\sqrt{2}$ are independently distributed according to (2.26), so that $Z/\sqrt{2}$ takes over the role of X' in the approaches of Hopkins and Holgate above, and $Z^2/2$ plays the same role as $W^2 - X^2$. Thus for sample pairs (X_i, Z_i), Besag and Gleaves [1973] propose two

"T-square" tests analogous to Holgate's h_B and h_N, namely

$$t_B = \Sigma X_i^2/(\Sigma X_i^2 + \tfrac{1}{2}\Sigma Z_i^2)$$

and

$$t_N = \frac{1}{s}\sum_i \left(\frac{X_i^2}{X_i^2 + \frac{1}{2}Z_i^2}\right).$$

When H is true t_B is $B(s, s)$ (or, equivalently, $\Sigma X_i^2/\Sigma \frac{1}{2} Z_i^2$ is distributed as $F_{2s,2s}$) and t_N is approximately $\mathfrak{N}(\frac{1}{2}, 1/(12s))$ $(s \geqslant 10)$. Diggle *et al.* [1976] compare the powers of B, h_B, h_N, t_B and t_N, and confirm the overall superiority of B (which can be used as norm). The T-square tests are better than their Holgate counterparts and t_N appears to be marginally superior to t_B, though it depends on the particular alternative hypothesis considered. The above tests, together with a new test, are compared by Hines and Hines [1979].

One objection to the use of distance methods as tests of randomness, voiced for example by Mead [1974], is that they "are concerned essentially with small-scale pattern". A particular problem is that of patterns which Diggle [1977a] describes as "locally random but globally heterogeneous", or "random-heterogeneous" for short. In general, extreme heterogeneity will be classified as aggregation using the t_N test, whereas random-heterogeneous patterns will tend to be missed because of the "local" nature of distance methods. However, Diggle [1977a] provides a supplementary procedure aimed specifically at distinguishing between the random-heterogeneous and the random-homogeneous (i.e. Poisson forest) patterns. The test statistic is given by

$$48s(s \log \bar{U} - \sum_i \log U_i)/(13s + 1) \tag{12.41}$$

which is approximately distributed as χ^2_{s-1} when H is true. Here $U_i = X_i^2 + \frac{1}{2}Z_i^2$, $\bar{U} = \Sigma U_i/s$, and the above statistic is a likelihood ratio test for constant D using (2.26) as the underlying null distribution for the distributions of X_i^2 and $\frac{1}{2}Z_i^2$. The statistic (12.41) is essentially a special case of Bartlett's statistic for testing constant variance and therefore uses his correction factor C; in this case $(13s + 1)/(12s)$.

Diggle [1977a] proposes a two-stage procedure in which the first stage is to calculate t_N: if t_N is significantly small a regular type pattern is suggested, while a significantly large t_N suggests aggregation. If t_N is not significant then the supplementary test (12.41) is carried out: significance would then imply random-heterogeneity. Thus the two tests lead to an approximate four-way characterisation of patterns. Although, in contrast to the work of Mead [1974] (discussed later in **12.6.2**), the two-stage procedure cannot be used to investigate pattern at different scales, it is much more readily carried out in the field and the data can also be used for density estimation.

Analogous to $\hat{\theta}$ and θ^* of (12.37) and (12.38), Diggle [1977b, 1979a] defines the estimators

$$\hat{\theta}_T = \pi\left(\sum_{i=1}^{s} X_i^2 + \sum_{i=1}^{s} \tfrac{1}{2}Z_i^2\right)\Big/(2s)$$

and

$$\theta_T^* = (\pi/s)(\Sigma X_i^2 \, \Sigma \tfrac{1}{2} Z_i^2)^{1/2}.$$

The simulation study reported by Diggle [1977b] shows that, except for regular patterns, θ_T^* is rather less robust than θ^*. However θ_T^* appears to be a useful estimate of $\theta (= 1/D)$ in that it is tolerably robust to a wide range of aggregated departures (modelled by a Poisson cluster process) from H.

12.5.2 Non-parametric methods

A potentially useful non-parametric estimate of D has been recently proposed by S.A. Patil *et al.* [1979]. Suppose $f(u)$ $(= g_1(y)$ of **2.1.4**) is the probability density function of $U = \pi X^2$, where X is the distance from a random point to the nearest object. The authors show that $f(0) = D$ for a very wide range of spatial distributions including regular lattices, cluster (generalised Poisson) models such as the Thomas, Pólya–Aeppli and negative binomial, and doubly stochastic (compound Poisson) models. For example, it is readily checked that this relationship holds for the exponential, binomial and negative binomial models of **2.1.4**.

Two methods have been proposed for estimating $f(0)$. The first is Eberhardt's [1978a] method described on p. 463. The second is based on the theory given by Rosenblatt [1956] and Loftsgaarden and Quesenberry [1965] and adapted to the present problem by S.A. Patil *et al.* [1979]. The method uses the order statistics $U_{(1)} < U_{(2)} < \ldots < U_{(s)}$ from the random sample $U_1, U_2, \ldots U_s$, and the estimate proposed is

$$\hat{D} = \hat{f}_s(0) = k(s)/(sU_{(k)}), \tag{12.42}$$

where $k(s)$ is a sequence of real numbers such that $k(s) \to \infty$ and $k(s)/s \to 0$ as $s \to \infty$, and $k = [k(s)]$ (the integral part of $k(s)$). If $f(u)$ is continuous on $(0, \infty)$, right continuous at $u = 0$ and $f(u) > 0$, the authors use the results of Moore and Yackel [1977] see also Devroye and Wagner [1977] with $p = 2$, i.e. ranked circles $U_{(i)}$ rather than ranked spheres) to show that $\sqrt{k(s)}\,[\hat{f}_s(0) - f(0)]$ is asymptotically $\mathfrak{N}(0, f^2(0))$ as $s \to \infty$. This result can be used to obtain a large-sample confidence interval for $f(0)$ $(= D)$. The authors suggest choosing $k(s) = \sqrt{s}$, but Burnham (personal communication) has indicated that other sequences may lead to more efficient estimates of $f(0)$.

If searching is restricted to a distance R, then the probability of finding an object in the circle of area πR^2 is

$$p_R = \Pr[U \leqslant \pi R^2] = F(\pi R^2).$$

Thus, given a sample of s points, the number of points s_1 which have a nearest neighbour closer than R is binomial with parameters s and p_R. The truncated distribution of U corresponding to this sampling scheme is

$$f_R(u) = f(u)/p_R \qquad 0 \leqslant u \leqslant \pi R^2,$$

so that setting $u = 0$ we have $D = f(0) = p_R f_R(0)$.

Thus an estimate of D is

$$\tilde{D}_R = \frac{s_1}{s} \tilde{f}_R(0),$$

where $\tilde{f}_R(0) = k(s_1)/(s_1 U_{(k_1)})$ and $k_1 = [k(s_1)]$. Using the δ-method we find that

$$V[\tilde{D}_R] \approx \frac{D^2}{k(s_1)} + \frac{D^2}{s}\left(\frac{1 - P_R}{P_R}\right)\left(1 + \frac{1}{k(s_1)}\right),$$

and this is estimated by

$$v[\tilde{D}_R] = (\tilde{D}_R)^2 \left[\frac{1}{k(s_1)} + \left(\frac{1}{s_1} - \frac{1}{s}\right)\left(1 + \frac{1}{k(s_1)}\right)\right].$$

The authors give a number of examples which suggest that the above estimates of D are satisfactory, provided that s is reasonably large ($s > 100$: Burnham, personal communication).

12.5.3. Other methods

There are several other methods which have been proposed for the study of plant and tree populations. These are the "aggregation circles" method (Aberdeen [1958]), the "angle-order" method (Morisita [1957]: this is a generalisation of the point-centred quarter method of Cottam and Curtis [1956]), and the "wandering quarter" method (Catana [1963]). The studies by Persson [1971] and Westman [1971] indicate that the estimates of density obtained by these methods are generally unsatisfactory and are non-robust to certain types of non-randomness.

12.6 PATTERN DETECTION(*)

In addition to estimating population density, ecologists are frequently interested in detecting any patterns in the population distributions of animals or their signs. The existence of pattern may shed light on such factors as social behaviour or habitat preference, and a variety of methods for detecting pattern have been developed; the type of method depending on whether we have full or only sample information on the location of the objects.

12.6.1 Use of random plots

The counts on random quadrats can be used for studying spatial pattern as well as estimating density. Iwao [1970], in discussing this approach, notes that counts on quadrats can often be described adequately by more than one model, and that the spatial distribution of even the same species may vary considerably from occasion to occasion to the extent that different models are warranted. Although such a fitted model could provide a useful empirical distribution of the data, it usually will not tell us anything about the underlying mechanisms which may give rise to the particular model. For example, the negative binomial model can be deduced from a variety of underlying biological processes (cf. Boswell and Patil [1970]).

(*) See the reviews by Cormack [1979] and Ord [1979] (Additional References).

1 Measures of dispersion

Iwao [1968, 1970] describes several measures of dispersion (aggregation) based on quadrat counts which can be used for indicating whether a population is random, overdispersed (clustered, contagious, aggregated), or underdispersed (regular). A common measure of dispersion is the so-called Poisson Index of Dispersion, $I = v/\bar{x}$, where $\bar{x}$ is the mean count on s plots and $v = \Sigma(x_i - \bar{x})^2/(s-1)$, the sample estimate of plot variance. When the underlying spatial distribution is Poisson, $E[I] = 1$ (Bartko *et al.* [1968], Dahiya and Gurland [1969]), while $E[I] > 1$ when there is clustering, and in this case I unfortunately varies with the mean density. The ratio σ^2/μ, the population version of I, is studied in detail by Stiteler and Patil [1971a] for regular patterns.

On p. 14 we saw that $(s-1)v/\bar{x}$, the Poisson Dispersion Test, provides a useful test of the Poisson model, and Gart [1970] gives the following graphical interpretation of the test (see also Ord [1972: 102]). Suppose we plot $\hat{\lambda}_i = if_i/f_{i-1}$ against i ($i = 1, 2, \ldots$), where f_i is the number of plots with exactly i objects (i.e. with $x = i$). If the model is Poisson then $E[\hat{\lambda}_i] \approx i\Pr[x = i]/\Pr[x = i-1] = \theta = E[x]$, so that the plot should be approximately linear with zero slope. Using weighted least squares with an appropriate choice of weights, an estimate of slope is

$$\hat{\beta} = 1 - [s/(s-1)](\bar{x}/v).$$

Thus a test of $\beta = 0$ is equivalent to using the Poisson Dispersion Test. The regression is still linear if the underlying model is negative binomial or binomial, but with positive and negative slopes, respectively. Gart also gives similar plotting methods for the logarithmic series distribution and various truncated distributions: further graphical methods are given by Grimm [1970]. Stiteler and Patil [1971b] review some of the above topics and introduce a vector of indices based on an extension of Lloyd's mean crowding (Lloyd [1967], Iwao [1976]) which they apply to a variety of discrete distributions.

Although the index $v/\bar{x}$ has been used extensively in the past as a means of described point processes as over-dispersed, random or under-dispersed, it has a number of drawbacks (Holgate [1972]), namely: (i) the index depends on the plot size; (ii) as already mentioned, for a fixed plot size there are many processes which can produce the same distribution of numbers per plot; (iii) the classification into overdispersed and underdispersed is not sufficiently fine; (iv) the method gives no indication as to whether overdispersion arises from heterogeneity in the density or from clustering. The same criticisms can be levelled at the use of k, the parameter of the negative binomial distribution (p. 25), as a measure of aggregation. In particular there are two basic models, one in which k is independent of quadrat size (e.g. Gérard and Berthet [1971]), and the other in which k is proportional to the size of the quadrat (cf. Cliff and Ord [1973: 59]).

Clearly what we would like is an index which is characteristic of a given species and does not vary with density. One such index has been proposed by Taylor [1961] who found that the empirical "power law", $v = a\bar{x}^b$, was satisfied by a number of species (cf. Southwood [1966: 9–12, 35]. Taylor [1971], Standen and Latter [1977: 217] and Taylor *et al.* [1978]). The constant a depends on the density, but b is often found to be independent of density and could be used as a species index. The appropriateness of the relationship can be checked by seeing whether the plot of log v (= log a + b log $\bar{x}$) versus log $\bar{x}$ is linear. Taylor *et al.* [1978] give examples where k is not constant and compare the power law approach with Iwao's [1968, 1979(*)] regression method.

Finally we note that Iwao's regression method can be used for determining sample sizes in sampling experiments (Iwao and Kuno [1971]) and for designing a sequential sampling scheme for determining whether a population density is above or below some threshold level, a common situation in pest control (cf. Iwao [1975]). An interesting application of several measures of dispersion to zooplankton is given by George [1974].

2 *Fitting distributions*

There is an extensive literature on the fitting of various discrete distributions to sample data. Rogers [1974] has a helpful discussion on the types of compound and generalised distributions that make up the range of distributions Poisson, Poisson–binomial, Poisson–Poisson (Neyman Type A), Poisson–negative binomial (Poisson–Pascal), negative binomial and perfect clustering (see also Ord [1972]: 126, table 6.1). He also discusses estimation for these distributions and applies his techniques to geographical data. Procedures for choosing and fitting one of these distributions are described by Gurland and Hinz [1971] and Ord [1972]: 132ff].

The so-called Thomas distribution, which is similar to the Neyman Type A distribution, has also been used. For both distributions the number of clusters is Poisson with constant mean: for the Neyman Type A the number of individuals per cluster is Poisson, while for the Thomas case the "parent" of the cluster is always present, the number of "offspring" being Poisson. Methods for estimating the parameters of these two distributions from quadrat samples are given by Gleeson and Douglas [1975].

The various distributions are also discussed by Douglas [1970] from the point of view of a mixing model in which impurities in a substance become broken up and distributed as specks. A further classification of discrete distributions is given by Kemp [1971]. Another distribution which shows some promise because of its flexibility is the log-zero Poisson distribution (Rao and Katti [1970], Katti and Rao [1970]). For an excellent survey of methods of selecting and fitting an appropriate discrete distribution the reader is referred to Ord [1972].

Frequently there is more than one level of clustering in a population in that clusters themselves are clustered, and hopefully counts from quadrats can be used to detect the presence of clustering at different scales. As mentioned on pages 23–25, the detection of non-randomness depends on the quadrat size (cf. Kershaw [1970, 1973: 136] and, for an interesting example, Schroder and Geluso [1975]) so that counts must be made from quadrats of different sizes.

(*)See Additional References.

However, apart from Prentice [1973] very little seems to have been done on analysing data from random plots of different sizes. The usual approach is to use contiguous (adjacent) rather than random plots for detecting levels of clustering. These methods are described below.

12.6.2 Contiguous plots

A common method for studying patterns, due to Greig-Smith [1952], is to count the number of objects on each quadrat of a 16×16 grid of 256 contiguous quadrats. The whole grid is then divided into two halves, each half into two quarters, each quarter into two eighths, and so on until the final division is of each pair of quadrats into two individual quadrats, and a hierarchical analysis of variance is then carried out (cf. Mead [1974: table 1]). Assuming a random (Poisson) distribution of objects, each mean square in the analysis of variance table is an estimate of the variance of the Poisson distribution for counts on an individual quadrat. The mean square, or I = (*mean square/overall quadrat mean*), for each quadrat grouping (block size) is then plotted against block size, and any peaks in this plot are generally interpreted as evidence of pattern at the corresponding scales. Various attempts at formalising this procedure have been made such as, for example, carrying out a test for randomness (similar to the Poisson Dispersion Test) for each block size by comparing I with the distribution of the appropriate chi-squared variable divided by its degrees of freedom. Thus if we find I to be significantly high for the mean square obtained for the sum of squares for sixteenths within eighths (or equivalently for blocks of 16 with blocks of 32), we supposedly infer that there is a pattern in the form of clumping at a scale of 16 quadrats. If I is significantly low we would then infer some regularity at a scale of 16 quadrats. Unfortunately such a procedure is not strictly correct as tests based on the I values will be dependent, and the rejection of spatial randomness at any particular scale logically invalidates all subsequent tests. However, as Mead [1974: 306] observes, Greig-Smith's method (which might be conveniently called the BQV or Block-Quadrat-Variance method) was only intended to provide an approximate empirical technique of assessing general patterns rather than a rigorous statistical procedure. It has been found useful by ecologists and it has also been used by geographers to determine the most important scale for study (e.g. Chorley *et al.* [1966], Moellering and Tobler [1972]). However, it seems that more care is needed in interpreting the plot of I versus the block size. For example, for a Poisson cluster process (e.g. Neyman Type A distribution), the plot of $E[I]$ against block size rises to a maximum which is then maintained with further increase in block size (Bartlett [1975: ch. 5]). If the parent process is itself clustered rather than Poisson, Bartlett suggests a further increase in I rather than a double peak corresponding to the two scales of clustering.

A modification of the above method, due to Kershaw [1957], consists of replacing the square grid of quadrats by a long grid or transect one quadrat wide and 256 quadrats long. One advantage of a long transect is the greater range of scales of pattern than can be investigated. However, the scale investigated is linear rather than areal, and the conclusion may depend very much on the direction of the transect. This and other difficulties were shown up by the Monte Carlo experiments of Errington [1973] and Usher [1975].

Ludwig [1979] regards the principal drawbacks of the above BQV method for a grid or a line as (i) the limitation of block-sizes to successive powers of 2, (ii) the dependence of the mean squares (variance estimates) on the starting position of the grid or transect if a regular pattern exists, and (iii) the lack of independence of the mean squares at the different block sizes. Several methods have been introduced to avoid these problems. To overcome the problem of starting position, Usher [1969] proposed a modification, called the stepped BQV or SBQV method by Ludwig, which effectively averages the BQV mean-squares over the starting position. The restriction to blocks of size 2^k can be avoided by taking sequential blocks of quadrats (1, 2), (2, 3), etc., rather than disjoint blocks (1, 2), (3, 4), etc.: this is the basis of Hill's [1973] Two Term Local Quadrat Variance (TTLQV) method. In an attempt to avoid dependence in the mean squares Goodall [1974] introduced a method in which pairs at a certain fixed distance apart are chosen at random without replacement, the method of Random Pairing of Quadrats for Variance (RPQV). The main disadvantage of this method is that in practice it is limited to relatively few spacings, since random pairing without replacement leads to only a few degrees of freedom for each mean square, unless a very large number of quadrats have been sampled—not the usual case. Also, as pointed out by Zahl [1977a], although the mean squares are independent conditionally on the particular random partition chosen (since these are based on disjoint sets of plots), they are not unconditionally independent since random selection without replacement from a finite population creates dependence. However, Ludwig [1979] suggests that accuracy of estimation is more important than lack of independence when searching for patterns and recommends using all the pairs at a given distance, his Paired-Quadrat Variance (PQV) method. Using simulation, he compares the above methods and concludes that TTLQV and PQV seem to provide the best evaluation of the scales and intensity of existing patterns. If two scales of pattern exist, the TTLQV method tends to emphasize the large-scale pattern while the PQV method tends to emphasize the small-scale pattern. Hypotheses suggested by the analysis can be tested by comparing the independent mean squares provided by the RPQV method. Ludwig's article should be consulted for an algebraic description of the above methods and a comparison of the blocked-quadrat methods (BQV, SBQV and TTLQV) versus the paired-quadrat methods (PQV and RPQV). In comparing this algebra with Table 1 of Mead [1974] it is helpful to note the identity

$$[x_1 - \tfrac{1}{2}(x_1 + x_2)]^2 + [x_2 - \tfrac{1}{2}(x_1 + x_2)]^2 = \tfrac{1}{2}(x_1 - x_2)^2.$$

Mead [1974] also considers some of the problems associated with the BQV method and presents a one-dimensional ("2 within 4" randomisation test which provides a valid scheme for simultaneously testing for non-randomness at different scales. In this case significance at one scale does not invalidate the test at a larger scale. One disadvantage of the method is that it is not independent of the scale of variation within each block of 4 quadrats. Mead also mentions briefly a two-dimensional "4's within 16" randomisation test in which the quadrats are successively partitioned into 1, 4, 16 ... blocks of 16 quadrats, each block consisting of a 4×4 grid of quadrats. For each partitioning the null hypothesis is that, within each block, the observed counts in the four 2×2 sub-blocks form a random sample from the $(16!)/(4!)^5 = 2{,}627{,}625$ possible combinations of four sets of four. Although the full randomisation procedure appears to be computationally prohibitive, one may compare the test statistic at each scale with a pseudo-random sample from its randomisation distribution – the so-called Monte Carlo test (cf. **1.3.10**).

Another valid test procedure, which is similar to Scheffé's S-method for simultaneous contrasts, has been proposed by Zahl [1974]. He compares his method (Zahl [1977a]) with BQV (using F-ratios rather than dispersion tests) and Goodall's RPQV, and concludes that BQV performs the best and RPQV the worst. Unfortunately Zahl's S-method requires considerable computing.

12.6.3 Map of locations

If a map of object locations is available so that sampling techniques are not required, then various point process models can be proposed for the pattern observed and tests of fit carried out. Hopefully the results of such an analysis may shed light on the underlying biological mechanisms that led to the observed pattern, though, as noted on p. 478 such an inference is not easy. Diggle [1979a, b] lists five types of model for consideration: (i) the Poisson forest, that is the two-dimensional, homogeneous Poisson Point process; (ii) Poisson cluster processes (leading to the generalised Poisson distributions) in which parents are randomly distributed according to a Poisson forest with each parent independently producing a random number of offspring, and each offspring being independently spatially distributed relative to the corresponding parent with a given bivariate density function (the final pattern may consist of either offspring only, or of parent and offspring, assumed indistinguishable); (iii) doubly stochastic (heterogeneous) Poisson process, leading to the so-called compound Poisson distribution, obtained by replacing the constant density D of a Poisson forest by a realisation of some stationary (isotropic) process $D(x, y)$; (iv) lattice processes in which individuals form a regular pattern such as a square or equilateral triangular lattice; and (v) inhibition processes in which no two individuals can be less than a certain distance apart.

For model (ii), general distance distributions are given by Bartlett [1975: ch. 1], and some explicit results for special cases are provided by Warren [1971] and Diggle [1975]. Unfortunately model (iii) ("apparent" contagion) cannot usually be distinguished from model (ii) ("true" contagion), as demonstrated by Bartlett [1964: 301] and discussed by Cliff and Ord [1973: 59] in relation to the negative binomial and Neyman Type A distributions. For example, in the notation of Gurland [1957] we have the generalised distributions *Poisson* v *logarithmic* ~ *negative binomial* and *Poisson* v *Poisson* ~ *Neyman Type A*; and the compound distributions *Poisson* $\wedge$ *gamma* ~ *negative binomial* and *Poisson* $\wedge$ *Poisson* ~ *Neyman Type A*. Although some progress has been made (cf. Cliff and Ord [1973: ch. 3]) in distinguishing between the different spatial processes generating the negative binomial distribution, relatively little work seems to have been done in distinguishing between models (ii) and (iii) in general. However, in spite of this problem, the concept of heterogeneity is a useful one in the description of empirical point patterns, and model (iii) and associated distance distributions are discussed by Matérn [1971: see also the discussion by Paloheimo] and Bartlett [1975].

Lattice processes represent an extreme form of regularity and, as Diggle [1979a] points out, their principal use seems to be as a yardstick for the assessment of inferential procedures applied to regular patterns in general. However, a pattern can be reduced to counts on a lattice if a rectangular grid is placed over the population area, and each quadrat thus formed is approximated by its centre. In some situations the quadrats can be made small enough so that only a few contain more than a single object, and it is then reasonable to reduce the data to a binary (presence/absence) form. This quadrat reduction method has been widely used in geographical problems in which a frequency distribution is constructed giving the number of quadrats with 0, 1, 2, ... objects (Cliff and Ord [1973: 58], Rogers [1974]).

There is an extensive literature on lattice processes and the reader is refererred to Bartlett [1975: ch. 2] for references. The distribution of the distance from a randomly located point to the nearest object (i.e. to the nearest lattice-node) is given by Persson [1964] for the square lattice, by Holgate [1965a] for the equilateral triangle, and by Eberhardt [1967] for a regular polygon. However, a more useful model may be obtained if we add to the lattice structure a Poisson forest with density ρ per unit area. The resulting patterns range from extreme regularity when $\rho = 0$, to a Poisson forest as $\rho \to \infty$ (the effect of the lattice becoming "swallowed up"). Diggle [1975] gives some results for distance distributions from this type of model.

Another way of constructing a stochastic model from a lattice is by random thinning, that is, by randomly removing individuals from the lattice. Distribution theory for this model is given by Brown and Holgate [1974]. In practice, models exhibiting regularity do not seem to have much application to animal population studies, though they could arise in a species exhibiting strong territorial instincts, e.g. banner-tailed kangaroo rat mounds (Schroder and Geluso [1975]) and sparrowhawk nests (Newton *et al.* [1977]). However, such models are particularly useful in geography and forestry: for example, mortality through competition can produce regularity in a mature stand of trees (Warren [1972: 803]).

Inhibition processes arise naturally, for example, in the modelling of centres of circular non-overlapping objects. Examples of such processes are given by Matérn [1960], Diggle *et al.* [1976], Bartlett [1975: ch. 3] and Ripley [1977: under the title of "hard-core" models]. A more general class of processes which include both inhibition (regular) and contagious patterns is the class of Markov point processes introduced by Ripley and Kelly [1977]. For general surveys of spatial point processes see Lewis [1972], Bartlett [1975], Cliff and Ord [1975: applications to human geography] and Ripley [1977].

Although there is a wide variety of models for describing point patterns, little has been done in the way of providing goodness of fit procedures. However, the reader should refer to Diggle [1979a: section 5] and the rather theoretical paper by Ripley [1977]. Both articles give a number of practical examples, and Monte Carlo methods (cf. **1.3.10**) can be used fruitfully here.

Tests for randomness based on distance methods have already been discussed in **12.5.1**. However, if one uses all the nearest neighbour distances rather than a sample, then there is the problem of dependence. This point is discussed by Cliff and Ord [1975: 305–6], Diggle [1975] and, in particular, Brown and Rothery [1978]. A quick test for randomness is given by Ripley and Silverman [1978]: see also Ripley [1979] (Additional References).

For wildlife populations, other models for patchiness can be developed. Rougharden [1977], for example, lists a number of references and considers a model in which patchiness "is an inevitable consequence simply of dispersal in a fluctuating environment". Also there is a need for greater cross-fertilisation between geography and ecology, as the methods of Cliff and Ord [1973] and Royaltey *et al.* [1975] have considerable potential in ecology (e.g. Jumars *et al.* [1977]).

12.7 SINGLE MARK RELEASE

12.7.1 Petersen method

The Petersen method is discussed extensively in Chapter 3 and there are few further theoretical developments to add to that chapter. However, a method of estimating N when there are no recaptures is described by Bell [1974]. Using the hypergeometric model of **3.1.1**, the probability of zero recaptures ($m_2 = 0$ in equation (3.1)) is

$$P = (N - n_1)!(N - n_2)!/[N!(N - n_1 - n_2)!], \tag{12.44}$$

and Bell suggests solving $P = 0{\cdot}50$ for an estimate of N, and $P = 0{\cdot}025$ and $P = 0{\cdot}975$ for a 95 percent confidence interval. In practice, however, a Petersen experiment should be designed so that $m_2 \geqslant 10$.

Gaskell and George [1972] give a Bayesian modification of the Petersen estimate of the form $(n_1 n_2 + AB)/(m_2 + B)$, where A is a prior estimate of N and $B \approx 2$, when m_2 is small (say $m_2 < 10$). Bayesian methods are also discussed by Johnson [1977]. However the main difficulty with such an approach is the specification of a sensible prior distribution for N which will combine tractably with the likelihood function.

Finally we note that in fisheries it is not always possible to count all the marked fish. Instead the proportion of marked fish is estimated from the proportion, $\hat{p}$ say, of marked fish in a sample from the catch, and the number of marked in the catch is estimated by multiplying $\hat{p}$ by the number in the catch. Summing over all catches associated with a given mark–release then gives an estimate of m_2, the total marked recovered (Kimura [1976]).

12.7.2 Inverse sampling

On p. 118 we review several Petersen-type estimates based on an inverse (sequential type) sampling scheme in which one continues to take the second sample until a prescribed number of marked (or unmarked) have been recovered. Using a beta-Pascal prior distribution for N, and making certain assumptions about costs, Freeman [1973a] discusses these sequential sampling plans for the case of sampling without replacement from a Bayesian viewpoint and compares them with the optimal solution. Freeman [1973b] also compares his optimal solution with the optimal solution of another procedure based on sequential tagging with samples of size one (cf. p. 189). He concludes that sequential recapture with constant cost of sampling is better than sequential tagging, and this in turn is better than sequential recapture with increasing costs. However, this theory is of limited application as the difficulties of obtaining a random sample generally outweigh considerations of cost, which, in any case, are difficult to evaluate.

Instead of taking the second sample until a prescribed number of marked or unmarked are captured, an alternative scheme is to continue the second sample until N is estimated with a prescribed coefficient of variation (Kuno [1977]). Now, conditional on n_1 and n_2 being fixed, the modified Petersen estimate N^* (p. 60) has an estimated coefficient of variation C given by

$$C^2 = \frac{v^*}{(N^*)^2} \approx \frac{v^*}{(N^* + 1)^2} = \frac{(n_1 - m_2)(n_2 - m_2)}{(n_1 + 1)(n_1 + 1)(m_2 + 2)}.$$

However, when it comes to variance estimation there is little difference between treating n_2 as fixed or random (p. 62), so that the above formula for C^2 can also be used to give the approximate relationship between m_2 and n_2 for the sequential scheme. Thus, solving the above equation for the smallest root of m_2 we have an expression for m_2 in terms of n_2, namely

$$\begin{aligned} 2m_2 = {} & [(n_1 + n_2) + C^2(n_1 + 1)(n_2 + 1)] - \{[n_1 + n_2 + C^2(n_1 + 1)(n_2 + 1)]^2 \\ & - 4[n_1 n_2 - 2C^2(n_1 + 1)(n_2 + 1)]\}^{1/2}. \end{aligned} \tag{12.45}$$

This is the boundary line for sequential estimation and represents a curve with a negative second derivative rising from the origin towards the horizontal asymptote

$$m_2 = \frac{n_1 - 2(n_1 + 1)C^2}{1 + (n_1 + 1)C^2}. \tag{12.46}$$

(The above equations differ slightly from those of Kuno [1977] who uses the biased Petersen estimate of N, $\hat{N}$, and a biased variance estimate). For a given n_1, the sequential procedure consists of plotting, cumulatively, m_2 against n_2 and continuing sampling until the boundary curve is crossed. In practice one could programme a calculator to give the value of m_2 for a given n_2 on the boundary curve: the sampling stops when the observed value of m_2 equals or exceeds the calculated value.

12.7.3 Underlying assumptions

1 Catchability

Variable catchability is one of the major headaches in the use of capture–mark–recapture methods. In addition to the references on pages 81–2 we note that catchability can vary with age and sex (Gliwicz [1970]), size (Joule and Cameron [1974]), type of trap and its location (Barbehenn [1973]), type of bait (Buchalczyk and Olszewski [1971]) and environmental conditions (Chapman and Trethewey [1972], Baümler [1975], Grunwald [1975], Perry *et al.* [1977]); methods of comparing capture rates are given by Linn and Downton [1975]. Roff [1973b: 49–51] gives an extensive survey of experiments which have variable catchabilities and Smith *et al.* [1975: 37] list some of the factors affecting the catchability of small mammals. In fact, as social behaviour has a bearing on catchability, capture–recapture experiments can be used for studying social structures (Slade [1976]).

Following the work of Cross and Stott [1975], Bohlin and Sundström [1977] studied the effect of variable catchability on the Petersen and removal methods for electro-fishing. For a trout population they found that the probability of capture was related to mark status and that marked fish had a higher probability of capture. They then carried out a further experiment on a population of known size to determine whether the difference was due to an inherent variation in catchability, so that the more catchable ones tended to be marked (violation of assumption (b), pages 59, 87), or that marking itself affected catchability (violation of assumption (c), pages 59, 86). They concluded that the former explanation was more likely. In fact, some individuals had a low probability of capture, and this can lead to severe underestimation of N (p. 87).

2 Effects of tagging

As mentioned above, the process of catching and tagging, and even the tag itself, can seriously affect the subsequent catching of an animal. In an attempt to overcome this problem there is an increasing interest in trap design, as seen for example in such journals as the *Journal of Mammalogy* and the *Journal of Wildlife Management*. Automatic tagging devices help to eliminate the effect of handling (e.g. Roussel and Pichette [1974: moose]), and sometimes territorial animals can be "tagged" (i.e. identified) without capture. For example, Thompson and Gidden [1972] used visual data such as length and basking location to identify alligators. Other techniques used for tagging are micronized dusts which fluoresce in characteristic colours under ultraviolet (Crumpacker and Williams [1973: wild flies]), chemiluminescent tags for tracking bats and other nocturnal animals (Buchler [1976]) and freezebranding (Hadow [1972: mammals], Daugherty [1976: anurans]). Radio transmitters are now being used extensively for studying the movements of many species (cf. **12.1.1**(*2*)). As the transmitters become smaller and cheaper they will no doubt be widely used for mark–recapture experiments.

The method of tagging is very critical with regard to fish populations as all tags have some effect on fish (p. 83). For example, the adverse effects of jaw tagging are well-known (e.g. Warner [1971]), and fin clipping appears to reduce the survival rate (e.g. Nicola and Cordone [1973], Mears and Hatch [1976]). Clearly methods are needed which (i) minimise handling, (ii) have a minimal effect on behaviour and survival, (iii) are not easily lost and, for commercial applications, (iv) which are quick, for mass tagging. For this reason paints and dyes have been receiving greater attention recently and examples are: injecting acrylic paints (Lotrich and Meredith [1974]), fluorescent marking with tetracycline antibiotics (e.g. Odense and Logan [1974]), external marking with dry pigment propelled by compressed air (Andrews [1972], Rinne and Deacon [1973], Phinney [1974] and Englehardt [1977]), and staining (Rinne and Deacon [1973], Sebens [1976]). Tags which release a dye can be used for studying fish movements on a small scale (Leaman [1976]). The lack of satisfactory tagging methods has hampered studies of the population biology of many important fishes, for example the engaulids which constitute nearly 20 percent of the world fish catch. One tag suitable for a small fish like the anchovy consists of a tiny, non-reactive, stainless steel wire which can be implanted automatically and instantaneously (using a gun) with minimal handling of the fish except to position it for tagging. It is also magnetised so that the tagged fish can be separated from the others by a magnetic sensor (Leary and Murphy [1975] and, for a similar method, Bearden and McKenzie [1972]). Clearly all tagging methods need to be checked for their effect on the fish by, for example, retaining some of the fish in tanks (Dominy and Myatt [1973], Rinne and Deacon [1973], Phinney [1974], Leary and Murphy [1975], Marullo *et al.* [1976]).

A review of some recent tagging studies for marine mammals is given by Eberhardt *et al.* [1979]. In the case of fur seal pups the marked and unmarked could not be expected to mix at this stage of their life, so that randomness needed to be achieved through a random allocation of marks and locations (Chapman and Johnson [1968]). Mark–recapture estimation of population numbers has not worked very well for whales, and Chapman [1974a] lists three contributing factors: (1) it is very difficult to determine the number of whales properly tagged: the tags may not penetrate, may miss or may cause immediate mortality; (2) tagging is difficult and expensive so that the number tagged and the number recaptured are small; (3) since the tags are internal they are often overlooked by the processing ships.

3 Tag loss

The loss or illegibility of tags and marks is a serious problem, particularly with birds and fish, and has not received sufficient attention in the literature. Double tagging seems to be the simplest method of checking on tag loss, and

Hubert *et al.* [1976] give an estimate of the probability of losing both tags (assuming indistinguishable tags) which can be shown to be identical to the one given by Chapman *et al.* [1965: 340], Cormack [1968: 463], Caughley [1971] and derived on pages 95–6. An application of this estimate is given by Best and Rand [1975: their P_s is the same as $\tilde{\pi}$]. Hubert *et al.* [1976] also relate tag loss to the length of period before recapture. However, any of the estimates can be plotted as a function of time (pp. 281ff). We note the following alternative expressions for $\tilde{m}_2$:

$$\begin{aligned} \tilde{m}_2 &= m_{AB}/(1-\tilde{\pi})^2 \\ &= (m_C + 2m_{AB})/2(1-\tilde{\pi}) \qquad (12.47) \\ &= m_C + m_{AB})/(1-\tilde{\pi}^2). \end{aligned}$$

4 Non-reporting of tags

As mentioned on p. 97 an incomplete tag return leads to an overestimate of N. This problem of the non-reporting, or mis-reporting, of tags arises particularly when agencies must depend on commercial fishermen, sportsmen and hunters for the return of tags. Some of the biases which can arise (cf. p. 360) have been studied in detail by Sen with regard to Canadian water-fowl surveys. For instance, he studied the effect of response errors (Sen [1972, 1973a, c]) and found that they can be large compared to the sampling and non-response errors combined. The tendency is for hunters to inflate their hunting successes through pride, prestige or memory lag, so that estimates based on mailed questionnaires can have large positive biases. Those hunters who bagged nothing were the worst offenders. Bias can also arise through an imperfect sampling frame (Sen [1970b, 1972]). Other problems associated with migrating bird surveys are also discussed in the review by Sen [1976]: he uses a sampling technique of combining data from successive surveys to increase precision (Sen [1971b, c, 1973b] and Sen *et al.* [1975]). Similar problems arise in trying to calculate harvest estimates for a sports fishery from a postal survey (Carline [1972]): again there is a tendency for numbers to become inflated and for unsuccessful fishermen not to return their cards.

12.8 MULTIPLE RECAPTURE METHODS

12.8.1 Conditional multinomial distributions

We recall from p. 131, equation (4.1), the following multinomial model for a Schnabel census, namely

$$f(\{a_w\}) = \frac{N!}{\prod_w a_w!(N-r)!} Q^{N-r} \prod_w P_w^{a_w}, \qquad (12.48)$$

where a_w is the number of animals with capture history w, P_w is the probability of being in that category, $Q = 1 - \Sigma_w P_w = 1 - P$, say and r is the total number of *different* animals caught in the experiment. The probability function for r is binomial with parameters N and P, so that the joint distribution of $\{a_w\}$, conditional on r (the sufficient statistic for N), is

$$f(\{a_w\}|r) = \frac{r!}{\prod_w a_w!} \prod_w \left(\frac{P_w}{P}\right)^{a_w}. \tag{12.49}$$

Suppose $P_w = P_w(\boldsymbol{\theta})$ is a function on an unknown vector parameter $\boldsymbol{\theta}$, and let $\hat{\boldsymbol{\theta}}_u$ and $\hat{N}_u$ maximise (12.48) (the unconditional maximum likelihood estimates). One method of finding maximum likelihood estimates is to use a conditional approach (Fienberg [1972: 593]; namely find $\hat{\boldsymbol{\theta}}_c$, the maximum likelihood estimate of $\boldsymbol{\theta}$ for the conditional distribution (12.49), and then set

$$\hat{N}_c = [r/\hat{P}_c],$$

where $\hat{P}_c = P(\hat{\boldsymbol{\theta}}_c)$, and $[r/\hat{P}_c]$ is the greatest integer not exceeding $r/\hat{P}_c$. For most practical purposes $\hat{N}_c$ and $\hat{\boldsymbol{\theta}}_c$ are almost identical to $\hat{N}_u$ and $\hat{\boldsymbol{\theta}}_u$ (Darroch [1958: 357]). In fact $\hat{N}_u \leqslant \hat{N}_c$, and the asymptotic distributions of $\hat{N}_c$ and $\hat{\boldsymbol{\theta}}_c$ are the same as for the unconditional estimates (Sanathanan [1972]). We can therefore make use of this asymptotic equivalence and find the asymptotic variance–covariance matrix of $\hat{\boldsymbol{\theta}}_u$. Thus, applying standard large-sample maximum likelihood theory to (12.49), we find the asymptotic variance–covariance matrix conditional on r, and then make this matrix unconditional by using the fact that r is binomial.

On p. 14 we noted that the conditional distribution (12.49) can also be used for testing the hypothesis $P_w = P_w(\boldsymbol{\theta})$.

12.8.2 Fixed or random sample sizes

For the model considered in section 4.1.2 i.e. $\boldsymbol{\theta}' = (p_1, p_2, \ldots, p_s)$ and fixed sample sizes n_i, the appropriate distribution for the Schnabel census is (equation (4.3))

$$f(\{a_w\}|\{n_i\}) = \frac{N!}{\prod_w a_w!(N-r)!} \prod_{i=1}^{s} \binom{N}{n_i}^{-1}, \tag{12.50}$$

and the maximum likelihood estimate, $\hat{N}$, is the appropriate root of equation (4.4). As r is sufficient for N, $\hat{N}$ can also be obtained by maximising $f(r|\{n_i\})$, which takes the same form as (12.50) with respect to the terms involving N. Berg [1976], using a theorem in a previous paper (Berg [1974]) concerning unbiased estimation for a factorial series distribution like (12.50), derives a ratio estimate of N, $\tilde{N}$ say, which is minimum variance unbiased provided $N \leqslant \sum_i n_i$. This estimate, which is the same as that given by Pathak [1964], is difficult to compute as the terms in the numerator and denominator tend to grow rapidly with increasing arguments. To get round this computational problem Berg [1976] derives a useful recurrence relation for $\tilde{N}$ which is independent of the rapidly growing terms. He also proves that $\tilde{N} \leqslant \hat{N}$, which is a strict inequality for $\max n_i < r < \sum_i n_i - 1$: the inequality becomes less sharp as r approaches $\sum_i n_i$. A variance estimate of $\tilde{N}$ is given by Berg [1974] which may be calculated at the same time as $\tilde{N}$. Berg also gives a numerical example based on the cricket frog data on p. 135. When all the n_i are unity, $\tilde{N}$ reduces to the ratio of two Stirling numbers of the second kind; a case discussed in detail by Berg [1975] along with an example. If $s = 2$, $\tilde{N}$ reduces to the unbiased version of the Petersen estimate (N^* on p. 60).

Lee [1972] discusses the above model and demonstrates that the approximation for the variance of r, namely $\sigma^2(N)$ on p. 134, is close to the exact variance $[N - \rho(N)]\,[\rho(N) - \rho(N-1)]$. In fact the approximation can be obtained by replacing the difference $\rho(N) - \rho(N-1)$ by the derivative $d\rho(N)/dN$. She also considers the problem of designing a Schnabel census and gives several useful graphs for determining the number of samples to give a prescribed accuracy for $\hat{N}$.

The maximum likelihood estimate of N is essentially unchanged if the n_i are now regarded as random variables. Thus the unconditional maximum likelihood estimates for (4.2) are close to $\hat{N}$ and $\hat{p}_i = n_i/\hat{N}$, where $\hat{N}$ is again the appropriate root of (4.4) (though according to Otis *et al.* [1978: 25] this root may differ by more than unity from the true maximum likelihood estimate and they suggest maximising the likelihood function directly). We note that Darroch [1958: 357–9] and Wittes [1974] use the conditional approach of **12.8.1** to find the asymptotic variance–covariance matrix of the $\hat{p}_i$.

12.8.3 Multiple-record systems

Wittes [1974] applied the above theory to epidemiological data in which the target population of size N is characterised by a certain rare trait. Sampling from the population at large is usually prohibitively expensive, so that one procedure is to merge existing, but incomplete lists, of the members of the target population. We thus have a Schnabel census in which "caught in the ith sample" corresponds to "being on list i", and n_i is the number on the ith list. In addition to estimating $\hat{N}$, Wittes is interested in the estimate $\hat{p}_i$ $(i = 1, 2, \ldots, s)$, which is a measure of the relative efficiency of the ith recording system, and $\hat{P} = r/\hat{N}$: she also gives their asymptotic variances. Wittes discusses design questions such as "How many lists should be merged?" or "Under what conditions is it appropriate to add in another list?". Further applications of the Schnabel or CMR (Capture–Mark–Recapture) technique to epidemiological data may be found in Wittes and Sidel [1968], Lewis and Hassanein [1969], and Wittes *et al.* [1974]. The examples given in these papers and Wittes [1974] deal with (i) the surveillance of infectious diseases reported by physicians, nurses and bacteriological laboratories, (ii) the estimation of the total number of patients receiving methicillin (a synthetic penicillin) as reported by nurses, medication sheets and pharmacists, and (iii) infants with a specific abnormality reported by obstetric clinics, hospitals, departments of public and mental health, and public and private schools.

The above technique of merging several lists can be conveniently called the Multiple-Record System or MRS (El-Khorazaty *et al.* [1976]). A special case, the DRS or Dual Record System, has evidently been used fairly extensively in the last

last 30 years to adjust for gaps in the recording of vital events and to estimate population growth rates (cf. Carver [1976]). Chandrasekar and Deming [1949] examined the problem theoretically and their estimate of N is simply $\hat{N}$, the Petersen estimate. Their approach was extended to situations involving three or more lists by Chakraborty [1963] and Das Gupta [1964]. Recently the above capture–recapture methods were used by El-Khorazaty [1975] and El-Khorazaty and Sen [1976] for estimating numbers of events like births and deaths in a human population from incomplete lists. They developed models for the two and three list situations which allowed for dependence between lists, for example in the case of two lists $p_{12} \neq p_1 p_2$. These models may have applications in ecology where unequal catchability, due for example to mark or size selectivity, is analogous to dependence between lists. For a detailed survey of this work and other methods for estimating N from a DRS or MRS, the reader is referred to El-Khorazaty *et al.* [1977; though their equation (4.9) is incorrect].

12.8.4 Loglinear models

Another method of estimating N from a Schnabel census which shows some promise is to use the theory of incomplete contingency tables (Fienberg [1972], Haberman [1974], Bishop *et al.* [1975], El-Khorazaty *et al.* [1976], Koch *et al.* [1976, 1977]). Since there are s samples, the number of different capture histories is 2^s which can be arranged in an s-dimensional contingency table. However, the number of animals in one of the cells, namely the $N - r$ animals not caught at all, is unobservable, so that the contingency table is incomplete. By fitting a suitable log-linear model to the rest of the table, N can be estimated. A variety of models can be fitted depending on how many interaction terms are included. These interaction terms correspond to dependencies between various samples, and for brief readable discussions of the problem see Bishop *et al.* [1975: ch. 6] and Cormack [1979]). When the samples are independent so that all the interactions are zero, the estimate of N obtained is simply $\hat{N}$ of (4.4).

12.8.5 Sequential Schnabel census

On p. 188 we discussed the sequential Schnabel census and mentioned that Goodman [1953] derived the minimum variance unbiased estimate of N as the ratio of two determinants which are generally of high order. Recently Berg and Ericson [1977] obtained a useful recurrence relationship for calculating this estimate and a non-negative unbiased estimate of its variance, which avoids the problem of large determinants. They also examined the large sample result given by Goodman that n^2/N is approximately χ^2_{2L}, where $n = \Sigma_i n_i$ and found that this approximation gave poor results for their simulated examples. Evidently the approximation should be used with great care, and only for very large populations.

When samples are of size 1 (e.g. $n_i = 1$) then the above estimate of N is also a ratio of two Stirling numbers of the second kind ($\hat{N}_6$ on p. 189). Although these numbers are also difficult to handle as they grow rapidly with increasing argument, a useful recurrence relationship which is essentially the same as the one mentioned above and which is not affected by rapid growth, has been given by Berg [1975]. If the animals have a variable catchability (called model M_h in the next section) so that P_j is the (conditional) probability of drawing the jth animal ($j = 1, 2, \ldots, N$), given that one animal is caught, then Holst [1971] proves that as $N \to \infty$, $A_N s^2/N$ is asymptotically distributed as χ^2_{2L} where $A_N = N \sum_{j=1}^{N} P_j^2$. When we have equicatchability, $P_j = 1/N$, $A_N = 1$, and we have Goodman's result, with $n = s$. Since $\Sigma_j P_j = 1$ and $\Sigma a_j^2 \; \Sigma b_j^2 \geqslant (\Sigma a_j b_j)^2$, we see that $A_N \geqslant 1$, so that variable catchability leads to the underestimation of N.

Samuel [1968, 1969] considered four different stopping rules for the above sequential scheme (pages 191–3) and these are compared from a Bayesian viewpoint by Freeman [1972].

12.8.6 Further models

1 Modelling catchability

More recently Otis *et al.* [1978], utilising the results of Burnham [1972] and Pollock [1974], presented a comprehensive package of models for finding an estimate of N, and its variance estimate, under a wide variety of experimental conditions (see also Pollock [1975a] for a helpful elementary discussion). They presented 6 models which allow for three types of variation in capture probabilities: model M_t given by (4.2) (variation with trapping occasion or time, e.g. unequal effort, trapping at different times of the day, weather changes), M_b (variation by behavioural responses, e.g. trap shyness or trap addiction), M_h (variation by individual response or heterogeneity, e.g. social dominance, size selectivity in electro-fishing), and various combinations M_{tb}, M_{bh}, M_{th} and M_{tbh}. They also include the model M_o in which there is no variation, and a generalised removal model in which removal corresponds to tagging (p. 323).

If p_{ij} is the probability that the ith animal ($i = 1, 2, \ldots, N$) is caught in the jth sample ($j = 1,2, \ldots, s$), and we can assume that the animals are independent of one another as far as catching is concerned, then the likelihood function is $\prod_i \prod_j p_{ij}^{x_{ij}} (1 - p_{ij})^{(1-x_{ij})}$, where $x_{ij} = 1$ if the ith animal is caught in the jth sample and $x_{ij} = 0$ otherwise. The various models can now be described mathematically as follows: M_o, $p_{ij} = p$; M_t, $p_{ij} = p_j$; M_b, $p_{ij} = p$ for any first capture and $p_{ij} = c$ for any recapture; M_h, $p_{ij} = p_i$ ($i = 1, 2, \ldots, N$) where the p_i are a random sample from some density $f(p)$ (this model is discussed in **12.8.7**(2)); M_{bt}, $p_{ij} = p_j$ ($j = 1, 2, \ldots, s$) for any first capture and $p_{ij} = c_j$ ($j = 2, 3, \ldots, s$) for any recapture; M_{th}, $p_{ij} = \phi_i \theta_j$; M_{bh}, $p_{ij} = p_i$ for any first capture and $p_{ij} = c_i$ for any recapture; and M_{tbh}, $p_{ij} = P_{ij}$ for any first capture and $p_{ij} = c_{ij}$ for any recapture. The authors use an algorithm described in Brent [1973] to calculate the unconditional maximum likelihood estimate $\hat{N}$ for the models M_o (p. 164), M_t (discussed in **4.1** and **12.8.2** above), M_b (Pollock [1974] and p. 312 with n_i replaced by u_i) and M_{bh}; the model M_{bh} being regarded as the most realistic and useful one of the models. The method used for analysing M_{bh}, which the authors call the generalised removal method (cf. **12.9.3**) and includes M_b as a special case, can also be applied to removal experiments as the estimate of N is based on just the u_j, the animals caught for the first time in the jth sample ($j = 1, 2, \ldots, s$). Suitable estimation procedures are not yet available for M_{bt}, M_{th} and M_{tbh}, but a generalised jackknife estimator due to Burnham [1972] (cf. Burnham and Overton [1978, 1979] is available for M_h: this estimator is described below (equation (12.53)). Where an estimation procedure for a model is available, Otis *et al.* [1978] give an approximate 95 percent confidence interval for N in the form $\hat{N} \pm 1{\cdot}96\hat{\sigma}_N$, where $\hat{\sigma}_N$ is the estimated large-sample standard deviation of $\hat{N}$. It appears that this confidence interval is quite satisfactory (i.e. has a true confidence level of at least 90 percent) when the correct model is used and the interval is not too wide. If the average probabilities of capture on each sampling occasion are less than 0·1 then $\hat{N}$ is generally significantly biased and the calculated confidence interval tends to be impracticably wide. In this case the interval still provides relevant information, though about the failure of the experiment rather than about the population size. It is generally recommended that the average probability be greater than 0·2 and $s \geqslant 5$ for reasonably short confidence intervals, unless N is very large. If the average probability of capture is too small, say less than 0·1, it may be a waste of time carrying out the experiment.

2 Comparing estimates

Using simulation, Otis *et al.* [1978] give a detailed discussion of the robustness of each estimate and its associated confidence interval. They show that $\hat{N}_0$ (the subscript for $\hat{N}$ denotes the model used) is very sensitive to departures from model M_0, in particular heterogeneity, and confirm Darroch's comment (p. 164) that there is little to be gained by using M_0 instead of M_t when only time-specific changes in the probability of capture are likely to be present. The estimate $\hat{N}_t$ is also non-robust, tending to be negatively biased when there is heterogeneity, and positively or negatively biased if the animals tend to become trap shy or trap-addicted, respectively. Burnham and Overton [1969: 53] showed that several alternatives to $\hat{N}_t$ such as the Schnabel estimate (N' of p. 139) are also sensitive to heterogeneity. The estimate $\hat{N}_b$ exhibits the same non-robustness to heterogeneity as $\hat{N}_t$ and also appears to be sensitive to significant changes in the probability of capture over time. The jackknife estimator $\hat{N}_h$ seems to be the most robust of the estimators presented, though at times the confidence interval may be wide and unreliable. The removal type estimate $\hat{N}_{bh}$ seems reasonably robust, but the confidence interval based on it is unreliable as the variance estimator appears ill-behaved and the distribution of $\hat{N}_{bh}$ is non-normal.

In addition to the various estimation procedures, Otis *et al.* [1978] provide a battery of 7 tests along with a discriminant analysis procedure for selecting the "best" model. However the problem remains of what to do when one of the models without an adequate estimation procedure is clearly better than the rest. In testing the goodness of fit of M_t, the authors prefer to use Leslie's [1958] test (p. 161) rather than the improved multivariate version due to Carothers [1971] (cf. 13.1.7) as the latter requires some arbitrary trimming of the data and is therefore difficult to automate-simulate. We note that Leslie's test is a test of M_t and not, as Roff [1973b] appears to suggest, of M_0. The sensitivity of Leslie's test needs further investigation.

The linearity of Schumacher and Eschmeyer's regression line (p. 141) will provide a check, but only a rough check, on the validity of model M_t. However, Roff [1973b] has found that the plot may appear linear even if N changes by 10 percent between each sampling occasion, that is by 50 percent over six occasions. The plot will also tolerate a certain amount of heterogeneity of catchability.

3 *Experimental design*

The first consideration in any planned Schnabel experiment is whether the population is closed or not. Some tests of closure have been given by Robson and Flick [1965] and Robson and Regier [1968] (utilising age or size data cf. pages 72–81), Pollock *et al.* [1974: tests for recruitment and or mortality based essentially on model M_t] and Otis *et al.* [1978: utilising the times between the first and last capture of animals caught at least twice]. However, as pointed out by Otis *et al.*, these tests will generally have little chance of rejecting closure unless samples are large and there is a marked departure from total closure. Also these tests are often confounded with behavioural response to capture (e.g. trap shyness). For example, the test that Otis *et al.* propose, due to Burnham and Overton [1978], does not seem to be affected by heterogeneity (M_h) or time variation (M_t) but it evidently rejects strongly when model M_b is true. Because of these problems of interpretation Otis *et al.* suggest that the best evidence for a closed population is biological rather than statistical and that the experiment should be designed so that the closure assumption is satisfied as closely as possible. For example, trapping should not be carried out during known migration times or when recruitment may occur due to, say, juveniles becoming catchable. Also variable weather and trapping conditions should be avoided where possible, and the time taken for the overall experiment should be as short as possible. It has been suggested in the past that trap locations should be randomised on each sampling occasion to reduce possible trap response, but this may be difficult to carry out in practice and further studies are needed on this question.

In planning an experiment there are a number of important questions relating to design such as the number of traps, trap spacing and the number of samples s. A helpful discussion on such practical matters is given by Otis *et al.* [1978] and should be read carefully. They also discuss data anomalies such as multiple captures in a trap, accidental deaths, etc., though the question of accidental deaths needs further investigation. Two of the recommendations which should be underlined are $N > 25$ and $\bar{p} > 0{\cdot}1$, where $\bar{p}$ is the average probability of capture in a sample. Although the authors do not guarantee that the data can be satisfactorily analysed if these conditions are met, they believe that precise estimation will not be achieved without meeting these conditions.

12.8.7 Frequency of capture of models

Models based on frequency of capture are discussed by Wilbur and Landwehr [1974], though their use of the terms "number of captures" of an animal (x, say) and the "number of recaptures" ($x - 1$) is confusing. Strictly speaking the models considered in **4.1.6** only apply to captures and not recaptures, though it is tempting to use the first captures of animals as simply a means of identifying a known population and then say that the number of "captures" for a member of this population is $x - 1$. However, this procedure is not valid: such a population with a known zero class should be established before the experiment is carried out, though it may be actually identified after the experiment (e.g. Wilbur and Landwehr [1974: table 1]. This point is also made on p. 162.

1 Truncated models

If the population size is unknown, and model M_0 holds, we can use the truncated binomial or Poisson models described on pages 169–70. However, if there is heterogeneity or trap response (models M_h, M_b and M_{bh}), several truncated models such as the geometric (p. 170), negative binomial (p. 174) and Skellam's model (p. 177) have been proposed. However, these models should be used with extreme care as estimates of $N - r$, the zero frequency-of-capture class, can be quite misleading for two reasons (Wilbur and Landwehr [1974]). Firstly, as was mentioned in **12.6.1** and on p. 174, many of the generalised distributions such as the negative binomial can be generated by a variety of biological mechanisms and it is impossible to determine from the distribution which is the correct mechanism. This means that the distributions are statistical descriptions rather than models and there is no reason why such a distribution should be extendable to an unobserved category. Secondly, and what is more serious, it is often possible to find several different distributions which fit the frequency data well and yet give rise to widely different estimates for the zero-frequency class. Examples of this problem are given by Carothers [1973b], Roff [1973b], Wilbur and Landwehr [1974] and Efron and Thisted [1976]. For example, Carothers [1973b: 146] gave data which were fitted satisfactorily by both the truncated binomial and geometric models, but both models gave poor estimates of the known population size. Roff [1973b] showed that the goodness of fit test for the truncated Poisson was unreliable since the model can still give a satisfactory fit even when there is variable catchability (varying according, for example, to a compound Poisson distribution) and changing population size. To a lesser extent he found that the truncated geometric distribution has the same ability to fit recapture data, and at low probabilities of capture (say 0·1) it was difficult to distinguish between the two models on the basis of a goodness of fit test. In his example the estimate from the geometric model was about twice that from the Poisson.

Efron and Thisted [1976], in considering the allied problem of determining how many different words an author knows, given a sample of the author's writing, showed convincingly that different models which fit the observed data equally well give totally different estimates of the population size, even when $r = 31\,534$ and $n = 884\,647$ $(= \Sigma\, n_i)$. The authors provided a rough lower bound for estimating a certain parameter $N(t)$ as $t \to \infty$ (this limiting case corresponding to our N above), and no upper bound.

Clearly the parametric truncated models of **4.1.6** are open to question and there is serious doubt as to whether these models should be used at all. What is needed is a more robust non-parametric model such as the following.

2 *Burnham's non-parametric model*

Models allowing for heterogeneity (M_h) can be generated by allowing the probability of capture p of the binomial model (or the parameter λ in the corresponding Poisson approximation) to vary according to some distribution with density function $f(p)$. Thus the ith animal ($i = 1, 2, \ldots, N$) has probability p_i of being caught in a sample and $\{p_1, p_2, \ldots, p_N\}$ represents a random sample from $f(p)$. If f_x ($x = 0, 1, \ldots, s$) is the number of animals caught x times, then Burnham and Overton [1978, 1979] show that the joint distribution of the f_x is

$$g(f_0, f_1, \ldots, f_s) = \frac{N!}{\prod_{x=0}^{s} f_x!} \prod_{x=0}^{s} (\pi_x^{f_x}) \tag{12.51}$$

where

$$\pi_x = \int_0^1 \binom{s}{x} p^x (1-p)^{s-x} f(p)\, dp.$$

Evidently when this model was first considered, the intent was to let $f(p)$ be a beta distribution with parameters α and β, say, and then use a standard parametric approach using maximum likelihood. In this case the probability distribution $\{\pi_x\}$ is simply Skellam's model (p. 177, equation (4.34)). However, this parametric approach was investigated by Burnham [1972] and found to be unsatisfactory. He also considered the case $\alpha = 1$ and, by equating the random variables $r \left(= \sum_{x=1}^{s} f_x\right)$ and $n \left(= \Sigma n_i = \sum_{x=0}^{s} x f_x\text{, the total number of captures}\right)$ to their expected values, obtained the moment estimate

$$\hat{N} = (1 - s^{-1})r/(1 - \bar{x}^{-1}), \tag{12.52}$$

where $\bar{x} = n/r$. Apart from the term $(1 - s^{-1})$ this is the same as the estimator based on the truncated geometric distribution (cf. p. 171, where the effect of truncation on the right can be ignored). In the Petersen situation, i.e. $s = 2$, $\hat{N}$ of (12.52) reduces to an estimate proposed by Eberhardt (denoted by $\hat{N}_\beta$ on p. 178). Unfortunately the beta distribution does not include the special case of p constant for all animals (i.e. $\alpha = \infty$, $\beta = \infty$). In this case we have, for large N, $E[\hat{N}_\beta] \approx N(2 - p)$, which is unsatisfactory.

If no assumption is made about $f(p)$, Burnham [1972] has shown that the maximum likelihood estimate of N is r, the number of different animals caught. Using r as a "naïve estimate, Burnham and Overton [1978] give a generalised jackknife statistic (cf. 1.3.9) based on r. Assuming that

$$E[r] - N = a_1 s^{-1} + a_2 s^{-2} + \dots ,$$

where $a_1, a_2, \dots$ do not depend on s, they give a "kth order" estimate of the form

$$\hat{N}_{Jk} = \sum_{x=1}^{s} a_{xk} f_x = r + \sum_{x=1}^{s} (a_{xk} - 1) f_x \qquad (12.53)$$

with large-sample variance estimator

$$v[\hat{N}_{Jk}] = \sum_{x=1}^{s} a_{xk}^2 f_x - \hat{N}_{Jk}.$$

The estimate $\hat{N}_{Jk}$ removes the bias terms up to and including $a_k s^{-k}$, and the authors give a table for the formula of $\hat{N}_{Jk}$ for $k = 1, 2, \dots, 5$. They also propose a sequential testing procedure for determining which value of k should be used. A goodness of fit test for M_h (i.e. $p_{ij} = p_i$) is given by the authors and is described in Otis *et al.* [1978: 117]. We note that $\hat{N}_{Jk}$ (previously denoted by $\hat{N}_h$) has already been compared with other estimates in **12.8.6**(*2*). It appears to be reasonably robust against time trends, provided s is sufficiently large (say $s \geqslant 5$) and very few members, if any, of the population are uncatchable. Further details, and the results of a simulation study, are given in Burnham and Overton [1979].

3 Waiting time models

Even if there is no heterogeneity of capture, frequency of capture methods are often plagued by two further problems: (1) the possibility that the probability of capture changes once an individual is caught for the first time (model M_b instead of M_0), and (2) the presence of transient or migrant animals passing through the area which are generally not caught more than once. If there are no migrants then (1) can be avoided by considering the times to first capture only. For example, if the probability of capture for the first time is constant and equal to $p(= 1 - q)$ we have the geometric model

$$f(y) = q^{y-1} p/(1 - q^s), \quad y = 1, 2, \dots, s, \qquad (12.54)$$

where $f(y)$ is the probability than an animal is caught for the first time in the yth sample, given that it is caught at least once in s samples. If $\bar{y}$ is the sample average, the maximum likelihood estimate $\hat{p}$ can be found using the method of p. 171. Also if r is the number of different animals caught then, since r is binomial,

$$E[r] = N(1 - q^s),$$

and N can be estimated by

$$\hat{N} = r/(1 - \hat{q}^s). \qquad (12.55)$$

Another way of avoiding (1) is to use the removal method for which animals are regarded as "removed" once they have been captured and tagged, provided the traps do not become too saturated with marked animals so that p is no longer constant.

The problem of migrants has been considered by several authors (cf. pages 181–2) but only for the case when the probability of capture remains unchanged when an animal is captured for the first time (model M_0). When both migrants and change in capture probability are possibilities, MacArthur and MacArthur [1974] has given a non-linear regression model, and Manly [1977a] uses the method of moments, for estimating the total number of residents and the number of transients. Manly's method, couched in the framework of mist netting birds, assumes that birds are caught and released almost continuously, so that the time of first capture of an individual is known accurately. Thus, assuming that uncaptured residents are captured at random at a constant mean rate θ, and the experiment lasts one unit of time, then the density function of the time to first capture, given that the individual is captured during the experiment, is

$$f(t) = \theta e^{-\theta t}/(1 - e^{-\theta T}), \quad 0 \leqslant t \leqslant T, \tag{12.56}$$

where $T = 1$. The first and second moments about the origin for this distribution are

$$\mu_1(\theta) = \theta^{-1} - e^{-\theta}/(1 - e^{-\theta})$$

and

$$\mu_2'(\theta) = 2\mu_1(\theta)/\theta - e^{-\theta}/(1 - e^{-\theta}).$$

As far as migrants are concerned, it may be reasonable to assume that they are equally likely to be captured at any time during the experiment. In this case the first capture time of a migrant is uniform on [0, 1] with mean $\frac{1}{2}$ and second moment about the origin of $\frac{1}{3}$. Now suppose there are r animals captured with first capture times $t_1, t_2, \dots, t_r$ and let π be the probability that an animal is a resident, given that it is caught. Then, if $m_1 = \bar{t}$ and $m_2 = \Sigma t_i^2/r$, we have

$$E[m_1] = \pi\mu_1(\theta) + (1 - \pi)/2$$

and

$$E[m_2] = \pi\mu_2'(\theta) + (1 - \pi)/3.$$

Replacing $E[m_i]$ by m_i and solving for θ and π gives us moment estimates $\hat{\theta}$ and $\hat{\pi}$, say. An estimate of R, the total number of residents, is then given by

$$\hat{R} = \hat{\pi}r/(1 - e^{-\hat{\theta}}).$$

Manly [1977a] gives a graphical method for finding $\hat{\theta}$, approximate variances and covariances of $\hat{\pi}$, $\hat{\theta}$ and $\hat{R}$, and a numerical example.

The above method can also be used if trapping is not continuous but takes place at times $t = 1, 2, \dots, s$ so that the time to first capture is given by the geometric model (12.54). The distribution of times for migrants is now the discrete uniform distribution given by

$$g(x) = \frac{1}{s}, \quad x = 1, 2, \dots, s,$$

which has mean $(s + 1)/2$ and second moment about the origin of $(s + 1)(2s + 1)/6$. Thus we now have the equations

$$E[m_1] = \pi\mu_1(p) + (1 - \pi)(s + 1)/2$$

and

$$E[m_2] = \pi\mu_2'(p) + (1 - \pi)(s + 1)(2s + 1)/6,$$

where $\mu_1(p)$ and $\mu_2'(p)$ are readily obtained from the probability generating function associated with (12.54). From these equations we can obtain moment estimates of p and π, and R is estimated by

$$\hat{R} = \hat{\pi}r/(1 - \hat{q}^s).$$

12.9 USING KNOWLEDGE OF SAMPLING EFFORT

12.9.1 Robustness of removal method

In 7.2 we discussed in some detail a sampling procedure, called the removal method, in which s samples of size n_i $(i = 1, 2, \ldots, s)$ are each removed from the population. The same trapping or catching effort is used for each sample and, on this basis, it is assumed that $p_i = p$ $(i = 1, 2, \ldots, s)$, where p_i is the probability of capture in the ith sample. However, for many populations the probability of capture varies from individual to individual, and this heterogeneity can lead to a serious underestimation of N the population size. Therefore suppose that the probability of capture of the jth animal P_j $(j = 1, 2, \ldots N)$ is the same for all s samples, and that the P_j represent a random sample from a catchability distribution with density function $g(p)$. Then arguing as on p. 316 it can be shown that the joint distribution of the n_i is

$$f(\{n_i\}) = \frac{N!}{\sum_{i=1}^{s} n_i!\,(N - n)!}\,\pi_1^{n_1}\,\pi_2^{n_2}\,\ldots\,\pi_s^{n_s}\,(1 - \pi)^{N-n} \tag{12.57}$$

where $\pi = \sum_i \pi_i$, $n = \sum_i n_i$ and

$$\pi_i = E[P(1 - P)^{i-1}]. \tag{12.58}$$

Using (12.57) and (12.58) for the cases $s = 2$ (p. 322) and $s = 3$ (p. 316), it was shown that the maximum likelihood estimate of N, based on the model $P_j = p$, is generally insensitive to variable P_j, provided there are no animals with a very low probability of capture. As s increases, the effect of these hard-to-catch animals becomes more serious. This problem of underestimation has been demonstrated, for example, by Cross and Stott [1975] and Bohlin and Sundström [1977] in the case of electro-fishing.

Variable catchability has a similar, though probably smaller, effect on Hayne's regression model (p. 325) $E[n_i|x_i] = p(N - x_i)$, where $x_i = \sum_{j=1}^{i-1} n_j$, the cumulative catch. Suppose that $\bar{p}_i$ is the average conditional probability of capture in the ith sample for those animals not previously captured. Then analogous to (7.1), we have

$$f(\{n_i\}) = \prod_{i=1}^{s} \binom{N - x_i}{n_i} \bar{p}_i^{n_i}(1 - \bar{p}_i)^{N-x_{i+1}}$$

which leads to (12.57) with $\pi_1 = p_1$ and

$$\pi_i = (1 - \bar{p}_1)(1 - \bar{p}_2) \ldots (1 - \bar{p}_{i-1})\bar{p}_i, \quad (i = 2, 3, \ldots, s). \tag{12.59}$$

Thus

$$E[n_i|x_i] = \bar{p}_i(N - x_i), \tag{12.60}$$

where $\bar{p}_i$ can be evaluated by solving (12.59) using (12.58). For example, if $g(p)$ is uniform on $[c, d]$ $(0 \leqslant c \leqslant d \leqslant 1)$ and $w = c/d$ (cf. p. 316) then, for $s = 3$,

$$\bar{p}_1 = \tfrac{1}{2}d(1 + w)$$

$$\bar{p}_2 = \frac{\tfrac{1}{2}d(1 + w) - \tfrac{1}{3}d^2(1 + w + w^2)}{1 - \tfrac{1}{2}d(1 + w)}$$

and

$$\bar{p}_3 = \frac{\tfrac{1}{2}d(1 + w) - \tfrac{2}{3}d^2(1 + w + w^2) + \tfrac{1}{4}d^3(1 + w)(1 + w^2)}{1 - d(1 + w) + \tfrac{1}{3}d^2(1 + w + w^2)}.$$

Thus if $d = 0{\cdot}5$ and $w = 0$, so that $[c, d] = [0, 0{\cdot}5]$, then $\bar{p}_1 = \frac{1}{4}$, $\bar{p}_2 = \frac{2}{9}$ and $\bar{p}_3 = \frac{11}{56}$, which are not as different as might be expected.

Suppose that the probability of capture is the same for all animals but there are fluctuations in the sampling effort, so that p_i, the probability of capture in the ith sample, is now a random variable with mean p. Then, setting $p_i = p + u_i$ (additive fluctuation) or $p_i = p(1 + v_i)$ (multiplicative fluctuation) in $n_i = p_i(N - x_i) + e_i$, where $E[e_i|x_i] = 0$ and e_i is independent of u_i or v_i, we obtain $n_i = p(N - x_i) + w_i$, where $E[w_i|x_i] = 0$. Thus we still have a straight line, but the error term is now inflated from e_i to w_i, so that regression estimates will have greater variances. Hayne's method, which consists of fitting the straight line $y = p(N - x)$ and extrapolating to $y = 0$, is widely used for investigating fresh-water fish populations (e.g. Mann [1971], Le Cren *et al.* [1977]). If the regression is markedly curved then a possible procedure is to fit a low-degree polynomial, say $y = \alpha + \beta_1 x + \beta_2 x^2$ and, if β_2 is significant, extrapolate the curve to $y = 0$. For example, if $\hat{\alpha}$, $\hat{\beta}_1$ and $\hat{\beta}_2$ are the least squares estimates of the regression parameters, then an estimate of N is the appropriate root of $\hat{\alpha} + \hat{\beta}_1 x + \hat{\beta}_2 x^2 = 0$.

The above methods are frequently applied to experiments in which $s = 3$ and even $s = 2$ (p. 315), though in the latter case it is not possible to test for a constant probability of capture. However, in designing a removal experiment, Otis *et al.* [1978] suggest that s should be at least 4. This makes good sense, though three samples might be satisfactory if a high proportion of the population is caught by the end of the experiment. In using Zippin's maximum likelihood estimate $\hat{N}$ there is one further question of design that must be considered, the question of experimental failure (p. 312). Otis *et al.* [1978] demonstrate that if $p < 0{\cdot}1$ there is a high probability (e.g. $0{\cdot}20$) that the experiment will fail. Even if the experiment does not fail in this case, $\hat{N}$ tends to be significantly biased (say 15–20 percent): when the failure criterion is close to being satisfied, $\hat{N}$ evidently tends to be biased high. For these reasons it is recommended that $p > 0{\cdot}2$.

In analysing catch–effort data Nelson and Clark [1973] suggest making a correction for trapping devices that become sprung during the experiment. They recommend subtracting half a trapping unit from the total trapping effort for each trap sprung, the assumption being that, on average, each trap is sprung for half the trapping interval. This assumption can perhaps be met experimentally by adjusting the middle of the trapping interval to coincide with the peak of activity of the animals being sought. Andersson [1976] gives a complicated model for dealing with the problem of trap saturation.

12.9.2 A Bayesian removal model

Assuming $p_i = p$ $(i = 1, 2, \ldots, s)$, the joint distribution of the n_i is given by (7.20) on p. 312. The maximum likelihood equations are obtained by differentiating the logarithm of (7.20) with respect to p and taking first differences with respect to N; namely

$$p = T/(sN - X) \tag{12.61}$$

and

$$N = T/(1 - q^s), \tag{12.62}$$

where

$$T = \sum_{i=1}^{s} n_i \; (= x_{s+1}) \quad \text{and} \quad X = \sum_{i=1}^{s} (s-i)n_i \left(= \sum_{i=1}^{s} x_i \right).$$

The procedure given on p. 312 is to substitute (12.62) in (12.61) and solve the resulting equation (7.22) for p. Substituting this value of p back into (12.62) gives $\hat{N}$. However, several authors (Otis *et al.* [1978: model M_b], Carle and Strub [1978]) have found that this method does not necessarily provide an answer which is within unity of the exact maximum likelihood estimate, $\hat{N}_E$ say. The reason for this is that $\hat{N}_E$ is an integer (the right-hand side of (12.62) suitably rounded off), and this should be taken into account when substituting (12.62) in (12.61) A more accurate method is to substitute (12.61) in (12.62) and solve the resulting polynomial in N. However, the best method is to substitute (12.61) in the likelihood function (7.20) and then maximise the logarithm of the resulting function of N directly.

Carle and Strub [1978] point out that the failure criterion given on p. 312, which can also be written in the form $X \leqslant T(s-1)/2$, applies only to the approximate procedure given there. Computer simulation indicates that the appropriate condition of failure for $\hat{N}_E$, rather than $\hat{N}$, is

$$X \leqslant [(T-1)(s-1)/2] - 1.$$

An alternative method of estimating N is given by Carle and Strub who suggest incorporating a prior beta distribution for p and then integrating out p. This leads to a likelihood function proportional to

$$\frac{N!\Gamma(\alpha+\beta)\Gamma(T+\alpha)\Gamma(sN-X-T+\beta)}{\prod_i n_i!(N-T)!\Gamma(\alpha)\Gamma(\beta)\Gamma(sN-X+\alpha+\beta)}, \tag{12.63}$$

where $\alpha \geqslant 1$ and $\beta \geqslant 1$. Assuming a uniform prior distribution ($\alpha = 1$, $\beta = 1$), (12.63) is maximised for the smallest integer greater than or equal to T which satisfies

$$\frac{N+1}{N-T+1}\prod_{i=1}^{s}\left\{\frac{sN-X-T+1+s-i}{sN-X+2+s-i}\right\} \leqslant 1.$$

The authors call this estimate of N, $\tilde{N}$ say, the maximum weighted likelihood (MWL) estimate. They show that $\tilde{N}$ has smaller bias and variance than $\hat{N}_E$, is more robust with regard to certain departures from the underlying assumptions, and never "fails". Also the asymptotic variance of $\tilde{N}$ is the same as that of $\hat{N}$, and (7.23) is a reasonable approximation for the variance of $\tilde{N}$ when $Np^3 \geqslant 7q^2(1+q)$. However, several graphs are provided for assessing the general usefulness of $\tilde{N}$ for $s = 3$ and different values of N and p. If the failure criterion for $\hat{N}$ is satisfied, $\tilde{N}$ is usually an unreliable estimate (cf. the discussion on $\hat{N}_b$ in Otis *et al.* [1978]).

12.9.3 Generalised removal method

Since variable catchability seems to be a fact of life, it would seem more appropriate to develop a model based on the distribution (12.57) with the π_i given by (12.58). However, because of difficulties encountered by Burnham [1972] with similar models in capture–recapture studies, it does not seem appropriate to model the distribution of P using, for example, a flexible distribution. For this reason Otis *et al.* [1978] suggest using the parametrisation (12.59). Since the individuals with a high probability of capture will tend to be removed first, we can expect $\bar{p}_1 > \bar{p}_2 > \ldots > \bar{p}_s$ and, in most cases, $\bar{p}_1 - \bar{p}_2 > \bar{p}_2 - \bar{p}_3 > \ldots > \bar{p}_{s-1} - \bar{p}_s$. Otis *et al.* [1978: Appendix H] suggest testing the sequence of hypotheses H_k: $\bar{p}_k = \bar{p}_{k+1} = \ldots \bar{p}_s$ for $k = 1, 2, \ldots, s-1$, and choosing the smallest k that gives a "suitable" fit to the data. Once k has been chosen, the maximum likelihood estimate corresponding to the model represented by H_k can be calculated iteratively. The authors call this method the generalised removal method. A sequence of exact tests for finding k are given by Skalski and Robson [1979].

Removals can also be effected by marking (p. 323) so that (Otis *et al.* [1978]) apply the generalised removal method to the unmarked animals in the capture–recapture model, M_{bh} (cf. p. 493). Although the information from the marked captures is ignored, the problem of trap response is avoided. However, it should be noted that marked animals occupy traps, and as the experiment progresses, fewer traps become available for catching unmarked animals. In this situation (12.58) is open to question but (12.59) still holds, so that the generalized removal method is still appropriate. For the same reason, namely trap response, some authors (e.g. Tanaka [1967, 1970], Tanaka and Murakami [1977]) suggest applying Hayne's method to unmarked animals. However, in this case, trap saturation by the marked will lead to $\bar{p}_1 > \bar{p}_2 > \ldots > \bar{p}_s$, so that (12.60) is no longer linear. A proportional trapping model is given by Good *et al.* [1970][*].

12.9.4 Effective trapping area

In applying the removal method, lines of traps are often used in conjunction with (e.g. Bobeck [1973]) or instead of a grid, and the lines are sometimes arranged in the form of an octagon. As mentioned in **12.1.1**(*3*), one method of estimating the effective trapping area of a grid or octagon is to use assessment lines placed across the boundary of the trapping network, usually after the removal trapping (Wheeler and Calhoun [1968]). The octagon removal census with an assessment line was investigated by Gentry, Smith and Chelton [1971] who found the method to be unreliable in their particular study, perhaps because of the low population density. Using a large octagon and more assessment lines, Kaufman *et al.* [1971] were more successful. Hansson [1974] considered varying the trap spacing along a trap line with a higher trap density in the centre of the line. However, this method did not lead to the Hayne regression line of catch versus accumulated catch along the line, usually associated with the use of assessment lines, and he concluded that the animals tended to move to the centre of the line. A major problem in using assessment lines after removal trapping is that animals tend to move into the study area when the population becomes depleted. For this reason O'Farrell *et al.* [1977] recommend that "removal" be done by marking. However, the above methods of estimating the effective trapping area are rather subjective and need further investigation.

When grid trapping is used, some authors try to avoid the edge effect by considering the captures on just an inner grid and calculating the effective trapping area as the area of the subgrid plus a strip of width equal to half the distance between the lines of traps. However, a more appropriate method is the method of nested subgrids described in **12.1.1**(*4*), though a note of warning should be sounded with regard to removal trapping: as the population is depleted the strip width W may tend to increase (Hansson [1969]). The subgrid or "inner square" method has been criticised by some of the proponents of the assessment line method (e.g. Smith *et al.* [1971]). However Barbehenn [1974: 224–5], in rejecting some of these criticisms, feels that as outside invasion is likely with removal trapping, he cannot see the point of "intentionally inducing an invasion by extended trapping and then attempting to measure the area of effect from a handful of animals taken on assessment lines". Taking everything into consideration it would appear, from a statistical viewpoint, that the best approach is to "remove" by marking and use the generalised removal model. Density is then estimated by the method of **12.1.1**(*4*).

Finally we note an interesting but time-consuming method of density estimation proposed by Sarrazin and Bider [1973]. They combined removal trapping with data on population activity obtained by checking fine sand transects for tracks every two hours. Basically their method consists of estimating how many animals would have to be removed in order to reduce "activity" to zero in the study area.

(*)See Additional References.

CHAPTER 13

RECENT DEVELOPMENTS: OPEN POPULATIONS

13.1 MARK RELEASES DURING SAMPLING PERIOD

13.1.1 Introduction

The multiple recapture method, or Schnabel census, of Chapter 5 has been widely used for the study of open populations, and recent applications of the Jolly–Seber (J-S) method include (see also p. 205) the studies by Brussard and Ehrlich [1970: butterflies], Manga [1972: beetles], Sonleitner [1973: fruit flies in field cages], Morgan [1974: a long extensive capture–recapture programme with rock lobsters], Dolbeer and Clark [1975: snowshoe hares, using the no recruitment model cf. p. 218], White [1975: 181, alpine grasshoppers with the model slightly modified to allow for inadequate dispersal], Conn [1976 narcissus bulb flies], Cook *et al.* [1976: tropical butterflies], Ericson [1977: beetles], Siniff *et al.* [1977: seals, with the model slightly modified to allow for the "incomplete" tagging of those released; ν_i and B_i are simply redefined] and Van Noordwijk [1978]. Examples of Ricker's two-release method for estimating the survival rate (p. 222) are given by Mathews and Buckley [1974: salmon], Nixon *et al.* [1974: 7, fox squirrels] and Howe *et al*. [1976: 655, flounder]. Chapman [1974a] also suggests the use of Ricker's method for whales because of problems such as the non-return of tags. Hopefully the proportion of tags overlooked would be the same for each release so, that the constant of proportionality would cancel out when ratios are taken.

Mention must also be made of an ingenious experiment carried out by Carothers [1973b] using a known population of taxi cabs (see also Bishop and Bradley [1972]). As noted by Cormack [1979], this experiment demonstrates "how difficult it is to devise a sampling scheme which eliminates heterogeneity of capture from such an organism which exhibits home-range behaviour".

In recent years further developments have been of three types. First, there has been an examination of the models described in Chapter 5 with regard to robustness (section **13.1.2**), and clearly more robust methods of variance estimation are required. Secondly, some special cases have been examined in more detail, and these are described in sections **13.1.3** to **13.1.5**. In particular the multi-sample single recapture census of **13.1.4** has proved to be a versatile model incorporating bird-banding and exploited fisheries. Further special models can always be obtained by imposing constraints like $p_i = p$ or $\phi_i = \phi$, and the maximum likelihood estimates will then have smaller variances as there are now fewer parameters to estimate. However, there is a penalty: the maximum likelihood equations now have to be solved iteratively and the (asymptotic) variances of the estimators are not easily found. For example, Jolly [1979b] assumes a constant ϕ and uses Poisson approximations to the multinomial distributions to obtain maximum likelihood equations for the p_i and ϕ: these equations must be solved iteratively. Further examples of constraining parameters are given in the bird-banding models of **13.1.5**.

A third area of research activity has been the development of a number of generalisations of the J-S method which allow survival, and possibly catchability, to depend on mark-status. These models are discussed in **13.1.6**.

In conclusion we mention a few developments in the computational aspects of the subject. A number of computer programmes for the Jolly–Seber method and various modifications have been developed by several authors including Davies [1971], White [1971a, b] and, for analysing bird-banding data, Anderson *et al.* [1974] and Brownie *et al.* [1978]. Recently Arnason and Baniuk [1978] extended the previous work of Arnason and Kreger [1973] and produced an extensive computer package, called POPAN-2, for handling multiple recapture data. The manual for this package contains a great deal of useful advice and discusses the important problem of pooling data from consecutive samples to increase the efficiency of estimation.

13.1.2 Robustness

The robustness of the estimates and their large sample variances discussed in Chapter 5 have been investigated by Manly [1970, 1971a], Gilbert [1973], Carothers [1973a], Roff [1973a] and Bishop and Sheppard [1973], and their results are summarised below.

Using Monte Carlo simulations and Taylor series expansions, Gilbert [1973] obtained approximations for the relative biases of the J-S estimates $\hat{N}_i$ and their "unbiased" modifications N_i^* (p. 204) for small samples. He considered the special case of a closed population, with $N = 50$ and 300 for the two models (cf. **12.8.6**): M_0 (constant probability of capture p for all animals and all trapping occasions) and M_h (heterogeneity, that is, catchability varying with the individual but constant over time). Following Carothers [1973a], he partitioned the relative bias of his estimates into two components, one which he called the relative small sample bias (RSSB) and the other which he calls the asymptotic relative bias (ARB) arising from heterogeneity. Gilbert concluded that serious negative bias can arise in practice using either $\hat{N}_i$ or N_i^* if the samples are too small, or if heterogeneity is present. As $N_i^* < \hat{N}_i$ the bias of the former is smaller and generally negative, while the bias of $\hat{N}_i$ can be positive or negative. Under model M_0 the small sample bias is severe if $p < 0{\cdot}20$, while if $p > 0{\cdot}20$ and $s > 10$ the negative bias of N_i^* is less than $0{\cdot}1N_i$. Under model M_h there is little bias due to small samples or heterogeneity if $\bar{p} > 0{\cdot}50$, where $\bar{p}$ is the average probability of capture. This suggests that an experimenter need not attempt to design an experiment so that all animals have the same probability of capture, but rather that nearly all the animals have a probability of capture greater than 0·5. Gilbert also provided confirmation of the rule that if $\bar{p} > 0{\cdot}20$ and $s \geqslant 10$, do not compute $\hat{N}_i$ or N_i^* unless m_i, the number of marked individuals in the ith sample, and r_i, the number released from the ith sample and subsequently recaptured, are both at least 3 if large biases are to be avoided.

However, even if $m_i \geqslant 3$ and $r_i \geqslant 3$ the variances for some trapping occasions may be quite large unless $\bar{p} \geqslant 0{\cdot}50$. It should also be noted that the small-sample bias decreases as N increases, but the bias due to heterogeneity could remain large, resulting in a large total bias. Although the J-S method is not appropriate for closed populations, Gilbert's study does at least highlight some of the problems involved.

The study by Carothers [1973a] was partly an extension of Gilbert's study to open populations rather than just models M_0 and M_h. However, he was not concerned with the question of small sample bias on the grounds that when the variances, and in particular estimated variances, are small enough to be of practical use, the small-sample biases are negligible: in this case the modifications N_i^*, etc., seem unnecessary. Carothers considered a model like M_h but with a constant survival probability ϕ, and a constant number of "births" or new additions, B, between samples. Assuming a large population, he obtained first-order expressions for the asymptotic relative biases, ARB. By making the assumption that deaths balance new additions ($B = (1-\phi)N$), these ARB's conveniently do not depend on population size, provided the population is large enough for first-order terms to be adequate. Carothers computed the theoretical ARB's for various parameter values and compared them with the results from (smaller) simulated populations with $N_1 = 100$ and 500. There was satisfactory agreement between theory and simulation (cf. ARB and MRB in his tables 4 and 5) and he came to the following conclusions: (i) $V[\hat{N}_i]$ is a maximum for $i = 2$ and $i = s$ and a minimum for the "central" value(s) of i (a fact also observed by Gilbert); (ii) $V[\hat{\phi}_i]$ generally increases as $i \to s$; (iii) as s increases, the maximum values of $V[\hat{N}_i]$ and $V[\hat{\phi}_i]$ remain almost unchanged while the minimum values decrease; (iv) as ϕ decreases both $V[\hat{N}_i]$ and $V[\hat{\phi}_i]$ increase, but the dependence of these variances on i and s becomes less marked; and (v) for a wide range of catchability distributions, the effect of variable catchability is to a large extent dependent only on the mean and coefficient of variation, γ, of the catchability distribution. For large N_1 (e.g. $N_1 = 500$) and a moderate variation in catchability (coefficient of variation of 0·2), the ARB's are small, roughly 5 percent for N_i and 0·5 percent for ϕ_i; for $N_1 = 100$ percentages are approximately double. Increasing the coefficient of variation to 0·8 led to ARB's of up to 50 percent and 5 percent respectively, irrespective of N_1. Looking at his tables, it would appear that a rough approximation for ARB(N_i) when $\gamma \leqslant 0{\cdot}05$ is γ^2; thus if $\gamma = 0{\cdot}2$, ARB is about 4 percent.

Manly [1970] investigated the effect of age-dependent mortality on the J-S, the M-P (Manly and Parr, p. 233), and the F-F (Fisher and Ford [1947]) estimates for an open population in which the probability of capture is constant. He concluded that the J-S estimates can tolerate a moderate amount of age-dependent mortality, while the M-P estimates, though not affected by age-dependent mortality, are more sensitive to the effects of small sample data. In the latter case the Fisher and Ford model provided better estimates when the mortality was approximately constant; this is not surprising as, with a single survival probability, there are fewer parameters to estimate.

Manly [1971a] examined the large-sample variances of the J-S estimates for the case of constant probability of capture and concluded that the usual large-sample confidence intervals, for example $\hat{N}_i \pm 1{\cdot}96\hat{\sigma}_i$, where $\hat{\sigma}_i = (\hat{V}[\hat{N}_i])^{1/2}$, are not very reliable. One of the reasons for this is that $V[\hat{N}_i|N_i]$ and $V[\hat{\phi}_i]$ are proportional to N_i^2 and ϕ_i^2 respectively, so that overestimates appear to be less accurate than they really are, while underestimates are not as accurate as they appear — a more serious problem. The effect is less with survival estimates, and Manly found that the log and inverse transformations were only partly successful in overcoming the correlation.

Using simulation, Roff [1973a] came to similar conclusions about the accuracy of variance estimates and the width of confidence intervals for population size. Clearly if a low coefficient of variation is required, the sampling intensity, i.e. the probability of capture usually needs to be high, though the number of samples and migration are contributing factors. Robson and Regier (cf. p. 64) recommended an accuracy of 10 percent for careful research into population dynamics. This corresponds to a coefficient of variation of about 5 percent which, as Roff [1973a] demonstrated, could only be attained if the population is large ($N > 500$) and the sampling intensity high ($p > 0{\cdot}50$).

Bishop and Sheppard [1973] compared the J-S and F-F methods by simulating a population which satisfied the conditions of M_0 except for a constant survival probability ϕ and the addition of new unmarked individuals to maintain a constant population size. In their simulations s was high (10 or 20) and the sampling intensity (proportion captured) low, namely 5, 9 and 12 percent. They found that the J-S method gave poor estimates when ϕ was low (say 0·5) and $p = 0{\cdot}05$; a not unexpected result in the light of previous comments made in **12.8.6** about low probabilities of capture in closed populations. In this situation, where few recaptures were made, Fisher and Ford's model provided better estimates, as expected. For the examples simulated, the J-S estimates of N_i and θ_i were almost always positively biased, a conclusion which appears to be in contradition with some of the above studies. However, the parameter values and the method of averaging estimates are different here, making comparisons with other work difficult.

Because of the general unreliability of the J-S confidence intervals Manly [1977b] proposed an alternative method based on the generalised jackknife approach of Gray and Schucany [1972]. Suppose that the sample data are divided into K comparable sized subsamples at random. Let $\hat{\theta}$ be the estimate of a parameter θ based on all the data and let $\hat{\theta}_{-i}$ be the estimate based on all the data, but with the ith subsample omitted ($i = 1, 2, \ldots, K$). Define the "pseudo-values"

$$G_i(\hat{\theta}) = (\hat{\theta} - R\hat{\theta}_{-i})/(1 - R), \quad i = 1, 2, \ldots, K,$$

where R is a function of K. The average of these pseudo-values, namely

$$G(\hat{\theta}) = \sum_{i=1}^{K} G_i(\hat{\theta})/K, \tag{13.1}$$

is called the generalised jackknife estimate of θ. When Quenouille [1956] proposed the jackknife technique he used $R = (K - 1)/K$ for which the pseudo-values are now

$$J_i(\hat{\theta}) = K\hat{\theta} - (K-1)\hat{\theta}_{-i},$$

with mean

$$J(\hat{\theta}) = \sum_{i=1}^{K} J_i(\hat{\theta})/K \tag{13.2}$$

and variance estimate

$$S_J^2 = \sum_{i=1}^{K} [J_i(\hat{\theta}) - J(\hat{\theta})]^2/(K-1). \tag{13.3}$$

Quenouille showed that $J(\hat{\theta})$ will often have less bias than $\hat{\theta}$, and Tukey [1958] conjectured that $\sqrt{K}\,[J(\hat{\theta}) - \theta]/S_J$ is approximately distributed as the t-distribution with $K-1$ degrees of freedom. This result has been generalized by Gray and Schucany [1972: 154] who prove that under fairly general conditions $\sqrt{K}\,[G(\hat{\theta}) - \theta]/S_J$ is approximately distributed as the t-distribution with $K - 1$ degrees of freedom. Using this result, with $R = (K-1)^2/K^2$ and an appropriate scale factor derived empirically, Manly [1977b] derived approximate confidence intervals for the various population parameters. Using simulation, Manly showed that the intervals appear to be generally satisfactory if 20–30 subsamples are used.

Finally we note one other simulation study carried out by Hilborn *et al.* [1976] who considered the use of the number of marked individuals known to be alive at a given time (having been caught both before and after that time) as an index (lower bound) of population size at that time. They showed that the index is insensitive to population size and robust, provided the catchabilities of marked and unmarked are not too different and the average probability of capture is at least 0·5.

13.1.3 Some special models

1 Releases made at age zero

Manly [1972–3] considers a model in which batches of individuals of (coded) age zero are released at times $t_1, t_2, \ldots, t_s$ and recapture samples are taken at times $t_1', t_2', \ldots, t_s'$, where $t_1 < t_1' \leqslant t_2 < \ldots \leqslant t_s < t_s'$. He assumes that (i) all recaptured individuals are released again and that the recapture of an animal does not affect its future survival; (ii) $t_i' - t_i$ and $t_i - t_{i-1}'$ are constant for all i, that is $t_{k+i}' - t_i$ $(= \Delta_k)$ does not depend on i; (iii) all animals alive at time t_i' have the same probability p_i of being caught at time t_i', and (iv) the probability ϕ_k of an animal surviving until at least age Δ_k is the same for all animals irrespective of when released. Equating certain random variables with their expected values, Manly obtains moment estimates of the p_i and ϕ_k together with appropriate expressions for variances and covariances; he also considers the case of $p_i = p$ $(i = 1, 2, \ldots, s)$. Plotting the estimate of ϕ_k against Δ_k gives an estimate of the so-called survival curve. Unfortunately the above method will have limited application as, for many populations, the probability of capture depends on age, at least as far as juvenile and adults are concerned. Manly applied the method to some insect data which seemed to satisfy the underlying assumptions.

2 *Death only model*

For the case of mortality only an alternative model has been proposed by Rafail [1971a]. He assumes that the parameters p_i and ϕ_i depend on whether the animals are tagged or untagged, so that an appropriate notation for these two groups is $\{p_{(t)i}, \phi_{(t)i}\}$ and $\{p_{(u)i}, \phi_{(u)i}\}$ respectively. It should be noted that the assumption $\phi_{(t)i} = \phi_{(u)i}$ is not required in any of the *general* capture–recapture models of Chapter 5 (cf. p. 196 where assumption (b) refers to marked animals only) except in the estimate of B_i (p. 200). What are estimated by the J-S and M-P models are $p_{(t)i}$ as all the information about capture and survival is provided by the tagged population (p. 198). Since, by redefining ϕ_i, we can include permanent immigrants as well as deaths, it is perhaps not surprising that Rafail's [1971a] estimates of $p_{(t)i}$ and $\phi_{(t)i}$ are the same as those of Manly–Parr (cf. $\tilde{p}_i$ and $\tilde{\phi}_i$ of pages 235–6).

Information about the unmarked population sizes U_i is given by the numbers $\{u_i\}$ of unmarked caught in each sample. Unfortunately we have only s items of information $\{u_i\}$ to estimate $2s$ parameters $\{U_i, p_{(u)i}\}$. However, if we assume $p_{(u)i} = p_{(t)i}$ it is then possible to estimate N_i by $n_i/\hat{p}_{(t)i}$. If migration is present then the numbers of new animals B_i joining the population can only be estimated if we can make the further assumption that $\phi_{(t)i} = \phi_{(u)i}$.

Returning to the death-only situation, we note that Darroch [1959] assumes both $p_{(t)i} = p_{(u)i}$ and $\phi_{(t)i} = \phi_{(u)i} = \phi_i$ to obtain more efficient estimates of each $N_1\phi_1\phi_2 \ldots \phi_{i-1}$ (called N_i on p. 218), the *expected* size of the population at the time of the ith sample. However, Rafail [1971a] allows $\phi_{(t)i}$ and $\phi_{(u)i}$ to differ and makes the weaker assumption about catchability that

$$p_{(u)i} = cp_{(t)i}. \tag{13.4}$$

Unfortunately there is a price to be paid as we have one too many parameters, namely c, and this cannot be estimated without a further constraint on the parameters. If $u_{i,i+1}$ is the number of unmarked caught in the ith sample which are also caught in the $(i + 1)$th, then

$$\begin{aligned} E[u_{i,i+1}|U_i] &= U_i p_{(u)i}\phi_{(t)i}p_{(t),i+1} \\ &= U_i p_{(t)i}\phi_{(t)i}p_{(t),i+1}c, \\ E[u_{i+1}|U_i] &= U_i q_{(u)i}\phi_{(u)i}p_{(u),i+1} \\ &= U_i q_{(u)i}\phi_{(u)i}p_{(t),i+1}c, \end{aligned}$$

and we have the estimating equations proposed by Rafail (his equation 6d), namely

$$\frac{1 - cp_{(t)i}}{p_{(t)i}} \cdot \frac{\phi_{u(i)}}{\phi_{t(i)}} = \frac{u_{i+1}}{u_{i,i+1}}.$$

By taking a consecutive pair of such equations (for $i - 1$ and i) we have an equation relating c and the ratio

$$r = [\phi_{(t),i-1}\phi_{(u)i}]/[\phi_{(t)i}\phi_{(u),i-1}]. \tag{13.5}$$

Here $\{\phi_{(t)i}\}$ and $\{p_{(t)i}\}$ can be estimated using the J-S or M-P methods, but c can only be estimated if estimates of the $\phi_{(u)i}$ are available, for example from age composition data in fisheries, or if a further assumption about the $\phi_{(u)i}$'s is made. Rafail suggests setting $\phi_{(u)i-1} = \phi_{(u)i}$ for some i, so that c can be estimated by $\hat{c}$, say. Then N_i is estimated by

$$\frac{u_i}{\hat{p}_{(u)i}} + \frac{m_i}{\hat{p}_{(t)i}} = \frac{1}{\hat{p}_{(t)i}}\left(\frac{u_i}{\hat{c}} + m_i\right).$$

No variance estimates are given. The usefulness of this approach depends on the validity of the assumption (13.4) which is very much open to question, particularly when the differences between $p_{(u)i}$ and $p_{(t)i}$ is due to heterogeneity of catchability. In this case the more catchable ones are caught first, so that $p_{(u)i}$ may vary relative to $p_{(t)i}$.

Rafail [1971b, 1972a, b, c] extends this method further to the case when effort data can be utilised, as for example in fisheries. He expresses r of (13.5) in terms of instantaneous mortality rates and catchability coefficients and also considers the case $r = 1$ (Rafail [1972a]). However, his model, which he calls Example II, has limited application to commercially exploited populations as it requires the untagged to be retained and the tagged to be released, a difficult requirement for most fishing fleets.

13.1.4 The multi-sample single recapture census

1 General model

Although the theory of 5.1.1 was derived on the basis that releases followed immediately after sampling, we saw on pages 208–17 that the estimates can be applied to a wide variety of models, including the situation where release and recapture are operated independently. An important special case is the so-called multi-sample single recapture census in which there is a 100 percent removal on capture. In this case the model can be generalised (Seber [1962]) to deal with the situation where releases and recaptures operate at different times. Although the general model of **5.1.1** can be adapted to deal with this modification, the notation there becomes inadequate as it requires calling the first release sample 1 and the ith recapture sample, sample $i + 1$: this leads to the problems of suffix notation mentioned on pages 212–14. We shall therefore restate this modified model using a more convenient notation.

Suppose that releases are made at times $t_1, t_2, \ldots, t_s$ and recapture samples are taken at times $t'_1, t'_2, \ldots, t'_s, \ldots, t'_{s+k}$ where

$$t_1 < t'_1 \leqslant t_2 < t'_2 \leqslant \ldots \leqslant t_s < t'_s < t'_{s+1} \ldots < t'_{s+k}.$$

We are allowing for the possibility of k further recapture samples at times

$t'_{s+1}, \ldots, t'_{s+k}$ without corresponding releases. The following assumptions are now made:

(a) Every marked individual has the same probability p_i ($= 1 - q_i$) of being caught in the ith sample at time t'_i, given that it is alive and in the population just before time t'_i ($i = 1, 2, \ldots, s + k$).
(b) For $i = 1, 2, \ldots, s$, every marked individual has the same probability $1 - \phi_i$ of dying or permanently emigrating in the time-interval t_i to t'_i, given that it it is alive and in the population just after the ith release at time t_i.
(c) For $i = s + 1, s + 2, \ldots s + k$, every marked individual has the same probability $1 - \phi_i$ of dying or permanently emigrating in the interval t'_{i-1} to t'_i, given that it is alive and in the population just after time t'_{i-1} (see Fig. 13.1).

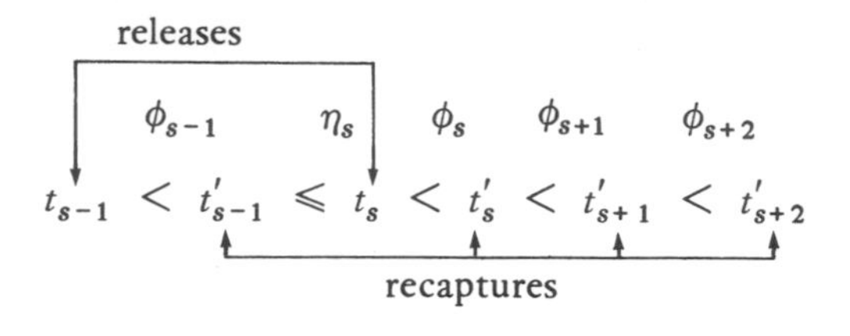

Fig. 13.1 Pattern of survival probabilities

(d) Every marked individual has the same probability $1 - \eta_i$ of dying or permanently emigrating in the interval t'_{i-1} to t_i, given that it is alive and in the population just after the sample at time t'_{i-1} ($i = 2, \ldots, s$).
(e) Either there is no emigration, or the emigration is permanent, so that emigrants can be regarded as being "dead".

Thus if m_{ij} is the number from release i caught in sample j, the joint distribution of the m_{ij} (cf. p. 213) is proportional to

$$\prod_{i=1}^{s} \{\beta_i^{m_{ii}} (\alpha_i\beta_{i+1})^{m_{i,i+1}} \ldots (\alpha_i\alpha_{i+1} \ldots \alpha_{t-1}\beta_t)^{m_{it}}(1-\theta_i)^{R_i - r_i}\}$$

$$= \prod_{i=1}^{s-1} \{\alpha_i^{T_i - m_i}\beta_i^{m_i}(1-\theta_i)^{R_i - r}\}\, \beta_s^{m_s}(1-\theta_s)^{R_s - r_s}\gamma_{s+1}^{m_{s+1}} \ldots \gamma_t^{m_t}, \tag{13.6}$$

where

$$t = s + k,$$

$$\beta_i = \phi_i p_i \quad (i = 1, 2, \ldots, t), \tag{13.7}$$

$$\alpha_i = \begin{cases} \phi_i q_i \eta_{i+1} & (i = 1, 2, \ldots, s-1) \\ \phi_i q_i & (i = s, \ldots, t), \end{cases} \tag{13.8}$$

$$\gamma_j = \alpha_s\alpha_{s+1} \ldots \alpha_{j-1}\beta_j \quad (j = s+1, \ldots, t),$$

and

$$\theta_i = \beta_i + \alpha_i\beta_{i+1} + \ldots + \alpha_i\alpha_{i+1}\ldots\alpha_{t-1}\beta_t,$$

where θ_i is the probability of recapture from the ith release. Here R_i is the number in the ith release, r_i is the number recovered from R_i, m_i is the number of recaptures (marked) in the ith sample, and T_i is the number of different individuals recaptured after the ith release from the first i releases. Thus $T_i - m_i$ $(= T_{i+1} - r_{i+1} = z_{i+1})$ is the number recovered after the ith sample from releases made prior to the ith sample or, expressed another way, the number recovered after the $(i+1)$th release from releases prior to the $(i+1)$th release. (For this latter reason I have denoted this number by z_{i+1} rather than z_i as the theory also applies to bird-banding models). The maximum likelihood estimates for the estimable parameters in (13.6) are:

$$\hat{\alpha}_i = \frac{T_i - m_i}{T_i}\cdot\frac{r_i}{R_i}\cdot\frac{R_{i+1}}{r_{i+1}}, \quad (i = 1, 2, \ldots, s-1), \tag{13.9}$$

$$\hat{\beta}_i = \frac{m_i}{T_i}\cdot\frac{r_i}{R_i}, \quad (i = 1, 2, \ldots, s) \tag{13.10}$$

$$\hat{\gamma}_j = \beta_s\frac{m_j}{m_s}, \quad (j = s+1, s+2, \ldots, t), \tag{13.11}$$

and

$$\hat{\theta}_i = \frac{r_i}{R_i}. \tag{13.12}$$

Now Robson and Youngs [1971] show that the conditional distribution of m_i, given r_i and T_i, is binomial, namely

$$f(m_i|r_i, T_i) = \binom{T_i}{m_i}\left(\frac{\beta_i}{\theta_i}\right)^{m_i}\left(1 - \frac{\beta_i}{\theta_i}\right)^{T_i - m_i}, \tag{13.13}$$

which indicates that m_i is independent of r_i. Thus, since $E[r_i] = R_i\theta_i$,

$$\begin{aligned} E[\hat{\beta}_i] &= \underset{r_i}{E}\, E[\hat{\beta}_i|r_i] \\ &= \underset{r_i}{E}\left[\frac{\beta_i}{\theta_i}\cdot\frac{r_i}{R_i}\right] \\ &= \beta_i, \end{aligned}$$

and $\hat{\beta}_i$ is unbiased. Similarly, writing

$$\hat{\alpha}_i = \left(\frac{r_i}{R_i} - \hat{\beta}_i\right)\Big/\left(\frac{r_{i+1}}{R_{i+1}}\right),$$

and noting that the numerator and denominator are statistically independent, leads to

$$E[\hat{\alpha}_i] = (\theta_i - \beta_i)R_{i+1}E[r_{i+1}^{-1}].$$

Since r_{i+1} is binomial with parameters R_{i+1} and θ_{i+1} we can make the usual adjustment for bias based on the binomial distribution, namely

$$\alpha_i^* = \frac{T_i - m_i}{T_i} \cdot \frac{r_i}{R_i} \cdot \frac{R_{i+1}+1}{r_{i+1}+1}. \tag{13.14}$$

Since $\theta_i - \beta_i = \alpha_i\theta_{i+1}$ we have (cf. p. 222)

$$\begin{aligned} E[\alpha_i^*] &\approx \alpha_i\{1-(1-E[r_{i+1}/R_{i+1}])^{R_{i+1}+1}\} \\ &= \alpha_i\{1-(1-\theta_{i+1})^{R_{i+1}+1}\}, \end{aligned} \tag{13.15}$$

and α_i^* is approximately unbiased. The large-sample variances and covariances of the α_i^* will be the same as those of the $\hat{\alpha}_i$ (p. 213).

A general goodness-of-fit test of the model is given by

$$T = \sum_{i=1}^{s}\sum_{j=1}^{t} (m_{ij} - E_{ij})^2/E_{ij},$$

which is approximately distributed as chi-squared with $(s-1)(t-1) - \frac{1}{2}s(s-1)$ degrees of freedom when the model is true. Here

$$E_{ii} = R_i\hat{\beta}_i \quad (i = 1, 2, \ldots, s),$$

$$E_{ij} = \begin{cases} R_i\hat{\alpha}_i\hat{\alpha}_{i+1} \ldots \hat{\alpha}_{j-1}\hat{\beta}_j & (j = i+1, \ldots, s) \\ R_i\hat{\alpha}_i\hat{\alpha}_{i+1} \ldots \hat{\alpha}_{s-1}\hat{\gamma}_j & (j = s+1, \ldots, t), \end{cases}$$

and the terms involving $R_i - r_i$ are not included in T as their contribution is zero (since $R_i - r_i - R_i(1-\hat{\theta}_i) = 0$).

Unfortunately the basic parameters ϕ_i, p_i and η_i (cf. 13.7) and (13.8) are not estimable unless some constraints are applied to reduce the number of free parameters. In Seber [1962] we assumed that releases are made immediately after recaptures so that $t_i' = t_{i+1}$ and $\eta_{i+1} = 1$ $(i = 1, 2, \ldots, s-1)$. Bearing in mind the difference in notation – that is, the release at time t_1 and recaptures at time t_i' are called samples 1 and $i+1$ respectively, so that m_{ij}, m_i and p_i become $m_{i,j+1}$, m_{i+1} and p_{i+1} in (5.24) – this model is best analysed using the general theory of 5.1.1.

Manly [1974b], however, assumed that the survival probability per unit time is constant for the interval between any two releases (though it may vary from interval to interval). This implies that for the time-interval (t_i, t_{i+1}) we have

$$\phi_i = \psi_i^{1-\Delta_i} \quad \text{and} \quad \eta_{i+1} = \psi_i^{\Delta_i}, \quad (i = 1, 2, \ldots, s-1)$$

where $\Delta_i = (t_{i+1} - t_i')/(t_{i+1} - t_i)$, and $\psi_i = \phi_i\eta_{i+1}$ may be interpreted as the probability of "natural" survival (that is ignoring the recapture sample). Since we now have a one-to-one correspondence between the parameters (α_i, β_i) and (ψ_i, p_i), the maximum likelihood estimates $\hat{\psi}_i$, $\hat{p}_i$ satisfy $\hat{\alpha}_i = \hat{\psi}_i\hat{q}_i$ and $\hat{\beta}_i = \hat{\psi}_i^{1-\Delta_i}\hat{p}_i$. Thus for $i = 1, 2, \ldots, s-1$

$$\hat{\psi}_i - \hat{\beta}_i\hat{\psi}_i^{\Delta_i} = \hat{\alpha}_i$$

and

$$\hat{p}_i = \hat{\beta}_i\hat{\psi}_i^{\Delta_i - 1}$$

which can be solved for $\hat{\psi}_i$ and $\hat{p}_i$. Using the delta method Manly shows that, asymptotically,

$$V[\hat{\psi}_i] = \psi_i^2 \left\{ \frac{p_i^2}{E[m_i]} + \frac{(1-p_i)^2}{E[z_{i+1}]} - \frac{1}{E[T_i]} + \frac{1}{E[r_i]} - \frac{1}{R_i} + (1-p_i)^2 \left(\frac{1}{E[r_{i+1}]} - \frac{1}{R_{i+1}} \right) \right\} \Big/ (1 - \Delta_i p_i)^2,$$

$$\text{cov}\,[\hat{\psi}_i, \hat{\psi}_{i+1}] = \frac{-\psi_i \psi_{i+1}(1-p_i)}{(1-\Delta_i p_i)(1-\Delta_{i+1}p_{i+1})} \left(\frac{1}{E[r_{i+1}]} - \frac{1}{R_{i+1}} \right)$$

and

$$\text{cov}\,[\hat{\psi}_i, \hat{\psi}_j] = 0, \quad j > i+1.$$

Estimates are obtained by replacing expectations by random variables. When $\Delta_i = 0$ it can be shown that, given the difference in notation, the above equations reduce to (5.15) to (5.17). Manly [1974b] describes an application of the above theory to a moth population where it was found convenient to make releases in the mornings and recapture samples in the evenings.

2 *Constant probability of capture*

When the numbers of recaptures are small, we may end up with a series of very unreliable estimates unless more constraints are imposed on the above model, thus reducing the number of unknown parameters. Manly [1975a] describes a model in which (i) the times between releases and recaptures are kept constant, that is $t_1 = 1$, $t_2 = 2$, $t_3 = 3$, etc. and $t_1' = 2 - \Delta$, $t_2' = 3 - \Delta$, etc., (ii) the various parameters are constant over time, namely $p_i = p$ $(i = 1, 2, \ldots, t)$, $\psi_i = \psi$ $(i = 1, 2, \ldots, s-1)$, etc., and (iii) enough recapture samples are taken after the last release to ensure that there are virtually no marked individuals left at the end of the experiment. The likelihood function is now given by (13.6) with $\alpha_i = \alpha$ and $\beta_i = \beta$ for all i, that is proportional to

$$\alpha^u \beta^{r.} \prod_{i=1}^{s} (1-\theta_i)^{R_i - r_i} \tag{13.16}$$

with $u = \sum_{i=1}^{s} \sum_{j=i}^{t} (j-i)m_{ij}$, $\quad r_{.} = \sum_{i=1}^{s} r_i \quad$ and

$$\theta_i = \beta(1 + \alpha + \ldots \alpha^{t-i}) = \beta(1-\alpha^{t-i+1})/(1-\alpha)$$

If assumption (iii) is true then $\alpha^{t-i+1} \approx 0$ $(i = 1, 2, \ldots, s)$ and $\theta_i \approx \beta/(1-\alpha)$. With this approximation the maximum likelihood estimates of α and β are

$$\hat{\alpha} = \frac{u}{r_{.} + u} \tag{13.17}$$

and

$$\hat{\beta} = \frac{r_{.}^2}{R_{.}(r_{.} + u)}, \tag{13.18}$$

where $R_{\bullet} = \sum_{i=1}^{s} R_i$. Thus the maximum likelihood estimates of ψ (survival probability per unit time) and p satisfy

$$\hat{\psi} - \hat{\beta}\hat{\psi}^{\Delta} = \hat{\alpha} \quad \text{and} \quad \hat{p} = \hat{\beta}\hat{\psi}^{\Delta-1}.$$

Manly [1975a] shows that for large r the variance and covariances are approximately

$$V[\hat{\psi}] = [\psi^2 p^2(1-\theta) + \psi q(1-\psi)^2]/[R_{\bullet}\theta(1-\Delta p)^2] \quad (= \sigma^2[\hat{\psi}]) \tag{13.19}$$

$$V[\hat{p}] = p^2 q[q(1-\theta) + (1-\Delta+\Delta\psi q)^2/\psi]/[R_{\bullet}\theta(1-\Delta p)^2]$$

and

$$\text{cov}[\hat{\psi}, \hat{p}] = pq\{\psi p(1-\theta) - (1-\psi)[1-\Delta+\Delta\psi q]\}/[R_{\bullet}\theta(1-\Delta p)^2],$$

where θ represents the "total" probability of capture, and can be estimated by $r_{\bullet}/R_{\bullet}$. An application of the above theory is given by Manly [1975b].

Manly [1975a] notes that the above theory still holds if Δ is now *negative* and an animal is not available for recapture until one unit of time has elapsed, that is, we now have $(t_2 = 2) < (t_1' = 2 - \Delta) < t_3 = 3$ etc. This is very convenient since it provides a means of overcoming the problem of low numbers of recaptures in the first recapture sample after release. If these numbers are small they can be simply ignored and the theory applied with negative Δ.

We note that $\sigma[\hat{\psi}]\sqrt{R_{\bullet}}$ (cf. (13.19)) is a function of just p, ψ and Δ, and is tabulated by Manly [1977d]. This table is reproduced here as Table 13.1 and we now give two examples of the use of this table.

Example 13.1

Suppose it is expected that $\psi \approx 0{\cdot}7$ per day and an initial lack of catchability of released animals means that they cannot be recaptured until $\frac{3}{4}$ day after their release. In this case Δ can be set anywhere between the limits $-\frac{3}{4}$ to $\frac{1}{4}$ (which correspond to the ith sample taken between $\frac{3}{4}$ day after the $(i+1)$th release and $\frac{1}{4}$ day before the $(i+1)$th release: for this range no individuals from the $(i+1)$th release will be caught in sample i). The best value of Δ which gives the smallest entry in Table 13.1 depends on p. If the number of traps can be determined so that the "optimal" value of $p = 0{\cdot}5$ can be anticipated, then Δ should be set at $-\frac{3}{4}$ which leads to a standard error of $\sigma[\hat{\psi}] = 0{\cdot}37/\sqrt{R}$. Thus releases of a total of 100 animals would be needed to make the standard error as small as 0·04.

Example 13.2

Suppose $\psi \approx 0{\cdot}9$ with animals being available for recapture almost immediately after their release. Then Δ can be set between 0 and 1, and Table 13.1 indicates $\Delta = 0$. With the optimal value of $p = 0{\cdot}1$, and $\Delta = 0$, we have $\sigma[\hat{\psi}] = 0{\cdot}16/\sqrt{R_{\bullet}}$. If the experiment is badly *designed* so that $p = 0{\cdot}9$ and $\Delta = 0{\cdot}5$, then $\sigma[\hat{\psi}]$ is more than doubled.

TABLE 13.1

Standard errors (per animal released) for survival estimators from multi-sample single recapture experiments. If $R_{\cdot}$ animals are released then the standard error is the tabulated value divided by $\sqrt{R_{\cdot}}$ (reproduced from Manly [1977d: Table 1])

	(a) $\psi = 0{\cdot}1$					(b) $\psi = 0{\cdot}3$				
Δ	$p = 0{\cdot}1$	0·3	0·5	0·7	0·9	$p = 0{\cdot}1$	0·3	0·5	0·7	0·9
$-\frac{3}{4}$	5·69	2·59	1·56	0·99	0·60	2·63	1·26	0·84	0·62	0·50
$-\frac{1}{2}$	4·37	2·07	1·29	0·84	0·51	2·32	1·16	0·79	0·60	0·49
$-\frac{1}{4}$	3·35	1·66	1·07	0·72	0·45	2·04	1·06	0·75	0·59	0·49
0	2·58	1·34	0·90	0·63	0·41	1·80	0·98	0·72	0·58	0·49
$\frac{1}{4}$	1·98	1·08	0·77	0·57	0·39	1·59	0·91	0·71	0·59	0·52
$\frac{1}{2}$	1·53	0·88	0·67	0·54	0·40	1·40	0·85	0·70	0·63	0·57
$\frac{3}{4}$	1·18	0·73	0·61	0·55	0·47	1·24	0·80	0·71	0·70	0·72

	(c) $\psi = 0{\cdot}5$					(d) $\psi = 0{\cdot}7$				
Δ	$p = 0{\cdot}1$	0·3	0·5	0·7	0·9	$p = 0{\cdot}1$	0·3	0·5	0·7	0·9
$-\frac{3}{4}$	1·36	0·72	0·55	0·48	0·45	0·60	0·40	0·37	0·36	0·37
$-\frac{1}{2}$	1·27	0·70	0·55	0·48	0·46	0·59	0·40	0·38	0·38	0·39
$-\frac{1}{4}$	1·20	0·69	0·55	0·50	0·47	0·58	0·41	0·39	0·40	0·41
0	1·12	0·67	0·56	0·52	0·50	0·57	0·41	0·40	0·42	0·45
$\frac{1}{4}$	1·06	0·67	0·58	0·55	0·55	0·56	0·42	0·43	0·46	0·50
$\frac{1}{2}$	0·99	0·66	0·60	0·60	0·63	0·54	0·43	0·46	0·51	0·58
$\frac{3}{4}$	0·94	0·66	0·64	0·69	0·80	0·53	0·45	0·50	0·59	0·75

	(e) $\psi = 0{\cdot}9$					
Δ	$p = 0{\cdot}05$	0·1	0·3	0·5	0·7	0·9
$-\frac{3}{4}$	0·18	0·16	0·17	0·19	0·21	0·23
$-\frac{1}{2}$	0·18	0·16	0·18	0·20	0·22	0·24
$-\frac{1}{4}$	0·18	0·16	0·18	0·21	0·24	0·26
0	0·18	0·16	0·19	0·23	0·26	0·29
$\frac{1}{4}$	0·18	0·16	0·20	0·24	0·28	0·32
$\frac{1}{2}$	0·18	0·16	0·20	0·26	0·32	0·38
$\frac{3}{4}$	0·18	0·17	0·21	0·28	0·37	0·50

Note: ψ = probability of survival per unit interval of time
p = probability of recapture in a particular sample
Δ = time between a recapture sample and the next release.

Manly [1977d] also considers the case where several types of animal are released and gives a table corresponding to Table 13.1 for comparing the survival rates for two types when the probability of capture is the same for both types.

3 Survival ratios

Mark releases can also be used for estimating survival ratios for two or more species. The ratio of the survival probability of species i to the survival probability of species 1, a "control" species, can be regarded as a relative "coefficient of selectivity" for the species i ($i > 1$)when most of the mortality is from predation. The estimation of such coefficients is important, for example, in studies associated with mimicry. Assuming that the probability of capture in a given sample is the same for all species, Manly [1972] shows how to estimate these coefficients from a single mark-release. Using simulation, Manly [1978] compares several methods that he has proposed (Manly [1973, 1975a, 1977d]) for bias, efficiency, robustness and normality of the estimates. His models are based on a multi-sample single recapture census for each species, with various constraints such as probability of capture being constant over time and (or) species.

4 Exploitation model

ESTIMATION. On p. 213 we showed that the multi-sample single recapture census could be applied to the situation where the sampling (exploitation) is continuous and tag releases are made at the beginning of each year. Here α_i is now the probability of an individual surviving the ith year, and, as it refers to survival from both natural and exploitation mortality, it is sometimes called the total survival rate for year i. The parameter β_i is the probability that an individual is caught and its tag recovered in year i; β_i is sometimes referred to as the observed exploitation rate for year i. As the model for this new situation is still given by (13.6), estimates of α_i and β_i are given by equations (13.9) and (13.10). Noting the equivalences $z_{i+1} = T_i - m_i$ and $r_i + z_i = T_i$, approximate expressions for the variances and covariances of the $\hat{\alpha}_i$ are given on p. 213. In fisheries the observed exploitation rate is an important parameter in its own right, so we include the following results:

$$V[\hat{\beta}_i] \approx \beta_i^2 \left\{ \frac{1}{E[r_i]} - \frac{1}{R_i} + \frac{1}{E[m_i]} - \frac{1}{E[T_i]} \right\}$$

and

$$\text{cov}\,[\hat{\beta}_i, \hat{\beta}_j] \approx 0 \quad \text{for } i \neq j.$$

If we use the notation $\alpha_i \to S_i$ and $\beta_i \to E_i^*$, and replace expected random variables by the random variables, the above variance and covariance expressions for the $\hat{\alpha}_i$ and $\hat{\beta}_i$ lead to the estimates given by Youngs and Robson [1975]. These authors also give a number of useful graphs for planning such studies and the following discussion is based on their paper, with the above changes in notation.

EXPERIMENTAL DESIGN. Assuming $\alpha_i = \alpha$, $\beta_i = \beta$ and $R_i = R$ for $i = 1, 2, \ldots, s$, it can be shown that

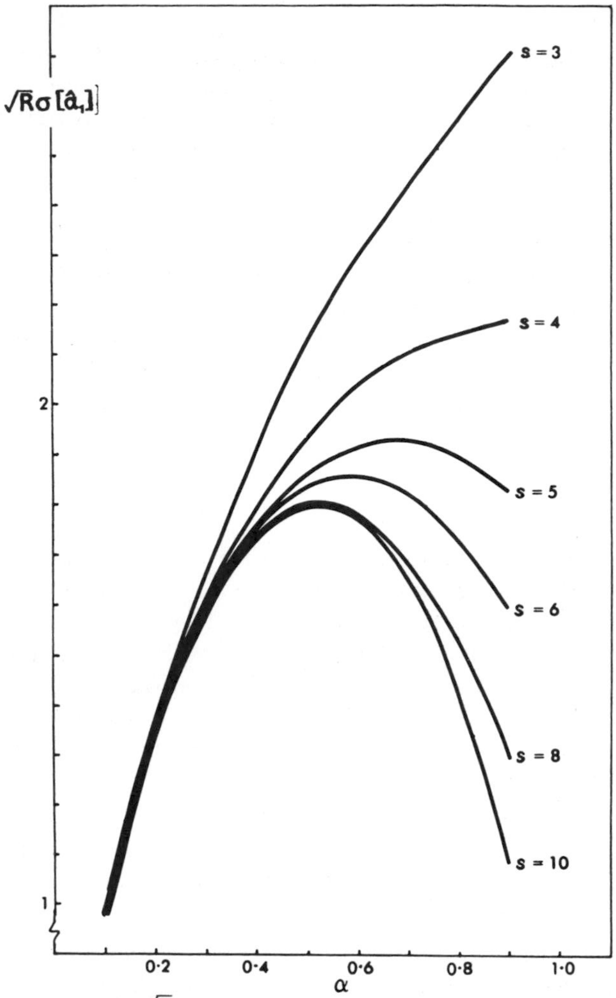

Fig. 13.2 Relationship of $\sqrt{R}\sigma[\hat{\alpha}_1]$ as a function of survival (α) and number of years of tag releases (s) with a reported exploitation rate (β) equal to 0·1: from Fig. 2, Youngs and Robson [1975].

$$RV[\hat{\alpha}_i] \approx \frac{\alpha^2}{\beta}\left[\frac{1-\alpha^2}{\alpha(1-\alpha^{s-i})} - \frac{(1-\alpha)^2(1-\alpha^{i-1})}{(1-\alpha^i)(1-\alpha^{s-i})(1-\alpha^{s-i+1})} - 2\beta\right] \quad (13.20)$$

$$(= R\sigma^2[\hat{\alpha}_i], \quad \text{say}),$$

and the authors give graphs of $\sqrt{R}\,\sigma[\hat{\sigma}_1]$ versus α for $\beta = 0{\cdot}01$, 0·1 and 0·3, and different s ($= k$ in their notation). Suppose that a fishery manager wishes to estimate the survival probability for the first year of study of a three-year release programme and be within ± 0·1 with 95 percent confidence. Then using the approximate confidence interval $\hat{\alpha}_1 \pm 2\sigma[\hat{\alpha}_1]$, we have that $2\sigma[\hat{\alpha}_1] = 0{\cdot}1$ or

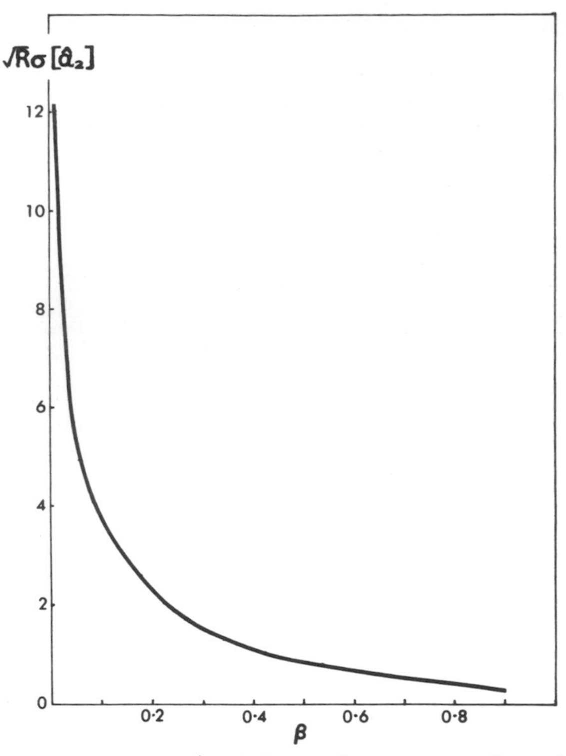

Fig. 13.3 Relationship between $\sqrt{R}\sigma[\hat{\alpha}_2]$ and β for $s = 3$ years of tag releases: from Fig. 5, Youngs and Robson [1975].

$\sigma[\hat{\alpha}_1] = 0{\cdot}05$. Suppose further that preliminary studies have indicated that $\alpha \approx 0{\cdot}6$ and $\beta \approx 0{\cdot}1$. Then entering Fig. 13.2 with $\alpha = 0{\cdot}6$ and $s = 3$, we have $\sqrt{R}\,\sigma[\hat{\alpha}_1] = 2{\cdot}3$ or $R = (2{\cdot}30/0{\cdot}05)^2 = 2116$, so that 2116 tags must be released at the beginning of each of the three years. If a graph is not available for a particular β then, for a conservative R (i.e. too many tags), we would choose the nearest value below β which is graphed. Now the maximum value of $\sigma[\hat{\alpha}_i]\sqrt{R}$ is a function of β and occurs when $i = 2$ and $s = 3$ with α at its maximum value. This relationship is shown graphically in Fig. 13.3 and gives an *upper bound* on the number of tags that need to be released each year in order to have at least the desired level of confidence.

HYPOTHESIS TESTING. Youngs and Robson [1975] also give a number of test procedures for testing the above model (13.6). As pointed out by Robson and Youngs [1971], a test of the model can be obtained from the distribution of the $\{m_{ij}\}$ conditional on the sufficient statistics $(r_1, r_2, \ldots, r_s, T_2, \ldots, T_s)$. This conditional distribution has "rank"

$$s(s+1)/2 - (2s-1) = (s-1)(s-2)/2 = 1 + 2 + \ldots + s - 2, \qquad (13.21)$$

and may be expressed as a product of multi-hypergeometric distributions of

TABLE 13.2

Contingency table for testing the goodness of fit of model (13.6): from Youngs and Robson [1975].

Year tagged	Year of recapture 2	3	...	s	Total
1	m_{12}	m_{13}	...	m_{1s}	$T_2 - r_2$
2	m_{22}	m_{23}	...	m_{2s}	r_2
Total	m_2	m_{23}^*	...	m_{2s}^*	T_2

successively smaller "ranks" $s-2, s-3, \ldots, 1$. Each distribution leads to an (asymptotically) independent chi-squared test based on a contingency table. The first table is Table 13.2 and the usual test statistic for homogeneity based on this table is approximately χ^2_{s-2}, when the model is true. The second table is Table 13.3 and the test statistic is asymptotically χ^2_{s-3}. This statistic is approximately independent of the previous test statistic so that the two chi-squared statistics are additive. The $(i-1)$th such table is given by Table 13.4 and the chi-squared statistic for this table is approximately independent of the previous chi-squared statistics. Proceeding in this manner we end up with $(s-2)$ independent chi-squared statistics which may be added to give one combined test of the model with $(s-1)(s-2)/2$ degrees of freedom (cf. (13.21)).

TABLE 13.3

Contingency table for testing the goodness of fit of model (13.6): fom Youngs and Robson [1975].

Year tagged	Year of recapture 3	4	...	s	Total
1 or 2	m_{23}^*	m_{24}^*	...	m_{2s}^*	$T_3 - r_3$
3	m_{33}	m_{34}	...	m_{3s}	r_3
Total	m_3	m_{34}^*	...	m_{3s}^*	T_3

Certain alternatives to the model can be tested by partitioning each of the above chi-squares. An important alternative is the possibility of a type I loss or initial tagging mortality (cf. p. 231). A test against this alternative is obtained by partitioning each $2 \times (s-i+1)$ table, Table 13.4 ($m_{ij}^* = m_{1j} + m_{2j} + \ldots + m_{ij}$), into the two tables, Tables 13.5 and 13.6. We thus have a partition of χ^2_{s-1} into χ^2_1 and χ^2_{s-i-1}, respectively. The sum of the $(s-2)$ χ^2_1 statistics provides a test against short-term tagging effects.

Youngs and Robson [1975] demonstrate the above theory and tests on some data for lake trout.

TABLE 13.4

Table to be partitioned for testing type I loss: from Youngs and Robson [1975].

Year tagged	Year of recapture i	$i+1$	$,\ldots$	s	Total
$1, 2, \ldots, i-1$	$m^*_{i-1,i}$	$m^*_{i-1,i+1}$	$\cdots$	$m^*_{i-1,s}$	$T_i - r_i$
i	m_{ii}	$m_{i,i+1}$	$\cdots$	$m_{i,s}$	r_i
Total	m_i	$m^*_{i,i+1}$	$\cdots$	$m^*_{i,s}$	T_i

TABLE 13.5

Contingency table for detecting type I losses; all the columns of Table 13.4, but the first, are pooled: from Youngs and Robson [1975].

$m^*_{i-1,i}$	$T_i - r_i - m^*_{i-1,i}$	$T_i - r_i$
m_{ii}	$r_i - m_{ii}$	r_i
m_i	$T_i - m_i$	T_i

TABLE 13.6

A partition of Table 13.4 involving all the columns but the first: from Youngs and Robson [1975].

$m^*_{i-1,i+1}$	$\cdots$	$m^*_{i-1,s}$	$T_i - r_i - m^*_{i-1,i}$
$m_{i,i+1}$	$\cdots$	$m_{i,s}$	$r_i - m_{ii}$
$m^*_{i,i+1}$	$\cdots$	m^*_{is}	$T_i - m_i$

INSTANTANEOUS MORTALITY RATES. Let μ_i, μ_{Ei} and $Z_i = \mu_i + \mu_{Ei}$ be the instantaneous natural, exploitation, and total mortality rates respectively in year i (p. 3). Suppose that a proportion $1 - \nu_i$ of those tagged and released at the beginning of year i die immediately after release, and a proportion ρ_i of the tags recaptured in year i are actually reported. Then, if the time unit is one year,

$$\alpha_i = e^{-Z_i} \quad (= S_i \text{ of } 6.2) \tag{13.22}$$

and, from p. 272, equation (6.2),

$$\beta_i = c_i\mu_{Ei}(1 - \alpha_i)/Z_i \quad (= c_i u_i, \text{ say}), \tag{13.23}$$

where $c_i = \rho_i\nu_i$ is usually assumed to be constant ($= c$, say). Then from the maximum likelihood estimates $\hat{\alpha}_i$ and $\hat{\beta}_i$ given by equations (13.9) and (13.10), we can use (13.22) and (13.23) to obtain the maximum likelihood estimates $\hat{Z}_i = -\log \hat{\alpha}_i$ of Z_i and $\hat{G}_i = -\hat{\beta}_i \log \hat{\alpha}_i/(1 - \hat{\alpha}_i)$ of $G_i = c_i\mu_{Ei}$. Also, using the delta method (p. 7), we have

$$V[\hat{Z}_i] \approx V[\hat{\alpha}_i]/\alpha_i^2 \tag{13.24}$$

and

$$\operatorname{cov}[\tilde{Z}_i, \tilde{Z}_j] \approx \operatorname{cov}[\hat{\alpha}_i, \hat{\alpha}_j]/\alpha_i\alpha_j. \tag{13.25}$$

If an estimate of c_i is available then μ_{Ei}, and hence μ_i, can be estimated. A number of special cases will now be considered.

Suppose that $c_i = c$ and $\mu_i = \mu$, then using the approximation (p. 330, equation (8.8))

$$(1 - e^{-w})/w \approx e^{-(1/2)w},$$

we have

$$\frac{e^{-\mu} - e^{-(\mu+\mu_{Ei})}}{1 - e^{-(\mu+\mu_{Ei})}} \cdot \frac{\mu_{Ei} + \mu}{\mu_{Ei}} \approx e^{-(1/2)\mu}$$

or

$$e^{-\mu} - e^{-Z_i} \approx e^{-(1/2)\mu}(1 - \alpha_i)\mu_{Ei}/Z_i$$

$$= e^{-(1/2)\mu}\beta_i/c.$$

Then rearranging the above equation, and setting $\phi = e^{-\mu}$ and $k = c^{-1}e^{-(1/2)\mu}$, we have from (13.22)

$$\alpha_i \approx \phi - k\beta_i. \tag{13.26}$$

Thus a regression of $\hat{\alpha}_i$ versus $\hat{\beta}_i$ will be approximately linear with the intercept estimating ϕ, the natural mortality rate. This model, (13.26), was proposed by Youngs [1972], though he used a less efficient estimate of β_i, namely $\beta_i^* = m_{ii}/R_i$ in his example. If the exploitation is seasonal then it transpires that the only change in (13.26) is in the constant k.

Now if just $c_i = c$ we have, for large releases,

$$E[\hat{G}_i] \approx G_i = c\mu_{Ei}$$

so that

$$E[\hat{Z}_i] \approx Z_i$$

$$= \mu_i + \mu_{Ei}$$

$$\approx \mu_i + \left(\frac{1}{c}\right)E[\hat{G}_i].$$

Thus if $\mu_i \approx \mu$ or $\mu_i = \mu + \epsilon_i$, where $E[\epsilon_i] = 0$, we have the following regression model

$$\hat{Z}_i = \mu + \left(\frac{1}{c}\right)\hat{G}_i + e_i, \tag{13.27}$$

where e_i is the "error" satisfying $E[e_i] \approx 0$, and c can be estimated from the slope of the fitted line. This model, together with an example, was proposed by Youngs [1974] with the following changes in notation: $\hat{G} \to R$, $\alpha \to S$ and $\beta \to f$. It should be noted that some care is needed in fitting (13.26) and (13.27) as they are, both, so-called, "functional relationships" (cf. Seber [1977: 210] and Ricker [1973]). Equation (13.23) is discussed further by Youngs [1976: see his equation (2)] from the point of view of μ_{Ei} varying throughout the year.

If $\mu_i = \mu$ and effort data are available then we have the model $E[\hat{Z}_i] \approx Z_i = \mu + k\mathring{f}_i$, where $\mathring{f}_i$ is the total effort in year i, and least squares estimates of μ and k, the Poisson catchability coefficient, can be readily calculated. Although the $\hat{Z}_i$ are correlated the (unweighted) least squares estimates will still be approximately unbiased. However the usual variance formulae for the least squares estimates will not be correct unless the covariances are small and the variances approximately equal. A more appropriate procedure would be to use a generalised or weighted least squares analysis (p. 11) using an estimated variance–covariance matrix of the $\hat{Z}_i$ obtained from (13.24) and (13.25).

If $\mu_i = \mu$, $\mu_{Ei} = \mu_E$ and $c_i = c$, then $\alpha_i = \alpha$ and $\beta_i = \beta$ and we are back to the model of (13.16). In this case μ and μ_E can be estimated if an estimate of c is available. Alternatively, we note that

$$E[m_{ij}] = R_i\alpha^{j-i}\beta = R_i e^{-Z(j-i)}cP_1, \quad \text{say},$$

where $P_1 = \beta/c$, and this leads to the model

$$E[\log(R_i/m_{ij})] \approx -[Z + \log(cP_1)] + Z(j - i + 1).$$

This model is considered on p. 239, equation (5.34), but with a change in notation for the m_{ij} so that $(j - i + 1)$ becomes j.

5 *Estimating tag loss*

Methods of estimating tag loss using double tagging are described in 6.3.3. Although not specifically stated there, the essence of the method is that the conditional distribution $f(m_A(t), m_B(t), m_{AB}(t)|m(t))$, from which $\pi_A(t)$, $\pi_B(t)$ and $m(t)$ are estimated, is the same as for the closed population case of p. 95, irrespective of the distribution of $m(t)$ $(= m_2)$. When the tags are indistinguishable and $\pi(t)$ is the probability of losing a tag by time t, then a common tagging model (cf. Bayliff and Mobrand [1972]) is to set $1 - \pi(t) = \rho \exp(-Lt)$, where $1 - \rho$ is the proportion of tags immediately lost on release and L is the instantaneous tag loss rate (two tags on the same individual are regarded as two separate tags). Thus we can fit the model (cf. p. 282: $m_c(t) = m_A(t) + m_B(t)$)

$$\begin{aligned} E\left[\log\left\{\frac{2m_{AB}(t_i)}{m_C(t_i) + 2m_{AB}(t_i)}\right\}\right] &\approx \log[1 - \pi(t_i)] \\ &= \log\rho - Lt_i, \end{aligned} \qquad (13.28)$$

which is a straight line. Lenarz *et al.* [1973] compared the loss rates for different types of tags and then pooled the data to give estimates of ρ and L based on two points $(i = 1, 2)$ corresponding to two years of returns. Similar models can be developed as follows if single-tagged animals are also released.

Suppose that M_S and M_D single- and double-tagged individuals are released at the beginning of a single release experiment, and let $m_S(t)$ be the number of single-tagged individuals caught at time t which still retain their tags.

Then assuming that both groups of individuals behave the same with respect to mortality, exploitation, etc., and that tags are indistinguishable, we have

$$\frac{E[m_{AB}(t)]}{E[m_S(t)]} = \frac{M_D[1-\pi(t)]^2}{M_S[1-\pi(t)]} \tag{13.29}$$

$$= \frac{M_D}{M_S}\rho \exp(-Lt).$$

This leads to the regression model

$$E\left[\log\left\{\frac{m_{AB}(t)M_S}{m_S(t)M_D}\right\}\right] \approx \log\rho - Lt. \tag{13.30}$$

Similarly

$$\frac{E[m_C(t)]}{E[m_S(t)]} = \frac{M_D 2\pi(t)[1-\pi(t)]}{M_S[1-\pi(t)]},$$

which leads to

$$E\left[\log\left\{1-\frac{m_C(t)M_S}{2m_S(t)M_D}\right\}\right] \approx \log\rho - Lt. \tag{13.31}$$

The three models (13.28), (13.30) and (13.31), with $\rho = 1$, were developed by Chapman *et al.* [1965] using the usual exponential model for mortality and exploitation (p. 272). However, we need only to assume that single- and double-tagged individuals behave identically, so that when taking ratios such as (13.29), the factors relating to mortality and exploitation cancel out. The three models, with general ρ, were also investigated by Bayliff and Mobrand [1972], who considered methods of combining the three estimates of L obtained from these models.

13.1.5 Bird-banding models(*)

1 Time-specific model

Band returns have been used extensively for estimating survival rates for birds for many years, but it is only recently (cf. Brownie *et al.* [1978]) that proper statistical models, together with their associated goodness-of-fit tests have been developed. The choice of model depends on whether the survival rate and band recovery rate are time- and(or) age-dependent. Invariably the band recovery rate is at least time-dependent, as the recovery of a band depends on such factors as the mortality rate and the probability that a band is returned, given that it is found. If survival is allowed to be completely age- and time-dependent, then there are too many parameters in the model, with a consequent lack of identification, and some constraints on the parameters are needed. These are provided by the fact that the survival rates (and hence the band recovery rates) of adult birds tend to become age-independent and depend only on the calendar year. Thus if just *adults* are banded then survival is time-dependent only. If we also assume that the band recovery rate does not depend on age then we arrive once again at the basic model (13.6) which can be best described by the equation

(*)See Nelson *et al.* [1978] (Additional References) for a band loss study.

$$E[m_{ij}] = R_i\alpha_i\alpha_{i+1} \ldots \alpha_{j-1}\beta_j \quad (i = 1, 2, \ldots, s; j = 1, 2, \ldots, t) \quad (13.32)$$

or, in the notation of Brownie *et al.* [1978] which is given in brackets below,

$$E[R_{ij}] = N_iS_iS_{i+1} \ldots S_{j-1}f_j.$$

Here $R_i(N_i)$ is the number released at the beginning of the ith calendar year, $m_{ij}(R_{ij})$ is the number from $R_i(N_i)$ recovered in the jth calendar, $\alpha_i(S_i)$ is the probability of survival in the ith year ("survival rate" in year i) and $\beta_j(f_j)$ is the probability that a band is recovered in year j (the so-called band recovery rate or reported exploitation rate in year j).

The above model was originally developed independently by Robson and Youngs [1971] and Seber [1970b] and is described in the context of bird-banding on pages 240–3. Further numerical examples of this model are given in Anderson and Sterling [1974], Anderson [1975] and Brownie *et al.* [1978: ch. 2, model 1]. In my approach on p. 240 I emphasised the case of non-game birds for which $\beta_j = (1 - \alpha_j)\lambda_j$ (with $\alpha_j = \phi_j$), λ_j being the probability that a banded bird is found and its band returned in year j. For game birds, where bands are returned by hunters, β_j can be expressed in the form $\beta_j = H_j\delta_j$, where H_j is the so-called harvest rate and δ_j is the band reporting rate, that is the probability that a hunter will report the band, given that he has killed and retrieved a banded bird in year j. Clearly H_j and δ_j cannot be separately estimated unless further information is available, for example δ_j is estimated using reward bands (Henny and Burnham [1976]). We note that $H_j < 1 - \alpha_j$ as $1 - \alpha_j$ also includes natural mortality as well as hunting mortality. The relationships between the various parameters are discussed in detail by Anderson and Burnham [1976] and Brownie *et al.* [1978: 14]. The estimate $\hat{\beta}_j$ has been used as an index of hunting pressure in preference to the proportions of bands recovered in the first season after a band release (Anderson and Sterling [1974]). The special cases of constant survival rate α and(or) constant recovery rate β are considered by Brownie *et al.* [1978: models (2) and (3)]. These models should be fitted where possible, as the estimation of fewer parameters leads to more efficient estimates (e.g. Raveling [1976]). However, explicit solutions of the maximum likelihood equations do not exist and the equations need to be solved iteratively. The case of constant α and β is discussed further in section 4 below.

An important problem which has not been adequately considered is the question of band loss or band illegibility. However, if double banding is used then the returns from each release can be corrected using the methods of pages 281–2. Thus if $m_C(t)$ and $m_{AB}(t)$ are returns for a given year t years after the initial release, then the corrected number of returns is $\tilde{m}(t)$ $(= m_{AB}(t)/[1 - \tilde{\pi}(t)]^2$, cf. p. 489). If $\pi(t)$ depends only on t, the period after release, and not on the year of release, then a sequence of $\pi(t)$ values can be calculated for each release and averaged over releases. This smoothed estimate can then be used in the calculation of $\tilde{m}(t)$.

An important use of banding data is to determine areas of breeding, wintering, migration and harvest, as well as degree of homing. Computer methods for plotting band recoveries on maps and comparing distributions are described by Cowardin [1977] and Davenport [1977]; computer methods have also been used for constructing suitable number and colour sequences for bands (Duncan [1971]).

2 Time- and age-specific models

For a number of reasons the recovery rates for newly banded birds may be different from those of birds banded in previous years. The above model (13.32) has been modified by Brownie *et al.* [1978: 30] to deal with this generalisation. Further adaptions have been given by Brownie [1973], Brownie and Robson [1976] and Brownie *et al.* [1978: ch. 3] which allow for the survival and recovery rates to be age-dependent with respect to the two groups young–adults (adults being birds older than one year) or the three groups young–subadults–adults (adults being more than two years old). These models assume that only two groups, young and adults, are distinguishable – the usual situation in practice – and that separate records are kept for the numbers released and recovered from the two groups. Models where the three age-groups are recognisable are given by Brownie [1973] and discussed in detail by Brownie *et al.* [1978: ch. 4]. A two-group model which imposes some constraints on the parameters for the young birds is given by Johnson [1974]: he assumes that two proportionality factors relating young to adult parameters are constant over time.

For many species it is easier to band the young than to band adults, so that there is a strong temptation to band only the young. However, Brownie *et al.* [1978: 112–3] take the view that this is a pointless practice as the parameters are no longer identifiable (that is uniquely determined from the model) and therefore cannot be estimated. This lack of identifiability is due to the fact that young generally have a lower survival rate than adults and that the first-year band recovery rate for young is typically higher than for adults. A different survival rate for young has been observed with other animals (e.g. Storm *et al.* [1976: foxes]). For a different approach see (*)North [1979] and North and Morgan [1979].

3 Age-specific model

One further model which merits special reference, because of its misuse rather than its usefulness, is the so-called age-specific model (Seber [1971]) discussed on p. 251. This model assumes that (i) the survival and recovery rates are age-dependent only, that is they are unaffected by year to year changes in the hunting regulations, habitat, weather, etc., and (ii) the age-specific recovery rate is a constant fraction of the age-specific mortality. Thus if α_j and β_j now refer to birds of age j rather than to calendar year j then, by assumption (ii), $\beta_j/(1-\alpha_j) = c$ $(j = 1, 2, \ldots, t-1)$, where c is a constant which may be identified with λ of section *1* in the non-game bird situation (cf. Anderson and Burnham [1976: Appendix B]). An analysis of this particular model, which

(*)In Additional References.

requires one constraint on the α_j's such as setting two of the α_j's equal for the purposes of identifiability, has been carried out by Seber [1971: $\alpha_j = \phi_j$ and $\beta_j = \lambda(1 - \phi_j)$][(*)]. A goodness of fit test of this model which, as might be expected, does not depend on the particular choice of identifiability constraint, is given by Burnham and Anderson [1979]. They also give a contingency table statistic to test whether the first year recovery rates are constant, that is independent of time.

However, the above assumptions are demonstrably invalid for game bird studies, e.g. mallard (Anderson [1975], Burnham and Anderson [1979]) and are questionable for non-game species. For instance, the recovery rates of nearly all waterfowl species vary significantly by both age and years (Anderson [1975: 13]). Apart from the problem of time-dependent survival, the harvest rate depends on the hunting regulations which may change with time (e.g. Hopper *et al.* [1975: 63]), and the band reporting rate tends to fall off with time as hunters lose interest in the programme (Henny and Burnham [1976]). Not only are the basic assumptions generally invalid, but the estimates obtained from the above model are very sensitive to departures from the underlying assumptions.

In addition to the conditional maximum likelihood method of Seber [1971] there is another time-hallowed method of analysing the above model, called the composite dynamic method (Hickey [1952], Geis and Taber [1963], Geis *et al.* [1971], Geis [1972a, b]). This method, which is frequently set out in the form of a life-table analysis (p. 252), makes the additional assumption (iii) that the experiment is continued for a sufficient number of years after the last release, so that virtually all the banded birds are dead by the end of the experiment, that is the identifiability constraint on the α_j takes the form $\alpha_t = 1$. However, such an assumption is not very helpful in the case of bird-banding studies where estimates are needed in the year following the last release rather than several years later. If this assumption is not met, which is often the case, then the survival estimates will be too low (Seber [1972: table 2]; see also Burnham and Anderson [1979]). Clearly the composite dynamic method along with the various *ad hoc* life table methods associated with it (pages 253–4) should be dropped. As far as bird-banding experiments are concerned we quote the following from Brownie *et al.* [1978: 115, see also 184]: "We cannot emphasise too strongly that, based on our current knowledge, there is no valid way to estimate age-specific survival rates from only the banding of young." In fact, anyone wishing to analyse bird-banding data is advised to study this comprehensive survey by Brownie *et al.* and in particular their Appendix A. This survey lists all the various models, together with appropriate goodness of fit tests and examples, and also discusses various practical aspects such as pooling (adult males and females, geographical areas), coding data, experiments with banding twice a year, the possible utilisation of live recapture data, unequal time-intervals, average survival estimates, estimation of mean life-span, and the design of banding studies.

(*)See also North and Cormack [1980], North and Morgan [1979] (Additional References).

4 Constant survival and recovery rates

The case of constant α and β has already been considered several times in this book under different guises. Apart from a constant term, the joint distribution of the $\{m_{ij}\}$ is given by (13.16), which is the same as the model on p. 245 with $\alpha = \phi$ and $\beta = (1-\phi)\lambda$. This particular model, which requires an iterative solution of the maximum likelihood equations, is considered by Brownie *et al.* [1978: model 3]. A check on this model is provided by the linearity of the regression model of p. 254, namely

$$\begin{aligned} Y_j &= \log[100\beta] + (j-1)\log\alpha + e_j \\ &= a + (j-1)\log\alpha + e_j, \text{ say.} \end{aligned} \tag{13.33}$$

Any non-linearity will indicate departures from the underlying assumptions, for example, decline in the recovery rate due to band loss (Henny [1972], Wakeley and Mendall [1976]). It should be noted that a linear model is still appropriate even if the α_j are not constant but are random variables. Setting

$$\begin{aligned} \log\alpha_j &= E[\log\alpha_j] + (\log\alpha_j - E[\log\alpha_j]) \\ &= b + u_j, \quad \text{say,} \end{aligned}$$

we have

$$\begin{aligned} Y_j &= a + (j-1)b + (j-1)u_j + e_j \\ &= a + (j-1)b + e_j', \quad \text{say,} \end{aligned}$$

which is still a straight line.

An alternative method of estimation is available if we use the conditional distribution of the m_{ij} given the r_i, that is $f(\{m_{ij}\}|\{r_i\})$. This leads to the generalisation of Haldane's method given on p. 246. However, this approach does not provide an estimate of β, and the estimate of α is not the asymptotically efficient maximum likelihood estimate. If $\alpha^{t-s+1} \approx 0$ (assumption (iii) of section *3*) so that there are no banded birds left by the end of the experiment, then the maximum likelihood estimates of α and β are given by $\hat{\alpha}$ and $\hat{\beta}$ of (13.17) and (13.18). Noting that

$$u = \sum_{k=1}^{s} (k-1)D_k,$$

where $D_k = \sum_{i=1}^{s} m_{i,i+k-1}$ is the number of bands returned from birds which die in their kth year after release, we find that $\hat{\alpha} = \Sigma_k(k-1)D_k/\Sigma_k kD_k$, which is Lack's estimate (p. 247, with $\alpha = \phi$). Unfortunately this estimate of α is frequently used when assumption (iii) is false.

If Lack's method is applicable, then the underlying model is the geometric distribution (5.43) on p. 247. This model has also been studied by Chapman and Robson [1960] and, as noted by Eberhardt [1972], the theory based on equation (10.15) of p. 414 applies here. Thus, substituting $n = \Sigma_j D_j$ and $X = \sum_{x=0}^{r} xn_x = \Sigma(j-1)D_j$, their estimate $X/(n+X)$ is simply Lack's estimate.

When there is a single release, as for example in fisheries, we find that the conditional and unconditional approaches give the same maximum likelihood estimate $\tilde{\alpha}$ say, the solution of (5.40) on p. 246 with $s = 1$ and $\phi = \alpha$. As the conditional likelihood function is the same as for a sample from a truncated geometric distribution (cf. (5.42) of p. 247 with $i = 1$), $\tilde{\alpha}$ is also the solution of (10.19) on p. 421 with $k + 1 = t$. For a similar reason we see that $\tilde{\alpha}$ is the solution of (6.11) on p. 275.

13.1.6 Generalised hypergeometric models

1 Introduction

The first models developed for studying mark–recapture experiments for closed populations were hypergeometric (cf. equations (3.1), (4.5) and (4.3)). These models are based on the assumption that we have independent simple random samples of predetermined size. However, the hypergeometric distribution and its generalisations can be difficult to handle mathematically and do not readily allow for modifications corresponding to changes in the catchability assumptions. In contrast to these distributions in which the random element in the sampling process is due to the activity of the experimenter, the multinomial distribution focuses attention on the activity of the animals. Assuming independent movement, the animals can be regarded as N independent multinomial "trials" with certain probabilities of belonging to different capture categories (equation 4.1)). This model can also be derived directly from the multi-hypergeometric model (4.3) by allowing the sample sizes to be random variables (the approach in **4.1.2** is the reverse of this for ease of exposition). We saw in **12.8.6** that the multinomial distribution is very flexible and can accommodate a variety of assumptions about the catchability of animals.

As pointed out by Cormack [1979] in his excellent overview of capture–recapture models, the historical development of stochastic models for open populations has been in the reverse order from those of closed populations. Models based on the multinomial distribution came first, with special cases (Darroch [1959], Seber [1962], Cormack [1964]) leading to the general models of Jolly [1965] and equation (5.5). These were then followed by the complex generalised hypergeometric models of Robson [1969] and Pollock [1975b]. Their approach has considerable potential and will lead to further models (e.g. Pollock [1979]).

The joint distribution for a multinomial-type model can be built up in several ways. In Chapter 5 (p. 198) we studied the fate of the u_i unmarked individuals in each subsequent sample, and each u_i led to a multinomial distribution. Unfortunately this approach, although notationally and mathematically convenient, does not readily lend itself to further generalisations in which the probabilities of capture and survival can vary with mark status. However, by appropriately specifying these probabilities, a model of the form (5.2) could always be written down, though some intuition is usually needed in identifying the appropriate data groups – the sufficient statistics – as for example in (5.3).

Another approach, utilised by Jolly [1965] and which has considerable flexibility, is to take a group of individuals with a common capture history and break the group down into subgroups according to their future fate. For example, a member of a particular group in the population before sample i is either caught or not caught in sample i. If caught it is either lost on capture or released, and if released it dies before sample $i + 1$ or survives until sample $i + 1$. Those not caught in sample i either die before sample $i + 1$ or survive until sample $i + 1$. The numbers in each category are binomially or trinomially distributed conditionally on their immediate ancestors, and the overall likelihood function is a product of such distributions.

A third approach which leads to a generalised hypergeometric model is to regard the size of the various groups which have the same capture and survival characteristics as fixed parameters, and assume that simple random samples are taken from these groups. For example, referring to pages 199–200, the basic groups are $M_i + U_i$ $(= N_i)$ and $M_i - m_i + R_i$ from which are taken samples of size n_i (fixed) and $r_i + z_i$ (random). Since the triples $\{m_i, r_i, T_i\}$, where $T_i = r_i + z_i$, are sufficient for the parameters $\{M_i, U_i\}$, we find that the relevant part of the likelihood function for estimating these parameters is

$$\prod_{i=2}^{s} \left\{ \frac{\binom{M_i}{m_i}\binom{U_i}{u_i}}{\binom{N_i}{n_i}} \right\} \prod_{i=2}^{s-1} \left\{ \frac{\binom{M_i - m_i}{z_i}\binom{R_i}{r_i}}{\binom{M_i - m_i + R_i}{T_i}} \right\} \tag{13.34}$$

$$= L_1 \times L_2, \quad \text{say}.$$

Here L_1 requires that the marked and unmarked have the same chance of being caught in the ith sample, while L_2 requires that marked individuals have the same furture survival and capture characteristics after sample i. By setting first differences of the logarithm of (13.34) equal to zero we readily find that the maximum likelihood estimates of M_i and N_i are given by (5.7) and (5.8). Other parameters can then be introduced deterministically, namely $p_i = n_i/N_i$, $\phi_i = M_{i+1}/(M_i - m_i + R_i)$ etc. and can be estimated as on p. 200.

This type of approach forms the basis of a number of very general models proposed by Robson [1969] and Pollock [1975b, 1979]. Cormack [1972, 1973] utilised similar ideas to develop intuitive estimates for certain "non-standard" situations. Robson [1969] showed that the assumption of equal survival rates could be weakened to allow animals with different capture histories to have different survival rates. Pollock [1974, 1975b] further generalised Robson's model to allow the catchability to also vary with capture history. In both models some constraints on the degree of unequal survival and catchability are required in order for all the parameters to be estimable (i.e. identifiable). It is assumed below that releases immediately follow samples.

2 Robson–Pollock model

After the $(i-1)$th sample, individuals can be grouped by their capture history in the first $i-1$ samples. We will denote members of such a group by the superscript v (v depends on i). Thus $M_i^{(v)}$ is the number of marked individuals belonging to group v which are in the population just prior to the ith sample (time t_i^-), and $M_i^{(0)} = U_i$ represents the number of unmarked in the population at time t_i^-. Robson and Pollock define the following random variables (with their notation given in brackets):

$u_i = m_i^{(0)}$, the number from U_i caught in sample i $(= n_i^{(0)})$;
$m_i^{(v)}$ = number of $M_i^{(v)}$ caught in sample i $(= n_i^{(v)})$;
$R_{i-1}^{(v)}$ = number released after sample $i-1$ belonging to group v $(= m_{i-1}^{(v)})$;
$r_{i-1}^{(v)}$ = number of different individuals subsequently recaptured from the $R_{i-1}^{(v)}$ released $(= R_{i-1}^{(v)})$;
$z_i^{(v)}$ = number belonging to $M_i^{(v)}$ which are not captured in sample i but are subsequently recaptured $(= Z_i^{(v)})$;
$T_{i-1}^{(v)}$ = number of different individuals belonging to group v after sample $i-1$ which are subsequently recaptured $(= T_{i-1}^{(v)})$;
$S_{i-1}^{(v)}$ = number of individuals which would have belonged to group v if they had not been captured in sample $i-1$ $(= S_{i-1}^{(v)})$;
$m_i = \sum_{v \neq 0} m_i^{(v)}$;
$n_i = m_i + u_i$.

Pollock derives the general likelihood function for the above situation but considers in detail the special case in which the first, but not subsequent, capture affects the individual's survival for L periods and the catchability for Q periods $(Q \leqslant L)$. In this case he defines $M_i^{(v)}$ as the number of marked first captured in sample $i-v$ which are alive at time t_i^- $(v = 1, 2, \ldots, L; i = v+1, \ldots, s)$ and $M_i^{(L+1)}$ is the number first captured in or before sample $i-L-1$ which are alive at time t_i^-. Thus $v = 0, 1, \ldots, L+1$ and the total population size at time t_i^- is

$$N_i = U_i + \sum_{v=1}^{L+1} M_i^{(v)} = \sum_{v=0}^{L+1} M_i^{(v)}.$$

We note that for $v \geqslant 1$, membership in group $v+1$ after sample i (i.e. before sample $i+1$) does not depend on whether an individual is caught or not caught in sample i. Therefore those caught in sample i from group v would have belonged to group $v+1$ after sample i if they had not been caught in sample i, that is

$$S_i^{(v+1)} = m_i^{(v)}, \quad v = 1, 2, \ldots, L-1.$$

Also $R_i^{(L+1)}$ is the number released from sample i which now belong to group $L+1$ and therefore belonged to group L *or* $L+1$ prior to sample i. The same description applies to $r_i^{(L+1)}$ and $z_i^{(L+1)}$, and forms the basis for the equation

$$S_i^{(L+1)} = m_i^{(L)} + m_i^{(L+1)}. \tag{13.35}$$

Finally we have the relationships

$$z_i^{(v)} = T_i^{(v+1)} - r_i^{(v+1)} = T_{i-1}^{(v)} - m_i^{(v)}, \quad v = 1, 2, \ldots, L,$$

$$z_i^{(L)} = T_{i-1}^{(L)} - m_i^{(L)},$$

$$z_i^{(L+1)} = T_{i-1}^{(L+1)} - m_i^{(L+1)},$$

and

$$z_i^{(L)} + z_i^{(L+1)} = T_i^{(L+1)} - r_i^{(L+1)}$$

$$= T_{i-1}^{(L)} + T_{i-1}^{(L+1)} - m_i^{(L)} - m_i^{(L+1)}.$$

With regard to estimation for the above model, the maximum likelihood estimates for the $M_i^{(v)}$ take simple forms only when $Q = L - 1$ (called model 1 by Pollock) or $Q = L$ (model 2). Pollock shows that in both cases, for $v = 1, 2, \ldots, L - 1$ and $i = v + 1, \ldots, s$, the maximum likelihood estimate of $M_i^{(v)}$ is given by

$$\hat{M}_i^{(v)} = m_i^{(v)} + \frac{R_i^{(v+1)} z_i^{(v)}}{r_i^{(v+1)}}. \tag{13.36}$$

As on p. 200 (equation 5.6)), this equation follows from the intuitive relationship

$$\frac{z_i^{(v)}}{M_i^{(v)} - m_i^{(v)}} \approx \frac{r_i^{(v+1)}}{R_i^{(v+1)}}. \tag{13.37}$$

Here, those $R_i^{(v+1)}$ caught and released from sample i and belonging to group $v + 1$, belonged to group v before they were caught in sample i. Since these individuals are affected only by their first capture (and not their capture in sample i), the two groups $M_i^{(v)} - m_i^{(v)}$ and $R_i^{(v+1)}$ have the same subsequent survival and capture probabilities, and can therefore be compared. The estimate $\hat{M}_i^{(v)}$ can also be obtained intuitively from the conditional distribution

$$\Pr[r_i^{(v+1)} | T_i^{(v+1)}, m_i^{(v)}] = \frac{\binom{R_i^{(v+1)}}{r_i^{(v+1)}} \binom{M_i^{(v)} - m_i^{(v)}}{T_i^{(v+1)} - r_i^{(v+1)}}}{\binom{M_i^{(v)} - m_i^{(v)} + R_i^{(v+1)}}{T_i^{(v+1)}}}$$

by equating $r_i^{(v+1)}$ to its (conditional) expected value. Pollock [1974] uses this distribution to show that, asymptotically,

$$V[\hat{M}_i^{(v)} | m_i^{(v)}] = (M_i^{(v)} - m_i^{(v)})(M_i^{(v)} - m_i^{(v)} + R_i^{(v+1)}) \left(\frac{1}{E[r_i^{(v+1)}]} - \frac{1}{R_i^{(v+1)}} \right). \tag{13.38}$$

We now consider the estimation of U, $M_i^{(L)}$ and $M_i^{(L+1)}$. For the above two cases, and in fact for the general model, U can be estimated only if there is another group $M_i^{(v)}$ of marked which has the same probability of capture in sample i. In the Jolly–Seber model *all* the marked were assumed to have the same probability p_i as the unmarked of being caught in sample i. The two cases are now considered separately, and the maximum likelihood estimates are derived below using intuitive arguments.

MODEL (1): $Q = L - 1$. When $Q = L - 1$ the effect of the first capture of individuals in the groups $M_i^{(L)}$ and $M_i^{(L+1)}$, which took place more than Q periods prior to sample i, will have worn off by sample i. Thus $M_i^{(L)} + M_i^{(L+1)}$ can be compared with U so that we have the intuitive equation

$$\frac{u_i}{\hat{U}_i} = \frac{m_i^{(L)} + m_i^{(L+1)}}{\hat{M}_i^{(L)} + \hat{M}_i^{(L+1)}}$$

or

$$\hat{U}_i = \frac{u_i(\hat{M}_i^{(L)} + \hat{M}_i^{(L+1)})}{m_i^{(L)} + m_i^{(L+1)}}, \quad i = 1, 2, \ldots, s. \tag{13.39}$$

As the two groups $M_i^{(L)}$ and $M_i^{(L+1)}$ can be effectively pooled, and $R_i^{(L+1)}$ represents marked individuals from groups L and $L + 1$ prior to sample i, we can argue along the lines of (13.37) and obtain the intuitive estimate

$$\hat{M}_i^{(L)} + \hat{M}_i^{(L+1)} = m_i^{(L)} + m_i^{(L+1)} + \frac{R_i^{(L+1)}(z_i^{(L)} + z_i^{(L+1)})}{r_i^{(L+1)}}$$

$$(i = L + 1, \ldots, s - 1). \tag{13.40}$$

The two groups have the same future survival and catchabilities so

$$\frac{T_{i-1}^{(L)}}{\hat{M}_i^{(L)}} = \frac{T_{i-1}^{(L+1)}}{\hat{M}_i^{(L+1)}} \left(= \frac{T_{i-1}^{(L+1)} + T_{i-1}^{(L+1)}}{\hat{M}_i^{(L)} + \hat{M}_i^{(L+1)}} \right), \tag{13.41}$$

which, in conjunction with (13.40), leads to $\hat{M}_i^{(L)}$ and $\hat{M}^{(L+1)}$. Combining the above estimates with (13.36) gives us

$$\begin{aligned} \hat{N}_i &= \sum_{v=1}^{L-1} \hat{M}_i^{(v)} + \hat{U}_i + \hat{M}_i^{(L)} + \hat{M}_i^{(L+1)} \\ &= \sum_{v=1}^{L-1} M_i^{(v)} + \frac{(u_i + m_i^{(L)} + m_i^{(L+1)})(\hat{M}_i^{(L)} + \hat{M}_i^{(L+1)})}{m_i^{(L)} + m_i^{(L+1)}}, \end{aligned}$$

$$i = L + 1, \ldots, s - 1. \tag{13.42}$$

From Pollock [1975b] we have that the $M_i^{(v)}$ $(v = 1, 2, \ldots, L - 1)$ and $M_i^{(L)} + M_i^{(L+1)}$ are asymptotically mutually independent with asymptotic variances given by (13.38) and (cf. (13.35))

$$\begin{aligned} V[\hat{M}_i^{(L)} + \hat{M}_i^{(L+1)}] &= (M_i^{(L)} + M_i^{(L+1)} - E[S_i^{(L+1)}]) \\ &\times (M_i^{(L)} + M_i^{(L+1)} - E[S_i^{(L+1)}] + R_i^{(L+1)}) \left\{ \frac{1}{E[r_i^{(L+1)}]} - \frac{1}{R_i^{(L+1)}} \right\}. \end{aligned}$$

Thus, from (13.42), we have asymptotically

$$V[\hat{N}_i] = \sum_{v=1}^{L-1} V[M_i^{(v)}] + V[A_i],$$

where

$$V[A_i] = V\left[\left(\frac{u_i + m_i^{(L)} + m_i^{(L+1)}}{m_i^{(L)} + m_i^{(L+1)}}\right)(\hat{M}_i^{(L)} + \hat{M}_i^{(L+1)})\right]$$

$$= N_i^*(N_i^* - n_i^*)\left\{\left(\frac{M_i^{(L)} + M_i^{(L+1)} - E[S_i^{(L+1)}] + R_i^{(L+1)}}{M_i^{(L)} + M_i^{(L+1)}}\right)\right.$$

$$\left. \times\left(\frac{1}{E[r_i^{(L+1)}]} - \frac{1}{R_i^{(L+1)}}\right) + \frac{1-\rho_i^*}{E[S_i^{(L+1)}]}\right\}, \tag{13.43}$$

and

$$N_i^* = U_i + M_i^{(L)} + M_i^{(L+1)},$$

$$n_i^* = u_i + m_i^{(L)} + m_i^{(L+1)} = u_i + S_i^{(L+1)}, \text{ and}$$

$$\rho_i^* = [M_i^{(L)} + M_i^{(L+1)}]/N_i^*.$$

It should be noted that, for mathematical convenience, Pollock treats the $R_i^{(v)}$, $m_i^{(v)}$ $(i = 1, 2, \ldots, L-1)$ and $m_i^{(L)} + m_i^{(L+1)} + u_i$ $(= S_i^{(L+1)} + u_i)$ as fixed parameters and $S_i^{(L+1)}$ and u_i as random variables. However, when it comes to the variance formulae given above, it does not matter, asymptotically, whether we write $m_i^{(v)}$ or $E[m_i^{(v)}]$ as in both cases we would use $m_i^{(v)}$ in the estimation of the variances.

MODEL (2): $Q = L$. In this case we can compare the two groups U_i and $M_i^{(L+1)}$ so that

$$\frac{u_i}{\hat{U}_i} = \frac{m_i^{(L+1)}}{\hat{M}_i^{(L+1)}}$$

or

$$\hat{U}_i = u_i \frac{\hat{M}_i^{(L+1)}}{m_i^{(L+1)}}. \tag{13.44}$$

It transpires that equation (13.36) now also holds for $v = L$, namely

$$\hat{M}_i^{(L)} = m_i^{(L)} + \frac{R_i^{(L+1)}z_i^{(L)}}{r_i^{(L+1)}}, \tag{13.45}$$

and, using a similar argument,

$$\hat{M}_i^{(L+1)} = m_i^{(L+1)} + \frac{R_i^{(L+1)}z_i^{(L+1)}}{r_i^{(L+1)}}. \tag{13.46}$$

Combining the above three estimates with (13.36) leads to

$$\hat{N}_i = \sum_{i=1}^{L} \hat{M}_i^{(v)} + \frac{(u_i + m_i^{(L+1)})\hat{M}_i^{(L+1)}}{m_i^{(L+1)}}, \quad i = L+2, \ldots, s-1, \tag{13.47}$$

and, using (13.38) (which now also holds for $v = L$) we have, asymptotically,

$$V[\hat{N}_i] = \sum_{v=1}^{L-1} V[M_i^{(v)}] + V[B_i], \tag{13.48}$$

where

$$\begin{aligned} V[B_i] &= V\left[\hat{M}_i^{(L)} + \frac{(u_i + m_i^{(L+1)})\hat{M}_i^{(L+1)}}{m_i^{(L+1)}}\right] \\ &= V[\hat{M}_i^{(L)}] + V\left[\frac{(u_i + m_i^{(L+1)})\hat{M}_i^{(L+1)}}{m_i^{(L+1)}}\right] \\ &\quad + 2\,\mathrm{cov}\left[\hat{M}_i^{(L)}, \frac{(u_i + m_i^{(L+1)})\hat{M}_i^{(L+1)}}{m_i^{(L+1)}}\right]. \end{aligned} \tag{13.49}$$

From Pollock [1975b] we have the asymptotic expressions

$$V[\hat{M}_i^{(L+1)}] =$$

$$(M_i^{(L+1)} - E[m_i^{(L+1)}])(M_i^{(L+1)} - E[m_i^{(L+1)}] + R_i^{(L+1)})\left\{\frac{1}{E[r_i^{(L+1)}]} - \frac{1}{R_i^{(L+1)}}\right\}, \tag{13.50}$$

$$\begin{aligned} V\left[\frac{(u_i + m_i^{(L+1)})\hat{M}_i^{(L+1)}}{m_i^{(L+1)}}\right] &= N_i'(N_i' - n_i')\left\{\frac{(M_i^{(L+1)} - E[m_i^{(L+1)}] + R_i^{(L+1)}}{M_i^{(L+1)}}\right. \\ &\quad \times \left.\left[\frac{1}{E[r_i^{(L+1)}]} - \frac{1}{R_i^{(L+1)}}\right] + \frac{1-\rho_i'}{E[m_i^{(L+1)}]}\right\}, \end{aligned} \tag{13.51}$$

and

$$\mathrm{cov}\left[\hat{M}_i^{(L)}, \frac{(u_i + m_i^{(L+1)})\hat{M}_i^{(L+1)}}{m_i^{(L+1)}}\right] = (M_i^{(L)} - m_i^{(L)})(N_i' - n_i')\left\{\frac{1}{E[r_i^{(L+1)}]} - \frac{1}{R_i^{(L+1)}}\right\} \tag{13.52}$$

with

$$N_i' = U_i + M_i^{(L+1)},$$

$$n_i' = u_i + m_i^{(L+1)}$$

and

$$\rho_i' = M_i^{(L+1)}/N_i'.$$

For this model Pollock assumes that the $m_i^{(v)}$ $(i = 1, 2, \ldots, L)$ and $m_i^{(L+1)} + u_i$ are fixed parameters, and $m_i^{(L+1)}$ and u_i random variables.

In comparing the above two models we note that the estimates of $M_i^{(v)}$ $(i = 1, 2, \ldots, L-1)$ and $M_i^{(L)} + M_i^{(L+1)}$ are the same for both models, but the individual estimates of $M_i^{(L)}$ and $M_i^{(L+1)}$ differ. Also, for model (1), N_i is estimable for $i = L+1, \ldots, s-1$, while for model (2) it is estimable for $i = L+2, \ldots, s-1$.

We shall now consider some special cases.

Example 13.1 $(L = 0, Q = 0)$

We note that the Jolly–Seber model corresponds to $L = 0$ and $Q = 0$ and is therefore a special case of model (2) above (called Example I in Robson [1969]). For this case we only have two groups U_i $(= M_i^{(0)})$ and $M_i = M_i^{(1)}$ $(= M_i^{(L+1)})$.

Thus from equation (13.44)

$$\hat{U}_i = u_i\hat{M}_i/m_i$$

or

$$\hat{N}_i = \hat{M}_i + \hat{U}_i$$

$$= (m_i + u_i)\hat{M}_i/m_i$$

$$= n_i\hat{M}_i/m_i, \tag{13.53}$$

as on p. 200. Also (13.46) is equivalent to (5.7) on p. 200. As far as variance estimation is concerned, the only term that contributes to $V[\hat{N}_i]$ of (13.48) is (13.51): this reduces to $V[\hat{N}_i|N_i]$ of (5.14), as expected, since $N_i' = N_i$ and $n_i' = n_i$.

Example 13.2 ($L = 1$, $Q = 0$)

Robson [1969: Example 11] considers the special case $L = 1$ and $Q = 0$ of model (1) in which the first capture (the tagging process) affects survival for one period but does not affect catchability. The basic parameters are now U_i, $M_i^{(1)}$ and $M_i^{(2)}$ representing, respectively, the unmarked, those newly captured in sample $i-1$ and those first captured prior to sample $i-1$, all of which are present just before sample i. Robson shows that (cf. 13.35)

$$\hat{N}_i = \frac{n_i}{S_2^{(2)}}(\hat{M}_i^{(1)} + \hat{M}_i^{(2)})$$

$$= \frac{u_i + m_i^{(1)} + m_i^{(2)}}{m_i^{(1)} + m_i^{(2)}}(\hat{M}_i^{(1)} + \hat{M}_i^{(2)}),$$

which is the same as the last term of (13.42). Setting $m_i = m_i^{(1)} + m_i^{(2)}$ and $\hat{M}_i = \hat{M}_i^{(1)} + \hat{M}_i^{(2)}$, we see that (13.53) holds once again, which is to be expected as both marked and unmarked have the same probability of capture, p_i say, in sample i (since $Q = 0$). Also the estimate $\hat{M}_i$ is given by (13.40), namely

$$\hat{M}_i = m_i + R_i^{(2)}z_i/r_i^{(2)}, \tag{13.54}$$

where $z_i = z_i^{(1)} + z_i^{(2)}$ is the number of marked not caught in sample i but subsequently recaptured, and $R_i^{(2)}$ is the number which belonged to $M_i^{(1)}$ or $M_i^{(2)}$, that is, to the marked population $M_i = M_i^{(1)} + M_i^{(2)}$, before being captured and released from sample i. Equation (13.54) differs from the corresponding equation for the case ($L = 0$, $Q = 0$) in that we have $R_i^{(2)}$ and $r_i^{(2)}$ instead of R_i and r_i. Here $R_i = R_i^{(1)} + R_i^{(2)}$, and the group of $R_i^{(1)}$ individuals released from sample i (which were unmarked prior to sample i) have a different probability of survival from sample i to sample $i+1$ than the group $M_i - m_i$. Finally, from (13.43) with $N_i^* = N_i$, $n_i^* = n_i$ and $\rho_i = M_i/N_i$, we have

$$V[\hat{N}_i] = V[A_i]$$

$$= N_i(N_i - n_i)\left\{\frac{(M_i - E[m_i] + R_i^{(2)})}{M_i}\left(\frac{1}{E[r_i^{(2)}]} - \frac{1}{R_i^{(2)}}\right) + \frac{1-\rho_i}{E[m_i]}\right\}.$$

Cormack [1972; 1973, no losses on capture, i.e. $R_i^{(1)} = u_i$, $R_i^{(2)} = m_i$] also considers the above case but from an intuitive viewpoint. Using his notation, in which "dash" and "star" represent the groups 1 and 2, we have

$$\begin{aligned}\hat{M}_i &= \hat{M}'_i + \hat{M}^*_i \\ &= m'_i + m^*_i + s^+_i(z'_i + z^*_i)/r^+_i \\ &= m^{(1)}_i + m^{(2)}_i + R^{(2)}_i(z^{(1)}_i + z^{(2)}_i)/r^{(2)}_i\end{aligned}$$

which is the same as (13.54). However, his separate estimates $\hat{M}'_i$ and $\hat{M}^*_i$ are not the maximum likelihood estimates of $M^{(1)}_i$ and $M^{(2)}_i$ (Pollock [1974: 34]). Cormack [1972] also gives the following estimates:

$$\hat{p}_i = m_i/\hat{M}_i, \quad \hat{\phi}^{(1)}_i = \hat{M}^{(1)}_{i+1}/R^{(1)}_i$$

and

$$\hat{\phi}^{(2)}_i = \hat{M}^{(2)}_{i+1}/(\hat{M}^{(2)}_i - m_i + R^{(2)}_i).$$

Since the two groups U_i and $M^{(2)}_i$ have the same survival rates, the effect of tagging having worn off the group $M^{(2)}_i$, $\hat{\phi}^{(2)}_i$ is also the survival estimate for the unmarked population; $\hat{\phi}^{(1)}_i$ is the survival estimate for the newly marked.

Example 13.3 ($L = 1, Q = 1$)

This particular case is considered intuitively by Cormack [1972, 1973]. His estimates $\hat{M}'_i$ and $\hat{M}^*_i$ are now the maximum likelihood estimates of $M^{(1)}_i$ and $M^{(2)}_i$, respectively, and he gives

$$\begin{aligned}\hat{N}_i &= M'_i + (n_i - m'_i)M^*_i/m^*_i \\ &= \hat{M}^{(1)}_i + \left(\frac{u_i + m^{(2)}_i}{m^{(2)}_i}\right)\hat{M}^{(2)}_i,\end{aligned}$$

which is (13.47) with $L = 1$.

The case ($L = 2, Q = 0$) is considered by Robson [1969] and involves the solution of quadratic equations.

HYPOTHESIS TESTING. Pollock [1975b] gives chi-squared tests for testing the three nested hypotheses: H_0 ($L = c - 1, Q = c - 1$), H_1 ($L = c, Q = c - 1$), and H_2 ($L = c, Q = c$). We can test H_1 versus H_2, a test of equal catchability of the $M^{(c)}_i$ and $M^{(c+1)}_i$ individuals, using a series of contingency tables of the form:

$m^{(c)}_i$	$m^{(c+1)}_i$
$T^{(c)}_{i-1} - m^{(c)}_i$	$T^{(c+1)}_{i-1} - m^{(c)}_i$
$T^{(c)}_{i-1}$	$T^{(c+1)}_{i-1}$

where $i = c + 2, \ldots, s - 1$. Each table gives an asymptotic chi-squared test with one degree of freedom and, because the tests are independent, an overall test of H_1 versus H_2 can be obtained by adding all the test statistics to give an overall asymptotic chi-squared test with $s - c - 2$ degrees of freedom.

To test H_0 versus H_1, a test of equal survival of the $M^{(c)}_i$ and $M^{(c+1)}_i$ individuals given their equal catchability, we have a similar set of tables:

$r^{(c)}_i$	$r^{(c+1)}_i$
$R^{(c)}_i - r^{(c)}_i$	$R^{(c+1)}_i - r^{(c+1)}_i$
$R^{(c)}_i$	$R^{(c+1)}_i$

where $i = c + 1, \ldots, s - 1$. Once again we can compute an overall chi-squared statistic, this time with $s - c - 1$ degrees of freedom. Pollock suggests that an appropriate procedure might be to test H_1 versus H_2 and then, if H_1 is accepted, test H_0 versus H_1. This process could be continued back through smaller values of L until a test was rejected or until $L = 0$, $Q = 0$.

13.1.7 Test for random sampling

On p. 161 we described a method due to Leslie for testing whether sampling from an identifiable group of G individuals is random. By an appropriate choice of G, the method can also be applied to open populations (pages 162, 226), provided all emigration is permanent. If all the individuals are given a unique identifying mark on first capture then, following Carothers [1971], all the recapture data can be presented in the form of a matrix $\mathbf{X} = [(X_{ij})]$ in which the data from the ith individual are placed in row i and the data from the jth sample in column j. If the ith individual is captured in the jth sample we set $X_{ij} = 1$. However, if the ith individual is not captured in the jth sample but is known to be in the population at that time, we set $X_{ij} = 0$. All other elements are left blank ($= b$). Thus for 4 marked individuals and 5 samples a possible array is

$$\mathbf{X} = \begin{bmatrix} 1 & 1 & 0 & 0 & 1 \\ 1 & 0 & 0 & 1 & b \\ b & 1 & 1 & 1 & b \\ b & b & b & 1 & b \end{bmatrix}. \tag{13.55}$$

Here the first individual is captured in samples 1, 2 and 5 only, the second individual in samples 1 and 4 only, the third individual in samples 1, 2 and 3 only; etc. The third individual may have been in the population right from the beginning or entered by birth or immigration prior to sample 2; it also may have died or left the population after sample 4. If the population is closed we simply replace all the blanks by zeros.

The appropriate data for testing consists of a sequence of 0's or 1's for each individual, but, in the case of open populations, excluding the initial capture and/or final recapture when either of these is used to determine the individual's presence in the population. Thus the test data for (13.55) are represented by

$$\begin{bmatrix} - & \boxed{\begin{matrix} 1 & 0 \\ 0 & 0 \end{matrix}} & \begin{matrix} 0 \\ - \end{matrix} & \begin{matrix} - \\ - \end{matrix} \\ - & - \;\; 1 & - & - \\ - & - \;\; - & - & - \end{bmatrix}. \tag{13.56}$$

(The boxed block is labelled **B**.)

To apply Leslie's test we select a rectangular block, **B** say, of the test data consisting of G rows and t columns as shown for example in (13.56) (though in practice we must have $G > 20$), and calculate the statistic

$$T = \sum_{i=1}^{G} (R_i - \bar{R})^2 \Big/ \sum_{j=1}^{t} P_j Q_j.$$

Here R_i is the number of 1's (recaptures) in row i of **B**, $\bar{R} = \sum_{i=1}^{G} R_i/G$, and P_j $(= 1 - Q_j)$ is the proportion of 1's in the jth column of **B**. Referring to the notation of p. 161, $P_j = g_j/G$, and it is readily seen that $\bar{R} = \mu$ and

$$\sum_{x=0}^{t} f_x(x - \mu)^2 = \sum_{i=1}^{G} (R_i - \bar{R})^2.$$

Although Leslie did not give a rigorous proof that T is approximately χ^2_{G-1} $(G > 20)$, Carothers [1971] points out that

$$Q = (1 - 1/G)T$$

is identical to Cochran's "Q" statistic. Cochran [1950] showed that under the null hypothesis, in this case the hypothesis of random sampling, Q is approximatley χ^2_{G-1} and gave a method for deriving the randomisation distribution Q for small values of G and t. Carothers [1971] proposes a modification appropriate to small data sets, namely

$$Q' = \frac{Q}{t(G-1)} = \frac{T}{tG}.$$

Here Q' is approximately distributed as a beta distribution with $\frac{1}{2}(G-1)$ and $\frac{1}{2}(G-1)(t-1)$ degrees of freedom, or equivalently

$$F' = Q'/(1 - Q')$$

which has approximately an F-distribution with $G - 1$ and $(G-1)(t-1)$ degrees of freedom, under the null hypothesis. This modification is motivated by simulation experiments and by regarding the block **B** as a two-way classification of binary data with columns (samples) as a fixed factor and rows (individuals) as a random factor.

On p. 161 we described a procedure for closed populations in which the unmarked individuals in each sample identified a separate block of data; the chi-squareds from each block are then pooled to give an overall chi-squared statistic. A similar procedure can be applied to open populations though there is some subjectivity in the choice of G and t for each block: large G will tend to lead to small t, and vice versa. For example, **B** in (13.56) could be either 1×3 or 2×2. As we require at least 20 rows in each block, some extensive trimming of the data is inevitable with a consequent loss of information. However, Carothers proposes a multivariate extension of the above test which uses a more efficient block construction as there can now be less than 20 individuals in each block. Instead we now require that there must be at least 20 entries with 1 or 0 in each column of **X**, and columns not satisfying this requirement are deleted before the blocks are chosen.

13.1.8 Loglinear models

In **12.8.4** we mentioned the possibility of using loglinear models for the analysis of capture–recapture data from closed populations. Cormack [1979] appears to be the only person to consider the possibility of extending the method to open populations. He points out that when mortality is operating, all the interactions in the loglinear model are non-zero except those involving the first sample. (For a different solution see Cormack [1981], Additional References).

13.2 CATCH–EFFORT MODELS

13.2.1 Mortality only

If the effort does not change too much from one unit of time (usually a year) to the next then, from (8.10) on p. 330,

$$\begin{aligned} E[y_i] &\approx \mu + k(f_i + f_{i+1})/2 \\ &\approx \mu + kf_i \\ &= Z_i', \quad \text{say,} \end{aligned} \tag{13.57}$$

where $y_i = \log(n_i f_{i+1}/n_{i+1} f_i)$ and f_i is the total effort in year i. Here n_i and n_{i+1} are the total catches or catches from a particular age-class in years i and $i+1$: the use of fully recruited age-class avoids problems relating to possible recruitment (p. 332). When exploitation takes place continuously at a uniform rate throughout the year then $Z_i' = Z_i$, the total instantaneous mortality rate. Alternatively, if exploitation is seasonal, say for a fraction τ of the year, then Z_i' is not the instantaneous mortality rate as this rate now depends on whether exploitation is taking place or not. However, from the bottom of p. 331,

$$\begin{aligned} E[N_{i+1}|N_i] &= N_i \exp\{-(\mu + k\mathring{f}_i\tau_i)\} \\ &= N_i \exp\{-(\mu + kf_i)\} \\ &= N_i \exp\{-Z_i'\}, \end{aligned}$$

which is of the same form as (8.4) (with $k_i = k$). Thus (13.57) still holds with Z_i' now regarded as the "annual" instantaneous mortality rate (although this is strictly a misuse of terms). Actually the above equation will hold even if the exploitation rate changes throughout the year: all we need to know is f_i, the *total* effort for the year.

The basis for equation (13.57) is the fundamental equation (8.6), namely $E[n_i/f_i|N_i] = k\bar{N}_i$, and this equation holds for any interval of time, and not just for intervals of one year. Thus n_i/f_i could refer to the catch per unit effort for the whole year or for just a month; and estimates of Z_i' can be obtained for different months based on monthly catches from the same age-class and one year apart. Such estimates can then be averaged over month and area.

Since we assume the f_i to vary slowly, the regression estimates of μ and k will be imprecise (p. 330). However, in many cases it is possible to estimate μ from age or age-length data (cf. **10.3**) and thus obtain an estimate of k from the equation $Z_i' - \mu = kf_i$ (= G_i, say); an average estimate $\bar{k}$ of k will lead to a smoothed estimate $\bar{k}f_i$ of G_i and hence to a smoothed estimate of Z_i'. These methods have been used for whales, in particular the baleen species (fin, blue and sei whales: Chapman [1974a]).

A model with variable natural mortality is given by Mathews and Buckley [1976]. They show that the weight of a fish at time t is given by $w(t) = a \exp(bt)$ and assume that the instantaneous rate of natural mortality is inversely proportional to weight, that is $\mu_t = c/w(t) = ca^{-1} \exp(-bt)$. This relationship can then be incorporated into the catch equations.

Alternative catch–effort models based on the population dynamics of a fishery not in equilibrium are available (cf. Wise [1972: introductory remarks], Francis [1974] and, in particular, Walter [1975]).

A major problem of large-scale fisheries is the measurement and standardisation of fishing effort (Pope [1975]). Also catchability may vary from area to area and a spatial catch effort model is considered by Caddy [1975].

13.2.2 Mortality and recruitment

1 Proportional recruitment

Another method which has been used for whales, due to Allen [1966], assumes that catching, natural mortality and recruitment proceed successively and not simultaneously. This assumption is not unreasonable for short fisheries such as whaling. Thus if N_i is the size of the *catchable* population (= $\rho_i N_i$ in **8.1.3**) at the beginning of year i (when the season begins) and w_i is the fraction of this population newly recruited at the beginning of the year, then we have the deterministic equation

$$N_i \approx (N_{i-1} - n_{i-1})e^{-\mu} + w_i N_i \tag{13.58}$$

or

$$N_i \approx (N_{i-1} - n_{i-1})e^{-\mu}/(1 - w_i). \tag{13.59}$$

Also

$$\begin{aligned} E[n_i] &= kf_i\bar{N}_i \\ &\approx kf_i(N_i - \tfrac{1}{2}n_i), \end{aligned} \tag{13.60}$$

which, on successive substitution using (13.58), leads to $E[n_i] \approx g_i(k, \mu, N_1)$ where g_i is a function of the unknown parameters k, μ and N_1; the w_i(and n_i) are assumed known. These parameters are then estimated by minimising $\Sigma_i [n_i - g_i(k, \mu, N_1)]^2$ with respect to k, μ and N_1. This minimisation is best done iteratively on a computer, though the computations are somewhat simpler if an independent estimate of μ is available. We can then either minimise with respect to just k and N_1 or else use this value of μ as a first step in the iterations. We note that

$$w_i = \sum_a \pi_{i,a} R_{i,a},$$

where $\pi_{i,a}$ is the proportion of N_i in the ath age-class, $R_{i,a}$ is the proportion of this group newly recruited at the beginning of the ith year and the summation is overall age-groups in which recruitment occurs (i.e. $R_{i,a} > 0$). Thus if $P_{i,a}$, the proportion of the ith catch in the ath class, is an unbiased estimate of $\pi_{i,a}$, then w_i is estimated by (Allen [1966])

$$\hat{w}_i = \sum_a P_{i,a}\hat{R}_{i,a}.$$

The estimates of $R_{i,a}$ can be calculated using the method of Allen [1966, 1968] described on pages 336–340, and in Chapman [1974a: his definition of B is different from that of Allen]. Allen's [1966] paper should be consulted for further details.

2 *Constant instantaneous recruitment rate*

Chapman [1970, 1974a] has developed a simple modification of the Leslie method which has been used with some success in the study of whales. Suppose that the instantaneous recruitment rate λ is constant (cf. p. 260) so that equation (13.58) is replaced by

$$N_i \approx (N_{i-1} - n_{i-1})\,e^{-(\mu-\lambda)}.$$

Then by successive substitution we have, for example,

$$N_3 \approx N_1\,e^{-2(\mu-\lambda)} - n_1\,e^{-2(\mu-\lambda)} - n_2\,e^{-(\mu-\lambda)}$$

$$\approx N_1[1 + 2(\lambda-\mu)] - n_1 - n_2$$

or, in general,

$$N_i \approx N_1[1 + (i-1)(\lambda-\mu)] - x_i'\}$$

where $x_i = \sum_{j=1}^{i-1} n_j$, the cumulative catch. Thus from (13.60) we have:

$$E[n_i/f_i] \approx k\{N_1[1 + (i-1)(\lambda-\mu)] - x_i'\}.$$

where $x_i' = x_i + \frac{1}{2}n_i$, the cumulative catch to the *middle* of the season. If the catch rate is approximately constant we have

$$E[x_i'] = a + (i - 1 + \tfrac{1}{2})b.$$

Let $\hat{a}$ and $\hat{b}$ be the least squares estimates of a and b, then

$$x_i' \approx (\hat{a} + \tfrac{1}{2}\hat{b}) + (i-1)\hat{b}$$

and

$$\begin{aligned} n_i/f_i &\approx k\{N_1[1 + (i-1)(\lambda-\mu)] - (\hat{a} + \tfrac{1}{2}\hat{b}) - (i-1)\hat{b}\} \\ &= k\{N_1 - \hat{a} - \tfrac{1}{2}\hat{b} + (i-1)[N_1(\lambda-\mu) - \hat{b}]\} \\ &= c + (i-1)d, \quad \text{say,} \end{aligned}$$

If $\hat{c}$ and d are the least squares estimates of c and d, then N_1 can be estimated by

$$\hat{N}_1 = \frac{\hat{d}(\hat{a} + \frac{1}{2}\hat{b}) - \hat{c}\hat{b}}{\hat{d} - (\lambda-\mu)\hat{c}}$$

Thus $\hat{N}_1$ can be calculated if an estimate of $\lambda - \mu$ (denoted by $r - M$ in Chapman's notation) is available.

3 Partial recruitment factors

Pope [1977] has generalised the theory of **8.1.1** to allow for the possibility of recruitment, and considers his approach as a possible alternative to Gulland's virtual population analysis (described briefly by Pope [1972]). His generalisation is made by introducing "partial recruitment factors" S_a so that the instantaneous fishing (exploitation) rate for year i and age-group a is $S_a\mu_{Ei}$; $S_a = 1$ for fully recruited age-classes. Thus, setting $\Delta_i = 1$ for all i and using the same arguments that led to (8.7), we have approximately

$$E[Y_{ia}] = \log\left[\frac{S_a\mu_{Ei}Z_{i+1}\{1-\exp(-Z_i)\}}{S_{a+1}\mu_{E,i+1}Z_i\{1-\exp(-Z_{i+1})\}}\right] + Z_i, \tag{13.61}$$

where $Y_{ia} = \log(n_{ia}/n_{i+1,a+1})$, $Z_i = S_a\mu_{Ei} + \mu$ and $Z_{i+1} = S_{a+1}\mu_{E,i+1} + \mu$. Equation (13.61) is a non-linear regression model, and estimates of S_a and μ_{Ei} (Pope assumes that μ is known) can be obtained iteratively by minimising $\Sigma_i\Sigma_a(Y_{ia} - E[Y_{ia}])^2$. This model does not require knowledge of effort, but there is a penalty: estimates may not be very precise because of the large number of parameters to be estimated. It should be noted that some of the difficulties on p. 332 that relate to assessing the data relating to a given age-group also apply here to the exploitation rate. Although μ_{Ei} is the exploitation rate in year i for fully recruited age-groups, it has to be "spread" over several age-groups.

Using the approximation (8.8), (13.61) can be linerarised to give

$$E[Y_{ia}] \approx \mu + \log(\mu_{Ei}/\mu_{E,i+1}) + \log(S_a/S_{a+1}) + \tfrac{1}{2}\mu_{Ei}S_a + \tfrac{1}{2}\mu_{E,i+1}S_{a+1}.$$

This equation resembles that obtained from a two-way analysis of variance model, and Pope discusses the above model from this viewpoint.

4 Constant recruitment

Equation (13.58) can be written as

$$N_i \approx (N_{i-1} - n_{i-1})e^{-\mu} + R_i, \quad (i = 1, 2, \ldots, s), \tag{13.62}$$

where R_i is the recruitment in year i and N_1 is the initial size of the *exploitable* population. Chapman [1974b] (see also Breiwick [1978a]) assumed that (i) the population was initially unexploited, (ii) the population was initially in equilibrium, so that the recruits entering the exploitable population equalled the number of natural deaths ($\approx N_1(1-e^{-\mu})$) in the exploitable population, and (iii) the age of recruitment was greater than the period of exploitation, so that all the individuals recruited during the exploitation period were present at the beginning. If these assumptions are tenable, as for example in some whale populations, then $R_i = N_1(1-e^{-\mu})$, which combined with (13.62) and (13.60) leads to

$$E[n_i/f_i] \approx k\left(N_1 - \sum_{k=1}^{i} n_k e^{-(i-k)\mu} - \tfrac{1}{2}n_i\right). \tag{13.63}$$

The unknown parameters k, N_1 and μ can be estimated using non-linear least squares. If μ is known then (13.63) is a simple modification of Leslie's method (p. 297). Breiwick [1978a] also considers the approximation $e^{-\mu} \approx 1 - \mu$. Unfortunately the above model is of limited application because of assumption (i). For further applications of (13.61) and previous models to whale populations see Breiwick [1977, 1978b].

13.3 USE OF AGE DATA

13.3.1 Life-table analysis of band data

Life-table methods are still being applied to populations which do not satisfy the stringent assumptions necessary for the valid application of these methods. In particular the so-called composite dynamic method, usually set out in the form of a life-table analysis, is still widely used, especially in the analysis of bird-band returns. The shortcomings of this method as applied to bird-band returns have already been discussed in section **13.1.5**(*3*) and my own feeling about this is summed up in the words of Eberhardt [1972: 167]; "There seems to be little advantage in continuing the practice of life-table analysis of banding data. However, the dynamic table is sometimes used as a means of calculating the Lack estimate." In fact, if life-table methods were no longer used for analysing bird-banding data there would be much less confusion over the legitimate use of life tables.

The approach to life-table analysis in the past has tended to be deterministic in that one talks about observed rates rather than estimates of probabilities. However, a stochastic approach, with possibly a life-table format, is preferred as it enables one to calculate variances of the various estimates (cf. p. 410). Life-tables can also be constructed from capture–recapture data (Manly and Seber [1973]).

13.3.2 Age-specific data

For the age-specific case a stochastic approach, due to Chiang, is given in detail on pages 408–413. When the probability of survival S is constant from year to year, then, using the usual life-table notation, the maximum likelihood estimate of $q = 1 - S$ is (p. 413)

$$
\begin{aligned}
q_{\text{pool}} &= \sum_{x=0}^{w-1} d_x \Big/ \sum_{x=0}^{w-1} l_x \\
&= \sum_{x=0}^{w-1} d_x \Big/ \left[\sum_{x=0}^{w-1} (x+1) d_x + w d_w \right] \qquad (13.64)
\end{aligned}
$$

where d_x is the number of deaths from the cohort of size l_0 in the age-class $[x, x+1)$, and w is the oldest age attained. If this formula is applied, or should I say misapplied, to bird-banding data, then, setting $D_{x+1} = d_x$ and $t = w + 1$ (since data are collected for $w + 1$ years to record d_w), we obtain

$$q_{\text{pool}} = \sum_{j=1}^{t-1} D_j \Big/ \left[\sum_{j=1}^{t-1} jD_j + (t-1)D_t \right], \tag{13.65}$$

which looks like Lack's estimate but is not as the last age-class D_t $(= d_w)$ is not added to the numerator and denominator. The similarity of these two estimates has caused some confusion in the literature because several authors have used the life-table approach simply as a means of calculating Lack's estimate. I recommend that this use of the life table be dropped.

The theory on p. 413 also applies to the situation where some members of the cohort are still alive at the end of the experiment. In this case w refers to the time taken by the experiment, so that there are l_w animals of age w alive at the end of the experiment and d_w $(= l_w)$ is the number dying *after* the experiment has finished and not just those dying in the period $[w, w + 1)$. Examples of this, in which the cohort is followed up using radio telemetry, are given by Trent and Rongstad [1974: cottontail rabbits], Brand *et al.* [1975: snowshoe hares] and Mech [1977: voles]. Trent and Rongstad worked with adult animals only and the time-unit was one day instead of one year, the experiment lasting n $(= w)$ days. Assuming a constant daily survival rate S, they used $S_{\text{pool}} = 1 - q_{\text{pool}}$ (cf. their equation (2) with $y_i = d_{i-1}, x_i = l_{i-1}$) as their estimate of S. As the d_x were usually 0 and occasionally 1, the daily mortality rate was small ($q_{\text{pool}} = 0{\cdot}0037$), so that Σd_x was approximately distributed as Poisson with mean $q\Sigma l_x$. Tables of the Poisson distribution were used to obtain a confidence interval for S and S^n (the survival rate for n days).

A plot of l_x versus x is called a survival curve (p. 395). Manly [1974a] gives a method for fitting such a curve based on a discrete analogue of the continuous gamma distribution.

13.3.3 Time-specific data

For time-specific data with age-dependent survival, the stochastic approach of Chapman and Robson (pages 429–430) is preferred. Underlying this approach there are two basic assumptions which need to be checked (Murphy and Whitten [1976]): (i) samples are representative of the population age structure, and (ii) the population is stationary. Assumption (i) may not hold because of variable catchability: for example, juveniles are sometimes underrepresented as they have a lower catchability (e.g. Wilbur [1975: turtles]). Gill-net selectivity in fisheries can also lead to a violation of (i) e.g. Healey [1975: 430]). Frequently assumption (ii) does not hold, or may hold only for certain years (e.g. Nellis and Keith [1976: 395]). Spinage [1972] constructed life tables for several African ungulates using age data from skulls. Unfortunately skulls of different sizes may deteriorate at different rates, though some-times correction factors can be made using data from live populations.

Although the life-table format is a convenient method of summarising time-specific data, the use of life-table methods in this area has again led to a great deal of confusion. This confusion is usually caused by practitioners being uncertain as to whether a sample of ages is an $\{l_x\}$ or a $\{d_x\}$ series (cf. pages 401–2). For example, when survival is age-dependent, the sample age-series

$\{n_x\}$ as used by Chapman and Robson [1960] can definitely be used as an $\{l_x\}$ series as the estimates of survival are (p. 429) $\hat{S}_x = n_{x+1}/n_x$. However, when survival is independent of age, several estimates of q (= $1 - S$) have been proposed (pages 414–428). For example, setting $n_x = l_x$ we have q_{pool} (see equation (13.64));

$$\tilde{q} = \frac{\Sigma l_x}{\Sigma(x+1)l_x} \quad \left(= \frac{n}{X+n}, \text{ say}\right);$$

$$\hat{q}_1 = \frac{l_0}{\Sigma l_x} = \frac{\Sigma d_x}{\Sigma l_x} = \frac{\Sigma d_x}{\Sigma(x+1)d_x}; \text{ and}$$

$$\hat{q}_2 = \frac{d_0}{l_0} = \frac{d_0}{\Sigma d_x},$$

where x ranges from 0 to r (r being the maximum *observed* age in the sample). Referring to p. 415 and noting that $d_r = l_r$, it is readily seen that

$$S_{\text{pool}} = 1 - q_{\text{pool}} \quad (= (n_1 + \ldots + n_r)/(n_0 + \ldots + n_{r-1})).$$

is simply Jackson's [1939] survival estimate, and $\tilde{S} = 1 - \tilde{q}$ is the maximum likelihood estimate of survival given by Chapman and Robson [1960]. The estimate $\hat{q}_1$ (= n_0/n), which is commonly used in life-table analysis, is in fact Heincke's estimate of mortality and differs from q_{pool} only in the inclusion of the age-class r. Finally $\hat{q}_2$ is a simple estimate of mortality based on the first entry in the d_x column.

It is interesting to note that if $\{n_x\}$ is incorrectly treated as a $\{d_x\}$ rather than an $\{l_x\}$ series, then the use of, for example, $\hat{q}_2$ and $\hat{q}_1$ is in fact equivalent to using $\hat{q}_1$ and $\tilde{q}$ with the correct series $\{l_x\}$. Also $\tilde{q}$ will still be a satisfactory estimate as $d_x \approx ql_x$. Thus when the survival rate is independent of age, ecologists can freely confuse l_x and d_x and still get a reasonable estimate of S. However when the geometric model (10.15) is applicable (that is truncation at w, the maximum *population* age, can be ignored, cf. p. 420), it is best to use the unbiased modification of $\tilde{S}$, namely $\hat{S} = X/(X + n - 1)$ as this has minimum variance. Examples using $\hat{S}$ are given by Hanson and Eberhardt [1971: 37, Canada geese], Havey [1974: salmon], Storm *et al.* [1976: 56, red foxes] and Rose [1977: rabbits].

When the population is stationary over several years, $\{n_x\}$ series for different years can be pooled, as, for example, in the study of Kimball and Wolfe [1974] for an elk population. In applying the methods of Chapman and Robson, they found that the survival rate was not constant for the older age-groups, so that the data were truncated at a certain age and S_{seg}, based on a truncated geometric distribution (pages 420–1), was calculated. Spinage [1970: 69, 75] argues that the $\{n_x\}$ series still gives a useful indication of the age structure even if the population is not stationary. Admittedly changes in the age structure will reflect population changes (e.g. Turner *et al.* [1970]) though care is needed in interpreting changes in age ratios (Caughley [1974b]).

As mentioned on p. 529 the same geometric models of **10.3** are obtained for a special case (constant survival and recovery rates) of the joint distribution of tag returns from a single release, conditional on the total number of returns. Thus Eberhardt's [1972] simulation results apply here, namely: (i) S_{seg} is generally biased and rather inefficient. (ii) $\hat{S}$, $\hat{S}_1$ $(= 1 - \hat{q}_1)$, S_{seg} and two regression estimates (equation (10.27) and its modification, called the adjusted regression by Eberhardt) all tend to estimate the first-year survival rate when the survival rate does not remain constant. We see this, for example, from the following:

$$\begin{aligned} E[\hat{S}] &= E\left[\frac{X}{n + X - 1}\right] \\ &\approx \frac{E[X]}{E[n + X]} \\ &= S_0\left[\frac{1 + 2S_1 + 3S_1S_2 + \ldots}{1 + 2S_0 + 3S_0S_1 + \ldots}\right] \\ &\approx S_0, \end{aligned}$$

when S_i is changing slowly with i. (iii) As expected, $\hat{S}$ is much more efficient than the other estimates which have variances that are usually at least twice as large as that of $\hat{S}$. (iv) The test proposed by Chapman and Robson (p. 416), based on $\hat{S} - \hat{S}_1$ and designed to test the consistency of the first-year recoveries with the remaining years, is robust to changing survival.

The regression or "catch curve" methods of **10.3.2** are widely used in fisheries (cf. Ricker [1975: ch. 2] for a detailed discussion) and could be applied to other animal populations. Further examples of catch curves are given by Davidoff *et al.* [1973: 1676–7, white fish] and Manooch and Huntsman [1977: red porgy]. Crowe [1975: bobcats] uses the regression method (with logs to the base 10) to estimate S, and then uses the straight line to reconstruct a theoretical life table with a constant survival rate. As mentioned on p. 422, it is often unrealistic to age the whole catch, and a simpler method is to stratify the sample by length and then age just a subsample from each length class. The sampler can either subsample a fixed number from each length category or else subsample proportionally to the number in each length category. Southward [1976] found that the second method is more efficient and this is confirmed theoretically by Kimura [1977]. Some of the subsampling problems associated with large trawl catches from research vessels are discussed by Hughes [1976]. Sen and Southward [1977] mention several methods of sampling landed catches and describe a double sampling procedure, proposed by Southward [1976], for estimating the age structure and total catch. The problem of subsampling is also considered by Tomlinson [1971].

Survival and growth rates can be estimated directly from size data, though fairly stringent assumptions are required (e.g. Ebert [1973]). However, Van Sickle [1977] gives a more general stationary model for fisheries using size-specific rather than age-specific mortality and growth rates. As a result mortalities can be estimated directly from observed size-frequency curves and growth data, without explicitly determining the population's age distribution. A model for sperm whales based on the changes in the mean size of whales in catches is given by Holt [1977].

ESTIMATING POPULATION SIZE: For a stationary population N_x, the number of individuals of age x, is constant from year to year. Suppose that individuals live no longer than $w + 1$ years, so that $N_{w+1} = N_{w+2} = \ldots = 0$. If we can observe in a given year *all* the animals that die at age x, say d_x of them, then $d_x = N_x - N_{x+1}$ $(x = 0, 1, \ldots, w)$. Now, for a stationary population, the number born each year equals the number that die, so that $N_0 = \Sigma d_x$. Thus, from the identity

$$\begin{aligned} N_y &= (N_y - N_{y+1}) + (N_{y+1} - N_{y+2}) + \ldots + N_w \\ &= \sum_{x=y}^{w} d_x, \end{aligned}$$

we have

$$\begin{aligned} N &= N_0 + N_1 + \ldots N_w \\ &= \sum_{x=0}^{w} d_x + \sum_{x=1}^{w} d_x + \ldots + d_w \\ &= \sum_{x=0}^{w} (x+1) d_x. \end{aligned} \tag{13.66}$$

Evidently this method of estimating N from the d_x has been widely used by zoologists, and its properties are discussed by Holgate [1973] using the methods of the next section. In Holgate's notation we have

$$d_x \to d_{x+1,t}, \quad w + 1 \to k \quad \text{so that} \quad N = \sum_{i=1}^{k} i d_{i,t}.$$

13.3.4 Matrix methods for stationary populations

Let $\mathbf{N}_t = (N_{t0}, N_{t1}, \ldots, N_{tw})'$, where N_{tx} is the number of females in a population at time t belonging to age-class x, and let $f_x (x = 0, 1, \ldots, w)$ represent the number of female offspring to an individual of age x. Then, from Leslie [1945, 1948] and Lewis [1942] (or more recently Emlen [1973: 240–260], Doubleday [1975], Hearon [1976]),

$$N_{t+1,0} = \sum_{x=0}^{w} f_x N_{tx}.$$

Also if S_x is the proportion of age-class x that survives to age class $x + 1$ then

$$N_{t+1,x} = S_{x-1} N_{t,x-1}, \quad x > 0.$$

Thus in matrix notation we have $\mathbf{N}_{t+1} = \mathbf{A}\mathbf{N}_t$, where

$$\mathbf{A} = \begin{bmatrix} f_0 & f_1 & \cdots & f_w \\ S_0 & 0 & \cdots & 0 \\ 0 & S_1 & & \vdots \\ 0 & \cdots\ 0 & S_{w-1} & 0 \end{bmatrix} \tag{13.67}$$

is commonly called the Leslie matrix. Here the f_x and S_x are called the age-specific fecundity and survival coefficients as they depend only on age and not on time t. Since the matrix $\mathbf{A}$ is non-negative, that is has non-negative elements, a number of theorems apply (cf. Gantmacher [1959: 53] or Lancaster [1969: 282]. In particular, if all the elements of $\mathbf{A}^k$ are positive for some positive integer k, then $\mathbf{A}$ is called primitive and has a positive eigenvalue λ_1 of algebraic multiplicity 1 such that (i) any other eigenvalue is smaller in absolute value and (ii) the eigenvector $\mathbf{v}_1$ corresponding to λ_1 has all positive elements. A sufficient condition for $\mathbf{A}$ to be primitive is that two consecutive f's, f_x and f_{x+1} say, are positive (Demetrius [1971]).

Now if $\mathbf{N}_t = k\mathbf{v}_1$, that is the population has an age distribution determined by $\mathbf{v}_1$, then $\mathbf{N}_{t+1} = \mathbf{A}\mathbf{N}_t = k\mathbf{A}\mathbf{v}_1 = k\lambda_1\mathbf{v}_1 = \lambda_1\mathbf{N}_t$. This means that once a population has an age distribution determined by $\mathbf{v}_1$ it maintains this distribution. Such a population is said to have a stable age distribution and is called a stable population. In fact, as λ_1 is the dominant eigenvalue we find that as $t \to \infty$ the age distribution tends to the stable age distribution with $\mathbf{N}_{t+1} = \lambda_1\mathbf{N}_t$, irrespective of the starting age distribution. This means that after an initial period the population attains a stable age structure given by $\mathbf{v}_1$ and then increases, decreases, or remains constant in size, depending on whether $\lambda_1 > 1$, $\lambda_1 < 1$ or $\lambda_1 = 1$. When $\lambda_1 = 1$ the population is said to be stationary or in equilibrium. The rate at which such a population approaches its stable state depends on its initial state and the ratio of λ_1 to the eigenvalue of $\mathbf{A}$ with the next largest magnitude. The parameter λ_1 is related to r_s, the intrinsic rate of increase (Caughley and Birch [1971]), by the equation $\log \lambda_1 = r_s$. Life tables have sometimes been used incorrectly for estimating r_s and the reader is referred to the above paper for a discussion of the problem.

Although the above matrix model has been expressed deterministically it can be readily converted into a stochastic model by writing $E[\mathbf{N}_t]$ instead of $\mathbf{N}_t$; in this case the elements of $\mathbf{A}$ can be described as probabilities (cf. Pollard [1966] who also considers the associated covariance structure and a number of extensions). If an estimate of $\mathbf{A}$ is available then the above theory can be used to study the dynamics and stability of the population (Paulik [1973: 312–315], Smith [1975], Smith [1977] and Beddington [1978: see also the references]) or to predict yields (Béland [1974]). Conversely, if we can assume $\lambda_1 = 1$ and we know the f_x then we can study the S_x. For example, Vaughan and Saila [1976] show how to estimate S_0, given estimates of the other S_x. The Leslie matrix can also be used for assessing environmental impact (Horst [1977]) and has been used in conjunction with size-class data, when growth rates are unknown, by Pip and Stewart [1975] who use a model of Lefkovitch [1965]. In conjunction with other population equations (e.g. for egg survival) it has also formed the basis of a number of fisheries models (e.g. Allen and Basasibwaki [1974]). Eberhardt and Siniff [1977] develop a simple model given by Leslie [1966] for determining management policies relating to marine mammals.

Using a more general matrix **A**, the above model has been extended by a number of authors (cf. Usher [1972]) and was applied to problems of harvesting by Lefkovitch [1967] and Williamson [1967]. If the population is stable and $\lambda_1 > 1$ it is possible to harvest the population, after it has reproduced, in such a way that the population remains stationary. Even if $\lambda_1 = 1$ a positive harvest is still possible under certain conditions. An important question considered by Beddington and Taylor [1973], Beddington [1974], Rorres and Fair [1975] and Doubleday [1975] is that of finding the optimal harvesting scheme, that is, a scheme which maintains stationarity but maximises the harvest, either in terms of numbers of animals or biomass. Doubleday [1975] also considers the problem of maximising a fisheries yield where nets of fixed mesh size take a constant proportion of all individuals larger, and no individuals smaller, than a certain minimum size. However, these approaches have certain shortcomings; for example, they refer to a population which has reached a stable age distribution, and this may not be the case. For a further analysis of this problem see Mendelssohn [1976].

13.3.5 Stage-specific survival rates

As insects pass through various stages of development (egg, larval instar, etc.) it would seem appropriate to estimate survival rates for each stage. If a cohort of eggs can be followed through the various stages then the stochastic model of **10.2** can be used to estimate the P_x, where P_x is now the probability that an insect entering stage number x survives to enter stage number $x + 1$. Examples where cohort and related information is available are given by Beaver [1966], Berryman [1968] and McLaren and Pottinger [1969]. However, in practice, information about the stages is usually obtained by taking a sequence of samples from the population and noting the proportions of insects in the different stages for each sample. Unfortunately such data are not easy to analyse as the time when an insect enters the cycle of stages is random, so that the insects get out of phase and the various stages overlap. A number of attempts have been made to set up an adequate model and obtain estimates of stage-specific survival, average duration of each stage, numbers entering each stage, and daily survival rates. Manly [1974d] compared five such methods due respectively to Richards and Waloff [1954], Richards *et al.* [1960] (though see Manly [1975c] for a stochastic justification), Dempster [1961], Kiritani and Nakasuji [1967] and Manly [1974c], under a variety of conditions using simulation. All the methods require fairly strong assumptions and are listed for comparison by Manly [1974d]. For example, four of the methods assume a constant daily survival rate throughout, and all the methods assume that the durations of the stages are the same for all insects. Kiritani and Nakasuji's (K–N) method requires the sampling to be carried out at equal intervals of time until almost all the insects are dead, while Manly's method assumes a normal distribution for entry times. Manly concluded that the K–N method should be used for estimating stage-specific survival, provided the conditions about sampling are satisfied. Otherwise Manly's method should be used, particularly if it is desired to estimate the actual number of insects entering each stage. In later papers Manly [1976, 1977c] extended the K–N method so that it could be used with populations that have been sampled at irregular intervals of time. He also gave procedures for estimating various other parameters in addition to age-specific survival rates and demonstrated that Tukey's jackknife method for variance estimation (p. 508) is reliable. As Manly's extension seems to be the best of the above methods, the basic equations are summarised as follows, using Manly's [1977c] notation. For a *given* stage let:

M = number of insects that enter the stage,
a = duration of the stage (measured, say, in days and assumed to be the same for all insects),
$e^{-\theta}$ = daily survival rate (the probability of surviving for one day),
$\phi(u)$ = probability density function for the distribution of time of entry to the stage ($\phi(u) \rightarrow f(x)$ in Manly's notation),
μ = mean time of entry to the stage, and
σ^2 = variance of the time of entry to the stage.

Now the expected number, $N(t)$ say, of the insects in this stage at time t will be those individuals that entered in the time-interval $(t - a, t)$ and survived to time t, that is

$$N(t) = M \int_{t-a}^{t} \phi(u)\, e^{-\theta(t-u)} du. \tag{13.68}$$

We shall require the following expressions:

$$A = \int_{-\infty}^{\infty} N(t)dt = M(1 - e^{-\theta a})/\theta,$$

$$B = \frac{1}{A}\int_{-\infty}^{\infty} tN(t)dt = \theta^{-1} + \mu - a e^{-\theta a}/(1 - e^{-\theta a}),$$

and

$$C = \frac{1}{A}\int_{-\infty}^{\infty} t^2 N(t)dt - B^2$$

$$= \theta^{-2} + \sigma^2 - a^2 e^{-\theta a}/(1 - e^{-\theta a})^2.$$

Suppose that the insects start to enter the stage being considered just after t_1 and that all insects have left the stage by some time t_n. Suppose also that the population is sampled at times $t_1, t_2, \ldots, t_n$ so that we have estimates $\hat{N}(t_i)$ of $N(t)$ at these times ($\hat{N}(t_1) = \hat{N}(t_n) = 0$). Then using the trapezoidal rule for estimating the areas under the curves $N(t)$, $tN(t)$ and $t^2N(t)$, estimates of A, B and C are

$$\hat{A} = \frac{1}{2} \sum_{j=1}^{n} (b_j + b_{j+1}) \hat{N}(t_j),$$

$$\hat{B} = \frac{1}{2} \sum_{j=1}^{n} (b_j + b_{j+1}) t_j \hat{N}(t_j)/\hat{A}, \quad \text{and}$$

$$\hat{C} = \frac{1}{2} \sum_{j=1}^{n} (b_j + b_{j+1}) t_j^2 [\hat{N}(t_j)]^2/\hat{A}^2 - \hat{B}^2,$$

where $b_j = t_j - t_{j-1}$. As we have five unknown parameters for each stage, namely M, θ, a, μ and σ^2, we cannot estimate these parameters by equating A, B and C to their estimates. However, by assuming θ to be constant for all stages, Manly shows how the data from the different stages can be pooled to provide "moment-type" estimates of all the parameters. Finally there is the important problem of finding estimates $\hat{N}(t_i)$ of $N(t_i)$. If $n(t_i)$ is the number sampled from a given stage at time t_i, and a proportion p_i of the population is sampled, then $\hat{N}(t_i) = n(t_i)/p_i$. Usually p_i is kept constant throughout the experiment and is equal to the proportion of the population area (or population volume, as in the case of trees) sampled.

BIRLEY'S METHOD. An alternative approach to stage-specific survival is given by Birley [1977] who uses the more realistic discrete version of (13.68) (since sampling is discrete), namely

$$N_t = M_t \sum_{j=0}^{a} \phi(t-j)\, \alpha_j,$$

where N_t is the expected number in a given stage on day t, $\phi(t-j)$ is the fraction of the population entering the stage on day $t-j$, and α_j is the probability of an insect in the stage surviving j days. Setting $l_0 = M$ and $l_j = M\alpha_j$ ($= s_j$ in Birley's notation), we have for that stage a life-table series $\{l_j\}$, where the time-unit is one day. If $\phi(u)$ can be estimated, for example from emergence traps, cage experiments, or from visible differences between old and new, and estimates $\hat{N}_{t_i}$ are available at times t_i, then the l_j can be estimated by minimising

$$\sum_i [\hat{N}_{t_i} - \sum_{j=0}^{a} \phi(t_i - j) l_j]^2,$$

subject to $l_0 \geqslant l_1 \geqslant \ldots \geqslant l_a \geqslant 0$. Basically this is a so-called "constrained multiple linear regression" and can be solved using quadratic programming. Alternatively the l_j could be modelled in terms of a smooth function with fewer parameters. For example, if the survival rate is constant we can assume $l_j = Me^{-\theta j} = M\phi^j$ say. One advantage of Birley's model is that it can be modified to take care of catastrophes, for example sudden changes in survival due to the application of an insecticide. However, an objective method of determining a is not given, though a general plot of l_i against i for a guessed a will give further information about a.

METHOD OF READ AND ASHFORD. A very general method, showing considerable potential, has been given by Read and Ashford [1968] and Ashford *et al.* [1970]. The main advantage of this approach over the above method is that it allows for a random stage-length. If X_i is the time that an individual is in stage i ($i = 1, 2, \ldots, k$), then $X_i = \min(Y_i, Z_i)$ where Y_i would be the time spent in stage i if there were no mortality and Z_i would be the time an individual lives in stage i, assuming no transition to stage $i+1$. Let f_i, g_i, h_i and F_i, G_i, H_i be the respective density and distribution functions of X_i, Y_i, Z_i. Then, assuming the independence of Y_i and Z_i, $\Pr[X_i > x] = \Pr[Y_i > x] \Pr[Z_i > x]$ or $1 - F_i(x) = [1 - G_i(x)][1 - H_i(x)]$. Thus, differentiating we have

$$f_i(x) = g_i(x)[1 - H_i(x)] + h_i(x)[1 - G_i(x)]. \tag{13.69}$$

Although $i = 1, 2, \ldots, k$, it is convenient to introduce a stage $i = 0$ corresponding to the existence of an individual immediately prior to birth. An individual in stage 0 is assumed either to pass to stage 1 (i.e. to be born) or to die in stage 0. For example, in the case of insects this would correspond to the hatching or non-hatching of eggs laid by a previous generation. Clearly birth into stage 1 can be treated as simply a transition from stage 0 to stage 1. If we know the time when births can begin (t_0 say), we can use this point in time as our time origin $t = 0$. In this case the time of birth will have a density function given by (13.69) with $i = 0$.

If $p_i(t)$ is the probability that an individual is alive in stage i at time t, it follows that

$$p_i(t) = \int_0^t \phi_{i-1}(u_{i-1})[1 - F_i(t - u_{i-1})]\, du_{i-1}, \tag{13.70}$$

where ϕ_{i-1} is the density function of the time of entry to stage i from stage $i-1$. Here $\phi_0(u_0) = k_0(u_0)$ and

$$\phi_i(u_i) = \int_0^{u_i} \phi_{i-1}(u_{i-1}) k_i(u_i - u_{i-1}) du_{i-1}, \tag{13.71}$$

where

$$k_i(u) = g_i(u)[1 - H_i(u)],$$

the probability density function of the time spent in stage i given entry to stage $i+1$. Ashford *et al.* [1970] assume that death is a Poisson process so that

$$h_i(z_i) = \theta_i e^{-\theta_i z_i}, \tag{13.72}$$

which leads to

$$k_i(u) = g_i(u) e^{-\theta_i u}.$$

They also consider modelling g_i by a one-parameter Erlangian distribution, namely a scaled Gamma distribution with 2 or 3 degrees of freedom and an unknown scale factor, and then estimate all the parameters using maximum likelihood estimation. Usually t_0 will not be known and it can be estimated along with the other parameters by simply replacing t by $t - t_0$ in the likelihood function. Unfortunately considerable computations are required and some of the equations are not given, which will unfortunately preclude the use of the method by some ecologists.

If the generation length is constant ($= a_i$) rather than random, and $h_i(z_i)$ is given by (13.72), then

$$1 - F_i(x) = \begin{cases} e^{-\theta_i x} & 0 \leqslant x < a_i \\ 0 & \text{otherwise,} \end{cases}$$

and

$$p_i(t) = \int_{t-a_i}^{t} \phi_{i-1}(u_{i-1})\, e^{-\theta_i(t-u_{i-1})} du_{i-1},$$

which is the same as the expression for $N(t)/M$ given by (13.68). Also $\phi_i(u_i) = \phi_{i-1}(u_i + a_i)$.

OTHER METHODS. One approach to the above type of problem is to model the population by a continuous-time Markov chain with transition states corresponding to the various stages and an absorbing state corresponding to death. Moon [1976] developed such a model for a mosquito population and obtained an expression for the expected number of adult mosquitoes, $N(T)$, alive at time T.

Finally, a number of rather specialised models have been considered by Ruesink [1975].

13.4 GEOGRAPHIC STRATIFICATION

13.4.1 Two strata

In Chapter 11 we considered the problem of estimating population parameters for several populations, or equivalently, a single population stratified geographically with s strata. That chapter is based largely on the papers of Chapman and Junge [1956] and Darroch [1961] who dealt with the case of a Petersen experiment carried out simultaneously in each of the population areas, using tags which also indicated the place of release. It is seen that for an open population with just natural mortality, and possibly permanent emigration out of the population, the parameters are not identifiable, so that it is possible to estimate only certain functions of the parameters (p. 439). It is not possible, for example, to estimate the probability of surviving from the time of the first sample to the time of the second. However, if a third sample is taken, it is now possible to estimate most of the population parameters. For instance, when there are just two strata, Iwao [1963] gave estimates for a restricted deterministic migration model in which animals which migrate from one stratum to another between samples 1 and 2 may not migrate again between samples 2 and 3. This restriction was removed by Arnason [1972a] who derived the moment estimates given below.

Let the suffices A and B denote the two areas and let $x, y = A$ or B. For $j = 1, 2, 3$ and $i = 1, \ldots, j-1$, let

m_{ijxy} = number of animals taken in sample j (at time j) from the area y that were last captured in sample i from area x,
n_{jy} = size of sample j from area y,
N_{jy} = size of population y at time j,

ϕ_{jxy} = probability that an individual, alive and in area x at time j, is alive and in area y at time $j+1$,

ϕ_{jx} = probability that an individual, alive and in area x at time j, is alive and in the population at time $j+1$,

$= \phi_{jxA} + \phi_{jxB}$,

p_{jx} = probability that an individual, alive and in area x at the time the jth sample is taken, is captured in the jth sample,

$$|\mathbf{M}_{ij}| = m_{ijAA}m_{ijBB} - m_{ijAB}m_{ijBA},$$
$$|\mathbf{N}_{ijA}| = n_{jA}m_{ijBB} - n_{jB}m_{ijBA},$$
$$|\mathbf{N}_{ijB}| = n_{jB}m_{ijAA} - n_{jA}m_{ijAB},$$
$$|\mathbf{D}_1| = m_{13AA}m_{23BB} - m_{13AB}m_{23BA}, \text{ and}$$
$$|\mathbf{D}_2| = m_{13AB}m_{23AA} - m_{13AA}m_{23AB}.$$

Equating the m_{ijxy} and n_{jy} to their expected values, Arnason [1972a] obtained the estimates

$$\hat{p}_{1A} = |\mathbf{M}_{12}|/|\mathbf{N}_{12A}|,$$
$$\hat{N}_{1A} = n_{1A}/\hat{p}_{1A},$$
$$\hat{\phi}_{1AA} = [m_{12AA} + (n_{2A}|\mathbf{D}_1|/|\mathbf{M}_{23}|)]/n_{1A},$$
$$\hat{\phi}_{1AB} = [m_{12AB} + (n_{2B}|\mathbf{D}_2|/|\mathbf{M}_{23}|)]/n_{1A},$$
$$\hat{\phi}_{1A} = \hat{\phi}_{1AA} + \hat{\phi}_{1AB},$$
$$\hat{p}_{2A} = |\mathbf{M}_{23}|/|\mathbf{N}_{23A}|,$$
$$\hat{N}_{2A} = n_{2A}/\hat{p}_{2A},$$
$$\widehat{\phi_{2AA}p_{3A}} = m_{23AA}/n_{2A},$$
$$\widehat{\phi_{2BA}p_{3A}} = m_{23BA}/n_{2B},$$

The estimates for p_{1B}, N_{1B}, ϕ_{1BB}, ϕ_{1BA}, etc. are given by the above equations with A and B interchanged. Arnason gives two rules of thumb for planning and assessing experiments. (i) For planning experiments use prior knowledge of the parameters $\{N_{1x}, \phi_{ixy}\}$ to find the p_{ix} which give all $E[m_{ijxy}] > 8$ (preferably greater than 10). (ii) Once an experiment is carried out we must have all $m_{ijxy} > 5$. These two rules will ensure that, provided the underlying model is correct, the estimates are approximately unbiased, have a coefficient of variation of less than 50 percent, and have a low probability of being inadmissible, e.g. $\hat{p}_{ix}$ lying outside [0, 1]. Inadmissible estimates of some of the parameters may occur if they are close to 0 or 1: if such an estimate occurs then it is replaced by its closest boundary value, 0 or 1. Arnason [1972b] also gives asymptotic variance formulae which, on the basis of a limited number of simulations, are reasonably accurate if all $m_{ijxy} \geqslant 12$, but which have poor sample estimates unless the m_{ijxy} are much larger.

13.4.2 More than two strata

In a further paper Arnason [1973] gives moment estimates for the case of three samples and more than two strata, though his method is readily generalised to handle k samples, as in the algebra below. Suppose then that the strata are denoted by $A, B, C, \ldots, Q$ so that x and y can now take any of these values. For $j = 1, 2, \ldots, k$ and $i = 1, 2, \ldots, j-1$, define the following matrices:

$$\mathbf{M}_{ij} = [(m_{ijxy})] = \begin{bmatrix} m_{ijAA} & m_{ijAB} & \cdots & m_{ijAQ} \\ m_{ijBA} & m_{ijBB} & \cdots & m_{ijBQ} \\ \cdots & \cdots & \cdots & \cdots \\ m_{ijQA} & m_{ijQB} & \cdots & m_{ijQQ} \end{bmatrix}$$

$$\mathbf{\Phi}_j = [(\phi_{jxy})] = \begin{bmatrix} \phi_{jAA} & \phi_{jAB} & \cdots & \phi_{jAQ} \\ \phi_{jBA} & \phi_{jBB} & \cdots & \phi_{jBQ} \\ \cdots & \cdots & \cdots & \cdots \\ \phi_{jQA} & \phi_{jQB} & \cdots & \phi_{jQQ} \end{bmatrix}$$

$$\mathbf{D}_{n_j} = \begin{bmatrix} n_{jA} & 0 & \cdots & 0 \\ 0 & n_{jB} & \cdots & 0 \\ \cdots & \cdots & \cdots & \cdots \\ 0 & 0 & \cdots & n_{jQ} \end{bmatrix}$$

$$\mathbf{N}_j' = (N_{jA}, N_{jB}, \ldots, N_{jQ});$$

the matrix $\mathbf{D}_{p_j}$ and the vector $\mathbf{n}_j$ are similarly defined. Equating random variables to their expected values leads to the equations

$$\begin{aligned} \hat{N}_{j+1,x} &= n_{j+1,x}/\hat{p}_{j+1,x}, \\ m_{j,j+1,xy} &= n_{jx}\,\hat{\phi}_{jxy}\,\hat{p}_{j+1,y}, \quad \text{and} \\ \hat{N}_{j+1,y} &= \sum_x \hat{N}_{jx}\,\hat{\phi}_{jxy}. \end{aligned} \tag{13.73}$$

These equations, in matrix form, are respectively

$$\hat{\mathbf{N}}_{j+1} = \hat{\mathbf{D}}_{p_{j+1}}^{-1}\,\mathbf{n}_{j+1}, \tag{13.74}$$

$$\hat{\mathbf{\Phi}}_j = \mathbf{D}_{n_j}^{-1}\,\mathbf{M}_{j,j+1}\,\hat{\mathbf{D}}_{p_{j+1}}^{-1} \tag{13.75}$$

and

$$\hat{\mathbf{N}}_{j+1} = \hat{\mathbf{\Phi}}_j'\hat{\mathbf{N}}_j. \tag{13.76}$$

Using (13.75) and (13.74) we have

$$\begin{aligned} \hat{\mathbf{N}}_j &= (\hat{\mathbf{\Phi}}_j')^{-1}\hat{\mathbf{N}}_{j+1} \\ &= \mathbf{D}_{n_j}(\mathbf{M}_{j,j+1}')^{-1}\hat{\mathbf{D}}_{p_{j+1}}\,\hat{\mathbf{D}}_{p_{j+1}}^{-1}\,\mathbf{n}_{j+1} \\ &= \mathbf{D}_{n_j}(\mathbf{M}_{j,j+1}')^{-1}\,\mathbf{n}_{j+1}, \end{aligned} \tag{13.77}$$

which are the same as those given by Arnason [1973], except that we have used column rather than row vectors. The above notation is similar to that used in Chapter 11 with the difference that the role of the suffices (i, j) is now taken over by (x, y). When $k = 3$ and there are just two areas A and B the above estimates reduce to those given earlier in this section.

We note from p. 434 and equations (11.8) and (11.5) that $\mathbf{U}^*$ can be estimated by

$$\begin{aligned}\hat{\mathbf{U}}^* &= (\hat{\mathbf{\Theta}}')^{-1}\hat{\mathbf{U}} \\ &= \mathbf{D}_a(\mathbf{M}')^{-1}\hat{\mathbf{D}}_p\hat{\mathbf{U}} \\ &= \mathbf{D}_a(\mathbf{M}')^{-1}\mathbf{u}. \end{aligned} \tag{13.78}$$

As this does not require the assumption that $\mathbf{\Theta 1} = \mathbf{1}$ (i.e. no mortality) which is used in the estimation of $\hat{\mathbf{U}}$, (13.78) still holds even when there is mortality. However, we can go one step further and write down a similar equation for total numbers rather than just numbers of unmarked, namely

$$\hat{\mathbf{N}}^* = (\hat{\mathbf{\Theta}}')^{-1}\hat{\mathbf{N}} = \mathbf{D}_a(\mathbf{M}')^{-1}\mathbf{n}, \tag{13.79}$$

where $\mathbf{n} = [(\mathrm{n}_j)]$ and $\hat{N}_j = n_j/\hat{p}_j$. When there is just one stratum the above estimate reduces to the Petersen estimate. Thus $\hat{\mathbf{N}}^*$ can be regarded as a generalised Petersen estimate which, like the one-dimensional case (p. 71), is not affected by mortality under fairly general conditions. Also $\hat{\mathbf{N}}^*$ is the estimate proposed by Chapman and Junge [1956]. We note that (13.77) is of the same form as $\hat{\mathbf{N}}^*$, so that Arnason's method is just the generalised Petersen method applied to pairs of samples: samples 1 and 2 for the estimation of $\mathbf{N}_1$, samples 2 and 3 for the estimation of $\mathbf{N}_2$, etc.

We noted above that survival probabilities cannot be estimated from just two samples because of non-identifiability of the parameters. However, when we have more than two samples we also have the equations

$$m_{j,j+2,xy} = \sum_z n_{jx}\,\phi_{jxz}\,(1-\hat{p}_{j+1,z})\,\hat{\phi}_{j+1,zy}\,\hat{p}_{j+2,y}$$

which, combined with the equations for $m_{j,j+1,xy}$ and $m_{j+1,j+2,xy}$ (cf. (13.73)), lead to

$$\begin{aligned} m_{j,j+2,xy} &= \sum_z n_{jx}\,\hat{\phi}_{jxz}\,m_{j+1,j+2,zy}/n_{j+1,z} \\ &\quad - \sum_z m_{j,j+1,xz}\,m_{j+1,j+2,zy}/n_{j+1,z}. \end{aligned}$$

Expressed in matrix form, these equations are

$$\mathbf{M}_{j,j+2} = \mathbf{D}_{n_j}\,\hat{\mathbf{\Phi}}_j\,\mathbf{D}_{n_{j+1}}^{-1}\,\mathbf{M}_{j+1,j+2} - \mathbf{M}_{j,j+1}\,\mathbf{D}_{n_{j+1}}^{-1}\,\mathbf{M}_{j+1,j+2}$$

so that

$$\hat{\mathbf{\Phi}}_j = \mathbf{D}_{n_j}^{-1}(\mathbf{M}_{j,j+2}\,\mathbf{M}_{j+1,j+2}^{-1}\,\mathbf{D}_{n_{j+1}} + \mathbf{M}_{j,j+1}). \tag{13.80}$$

Estimates of survival, $\phi_{jx} = \sum_y \phi_{jxy}$, are obtained from the row sums of $\hat{\mathbf{\Phi}}_j$. Also we have the estimates $\hat{p}_{jx} = n_{jx}/\hat{N}_{jx}$ $(j = 1, 2, \ldots, k-1)$. However, estimates are not available for N_{kx} or $\mathbf{\Phi}_k$ unless some form of extrapolation is used.

The assumptions underlying the above method are (Arnason [1973: 3]):

(i) All animals behave independently with respect to capture, survival and migration, and that all individuals in a particular stratum at any time have the same probability of being subject to these events (regardless of age, size, capture history, etc.), though these probabilities may differ among areas and sampling times. (ii) Animals may migrate freely among areas between sampling times, but if they emigrate beyond the study areas *A*, *B*, .., , *Q*, they do so permanently. (iii) There are no births or immigrants. (iv) There are no losses on capture. However, as in the Jolly–Seber method, assumption (iv) can be relaxed to allow for the possibility that R_{jx} marked individuals are released from stratum x immediately after the jth sample. Provided the R_{jx} behave the same with respect to survival, migration and capture, then moment estimates can be readily calculated as before, giving (cf. (13.75) and (13.77)

$$\hat{\mathbf{N}}_j - \mathbf{n}_j + \mathbf{R}_j = \mathbf{D}_{R_j}(\mathbf{M}'_{j,j+1})^{-1}\mathbf{n}_{j+1}$$

and

$$\hat{\mathbf{\Phi}}_j = \mathbf{D}_{R_j}^{-1}(\mathbf{M}_{j,j+2}\,\mathbf{M}_{j+1,j+2}^{-1}\,\mathbf{D}_{R_{j+1}} + \mathbf{M}_{j,j+1}).$$

where $\mathbf{R}'_j = (R_{jA}, R_{jB}, \ldots, R_{jQ})$.

CHAPTER 14

CONCLUSION

14.1 COMPARISON OF METHODS

14.1.1 General Comments

A wide variety of models have been discussed in this book, and the question now arises as to which method should be used in a given situation. Obviously the choice of method will depend very much on the nature of the population, its distribution over the population area, and the method of sampling the population. Where possible, the experiment should be designed so that more than one method of estimation can be used. For example, the distance methods of **2.1.4** and **12.5** can sometimes be used along with quadrat sampling: the Petersen method can be incorporated in the CIR method of Chapter 9; and if the population is stationary, the time-specific methods of Chapter 10, using an age analysis of the removal, can be used with the CIR method. Since the generalised removal method of **12.9.3** can be applied to the unmarked animals in a capture–recapture experiment, live trapping should be used where possible, so that more options are available for data analysis. A common procedure after a multiple-recapture experiment in a closed population is to use kill-trapping or hunting and obtain an independent Petersen estimate based on r, the number of marked animals in the population at the end of the experiment (cf. **4.1.2**). Since the method of sampling the animals for marking, namely live trapping, will be very different from the method of sampling for recaptures, namely kill-trapping, the Petersen estimate will hopefully be unbiased (p. 86). The following is a selection of articles which obtain estimates by more than one method: Dunnet [1957, rabbits; Hayne's regression method and variations of the Petersen method], Flyger [1959, squirrels: Schnabel method using trapping and sighting records], Wood [1959, foxes: relative density estimates], Huber [1962, rabbits: Petersen and Schnabel methods], Ammann and Ryel [1963, ruffed grouse: relative density estimates], Eberhardt *et al.* [1963] rabbits: Petersen, CIR, frequency of capture, catch–effort, and other miscellaneous methods], Muir [1964, fish: catch–effort, Schumacher–Eschmeyer and age-composition methods], Chapman [1964, fur seals: age-composition, CIR, and tag release methods], Fischler [1965, crabs: catch–effort and single tag release methods], Bergerud and Mercer [1966, ptarmigan: direct count, King, Petersen, relative density and aerial methods], Mosby [1969, squirrels: Petersen, Schnabel, frequency of capture, age–composition methods], Phillips and Campbell [1970, whelks: Schnabel, Schumacher and Eschmeyer, Jolly–Seber, frequency of capture methods], Kemp and Keith [1970, squirrels: Petersen, frequency of capture and age-composition methods], Rusch and Keith [1971, ruffed grouse: transect counts, drumming count index, Petersen method, King strip method], Gromadzki and Trojan [1971, rabbits: Petersen and removal methods, digging out burrows], Lidicker [1973, voles: Petersen and Jolly–Seber methods, runway transect counts], and Trent and Rongstad [1974, rabbits: Petersen, frequency of capture and bounded counts method].

In the past, little attention has been devoted to the problem of designing an experiment to yield an estimate with a given minimum accuracy or precision. For example, in many of the early applications of the Petersen method, too few individuals were tagged, so that the number of recaptures was too small and the resulting confidence intervals too wide. However, for many of the models the variance formulae are complicated, so that it is not easy to plan for a given precision. Clearly more research is needed on the question of design and useful graphs for planning are needed for some methods. Examples of such graphs are given on pages 67–9, 368–70, 389–91, 517, 519–20 and in A2.

Where possible, the robust but less efficient regression estimates should be calculated along with the more efficient maximum-likelihood estimates. The regression method is particularly useful when expected values appear to be correct (as indicated by the appropriate graph(s)), but the variances predicted by the model underlying the maximum-likelihood theory are open to question because of possible departures from the underlying assumptions, e.g. sampling is not strictly random (cf. 4.1.3(*1*)). However, in all cases, the assumptions underlying a particular model should be studied carefully and, where possible, appropriate tests carried out. If there is likely to be any question about the validity of the underlying assumptions, the sample data should be collected in such a way that empirical variance estimates are available from replicated samples. A comparison of the sampling variance with the estimated theoretical variance predicted by the model will often throw some light on the validity of the model.

When samples are large enough, it is sometimes possible to use inter-penetrating subsamples. Here the total sample is split up randomly into subsamples from which separate estimates of population size, etc. can be calculated and then averaged (e.g. 3.4.3). But this technique has so far received little attention in the literature. Another technique which shows considerable promise in ecology is the jackknife method. It can be used for bias reduction and variance estimation.

Davis [1963] has stated that "the failure of wildlife investigators to check population estimates against known number is a deplorable situation". In spite of the fact that some methods have been widely used for estimating population size, these methods are seldom checked against a known population.

14.1.2 Individual methods

1 Absolute density

SAMPLE PLOTS. Total counts over the whole population are rarely possible, and one must usually resort to counting on sample plots. This method, however has a limited usefulness for animal populations because of (i) the problem of locating and marking out the sample plots, particularly if the terrain is difficult and the animals mobile, and (ii) the problem of finding all the animals on the plot. When the terrain is difficult and the population area large, strip transects will generally be more appropriate than quadrats. This seems to be the case in aerial censusing (12.3).

We recall from Chapter 2 that accuracy of a population-density estimate depends very much on the proportion p of the total area sampled, and some care should be given to the design of such experiments. It was shown in 2.1.1(*1*) that when $p < 0{\cdot}2$, $C^2 \approx 1/n$, and the accuracy depends only on the total number n of animals seen when less than 20 per cent of the population area is sampled. However, the validity and usefulness of this formula when the population is not randomly distributed needs further investigation; this could be done using simulation or by working with populations of known size. When non-randomly distributed populations are to be compared, the data can sometimes be transformed to stabilise the variance and achieve some measure of normality (cf. Southwood [1966: 8–12]). For example, the logarithmic transformation (cf. Koch [1969]) is often used for contagious populations; this leads to a study of ratios of population densities rather than of differences. If a rough estimate of k is available, then the negative binomial model (p. 25) is useful for planning purposes.

If the population density is known to vary considerably over different but well defined parts of the population area then stratified sampling should be used. As considerable gains in precision can be achieved if the allocation of sampling effort to each stratum or domain is optimal (i.e. the fraction of each domain area sampled is proportional to the square root of the domain density), one should endeavour to obtain rough estimates of domain densities before the experiment is carried out. Even very approximate estimates will lead to more precise estimates than those obtained using proportional allocation (i.e. the same fraction of each domain area sampled).

LINE TRANSECTS. This method is particularly useful when the animals are too mobile to be sampled using sample plots, or when the animals are difficult to locate and must be flushed into the open. But the particular density estimate to be used for a given species depends very much on the nature of the species and its "flushing" behaviour. For this reason the line transect method is still in its infancy as far as animal populations are concerned, and it should be used with caution until we know more about the behaviour of the various estimates described in 2.1.3 and 12.4.3. In this respect further research is needed on known populations: the simulated experiments of Robinette *et al.* [1954, 1956, 1974], Gates [1969] and Burnham *et al.* [1980] are a step in the right direction.

It is not easy to make a direct comparison between the line transect method and quadrat sampling as theoretical variances depend on the spatial distribution of the animals. However if we can assume a random distribution of animals, and the exponential model given in 2.1.3(*2*) leading to the estimate $\hat{N}_5$ is valid, some general comments can be made. To begin with, it was noted in that section that $C^2 \approx 2/n$ for a random population, so that approximately twice the number of animals must be seen to give the same coefficient of variation C as the sample plot method. To compare the costs of the two methods, we note from (2.12) that

$$C^2 = V[\hat{N}_5]/N^2$$

$$\approx \frac{1}{NP}\left(1 - P + NP\, E\left[\frac{1}{n-2}\right]\right)$$

$$\approx \frac{2-P}{NP},$$

and since $P = 2L/(A\lambda_1)$, where L is the length of line transect, A the population area, and $1/\lambda_1$ the average distance at which animals are seen from the transect, we find that

$$L \approx A\lambda_1/(1 + NC^2).$$

Hence the cost of obtaining an estimate with a given C is approximately Lc_L, where c_L is the cost per unit length of transect. On the other hand, for the sample plot method, the proportion p of the population area which must be sampled for a given C is (cf. (2.2)) $1/(1 + NC^2)$, and if quadrats of area a are used, each costing c_q, then the total cost is $c_q Ap/a$. Hence the ratio of cost of the line transect method to the sample plot method is

$$c_L\lambda_1/c_q'$$

where $c_q' = c_q/a$, the quadrat cost on a per-unit-area basis. Therefore, as one would expect $c_L < c_q'$ and $\lambda_1 < 1$, the above ratio will generally be less than unity, so that the line transect method will give more value for money, provided the underlying assumptions are satisfied.

DISTANCE METHODS. These methods can be used for animals which are relatively immobile and readily seen. In **2.1.4**(*2*) a promising estimate $\tilde{D}$ (cf. (2.31)) of the population density D $(= N/A)$ was considered which is unbiased, even when the population is not randomly distributed according to a binomial or Poisson law but rather follows a negative-binomial law, as in the case of contagious populations. Another promising method is the non-parametric technique of **12.5.2**.

When the main effort involved is in searching for the animals rather than in locating sample points or quadrat boundaries it was shown on p. 44 that the closest-individual technique using $\hat{D}$ has approximately the same efficiency as quadrat sampling. Otherwise, suppose that c_q is the cost of locating and setting out a quadrat of area a and let c_p be the cost of locating a sampling point and measuring the rth closest individual. Then, from (2.30),

$$C[\hat{D}] = 1/\sqrt{(sr-2)}$$

or

$$s = \frac{1}{r}\left(\frac{1}{C^2} + 2\right).$$

Hence, for a given C, the closest-individual method costs

$$\frac{c_p}{r}\left(\frac{1}{C^2}+2\right),$$

which may be compared with

$$\frac{c_q A}{a(1+NC^2)},$$

the total cost for the sample plot method.

2 Relative density

The direct methods discussed in **2.2.1** for estimating relative density are particularly useful when one is interested either in detecting changes in population density with time or in comparing populations in different areas. However, it was stressed there that if any comparisons are to be made, then the censusing should be carried out under as nearly identical conditions as possible. As changing conditions have different effects on different density indices, one should endeavour to obtain more than one type of index. For example, Ammann and Ryel [1963] used correlation coefficients to compare a number of indices for ruffed grouse based on mail-carrier counts along their routes, drumming counts, opinion surveys, brood counts, and kill records.

Where possible, replicated samples should be used for determining indices, so that sampling estimates of variance can be calculated (cf. (1.9)). For example, instead of just counting the total number n of animals seen on a roadside count for a road of length l units, one should count the number n_i $(i = 1, 2, \ldots, l)$ for each unit of road-length; although the index $\bar{n} = \Sigma\, n_i/l = n/l$ is still based on the total n, an estimate of $V[\bar{n}]$ is now available, namely $\Sigma\,(n_i - \bar{n})^2/l(l-1)$. In this example we note that the general comments and theory of **2.1.2**(*1*) still apply, provided we think in terms of sample length rather than sample plot. For example, in choosing the road-length unit we want l to be as large as possible, so that by the central limit theorem $\bar{n}$ is approximately normal; however, the shorter the unit the more crucial is the "end effect" where animals may be allocated to the wrong unit.

Catches from trap-lines have been widely used as an index of population density, particularly for small-mammal populations. The basis of the method is that the catch is always proportional to population density, provided the trapping is standardised (Calhoun [1948]). Hansson [1967, 1968], however, has reviewed this method and concludes that the proportion caught may vary considerably, thus suggesting that the trap-line method can be very unreliable (particularly when there is a wide variation in home range: cf. Stickel [1948]). On the other hand, Southern [1965] points out that the method is simple and very suitable for naturalists who have little time to spare; both he and Ashby [1967: 411 ff.] demonstrate that the method can be a useful one for detecting population changes (see also Myllymäki *et al.* [1971] and Hansson [1972, 1975]).

3 Mark–recapture methods: closed population

PETERSEN METHOD. Of all the methods considered in this book the Petersen method appears to be the most useful, provided that the assumptions underlying the method are satisfied and there are sufficient recaptures in the second sample. With regard to the latter proviso it was noted in **3.1.1** that the coefficient of variation C of the Petersen estimate is given by

$$C^2 \approx N/(n_1 n_2) \approx 1/m_2, \qquad (14.1)$$

so that for $C < 0{\cdot}25$ we must have at least 16 recaptures. Many of the early applications of the Petersen method fell down on this score, as too few individuals were marked, so that less than 10 were recaptured. By studying Fig. 3.1–3.6 on pages 65–69 we see that for population sizes commonly encountered in unexploited populations a considerable proportion of the population must be marked for a reasonable accuracy. For example, setting $n_1 = n_2 = n$, say, and defining the accuracy A as on p. 64, we obtain the following values of $100n/N$, the percentage of the population that must be marked for a given A:

A	0·5	0·25	0·1	0·5	0·25	0·1	0·5	0·25	0·1
N	50	50	50	100	100	100	1000	1000	1000
$100n/N$	40	54	74	32	43	64	13	20	40

A number of authors have commented on the need to mark at least 50 per cent of the population, e.g. Huber [1962: 185], Strandgaard [1967: 650] and Mosby [1969: 61].

The main assumption underlying the Petersen estimate is that marked and unmarked have the same probability of being caught in the second sample. Unfortunately it is not always easy to detect departures from this assumption, so that even when all precautions are taken and the assumptions appear to be satisfied, the Petersen estimate may be biased. For example, Buck and Thoits [1965] carried out extensive Petersen experiments in large fishponds, and after draining the ponds found that the Petersen estimates were considerably biased, with errors much larger than one would expect from chance. Thus "in spite of normally accepted indications to the contrary", the assumptions underlying the Petersen method were not valid: recruitment and mortality were ruled out, so that the errors were apparently due to a variable catchability amongst the fish. Further examples from Buck and Thoits [1965: Table 1], Cormack [1968, 470–3], Robinette *et al.* [1954], Dunnet [1957], Huber [1962], Smith [1968], Mosby [1969], Eisenberg [1972], Rose and Armentrout [1974] in which estimates are compared with known population sizes, indicate that in some cases the Petersen estimate is satisfactory while in others it is most unreliable. If the Petersen estimate is to be used extensively for a given species then it should be compared with other estimates and, where possible tested on a known population.

If the second sample can be taken in stages then the regression method of **3.7** (p. 125) can be used for testing the assumption that marked and unmarked have the same probability of capture.

BINOMIAL MODEL. Various methods for overcoming the problem of variable catchability, such as pre-baiting, using different sampling methods for the two samples, changing trap positions, etc., were discussed in **3.2.2**. However, the most promising approach to the problem is to avoid recapturing altogether and to obtain an estimate of the proportion marked by simply observing the animals or, if toe-clipping is used, by counting their tracks (Marten [1970b]); tagged animals can also be detected using remote sensing techniques (Marten [1972a, b, 1973]).

If sight records are used, as in Example 3.9 (p. 110), then the second sample is obtained by sampling *with* replacement and the binomial model (3.3) is applicable. In this case we have from Bailey [1951]

$$C^2 \approx (N - n_1)/(n_1 n_2), \tag{14.2}$$

and since animals can be seen more than once, n_2, the number of animals sighted in the second sample, can even be larger than N, thereby increasing the precision of the experiment. Therefore, given a rough estimate of N, and given n_1, n_2 can be determined from (14.2) for a given C with no restriction on n_2 other than that of time and cost of sampling. If the experiment is limited by funds then we can determine the optimum allocation of effort as follows.

Let c_m be the cost of marking an animal in the first sample and let c_0 be the cost of observing an animal in the second sample. Then B, the total cost, is given by

$$B = n_1 c_m + n_2 c_0,$$

and minimising C^2 in (14.2), subject to the above constraint, leads to

$$n_1 = N\left[1 - \left(1 - \frac{B}{Nc_m}\right)^{\frac{1}{2}}\right]$$

or, when n_1/N is small,

$$n_1 \approx B/(2c_m).$$

In using the sighting method for the second sample, some animals will be more active than others, so that they will be more likely to be seen. However, this will not affect the Petersen estimate or its modified version $\hat{N}_1$, provided the probability of sighting a given animal is independent of its mark status.

If tracking is used to "observe" the animals, then, from Marten [1972a], the Petersen estimate for the unmarked population size is given by

$$\hat{U} = \hat{N} - n_1 = n_1 u_2/m_2 = u_2/\hat{k},$$

where $\hat{k}$ is the number of tracks per marked individual and u_2 is the number of tracks from unmarked individuals. Therefore, since u_2 and $\hat{k}$ are independent, we have, using the delta method,

$$V[\hat{U}] \approx U^2\left[\frac{V[u_2]}{(E[u_2])^2} + \frac{V[\hat{k}]}{(E[\hat{k}])^2}\right].$$

Following Marten, it is not unreasonable to assume that the number of tracks per individual is Poisson, so that

$$V[\hat{U}] \approx U^2\left[\frac{1}{E[u_2]} + \frac{1}{E[m_2]}\right]$$

and

$$C^2[\hat{U}] \approx \frac{1}{E[u_2]} + \frac{1}{E[m_2]}.$$

Marten points out that if animals vary in their activity, the above estimate of U will not be affected, provided activity is independent of mark status. Further extensions of the above tracking method are given in Marten [1972a].

SCHNABEL METHOD. Sometimes it is not possible to catch enough individuals on the first sampling occasion for a satisfactory application of the Petersen estimate, so that the Schnabel method must be used (e.g. Williams [1965]). In any case the latter method should be used if variable catchability is a problem as the various models in **12.8.6** are available.

To compare the Petersen and Schnabel sampling schemes from the point of view of efficiency it is convenient to look at the Petersen and Schnabel estimates as they have simple coefficients of variation, C_1 and C_2 respectively, of the form $C_i = \sqrt{(N/\lambda_i)}$, where $\lambda_1 = n_1 n_2$ and $\lambda_2 = \sum_{i=2}^{s} n_i M_i$ (cf. p. 130 for notation). We shall assume that for each method the probability of capture is constant, and we shall denote these probabilities by P_1 and P_2 respectively. Then, replacing the n_i and M_i by their expected values leads to

$$\begin{aligned} R^2 &= C_1^2/C_2^2 \\ &= \lambda_2/\lambda_1 \\ &\approx NP_2 \sum_{i=2}^{s} N(1 - Q_2^{i-1})/N^2P_1^2 \\ &= (sP_2 - 1 + Q_2^s)/P_1^2, \end{aligned}$$

where $Q_2 = 1 - P_2$. One basis for comparison is to assume that both methods have the same expected total sample size, i.e. $E[n_1 + n_2]$ for the Petersen methods equal $E[n_1 + n_2 + \ldots + n_s]$ for the Schnabel method; then $2NP_1 = sNP_2$ and

$$R = \sqrt{4(sP_2 - 1 + Q_2^s)/(s^2P_2^2)}. \qquad (14.3)$$

From Table 14.1 we see that for the values of P_2 and s commonly used in practice, the Schnabel estimate is more efficient than the Petersen estimate, though the difference is not great. For small P_2, R increases with s up to a maximum value and then decreases to zero as $s \to \infty$; the maximum is reached more quickly for larger values of P_2.

If one wishes to compare the costs of the two methods for the same efficiency (i.e. $R = 1$), suppose that c_T is the cost of setting up a single trap each day and c_m is the cost of marking an animal. Assuming that $P_i = aT_i$, where T_i is the number of traps used, and using individual marks or tags (i.e. marked individuals are not remarked), the ratio of the cost of the Petersen method to the Schnabel method is

$$\frac{2T_1 c_T + NP_1 c_m}{sT_2 c_T + N(1 - Q_2^{s-1})c_m}.$$

If rough estimates of N and a are available then the above ratio can be calculated for different values of s and T_2; T_1 is calculated from

$$P_1^2 = sP_2 - 1 + Q_2^s.$$

TABLE 14.1

Ratio (R) of the efficiencies of the Petersen and Schnabel estimates for different values of s (the number of samples in the Schnabel census) and P_2 (the probability of capture in each sample of the Schnabel census).

	Ratio (R)				
P_2	$s = 3$	4	5	6	12
0·05	1·14	1·20	1·23	1·25	1·25
0·1	1·13	1·18	1·20	1·21	1·16
0·2	1·11	1·14	1·14	1·13	1·01
0·3	1·10	1·11	1·09	1·06	0·89
0·5	1·05	1·03	0·99	0·95	0·75

REGRESSION METHODS. The basic equation underlying the regression method of Schumacher and Eschmeyer (**4.1.3**(*1*)) is $E[m_i/n_i|M_i] = M_i/N$, which holds when marked and unmarked have the same probability of capture in the ith sample. However, another useful equation is $E[u_i] = p_i U_i$, where p_i is the probability than an unmarked individual is caught in the ith sample. If p_i is the same for marked individuals, then $1/p_i$ can be estimated by $(M_i + 1)/(m_i + 1)$, and leads to the regression model

$$E\left[\frac{u_i(M_i + 1)}{(m_i + 1)} \,|M_i\right] \approx U_i \approx N - M_i.$$

Alternatively, if p_i is k times the probability of catching a marked individual, then $1/p_i$ can be replaced by $(M_i + 1)/k(m_i + 1)$ and

$$E\left[\frac{u_i(M_i + 1)}{(m_i + 1)} \,|M_i\right] \approx k(N - M_i), \tag{14.4}$$

which is Marten's model (4.18). However, departures from the underlying assumption that marked and unmarked have the same probability of capture take different forms. For example, Tanaka and Kanamori [1967] suggest that, with trap addiction, the traps will first be taken up by marked animals so that p_i will be proportional to some function g of the traps available for unmarked individuals, i.e. if T is the number of traps and T_i $(= T - m_i)$ is the number of traps not occupied by marked when the ith sample is complete, then $p_i = a\,g(T_i)$. Therefore, since $m_1 = 0$,

$$p_i = p_1\, g(T_i)/g(T)$$

and, given an estimate ρ_i of $g(T)/g(T_i)$, we have the regression model

$$E[\rho_i u_i | M_i] = p_1(N - M_i)$$

which is of the same form as (12.4). Tanaka gives an example where, on the basis of empirical information, he chooses $g(T) = T^x$. Leslie and Davis [1939] and Tanaka [1963b], using the analogy of the random movement of molecules suggest an alternative model

$$T\{\log T_i - \log (T_i - u_i)\} = p_1(N - M_i) + e_i,$$

which is again of the same form as (14.4).

4 Mark–recapture methods: open population

The single tag–release models of Chapter 6 are mainly applicable to commercially exploited populations where numbers are large and tagging experiments costly. However, since the information from a single release is minimal, strong assumptions have to be built into the models in order that the population size may be estimated. For this reason it is preferable to release several small batches of tagged individuals at suitable intervals rather than one large batch at the beginning, and then use the "multi-sample single recapture" models of **13.1.4**. Methods are described there for planning such experiments and the reader is also referred to Brownie *et al.* [1978] for a careful consideration of the design of bird-banding experiments.

For the small unexploited population where multiple recaptures are possible, the various methods described in **5.1.1** and **13.1.6** can be used. Clearly any variation in catchability will have a greater effect on this model than on the multi-sample single recapture model where animals are recaptured once only. The various special cases discussed in **5.1.3**, sections *1* and *2*, should not be used unless there is definite biological evidence that the conditions for their applicability are satisfied. When the conditions are satisfied, the more restrictive models will yield more efficient estimates. Carothers [1973a: 146] found a considerable gain in efficiency in using one of the restricted models, particularly at low sampling intensities.

5 Catch–effort methods

These are widely used for commercial fisheries and are particularly useful when tagging experiments are unsatisfactory. For example, if the population is very big and the fishing grounds extensive, the recaptures in each sample may be too few, even in spite of large tag–releases; also, when recaptures are few, the non-reporting of tags can be a major problem. Volume 155 of *Rapports et Procès-Verbaux des Réunions du Conseil International pour L'Exploration de la Mer* contains reviews of experience with catch–effort methods in marine fisheries: see especially Gulland [1964] and Paloheimo and Dickie [1964].

For small closed populations, the so-called removal method of **7.2** has been used with mixed success. For example, Gentry *et al.* [1968] found that the probability of capture p did not remain constant from sample to sample as required, while Grodzinksi *et al.* [1966] found that the method can be satisfactory for some species, particularly with prebaiting followed by intensive trapping (see also Babinska and Bock [1969]). The main departures from the assumption of constant p are (i) variation in trap efficiency from sample to sample due to environmental changes such as weather, etc., and (ii) variation in the trap response of individuals: both departures are discussed in some detail in **7.2.2**(*1*) and **12.9**.

To compare the two-sample removal method with the Petersen method we shall assume that for the Petersen method the probability of capture in each sample is p. Then, from p. 324, the ratio of the coefficient of variation of the removal estimate to that of the Petersen estimate is $\sqrt{(1+q)/p}$. Values of this ratio are:

p	0·1	0·2	0·3	0·4	0·5	0·6	0·9
ratio	4·36	3·00	2·39	2·00	1·73	1·53	1·10

thus showing that the Petersen estimate is much more efficient (though of course more costly).

For the three-sample removal method we find, using (7.23), that the corresponding ratio is

$$\sqrt{\left[\frac{(1-q^3)qp^2}{(1-q^3)^2-9p^2q^2}\right]}$$

which takes values:

p	0·1	0·2	0·3	0·4	0·5
ratio	2·14	1·42	1·09	0·88	0·73

Thus, for $p = 0{\cdot}3$ there is little difference between the two methods as far as precision is concerned. However, for this value of p the removal method would remove about 66 percent of the population, while the Petersen method would lead to a marking of 30 percent of the population.

6 *Change in ratio method*

To compare the CIR method with the Petersen method, suppose that n animals are classified according to x- and y-type before and after a removal of R_x x-type animals (see p. 354 for notation; here $n_1 = n_2 = n$, $R_y = 0$ and $R = R_x$). Then, if θ is the ratio of the cost of tagging to the cost of classification, Chapman [1955: $\theta = 1/\lambda$] considers the Petersen estimate that could be made if $2n/\theta$ animals were tagged and a second sample of R animals was inspected for tags. For this situation we find from Chapman that the ratio of the coefficient of variation of the Petersen estimate to that of the CIR method is approximately

$$\sqrt{\left[\left(\frac{1-P_1}{2P_1-u}\right)\left(\frac{u}{1-u}\right)\cdot\frac{\theta}{2}\right]},$$

where $P_1 = X_1/N_1$ (the initial proportion of x-types) and $u = R/N_1$. The values of θ required to make the above ratio unity are given in Table 14.2 for different values of P_1 and u. Since the comparison of the two methods is made under the conditions most favourable to the CIR method, i.e. the allocation of sampling effort is almost optimal (p. 363), this table suggests that unless tagging is prohibitively costly, the Petersen method is better value for money.

TABLE 14.2

The value of θ, the relative cost of tagging to cost of classification, required to give the same coefficients of variation for the Petersen and CIR estimates: from Chapman [1955: Table 3].

X_1/N_1 $(= P_1)$	$R/N_1 = u$ 0·25	0·20	0·15	0·10	0·05	0·02	0·01
0.1	–	–	–	–	6·34	19·60	41·80
0·2	–	–	3·54	6·76	16·62	46·55	96·53
0·3	3·00	4·58	7·28	12·86	29·85	81·20	166·90
0·4	5·51	8·00	12·28	21·00	47·49	127·40	260·70
0·5	9·00	12·80	19·27	32·40	71·20	192·08	392·04
0·6	14·25	20·00	29·75	49·50	109·25	289·10	589·05
0·7	23·00	32·00	47·22	78·00	171·00	450·80	917·33
0·8	40·51	56·00	81·16	135·00	294·51	774·20	1574·10
0·9	93·00	128·00	87·00	306·00	332·51	1744·40	3544·20

If considerations of cost are not important then $\sqrt{\theta}$ will represent the ratio of the coefficient of variation of the CIR estimate to that of the Petersen estimate. Therefore, since the Petersen estimate itself is not very precise unless a large fraction of the population is tagged, we find that CIR estimates are generally not very accurate unless a large fraction of the population is classified, or there is a big change in the proportion of x-types through the removal R; from p. 370 we recall that $\Delta P = P_1 - P_2$ should be at least greater than 0·10 (or possibly 0·05 if we have full confidence in the underlying assumptions).

7 *Age–composition methods*

Although life tables are widely used, the assumptions behind the construction of such tables are not always appreciated. For example, the time-specific method of **10.1.2**(*2*) is only valid when the population is stationary, so that any such an analysis should be accompanied by biological and statistical evidence for stationarity.

Because of the misuse of life tables by many ecologists I wish to emphasise again that life-table methods should not be used for the analysis of tag returns from dead animals (e.g. bird-band returns). The life-table approach only leads to confusion and, frequently, erroneous estimates.

14.2 OTHER TOPICS

When I first planned this book I had hoped to cover a wider range of topics. However, the more I searched the literature the more I realised that I had to restrict myself to just the problem of estimating animal abundance, without considering the nature of the factors underlying population change. This means that a wide range of topics relating to population dynamics have not been covered. For example, there have been many applications of the theory of stochastic processes to population systems (cf. Bartlett [1960], Keyfitz [1968], Pielou [1969]), while in the study of fishery dynamics numerous deterministic models have been developed (cf. Beverton and Holt [1957], Ricker [1958, 1975], Cushing [1968]); in subsequent years simulation (e.g. Eleveback and Varma [1965], King and Paulik [1967], Walters [1969]) and systems analysis (e.g. Watt [1966], Dale [1970]) have provided useful tools for population analysis. Also, such topics as density-dependence (Andrewartha [1961], Varley and Gradwell [1970]), population cycles (Lack [1954]), the natural regulation of population numbers (Lack [1954]), association between species (Greig-Smith [1964], Southwood [1966], Pielou [1969]), and estimating numbers of species (Williams [1964], Holgate [1969], Pielou [1969]), etc., have generated much research. However, I hope that this book will encourage statisticians to become increasingly interested in ecology, with its broad spectrum of challenging research problems.

APPENDIX

A1 Shortest 95% confidence interval for N/λ based on the Poisson distribution

		Entering variable m_2 (or m)			
m_2	Lower limit	Upper limit	m_2	Lower limit	Upper limit
0	0·088 5				
1	0·072 0	19·489	26	0·024 78	0·056 3
2	0·076 7	2·821	27	0·024 08	0·053 9
3	0·073 6	1·230	28	0·023 42	0·051 6
4	0·069 0	0·738	29	0·022 79	0·049 5
5	0·064 4	0·513	30	0·022 21	0·047 5
6	0·060 0	0·388	31	0·021 65	0·045 7
7	0·056 1	0·309	32	0·021 12	0·044 0
8	0·052 6	0·256	33	0·020 61	0·042 5
9	0·049 5	0·217	34	0·020 14	0·041 0
10	0·046 8	0·188	35	0·019 68	0·039 6
11	0·044 3	0·165	36	0·019 25	0·038 4
12	0·042 0	0·147	37	0·018 83	0·037 2
13	0·040 0	0·133	38	0·018 43	0·036 0
14	0·038 2	0·121	39	0·018 05	0·035 0
15	0·036 5	0·111	40	0·017 69	0·033 96
16	0·035 0	0·1020	41	0·017 33	0·033 00
17	0·033 62	0·0945	42	0·017 00	0·032 10
18	0·032 33	0·0880	43	0·016 68	0·031 24
19	0·031 14	0·0823	44	0·016 36	0·030 43
20	0·030 04	0·0773	45	0·016 06	0·029 66
21	0·029 01	0·0729	46	0·015 78	0·028 92
22	0·028 06	0·0689	47	0·015 50	0·028 22
23	0·027 16	0·0653	48	0·015 23	0·027 55
24	0·026 32	0·0620	49	0·014 98	0·026 91
25	0·025 52	0·0591	50	0·014 75	0·026 25

(Reproduced from Chapman [1948].)

Applications of the above table are given on p. 63 and pp. 139–40.

A2 Tag recoveries needed for prescribed probabilities of detecting incomplete tag-reporting with various levels of catch inspection

The parameters are defined in **3.2.4.**

$\alpha = 0.10$

ρ	$1-\beta$	p_0									
		.05	.10	.15	.20	.25	.30	.40	.50	.70	.90
.25	.50	6	4	4	3	3	4	4	5	9	28
	.80	24	14	11	10	10	9	10	11	17	51
	.90	39	23	18	15	14	14	14	15	23	66
	.95	54	31	24	20	19	18	18	19	28	80
	.99	88	50	37	32	29	27	26	28	40	109
.50	.50	39	23	18	15	14	14	14	15	23	66
	.80	136	76	56	47	42	39	38	39	55	147
	.90	210	116	85	71	63	58	55	56	76	201
	.95	284	156	114	94	83	76	71	73	97	252
	.99	453	246	178	146	128	117	108	109	143	365
.60	.50	84	47	36	30	27	26	25	27	38	106
	.80	276	151	110	91	80	74	69	71	95	247
	.90	421	229	166	136	120	110	101	102	134	344
	.95	564	305	221	180	158	144	132	133	172	436
	.99	888	478	344	280	243	221	201	200	256	639
.70	.50	197	109	80	66	59	55	52	53	72	191
	.80	615	332	240	196	171	156	143	143	186	469
	.90	925	498	359	291	253	230	209	208	266	662
	.95	1230	660	474	384	333	302	273	271	342	847
	.99	1917	1026	734	593	513	464	416	410	513	1256
.75	.50	322	176	128	106	93	86	80	81	108	278
	.80	981	527	380	308	268	244	221	220	280	696
	.90	1468	787	564	456	395	358	323	319	402	990
	.95	1943	1040	744	601	519	470	422	415	519	1272
	.99	3016	1609	1148	925	797	720	643	630	781	1895
.80	.50	568	307	222	182	159	145	133	134	173	439
	.80	1688	904	647	523	453	410	369	364	457	1121
	.90	2512	1341	958	772	667	602	539	529	658	1603
	.95	3314	1767	1260	1015	875	789	704	689	853	2065
	.99	5120	2725	1939	1558	1340	1207	1073	1046	1285	3093
.85	.50	1130	607	436	354	307	279	252	250	317	787
	.80	3290	1754	1251	1007	868	783	699	684	847	2052
	.90	4866	2590	1844	1482	1275	1148	1021	997	1225	2950
	.95	6401	3403	2420	1943	1670	1502	1333	1298	1590	3814
	.99	9845	5227	3712	2975	2554	2294	2031	1972	2402	5733
.90	.50	2832	1512	1079	869	750	677	605	593	736	1789
	.80	8074	4289	3048	2444	2100	1887	1672	1626	1985	4748
	.90	11881	6305	4475	3585	3076	2761	2442	2369	2880	6859
	.95	15579	8261	5859	4691	4022	3608	3186	3086	3743	8894
	.99	23864	12643	8959	7166	6138	5501	4850	4691	5669	13423
.95	.50	12548	6658	4724	3785	3246	2914	2576	2499	3036	7227
	.80	35080	18573	13151	10513	8999	8060	7097	6855	8262	19515
	.90	51367	27181	19237	15369	13149	11770	10352	9989	12013	28320
	.95	67153	35522	25132	20072	17167	15362	13503	13020	15641	36822
	.99	102459	54175	38312	30586	26148	23389	20540	19791	23732	55778

Table A.2 (*continued*)

$\alpha = 0.05$

ρ	$1-\beta$	p_0									
		.05	.10	.15	.20	.25	.30	.40	.50	.70	.90
.25	.50	9	6	5	5	5	6	7	8	14	46
	.80	30	19	15	13	13	13	14	15	25	75
	.90	47	28	22	19	18	18	18	20	32	93
	.95	63	37	29	25	23	22	23	25	38	109
	.99	101	57	44	37	34	33	32	35	51	143
.50	.50	63	37	29	25	23	22	23	25	38	109
	.80	180	100	75	63	57	53	51	54	76	207
	.90	264	146	108	90	80	75	71	74	102	271
	.95	346	191	140	116	103	95	90	92	125	331
	.99	530	289	211	174	153	141	131	133	177	458
.60	.50	137	77	58	49	45	42	41	44	63	174
	.80	368	202	148	123	109	101	95	97	132	346
	.90	533	291	212	175	154	142	131	134	178	460
	.95	693	376	273	224	197	180	166	168	221	566
	.99	1048	566	409	334	291	266	243	244	315	794
.70	.50	324	178	131	109	97	90	85	87	119	315
	.80	827	448	325	266	233	213	196	197	257	653
	.90	1183	638	461	375	327	298	272	272	350	879
	.95	1524	820	590	479	417	379	344	343	437	1091
	.99	2281	1223	877	710	615	558	503	498	628	1549
.75	.50	530	289	211	174	153	141	131	133	177	458
	.80	1325	714	515	419	364	332	302	301	386	968
	.90	1883	1011	726	589	511	464	420	416	528	1310
	.95	2417	1296	929	751	651	590	532	525	662	1631
	.99	3600	1924	1375	1110	959	867	777	764	953	2328
.80	.50	935	506	366	299	261	239	219	220	285	722
	.80	2289	1227	880	712	617	560	505	499	630	1554
	.90	3234	1730	1237	999	864	781	701	690	863	2114
	.95	4137	2209	1578	1272	1098	992	888	872	1084	2641
	.99	6132	3268	2329	1874	1614	1455	1297	1269	1567	3788
.85	.50	1862	1000	718	582	506	459	415	412	523	1297
	.80	4473	2388	1705	1374	1185	1070	957	939	1166	2836
	.90	6287	3350	2387	1920	1654	1491	1329	1299	1604	3877
	.95	8016	4267	3037	2441	2100	1891	1682	1641	2018	4859
	.99	11827	6286	4468	3585	3080	2769	2456	2390	2922	7000
.90	.50	4666	2490	1777	1432	1235	1115	997	977	1213	2947
	.80	11012	5854	4162	3340	2871	2582	2291	2230	2730	6544
	.90	15398	8176	5807	4655	3997	3590	3179	3088	3764	8989
	.95	19572	10386	7371	5905	5066	4548	4022	3902	4743	11299
	.99	28752	15243	10808	8651	7415	6649	5870	5684	6886	16345
.95	.50	20671	10967	7782	6234	5348	4799	4243	4116	5000	11905
	.80	47983	25412	18000	14393	12324	11041	9727	9402	11345	26830
	.90	66770	35343	25021	19997	17114	15324	13487	13022	15682	27011
	.95	84619	44776	31689	25317	21660	19389	17053	16456	19791	46652
	.99	123797	65478	46319	36989	31632	28303	24871	23978	28788	67742

Table A.2 (*concluded*)

$\alpha = 0.01$

ρ	$1-\beta$	p_0									
		.05	.10	.15	.20	.25	.30	.40	.50	.70	.90
.25	.50	17	12	10	10	10	11	13	16	28	92
	.80	45	28	23	21	20	20	22	25	42	131
	.90	65	39	31	28	27	26	28	32	51	155
	.95	84	50	39	35	33	32	33	37	59	175
	.99	126	73	57	49	46	44	45	49	75	218
.50	.50	126	73	57	49	46	44	45	49	75	218
	.80	278	157	118	100	90	85	83	88	127	350
	.90	382	213	158	133	119	112	108	113	159	432
	.95	479	266	197	164	147	137	130	136	188	506
	.99	692	381	280	232	205	190	179	184	250	661
.60	.50	274	154	116	98	89	84	82	87	125	347
	.80	578	319	235	195	174	162	153	158	217	579
	.90	781	428	314	259	230	212	199	204	276	723
	.95	971	531	387	319	281	259	242	246	329	855
	.99	1385	753	546	448	393	360	332	336	441	1131
.70	.50	648	357	262	218	193	179	169	174	238	629
	.80	1312	713	518	425	373	342	316	320	422	1083
	.90	1751	948	686	561	490	449	412	414	539	1368
	.95	2162	1167	843	687	599	547	500	500	645	1629
	.99	3048	1640	1180	959	834	758	688	685	874	2181
.75	.50	1060	578	422	347	305	281	261	266	353	914
	.80	2110	1140	823	671	586	535	489	490	633	1597
	.90	2802	1509	1087	883	769	700	636	634	811	2029
	.95	3446	1852	1331	1080	938	852	772	767	975	2424
	.99	4835	2591	1857	1503	1302	1180	1063	1050	1323	3261
.80	.50	1869	1011	732	597	522	477	437	439	570	1444
	.80	3661	1966	1413	1145	994	903	817	811	1029	2555
	.90	4834	2591	1857	1502	1301	1179	1063	1050	1323	3261
	.95	5927	3172	2270	1834	1586	1435	1290	1271	1593	3909
	.99	8775	4419	3156	2544	2196	1984	1775	1743	2169	5282
.85	.50	3724	2000	1437	1165	1011	918	830	824	1045	2593
	.80	7184	3840	2744	2214	1913	1729	1550	1524	1902	4647
	.90	9441	5038	3595	2896	2498	2255	2015	1976	2452	5957
	.95	11537	6150	4383	3528	3040	2741	2445	2393	2958	7161
	.99	16035	8534	6074	4882	4201	3782	3364	3283	4036	9719
.90	.50	9332	4980	3554	2863	2470	2229	1993	1954	2424	5894
	.80	17752	9444	6720	5398	4643	4178	3714	3621	4445	10689
	.90	23221	12342	8773	7040	6049	5438	4824	4695	5742	13759
	.95	28291	15027	10674	8560	7351	6604	5851	5687	6938	16586
	.99	39150	20774	14743	11812	10134	9096	8044	7804	9488	22601
.95	.50	41348	21937	15566	12469	10696	9600	8487	8232	10002	23813
	.80	77628	41128	29142	23312	19969	17897	15779	15263	18444	43682
	.90	101102	53539	37919	30318	25958	23254	20483	19795	23878	56448
	.95	122823	65021	46037	36797	31496	28206	24830	23981	28893	68223
	.99	169262	89565	63386	50642	43326	38784	34111	32916	39588	93317

(Reproduced from Paulik [1961].)

A3 Solution of $\exp(-x) + ax = 1$

Note: For $a < 0{\cdot}05$ take $x = 1/a$.

a	x	a	x	a	x	a	x	a	x
0·050	20·0000000	0·100	9·9995458	0·150	6·6581095	0·200	4·9651142	0·250	3·9206904
·051	19·6078431	·101	9·9004937	·151	6·6136295	·201	4·9395120	·251	3·9037180
·052	19·2307691	·102	9·8033798	·152	6·5697225	·202	4·9141487	·252	3·8868671
·053	18·8679244	·103	9·7081477	·153	6·5263771	·203	4·8890207	·253	3·8701363
·054	18·5185184	·104	9·6147429	·154	6·4835823	·204	4·8641246	·254	3·8535241
0·055	18·1818180	0·105	9·5231129	0·155	6·4413272	0·205	4·8394568	0·255	3·8370292
·056	17·8571425	·106	9·4332073	·156	6·3996012	·206	4·8150140	·256	3·8206502
·057	17·5438592	·107	9·3449775	·157	6·3583940	·207	4·7907929	·257	3·8043857
·058	17·2413787	·108	9·2583768	·158	6·3176956	·208	4·7667902	·258	3·7882344
·059	16·9491518	·109	9·1733599	·159	6·2774962	·209	4·7430026	·259	3·7721949
0·060	16·6666657	0·110	9·0898836	0·160	6·2377864	0·210	4·7194272	0·260	3·7562660
·061	16·3934414	·111	9·0079060	·161	6·1985568	·211	4·6960608	·261	3·7404463
·062	16·1290307	·112	8·9273866	·162	6·1597983	·212	4·6729003	·262	3·7247347
·063	15·8730138	·113	8·8482865	·163	6·1215021	·213	4·6499430	·263	3·7091299
·064	15·6249974	·114	8·7705681	·164	6·0836596	·214	4·6271857	·264	3·6936307
0·065	15·3846122	0·115	8·6941952	0·165	6·0462625	0·215	4·6046258	0·265	3·6782358
·066	15·1515112	·116	8·6191326	·166	6·0093024	·216	4·5822605	·266	3·6629441
·067	14·9253682	·117	8·5453466	·167	5·9727714	·217	4·5600869	·267	3·6477545
·068	14·7058763	·118	8·4728044	·168	5·9366617	·218	4·5381024	·268	3·6326657
·069	14·4927463	·119	8·4014745	·169	5·9009657	·219	4·5163044	·269	3·6176767
0·070	14·2857054	0·120	8·3313262	0·170	5·8656758	0·220	4·4946903	0·270	3·6027863
·071	14·0844963	·121	8·2623301	·171	5·8307850	·221	4·4732575	·271	3·5879935
·072	13·8888760	·122	8·1944575	·172	5·7962859	·222	4·4520036	·272	3·5732972
·073	13·6986147	·123	8·1276809	·173	5·7621718	·223	4·4309262	·273	3·5586962
·074	13·5134952	·124	8·0619734	·174	5·7284359	·224	4·4100228	·274	3·5441897
0·075	13·3333117	0·125	7·9973091	0·175	5·6950714	0·225	4·3892910	0·275	3·5297765
·076	13·1578693	·126	7·9336629	·176	5·6620721	·226	4·3687286	·276	3·5154556
·077	12·9869832	·127	7·8710106	·177	5·6294315	·227	4·3483333	·277	3·5012261
·078	12·8204781	·128	7·8093287	·178	5·5971436	·228	4·3281028	·278	3·4870869
·079	12·6581876	·129	7·7485942	·179	5·5652022	·229	4·3080351	·279	3·4730370
0·080	12·4999534	0·130	7·6887851	0·180	5·5336015	0·230	4·2881278	0·280	3·4590756
·081	12·3456253	·131	7·6298800	·181	5·5023357	·231	4·2683789	·281	3·4452017
·082	12·1950603	·132	7·5718581	·182	5·4713992	·232	4·2487864	·282	3·4314143
·083	12·0481222	·133	7·5146991	·183	5·4407865	·233	4·2293482	·283	3·4177125
·084	11·9046814	·134	7·4583837	·184	5·4104922	·234	4·2100622	·284	3·4040955
0·085	11·7646144	0·135	7·4028927	0·185	5·3805110	0·235	4·1909266	0·285	3·3905624
·086	11·6278033	·136	7·3482078	·186	5·3508378	·236	4·1719393	·286	3·3771122
·087	11·4941358	·137	7·2943110	·187	5·3214675	·237	4·1530985	·287	3·3637441
·088	11·3635044	·138	7·2411851	·188	5·2923952	·238	4·1344024	·288	3·3504572
·089	11·2358068	·139	7·1888131	·189	5·2636161	·239	4·1158490	·289	3·3372508
0·090	11·1109450	0·140	7·1371786	0·190	5·2351255	0·240	4·0974366	0·290	3·3241240
·091	10·9888254	·141	7·0862658	·191	5·2069187	·241	4·0791634	·291	3·3110759
·092	10·8693583	·142	7·0360593	·192	5·1789912	·242	4·0610277	·292	3·2981058
·093	10·7524581	·143	6·9865438	·193	5·1513386	·243	4·0430277	·293	3·2852129
·094	10·6380427	·144	6·9377049	·194	5·1239566	·244	4·0251617	·294	3·2723964
0·095	10·5260334	0·145	6·8895283	0·195	5·0968408	0·245	4·0074282	0·295	3·2596555
·096	10·4163548	·146	6·8420002	·196	5·0699872	·246	3·9898254	·296	3·2469895
·097	10·3089347	·147	6·7951072	·197	5·0433917	·247	3·9723518	·297	3·2343976
·098	10·2037037	·148	6·7488362	·198	5·0170503	·248	3·9550058	·298	3·2218790
·099	10·1005954	·149	6·7031744	·199	4·9909591	·249	3·9377858	·299	3·2094331
0·100	9·9995458	0·150	6·6581095	0·200	4·9651142	0·260	3·9206904	0·300	3·1970591

Table A.3 (*continued*)

a	*x*	*a*	*x*	*a*	*x*	*a*	*x*	*a*	*x*
0·300	3·1970591	0·350	2·6566127	0·400	2·2316119	0·450	1·8847348	0·500	1·5936243
·301	3·1847564	·351	2·6471411	·401	2·2240016	·451	1·8784240	·501	1·5882633
·302	3·1725241	·352	2·6377143	·402	2·2164218	·452	1·8721351	·502	1·5829188
·303	3·1603617	·353	2·6283321	·403	2·2088724	·453	1·8658679	·503	1·5775904
·304	3·1482684	·354	2·6189939	·404	2·2013532	·454	1·8596224	·504	1·5722783
0·305	3·1362436	0·355	2·6096996	0·405	2·1938638	0·455	1·8533984	0·505	1·5669823
·306	3·1242866	·356	2·6004487	·406	2·1864042	·456	1·8471958	·506	1·5617022
·307	3·1123968	·357	2·5912409	·407	2·1789740	·457	1·8410145	·507	1·5564381
·308	3·1005735	·358	2·5820758	·408	2·1715732	·458	1·8348542	·508	1·5511898
·309	3·0888160	·359	2·5729530	·409	2·1642015	·459	1·8287149	·509	1·5459572
0·310	3·0771238	0·360	2·5638723	0·410	2·1568586	0·460	1·8225965	0·510	1·5407403
·311	3·0654961	·361	2·5548333	·411	2·1495444	·461	1·8164989	·511	1·5355389
·312	3·0539325	·362	2·5458356	·412	2·1422588	·462	1·8104218	·512	1·5303531
·313	3·0424323	·363	2·5368790	·413	2·1350014	·463	1·8043652	·513	1·5251826
·314	3·0309949	·364	2·5279631	·414	2·1277721	·464	1·7983290	·514	1·5200275
0·315	3·0196198	0·365	2·5190875	0·415	2·1205707	0·465	1·7923131	0·515	1·5148876
·316	3·0083062	·366	2·5102520	·416	2·1133971	·466	1·7863172	·516	1·5097628
·317	2·9970537	·367	2·5014563	·417	2·1062510	·467	1·7803413	·517	1·5046532
·318	2·9858617	·368	2·4927000	·418	2·0991322	·468	1·7743854	·518	1·4995585
·319	2·9747297	·369	2·4839828	·419	2·0920406	·469	1·7684491	·519	1·4944788
0·320	2·9636570	0·370	2·4753044	0·420	2·0849759	0·470	1·7625325	0·520	1·4894139
·321	2·9526432	·371	2·4666645	·421	2·0779381	·471	1·7566355	·521	1·4843637
·322	2·9416876	·372	2·4580629	·422	2·0709268	·472	1·7507578	·522	1·4793282
·323	2·9307898	·373	2·4494991	·423	2·0639420	·473	1·7448995	·523	1·4743074
·324	2·9199493	·374	2·4409730	·424	2·0569834	·474	1·7390603	·524	1·4693011
0·325	2·9091655	0·375	2·4324843	0·425	2·0500510	0·475	1·7332402	0·525	1·4643092
·326	2·8984379	·376	2·4240326	·426	2·0431444	·476	1·7274391	·526	1·4593317
·327	2·8877660	·377	2·4156176	·427	2·0362635	·477	1·7216568	·527	1·4543685
·328	2·8771494	·378	2·4072391	·428	2·0294083	·478	1·7158932	·528	1·4494195
·329	2·8665874	·379	2·3988969	·429	2·0225784	·479	1·7101483	·529	1·4444847
0·330	2·8560797	0·380	2·3905906	0·430	2·0157738	0·480	1·7044219	0·530	1·4395640
·331	2·8456257	·381	2·3823199	·431	2·0089942	·481	1·6987139	·531	1·4346573
·332	2·8352249	·382	2·3740846	·432	2·0022395	·482	1·6930242	·532	1·4297645
·333	2·8248770	·383	2·3658845	·433	1·9955095	·483	1·6873527	·533	1·4248856
·334	2·8145814	·384	2·3577192	·434	1·9888042	·484	1·6816993	·534	1·4200205
0·335	2·8043377	0·385	2·3495885	0·435	1·9821232	0·485	1·6760639	0·535	1·4151691
·336	2·7941455	·386	2·3414922	·436	1·9754665	·486	1·6704464	·536	1·4103313
·337	2·7840042	·387	2·3334300	·437	1·9688339	·487	1·6648467	·537	1·4055072
·338	2·7739134	·388	2·3254016	·438	1·9622253	·488	1·6592647	·538	1·4006965
·339	2·7638728	·389	2·3174068	·439	1·9556405	·489	1·6537002	·539	1·3958994
0·340	2·7538818	0·390	2·3094453	0·440	1·9490792	0·490	1·6481533	0·540	1·3911155
·341	2·7439401	·391	2·3015170	·441	1·9425415	·491	1·6426237	·541	1·3863450
·342	2·7340472	·392	2·2936214	·442	1·9360271	·492	1·6371114	·542	1·3815878
·343	2·7242027	·393	2·2857585	·443	1·9295360	·493	1·6316164	·543	1·3768437
·344	2·7144062	·394	2·2779280	·444	1·9230678	·494	1·6261384	·544	1·3721128
0·345	2·7046573	0·395	2·2701297	0·445	1·9166226	0·495	1·6206774	0·545	1·3673948
·346	2·6949555	·396	2·2623632	·446	1·9102001	·496	1·6152333	·546	1·3626899
·347	2·6853006	·397	2·2546284	·447	1·9038002	·497	1·6098061	·547	1·3579979
·348	2·6756921	·398	2·2469251	·448	1·8974228	·498	1·6043955	·548	1·3533187
·349	2·6661296	·399	2·2392530	·449	1·8910677	·499	1·5990016	·549	1·3486523
0·350	2·6566127	0·400	2·2316119	0·450	1·8847348	0·500	1·5936243	0·550	1·3439987

Table A.3 (*continued*)

a	*x*	*a*	*x*	*a*	*x*	*a*	*x*	*a*	*x*
0·550	1·3439 987	0·600	1·1262 612	0·650	0·9336 939	0·700	0·7614 337	0·750	0·6058 600
·551	1·3393 577	·601	1·1221 820	·651	·9300 635	·701	·7581 691	·751	·6028 986
·552	1·3347 293	·602	1·1181 127	·652	·9264 412	·702	·7549 111	·752	·5999 427
·553	1·3301 135	·603	1·1140 533	·653	·9228 268	·703	·7516 598	·753	·5969 924
·554	1·3255 101	·604	1·1100 038	·654	·9192 204	·704	·7484 150	·754	·5940 475
0·555	1·3209 191	0·605	1·1059 641	0·655	0·9156 220	0·705	0·7451 767	0·755	0·5911 081
·556	1·3163 405	·606	1·1019 342	·656	·9120 315	·706	·7419 450	·756	·5881 742
·557	1·3117 742	·607	1·0979 141	·657	·9084 489	·707	·7387 198	·757	·5852 457
·558	1·3072 201	·608	1·0939 037	·658	·9048 742	·708	·7355 010	·758	·5823 227
·559	1·3026 782	·609	1·0899 029	·659	·9013 073	·709	·7322 887	·759	·5794 050
0·560	1·2981 485	0·610	1·0859 117	0·660	0·8977 481	0·710	0·7290 829	0·760	0·5764 927
·561	1·2936 307	·611	1·0819 301	·661	·8941 968	·711	·7258 834	·761	·5735 858
·562	1·2891 250	·612	1·0779 581	·662	·8906 531	·712	·7226 903	·762	·5706 842
·563	1·2846 312	·613	1·0739 955	·663	·8871 172	·713	·7195 036	·763	·5677 880
·564	1·2801 493	·614	1·0700 424	·664	·8835 889	·714	·7163 232	·764	·5648 970
0·565	1·2756 792	0·615	1·0660 987	0·665	0·8800 683	0·715	0·7131 491	0·765	0·5620 114
·566	1·2712 209	·616	1·0621 643	·666	·8765 553	·716	·7099 813	·766	·5591 310
·567	1·2667 743	·617	1·0582 393	·667	·8730 498	·717	·7068 198	·767	·5562 558
·568	1·2623 394	·618	1·0543 236	·668	·8695 519	·718	·7036 645	·768	·5533 859
·569	1·2579 161	·619	1·0504 171	·669	·8660 616	·719	·7005 154	·769	·5505 213
0·570	1·2535 043	0·620	1·0465 198	0·670	0·8625 787	0·720	0·6973 725	0·770	0·5476 618
·571	1·2491 040	·621	1·0426 317	·671	·8591 032	·721	·6942 358	·771	·5448 075
·572	1·2447 151	·622	1·0387 527	·672	·8556 352	·722	·6911 053	·772	·5419 584
·573	1·2403 377	·623	1·0348 828	·673	·8521 746	·723	·6879 808	·773	·5391 144
·574	1·2359 715	·624	1·0310 219	·674	·8487 214	·724	·6848 625	·774	·5362 756
0·575	1·2316 167	0·625	1·0271 701	0·675	0·8452 755	0·725	0·6817 503	0·775	0·5334 419
·576	1·2272 730	·626	1·0233 272	·676	·8418 370	·726	·6786 441	·776	·5306 132
·577	1·2229 405	·627	1·0194 933	·677	·8384 057	·727	·6755 440	·777	·5277 897
·578	1·2186 192	·628	1·0156 683	·678	·8349 816	·728	·6724 499	·778	·5249 712
·579	1·2143 089	·629	1·0118 521	·679	·8315 648	·729	·6693 618	·779	·5221 578
0·580	1·2100 097	0·630	1·0080 447	0·680	0·8281 552	0·730	0·6662 796	0·780	0·5193 494
·581	1·2057 214	·631	1·0042 461	·681	·8247 528	·731	·6632 035	·781	·5165 460
·582	1·2014 440	·632	1·0004 563	·682	·8213 575	·732	·6601 332	·782	·5137 476
·583	1·1971 774	·633	0·9966 752	·683	·8179 694	·733	·6570 689	·783	·5109 542
·584	1·1929 217	·634	0·9929 027	·684	·8145 883	·734	·6540 104	·784	·5081 658
0·585	1·1886 768	0·635	0·9891 389	0·685	0·8112 143	0·735	0·6509 579	0·785	0·5053 823
·586	1·1844 426	·636	·9853 837	·686	·8078 473	·736	·6479 112	·786	·5026 037
·587	1·1802 190	·637	·9816 371	·687	·8044 874	·737	·6448 703	·787	·4998 301
·588	1·1760 060	·638	·9778 989	·688	·8011 344	·738	·6418 352	·788	·4970 613
·589	1·1718 036	·639	·9741 693	·689	·7977 884	·739	·6388 059	·789	·4942 975
0·590	1·1676 118	0·640	0·9704 481	0·690	0·7944 493	0·740	0·6357 824	0·790	0·4915 385
·591	1·1634 304	·641	·9667 354	·691	·7911 171	·741	·6327 646	·791	·4887 843
·592	1·1592 594	·642	·9630 310	·692	·7877 918	·742	·6297 526	·792	·4860 350
·593	1·1550 988	·643	·9593 350	·693	·7844 733	·743	·6267 462	·793	·4832 905
·594	1·1509 485	·644	·9556 473	·694	·7811 617	·744	·6237 456	·794	·4805 508
0·595	1·1468 085	0·645	0·9519 679	0·695	0·7778 568	0·745	0·6207 506	0·795	0·4778 159
·596	1·1426 788	·646	·9482 967	·696	·7745 587	·746	·6177 612	·796	·4750 858
·597	1·1385 592	·647	·9446 338	·697	·7712 674	·747	·6147 775	·797	·4723 604
·598	1·1344 498	·648	·9409 790	·698	·7679 828	·748	·6117 994	·798	·4696 398
·599	1·1303 505	·649	·9373 324	·699	·7647 049	·749	·6088 269	·799	·4669 239
0·600	1·1262 612	0·650	·9336 939	0·700	0·7614 337	0·750	0·6058 600	0·800	0·4642 129

Table A.3 (*concluded*)

a	*x*	*a*	*x*	*a*	*x*	*a*	*x*
0·800	0·4642128	0·850	0·3343447	0·900	0·2145557	0·950	0·1034788
·801	·4615063	·851	·3318552	·901	·2122529	·951	·1013381
·802	·4588045	·852	·3293697	·902	·2099535	·952	·0992005
·803	·4561073	·853	·3268862	·903	·2076576	·953	·0970658
·804	·4534148	·854	·3244107	·904	·2053652	·954	·0949342
0·805	0·4507270	0·855	0·3219371	0·905	0·2030762	0·955	0·0928056
·806	·4480438	·856	·3194675	·906	·2007906	·956	·0906799
·807	·4453652	·857	·3170019	·907	·1985084	·957	·0885573
·808	·4426911	·858	·3145402	·908	·1962296	·958	·0864376
·809	·4400217	·859	·3120824	·909	·1939542	·959	·0843209
0·810	0·4373568	0·860	0·3096286	0·910	0·1916824	0·960	0·0822071
·811	·4346965	·861	·3071786	·911	·1894137	·961	·0800963
·812	·4320407	·862	·3047325	·912	·1871484	·962	·0779885
·813	·4293894	·863	·3022903	·913	·1848866	·963	·0758836
·814	·4267426	·864	·2998520	·914	·1826281	·964	·0737816
0·815	0·4241004	0·865	0·2974176	0·915	0·1803729	0·965	0·0716825
·816	·4214626	·866	·2949869	·916	·1781211	·966	·0695864
·817	·4188292	·867	·2925601	·917	·1758725	·967	·0674932
·818	·4162003	·868	·2901372	·918	·1736274	·968	·0654028
·819	·4135759	·869	·2877180	·919	·1713855	·969	·0633154
0·820	0·4109558	0·870	0·2853027	0·920	0·1691469	0·970	0·0612308
·821	·4083402	·871	·2828911	·921	·1669116	·971	·0591492
·822	·4057290	·872	·2804833	·922	·1646795	·972	·0570704
·823	·4031222	·873	·2780793	·923	·1624508	·973	·0549944
·824	·4005197	·874	·2756790	·924	·1602253	·974	·0529213
0·825	0·3979215	0·875	0·2732825	0·925	0·1580031	0·975	0·0508511
·826	·3953278	·876	·2708897	·926	·1557841	·976	·0487837
·827	·3927383	·877	·2685007	·927	·1535683	·977	·0467191
·828	·3901532	·878	·2661153	·928	·1513558	·978	·0446574
·829	·3875723	·879	·2637337	·929	·1491465	·979	·0425985
0·830	0·3849957	0·880	0·2613557	0·930	0·1469404	0·980	0·0405424
·831	·3824235	·881	·2589815	·931	·1447374	·981	·0384891
·832	·3798554	·882	·2566109	·932	·1425377	·982	·0364386
·833	·3772916	·883	·2542439	·933	·1403412	·983	·0343909
·834	·3747321	·884	·2518806	·934	·1381478	·984	·0323460
0·835	0·3721768	0·885	0·2495210	0·935	0·1359576	0·985	0·0303038
·836	·3696257	·886	·2471649	·936	·1337706	·986	·0282644
·837	·3670787	·887	·2448125	·937	·1315867	·987	·0262278
·838	·3645360	·888	·2424637	·938	·1294060	·988	·0241939
·839	·3619974	·889	·2401185	·939	·1272283	·989	·0221628
0·840	0·3594630	0·890	0·2377769	0·940	0·1250538	0·990	0·0201345
·841	·3569327	·891	·2354389	·941	·1228825	·991	·0181088
·842	·3544066	·892	·2331044	·942	·1207142	·992	·0160859
·843	·3518846	·893	·2307735	·943	·1185490	·993	·0140657
·844	·3493667	·894	·2284461	·944	·1163869	·994	·0120482
0·845	0·3468528	0·895	0·2261222	0·945	0·1142279	0·995	0·0100335
·846	·3443431	·896	·2238019	·946	·1120720	·996	·0080214
·847	·3418374	·897	·2214851	·947	·1099191	·997	·0060120
·848	·3393358	·898	·2191718	·948	·1077693	·998	·0040053
·849	·3368383	·899	·2168620	·949	·1056226	·999	·0020013
0·850	0·3343447	0·900	0·2145557	0·950	0·1034788	1·000	0·0000000

(Reproduced from Barton, David and Merrington [1960]. The above table is also given by David, Kendall and Barton, *Symmetric Functions and Allied Tables* (Cambridge University Press, for the Biometrika Trustees), 1966, Table 8.1.)

A4 Table for finding the maximum-likelihood estimate of P, the parameter of a truncated geometric distribution

In the following table the function

$$f(Q) = \frac{sQ^{s+1} - (s+1)Q^s + 1}{Q^{s+1} - Q^s - Q + 1}$$

is evaluated for different values of P ($= 1 - Q$) and s. Using linear interpolation, this table can be used for solving the equation $f(Q) = \bar{x}$ as described in **4.1.6** (*4*).

	P										
s	0·001	0·10	0·20	0·30	0·40	0·50	0·60	0·70	0·80	0·90	0·999
2	1·500	1·474	1·444	1·412	1·375	1·333	1·286	1·231	1·167	1·091	1·001
3	1·999	1·930	1·852	1·767	1·673	1·571	1·462	1·345	1·226	1·108	1·001
4	2·499	2·369	2·225	2·069	1·904	1·733	1·562	1·396	1·244	1·111	1·001
5	2·998	2·790	2·563	2·323	2·078	1·839	1·615	1·416	1·248	1·111	1·001
6	3·497	3·195	2·868	2·533	2·206	1·905	1·642	1·424	1·250	1·111	1·001
7	3·996	3·582	3·142	2·705	2·298	1·945	1·655	1·427	1·250	1·111	1·001
8	4·495	3·953	3·387	2·844	2·363	1·969	1·661	1·428	1·250	1·111	1·001
9	4·993	4·308	3·605	2·955	2·408	1·982	1·664	1·428	1·250	1·111	1·001
10	5·492	4·647	3·797	3·043	2·439	1·990	1·666	1·429	1·250	1·111	1·001
11	5·990	4·969	3·966	3·111	2·460	1·995	1·666	1·429	1·250	1·111	1·001
12	6·488	5·277	4·115	3·165	2·474	1·997	1·666	1·429	1·250	1·111	1·001
13	6·986	5·569	4·244	3·206	2·483	1·998	1·667	1·429	1·250	1·111	1·001
14	7·484	5·847	4·356	3·238	2·489	1·999	1·667	1·429	1·250	1·111	1·001
15	7·981	6·111	4·453	3·262	2·493	2·000	1·667	1·429	1·250	1·111	1·001
16	8·479	6·360	4·537	3·280	2·495	2·000	1·667	1·429	1·250	1·111	1·001
17	8·976	6·597	4·608	3·294	2·497	2·000	1·667	1·429	1·250	1·111	1·001
18	9·473	6·821	4·670	3·304	2·498	2·000	1·667	1·429	1·250	1·111	1·001
19	9·970	7·033	4·722	3·312	2·499	2·000	1·667	1·429	1·250	1·111	1·001
20	10·467	7·232	4·767	3·317	2·499	2·000	1·667	1·429	1·250	1·111	1·001
21	10·963	7·429	4·805	3·322	2·500	2·000	1·667	1·429	1·250	1·111	1·001
22	11·460	7·597	4·836	3·325	2·500	2·000	1·667	1·429	1·250	1·111	1·001
23	11·956	7·763	4·863	3·327	2·500	2·000	1·667	1·429	1·250	1·111	1·001
24	12·452	7·920	4·886	3·329	2·500	2·000	1·667	1·429	1·250	1·111	1·001
25	12·948	8·066	4·905	3·330	2·500	2·000	1·667	1·429	1·250	1·111	1·001
26	13·444	8·204	4·921	3·331	2·500	2·000	1·667	1·429	1·250	1·111	1·001

(Reproduced from Thomasson and Kapadia [1968].)

A5 Tabulation of

$$f(x) = \frac{1}{x} - \frac{1}{\exp(x) - 1}$$

Using linear interpolation, the following table can be used for solving the equation $f(x) = a$ (e.g. see equation (6.5)).

x	·0	·1	·2	·3	·4	·5	·6	·7	·8	·9
0		·4916	·4832	·4750	·4668	·4584	·4504	·4422	·4340	·4260
1	·4180	·4102	·4024	·3946	·3870	·3794	·3720	·3648	·3576	·3504
2	·3434	·3366	·3300	·3234	·3168	·3106	·3044	·2984	·2924	·2866
3	·2810	·2754	·2700	·2648	·2596	·2546	·2496	·2450	·2402	·2358
4	·2314	·2270	·2228	·2188	·2148	·2110	·2072	·2036	·2000	·1966
5	·1932	·1900	·1868	·1836	·1806	·1778	·1748	·1720	·1694	·1668
6	·1642	·1616	·1592	·1568	·1546	·1524	·1502	·1480	·1460	·1440
7	·1420	·1400	·1382	·1364	·1346	·1328	·1310	·1294	·1278	·1262
8	·1246	·1232	·1216	·1202	·1188	·1174	·1160	·1148	·1134	·1122

(Reproduced from Deemer and Votaw [1955].)

A6 Tabulation of $A_K(S) = \sum_{k=0}^{K} kS^k \Big/ \sum_{k=0}^{K} S^k$

S \ K =	2	3	4	5	6	7	8
.01	.0101	.0110	.0110	.0110	.0110	.0110	.0110
.02	.0204	.0204	.0204	.0204	.0204	.0204	.0204
.03	.0308	.0309	.0309	.0309	.0309	.0309	.0309
.04	.0415	.0416	.0416	.0416	.0416	.0416	.0416
.05	.0523	.0526	.0527	.0527	.0527	.0527	.0527
.06	.0632	.0638	.0638	.0638	.0638	.0638	.0638
.07	.0742	.0742	.0752	.0752	.0752	.0752	.0752
.08	.0854	.0868	.0869	.0869	.0869	.0869	.0869
.09	.0967	.0986	.0989	.0989	.0989	.0989	.0989
.10	.1081	.1107	.1111	.1111	.1111	.1111	.1111
.11	.1196	.1230	.1235	.1236	.1236	.1236	.1236
.12	.1312	.1355	.1363	.1364	.1364	.1364	.1364
.13	.1428	.1483	.1493	.1494	.1495	.1495	.1495
.14	.1545	.1612	.1625	.1627	.1628	.1628	.1628
.15	.1663	.1744	.1761	.1764	.1765	.1765	.1765
.16	.1781	.1879	.1900	.1904	.1905	.1905	.1905
.17	.1900	.2015	.2041	.2047	.2047	.2047	.2047
.18	.2019	.2153	.2186	.2193	.2195	.2195	.2195
.19	.2138	.2294	.2333	.2343	.2345	.2346	.2346
.20	.2258	.2436	.2484	.2496	.2499	.2499	.2499
.21	.2378	.2580	.2638	.2653	.2657	.2658	.2658
.22	.2498	.2727	.2795	.2814	.2818	.2819	.2819
.23	.2618	.2875	.2955	.2978	.2985	.2986	.2987
.24	.2737	.3025	.3118	.3147	.3155	.3158	.3158
.25	.2857	.3177	.3285	.3319	.3329	.3332	.3333
.26	.2977	.3330	.3454	.3495	.3508	.3512	.3513
.27	.3096	.3485	.3627	.3675	.3691	.3696	.3698
.28	.3216	.3641	.3803	.3860	.3879	.3886	.3888
.29	.3335	.3800	.3982	.4049	.4072	.4080	.4083
.30	.3453	.3959	.4164	.4242	.4270	.4281	.4284
.31	.3572	.4120	.4349	.4439	.4474	.4486	.4491
.32	.3690	.4282	.4538	.4642	.4682	.4698	.4704
.33	.3807	.4445	.4729	.4848	.4895	.4914	.4921
.34	.3924	.4610	.4923	.5058	.5114	.5137	.5145
.35	.4041	.4775	.5121	.5274	.5340	.5367	.5377
.36	.4157	.4942	.5321	.5494	.5570	.5603	.5616
.37	.4272	.5109	.5524	.5718	.5806	.5845	.5861
.38	.4387	.5277	.5730	.5948	.6049	.6094	.6114
.39	.4502	.5446	.5938	.6181	.6297	.6350	.6374
.40	.4615	.5616	.6149	.6420	.6552	.6614	.6643
.41	.4728	.5786	.6363	.6663	.6813	.6885	.6920
.42	.4841	.5957	.6579	.6910	.7080	.7164	.7205
.43	.4953	.6128	.6798	.7162	.7353	.7450	.7499
.44	.5064	.6299	.7019	.7418	.7633	.7745	.7801
.45	.5174	.6471	.7242	.7679	.7919	.8047	.8114
.46	.5284	.6644	.7467	.7945	.8212	.8358	.8435
.47	.5392	.6816	.7694	.8214	.8512	.8677	.8767
.48	.5500	.6988	.7923	.8488	.8817	.9005	.9109

.49	.5608	.7161	.8154	.8766	.9130	.9341	.9461
.50	.5714	.7333	.8387	.9048	.9449	.9686	.9824
.51	.5820	.7506	.8621	.9333	.9774	1.0041	1.0198
.52	.5925	.7678	.8857	.9623	1.0106	1.0403	1.0583
.53	.6029	.7850	.9094	.9917	1.0445	1.0775	1.0978
.54	.6132	.8022	.9333	1.0214	1.0789	1.1157	1.1387
.55	.6235	.8193	.9573	1.0514	1.1140	1.1547	1.1806
.56	.6336	.8365	.9813	1.0818	1.1497	1.1946	1.2237
.57	.6437	.8535	1.0055	1.1125	1.1860	1.2354	1.2681
.58	.6537	.8705	1.0297	1.1435	1.2229	1.2772	1.3136
.59	.6636	.8875	1.0540	1.1748	1.2604	1.3198	1.3604
.60	.6735	.9044	1.0784	1.2064	1.2984	1.3634	1.4084
.61	.6832	.9213	1.1028	1.2382	1.3370	1.4077	1.4576
.62	.6929	.9380	1.1273	1.2702	1.3761	1.4530	1.5081
.63	.7025	.9548	1.1518	1.3025	1.4157	1.4991	1.5598
.64	.7119	.9714	1.1763	1.3350	1.4557	1.5461	1.6127
.65	.7214	.9880	1.2009	1.3677	1.4963	1.5938	1.6668
.66	.7307	1.0044	1.2254	1.4006	1.5373	1.6423	1.7220
.67	.7399	1.0208	1.2499	1.4336	1.5787	1.6917	1.7786
.68	.7491	1.0371	1.2743	1.4667	1.6205	1.7417	1.8362
.69	.7581	1.0534	1.2988	1.5000	1.6626	1.7925	1.8950
.70	.7671	1.0695	1.3232	1.5333	1.7051	1.8439	1.9549
.71	.7760	1.0855	1.3476	1.5667	1.7479	1.8960	2.0158
.72	.7848	1.1014	1.3719	1.6002	1.7910	1.9487	2.0778
.73	.7936	1.1173	1.3961	1.6338	1.8343	2.0019	2.1407
.74	.8022	1.1330	1.4202	1.6773	1.8779	2.0557	2.2046
.75	.8108	1.1486	1.4443	1.7009	1.9217	2.1100	2.2694
.76	.8193	1.1641	1.4683	1.7345	1.9656	2.1647	2.3350
.77	.8277	1.1795	1.4921	1.7680	2.0097	2.2199	2.4014
.78	.8360	1.1947	1.5159	1.8015	2.0539	2.2754	2.2486
.79	.8443	1.2099	1.5395	1.8350	2.0981	2.3312	2.5364
.80	.8525	1.2249	1.5631	1.8683	2.1424	2.3872	2.6048
.81	.8605	1.2399	1.5865	1.9016	2.1868	2.4436	2.6737
.82	.8686	1.2547	1.6097	1.9348	2.2311	2.5001	2.7432
.83	.8765	1.2694	1.6328	1.9678	2.2754	2.5567	2.8131
.84	.8843	1.2839	1.6558	2.0007	2.3197	2.6135	2.8833
.85	.8921	1.2984	1.6786	2.0335	2.3638	2.6702	2.9538
.86	.8998	1.3127	1.7013	2.0662	2.4079	2.7270	3.0245
.87	.9075	1.3269	1.7238	2.0986	2.4518	2.7838	3.0953
.88	.9150	1.3409	1.7461	2.1309	2.4955	2.8405	3.1662
.89	.9225	1.3549	1.7683	2.1629	2.5391	2.8970	3.2372
.90	.9299	1.3687	1.7903	2.1948	2.5824	2.9534	3.3080
.91	.9372	1.3824	1.8121	2.2264	2.6255	3.0096	3.3788
.92	.9445	1.3960	1.8337	2.2578	2.6684	3.0655	3.4493
.93	.9517	1.4094	1.8552	2.2890	2.7110	3.1212	3.5196
.94	.9588	1.4227	1.8765	2.3200	2.7533	3.1765	3.5896
.95	.9658	1.4359	1.8975	2.3506	2.7953	3.2315	3.6593
.96	.9728	1.4490	1.9184	2.3811	2.8369	3.2861	3.7285
.97	.9797	1.4619	1.9391	2.4112	2.8783	3.3402	3.7972
.98	.9865	1.4747	1.9596	2.4411	2.9192	3.3940	3.8654
.99	.9933	1.4874	1.9799	2.4707	2.9598	3.4472	3.9332
1.00	1.0000	1.5000	2.0000	2.5000	3.0000	3.5000	4.0000

(Reproduced from Robson and Chapman [1961].)

The equation

$$A_K(S) = a \tag{1}$$

can be solved by using linear interpolation in the above table. However, when $K = 1$, $S = a/(1-a)$, and when $K = 2$:

$$S = \frac{-(1-a) + \sqrt{1 + 6a - 3a^2}}{2(2-a)}.$$

If greater accuracy is needed, or for $K > 8$, (1) must be solved iteratively. In this case $A_K(S)$ can be more conveniently written in the form

$$A_K(S) = \frac{S}{1-S} - \frac{(K+1)S^{K+1}}{1-S^{K+1}},$$

and (1) solved by repeated substitution in the right-hand side of

$$S = a(1-S) + \frac{(K+1)S^{K+1}(1-S)}{1-S^{K+1}}.$$

A7 Uniqueness of a certain polynomial root

If r and n_i $(i = 1, 2, \ldots, s)$ are positive integers such that r is greater than each n_i and $r < \sum_{i=1}^{s} n_i$, then the $(s-1)$th degree polynomial

$$N^{s-1}(N-r) = \prod_{i=1}^{s} (N-n_i)$$

has a unique root greater than r.

Proof:

Let

$$\phi(N) = \frac{N}{N-r} \prod_{i=1}^{s} \left(1 - \frac{n_i}{N}\right),$$

then we wish to prove that $\phi(N) = 1$ has a unique root greater than r. Now $\phi(N)$ is continuous for $N > r$, $\phi(r+0) = \infty$ and

$$\begin{aligned} \phi(N) &= 1 - (\Sigma\, n_i - r)/N + O(1/N^2) \\ &< 1 \end{aligned}$$

for large N. Hence $\phi(N) = 1$ has at least one root greater than r. If we can now show that $\phi'(N) = 0$ for only *one* value of N greater than r then this root is unique.

By taking logarithms we see that

$$\begin{aligned} \phi'(N) &= \frac{\phi(N)}{N} \left[\sum_{i=1}^{s} \frac{n_i}{N-n_i} - \frac{r}{N-r} \right] \\ &= \phi(N)\,\theta(N), \text{ say,} \end{aligned}$$

and we wish to prove that $\theta(N) = 0$ has a unique root greater than r. To do this we use a similar argument to that given above in considering $\phi(N) = 1$.

Thus $\theta(N)$ is continuous for $N > r$, $\theta(r+0) = -\infty$,

$$\theta(N) = \frac{1}{N}\left[\frac{(\Sigma\, n_i - r)}{N} + O\left(\frac{1}{N^2}\right)\right] = 0+$$

for large N and $\theta(N) = 0$ has at least one root N_0, say, greater than r. Now

$$\begin{aligned}\theta'(N_0) &= \frac{r}{(N_0 - r)^2} - \sum \frac{n_i}{(N_0 - n_i)^2} \\ &= \sum \frac{n_i}{(N_0 - n_i)(N_0 - r)} - \sum \frac{n_i}{(N_0 - n_i)^2} \\ &> 0,\end{aligned}$$

since $r > n_i$ and

$$\sum \frac{n_i}{N_0 - n_i} = \frac{r}{N_0 - r}.$$

Therefore $\theta'(N)$ is positive at every root of $\theta(N) = 0$ greater than r, and this can only be true if there is just one root. This in turn implies that $\phi(N) = 1$ has a unique root greater than r.

(The above proof was motivated by some unpublished work of Professor J.N. Darroch.)

REFERENCES

Aberdeen, J.E.C. (1958). The effect of quadrat size, plant size, and plant distribution on frequency estimates in plant ecology. *Aust. J. Bot.* **6**, 47–58. [478]

Abrahamsen, G. and Strand, L. (1970). Statistical analysis of population density data of soil animals, with particular reference to Enchytraeidae (Oligochaeta). *Oikos* **21**, 276–84. [451]

Adams, L. (1951). Confidence limits for the Petersen or Lincoln index used in animal population studies. *J. Wildl. Manag.* **15**, 13–19. [63–4]

Adams, L. (1959). An analysis of a population of snowshoe hares in northwestern Montana. *Ecol. Monogr.* **29**, 141–70. [54]

Adams, L. and Davis. S.D. (1967). The internal anatomy of home range. *J. Mammal.* **48**, 529–36. [20, 94]

Albers, P.H. (1976). Determining population size of territorial red-winged blackbirds. *J. Wildl. Manag.* **40**, 761–8. [453]

Allee, W.C., Emerson, A.E., Park, O., Park, T. and Schmidt, K.P. (1949). *Principles of Animal Ecology*. W.B. Saunders Co.: Philadelphia. [393]

Allen, D.L. (1942). A pheasant inventory method based upon kill records and sex ratios. *Trans. Nth Amer. Wildl. Conf.* **7**, 329–33. [353]

Allen, K.R. (1966). Some methods for estimating exploited populations. *J. Fish. Res. Bd. Canada* **23**, 1553–74. [336–40, 542]

Allen, K.R. (1968). Simplification of a method of computing recruitment rates. *J. Fish Res. Bd. Canada* **25**, 2701–2. [336–40, 543]

Allen, R.E. and McCullough, D.R. (1976). Deer-car accidents in Southern Michigan. *J. Wildl. Manag.* **40**, 317–25. [453]

Allen, R.L. and Basasibwaki, P. (1974). Properties of age structure models for fish populations. *J. Fish. Res. Bd. Canada* **31**, 1119–25. [551]

Ambrose, H.W., III (1969). A comparison of *Microtus pennsylvanicus* home ranges as determined by isotope and live trap methods. *Amer. Midl. Nat.* **81**, 535–55. [447]

Ambrose, H.W., III (1973). An experimental study of some factors affecting the spatial and temporal activity of *Microtus pennsylvanicus. J. Mammal.* **54**, 79–110. [447]

Amman, G.D. and Baldwin, P.H. (1960). A comparison of methods for censusing woodpeckers in spruce-fir forests of Colorado. *Ecology* **41**, 699–706. [21, 35]

Ammann, G.A. and Ryel, L.A. (1963). Extensive methods of inventorying ruffed grouse in Michigan. *J. Wildl. Manag.* **27**, 617–33. [20, 560, 564]

Anderson, J. (1962). Roe-deer census and population analysis by means of modified marking and release technique. *In* E.D. Le Cren and M.W. Holdgate (eds), *The Exploitation of Natural Animal Populations*, 72–82. Blackwell: Oxford. [110]

Anderson, D.R. (1975). Population ecology of the mallard: V. Temporal and geographic estimates of survival, recovery and harvest rates. *U.S. Fish Wildl. Serv., Resour. Publ. No. 125*, 110 pp. [526–8]

Anderson, D.R. and Burnham, K.P. (1976). Population ecology of the mallard: VI. The effect of exploitation on survival. *U.S. Fish Wildl. Serv., Resour. Publ. No. 128*, 66 pp. [526–7]

REFERENCES

Anderson, D.R., Burnham, K.P. and Crain, B.R. (1978). A log-linear model approach to estimation of population size using the line-transect sampling method. *Ecology* **59**, 190–3. [463, 465]

Anderson, D.R., Burnham, K.P. and Crain, B.R. (1979). Line transect estimation of population size: the exponential case with grouped data. *Commun. Statist.-Theor. Method A* **8**, 487–507. [462]

Anderson, D.R., Kimball, C.F. and Fieher, F.R. (1974). A computer program for estimating survival and recovery rates. *J. Wildl. Manag.* **38**, 369–70. [506]

Anderson, D.R., Laake, J.L., Crain, B.R., and Burnham, K.P. (1979). Guidelines for line transect sampling of biological populations. *J. Wildl. Manag.* **43**, 70–8. [467]

Anderson, D.R. and Pospahala, R.S. (1970). Correction of bias in belt transect studies of immotile objects. *J. Wildl. Manag.* **34**, 141–6. [463, 465]

Anderson, D.R. and Sterling, R.T. (1974). Population dynamics of molting pintail drakes banded in south-central Saskatchewan. *J. Wildl. Manag.* **38**, 266–74. [526]

Andersson, M. (1976). Influence of trap saturation on estimates of animal abundance based on catch per unit effort. *Oikos* **27**, 316–19. [502]

Andrewartha, H.G. (1961). *Introduction of Animal Populations.* Methuen: London. [572]

Andrews, A.K. (1972). Survival and mark retention of a small cyprinid marked with fluorescent pigments. *Trans. Amer. Fish. Soc.* **101**, 128–33. [488]

Andrzejewski, R. (1963). Processes of incoming settlement and disappearance of individuals and variations in the numbers of small rodents. *Acta theriologica* **7**, 169–213. [19]

Andrzejewski, R. (1967). Estimation of the abundance of small rodent populations for the use of biological productivity investigations. *In* K. Petrusewicz (ed.), *Secondary Productivity of Terrestrial Ecosystems,* vol. 1, 275–81. Institute of Ecology, Polish Academy of Sciences. [19]

Andrzejewski, R., Bujalska, G., Ryszkowski, L. and Ustyniuk, J. (1966). On a relation between the number of traps in a point of catch and trappability of small rodents. *Acta theiologica* **11**, 343–9. [83]

Andrzejewski, R., Fejgin, H. and Liro, A. (1971). Trappability of trap-prone and trap-shy bank voles. *Acta theriologica* **16**, 401–12.

Andrzejewski, R. and Jezierski, W. (1966). Studies on the European hare. XI: Estimation of population density and attempt to plan the yearly take of hares. *Acta theriologica* **11**, 433–48. [327]

Andrzejewski, R. and Wierzbowska, T. (1961). An attempt at assessing the duration of residence of small rodents in a defined forest area and the rate of interchange between individuals. *Acta theriologica* **5**, 153–72. [181]

Andrzejewski, R. and Wierzbowska, T. (1970). Estimate of the number of traps visited by small mammals based on a probabilistic model. *Acta theriologica* **15**, 1–14. [447]

Anscombe, F.J. (1949). The statistical analysis of insect counts based on the negative binomial distribution. *Biometrics* **5**, 165–73. [451]

Arnason, A.N. (1972a). Parameter estimates from mark–recapture experiments on two populations subject to migration and death. *Res. Popul. Ecol.* **13**, 97–113. [555]

Arnason, A.N. (1972b). Prediction methods and variance estimates for the parameters of the triple catch-two population model with migration and death. *Univ. of Manitoba, Dept. Comp. Science, Sci. Rep. No. 76,* 30 pp. [556]

Arnason, A.N. (1973). The estimation of population size, migration rates and survival in the stratified population. *Res. Popul. Ecol.* **15**, 1–8. [556–8]

Arnason, A.N. and Baniuk, L. (1978). *POPAN-2, A Data Maintenance and Analysis System for Recapture Data, release 3*, Charles Babbage Reseach Centre, Box 370, St. Pierre, Manitoba, Canada. [506]

REFERENCES

Arnason, A.N. and Kreger, N.S. (1973). User's Manual, POPAN-1, A program to calculate Jolly–Seber estimates from mark–recapture data with pooling methods to increase the precision. *Univ. of Manitoab, Dept. Comp. Science, Sci. Rep. No. 54,* 31 pp. [506]

Artmann, J.W. (1975). Woodcock status report, 1974. *U.S. Fish Wild. Serv., Spec. Sci. Rep. Wildl. No. 189,* 39 pp. [453]

Artmann, J.W. (1977). Woodcock status report, 1975. *U.S. Fish Wildl. Serv., Spec. Sci. Rep. Wildl. No. 200,* 36 pp. [453]

Ashby, K.R. (1967). Studies in the ecology of field mice and voles (*Apodemus sylvaticus, Clethrionomys glareolus and Microtus agrestis*) in Houghall Wood, Durham. *J. Zool., Lond.* **152**, 389–513. [564]

Ashford, J.R., Read, K.L.Q. and Vickers, G.G. (1970). A system of stocahastic models applicable to studies of animal population dynamics. *J. Animal Ecol.* **39**, 29–50. [399, 554]

Atwood, E.L. (1956). Validity of mail survey data on bagged waterfowl. *J. Wildl. Manag.* **20**, 1–16. [360]

Atwood, E.L. and Geis, A.D. (1960). Problems associated with practices that increase the reported recoveries of waterfowl bands. *J. Wildl. Manag.* **24**, 272–9. [97]

Ayre, G.L. (1962). Problems in using the Lincoln Index for estimating the size of ant colonies (*Hymenoptera: Formicidae*). *J.N.Y. Ent. Soc.* **70**, 159–66. [72]

Babińska, J. and Bock, E. (1969). The effect of pre-baiting on captures of rodents. *Acta theriologica* **14**, 267–70. [570]

Backiel, T. (1964). Tag detachment in *Vimba vimba* L. (in Polish). *Roczniki Nauk Rolniczych* **84–B**, 241–53. [281]

Bailey, G.N.A. (1968). Trap-shyness in a woodland population of Bank voles (*Clethrionomys glareolus*). *J. Zool., Lond.* **156**, 517–21. [20, 84]

Bailey, J.A. (1969). Trap responses of wild cottontails. *J. Wildl. Manag.* **33**, 48–58. [84]

Bailey, N.T.J. (1951). On estimating the size of mobile populations from capture–recapture data. *Biometrika* **38**, 293–306. [61, 118, 220, 566]

Bailey, N.T.J. (1952). Improvements in the interpretation of recapture data. *J. Animal Ecol.* **21**, 120–7. [61, 220]

Bailey, P.T. (1971). The red kangaroo, *Megaleia rufa* (Desmarest), in north-western New South Wales. I. Movements. *CSIRO Wildl. Res.* **16**, 11–28. [455, 458]

Balph, D.F. (1968). Behavioural responses of unconfined Uinta ground squirrels to trapping. *J. Wildl. Manag.* **32**, 778–94. [84]

Banks, C.J. and Brown, E.S. (1962). A comparison of methods of estimating population density of adult sunn pest, *Eurygaster integriceps* Put. (Hemiptera, Scutelleridae) in wheat fields. *Ent. Exp. and Appl.* **5**, 255–60. [83, 114–115]

Banks, E.M., Brooks, R.J. and Schnell, J. (1975). A radiotracking study of home range and activity of the brown lemming (*Lemmus trimucronatus*). *J. Mammal.* **56**, 888–901. [447]

Baranov, T.I. (1918). On the question of the biological basis of fisheries. *Rep. Div. Fish Management and Scientific Study of the Fishing Industry* **1**, 81–128. [329]

Barbehenn, K.R. (1958). Spatial and population relationships between *Microtus* and *Blarina. Ecology* **39**, 293–304. [327]

Barbehenn, K.R. (1973). The use of stratified traps in estimating density: *Peromyscus* and *Blarina. Acta theiologica* **18**, 395–402. [487]

Barbehenn, K.R. (1974). Estimating density and home range size with removal grids: the rodents and shrews of Guam. *Acta theriologica* **19**, 191–234. [449, 504]

Barkalow, F.S. Jr, Hamilton, R.B. and Soots, R.F. Jr (1970). The vital statistics of an unexploited gray squirrel population. *J. Wildl. Manag.* **34**, 489–500. [383]

REFERENCES

Barnard, G.A. (1963). Contribution to the discussion on Professor Bartlett's paper. *J. Roy. Statist. Soc., Series B* **25**, 294 [18]

Barnett, V.D. (1966). Evaluation of the maximum-likelihood estimator when the likelihood equation has multiple roots. *Biometrika* **53**, 151–65. [17]

Bartko, J.J., Greenhouse, S.W. and Patlak, C.S. (1968). On expectations of some functions of Poisson variates. *Biometrics* **24**, 97–102. [479]

Bartlett, M.S. (1960). *Stochastic Population Models in Ecology and Epidemiology.* Methuen's Monographs on Applied Probability and Statistics: London. [572]

Bartlett, M.S. (1964). The spectral analysis of two-dimensional point processes. *Biometrika* **51**, 299–311. [483]

Bartlett, M.S. (1975). *The Statistical Analysis of Spatial Pattern.* Chapman and Hall: London [481, 483–5]

Barton, D.E., David, F.N. and Merrington, M. (1960). Tables for the solution of the exponential equation, exp $(-a) + ka = 1$. *Biometrika* **47**, 439–45. [581]

Batcheler, C.L. (1971). Estimation of density from a sample of joint point and nearest-neighbor distances. *Ecology* **52**, 703–9. [473]

Batcheler, C.L. (1973). Estimating density and dispersion from truncated or unrestricted joint point-distance nearest-neighbour distances. *Proc. N.Z. Ecol. Soc.* **20**, 131–47. [473]

Batcheler, C.L. (1975a). Probable limits of error of the point distance–neighbour distance estimate of density. *Proc. N.Z. Ecol. Soc.* **22**, 28–33. [473]

Batcheler, C.L. (1975b). Development of a distance method for deer census from pellet groups. *J. Wildl. Manag.* **39**, 641–52. [473]

Batcheler, C.L. and Bell, D.J. (1970). Experiments in estimating density from joint point- and nearest-neighbour distance samples. *Proc. N.Z. Ecol. Soc.* **17**, 111–7. [473]

Batcheler, C.L. and Hodder, R.A.C. (1975). Tests of a distance technique for inventory of pine plantations. *N.Z.J. Forestry Science* **5**, 3–17. [473]

Baümler, W. (1975). Activity of some small mammals in the field. *Acta theriologica* **20**, 365–77 [487]

Bayliff, W.H. and Mobrand, L.M. (1972). Estimates of the rates of shedding of dart tags from yellowfin tuna. *Bull. Inter-Amer. Trop. Tuna Comm.* **15**, 441–52. [524–5]

Bazigos, G.P. (1975). The statistical efficiency of echo surveys with special reference to Lake Tanganyika. *FAO–Fisheries Technical Paper, No. 139,* 52 pp. [454]

Bazigos, G.P. (1976). The design of fisheries statistical surveys. *FAO–Fisheries Technical Paper, No. 133, Supplement 1,* 46 pp. [454]

Bearden, C.M. and McKenzie, M.D. (1972). Results of a pilot shrimp tagging project using internal anchor tags. *Trans. Amer. Fish. Soc.* **101**, 358–62. [488]

Beaver, R.A. (1966). The development and expression of population tables for the bark beetle. *Scolytus scolytus* (F.). *J. Animal Ecol.* **35**, 27–41. [398, 551]

Becker, G.F. and Van Orstrand, C.E. (1924). *Smithsonian Mathematical Tables: Hyperbolic Functions.* Smithsonian Institute: Washington. [136]

Beddington, J.R. (1974). Age structure, sex ratio and population density in the harvesting of natural animal populations. *J. appl. Ecol.* **11**, 915–24. [551]

Beddington, J.R. (1978). On the dynamics of sei whales under exploitation. *Rep. int. Whal. Commn.* **28**, 169–72. [550]

Beddington, J.R. and Taylor, D.B. (1973). Optimum age specific harvesting of a population, *Biometrics* **29**, 801–9. [551]

Béland, P. (1974). On predicting the yield from brook trout populations. *Trans. Amer. Fish. Soc.* **103**, 353–5. [550]

Bell. G. (1974). Population estimates from recapture studies in which no recaptures have been made. *Nature* **248**, 616. [485]

Bell, R.H.V. Grimsdell, J.J.R., Van Lavieren, L.P. and Sayer, J.A. (1973). Census of the Kafue lechwe by aerial stratified sampling. *E. Afr. Wildl. J.* **11**, 55–74. [455]

Bellrose, F.C. (1955). A comparison of recoveries from reward and standard bands. *J. Wildl. Manag.* **19**, 71–5. [97, 250]

Bellrose, F.C. and Chase, E.B. (1950). Population losses in the mallard, black duck. and blue-winged teal. *Illinois Nat. Hist. Survey Biol. Notes* **22**, 1–27. [253]

Berg, S. (1974). Factorial series distributions, with applications to capture–recapture problems. *Scand. J. Statist.* **1**, 145–52. [490–1]

Berg, S. (1975). Some properties and applications of a ratio of Stirling numbers of the second kind. *Scand. J. Statist.* **2**, 91–4. [491, 493]

Berg, S. (1976). A note on the UMVU estimate in a multiple–recapture census. *Scand. J. Statist.* **3**, 86–8. [490]

Berg, S. and Ericson, R. (1977). Goodman's sampling tagging experiment revisited. *Research Report No. 2, Dept. Statistics, University of Lund,* Sweden. [492]

Bergerud, A.T. (1971). The population dynamics of Newfoundland caribou. *Wildl. Monogr. No. 25,* 53 pp. [456, 458]

Bergerud, A.T. (1972). Changes in the vulnerability of ptarmigan to hunting in Newfoundland *J. Wildl. Manag.* **36**, 104–9. [452]

Bergerud, A.T. and Mercer, W.E. (1966). Census of the willow ptarmigan in Newfoundland. *J. Wildl. Manag.* **30**, 101–13. [560]

Berryman, A.A. (1968). Development of sampling techniques and life tables for the fir engraver *Scolytus ventralis. Canadian Entomologist* **100**, 1138–47. [398, 551]

Besag, J.E. and Diggle, P.J. (1977). Simple Monte Carlo tests for spatial pattern. *Appl. Statist.* **26**, 327–33. [18]

Besag, J.E. and Gleaves, J.T. (1973). On the detection of spatial pattern in plant communities *Bull. Inst. Int. Statist.* **45**, 153–8. [475]

Best, P.B. and Rand, R.W. (1975). Results of a pup-tagging experiment on the *Arctocephalus pusillus* rookery at Seal Island, False Bay, South Africa. *Rapp. P.-v. Cons. int. Explor. Mer.* **169**, 267–73. [114, 489]

Beverton, R.J.H. (1954). *Notes on the use of theoretical models in the study of exploited fish populations.* Misc. Contr. No. 2, U.S. Fishery Lab., Beaufort, N.C., 1–159. [276]

Beverton, R.J.H. and Holt, S.J. (1957). *On the Dynamics of Exploited Fish Populations. London*: Her Majesty's Stationery Office. [2, 94, 281, 306, 329, 431, 572]

Birch, M.W. (1964). A new proof of the Pearson–Fisher theorem. *Ann. Math. Statist.* **35**, 817–24. [15]

Birley, M. (1977). The estimation of insect density and instar survivorship functions from census data. *J. Animal Ecol.* **46**, 497–510. [553]

Bishop, J.A. and Bradley, J.S. (1972). Taxi-cabs as subjects for a population study. *J. Biol. Educ.* **6**, 227–31. [505]

Bishop, J.A. and Sheppard, P.M. (1973). An evaluation of two capture–repapture models using the technique of computer simulation. *In* M.S. Bartlett and R.W. Hiorns (eds.), *The Mathematical Theory of the Dynamics of Biological Populations,* 235–52. Academic Press: New York. [506–8]

Bishop, Y.M.M., Fienberg, S.E. and Holland, P.W. (1975). *Discrete Multivariate Analysis: Theory and Practice.* The MIT Press, Cambridge, Massachusetts. [492]

Bissel, A.F. and Ferguson, R.A. (1975). The jackknife–toy, tool or two-edged weapon? *The Statistician* **24**, 79–100. [18]

Blackith, R.E. (1978). Nearest-neighbour distance measurements for the estimation of animal populations. *Ecology* **39**, 147–50. [41, 48, 50]

REFERENCES

Blackith, R.E., Siddorn, J.W., Waloff, N. and Emden, H.F. Van (1963). Mound nests of the yellow ant, *Lasius flavus* L., on waterlogged pasture in Devonshire. *Ent. Mon. Mag.* **99**, 48–9. [41]

Blankenship, L.H., Humphrey, A.B. and MacDonald, D. (1971). A new stratification for mourning dove call-count routes. *J. Wildl. Manag.* **35**, 319–26. [451]

Bliss, C.I. (1971). The aggregation of species within spatial units. *In* G.P. Patil, E.C. Pielou and W.E. Waters (eds.), *Spatial Patterns and Statistical Distributions.* Statistical Ecology Series Vol. 1, 311–35. Penn State Univ. Press, University Park, Penn. [451]

Bliss, C.I. and Fisher, R.A. (1953). Fitting the negative binomial distribution to biological data; note on the efficient fitting of the negative binomial. *Biometrics* **9**, 176–96; 197–9, [175]

Boag, D.A. (1972). Effect of radio packages on behavior of captive red grouse, *J. Wildl. Manag.* **36**, 511–18. [448]

Boag, D.A., Watson, A. and Parr, R. (1973). Radio-marking versus back-tabbing red grouse. *J. Wildl. Manag.* **37**, 410–12. [448]

Bobek, B. (1973). Net production of small rodents in a deciduous forest. *Acta theriologica* **18**, 403–34. [504]

Boeker, E.L. and Bolen, E.B. (1972). Winter golden eagle populations in the southwest. *J. Wildl. Manag.* **36**, 477–84. [458]

Boguslavsky, G.W. (1955). A mathematical model for conditioning. *Psychometrika* **20**, 125–38. [190]

Boguslavsky, G.W. (1956). Statistical estimation of the size of a small population. *Science* **124**, 317–18. [191–2]

Bohlin, T. and Sundström, B. (1977). Influence of unequal catchability on population estimates using the Lincoln index and the removal method applied to electro-fishing. *Oikos* **28**, 123–9. [487, 500]

Bohrnstedt, G.W. and Goldberger, A.S. (1969). On the exact covariance of products of random variables. *J. Amer. Statist. Assoc.* **64**, 1439–42. [9]

Bole, B.P. (1939). The quadrat method of studying small mammal populations. *Cleveland Mus. Nat. Hist. Sci. Publ.* No. 5. [21–2]

Boswell, M.T. and Patil, G.P. (1970). Chance mechanisms generating the negative binomial distributions. *In* G.P. Patil (ed.), *Random Counts in Scientific Work,* Vol. 1, 3–22. Penn. State Univ. Press, University Park, Penn. [478]

Bouck, G.R. and Ball, R.C. (1966). Influence of capture methods on blood characteristics and mortality of the rainbow trout (*Salmo gairdneri*). *Trans. Amer. Fish. Soc.* **95**, 170–6. [83]

Bowden, D.C., Anderson, A.E. and Medin, D.E. (1969). Frequency distributions of male deer fecal group counts. *J. Wildl. Manag.* **33**, 895–905. [451]

Boyd, H. (1956). Statistics of the British population of the pink-footed goose. *J. Animal Ecol.* **25**, 253–73. [244–5]

Braaten, D.O. (1969). Robustness of the De Lury population estimator. *J. Fish. Res. Bd. Canada* **26**, 339–55. [304]

Brand, C.J., Vowles, R.H. and Keith, L.B. (1975). Snowshoe hare mortality monitored by telemetry. *J. Wildl. Manag.* **39**, 741–7. [448, 546]

Brander, R.B. and Cochran, W.W. (1969). Radio location telemetry. *In* R.H. Giles, Jr. (ed.), *Wildlife Management Techniques,* 3rd edn, 95–103. The Wildlife Society: Washington. [94]

Brass, W. (1958). Simplified methods of fitting the truncated negative binomial distribution. *Biometrika* **45**, 59–68. [175]

Bray, O.E. and Corner, G.W. (1972). A tail clip for attaching transmitters to birds. *J. Wildl. Manag.* **36**, 640–2. [448]

Breakey, D.R. (1963). The breeding season and age structure of feral mouse populations near San Francisco Bay, California. *J. Mammal.* **44**, 153–68. [403]

Breiwick, J.M. (1977). Analysis of the Antarctic fin whale stock in Area I. *Rep. int. Whal. Commn* **27**, 124–7. [545]

Breiwick, J.M. (1978a). Reanalysis of Antartic sei whale stocks. *Rep. int. Whal. Commn* **28**, 345–67. [544–5]

Breiwick, J.M. (1978b). Southern hemisphere sei whale stock sizes prior to 1960. *Rep. int. Whal. Commn* **28**, 179–82. [545]

Brent, R.P. (1973). *Algorithms for Minimization Without Derivatives.* Prentice-Hall, Inc.: Englewood Cliffs, N.J. [493]

Briese, L.A. and Smith, M.H. (1974). Seasonal abundance and movement of nine species of small mammals. *J. Mammal.* **55**, 615–29. [446]

Brock, V.E. (1954). A preliminary report on a method of estimating reef fish populations. *J. Wildl. Manag.* **18**, 297–308. [28]

Brooks, G.R. Jr (1967). Population ecology of the ground skink, *Lygosoma laterale* (Say). *Ecol. Monogr.* **37**, 71–87. [400]

Brown, D. and Rothery, P. (1978). Randomness and local regularity of points in a plane. *Biometrika* **65**, 115–22. [472, 485]

Brown, D.E. and Smith, R.H. (1976). Predicting hunting success from call counts of mourning and white-winged doves. *J. Wildl. Manag.* **40**, 743–9. [452]

Brown, L.E. (1954). Small mammal populations at Silwood Park Field Centre, Berkshire, England. *J. Mammal.* **35**, 161–76. [84]

Brown, S. and Holgate, P. (1974). The thinned plantation. *Biometrika* **61**, 253–61. [484]

Brownie, C. (1973). Stochastic models allowing age-dependent survival rates for banding experiments on exploited bird populations. Ph.D. Thesis, Biometrics Unit, Cornell Univ., Ithaka, New York. [527]

Brownie, C., Anderson, D.R., Burnham, K.P. and Robson, D.S. (1978). Statistical inference from band recovery data: A handbook. *U.S. Fish and Wildl. Serv., Resour. Publ. No. 131*, 212 pp. [506, 525–7, 569]

Brownie, C. and Robson, D.S. (1976). Models allowing for age-dependent survival rates for band-return data. *Biometrics* **32**, 305–23. [527]

Brunk, H.D. (1965). *An Introduction to Mathematical Statistics*, 2nd edn. Blaisdell: Waltham, Massachusetts. [186]

Brussard, P.F. and Ehrlich, P.R. (1970). The population structure of *Erebia epipsodea* (Lepidoptera: Satyrinae). *Ecology* **51**, 119–29. [505]

Buchalczyk, T. and Olszewski, J.L. (1971). Behavioural response of forest rodents against trap and bait. *Acta theriologica* **16**, 277–92. [487]

Buchalczyk, T. and Pucek, Z. (1968). Estimation of the numbers of *Microtus oeconomus* using the Standard Minimum method. *Acta theriologica* **14**, 461–82. [327]

Buchler, E.R. (1976). A chemiluminescent tag for tracking bats and other small nocturnal animals. *J. Mammal.* **57**, 173–6. [487]

Buck, D.H. and Thoits, C.F. 3rd (1965). An evaluation of Petersen estimation procedures employing seines in 1-acre ponds. *J. Wildl. Manag.* **29**, 598–621. [19, 86, 000]

Buckland, W.R. (1964). *Statistical Assessment of the Life Characteristic: a Bibliographic Guide.* Griffin's Statistical Monographs and Courses No. 13, Griffin: London. [287]

Buckner, C.H. (1957). Population studies on small mammals of southeastern Manitoba. *J. Mammal.* **38**, 87–97. [83]

Burge, J.R. and Jorgensen, C.D. (1973). Home range of small mammals: a reliable estimate. *J. Mammal.* **54**, 483–8. [446]

Burkitt, J.P. (1926). A study of the robin by means of marked birds. *Brit. Birds* **20**, 91–101. [397]

REFERENCES

Burnham, K.P. (1972). Estimation of population size in multiple capture studies when capture probabilities vary among animals. Ph.D. Dissertation, Ore, St. Univ. Corvallis, [493, 497, 503]

Burnham, K.P. (1979). A parametric generalization of the Hayne estimator for line transect sampling. *Biometrics* **35**, 587–960. [470]

Burnham, K.P. and Anderson, D.R. (1976). Mathematical models for nonparametric inferences from line transect data. *Biometrics* **32**, 325–36. [461–3, 470–1]

Burnham, K.P. and Anderson, D.R. (1979). The composite dynamic method as evidence for age-specific waterfowl mortality. *J. Wildl. Manag.* **43**, 356–66. [528]

Burnham, K.P., Anderson, D.R. and Laake, J.L. (1980). Estimation of density for line transect sampling of biological populations *Wildl. Managr. No. 72*, 202 pp. [467]

Burnham, K.P. and Overton, W.S. (1969). A simulation study of live-trapping and estimation of population size. *Oregon. State Univ., Dept. of Stat., Tech. Rep. No. 14.* [494]

Burnham, K.P. and Overton, W.S. (1978). Estimation of the size of a closed population when capture probabilities vary among animals. *Biometrika* **65**, 625–33. [494, 498]

Burnham, K.P. and Overton, W.S. (1979). Robust estimation of population size when capture probabilities vary among animals. *Ecology* **60**, 927–36. [494]

Caddy, J.F. (1975). Spatial model for an exploited shellfish population, and its application to the Georges Bank scallop fishery. *J. Fish. Res. Bd. Canada* **32**, 1305–28. [542]

Calef, G.W. (1973). Natural mortality of tadpoles in a population of *Rana Aurora. Ecology* **54**, 741–48. [452]

Calhoun, J.B. (1948). Announcement of program. North American census of small mammals. Release No. 1 (Rodent Ecology Project, Johns Hopkins University), 8 pp. [564]

Calhoun, J.B. and Casby, J.U. (1958). Calculation of home range and density of small mammals. U.S. Dept. of Health, Education, and Welfare, *Public Health Monogr.* **55**, 1–24. [171, 447]

Carle, F.L. and Strub, M.R. (1978). A new method for estimating population size from removal data. *Biometrics* **34**, 621–30. [315, 502]

Carline, R.F. (1972). Biased harvest estimates from a postal survey of sport fishery. *Trans. Amer. Fish. Soc.* **101**, 262–6. [489]

Carney, S.M. and Petrides, G.A. (1957). Analysis of variation among participants in pheasant cock-crowing censuses. *J. Wildl. Manag.* **21**, 392–7. [54]

Carothers, A.D. (1971). An examination and extension of Leslie's test of equal catchability. *Biometrics* **27**, 615–30. [161, 162, 228, 495, 539]

Carothers, A.D. (1973a). The effects of unequal catchability on Jolly–Seber estimates. *Biometrics* **29**, 79–100. [506, 569]

Carothers, A.D. (1973b). Capture–recapture methods applied to a population with known parameters. *J. Anim. Ecol.* **42**, 125–46. [496, 505]

Carroll, D. and Getz, L.L. (1976). Runway use and population density in *Microtus ochrogaster. J. Mammal.* **57**, 772–6. [54]

Carver, J.S. (1976), *Bibliography of the Dual Record System.* An Occasional Publication, International Program of Laboratories for Population Statistics, Univ. of North Carolina at Chapel Hill. [492]

Catana, A.J., Jr (1963). The wandering quarter method of estimating population density. *Ecology* **44**, 349–60. [478]

Caughley, G. (1966). Mortality patterns in mammals. *Ecology* **47**, 906–18. [393, 400–7]

Caughley, G. (1970). Eruption of ungulate populations, with emphasis on the Himalayan thar in New Zealand. Ecology **51**, 53–72. [407]

Caughley, G. (1971). Correction for band loss, *Bird-Banding* **42**, 220–1. [489]

Caughley, G. (1974a). Bias in aerial survey. *J. Wildl. Manag.* **38**, 921–33. [456–7]

Caughley, G. (1974b). Interpretation of age ratios. *J. Wildl. Manag.* **38**, 557–62. [548]

Caughley, G. (1977a). Sampling in aerial survey. *J. Wildl. Manag.* **41**, 605–15. [454–6]

Caughley, G. (1977b). *Analysis of Vertebrate Populations.* Wiley: London. [455–6]

Caughley, G. and Birch, L.C. (1971). Rate of increase. *J. Wildl. Manag.* **35**, 658–63. [550]

Caughley, G. and Goddard, J. (1972). Improving the estimates from inaccurate censuses. *J. Wildl. Manag.* **36**, 135–40. [457]

Caughley, G. and Goddard, J. (1975). Abundance and distribution of elephants in the Luangwa Valley, Zambia. *E. Afr. Wildl. J.* **13**, 39–48. [455]

Caughley, G., Sinclair, R. and Scott-Kemmis, D. (1976). Experiments in aerial survey. *J. Wildl. Manag.* **40**, 290–300. [456–7]

Caughley, G., Sinclair, R.G. and Wilson, G.R. (1977). Numbers, distribution and harvesting rate of kangaroos on the inland plains of New South Wales. *Aust. Wildl. Res.* **4**, 99–108. [457]

Chakraborty, P.N. (1963). On a method of estimating birth and death rates and the extent of registration. *Calcutta Stat. Assoc.* **12**, 106–12. [492]

Chakravarti, I.M. and Rao, C.R. (1959). Tables for some small sample tests of significance for Poisson distributions and 2 × 3 contingency tables. *Sankhyā* **21**, 315–26. [170]

Chandrasekar, C. and Deming, W.E. (1949). On a method of estimating birth and death rates and the extent of registration. *J. Amer. Statist. Assoc.* **44**, 101–15. [492]

Chapman, D.G. (1948). A mathematical study of confidence limits of salmon populations calculated from sample tag ratios. *Internat. Pac. Salmon Fisheries Comm. Bull.* **2**, 69–85. [63–4, 574]

Chapman, D.G. (1951). Some properties of the hypergeometric distribution with applications to zoological censuses. *Univ. Calif. Public. Stat.* **1**, 131–60. [60, 100, 122, 125]

Chapman, D.G. (1952). Inverse multiple and sequential sample censuses. *Biometrics* **8**, 286–306. [71, 119–21, 131–3, 138–9, 157–8, 187]

Chapman, D.G. (1954). The estimation of biological populations. *Ann. Math. Statist.* **25**, 1–15. [189, 238, 298, 309, 353, 355–6]

Chapman, D.G. (1955). Population estimation based on change of composition caused by selective removal. *Biometrika* **42**, 279–90. [353, 358, 362–3, 372–3, 377–9, 570–1]

Chapman, D.G. (1956). Estimating the parameters of a truncated Gamma distribution. *Ann. Math. Statist.* **27**, 498–506 [277]

Chapman, D.G. (1961). Statistical problems in the dynamics of exploited fish populations *Proc. 4th Berkeley Symp. 1960*, vol 4 153–68. [272–3, 304, 328–31, 344–5]

Chapman, D.G. (1964). A critical study of Pribilof fur seal population estimates. *Fishery Bulletin* **63**, 657–69. [560]

Chapman D.G. (1965). The estimation of mortality and recruitment from a single-tagging experiment. *Biometrics* **21**, 529–42 [266–72, 347–52]

Chapman, D.G. (1970). Reanalysis of Antartic fin whale population data. *Rep. int. Whal. Commn* **20**, 54–9. [543]

Chapman, D.G. (1974a). Estimation of population parameters of Antarctic baleen whales. *In* W.E. Schevill (ed.), *The Whale Problem: a Status Report*, 336–51. Harvard University Press, Mass. [488, 505, 542]

Chapman, D.G. (1974b). Estimation of population size and sustainable yield of whales in the Antarctic. *Rep. int. Whal. Commn* **24**, 82–90. [544]

Chapman, D.G. Fink, B.D. and Bennett, E.B. (1965). A method for estimating the rate of shedding of tags from yellowfin tuna. *Bull. Inter-Amer. Trop. Tuna. Comm.* **10**, 335–42. [489, 525]

Chapman, D.G. and Johnson, A.M. (1968). Estimation of fur seal pup populations by randomised sampling. *Trans. Amer. Fish. Soc.* **97**, 264–70. [117, 488]

Chapman, D.G. and Junge, C.O. (1956). The estimation of the size of a stratified animal population. *Ann. Math. Statist.* **27**, 375–89. [431–8, 555, 558]

Chapman, D.G. and Murphy, G.I. (1965). Estimates of mortality and population from survey–removal records. *Biometrics* **21**, 921–35. [353, 356, 376–7, 348–92]

Chapman, D.G. and Overton, W.S. (1966). Estimating and testing differences between population levels by the Schnabel estimation method. *J. Wildl. Manag.* **30**, 173–80. [122–5, 140]

Chapman, D.G. and Robson, D.S. (1960). The analysis of a catch curve. *Biometrics* **16**, 354–68. [171, 414–30, 529, 547]

Chapman, D.W. (1976). Acoustic estimates of pelagic ichthyomass in Lake Tanganyika with an inexpensive echo sounder. *Trans. Amer. Fish. Soc.* **105**, 581–7. [454]

Chapman, J.A., Henny, C.J. and Wight, H.M. (1969). The status, population dynamics, and harvest of the dusky Canada goose. *Wildl. Monogr. No. 18*, 48 pp.

Chapman, J.A. and Trethewey, D.E.C. (1972). Factors affecting trap responses of introduced eastern cottontail rabbits. *J. Wildl. Manag.* **36**, 1221–26. [487]

Chesness, R.A. and Nelson, W.M. (1964). Illegal kill of hen phasants in Minnesota. *J. Wildl. Manag.* **28**, 249–53. [361]

Chew, R.M. and Butterworth, B.B. (1964). Ecology of rodents in Indian Cove (Mojave Desert), Joshua Tree National Monument, California. *J. Mammal.* **45**, 203–25. [314]

Chiang, C.L. (1960a). A stochastic study of the life table and its applications. I: Probability distributions of the biometric functions. *Biometrics* **16**, 618–35. [408–12]

Chiang, C.L. (1960b). A stochastic study of the life table and its applications. II: Sample variance of the observed expectation of life and other biometric functions. *Human Biology* **32**, 221–38. [408–12]

Chitty, D. and Kempson, D.A. (1949). Prebaiting small mammals and a new design of live-trap. *Ecology* **30**, 536–42. [84]

Chitty, D. and Shorten, M. (1946). Techniques for the study of Norway rat (*Rattus norvegicus*). *J. Mammal.* **27**, 63–78. [82–3]

Chorley, R.J., Stoddart, D.R., Haggett, P. and Slaymaker, H.O. (1966). Regional and local components in the areal distribution of surface sand facies in the Breckland, Eastern England. *J. Sediment. Petrol.* **36**, 209–20. [481]

Chung, J.H. and De Lury, D.B. (1950). *Confidence Limits for the Hypergeometric Distribution*, University of Toronto Press. [62]

Clancy, D.W. (1963). The effect of tagging with Petersen disc tags on the swimming ability of fingerling Steelhead Trout (*Salmo gairdneri*). *J. Fish. Res. Bd. Canada* **20**, 969–81. [83]

Clapham, A.R. (1932). The form of the observational unit in quantitative ecology. *J. Ecology* **20**, 192–7. [21]

Clark, P.J. and Evans F.C. (1954). Distance to nearest neighbor as a measure of spatial relationships in populations. *Ecology* **35**, 445–53. [43, 45, 49, 50]

Clark, P.J. and Evans, F.C. (1955). Some aspects of spatial pattern in biological populations. *Science* **121**, 397–8. [43]

Cliff, A.D. and Ord, J.K. (1973). *Spatial Autocorrelation*. Pion Ltd: London. [18, 479, 483–5]

Cliff, A.D. and Ord, J.K. (1975). Model building and the analysis of spatial pattern in human geography (with Discussion). *J. Roy. Statist. Soc. B* **37**, 297–348. [18, 485]

Cochran, W.G. (1950). The comparison of percentages in matched samples. *Biometrika* **37**, 256–66. [540]

Cochran, W.G. (1954). Some methods for strengthening the common chi-squared tests. *Biometrics* **10**, 417–51. [163]

Cochran, W.G. (1977). *Sampling Techniques,* 3rd edn. John Wiley and Sons: New York. [5, 20, 23, 27, 55, 63, 64, 112–15, 134, 425, 426, 454]

Cohen, A., Peters, H.S. and Foote, L.E. (1960). Calling behaviour of mourning doves in two midwest life zones. *J. Wildl. Manag.* **24**, 203–12. [54]

Comfort, A. (1957). Survival curves of mammals in captivity. *Proc. Zool. Soc. London* **128**, 349–64. [396, 400, 409]

Comre, L.J. (1959). *Chambers' Six-figure Mathematical Tables,* vol. 2 – *Natural Values.* Chambers Ltd.: London. [136]

Conn, D.L.T. (1976). Estimates of population size and longevity of adult narcissus bulb fly *Merodon equestris* Fab. (Diptera: Syrphidae). *J. app. Ecol.* **13**, 429–34. [505]

Cook, L.M., Thomason, E.W. and Young, A.M. (1976). Population structure, dynamics and dispersal of the tropical butterfly *Heliconius charitonius. J. Animal Ecol.* **45**, 851–63. [505]

Cook, R.D. and Martin, F.B. (1974). A model for quadrat sampling with "visibility bias", *J. Amer. Statist. Assoc.* **69**, 345–9. [458]

Corbet, P.S. (1952). An adult population study of *Pyrrhosoma Nymphula* (Sulzer): (Odonata: Coenagrionidae). *J. Animal Ecol.* **21**, 206–22. [82]

Cormack, R.M. (1964). Estimates of survival from the sighting of marked animals. *Biometrika* **51**, 429–38. [128, 214–17, 530]

Cormack, R.M. (1966). A test for equal catchability. *Biometrics* **22**, 330–42. [88–93]

Cormack, R.M. (1968). The statistics of capture–recapture methods. *Oceanogr. Mar. Bio. Ann. Rev.* **6**, 455–506. [2, 187, 204, 208, 489, 565]

Cormack, R.M. (1972). The logic of capture–recapture estimates. *Biometrics* **28**, 337–43. [531, 537]

Cormack, R.M. (1973). Commonsense estimates from capture–recapture studies. *In* M.S. Bartlett and R.W. Hiorns (eds.), *The Mathematical Theory of the Dynamics of Biological Populations*, 225–34. Academic Press: New York. [531, 537]

Cormack, R.M. (1977). The invariance of Cox and Lewis's statistic for the analysis of spatial patterns. *Biometrika* **64**, 143–4. [475]

Cormack, R.M. (1979). Models for capture–recapture. *In* R.M. Cormack, G.P. Patil and D.S. Robson (eds.), *Sampling Biological Populations,* Statistical Ecology Series, Vol. 5, 217–55. International Co-operative Publishing House, Burtonsville, Maryland. [492, 505, 530, 541]

Corner, G.W. and Pearson, E.W. (1972). A miniature 30-MHz collar transmitter for small animals. *J. Wildl. Manag.* **36**, 657–61. [448]

Cottam, C. (1956). Uses of marking animals in ecological studies: marking birds for scientific purposes. *Ecology* **37**, 674–81. [93]

Cottam, G. and Curtis, J.T. (1949). A method for making rapid surveys of woodlands by means of pairs of randomly selected trees. *Ecology* **30**, 101–4. [50]

Cottam, G. and Curtis, J.T. (1955). Correction for various exclusion angles in the random pairs methods. *Ecology* **36**, 767. [50]

Cottam, G. and Curtis, J.T. (1956). The use of distance measures in phytosociological sampling. *Ecology* **37**, 451–60. [44, 50, 478]

Cottam, G., Curtis, J.T. and Hale, B.W. (1953). Some sampling characteristics of a population of randomly dispersed individuals. *Ecology* **34**, 741–57. [44]

REFERENCES

Coulson, J.C. (1960). A study of the mortality of the starling based on ringing recoveries. *J. Animal Ecol.* **29**, 251–71. [245, 250]

Cowardin, L.M. (1977). Analysis and machine mapping of the distribution of band recoveries. *U.S. Fish Wildl. Serv., Spec. Sci. Rep. Wildl. No. 198,* 8 pp. [527]

Cowardin, L.M. and Myers, V.I. (1974). Remote sensing for identification and classification of wetland vegetation. *J. Wildl. Manag.,* **38**, 308–314. [459]

Cox, D.R. (1969). Some sampling problems in technology. *In* N.L. Johnson and H. Smith Jr (eds.), *New Developments in Survey Sampling,* 506–7. Wiley-Interscience: Wiley, New York. [460, 463]

Cox, D.R. and Lewis, P.A.W. (1966). *The Statistical Analysis of Series of Events.* Methuen: London. [46]

Cox, T.F. (1976). The robust estimation of a forest stand using a new conditional distance method. *Biometrika* **63**, 493–9. [475]

Cox, T.F. and Lewis, T. (1976). A conditioned distance ratio method for analyzing spatial patterns. *Biometrika* **63**, 483–91. [475]

Craig, C.C. (1953a). On a method of estimating biological populations in the field. *Biometrika* **40**, 216–18. [47–9, 472]

Craig, C.C. (1953b). Use of marked specimens in estimating populations. *Biometrika* **40**, 170–6. [136–7, 170]

Crain, B.R., Burnham, K.P., Anderson, D.R. and Laake, J.L. (1978). *A Fouries Series Estimator of Population Density for Line Transect Sampling.* Utah State University Press. [463, 467]

Cross, D.G. and Stott, B. (1975). The effect of electric fishing on the subsequent capture of fish. *J. Fish. Biol.* **7**, 349–57. [487, 500]

Crow, E.L. (1956). Confidence limits for a proportion. *Biometrika* **43**, 423–35. [123]

Crow, E.L. and Gardner, R.S. (1959). Confidence intervals for the expectation of a Poisson variable. *Biometrika* **46**, 441–53. [63, 123]

Crowcroft, P. and Jeffers, J.N.R. (1961). Variability in the behaviour of wild house mice (*Mus musculus L.*) towards live-traps. *Proc. Zool. Soc. London* **137**, 573–82. [84, 162, 165]

Crowe, D.M. (1975). A model for exploited bobcat populations in Wyoming. *J. Wildl. Manag.* **39**, 408–15. [548]

Crumpacker, D.W. and Williams, J.S. (1973). Density, dispersion, and population structure in *Drosophila pseudoobscura. Ecol. Monogr.* **43**, 499–538. [487]

Cucin, D. and Regier, H.A. (1966). Dynamics and expoloitation of lake whitefish in Southern Georgian Bay. *J. Fish. Res. Bd. Canada* **23**, 221–74. [283]

Cushing, D.H. (1968). *Fisheries Biology: a Study in Population Dynamics.* University of Wisconsin Press. [572]

Czen Pin (1962). O minimaksowym estymatorze liczności populacji. *Zastosow. Mat.* **6**, 137–48. [121]

Dacey, M.F. (1963). Order neighbor statistics for a class of random patterns in multi-dimensional space. *Ann. Assoc. Amer. Geographers,* **53**, 505–15. [42]

Dacey, M.F. (1964a). Two-dimensional random point pattern: a review and an interpretation. *Papers, The Regional Science Association,* **13**, 41–55. [24, 42]

Dacey, M.F. (1964b). Modified Poisson probability law for point pattern more regular than random. *Ann. Assoc. Amer. Geographers,* **54**, 559–65. [41]

Dacey, M.F. (1965). Order distance in an inhomogeneous random point pattern. *The Canadian Geographer* **9**, 144–53. [41]

Dacey, M.F. (1966). A compound probability law for a pattern more dispersed than random and with areal inhomogeneity. *Economic Geography* **42**, 172–9. [41]

Dahiya, R.C. and Gurland, J. (1969). Functions of the sample mean and sample variance of a Poisson variate. *Biometrics* **25**, 171–3. [479]

Dale, M.B. (1970). Systems analysis and ecology. *Ecology* **51**, 2–16. [572]

Darling, D.A. and Robbins, H. (1967). Finding the size of a finite population. *Ann. Math. Statist.* **38**, 1392–8. [191–3]

Darroch, J.N. (1958). The multiple-recapture census. I: Estimation of a closed population. *Biometrika* **45**, 343–59. [131–4, 136, 139, 154, 164, 189–91, 234, 490]

Darroch, J.N. (1959). The multiple-recapture census. II: Estimation when there is immigration or death. *Biometrika* **46**, 336–51. [15, 218–19, 510, 530]

Darroch, J.N. (1961). The two-sample cature-recapture census when tagging and sampling are stratified. *Biometrika* **48**, 241–60. [431–45, 555]

Das Gupta, P. (1964). On the estimation of the total number of events and of the probabilities of detecting an event from information supplied by several agencies. *Calcutta Stat. Assoc.* **13**, 89–100. [492]

Dasmann, R.F. (1952). Methods for estimating deer populations from kill data. *Calif. Fish and Game* **38**(2), 225–33. [353]

Daugherty, C.H. (1976). Freeze branding as a technique for marking anurans. *Copeia 1976* 836–8. [487]

Davenport, D.A. (1977). Computerized tabulation and display of band recovery data. *U.S. Fish Wildl. Serv., Spec. Sci. Rep. Wildl. No. 199,* 7 pp. [527]

David, F.N. and Johnson, N.L. (1952). The truncated Poisson. *Biometrics* **8**, 275–85. [170]

Davidoff, E.B., Rybicki, R.W. and Doan, K.H. (1973). Changes in the population of Lake Whitefish (*Coregonus clupeaformis*) in Lake Winnipeg from 1944 to 1969. *J. Fish. Res. Bd. Canada* **30**, 1667–82. [548]

Davies, R.G. (1971). *Computer Programming in Qualitative Biology.* Academic Press: London and New York. [506]

Davis, D.E. (1957). Observations on the abundance of Korean mice. *J. Mammal.* **38**, 374–7. [315]

Davis, D.E. (1963). Estimating the numbers of game populations. *In* H.S. Mosby (ed.), *Wildlife Investigational Techniques,* 2nd edn, 89–118. The Wildlife Society: Washington. [19, 53–5, 57, 146, 376, 561]

Davis, D.E., Christian, J.J. and Bronson, F. (1964). Effect of exploitation on birth, mortality and movement rates in a woodchuck population. *J. Wildl. Manag.* **28**, 1–9. [233]

Davis, W.S. (1964). Graphic representation of confidence intervals for Petersen population estimates. *Trans. Amer. Fish. Soc.* **93**, 227–32. [64]

Debauche, H.R. (1962). The structural analysis of animal communities of the soil. *In* P.W. Murphy (ed.), *Progress in Soil Zoology,* 10–25. Butterworths: London, [25]

Deemer, W.L. and Votaw, D.V. Jr. (1955). Estimation of parameters of truncated or censored exponential distributions. *Ann. Math. Statist.* **26**, 498–504. [273, 345, 583]

Deevey, E.S. Jr. (1947). Life tables for natural populations of animals. *Quart. Rev. Biol.* **22**, 283–314. [393, 395]

Delong, K.T. (1966). Population ecology of feral house mice: interference by *Microtus. Ecology* **47**, 481–4. [160, 205]

De Lury, D.B. (1947). On the estimation of biological populations. *Biometrics* **3**, 145–67. [298–300, 303]

De Lury, D.B. (1951). On the planning of experiments for the estimation of fish pop-populations. *J. Fish. Res. Bd. Canada* **8**, 281–307. [298, 305, 307]

De Lury, D.B. (1954). On the assumptions underlying estimates of mobile populations. *In* O. Kempthorne (ed.), *Statistics and Mathematics in Biology.* Iowa State College Press: Ames, 287–93. [2]

De Lury, D.B. (1958). The estimation of population size by marking and recapture procedure. *J. Fish. Res. Bd. Canada* **15**, 19–25. [126, 142]

Demetrius, L. (1971). Primitivity conditions for growth matrices. *Math. Biosci.* **12**, 53–8. [550]

Dempster, J.P. (1961). The analysis of data obtained by regular sampling of an insect population. *J. Animal Ecol.* **30**, 429–32. [551]

De Vries, P.G. (1973). A general theory on line intersect sampling with applications to logging residue inventory. *Mededelingen Landbouwhogeschool Wageningen, Nederland* **73**, 1–23. [460]

De Vries, P.G. (1974). Multistage line intersect sampling. *Forest Science* **20**, 129–33. [460]

De Vries, P.G. (1979). Line intersect sampling – statistical theory. applications and suggestions for extended sue in ecological inventory. *In* R.M. Cormack, G.P. Patil and D.S. Robson (eds.), *Sampling Biological Populations*, Statistical Ecology Series, Vol. 5, 1–70. International Co-operative publishing House, Burtonsville, Maryland, [446]

Devroye, L.P., and Wagner, J. (1977). The strong uniform consistency of nearest neighbor density estimates. *Ann. Statist.* **5**, 536–40. [477]

Dice, L.R. (1938). Some census methods for mammals. *J. Wildl. Manag.* **2**, 119–30. [19, 51, 446]

Dice, L.R. (1952). *Natural Communities.* University of Michigan Press: Ann Arbor. [22, 52–6]

Dickie, L.M. (1955). Fluctuations in abundance of the giant scallop, *Placopecten magellanicus* (Gmelin), in the Digby area of the Bay of Fundy, *J. Fish Res. Bd. Canada* **12**, 797–857. [307–8]

Diggle, P.J. (1975). Robust density estimation using distance methods. *Biometrika* **62**, 39–48. [474, 483–5]

Diggle, P.J. (1977a). The detection of random heterogeneity in plant populations. *Biometrics* **33**, 390–4. [476]

Diggle, P.J. (1977b). A note on robust density estimation for spatial point patterns. *Biometrika* **64**, 91–5. [476]

Diggle, P.J. (1979a). Statistical methods for spatial point patterns in ecology. *In* R.M. Cormack and J.K. Ord (eds.), *Spatial and Temporal Analysis in Ecology*, Statistical Ecology Series, Vol. 8, 95–150. International Co-Operative Publishing House, Burtonsville, Maryland. [18, 472, 476, 483]

Diggle, P.J. (1979b). On parameter estimation and goodness-of-fit testing for spatial point patterns. *Biometrics* **35**, 87–101. [472, 483]

Diggle, P.J., Besag, J. and Gleaves, J.T. (1976). Statistical analysis of spatial point patterns by means of distance methods. *Biometrics* **32**, 659–67. [474–6, 484]

Dobson, R.M. (1962). Marking techniques and their application to the study of small terrestrial animals. *In P.W.* Mruphy (ed.), *Progress in Soil Zoology*, 228–39. Butterworths: London. [93–4]

Dobzhansky, T., Cooper, D.M., Phaff, H.J., Knapp, E.P. and Carson, H.L. (1956). Studies on the ecology of Drosophila in the Yosemite region of California. IV: Differential attraction of species of Drosophila to different species of yeast. *Ecology*, **37**, 544–50. [82]

Doebel, J.H. and McGinnes, B.S. (1974). Home range and activity of a gray squirrel population. *J. Wildl. Manag.* **38**, 860–7. [447]

Doi, T. (1974). Further development of whale sighting theory, *In* W.E. Schevill (ed.), *The Whale Problem: A Status Report*, 359–68, Harvard Univ. Press, Cambridge, Mass. [459]

Dolbeer, R.A. and Clark, W.R. (1975). Population ecology of Snowshoe hares in the central Rocky Mountains. *J. Wildl. Manag.* **39**, 535–49. [505]

Dole, J.W. (1965). Summer movements of adult leopard frogs, *Rana pipiens* (Schreber), in Northern Michigan. *Ecology* **46**, 236–55. [20]

Dominy, C.L. and Myatt, G.L. (1973). Short-term effects of three different marks on the recapture and survival of the alewife, *Alosa pseudoharengus. Trans. Amer. Fish. Soc.* **102**, 633–7. [488]

Dorney, R.S. (1958). Ruffed grouse roosts as a spring-census technique. *J. Wildl. Manag.* **22**, 97–9. [54]

Dorney, R.S., Thompson, D.R., Hale, J.B. and Wendt, R.F. (1958). An evaluation of ruffed grouse drumming counts. *J. Wildl. Manag.* **22**, 35–40. [54]

Doubleday, W.G. (1975). Harvesting in matrix population models. *Biometrics* **31**, 189–200. [549–51]

Douglas, J.B. (1970). Statistical models in discrete distributions. *In* G.P. Patil (ed.), *Random Counts in Scientific Work*, Vol. 3, 203–32. Penn. State University Press, University Park, Penn. [480]

Draper, N.R. and Smith, H. (1966). *Applied Regression Analysis.* Wiley: New York. [268, 305]

Duncan, K.W. (1971). A generalized computer program in Fortran IV for listing all possible color band permutations. *Bird-Banding* **42**, 279–87. [527]

Dunn, J.E. and Gipson, P.S. (1977), Analysis of radio telemetry data in studies of home range. *Biometrics* **33**, 85–101. [447]

Dunnet, G.M. (1957). A test of the recapture method of estimating the number of rabbits, *Oryctolagus cuniculus* (L). *C.S.I.R.O. Wild,. Res.* **2**, 90–100. [84, 560, 565]

Dunnet, G.M. (1963). A population study of the quokka, *Setonix brachyurus* Quoy and Gaimard (Marsupialia). *C.S.I.R.O. Wildl. Res.* **8**, 78–117. [71, 160]

Dyer, M.I. (1967). Photo-electric cell technique for analyzing radar film. *J. Wildl. Manag.* **31**, 484–91. [19]

Dzieciolowski, R. (1976a). Estimating ungulate numbers in a forest by track counts. *Acta theriologica* **21**, 217–22. [452]

Dzieciolowski, R. (1976b). Roe deer census by pellet-group counts. *Acta theriologica* **21**, 351–8. [453]

Eberhardt, L.L. (1967). Some developments in 'distance sampling'. *Biometrics* **23**, 207–16. [41–9, 484]

Eberhardt, L.L. (1968). A preliminary appraisal of line transects. *J. Wildl. Manag.* **32**, 82–8. [29, 34–5]

Eberhardt, L.L. (1969a). Population estimates from recapture frequencies. *J. Wildl. Manag.* **33**, 28–39. [62, 88, 171, 174, 178]

Eberhardt, L.L. (1969b). Population analysis. *In* R.H. Giles, Jr (ed.), *Wildlife Management Techniques,* 3rd edn. 457–95. The Wildlife Society: Washington. [393]

Eberhardt, L.L. (1972). Some problems in estimating survival from banding data. *In* Population Ecology of Migratory Birds. *U.S. Fish Wildl. Serv., Wildl. Res. Rep. No. 2,* 153–71. [529, 545, 548]

Eberhardt, L.L. (1976). Quantitative ecology and impact assessment. *J. Environ. Manag.* **4**, 27–70. [451]

Eberhardt, L.L. (1978a). Transect methods for population studies. *J. Wildl. Manag.* **42**, 1–31. [458–60, 462–71, 477]

Eberhardt, L.L. (1978b). Appraising variability in population studies. *J. Wildl. Manag.* **42**, 207–38, [452]

Eberhardt, L.L. (1978c). Line transects based on right-angle distances (in preparation). [458, 464]

Eberhardt, D.L., Chapman, D.G. and Gilbert, J.R. (1979). A review of marine mammal census methods. *Wildl. Monogr. No. 63,* 46 pp. [453, 455–6, 459, 488]

Eberhardt, L.L., Peterle, T.J. and Schofield, R. (1963). Problems in a rabbit population study. *Wildl. Monogr.* **10,** 1–51. [84, 170, 174, 299, 383, 560]

Eberhardt, L.L. and Siniff, D.B. (1977). Population dynamics and marine mammal management policies. *J. Fish. Res. Bd. Canada* **34,** 183–90. [551]

Ebert, T.A. (1973). Estimating growth and mortality rates from size data. *Oecologia* **11,** 281–98. [549]

Edwards, W.R. and Eberhardt, L.L. (1967). Estimating cottontail abundance from live-trapping data. *J. Wildl. Manag.* **31,** 87–96. [84, 165, 171–2]

Efron, B. (1979). Bootstrap methods: another look at the jackknife. *Ann. Statist.* **7,** 1–26. [18]

Efron, B. and Thisted, R. (1976). Estimating the number of unseen species: How many words did Shakespeare know? *Biometrika* **63,** 435–47. [496]

Einarsen, A.S. (1945). Quadrat inventory of pheasant trends in Oregon. *J. Wldl. Manag.* **9,** 121–31. [21]

Eisenberg, R.M. (1972). A critical look at the use of Lincoln index-type model for estimating population densities. *Tex. J. Sci.* **23,** 511–17. [152, 565]

Eleveback, L. and Varma, A. (1965). A selected bibliography on simulation and empirical sampling. *N.Y. Statistician* **16,** No. VI. [572]

El-Khorazaty, M.N. (1975). Methodological strategies for the analysis of categorical data from multiple-record systems. Univ. of North Carolina Institute of Statistics, Mimeo Series No. 1019, Chapel Hill, N.C. [492]

El-Khorazaty, M.N., Imrey, P.B., Koch, G.G. and Lewis, A.L. (1976). On log-linear models for multiple-record systems. *Proceedings of the Social Statistics Section, American Statistical Assoc.* [491]

El-Khorazaty, M.N., Imrey, P.B., Koch, G.G. and Wells, H.B. (1977). A review of methodological strategies for estimating the total number of events with data from multiple-record systems. *Internat. Statist. Rev.* **45,** 129–57. [492]

El-Khorazaty, M.N. and Sen, P.K. (1976). The capture–mark–recapture strategy as a method for estimating the number of events in a human population with data from dependent sources. Paper presented at the 9th International Biometric Conf., August 22–27, 1976, Boston, Mass. [492]

Elling, C.H. and Macy, P.T. (1955). Pink salmon tagging experiments in Icy Strait and Upper Chatham Strait, 1950. *U.S. Dept. Int. Fish. and Wildlife Ser. Fish. Bull.* **100,** 331–71. [290, 294]

Ellis, J.A., Westemeier, R.L., Thomas, K.P. and Norton, H.W. (1969). Spatial relationships among quail coveys. *J. Wildl. Manag.* **33,** 249–54. [41]

Emlen, J.T. (1971). Population densities of birds derived from transect counts. *The Auk* **88,** 323–42. [463]

Emlen, J.M. (1973). *Ecology: an Evolutionary Approach.* Addison-Wesley Publishing Co., Reading, Massachusetts. [549]

Emlen, J.T. Jr, Hine, R.L., Fuller, W.A. and Alfonso, P. (1957). Dropping boards for population studies of small mammals. *J. Wildl. Manag.* **21,** 300–14. [54]

Engelhardt, W.H. (1977). Retention of fluorescent pigment marks by two strains of large-mouth bass. *Trans. Amer. Fish. Soc.* **106,** 64–6. [488]

Epstein, B. (1954). Truncated life tests in the exponential case. *Math. Ann. Statist.* **25,** 555–64. [287]

Epstein, B. (1960a). Tests for the validity of the assumption that the underlying distribution of life is exponential: Part I. *Technometrics* **2**, 83–101. [46, 287]

Epstein, B. (1960b). Tests for the validity of the assumption that the underlying distribution of life is exponential: Part II. *Technometrics* **2**, 167–83. [46, 237]

Epstein, B. (1960c). Statistical life test acceptance procedures. *Technometrics* **2**, 435–6. [287]

Epstein, B. (1960d). Estimation from life test data. *Technometrics* **2**, 447–54. [287]

Epstein, B. and Sobel, M. (1953). Life testing. *J. Amer. Statist. Assoc.* **48**, 486–502. [287]

Epstein, B. and Sobel, M. (1954). Some theorems relevant to life testing from an exponential distribution. *Ann. Math. Statist.* **25**, 373–81. [287]

Ericson, D. (1977). Estimating population parameters of *Pterostichus cupreus* and *P. melanarius* (Carabidae) in arable fields by means of capture–recapture. *Oikos* **29**, 407–17. [231, 505]

Errington, J.C. (1973). The effect of regular and random distributions on the analysis of pattern. *J. Ecol.* **61**, 99–105. [482]

Evans, C.D., Troyer, W.A. and Lensink, C.J. (1966). Aerial census of moose by quadrat sampling units. *J. Wildl. Manag.* **30**, 767–76. [28, 456]

Evans, F.C. (1951). Notes on a population of the striped ground squirrel (*Citellus tridecemlineatus*) in an abandoned field in southeastern Michigan. *J. Mammal.* **32**, 437–49. [84]

Evans, W.E. (1974). Radio-telemetric studies of two species of small odontocete cetacenas. *In* W.E. Schevill (ed.), *The Whale Problem: a Status Report,* 385–94. Harvard Univ. Press, Cambridge, Mass. [459]

Facey, D.E., McCleave, J.D. and Doyon, G.E. (1977). Responses of atlantic salmon parr to output of pulsed ultrasonic transmitters. *Trans. Amer. Fish. Soc.* **106**, 489–96. [448]

Farner, D.S. (1945). Age groups and longevity in the American robin. *Wilson Bull.* **57**, 56–74. [247]

Farner, D.S. (1949). Age groups and longevity in the American robin: comments, further discussion and certain revisions. *Wilson Bull.* **61**, 68–81. [247]

Farner, D.S. (1955). Birdbanding in the study of population dynamics. *In* A. Wolfson (ed.), *Recent Studies in Avian Biology,* 397–449. Univ. of Illinois Press, Urbana. [247, 251, 397]

Faust, B.F., Smith, M.H. and Wray, W.B. (1971). Distances moved by small mammals as an apparent function of grid size. *Acta theriologica* **16**, 161–77. [446]

Feller, W. (1968). *An Introduction to Probability Theory and its Applications,* vol. 1, 3rd edn. Wiley: New York. [3, 15, 260]

Fieller, E.C. (1940). The biological standardisation of insulin. *J. Roy. Statist. Soc. Suppl.* **7**, 1–65. [11]

Fienberg, S.E. (1972). The multiple recapture census for closed populations and incomplete 2^k contingency tables. *Biometrika* **59**, 591–603. [490, 492]

Fischer, C.A. and Keith, L.B. (1974). Population responses of central Alberta ruffed grouse to hunting. *J. Wildl. Manag.* **38**, 585–600. [452]

Fischler, K.J. (1965). The use of catch–effort, catch–sampling, and tagging data to estimate a population of blue crabs. *Trans. Amer. Fish. Soc.* **94**, 287–310. [261–3, 300–2, 335–6, 560]

Fisher, H.I., Hiatt, R.W. and Bergerson, W. (1947). The validity of the roadside census as applied to pheasants. *J. Wildl. Manag.* **11**, 205–31. [53]

Fisher, R.A. (1970). *Statistical Methods for Research Workers,* 14th edn. Oliver and Boyd: Edinburgh. [55]

REFERENCES

Fisher, R.A. and Ford, E.B. (1947). The spread of a gene in natural conditions in a colony of the moth *Panaxia dominula* L. *Heredity, Lond.* **1**, 143–74. [98, 507]

Flyger, V.F. (1959). A comparison of methods for estimating squirrel populations. *J. Wildl. Manag.* **23**, 220–3, [84, 172, 560]

Flyger, V.F. (1960). Movements and home range of the gray squirrel *Sciurus Carolinensis*, in two Maryland woodlots. *Ecology* **41**, 365–9. [85]

Foote, L.E., Peters, H.S. and Finkner, A.L. (1958). Design tests for mourning dove call-count sampling in seven southeastern states. *J. Wildl. Manag.* **22**, 402–8. [28, 53, 54]

Ford, J.S. (1974). Echo integrator absolute constants put into real numbers. *Fish. Res. Board Can. Tech. Rep. No. 467,* 33 pp. [453]

Fordham, R.A. (1970). Mortality and population change of dominican gulls in Wellington, New Zealand; with a statistical appendix by R.M. Cormack. *J. Animal Ecol.* **39**, 13–27. [252]

Francis, R.C. (1974). Relationship of fishing mortality to natural mortality at the level of maximum sustainable yield under the logistic stock production model. *J. Fish. Res. Bd. Canada* **31**, 1539–42. [542]

Francis, W.J. (1973). Accuracy of census methods of territorial red-winged blackbirds. *J. Wildl. Manag.* **37**, 98–102. [58, 453]

Frantz, S.C. (1972). Fluorescent pigments for studying movements and home ranges of small mammals. *J. Mammal.* **53**, 218–23. [448]

Freeman, P.R. (1972). Sequential estimation of the size of the population. *Biometrika* **59**, 9–17. [493]

Freeman, P.R. (1973a). Sequential recapture. *Biometrika* **60**, 141–53. [486]

Freeman, P.R. (1973b). A numerical comparison between sequential tagging and sequential recapture. *Biometrika* **60**, 499–508. [486]

Frith, H.J. (1964). Mobility of the red kangaroo, *Megaleia rufa. CSIRO Wildl. Res.* **9**, 1–19. [455]

Gangwere, S.K. Chavin, W. and Evans, F.C. (1964). Methods of marking insects, with special reference to Orthoptera (Sens. Lat.). *Ann. Entomol. Soc. America* **57**, 662–0. [93]

Gantmacher, F.R. (1959). *The Theory of Matrices,* vol. 2, Chelsea Publishing Co.: New York. [550]

Gart, J.J. (1970). Some simple graphically oriented statistical methods for discrete data. *In* G.P. Patil (ed.), *Random Counts in Scientific Work,* Vol. 1, 171–91. Penn. State Univ. Press, University Park, Penn. [479]

Gaskell, T.J. and George, B.J. (1972). A Bayesian modification of the Lincoln index. *J. Appl. Ecol.* **41**, 377–84. [485]

Gates, C.E. (1969). Simulation study of estimators for the line transect sampling method *Biometrics* **25**, 317–28. [29, 32, 36–8, 562]

Gates, C.E. (1979). Line transect and related issues. *In* R.M. Cormack, G.P. Patil and D.S. Robson (eds.), *Sampling Biological Populations*, Statistical Ecology Series Vol. 5, 71–154. International Co-operative Publishing House, Burtonsville, Maryland. [467]

Gates, C.E., Clark, T.L. and Gamble, K.E. (1975). Optimizing mourning dove breeding population surveys. *J. Wildl. Manag.* **39**, 237–42. [453]

Gates, C.E., Gamble, K.E. and Clark, T.L. (1974). On fitting frequency distributions of calling mourning doves. *J. Wildl. Manag.* **38**, 370–1. [453]

Gates, C.E., Marshall, W.H. and Olson, D.P. (1968). Line transect method of estimating grouse population densities. *Biometrics* **24**, 135–45. [29–32, 40]

Gates, C.E. and Smith, W.B. (1972). Estimation of density of mourning doves from aural information. *Biometrics* **28**, 345–9. [53–4, 453]

Gates, J.M. (1966). Crowing counts as indices to cock pheasant populations in Wisconsin. *J. Wildl. Manag.* **30**, 735–44. [54]

Geis, A.D. (1955). Trap response of the cottontail rabbit and its effect on censusing. *J. Wildl. Manag.* **19**, 466–72. [84, 162, 165, 167–8]

Geis, A.D. (1972a). Use of banding data in migratory game bird research and management. *U.S. Fish Wild. Serv., Spec. Sci. Rep. Wildl. No. 154*, 47 pp. [528]

Geis, A.D. (1972b). Role of banding data in migrating bird population studies. *In* Population ecology of migratory birds. *U.S. Fish. Wildl. Serv., Wildl. Res. Rep.* **2**, 213–28. [528]

Geis, A.D. and Atwood, E.L. (1961). Proportion of recovered waterfowl bands reported. *J. Wildl. Manag.* **25**, 154–9. [250]

Geis, A.D., Smith, R.I. and Rogers, J.P. (1971). Black duck distribution, harvest characteristics, and survival. *U.S. Fish Wildl. Serv. Spec. Sci. Rep. Wildl. No. 139*, 241 pp. [528]

Geis, A.D. and Tabler, R.D. (1963). Measuring hunting and other mortality. *In* H.S. Mosby (ed.), *Wildlife Investigational Techniques*, 2nd edn, 284–98. The Wildlife Society: Washington. [361, 383, 528]

Gentry, J.B., Golley, F.B. and Smith, M.H. (1968). An evaluation of the proposed International Biological Program census method for estimating small mammal populations. *Acta theriologica* **13**, 313–27. [569]

Gentry, J.B. and Odum, E.P. (1957). The effect of weather on the winter activity of old-field rodents. *J. Mammal.* **38**, 72–7. [318]

Gentry, J.B., Smith, M.H. and Beyers, R.J. (1971). Use of radioactive isotopes to study movement patterns in small mammal populations. *Ann. Zool. Fennici* **8**, 17–21. [447]

Gentry, J.B., Smith, M.H. and Chelton, J.G. (1971). An evaluation of the octagon census method for estimating small mammal populations. *Acta theriologica* **16**, 149–59. [504]

George, D.G. (1974). Dispersion patterns in the zooplankton populations of a eutrophic reservoir. *J. Animal Ecol.* **43**, 537–51. [480]

Gérard, G. and Berthet, P. (1971). Sampling strategy in censusing patchy populations. *In* G.P. Patil, E.C. Pielou and W.E. Waters (eds.), *Spatial Patterns and Statistical Distributions*, Statistical Ecology Series Vol. 1, 59–67. Penn. State Univ. Press, University Park, Penn. [479]

Gerrard, P.J. and Chiang, H.C. (1970). Density estimation of corn rootworm egg populations based on frequency of occurrence. *Ecology* **51**, 237–45. [55–6]

Getz, L.L. (1961). Responses of small mammals to live-traps and weather conditions. *Amer. Midl. Nat.* **66**, 160–70. [84]

Gilbert, J.R., Quinn, T.J., II and Eberhardt, L.L. (1976). *An Annotated Bibliography of Census Procedures for Marine Mammals.* Prepared under the auspices of the U.S. Marine Mammal Commission. [1, 459]

Gilbert, R.O. (1973). Approximations of the bias in the Jolly–Seber capture–recapture model. *Biometrics* **29**, 501–26. [506]

Giles, R.H. Jr. (ed.) (1969). *Wildlife Management Techniques,* 3rd edn. The Wildlife Society: Washington. [2]

Gilmer, D.S., Ball, I.J., Cowardin, L.M. and Riechmann, J.H. (1974). Effects of radio packages on wild ducks. *J. Wildl. Manag.* **38**, 243–52. [448]

Gilmer, D.S., Klett, A.T. and Work, E.A. (1975). Evaluation of LANDSAT-1 system for monitoring waterfowl habitat in North Dakota. *Proceedings of the Workshop of Remote Sensing of Wildlife,* 77–98. Gouvernement du Québec, Service de la Recherche Biologique. [459]

Gilmer, D.S., Kuechle, V.B. and Ball, I.J., Jr. (1971). A device for monitoring radio-marked animals. *J. Wildl. Manag.* **35**, 829–32. [448]

Gilmer, D.S., Miller, S.E. and Cowardin, L.M. (1973). Analysis of radio-tracking data using digitized habitat maps. *J. Wildl. Manag.* **37**, 404–9. [448]

Gleeson, A.C. and Douglas, J.B. (1975). Quadrat sampling and the estimation of Neyman Type A and Thomas distributional parameters. *Austral. J. Statist.* **17**, 103–13. [480]

Gliwicz, J. (1970). Relation between trappability and age of individuals in a population of the bank vole. *Acta theriologica* **15**, 15–23. [487]

Goddard, J. (1967). The validity of censusing black rhinoceros populations from the air. *E. Afr. Wildl. J.* **5**, 18–23. [456]

Goddard, J. (1969). Aerial census of black rhinoceros using stratified random sampling. *E. Afr. Wildl. J.* **7**, 105–14. [456]

Golley, F.B., Gentry, J.B., Caldwell, L.D. and Davenport, L.B. Jr. (1965). Number and variety of small mammals on the A.E.C. Savannah River Plant. *J. Mammal.* **46**, 1–18. [327]

Goodall, D.W. (1974). A new method for the analysis of spatial pattern by random pairings of quadrats. *Vegatatio* **29**, 135–46. [482]

Goodman, L.A. (1953). Sequential sampling tagging for population size problems. *Ann. Math. Statist.* **24**, 56–69. [120, 188–9, 492]

Goodman, L.A. (1960). On the exact variance of products. *J. Amer. Statist. Assoc.* **55**, 708–13. [9]

Goodman, L.A. (1969). The analysis of population growth when the birth and death rates depend on several factors. *Biometrics* **25**, 659–81. [393]

Graham, A. and Bell, R. (1969). Factors influencing the countability of animals. *E. Afr. Agric. For. J.* **34**, (special issue), 38–43. [456]

Graves, H.B., Bellis, E.D. and Knuth, W.M. (1972). Censusing white-tailed deer by air-borne thermal infrared imagery. *J. Wildl. Manag.* **36**, 875–84. [456]

Gray, H.L. and Schucany, W.R. (1972). *The Generalized Jackknife Statistic.* Marcel Dekker: New York. [18, 508]

Gray, H.L., Schucany, W.R. and Watkins, T.A. (1975). On the generalised jackknife and its relation to statistical differentials. *Biometrika* **62**, 637–42. [18]

Green, R.G. and Evans, C.A. (1940). Studies on a population cycle of snowshoe hares on the Lake Alexander area. I: Gross annual census, 1932–1939. *J. Wildl. Manag.* **4**, 220–38. [106]

Greig-Smith, P. (1952). The use of random and contiguous quadrats in the study of the structure of plant communities. *Ann. Bot. N.S.* **16**, 293–316. [481]

Greig-Smith, P. (1964). *Quantitative Plant Ecology*, 2nd edn. Butterworths: London. [22–5, 41, 56, 572]

Grieb, J.R. (1970). The shortgrass prairie Canada goose population. *Wildl. Monogr. No. 22*, 49 pp. [452]

Grier, J.W. (1977). Quadrat sampling of a nesting population of bald eagles. *J. Wildl. Manag.* **41**, 438–43. [451, 456]

Grimm, H. (1970). Graphical methods for the determination of type and parameters of some discrete distributions. *In* G.P. Patil (ed.), *Random Counts in Scientific Work,* Vol. 1, 193–206. Penn. State Univ. Press, University Park, Penn. [479]

Grodzinski, W., Pucek, Z. and Ryszkowski, L. (1966). Estimation of rodent numbers by means of prebaiting and intensive removal. *Acta theriologica* **11**, 297–314. [315, 570]

Gromadzki, M. and Trojan, P. (1971). Estimation of population density in *Microtus arvalis* (Pall.) by three different methods. *Ann. Zool. Fennici* **8**, 54–9. [561]

Gross, J.E., Stoddart, L.C. and Wagner, F.H. (1974), Deomgraphic analysis of a northern Utah jackrabbit population. *Wildl. Monogr. No. 40,* 68 pp. [452, 468]

Grunwald, H. (1975). Changes in trappability of common vole. *Acta theriologica* **20,** 333–41. [487]

Gulland, J.A. (1955a). Estimation of growth and mortality in commercial fish populations. *U.K. Ministry Agric. and Fish, Fish. Invest., Ser. 2,* **18,** 1–46. [306]

Gulland, J.A. (1955b). On the estimation of population parameters from marked members. *Biometrika* **42,** 269–70. [272]

Gulland, J.A. (1963). On the analysis of double-tagging experiments. *In* North Atlantic Fish Marking Symposium, *I.C.N.A.F. Special Publication No. 4,* 228–9, [94, 281]

Gulland, J.A. (1964). Catch per unit effort as a measure of abundance. *Rapp. P.-v. Réun. Cons. perm. int. Explor. Mer* **155,** 8–14, [569]

Gullion, G.W. (1966). The use of drumming behaviour in ruffed grouse population studies. *J. Wildl. Manag.* **30,** 717–29. [54]

Gunderson, D.R. (1968). Foodplain use related to stream morphology and fish populations. *J. Wildl. Manag.* **32,** 507–14. [82]

Gurland, J. (1957). Some interrelations among compound and generalised distributions. *Biometrika* **44,** 265–8. [483]

Gurland, J. and Hinz, P. (1971). Estimating parameters, testing fit, and analyzing untransformed data pertaining to the negative binomial and other distributions. *In* G.P. Patil, E.C. Pielou and W.E. Waters (eds.), *Spatial Patterns and Statistical Distributions,* Statistical Ecology Series Vol. 1. 143–78. Penn. State Univ. Press, University Park, Penn. [480]

Guthrie, D.R., Osborne, J.C. and Mosby, H.S. (1967). Physiological change associated with shock in confined gray squirrels. *J. Wildl. Manag.* **31,** 102–8. [83]

Haberman, S.J. (1974). *The Analysis of Frequency Data.* Univ. of Chicago Press, Chicago. [492]

Hadow, H.H. (1972). Freeze-branding: a permanent marking technique for pigmented mammals. *J. Wildl. Manag.* **36,** 649–51. [487]

Hagen, A. Östbye, E. and Skar, H.J. (1973). A method for calculating the size of the trapping area in a catch–mark–release censusing of small rodents. *Norwegian J. of Zoology* **21,** 59–61. [449]

Hald, A.H. (1952). *Statistical Tables and Formulas.* Wiley: New York. [263]

Haldane, J.B.S. (1945). On a method of estimating frequencies. *Biometrika* **33,** 222–5. [121]

Haldane, J.B.S. (1953). Some animal life tables. *J. Inst. Actuaries* **79,** 83–9. [247]

Haldane, J.B.S. (1955). The calculation of mortality rates from ringing data. *Proc. XIth Int. Orn. Congr. Basel,* 454–8. [246, 248–9]

Hamley, J.M. (1975). Review of gillnet selectivity. *J. Fish. Res. Bd. Canada* **32,** 1943–69. [82, 305]

Hamley, J.M. and Regier, H.A. (1973). Direct estimates of gillnet selectivity to walleye (*Stizostedion vitreum vitreum*). *J. Fish. Res. Bd. Canada* **30,** 817–30. [82]

Hancock, D.A. (1963). Marking experiments with the commercial whelk (*Buccinum undatum*). *In* North Atlantic Fish Marking Symposium, *I.C.N.A.F., Special Publication No. 4,* 167–87. [82]

Hanson, W.C. and Eberhardt, L.L. (1971). A Columbia River Canada grose population. 1950–1970. *Wildl. Monogr. No. 28,* 61 pp. [547]

Hanson, W.R. (1963). Calculation of productivity, survival and abundance of selected vertebrates from sex and age ratios. *Wildl. Monogr.* **9,** 1–60. [353, 375, 381]

Hansson, L. (1967). Index line catches as a basis of population studies on small mammals. *Oikos* **18**, 261–76. [564]

Hansson, L. (1968). Population densities of small mammals in open field habitats in South Sweden in 1964–1967. *Oikos* **19**, 53–60. [564]

Hansson, L. (1969). Home range, population structure and density estimates at removal catches with edge effect. *Acta theriologica* **14**, 153–60. [449, 504]

Hansson, L. (1972). Evaluation of the small quadrat method of censusing small mammals. *Ann. Zool. Fennici* **9**, 184–90. [453, 564]

Hansson, L. (1974). Influence areas of trap stations as a function of number of small mammals exposed per trap. *Acta Theriologica* **19**, 19–25. [504]

Hansson, L. (1975). Comparison between small mammal sampling with small and large removal quadrats. *Oikos* **26**, 398–404. [453, 564]

Harke, D.T. and Stickley, A.R. Jr. (1968). Sensitive resettable odometer aids road-side census of red-winged blackbirds. *J. Wildl. Manag.* **32**, 635–6. [54]

Hartley, H.O. (1958). Maximum likelihood estimation from incomplete data. *Biometrics* **14**, 174–94. [169, 175]

Hartley, H.O., Homeyer, P.G. and Kozicky, E.L. (1955). The use of log transformations in analysing fall roadside pheasant counts. *J. Wildl. Manag.* **19**, 495–6. [53]

Harvard Computation Laboratory (1955). *Tables of the Cumulative Binomial Probability Distribution.* Harvard Univ. Press. [64, 122]

Harvey, J.M. and Barbour, R.W. (1965). Home range of *Microtus Ochrogaster* as determined by a modified minimum area method. *J. Mammal.* **46**, 398–402. [20]

Havey, K.A. (1960). Recovery, growth and movement of hatchery-reared Lake Atlantic salmon at Long Pond, Maine. *Trans. Amer. Fish. Soc.* **89**, 212–17. [304]

Havey, K.A. (1974). Population dynamics of landlocked salmon, *Solmo salar*, in Love Lake, Maine. *Trans. Amer. Fish. Soc.* **103**, 448–56. [547]

Hawes, M.L. (1977). Home range, territoriality, and ecological separation in sympatric shrews, *Sorex Vagrans* and *Sorex obscurus. J. Mammal.* **58**, 354–67. [447]

Hayne, D.W. (1949a). An examination of the strip census method for estimating animal populations. *J. Wildl. Manag.* **13**, 145–57. [36, 38, 468]

Hayne, D.W. (1949b). Two methods for estimating populations from trapping records. *J. mammal.* **30**, 399–411. [142, 146–50, 325, 327]

Hayne, D.W. (1949c). Calculation of size of home range. *J. Mammal.* **30**, 1–18. [000]

Hazard, J.W. and Pickford, S.G. (1979). Line intersect sampling of forest residue. *In* G.P. Patil and M.L. Rosenzweig (eds.), *Contemporary Quantitative Ecology and Related Ecometrics*, Statistical Ecology Series vol. 12, 493–503. International Co-operative Publishing House, Burtonsville, Maryland. [460]

Healey, M.C. (1975). Dynamics of exploited whitefish populations and their management with special reference to the Northwest Territories. *J. Fish. Res. Bd. Canada* **32**, 427–48. [426, 546]

Healy, M.J.R. (1962). Some basic statistical techniques in soil zoology. *In* P.W. Murphy (ed.), *Progress in Soil Zoology,* 3–9, Butterworths: London. [20]

Hearon, J.Z. (1976). Properties of the Leslie population matrix. *Bull. Math. Biol.* **38**, 199–203. [549]

Heezen, K.L. and Tester, J.R. (1967). Evaluation of radio-tracking by triangulation with special reference to deer movements. *J. Wildl. Manag.* **31**, 124–41. [94]

Heincke, F. (1913). Investigations on the plaice. General report. 1. The plaice fishery and protective measures. Preliminary brief summary of the most important points of the report. *Cons. int. Explor. Mer. Rapp. et P.-v.* **16**, 1–67. [415]

Hemingway, P. (1971). Field trials of the line transect method of sampling large populations of herbivores. *In* E. Duffy and A.S. Watts (eds.), *The Scientific Management of Animal and Plant Communities for Conservation*, 405–11. Blackwell: Oxford. [462]

Henderson, H.F., Hasler, A.D. and Chipman, G.G. (1966). An ultrasonic transmitter for use in studies of movements of fishes. *Trans. Amer. Fish. Soc.* **95**, 350–6. [94]

Henny, C.J. (1967). Estimating band-reporting rates from banding and crippling loss data. *J. Wildl. Manag.* **31**, 533–8. [250]

Henny, C.J. (1972). An analysis of the population dynamics of selected avian species: with special references to changes during the modern pesticide era. *U.S. Fish Wildl. Serv., Wildl. Res. Rep. No. 1,* 99 pp. [529]

Henny, D.J., Anderson, D.R. and Pospahala, R.S. (1972). Aerial surveys of waterfowl production in North America, 1955–71. *U.S. Fish Wildl. Serv., Spec. Sci. Rep. Wildl. No. 160,* 48 pp. [456]

Henny, C.J. and Burnham, K.P. (1976). A mallard reward band study to estimate band reporting rates. *J. Wildl. Manag.* **40**, 1–14. [250, 526–8]

Hessler, E., Tester, J.R., Siniff, D.B. and Nelson M.M. (1970). A biotelemetry study of survival of pen-reared pheasants released in selected habitats. *J. Wildl. Manag.* **34**, 267–74. [94]

Hewitt, O.H. (1967). A road-count index to breeding populations of red-winged blackbirds. *J. Wildl. Manag.* **31**, 39–47. [53]

Hickey, F. (1960). Death and reproductive rate of sheep in relation to flock culling and selection. *New Zeal. J. Agric. Res.* **3**, 332–44. [401]

Hickey, J.J. (1952). Survival studies of banded birds. *U.S. Fish Wildl. Serv., Spec. Sci. Rep. Wildl. No. 15,* 177 pp. (reprinted in 1972 with minor corrections). [245, 247, 528]

Hickman, G.L. (1972). Aerial determination of golden eagle nesting status. *J. Wildl. Manag.* **36**, 1289–92. [459]

Hilborn, R., Redfield, J.A. and Krebs, C.J. (1976). Reliability of enumeration for mark and recapture census of voles. *Can. J. Zool.* **54**, 1019–24. [509]

Hill, M.O. (1973). The intensity of spatial pattern in plant communities. *J. Ecol.* **61**, 225–35. [482]

Hines, W.G.S. and Hines, R.J.O. (1979). The Eberhardt statistic and the detection of nonrandomness of spatial point distributions. *Biometrika* **66**, 73–9. [476]

Hinkley, D.V. (1977). Jackknife confidence limits using Student t approximations. *Biometrika* **64**, 21–8. [18]

Hjort, J.G. and Ottestad, P. (1933). The optimum catch. *Hvalradets Skrifter, Olso* **7**, 92–127. [296]

Holdgate, M.W. (ed.), (1970). *Antarctic Ecology,* Vol. 1. Academic Press: London and New York. [459]

Holgate, P. (1964a). The efficiency of nearest neighbour estimators. *Biometrics* **20**, 647–9. [42–3]

Holgate, P. (1964b). A modified geometric distribution arising in trapping studies. *Acta theriologica* **9**, 353–6. [180–2]

Holgate, P. (1965a). The distance from a random point to the nearest point of a closely packed lattice. *Biometrika* **52**, 261–3. [484]

Holgate, P. (1965b). Tests of randomness based on distance methods. *Biometrika* **52**, 345–53. [41, 472]

Holgate, P. (1965c). Some new tests of randomness. *J. Ecol.* **53**, 261–6. [474]

Holgate, P. (1966). Contributions to the mathematics of animal trapping. *Biometrics* **22**, 925–36. [182–4]

Holgate, P. (1969). Species frequency distributions. *Biometrika* **56**, 651–60. [572]

Holgate, P. (1972). The use of distance methods for the analysis of spatial distributions of points. *In* P.A.W. Lewis (ed.), *Stochastic Point Processes*, 122–35. Wiley, New York. [473, 479]

Holgate, P. (1973). An estimator for the size of an animal population. *Biometrika* **60**, 135–40 135–40. [549]

Holst, L. (1971). A note on finding the size of a finite population. *Biometrika* **58**, 228–9. [493]

Holt, S.J. (1977). Estimation of sperm whale population sizes from changes in the mean size of whales in catches. *Rep. int. Whal. Commn* **27**, 363–7. [549]

Hope, A.C.A. (1968). A simplified Monte Carlo significance test procedure. *J. Roy. Statist. Soc., Series B* **30**, 582–98. [18]

Hopkins, B. (1954). A new method for determining the type of distribution of plant individuals. *Annals of Botany* **18**, 213–27. [43, 472]

Hopper, R.M., Geis, A.D., Grieb, J.R. and Nelson, L. Jr, (1975). Experimental duck hunting seasons, San Luis Valley, Colorado, 1963–1970. *Wildl. Monogr. No. 46*, 68 pp. [456, 528]

Hornocker, M.G. (1970). An analysis of mountain lion predation upon mule deer and elk in the Idaho primitive area. *Wildl. Monogr No. 21*, 39 pp. [456]

Horst, T.J. (1977). Use of the Leslie matrix for assessing environmental impact with an example for a fish population. *Trans. Amer. Fish. Soc.* **106**, 253–7. [550]

Hoskinson, R.L. (1976). The effect of different pilots on aerial telemetry error. *J. Wildl. Manag.* **40**, 137–9. [448]

Houser, A. and Dunn, J.E. (1967). Estimating the size of the threadfin shad population in Bull Shoals Reservoir from midwater trawl catches. *Trans. Amer. Fish. Soc.* **96**, 176–84. [25]

Howard, J.A. (1970). *Aerial Photo-ecology.* Faber and Faber: London. [459]

Howe, A.B., Coates, P.G. and Pierce, D.E. (1976). Winter flounder estuarine year-class abundance, mortality and recruitment. *Trans. Amer. Fish. Soc.* **105**, 647–57. [505]

Howell, J.C. (1951). The roadside census as a method of measuring bird populations. *Auk* **86**, 334–57. [53]

Huber, J.J. (1962). Trap response of confined cottontail populations. *J. Wildl. Manag.* **26**, 177–85. [84, 162, 560, 565]

Hubert, G.F., Jr, Storm, G.L., Phillips, R.L. and Andrews, R.D. (1976). Ear tag loss in red foxes. *J. Wildl. Manag.* **40**, 164–7. [489]

Hughes, R.D. (1965). On the composition of a small sample of individuals from a population of the banded hare wallaby, *Lagostrophus fasciatus* (Peron and Lesneur). *Austral. J. Zool.* **13**, 75–95. [403]

Hughes, R.D. and Gilbert, N. (1968). A model of an aphid population – a general statement. *J. Animal Ecol.* **37**, 553–63. [399]

Hughes, R.N. (1970). Population dynamics of the bivalve *Scrobicularia plana* (Da Costa) on an intertidal mud-flat in North Wales. *J. Animal Ecol.* **39**, 333–55. [24]

Hughes, S.E. (1976). System for sampling large trawl catches of research vessels. *J. Fish. Res. Bd. Canada* **33**, 833–9. [548]

Hunter, A.J. and Griffiths, H.J. (1978). Bayesian approach to estimation of insect population size. *Technometrics* **20**, 231–4. [451]

Hunter, W.R. and Grant, D.C. (1966). Estimates of population density and dispersal in the Natacid Gastropod, *Polinices duplicatus*, with a discussion of computational methods. *Biolog. Bull.* **131**, 292–307. [152]

Hutton, T.A., Hatfield, R.E. and Watt, C.C. (1976). A method for orienting a mobile radio-tracking unit. *J. Wildl. Manag.* **40**, 192–3. [448]

Imber, M.J. and Williams, G.R. (1968). Mortality rates of a Canada population in New Zealand. *J. Wildl. Manag.* **32**, 256–67. [253]

International Commission for the Northwest Atlantic Fisheries (1963). The selectivity of fishing gear. *I.C.N.A.F., Special Publication No. 5*, 1–225. [82]

International Pacific Halibut Commission (1960). *Utilisation of Pacific Halibut Stocks; Yield per Recruitment.* Seattle University. [351]

Iwao, S. (1963). On a method for estimating the rate of population interchange between two areas. *Res. Popul. Ecol.* **5**, 44–50. [555]

Iwao, S. (1968). A new regression method for analyzing the aggregation pattern of animal populations. *Res. Popul. Ecol.* **10**, 1–20. [479–80]

Iwao, S. (1970). Problems of spatial distribution in animal population ecology. *In* G.P. Patil (ed.), *Random Counts in Scientific Work,* Vol. 2, 117–49, The Pennsylvania State Univ. Press, University Park, Penn. [478–9]

Iwao, S. (1975). A new method of sequential sampling to classify populations relative to a critical denisty. *Res. Popul. Ecol.* **16**, 281–8. [480]

Iwao, S. (1976). A note on the related concepts 'mean crowding' and 'mean concentration'. *Res. Popul. Ecol.* **17**, 240–2. [479]

Iwao, S. and Kuno, E. (1971). An approach to the analysis of aggregation pattern in biological populations. *In* G.P. Patil, E.C. Pielou and W.E. Waters (eds.), *Spatial Patterns and Statistical Distributions,* Statistical Ecology Series Vol. 1, 461–512. Penn. State Univ. Press, University Park, Penn. [480]

Jackson, C.H.N. (1933). On the true density of tsetse flies. *J. Animal Ecol.* **2**, 204–9. [1]

Jackson, C.H.N. (1937). Some new methods in the study of *Glossina morsitans. Proc. Zool. Soc. London 1936,* 811–96. [1]

Jackson, C.H.N. (1939). The analysis of an animal population. *J. Animal Ecol.* **8**, 238–46. [1, 415, 547]

Jackson, C.H.N. (1940). The analysis of a tsetse-fly population: I. *Ann. Eugen. London,* **10**, 332–69. [1]

Jackson, C.H.N. (1944). The analysis of a tsetse-fly population: II. *Ann. Eugen. London* **12**, 176–205. [1]

Jackson, C.H.N. (1948). The analysis of a tsetse-fly population: III. *Ann. Eugen. London* **14**, 91–108. [1]

Järvinen, O. and Väisänen, R.A. (1975). Estimating relative densities of breeding birds by the line transect method. *Oikos* **26**, 316–22. [462]

Jenkins, D., Watson, A. and Miller, G.R. (1967). Population fluctuations in the red grouse (*Lagopus lagopus scoticus*). *J. Animal Ecol.* **36**, 97–122. [225]

Jennrich, R.I. and Turner, F.B. (1969). Measurement of non-circular home range. *J. Theoret. Biol.* **22**, 227–37. [20, 446]

Jester, D.B. (1977). Effects of color, mesh size, fishing in seasonal concentrations, and baiting on catch rates of fishes in gill nets. *Trans. Amer. Fish. Soc.* **106**, 43–56. [82]

Johnson, A.M. (1968). Annual mortality of territorial male fur seals and its management significance. *J. Wildl. Manag.* **32**, 94–9. [418]

Johnson, D.H. (1974). Estimating survival rates from banding of adult and juvenile birds. *J. Wildl. Manag.* **38**, 290–7. [527]

Johnson, D.H. (1977). Some Bayesian statistical techniques useful in estimating frequency and density *U.S. Fish. Wildl. Serv., Spec. Sci. Rep. Wildl. No. 203,* 10 pp. [485]

Johnson, M.G. (1965). Estimates of fish populations in warmwater streams by the removal method. *Trans. Amer. Fish. Soc.* **94**, 350–7. [315]

Jolly, G.M. (1965). Explicit estimates from capture–recapture data with both death and immigration–stochastic model. *Biometrika* **52**, 225–47. [196–205, 225, 530]

Jolly, G.M. (1969a). Sampling methods for aerial censuses of wildlife populations. *E. Afr. Agric. For. J.* **34**, (special issue), 46–9. [454]

Jolly, G.M. (1969b). The treatment of errors in aerial counts of wildlife populations. *E. Afr. For. J.* **34** (special issue), 50–5. [454, 456]

Jolly, G.M. (1979a), Sampling of large objects. *In* R.M. Cormack, G.P. Patil and D.S. Robson (eds.), *Sampling Biological Populations*, Statistical Ecology Series, Vol. 5, 193–201. International Co-operative Publishing House, Burtonsville, Maryland. [458, 460]

Jolly, G.M. (1979b). A unified approach to mark–recapture stochastic models, exemplified by a constant survival rate model. *In* R.M. Cormack, G.P. Patil and D.S. Robson (eds.), *Sampling Biological Populations,* Statistical Ecology Series, Vol. 5, 277–82. International Co-operative Publishing House, Burtonsville, Maryland. [505]

Jolly, G.M. and Watson, R.M. (1979). Aerial sample survey methods in the quantitative assessment of ecological resources. *In* R.M. Cormack, G.P. Patil and D.S. Robson (eds.), *Sampling Biological Populations*, Statistical Ecology Series, Vol. 5, 203–16. International Co-operative Publishing House, Burtonsville, Maryland. [454–6, 458]

Jones, J.W. (1959). *The Salmon.* Collins: London. [443]

Jones, R. (1956). The analysis of trawl haul statistics with particular reference to the estimation of survival rates. *Cons. Int. Explor. Mer, Rapp. et P.-v.* **140(*1*)**, 30–9. [426]

Jorgensen, C.D., Smith, H.D. and Scott, D.T. (1975). Small mammal estimates using recapture methods, with variables partitioned. *Acta theriologica* **20**, 303–18. [449]

Joule, J. and Cameron, G.N. (1974). Field estimation of demographic parameters: influence of *Sigmodon hispidus* population structure. *J. Mammal.* **55**, 309–18. [487]

Jumars, P.A., Thistle, D. and Jones, M.L. (1977). Detecting two-dimensional spatial structure in biological data. *Oecologia* **28**, 109–23. [485]

Junge, C.O. (1963). A quantitative evaluation of the bias in population estimates based on selective samples. *In* North Atlantic Fish Marking Symposium, *I.C.N.A.F., Special Publication No. 4,* 26–8. [86]

Junge, C.O. and Libosvárský, J. (1965). Effects of size selectivity on population estimates based on successive removals with electrical fishing gear. *Zool. Listy* **14**, 171–8. [315]

Justice, K.E. (1961). A new method for measuring home ranges of small mammals. *J. Mammal.* **42**, 462–70. [20, 54]

Kadlec, J.A. and Drury, W.H. (1968). Aerial estimation of the size of gull breeding colonies. *J. Wildl. Manag.* **32**, 287–93. [458]

Kale, B.K. (1961). On the solution of the likelihood equation by iteration processes. *Biometrika* **48**, 452–6. [17]

Kale, B.K. (1962). On the solution of the likelihood equations by iteration processes: the multiparametric case. *Biometrika* **49**, 479–86. [17]

Kathirgamatamby, N. (1953). Note on the Poisson index of dispersion. *Biometrika* **40**, 225–8. [14, 45]

Katti, S.K. and Rao, V. (1970). The log-zero-Poisson distribution. *Biometrics* **26**, 801–13. [480]

Kaufman, D.W., Smith, G.C., Jones, R.M., Gentry, J.B. and Smith, M.H. (1971). Use of assessment lines to estimate density of small mammals. *Acta theriologica* **16**, 127–47. [504]

Kaye, S.V. (1961). Movements of harvest mice tagged with gold-198. *J. Mammal.* **42**, 323–37. [20]

Keeping, E.S. (1962). *Introduction to Statistical Inference.* D. Van Nostrand: Princeton, New Jersey. [38, 82, 87, 98, 268]

Keith, L.B., Meslow, E.C. and Rongstad, O.J. (1968). Techniques for snowshoe hare 32, 795–801. [165, 170]

Keith, L.B., Meslow, E.C. and Rongstad, O.J. (1968). Techniques for snowshoe hare population studies. *J. Wildl. Manag.* **32**, 801–12. [83]

Kelker, G.H. (1940). Estimating deer populations by a differential hunting loss in the sexes. *Proc. Utah Acad. Sci., Arts and Letters.* **17**, 6–69. [353]

Kelker, G.H. (1944). Sex-ratio equations and formulas for determining wildlife populations. *Proc. Utah Acad. Sci., Arts and Letters,* **19–20**, 189–98. [353]

Kelker, G.H. and Hanson, W.R. (1964). Simplifying the calculation of differential survival of age classes. *J. Wildl. Manag.* **28**, 411. [353]

Kelley, T.L. (1948). *The Kelley Statistical Tables.* (Revised 1948.) Harvard Univ. Press: Cambridge, Massachusetts. [365]

Kelly, G.F. and Barker, A.M. (1963). Estimation of population size and mortality rates from tagged redfish, *Sebastes marinus* L., at Eastport, Maine. *In* North Atlantic Fish Marking Symposium, *I.C.N.A.F., Special Publication No. 4,* 204–9. [172]

Kelso, J.R.M., Pickett, E.E. and Dowd, R.G. (1974). A digital echo-sounding system used in determining abundance of freshwater pelagic fish in relation to depth. *J. Fish. Res. Bd. Canada* **31**, 1101–4. [453]

Kemp, C.D. (1971). Properties of some discrete ecological distributions. *In* G.P. Patil, E.C. Pielou and W.E. Waters (eds.), *Spatial Patterns and Statistical Distributions*, Statistical Ecology Series Vol. 1, 1–22. Penn. State Univ. Press, University Park, Penn. [480]

Kemp, G.A. and Keith, L.B. (1970). Dynamics and regulation of red squirrel (*Tamiasciurus hudsonicus*) populations. *Ecology* **51**, 763–79. [94, 560]

Kendall, M.G. and Moran, P.A.P. (1963). *Geometrical Probability.* Griffin's Statistical Monographs and Courses No. 10. Griffin: London. [41, 44, 50, 474]

Kendall, M.G. and Stuart, A. (1977). *The Advanced Theory of Statistics,* vol. 1, 4th edn. Griffin: London. [32,36, 161, 174]

Kendall, M.G. and Stuart, A. (1979). *The Advanced Theory of Statistics,* vol. 2, 4th edn, Griffin: London. [14, 38, 46, 286]

Kendall, M.G. and Stuart, A. (1976). *The Advanced Theory of Statistics,* vol. 3, 3rd edn. Griffin: London. [9]

Kendeigh, S.C. (1944). Measurement of bird populations. *Ecol. Monogr.* **14**, No. 1, 67–106. [21]

Kennedy, W.A. (1954). Tagging returns, age studies and fluctuations in abundance of Lake Winnipeg whitefish, 1931–1951. *J. Fish. Res. Bd. Canada* **11**, 284–309. [426]

Kenyon, W.W., Scheffer, V.B. and Chapman, D.G. (1954). A population study of the Alaska fur seal herd. *U.S. Fish and Wildl. Serv., Spec. Sci. Report: Wildl. No. 12,* 1–77. [433]

Kerbes, R.H. (1975). Lesser snow geese in the eastern Canadian Arctic. *Canadian Wildlife Service Report Series, No. 35.* [456]

Kerbes, R.H. and Moore, H.D. (1975). Use of current satellite imagery to predict the nesting success of Lesser Snow Geese. *Proceedings of the Workshop on Remoting Sensing of Wildlife*, 99–112. Gouvernement du Quebec, Service de la Recherche Biologique. [459]

Kershaw, K.A. (1957). The use of cover and frequency in the detection of pattern in plant communities. *Ecology* **38**, 291–9. [481]

Kershaw, K.A. (1970). An empirical approach to the estimation of pattern intensity from density and cover data. *Ecology* **51**, 729–34. [480]

Kershaw, K.A. (1973). *Quantitative and Dynamic Plant Ecology* (2nd edn). Edward Arnold Ltd.: London [25, 480]

Ketchen, K.S. (1950). Stratified subsampling for determining age distributions. *Trans. Amer. Fish. Soc.* **79**, 205–12. [422]

Ketchen, K.S. (1953). The use of catch–effort and tagging data in estimating a flatfish population. *J. Fish. Res. Bd. Canada* **10**, 459–83. [298, 308]

Keuls, M., Over, H.J. and De Wit, C.T. (1963). The distance method for estimating densities. *Statistica Neerlandica* **17**, 71–91. [41]

Keyfitz, N. (1968). *Introduction to the Mathematics of Population.* University of Chicago. [393, 572]

Kikkawa, J. (1964). Movement, activity and distribution of the small rodents *Clethrionomys glareolus and Apodemus sylvaticus* in woodland. *J. Animal Ecol.* **33**, 259–99. [20, 81, 205, 315]

Kimball, A. (1960). Estimation of mortality intensities in animal experiments. *Biometrics* **16**, 505–21. [410–11]

Kimball, J.F. Jr. and Wolfe, M.L. (1974). Population analysis of a Northern utah elk herd. *J. Wildl. Manag.* **38**, 161–74. [430, 458, 547]

Kimball, J.W. (1948). Pheasant population characteristics and trends in the Dakotas. *Trans. Nth Amer. Wildl. Conf.* **13**, 291–311. [375]

Kimura, D.K. (1976). Estimating the total number of marked fish present in a catch. *Trans. Amer. Fish. Soc.* **105**, 664–8. [486]

Kimura, D.K. (1977). Statistical assessment of the age-length key. *J. Fish. Res. Bd. Canada* **34**, 317–24. [548]

King, C.E. and Paulik, G.J. (1967). Dynamic models and the simulation of ecological systems. *J. Theoret. Biol.* **16**, 251–67. [572]

King, J.G., Robards, F.C. and Lensink, C.J. (1972). Census of the bald eagle breeding population in southeast Alaska. *J. Wildl. Manag.* **36**, 1292–5. [459]

King, L.J. (1969). *Statistical Analysis in Geography.* Prentice-Hall: Englewood Cliffs, New Jersey. [25]

Kiritani, K., Hokyo, N., Sasaba, T. and Nakasuji, F. (1970). Studies on population dynamics of the green rice leafhopper. *Res. Popul. Ecol.* **12**, 137–53. [399]

Kiritani, K. and Nakasuji, F. (1967). Estimation of the stage-specific survival rate in the insect population with overlapping stages. *Res. Popul. Ecol.* **9**, 143–52. [551]

Kline, P.D. (1965). Factors influencing roadside counts of cottontails. *J. Wildl. Manag.* **29**, 665–71. [53]

Knowlton, F.F., Martin, P.E. and Haug, J.C. (1968). A telemetric monitor for determining animal activity. *J. Wildl. Manag.* **32**, 943–8. [94]

Koch, A.L. (1969). The logarithm in biology. II: Distributions simulating the log-normal *J. Theoret. Biol.* **23**, 251–68. [562]

Koch, G.G., El-Khorazaty, M.N. and Lewis, A.L. (1976). The asymptotic covariance structure of log-linear model estimated parameters for the multiple recapture census. *Commun. Statist.–Theor. Meth. A*, **5**(14), 1425–45. [492]

Koch, G.G., Imrey, P.B., Freeman, D.H., Jr. and Tolley, H.D. (1977). The asymptotic covariance structure of estimated parameters from contingency table log-linear models. *Proceedings of the 9th International Biometric Conference*, 317–36. [492]

Koeppl, J.W., Slade, N.A., Harris, K.S. and Hoffmann, R.S. (1977). A three-dimensional home range model. *J. Mammal.* **58**, 213–20. [447]

Koeppl, J.W., Slade, N.A. and Hoffmann, R.S. (1975). A bivariate home range model with possible application to ethological data analysis. *J. Mammal.* **56**, 81–90. [446]

Kolz, A.L., Corner, G.W. and Johnson, R.E. (1973). A multiple-use wildlife transmitter. *U.S. Fish Wildl. Serv., Spec. Sci. Rep. Wildl. No. 163,* 11 pp. [448]

Kolz, A.K., Corner, G.W. and Tietjen, H.P. (1972). A radio-frequency beacon transmitter for small mammals. *J. Wildl. Manag.* **36**, 177–9. [448]

Kovner, J.L. and Patil, S.A. (1974). Properties of estimators of wildlife population density for the line transect method. *Biometrics* **30**, 225–30. [471]

Kozicky, E.L. (1952). Variations in two spring indices of male ring-necked pheasant populations. *J. Wildl. Manag.* **16**, 429–37. [54]

Kozicky, E.L., Bancroft, T.A. and Homeyer, P.G. (1954). An analysis of woodcock singing ground counts, 1948–1952. *J. Wildl. Manag.* **18**, 259–66. [53–4]

Kozicky, E.L., Hendrickson, G.O., Homeyer, P.G. and Speaker, E.B. (1952). The adequacy of the fall roadside pheasant census in Iowa. *Trans. Nth Amer. Wildl. Conf.* **17**, 293–305. [53]

Krebs, C.J. (1966). Demographic changes in fluctuating populations of *Microtus californicus. Ecol. Monogr.* **36**, 239–73. [160]

Kruk-De Bruin, M., Röst, L.C.M. and Draisma, F.G.A.M. (1977). Estimates of the number of foraging ants with the Lincoln-index method in relation to the colony size of *Formica polyctena. J. Animal Ecol.* **46**, 457–70. [106]

Kuno, E. (1976). Multi-stage sampling for population estimation. *Res. Popul. Ecol.* **18**, 39–56. [453]

Kuno, E. (1977). A sequential estimation technique for capture–recapture censuses. *Res. Popul. Ecol.* **18**, 187–94. [486]

Kurtén, B. (1953). On the variation and population dynamics of fossil and recent mammal populations. *Acta Zool. Fennica* **76**, 1–122. [403]

Lack, D. (1943). The age of the blackbird. *Brit. Birds* **36**, 166–75. [247]

Lack, D. (1951). Population ecology in birds. *Proc. Xth Int. Orn. Congr. Uppsala,* 409–48. [247]

Lack, D. (1954). *The Natural Regulation of Animal Numbers.* Clarendon Press: Oxford. [572]

Lancaster, P. (1969). *Theory of Matrices.* Academic Press: New York. [550]

Lance, A.N. and Watson, A. (1977). Further tests of radio-marking on red grouse. *J. Wildl. Manag.* **41**, 579–82. [448]

Lander, R.H. (1962). A method of estimating mortality rates from change in composition. *J. Fish. Res. Bd. Canada* **19**, 159–68. [353, 387]

Langman, V.A. (1973). A radio-biotelemetry system for monitoring body temperature and activity levels in the Zebra Finch. *The Auk* **90**, 375–83. [448]

Laplace, P.S. (1786). Sur les naissances, les mariages et les morts. *Histoire de l'Académie Royale des Sciences, Année 1783*, Paris, p. 693. [104]

Larsen, T. (1972). Air and ship census of polar bears in Svalbard (Spitsbergen). *J. Wildl. Manag.* **36**, 562–70. [455–7]

Lauckhart, J.B. (1950). Determining the big-game population from the kill. *Trans. Nth Amer. Wildl. Conf.* **15**, 644–9. [353]

Laughlin, R. (1965). Capacity for increase: a useful population statistic. *J. Animal Ecòl.* **34**, 77–91. [396]

Lavigne, D.M., Innes, S., Kalpakis, K. and Ronald, K. (1975). An aerial census of western Atlantic harp seals (Pagophilus groenlandicus) using ultraviolet photography. *Int. Comm. Northwest Atl. Fish. Res. Doc.* 75/XII/144. [456]

Lavigne, D.M. and Oritsland, N.A. (1974a). Black polar bears, *Nature* **251**, 218–9. [456]

Lavigne, D.M. and Oritsland, N.A. (1974b). Ultraviolet photography: a new application for remote sensing of mammals. *J. Fish. Res. Bd. Canada* **52**, 939–51. [456]

Laws, R.M., Parker, I.S.C. and Johnstone, R.C.B. (1975). *Elephants and their Habitats.* Clarendon Press, Oxford. [456]

Laycock, W.A. and Batcheler, C.L. (1975). Comparison of distance-measurement techniques for sampling tussock grassland species in New Zealand. *J. Range Manag.* **28**, 235–9. [473]

Leaman, B.M. (1976). An inexpensive tag for short-term visual tracking studies. *J. Fish. Res. Bd. Canada* **33**, 1628–9. [488]

Leary, D.F. and Murphy, G.I. (1975). A successful method for tagging the small, fragile engraulid, *Stolephorus purpureus. Trans. Amer. Fish. Soc.* **104**, 53–5. [488]

Le Cren, E.D. (1965). A note on the history of mark–recapture population estimates. *J. Animal Ecol.* **34**, 453–4. [59]

Le Cren, E.D., Kipling, C. and McCormack, J.C. (1977). A study of the numbers, biomass and year-class strengths of perch (*Perca fluviatilis* L.) in Windermere from 1941 to 1966. *J. Animal Ecol.* **46**, 281–281–307. [501]

Lee, F.S. (1972). Sample size determination for mark–recapture experiments. M.S. Thesis, Cornell University, Ithaka. [134, 491]

Leedy, D.L. (1949). Ohio pheasant nesting surveys based on farmer interviews. *J. Wildl. Manag.* **13**, 274–86. [54]

Lefkovitch, L.P. (1965). The study of population growth in organisms grouped by stages. *Biometrics* **21**, 1–18. [550]

Lefkovitch, L.P. (1967). A theoretical evaluation of population growth after removing individuals from some age groups. *Bull. Ent. Res.* **57**, 437–45. [551]

Lenarz, W.H., Matter, F.J., III, Beckett, J.S. Jones, A.C. and Mason, J.M. Jr (1973). Estimation of rates of tag shedding by Northwest Atlantic bluefin tuna. *U.S. Natl Mar. Fish. Serv. Fish. Bull.* **71**, 1103–5. [524]

Leopold, A. (1933). *Game Management.* Charles Scribner's Sons: New York. [36]

Le Resche, R.E., and Rausch, R.A. (1974). Accuracy and precision of aerial moose censusing. *J. Wildl. Manag.* **38**, 175–82. [456]

Leslie, P.H. (1945). On the use of matrices in certain population mathematics. *Biometrika* **33**, 183–212. [549]

Leslie, P.H. (1948). Some further notes on the use of matrices in population mathematics. *Biometrika* **35**, 213–45. [549]

Leslie, P.H. (1952). The estimation of population parameters from data obtained by means of the capture–recapture method. II: the estimation of total numbers. *Biometrika* **39**, 363–88. [160]

Leslie, P.H. (1958). Statistical appendix. *J. Animal Ecol.* **27**, 84–6. [161, 494]

Leslie, P.H. (1966). The intrinsic rate of increase and the overlap of successive generations in a population of guillemots (*Uria aalge* Pont.). *J. Animal Ecol.* **35**, 291–301. [551]

Leslie, P.H. and Chitty, D. (1951). The estimation of population parameters from data obtained by means of the capture–recapture method. I: The maximum likelihood equations for estimating the death rate. *Biometrika* **38**, 269–92. [201, 205]

Leslie, P.H., Chitty, D. and Chitty, H. (1953). The estimation of population parameters from data obtained by means of the capture–recapture method. III: An example of the practical applications of the method. *Biometrika* **40**, 137–69. [81, 160, 201. 224–5]

Leslie, P.H. and Davis, D.H.S. (1939). An attempt to determine the absolute number of rats on a given area. *J. Animal Ecol.* **8**, 94–113. [298, 569]

Leslie, P.H. and Ranson, R.M. (1940). The mortality, fertility and rate of natural increase of the vole (*Microtus agrestis*) as observed in the laboratory. *J. Animal Ecol.* **9**, 27–52. [400]

Leslie, P.H., Tener, T.S., Vizoso, M. and Chitty, H. (1955). The longevity and fertility of the Orkney vole, *Microtus orcadensis*, as observed in the laboratory. *Proc. Zool. Soc. Lond.* **125**, 115–25. [400]

Lewis, C.E. and Hassanein, K.M. (1969). The relative effectiveness of different approaches to the surveillance of infections among hospitalized patients. *Medical Care* **7**, 379–84. [491]

Lewis, E.G. (1942). On the generation and growth of a population. *Sankhyā* **6**, 93–6. [549]

Lewis, J.C. and Farrar, J.W. (1968). An attempt to use the Leslie census method on deer *J. Wildl. Manag.* **32**, 760–4. [299]

Lewis, P.A.W. (ed.) (1972). *Stochastic Point Processes.* Wiley: New York. [485]

Libosvárský, J. (1962). Application of De Lury method in estimating the weight of fish stock in small streams. *Int. Revue ges. Hydrobiol.* **47**, 515–27. [303, 314–15]

Lidicker, W.Z. Jr (1966). Ecological observations on a feral house mouse population declining to extinction. *Ecol. Monogr.* **36**, 27–50. [205]

Lidicker, W.Z., Jr.(1973). Regulations of numbers in an island population of the California vole, a problem in community dynamics. *Ecol. Monogr.* **43**, 271–302. [561]

Lieberman, G.J. and Owen, D.B. (1961). *Tables of the Hypergeometric Distribution.* Stanford University Press. [63, 66]

Lincoln, F.C. (1930). Calculating waterfowl abundance on the basis of banding returns. *U.S. Dept. Agric. Circ. No. 118,* 1–4. [1, 104]

Lindzey, F.G., Thompson, S.K. and Hodges, J.I. (1977). Scent station index of black bear abundance. *J. Wildl. Manag.* **41**, 151–3. [453]

Linhart, S.B. and Knowlton, F.F. (1975). Determining the relative abundance of coyotes by scent station lines. *Wildl. Soc. Bull.* **3**, 119–24. [453]

Linn, I.J. and Downton, F. (1975). The analysis of data obtained from small mammal index trappings. *Acta theriologica* **20**, 319–31. [487]

Lloyd, M. (1967). Mean crowding. *J. Animal Ecol.* **36**, 1–29. [21, 479]

Loftsgaarden, D.O. and Quesenberry, C.P. (1965). A non-parametric estimate of a multivariate density function. *Ann. Math. Statist.* **36**, 1049–51. [487]

Lonsdale, E.M., Bradach, B., Thorne, E.T. (1971). A telemetry system to determine body temperature in pronghorn antelope. *J. Wildl. Manag.* **35**, 747–51. [448]

Lord, G.E. (1973). Population and parameter estimation in the acoustic enumeration of a migrating fish population. *Biometrics* **29**, 713–25. [453]

Lotrich, V.A. and Meredith, W.H. (1974). A technique and the effectiveness of various acrylic colors for subcutaneous marking of fish. *Trans. Amer. Fish. Soc.* **103**, 140–2. [448]

Lowe, J.I. (1956). Breeding density and productivity of mourning doves on a county-wide basis in Georgia. *J. Wildl. Manag.* **20**, 428–33. [54]

Lucas, H.A. and Seber, G.A.F. (1977). Estimating coverage and particle density using the line intersect method. *Biometrika* **64**, 618–22. [460, 467]

Ludwig, J.P. (1967). Band loss – its effect on banding data and apparent survivorship in the ringed-billed gull populations of the Great Lakes. *Bird-Banding* **38**, 309–23. [251]

Ludwig, J. (1979). A new quadrat variance method for the analysis of spatial pattern. *In* R.M. Cormack and J.K. Ord (eds.), *Spatial and Temporal Analysis in Ecology,* Statistical Ecology Series, Vol. 8, 289–303. International Co-operative Publishing House, Burtonsville, Maryland. [482]

Luke, D. McG., Pincock, D.G. and Stasko, A.B. (1973). Pressure-sensing ultrasonic transmitter for tracking aquatic animals. *J. Fish. Res. Bd. Canada* **30**, 1402–4. [448]

Lund, G.F. (1974). A design for receiving antennas for laboratory radio transmitters, with field application. *J. Mammal.* **55**, 237–8. [448]

MacArthur, R.H. and MacArthur, A.T. (1974). On the use of mist nets for population studies of birds. *Proc. National Academy of Sciences, Washington* **71**, 3230–3. [499]

McCaffery, K.R. (1973). Road-kills show trends in Wisconsin deer populations. *J. Wildl. Manag.* **37**, 212–6. [452]

McCaffery, K.R. (1976). Deer trails counts as an index to populations and habitat use. *J. Wildl. Manag.* **40**, 308–16. [452]

McCleave, J.D. and Stred, K.A. (1975). Effect of dummy telemetry transmitters on stamina of Atlantic salmon (*Salmo salar*) smolts. *J. Fish. Res. Bd. Canada* **32**, 559–63. [448]

McClure, H.E. (1945). Comparison of census methods for pheasants in Nebraska. *J. Wildl. Manag.* **9**, 38–45. [54]

MacDonald, D. and Dillman, E.G. (1968). Techniques for estimating non-statistical bias in big game harvest surveys. *J. Wildl. Manag.* **32**, 119–29. [360]

McLaren, A.D. (1967). Appendix: statistical analysis of the spacing of the settled aphids over the leaf and over the glass. *J. Animal Ecol.* **36**, 163–70. [41]

McLaren, G.F. and Pottinger, R.P. (1969). A technique for studying the population dynamics of the cabbage aphid *Brevicoryne brassicae* (L.). *N.Z.J. agric. Res.* **12**, 757–70. [551]

McLaren, I.A. (1961). Methods of determining the numbers and availability of ring seals in the eastern Canadian Arctic. *Arctic* **14**, 162–75. [459]

McLaren, I.A. (1966). Analysis of an aerial census of ringed seals. *J. Fish. Res. Bd. Canada* **23**, 769–73. [459]

McLeese, D.W. and Wilder, D.G. (1958). The activity and catchability of the lobster (*Homarus americanus*) in relation to temperature. *J. Fish. Res. Bd. Canada* **15**, 1345–54. [305]

MacLulich, D.A. (1951). A new technique of animal census, with examples. *J. Mammal.* **32**, 318–28. [20, 51, 449]

MacNeill, I.B. (1971a). Quick statistical methods for analysing the sequences of fish counts provided by digital echo counters. *J. Fish. Res. Bd. Canada* **28**, 1035–42. [453]

MacNeill, I.B. (1971b). On the estimation of fish population distributions using acoustic methods. *In* G.P. Patil, E.C. Pielou and W.E. Waters (eds.), *Spatial Patterns and Statistical Distributions,* Statistical Ecology Series, Vol. 1, 553–82. Penn. State Univ. Press, University Park, Penn. [453]

MacNeill, I.B. (1972). Censored overlapping Poisson counts. *Biometrika* **59**, 427–34. [453]

Magnusson, W.E., Caughley, G.J. and Grigg, G.C. (1978). A double-survey estimate of population size from incomplete counts. *J. Wildl. Manag.* **42**, 174–6. [458]

Manga, N. (1972). Population metabolism of *Nebria brevicollis* (F.) (*Col. Carabidae*). *Oecologia* **10**, 223–42. [505]

Manly, B.F.J. (1969). Some properties of a method of estimating the size of mobile animal populations. *Biometrika* **56**, 407–10. [233]

Manly, B.F.J. (1970). A simulation study of animal population estimation using the capture–recapture method. *J. appl. Ecol.* **7**, 13–39. [237, 506–7]

Manly, B.F.J. (1971a). A simulation study of Jolly's method for analysing capture–recapture data. *Biometrics* **27**, 415–24. [204, 506–8]

Manly, B.F.J. (1971b). Estimates of a marking effect with capture–recapture sampling *J. appl. Ecol.* **8**, 181–9. [231]

Manly, B.F.J. (1972). Estimating selective values from field data. *Biometrics* **28**, 1115–25. [518]

Manly, B.J.F. (1972–3). Estimating survivorship from the recaptures of animals first released at age zero. *Biometrie-Praximetrie* **13**, 1–14. [509]

Manly, B.F.J. (1973). A note on the estimation of selective values from recaptures of marked animals when selection pressures remain constant over time. *Res. Popul. Ecol.* **14**, 151–8. [518]

Manly, B.F.J. (1974a). A discrete analogue of the Gamma distribution applied to data on the survival of birds. *Biometrie-Praximetrie* **14**, 1–14. [546]

Manly, B.F.J. (1974b). Estimating survival from a multi-sample single recapture census where recaptures are not made at release times. *Biom. Zeitschr.* **16**, 185–90. [514]

Manly, B.F.J. (1974c). Estimation of stage-specific survival rates and other parameters for insect populations developing through several stages. *Oecologia* **15**, 227–85. [551]

Manly, B.F.J. (1974d). A comparison of methods for the analysis of insect stage-frequency data. *Oecologia* **17**, 335–48. [551]

Manly, B.F.J. (1975a) Estimating survival from a multi-sample single recapture census – the case of constant survival and recapture probabilities. *Biom. Zeitschr.* **17**, 431–5. [515–8]

Manly, B.F.J. (1975b). A second look at some data on a cline. *Heredity* **34**, 423–37. [516]

Manly, B.F.J. (1975c). A note on the Richards, Waloff and Spradberry method for estimating stage-specific rates for insect populations. *Biom. Zeitschr.* **17**, 77–83. [551]

Manly, B.F.J. (1976). Extensions to Kiritani and Nakasuji's method for analysing insect stage-frequency data. *Res. Popul. Ecol.* **17**, 191–9. [552]

Manly, B.F.J. (1977a). The analysis of trapping records for birds trapped in mist nests. *Biometrics* **33**, 404–10. [499]

Manly, B.F.J. (1977b). A simulation experiment on the application of the jackknife with Jolly's method for the analysis of capture–recapture data. *Acta theriologica* **22**, 215–23. [18, 508]

Manly, B.F.J. (1977c). A further note on Kiritani and Nakasuji's model for stage-frequency data including comments on the use of Tukey's jackknife technique for estimating variances. *Res. Popul. Ecol.* **18**, 177–86. [18, 552]

Manly, B.F.J. (1977d). A note on the design of experiments to estimate survival and relative survival. *Biometrical Journal* (formerly *Biom. Zeitschr.*) **19**, 687–92. [516–8]

Manly, B.F.J. (1978). A simulation study of three methods for estimating selective values and survival rates from recapture data. *Res. Popul. Ecol.* **20**, 15–22. [518]

Manly, B.F.J. and Parr, M.J. (1968). A new method of estimating population size, survivorship and birth rate from capture–recapture data. *Trans. Soc. Brit. Ent.* **18**, 81–9. [226, 233–5]

Manly, B.F.J. and Seber, G.A.F. (1973). Animal life tables from capture–recapture data. *Biometrics* **29**, 487–500. [545]

Mann, R.H.K. (1971). The populations, growth and production of fish in four small streams in Southern England. *J. Animal Ecol.* **40**, 155–90. [325, 501]

Manooch III, C.S. and Huntsman, G.R. (1977). Age, growth, and mortality of the red porgy, *Pagrus pagrus*. *Trans. Amer. Fish. Soc.* **106**, 26–33. [548]

Margetts, A.R. (1963). Measurement of the efficiency of recovery and reporting of tags from recaptured fish. *In* North Atlantic Fish Marking Symposium, *I.C.N.A.F., Special Publication No. 4*, 255–7. [100]

Marriott, F.H.C. (1979). Barnard's Monte Carlo tests: how many simulations? *Appl. Statist.* **28**, 75–7. [18]

Marsden, H.M. and Baskett, T.S. (1958). Annual mortality in a banded bobwhite population. *J. Wildl. Manag.* **22**, 414–19. [400]

Marten, G.G. (1970a). A regression method for mark–recapture estimates with unequal catchability. *Ecology* **51**, 291–5. [128, 150–2]

Marten, G.G. (1970b). Censusing mouse populations by means of smoked-paper tracking. Small Mammal Newsl. (I.B.P. working group on small mammals, Warsaw, Poland) **4**, 45–73. [54, 566–7]

Marten, G.G. (1972a). Censusing mouse populations by means of tracking. *Ecology* **53**, 859–76. [566–7]

Marten, G.G. (1972b), The remote sensing approach to censusing. *Res. Popul. Ecol.* **14**, 36–57. [447, 566]

Marten, G.G. (1973). Time patterns of *Peromyscus* activity and their correlations with weather. *J. Mammal.* **54**, 169–88. [447, 566]

Martin, J.T. and Zickefoose, J. (1976). The effectiveness of aerial survey for determining the distribution of rabbit warrens in a semi-arid environment. *Aust. Wildl. Res.* **3**, 79–84. [459]

Martinson, R.K. (1966). Proportion of recovered duck bands that are reported. *J. Wildl. Manag.* **30**, 264–8. [250]

Martinson, R.K. and Grondahl, C.R. (1966). Weather and pheasant populations in Southwestern Dakota. *J. Wildl. Manag.* **30**, 74–81. [54]

Martinson, R.K. and McCann, J.A. (1966). Proportion of recovered goose and brant brands that are reported. *J. Wildl. Manag.* **30**, 856–8. [250]

Marullo, F., Emiliani, D.A., Caillouet, C.W.and Clark, S.H. (1976). A vinyl streamer tag for shrimp (*Penaeus* spp.). *Trans. Amer. Fish. Soc.* **105**, 658–63. [488]

Matérn, B. (1960). Spatial variation. *Meddelanden från Statens Skogsforskningsinstitut* **49**, No. 5. [484]

Matérn, B. (1971). Doubly stochastic Poisson processes in the plane. *In* G.P. Patil, E.C. Pielou and W.E. Waters (eds.), *Spatial Patterns and Statistical Distributions,* Statistical Ecology Series, Vol. 1, 195–213. Penn. State Univ. Press, University Park, Penn. [484]

Mathews, S.B. and Buckley, R. (1974). Natural mortality rate in last winter of life of coho salmon (*Oncorhynchus kisutch*) resident in Puget Sound. *J. Fish. Res. Bd. Canada* **31**, 1158–60. [505]

Mathews, S.B. and Buckley, R. (1976). Marine mortality of Puget Sound coho salmon (*Oncorhynchus kisutch*). *J. Fish. Res. Bd. Canada* **33**, 1677–84. [542]

Mathisen, O.A., Croker, T.R. and Nunnallee, E.P. (1977). Acoustic estimation of juvenile sockeye salmon. *Rapp. P.-V. Réun. Cons. Int. Explor. Mer* **170**, 279–86. [454]

Mathisen, O.A., Ostvedt, O.J. and Vestnes, G. (1974). Some variance components in acoustic estimation of nekton. *TETHYS* **6**, 303–12. [454]

Maza, B.G. French, N.R. and Aschwanden, A.P. (1973). Home range dynamics in a population of heteromyid rodents. *J. Mammal.* **54**, 405–25. [446]

Mazurkiewicz, M. (1971). Shape, size and distribution of home ranges of *Clethrionomys glareolus* (Schreber, 1780). *Acta theriologica* **16**, 23–60. [446]

Mead, R. (1974). A test for spatial pattern at several scales using data from a grid of contiguous quadrats. *Biometrics* **30**, 295–307. [476, 481, 482]

Mears, H.C. and Hatch, R.W. (1976). Overwinter survival of fingerling brook trout with single and multiple fin clips. *Trans. Amer. Fish. Soc.* **105**, 669–74. [488]

Mech, L.D. (1967). Telemetry as a technique in the study of predation. *J. Wildl. Manag.* **31**, 492–6. [94]

Mech, L.D. (1977). Productivity, mortality, and population trends of wolves in Northeastern Minnesota. *J. Mammal* **58**, 559–74. [546]

Mendelssohn, R. (1976). Optimization problems associated with a Leslie matrix. *Amer. Natur.* **110**, 339–40. [551]

Menhinick, E.F. (1963). Estimation of insect population density in herbaceous vegetation, with emphasis on removal sweeping. *Ecology* **44**, 617–21. [315, 327]

Mercer, M.C. (1975). Modified Leslie–De Lury population models of the long-finned pilot whale (*Globicephala melaena*) and annual production of the short-finned squid (*Illex illecebrosus*) based upon their interaction at Newfoundland. *J. Fish. Res. Bd. Canada* **32**, 1145–54. [305]

Mertz, D.B. (1971). Life history phenomena in increasing and decreasing populations. *In* G.P. Patil, E.C. Pielou and W.E. Waters (eds.), *Sampling and Modeling Biological Populations Dynamics*, Statistical Ecology Series, Vol. 2, 361–99. Penn. State Univ. Press, University Park, Penn. [397]

Meserve, P.L. (1977). Three-dimensional home ranges of cricetid rodents. *J. Mammal* **58**, 549–58. [447]

Meslow, E.C. and Keith, L.B. (1968). Demographic parameters of a snowshoe hare population. *J. Wildl. Manag.* **32**, 812–34. [233, 383]

Metzgar, L.H. (1972). The measurement of home range shape. *J. Wildl. Manag.* **36**, 643–5. [447]

Metzgar, L.H. (1973a). Home range shape and activity in *Peromyscus leucopus*. *J. Mammal* **54**, 383–90. [446–8]

Metzgar, L.H. (1973b). A comparison of trap- and track-revealed home ranges in *Peromyscus*. *J. Mammal* **54**, 513–5. [448]

Metzgar, L.H. (1973c). Exploratory and feeding home ranges in Peromyscus. *J. Mammal.* **54**, 760–3. [448]

Metzgar, L.H. and Sheldon, A.L. (1974). An index of home range size. *J. Wildl. Manag.* **38**, 546–51. [448]

Michner, C.D., Cross, E.A., Daly, H.V., Rettermeyer, C.W. and Wille, A. (1955). Additional techniques for studying the behaviour of wild bees. *Insectes Sociaux* **2**, 237–46. [96]

Mikulski, P.W. and Smith, P.J. (1975). A variance bound for unbiased estimation in inverse sampling. *Biometrika* **62**, 216–7. [121]

Miller, L.S. (1957). Tracing vole movements by radioactive excretory products. *Ecology* **38**, 132–6. [20]

Miller, R.G. Jr. (1966). *Simultaneous Statistical Inference*. McGraw-Hill: New York. [11]

Miller, R.G. (1974). The jackknife– a review. *Biometrika* **61**, 1–15. [18]

Mitra, S.K. (1958). On the limiting power function of the frequency chi-squared test. *Ann. Math. Statist.* **29**, 1221–33. [195, 243]

Moellering, H. and Tobler, W.R. (1972). Geographical variances. *Geograph. Anal.* **4**, 34–50. [481]

Mohr, C.O. and Stumpf, W.A. (1966). Comparison of methods for calculating areas of animal activity. *J. Wildl. Manag.* **30**, 293–304. [20]

Molenaar, W. (1973). Simple approximations to the Poisson, binomial and hypergeometric distributions. *Biometrics* **29**, 403–7. [63]

Mood, A.M. (1940). The distribution theory of runs. *Ann. Math. Statist.* **11**, 367–92. [185]

Moon, T.E. (1976). A statistical model of the dynamics of a mosquito vector (*Culex tarsalis*) population. *Biometrics* **32**, 355–68. [555]

Moore, D.S. and Yackel, J.W. (1977). Large sample properties of nearest neighbour density function estimators. *In* S.S. Gupta and D.S. Moore (eds.), *Statistical Decision Theory and Related Topics,* II, Academic Press: New York. [477]

Moore, G. and Wallis, W.A. (1943). Time series significance tests based on signs of differences. *J. Amer. Statist. Assoc.* **38,** 153–64. [158]

Moore, P.G. (1954). Spacing in plant populations. *Ecology* **35,** 222–7. [42–3, 472]

Moore, W.H. (1954). A new type of electrical fish-catcher. *J. Animal Ecol.* **23,** 373–5. [315]

Moose, P.H. and Ehrenberg, J.E. (1971). An expression for the variance of abundance estimates using a fish echo integrator. *J. Fish. Res. Bd. Canada* **28,** 1293–1301. [453]

Moran, P.A.P. (1951). A mathematical theory of animal trapping. *Biometrika* **38,** 307–11. [312]

Moran, P.A.P. (1971). Estimating structural and functional relationships. *J. Multivariate Anal.* **1,** 232–55. [453]

Morgan, G.R. (1974). Aspects of the population dynamics of the western rock lobster, *Panulirus cygnus,* George. I. Estimation of population density. *Aust. J. Mar. Freshwater Res.* **25,** 235–48. [505]

Morisita, M. (1954). Estimation of population density by spacing method. *Mem. Fac. Sci. Kyushu Univ.* **E1,** 187–97. [42, 44]

Morisita, M. (1957). A new method for the estimation of density by the spacing method applicable to non-randomly distributed populations. *Physiology and Ecology* **7,** 134–44 (Japanese): *U.S.D.A.* Forest Service translation no. 11116. [42, 478]

Morris, R.F. (1955). Population studies on some small forest mammals in Eastern Canada. *J. Mammal.* **36,** 21–35. [84]

Morris, R.F. and Miller, C.A. (1954). The development of life tables for the spruce budworm. *Canadian J. Zool.* **32,** 283–301. [399]

Mosby, H.S. (1967). Population dynamics. *In* O.H. Hewitt (ed.), *The Wild Turkey and its Management,* 113–36. The Wildlife Society: Washington. [398]

Mosby, H.S. (1969). The influence of hunting on the population dynamics of a woodlot gray squirrel population. *J. Wildl. Manag.* **33,** 59–73. [560, 565]

Moyle, J.B. and Lound, R. (1960). Confidence limits associated with means and medians of series of net catches. *Trans. Amer. Fish. Soc.* **89,** 53–8. [25]

Muir, B.S. (1963). Vital statistics of *Esox masquinongy* in Nogies Creek, Ontario. I: Tag loss, mortality due to tagging, and the estimate of exploitation. *J. Fish. Res. Bd. Canada* **20,** 1213–30. [333]

Muir, B.S. (1964). Vital statistics of *Esox masquinongy* in Nogies Creek, Ontario. II: Population size, natural mortality and effect of fishing. *J. Fish. Res. Bd. Canada* **21,** 727–46. [333–4, 560]

Muir, B.S. and White, H. (1963). Application of the Paloheimo linear equation for estimating mortalities to a seasonal fishery. *J. Fish. Res. Bd. Canada* **20,** 839–40. [331–2]

Murphy, E.C. and Whitten, K.R. (1976). Dall sheep demography in McKinley Park and a reevaluation of Murie's data. *J. Wildl. Manag.* **40,** 597–609. [458, 546]

Murphy, G.I. (1952). An analysis of silver salmon counts at Benkow Dam, South Fork of Eel River, California. *Calif. Fish and Game* **38,** 105–12. [358]

Murphy, G.I. (1960). Estimating abundance from longline catches. *J. Fish. Res. Bd. Canada* **17,** 33–40. [306]

Murphy, G.I. (1965). A solution of the catch equation. *J. Fish. Res. Bd. Canada* **22,** 191–202. [333]

Murphy, P.W. (ed.), (1962). *Progress in Soil Zoology.* Butterworths: London. [21]

Murton, R.K. (1966). A statistical evaluation of the effect of wood-pigeon shooting as evidenced by the recoveries of ringed birds. *The Statistician* **16**, 183–202. [247, 249]

Murton, R.K., Thearle, R.J.P. and Thompson, J. (1972). Ecological studies of the feral pigeon *Columbia livia* Var. I. Population, breeding biology and methods of control. *J. appl. Ecol.* **9**, 835–74. [228]

Myers, K. and Parker, B.S. (1975). A study of the biology of the wild rabbit in climatically different regions of eastern Australia. VI. Changes in numbers and distribution related to climate and land systems in semi-arid north-western New South Wales. *Aust. Wildl. Res.* **2**, 11–32. [459]

Myllymäki, A., Paasikallio, A., Pankakoshi, E. and Kanervo, V. (1971). Removal experiments on small quadrats as a means of rapid assessment of the abundance of small animals. *Ann. Zool. Fennici* **8**, 177–85. [453, 564]

Myres, M.T. (1969). Uses of radar in wildlife management. *In* R.H. Giles, Jr. (ed.), *Wildlife Management Techniques*, 3rd edn, 105–8. The Wildlife Society: Washington. [19]

Nakamura, K., Itô, Y., Nakamura, M., Matsumoto, T. and Hayakawa, K. (1971). Estimation of population productivity of *Parapleurus alliacens* Germar (Orthoptera: Acridiidae) on a *Miscanthus sinensis* Anders. Grassland. I. Estimation of population parameters. *Oecologia* **7**, 1–15. [228]

Neff, D.J. (1968). The pellet-group count technique for big game trend, census, and distribution: a review. *J. Wildl. Manag.* **32**, 597–614. [54]

Nellis, C.H. and Keith, L.B. (1976). Population dynamics of coyotes in central Alberta, 1964–68. *J. Wildl. Manag.* **40**, 389–99. [456, 459, 546]

Nelson, L., Jr and Clark, F.W. (1973). Correction for sprung traps in catch–effort calculations of trapping results. *J. Mammal.* **54**, 295–8. [502]

Nelson, R.D., Buss, I.O. and Baines, G.A. (1962). Daily and seasonal crowing frequency of ring-necked pheasants. *J. Wildl. Manag.* **26**, 269–72. [54]

New, J.G. (1958). Dyes for studying the movements of small mammals. *J. Mammal.* **39**, 416–29. [20, 93]

New, J.G. (1959). Additional uses of dyes for studying the movements of small mammals. *J. Wildl. Manag.* **23**, 348–51. [93]

Newman, D.E. (1959). Factors influencing the winter roadside count of cottontails. *J. Wildl. Manag.* **23**, 290–4. [53]

Newton, I., Marquiss, M., Weir, D.N. and Moss, D. (1977). Spacing of sparrowhawk nesting territories. *J. Animal Ecol.* **46**, 425–41. [484]

Nicholson, M.D. and Pope, J.A. (1977). The estimation of mortality from capture–recapture experiments. *In* J.H. Steele (ed.), *Fisheries Mathematics*, 77–85. The proceedings of a conference organised by the Institute of Mathematics and its Applications, Aberdeen, 1975. Academic Press: London. [273]

Nicola, S.J. and Cordone, A.J. (1973). Effects of fin removal on survival and growth of rainbow trout (*Salmo gairdneri*) in a natural environment. *Trans. Amer. Fish. Soc.* **102**, 753–8. [488]

Nixon, C.M., Donohoe, R.W. and Nash, T. (1974). Overharvest of fox squirrels from two woodlots in Western Ohio. *J. Wildl. Manag.* **38**, 67–80. [505]

Nixon, C.M., Edwards, W.R. and Eberhardt, L.L. (1967). Estimating squirrel abundance from live-trapping data. *J. Wildl. Manag.* **31**, 96–101. [84, 172–3]

Nixon, C.M., McClain, M.W. and Donohoe, R.W. (1975). Effects of hunting and mast crops on a squirrel population. *J. Wildl. Manag.* **39**, 1–25. [84, 174]

Norden. R.H. (1972). A survey of maximum likelihood estimation. *Internat. Statist. Rev.* **40**, 329–54. [4]

Norden, R.H. (1973). A survey of maximum likelihood estimation. Part 2. *Internat. Statist. Rev.* **41**, 39–58. [4]

North Atlantic Fish Marking Symposium (1963). *International Commission for Northwest Atlantic Fisheries, Special Publication No. 4,* 1–370. [83, 85]

Northcote, T.G. and Wilkie, D.W. (1963). Underwater census of stream fish populations. *Trans. Amer. Fish. Soc.* **92**, 146–51. [58]

Norton-Griffiths, M. (1973). Counting the Serengeti migratory wildebeest using two-stage sampling. *E. Afr. Wildl. J.* **11**, 135–49. [455–6]

Norton-Griffiths, M. (1975a). *Counting Animals.* African Wildlife Leadership Foundation, Nairobi. [455]

Norton-Griffiths, M. (1975b). The numbers and distribution of large mammals in Ruaha National Park, Tanzania. *E. Afr. Wildl. J.* **13**, 121–40. [455]

Norton-Griffiths, M. (1976). Further aspects of bias in aerial census of large mammals. *J. Wildl. Manag.* **40**, 368–71. [457]

Nunneley, S.A. (1964). Analysis of banding records of local population of Blue Jays and Redpolls at Granby, Mass. *Bird-Banding* **35**, 8–22. [109]

Odense, P.H. and Logan, V.H. (1974). Marking Atlantic salmon (*Salmo salar*) with oxytetracycline. *J. Fish. Res. Bd. Canada* **31**, 348–50. [488]

Odum, E.P. and Pontin, A.J. (1961). Population density of the underground ant *Lasius flavus,* as determined by tagging with P^{32}. *Ecology* **42**, 186–8. [105]

O'Farrell, M.J., Kaufman, D.W. and Lundahl, D.W. (1977). Use of live-trapping with the assessment line method for density estimation. *J. Mammal.* **58**, 575–82. [448, 504]

Omand, D.N. (1951). A study of populations of fish based on catch–effort statistics. *J. Wildl. Manag.* **15**, 88–98. [299, 304]

Ord, J.K. (1972). *Families of Frequency Distributions.* Griffin's Statistical Monographs and Courses, No. 30. Griffin: London. [479–80]

Orians, G.H. (1958). A capture–recapture analysis of a shearwater population. *J. Animal Ecol.* **27**, 71–84. [205, 228]

Otis, D.L., Burnham, K.P., White, G.C. and Anderson, D.R. (1978). Statistical inference for capture data from closed populations. *Wildl. Monogr. No. 62,* 135 pp. [160, 449, 491–8, 501–3]

Overton, W.S. (1965). A modification of the Schnabel estimator to account for removal of animals from the population. *J. Wildl. Manag.* **29**, 392–5. [154–6]

Overton, W.S. and Davis, D.E. (1969). Estimating the numbers of animals in wildlife populations. *In* R.H. Giles, Jr (ed.), *Wildlife Management Techniques,* 3rd edn, 403–55. The Wildlife Society: Washington. [36, 53–4, 58, 118, 170, 431]

Owen, D.B. (1962). *Handbook of Statistical Tables.* Addison-Wesley: Reading, Massachusetts. [123]

Pahl, P.J. (1969). On testing the goodness-of-fit of the negative binomial distribution when expectations are small. *Biometrics* **25**, 143–51. [25]

Paloheimo, J.E. (1961). Studies on estimation of mortalities. 1: Comparison of a method described by Beverton and Holt and a new linear formula. *J. Fish. Res. Bd. Canada* **18**, 645–62. [304, 330]

Paloheimo, J.E. (1963). Estimation of catchabilities and population sizes of lobsters. *J. Fish. Res. Bd. Canada* **20**, 59–88. [125–8, 305]

Paloheimo, J.E. (1972). A spatial bivariate Poisson distribution. *Biometrika* **59**, 489–92. [473]

Paloheimo, J.E. and Dickie, L.M. (1964). Abundance and fishing success. *Rapp. P.-v. Réun. Cons. Perm. int. Extor. Mer* **155**, 152–63. [569]

Paloheimo, J.E. and Kohler, A.C. (1968). Analysis of the Southern Gulf of St. Lawrence cod population. *J. Fish. Res. Bd. Canada* **25**, 555–78. [334]

Paludan, K. (1951). Contributions to tbe breeding biology of *Larus argentatus* and *Larus fuscus. Dansk naturh. Foren. Vidensk. Medd.* **114**, 1–128. [253]

Parker, B.S. and Myers, K. (1974). A study of the biology of the wild rabbit in climatically different regions of eastern Australia. V. The use of aerial surveys to map the distribution of rabbit warrens over large areas in semi-arid north-western New South Wales. *Aust. Wildl. Res.* **1**, 17–26. [459]

Parker, G.R. (1972). Biology of the Kaminuriak population of the barren-ground Caribou, Part I. *Canadian Wildlife Report Series, No. 20.* [456]

Parker, R.A. (1955). A method for removing the effect of recruitment on Petersen-type population estimates. *J. Fish. Res. Bd. Canada* **12**, 447–50. [263–6]

Parker, R.A. (1963). On the estimation of population size, mortality and recruitment. *Biometrics* **19**, 318–23. [256–60]

Parker, R.R. (1965). Estimation of sea mortality rates for the 1961 brood-year pink salmon of the Bella Coola Area, British Columbia. *J. Fish. Res. Bd. Canada* **22**, 1523–54. [383]

Parker, R.R. (1968). Marine mortality schedules of pink salmon of the Bella Coola River, Central British Columbia. *J. Fish. Res. Bd. Canada* **25**, 757–94. [205]

Parr, M.J. (1965). A population study of a colony of imaginal *Ischnura elegaris* (van der Linden) (Odonata: Coenagriidae) at Dale, Pembrokeshire. *Field Studies* **2**, 237–82. [204–5, 208]

Parr, M.J., Gaskell, T.J. and George, B.J. (1968). Capture–recapture methods of estimating animal numbers. *J. Biol. Educ.* **2**, 95–117. [205, 221, 229]

Parrish, B.B. (1962). Problems concerning the population dynamics of the Atlantic herring (*Clupea Harengus L.*) with special reference to the North Sea. *In* E.D. Le Cren and M.W. Holdgate (Editors), *The Exploitation of Natural Animal Populations*, 3–28, Blackwell: Oxford. [306]

Pathak, P.K. (1964). On estimating the size of a population and its inverse by capture mark method. *Sankhyā A* **26**, 75–80. [490]

Patil, G.P. and Ord, J.K. (1976). On size-biased sampling and related form-invariant weighted distributions. *Sankhyā B* **33**, 48–61. [460]

Patil, G.P. and Rao, C.R. (1978). Weighted distributions and size biased sampling with application to wildlife populations and human families. *Biometrics* **34**, 179–89. [460]

Patil, G.P., Taillie, C. and Wigley, R.L. (1979). Transect sampling methods and their application to the deep-sea red crab. *In* J. Cairns, Jr, G.P. Patil and W.E. Waters (eds.), *Environmental Biomonitoring, Assessment, Prediction and Management – Certain Studies and Related Quantitative Issues*, Statistical Ecology Series, Vol. 11, 51–75. International Co-operative Publishing House, Burtonsville, Maryland. [462–4, 467]

Patil, S.A., Burnham, K.P. and Kovner, J.L. (1979). Nonparametric estimation of plant density by the distance method. *Biometrics* **35**, 597–604. [464, 477]

Paulik, G.J. (1961). Detection of incomplete reporting of tags. *J. Fish. Res. Bd. Canada* **18**, 817–29. [97–100, 346–7, 000–0]

Paulik, G.J. (1962). Use of the Chapman–Robson survival estimate for single- and multi-release tagging experiments. *Trans. Amer. Fish. Soc.* **91**, 95–8. [275, 290–2]

Paulik, G.J. (1963a). Estimates of mortality rates from tag recoveries. *Biometrics* **19**, 28–57. [83, 238, 272–80, 286, 288, 293–5, 386]

Paulik, G.J. (1963b). Exponential rates of decline and type (1) losses for populations of tagged pink salmon. *In* North Atlantic Fish Marking Symposium, *I.C.N.A.F., Special Publication No. 4,* 230–7. [279, 294–5]

Paulik, G.J. (1973). Studies of the possible form of the stock-recruitment curve. *Rapp. P.-v. Réun. Cons. int. Explor. Mer* **164,** 302–15. [550]

Paulik, G.J. and Robson, D.S. (1969). Statistical calculations for change-in-ratio estimators of population parameters. *J. Wildl. Manag.* **33,** 1–27. [103, 353–84]

Paynter, R.A. (1966). A new attempt to construct life tables for Kent Island Herring Gulls. *Bull. Mus. of Comp. Zool., Harvard Univ.* **133,** (11), 489–528. [251]

Pearson, E.S. and Hartley, H.O. (1966). *Biometrika Tables for Statisticians,* vol. 1 (3rd edn). Cambridge University Press/Charles Griffin & Co. Ltd. [63–4, 123]

Pearson, P.G. (1955). Population ecology of the spade foot toad, *Scaphiopus h. holbrooki* (Harlan). *Ecol. Monogr.* **25,** 233–67. [150]

Pedersen, R.J. (1977). Big game collar-transmitter package. *J. Wildl. Manag.* **41,** 578–9. [448]

Peek, J.M., Urich, D.L. and Mackie, R.J. (1976). Moose habitat selection and relationships to forest management in Northeastern Minnesota. *Wildl. Monogr. No. 48,* 65 pp. [456]

Pelikán, J., Zejda, J. and Holišová, V. (1964). On the question of investigating small mammal populations by the quadrat method. *Acta theriologica* **9,** 1–24. [21]

Pennycuick, C.J. and Western, D. (1972). An investigation of some sources of bias in aerial transect sampling of large mammal populations. *E. Afr. Wildl. J.* **10,** 175–91. [456]

Perry, H.R., Pardue, G.B., Barkalow, F.S., Jr. and Monroe, R.J. (1977). Factors affecting trap responses of the gray squirrel. *J. Wildl. Manag.* **41,** 135–43. [487]

Persson, O. (1964). Distance methods. *Studia Forestalia Suecica No. 15,* 68 pp. [473, 484]

Persson, O. (1965). Distance methods II. *Forest Research Inst. of Sweden, Dept of Forest Biometry, Research Report No. 6,* 23 pp. [473]

Persson, O. (1971). The robustness of estimating density by distance measurements. *In* G.P. Patil, E.C. Pielou and W.E. Waters (eds.), *Sampling and Modeling Biological Populations and Population Dynamics.* Statistical Ecology Series Vol. 2. 175–87. Penn. State Univ. Press, University Park, Penn. [473, 478]

Peterle, T.J. (1969). Radio isotopes and their use in wildlife research. *In* R.H. Giles, Jr (ed.), *Wildlife Management Techniques,* 3rd edn, 109–18. The Wildlife Society: Washington. [94]

Petraborg, W.H., Wellein, E.G. and Gunvalson, V.E. (1953). Roadside drumming counts: a spring census method for ruffed grouse. *J. Wildl. Manag.* **17,** 292–5. [54]

Petrides, G.A. (1949). View points on the analysis of open season sex and age ratios. *Trans. Nth Amer. Wildl. Conf.* **14,** 391–410. [353, 376]

Petrides, G.A. (1954). Estimating the percentage kill in ringnecked pheasants and other game species. *J. Wildl. Manag.* **18,** 294–7. [380]

Petrusewicz, K. and Andrzejewski, R. (1962). Natural history of a free-living population of house mice (*Mus musculus* Linnaeus) with particular reference to groupings within the population. *Ekol. Pol.* (*A*) **10,** 1–122. [19]

Phillips, B.F. and Campbell, N.A. (1970). Comparison of methods of estimating population size using data on the whelk. *J. Animal Ecol.* **39,** 753–9. [211, 560]

Phinney, D.E. (1974). Growth and survival of fluorescent-pigment-marked and finclipped salmon. *J. Wildl. Manag.* **38,** 132–7. [383, 488]

Pielou, E.C..(1969). *An Introduction to Mathematical Ecology,* Wiley-Interscience: New York [25, 41, 50, 56, 572]

Pielowski, Z. (1969). Belt assessment as a reliable method of determining the number of hares. *Acta theriologica* **14**, 133–40. [28]

Pip, E. and Stewart, J.M. (1975). A method of fitting population data to a matrix model when the growth rate is unknown. *Int. Revue ges. Hydrobiol.* **60**, 669–72. [550]

Pollard, J.H. (1966). On the use of the direct matrix product in analysing certain stochastic population models. *Biometrika* **53**, 397–415. [550]

Pollard, J.H. (1971). On distance estimators of density in randomly distributed forests. *Biometrics* **27**, 991–1002. [472]

Pollock, K.H. (1974). The assumption of equal catchability of animals in tag–recapture experiments. Ph.D. Dissertation, Cornell Univ., Ithaca, N.Y. [493, 531–3, 538]

Pollock, K.H. (1975a). Building models of capture–recapture experiments. *The Statistician* **25**, 253–60. [493]

Pollock, K.H. (1975b). A *K*-sample tag–recapture model allowing for unequal survival and catchability. *Biometrika* **62**, 577–83. [205, 530–4, 536–8]

Pollock, K.H. (1978). A family of density estimators for line transect sampling. *Biometrics* **34**, 475–8. [462]

Pollock, K.H. (1979). A *K*-sample capture–recapture model allowing for age-dependent survival and capture rates. (In preparation). [530–1]

Pollock, K.H., Solomon, D.L. and Robson, D.S. (1974). Tests for mortality and recruitment in a *K*-sample tag–recapture experiment. *Biometrics* **30**, 77–87. [495]

Pope, J.A. (ed.) (1975). Measurement of fishing effort. *Rapp. P.-v. Réun. Cons. int. Explor. Mer* **168**. [306, 542]

Pope, J.G. (1972). An investigation of the accuracy of virtual population analysis using cohort analysis. *Res. Bull. Int. Comm. NW. Atlant. Fish.* **9**, 65–74. [542]

Pope, J.G. (1977). Estimation of fishing mortality, its precision and implications for the management of fisheries. *In* J.H. Steele (ed.), *Fisheries Mathematics*, 63–76. The proceedings of a conference organised by the Institute of Mathematics and its Applications, Aberdeen, 1975. Academic Press: London. [544]

Pottinger, R.P. and Le Roux, E.J. (1971). The biology and dynamics of *Lithocolletis blancardella* on apple in Quebec. *Mem. Ent. Soc. Can.* **77**, 1–437. [399]

Prentice, R.L. (1973). A design for studying the clustering of plant or animal species using quadrat sizes in geometric progression. *J. theor. Biol.* **39**, 601–8. [481]

Progulske, D.R. and Duerre, D.C. (1964). Factors influencing spotlighting counts of deer. *J. Wildl. Manag.* **28**, 27–34. [54]

Pucek, Z. (1969). Trap response and estimation of numbers of shrews in removal catches. *Acta theriologica* **14**, 403–26. [81, 83–4, 327]

Pyburn, W.F. (1958). Size and movements of a local population of cricket frogs (*Acris crepitans*). *Tex. J. Sci.* **10**, 325–42. [144]

Quenouille, M.H. (1956). Notes on bias reduction. *Biometrika* **43**, 353–60. [18, 509]

Quick, H.F. (1963). Animal population analysis. *In* H.S. Mosby (ed.), *Wildlife Investigational Techniques*, 2nd edn, 190–228. The Wildlife Society: Washington. [393, 397–9, 402–3]

Quinn, T.J., II and Gallucci, V.F. (1979). Parametric models for line transect estimates of abundance. (Submitted to *Ecology*). [462]

Rafail, S.Z. (1971a). Estimation of abundance of fish populations by capture–recapture experiments. *Marine Biology* **10**, 1–7. [510]

Rafail, S.Z. (1971b). Estimation of some parameters of large fish populations by capture–recapture experiments. *Marine Biology* **10**, 8–12. [511]

Rafail, S.Z. (1972a). A further contribution to the study of fish populations by capture–recapture experiments. *Marine Biology* **14**, 338–40. [511]

Rafail, S.Z. (1972b). The application of a multiple–recapture technique for the study of relatively small fish populations. *Marine Biology* **16**, 253–60. [511]

Rafail, S.Z. (1972c). The application of a multiple-recapture technique for the study of some aspects of dynamics of large fish populations. *Marine Biology* **17**, 251–5. [511]

Raff, M.S. (1956). On approximating the point binomial. *J. Amer. Statist. Assoc.* **51**, 293–303. [98, 122, 140]

Ramsey, F. and Scott, J.M. (1979). Estimating population densities form variable plot surveys. *In* R.M. Cormack, G.P. Patil and D.S. Robson, *Sampling Biological Populations*, Statistical Ecology Series, Vol. 5, 155–82. International Co-operative Publishing House, Burtonsville, Maryland. [468]

Randolph, S.E. (1977). Changing spatial relationships in a population of *Apodemus sylvaticus* with the onset of breeding. *J. Animal Ecol.* **46**, 653–76. [447]

Rao, A.V. and Katti, S.K. (1970). Minimum chi-square estimation for the log-zero-Poisson distribution. *In* G.P. Patil (ed.), *Random Counts in Scientific Work*, Vol. 2, 165–88. Penn. State Univ. Press, University Park, Penn. [480]

Rao, C.R. and Chakravarti, I.M. (1956). Some sample tests of significance for a Poisson distribution. *Biometrics* **12**, 264–82. [14, 170]

Rasmussen, D.I. and Doman, E.R. (1943). Census methods and their applications in the management of mule deer. *Trans. Nth Amer. Wildl. Conf.* **8**, 369–79. [353–4]

Raveling, D.G. (1976). Status of giant Canada Geese nesting in Southeast Manitoba. *J. Wildl. Manag.* **40**, 214–26. [526]

Read, K.L.Q. and Ashford, J.R. (1968). A system of models for the life cycle of a biological organism. *Biometrika* **55**, 211–21. [399, 554]

Reeves, H.M., Cooch, F.G. and Munro, R.E. (1976). Monitoring arctic habitat and goose production by satellite imagery. *J. Wildl. Manag.* **40**, 532–41. [459]

Reeves, H.M. and Marmelstein, A.D. (1975). U.S.F.W. activities in application of satellite imagery to goose management. *Proceedings of the Workshop on Remote Sensing of Wildlife*, 113–26. Gouvernement du Québec, Service de la Recherche Biologique. [459]

Regier, H.A. and Robson, D.S. (1967). Estimating population number and mortality rates. *In* S.D. Gerking (ed.), *The Biological Basis of Freshwater Fish Production*, 31–66. Blackwell Scientific Publications: Oxford. [58]

Reid, V.H., Hansen, R.M. and Ward, A.L. (1966). Counting mounds and earth plugs to census mountain pocket gophers. *J. Wildl. Manag.* **30**, 327–34. [54]

Rice, L.A. and Lovrien, H. (1974). Analysis of mourning dove banding in South Dakota. *J. Wildl. Manag.* **38**, 743–50.

Rice, W.R. and Harder, J.D. (1977). Application of multiple aerial sampling to a mark–recapture census of white-tailed deer. *J. Wildl. Manag.* **41**, 197–206. [456]

Richards, O.W. and Waloff, N. (1954). Studies on the biology and population dynamics of British grasshoppers. *Anti-Locust Bull.* **17**, 1–182. [551]

Richards, O.W., Waloff, N. and Spradberry, J.P. (1960). The measurement of mortality in an insect population in which recruitment and mortality widely overlap. *Oikos* **11**, 306–10. [551]

Ricker, W.E. (1940). Relation of catch per unit effort to abundance and rate of exploitation. *J. Fish. Res. Bd. Canada* **5**, 43–70. [329]

Ricker, W.E. (1944). Further notes on fishing mortality and effort. *Copeia* **1**, 23–44. [329]

Ricker, W.E. (1956). Uses of marking animals in ecological studies: the marking of fish. *Ecology* **37**, 665–70. [93]

Ricker, W.E. (1958). Handbook of computations for biological statistics of fish populations. *Bull. Fish. Bd. Canada* **119**, 1–300. [2, 62, 83–4, 86, 127, 142–3, 222–3, 298, 302–6, 350, 380, 383, 423, 428, 572]

Ricker, W.E. (1973). Linear regressions in fishery research. *J. Fish. Res. Bd. Canada* **30**, 409–34. [523]

Ricker, W.E. (1975). Computation and interpretation of biological statistics of fish populations. *Bull. Fish. Res. Bd. Canada No. 191*, 382 pp. [548, 572]

Rinne, W.E. and Deacon, J.E. (1973). Fluorescent pigment and immersion stain marking techniques for *Lepidomeda mollispinis* and *Cyprinodon nevadensis*. *Trans. Amer. Fish. Soc.* **102**, 459–62. [488]

Riordan, L.E. (1948). The sexing of deer and elk by airplane in Colorado. *Trans. Nth Amer. Wildl. Conf.* **13**, 409–28. [21, 353, 359–60]

Ripley, B.D. (1977). Modelling spatial patterns (with Discussion). *J. Roy. Statist. Soc., Series B* **39**, 172–212. [18, 484]

Ripley, B.D. and Kelly, F.P. (1977). Markov point processes. *J. London Math. Soc.* **15**, 188, 92. [485]

Ripley, B.D. and Silverman, B.W. (1978). Quick tests for spatial interaction. *Biometrika* **65**, 641–2. [485]

Robel, R.J., Henderson, F.R. and Jackson, W. (1972). Some sharp-tailed grouse population statistics from South Dakota. *J. Wildl. Manag.* **36**, 87–98. [84]

Robinette, W.L. (1949). Winter mortality among mule deer in the Fishlake National Forest, Utah. *U.S. Fish and Wildl. Serv., Spec. Sci. Report No. 65*, 1–15. [380]

Robinette, W.L., Jones, D.A., Gashwiler, J.S. and Aldous, C.M. (1954). Methods of censusing winter-lost deer. *Trans. Nth Amer. Wildl. Conf.* **19**, 511–25. [361, 562, 565]

Robinette, W.L., Jones, D.A., Gashwiler, J.S. and Aldous, C.M. (1956). Further analysis of methods of censusing winter-lost deer. *J. Wildl. Manag.* **20**, 75–8. [40, 361, 562]

Robinette, W.L., Loveless, C.M. and Jones, D.A. (1974). Field tests on strip census methods. *J. Wildl. Manag.* **38**, 81–96. [462, 468, 562]

Robson, D.S. (1960). An unbiased sampling and estimation procedure for creel censuses of fishermen. *Biometrics* **16**, 261–77. [304]

Robson, D.S. (1961). On the statistical theory of a roving creel census of fishermen. *Biometrics* **17**, 415–37. [304]

Robson, D.S. (1963). Maximum likelihood estimation of a sequence of annual survival rates from a capture–recapture series. *In* North Atlantic Fish Marking Symposium, *I.C.N.A.F., Special Publication No. 4*, 330–5. [214]

Robson, D.S. (1969). Mark–recapture methods of population estimation. *In* N.L. Johnson and H. Smith Jr. (Editors), *New Developments in Survey Sampling*, 120–40. Wiley-Interscience, Wiley and Sons: New York. [71, 82, 87, 205, 231, 530–1, 536–8]

Robson. D.S. (1979). Approximations to some mark–recapture sampling distributions. *In* R.M. Cormack, G.P. Patil and D.S. Robson (eds.), *Sampling Biological Populations*, Statistical Ecology Series Vol. 5, 257–76. International Co-operative Publishing House, Burtonsville, Maryland. [134, 218, 234]

Robson, D.S. and Chapman, D.G. (1961). Catch curves and mortality rates. *Trans. Amer. Fish. Soc.* **90**, 181–9. [414–30, 585]

Robson, D.S. and Flick, W.A. (1965). A non-parametric statistical method for culling recruits from a mark–recapture experiment. *Biometrics* **21**, 936–47. [74–81, 127, 495]

Robson, D.S. and Regier, H.A. (1964). Sample size in Petersen mark–recapture experiments. *Trans. Amer. Fish. Soc.* **93**, 215–26. [60, 63, 64–70, 119]

Robson, D.S. and Regier, H.A. (1966). Estimates of tag loss from recoveries of fish tagged and permanently marked. *Trans. Amer. Fish. Soc.* **95**, 56–9. [282–6]

Robson, D.S. and Regier, H.A. (1968). Estimation of population number and mortality rates. *In* W.E. Ricker (ed.), *Methods for Assessment of Fish Production in Fresh Waters*, IBP Handbook No. 3, 124–58. Blackwell Scientific Publications: Oxford. [16, 71–3, 153–5, 160, 319]

Robson, D.S. and Whitlock, J.H. (1964). Estimation of a truncation point. *Biometrika* **51**, 33–9. [58]

Robson, D.S. and Youngs, W.D. (1971). Statistical analysis of reported tag–recaptures in the harvest from an exploited population. Biometrics Unit Rep. BU-369-M, Cornell Univ. 15 pp. [513, 526]

Roessler, M. (1965). An analysis of the variability of fish populations taken by otter trawl in Biscayne Bay, Florida. *Trans. Amer. Fish. Soc.* **94**, 311–18. [25]

Roff, D.A. (1973a). On the accuracy of some mark–recapture estimators. *Oecologia* **12**, 15–34. [506–8]

Roff, D.A. (1973b). An examination of some statistical tests used in the analysis of mark–recapture data. *Oecologia* **12**, 35–54. [487, 495–6]

Rogers, A. (1974). *Statistical Analysis of Spatial Dispersion.* Pion Ltd: London. [480, 484]

Rogers, G., Julander, P. and Robinette, W.L. (1958). Pellet-group counts for deer census and range-use index. *J. Wildl. Manag.* **22**, 193–9. [54]

Rorres, C. and Fair, W. (1975). Optimal harvesting policy for an age-specific population. *Math. Biosci.* **24**, 31–47. [551]

Rose, C.D. and Hassler, W.W. (1969). Application of survey techniques to the dolphin, *Coryphaena hippurus*, fishery of North Carolina. *Trans. Amer. Fish. Soc.* **98**, 94–103. [304]

Rose, F.L. and Armentrout, D. (1974). Population estimates of *Ambystoma tigrinum* inhabiting two playa lakes. *J. Animal Ecol.* **43**, 671–9. [565]

Rose, G.B. (1977). Mortality rates of tagged adult cottontail rabbits. *J. Wildl. Manag.* **41**, 511–4. [547]

Rosenblatt, M. (1956). Remarks on some nonparametric estimates of a density function. *Ann. Math. Statist.* **27**, 832–7. [477]

Rougharden, J.D. (1977). Patchiness in the spatial distribution of a population caused by stochastic fluctuations in resources. *Oikos* **29**, 52–9. [485]

Roussel, Y.E. and Pichette, C. (1974). Comparison of techniques used to restrain and mark moose. *J. Wildl. Manag.* **38**, 783–8. [487]

Royaltey, H.H., Astrachan, E. and Sokal, R.R. (1975). Tests for patterns in geographic variation. *Geogr. Analysis* **7**, 369–95. [485]

Ruesink, W.G. (1975). Estimating time-varying survival of arthropod life stages from population density. *Ecology* **56**, 244–7. [555]

Ruos, J.L. (1974). Mourning dove status report, 1973. *U.S. Fish. Wild. Serv., Spec. Sci. Rep. Wildl. No. 186*, 36 pp [453]

Ruos, J.L. (1977). Mourning dove status report, 1974. *U.S. Fish Wildl. Serv., Spec. Sci. Rep. Wildl. No. 207*, 26 pp. [453]

Rupp, R.S. (1966). Generalised equation for the ratio method of estimating population abundance. *J. Wildl. Manag.* **30**, 523–6. [354–5, 372]

Rusch, D.H. and Keith, L.B. (1971). Seasonal and annual trends in numbers of Alberta ruffed grouse. *J. Wildl. Manag.* **35**, 804–22. [560]

Ryszkowski, L. (1971). Estimation of small rodent density with the aid of coloured bait. *Ann. Zool. Fennici* **8**, 8–13. [448]

Ryszkowski, L., Andrzejewski, R. and Petrusewicz, K. (1966). Comparison of estimates of numbers obtained by the methods of release of marked individuals and complete removal of rodents. *Acta theriologica* **11**, 329–41. [19]

Sadleir, R.M.F.S. (1965). The relationship between agonistic behaviour and population changes in the deermouse, *Peromyscus maniculatus* (Wagner). *J. Animal Ecol.* **34**, 331–53. [205]

Sampford, M.R. (1955). The truncated negative binomial distribution. *Biometrika* **42**, 58–69. [174]

Sampford, M.R. (1962). *An Introduction to Sampling Theory: with Applications to Agriculture.* Oliver and Boyd: London. [22, 43, 174]

Samuel, E. (1968). Sequential maximum likelihood estimation of the size of a population. *Ann. Math. Statist.* **39**, 1057–68. [189, 191, 193, 493]

Samuel, E. (1969). Comparison of sequential rules for estimation of the size of a population. *Biometrics* **25**, 517–35. [136, 193, 493]

Sanathanan, L. (1972). Estimating the size of a multinomial population. *Ann. Math. Statist.* **43**, 142–52. [490]

Sanderson, G.C. (1950). Small-mammal population of a prairie grove. *J. Mammal.* **31**, 17–25. [51]

Sanderson, G.C. (1966). The study of mammal movements – a review. *J. Wildl. Manag.* **30**, 215–35. [20, 94]

Sands, W.A. (1965). Termite distribution in man-modified habitats in West Africa, with special reference to species segregation in the genus *Trinervitermes* (Isoptera, Termitidae, Nasutitermitinae). *J. Animal Ecol.* **34**, 557–71. [21, 41]

Sandvik, L. (1978). Correspondence: A note on direct multiple sample censuses. *Biometrics* **34**, 523. [133]

Sargeant, A.B., Pfeifer, W.K. and Allen, S.H. (1975). A spring aerial census of red foxes in North Dakota. *J. Wildl. Manag.* **39**, 30–9. [459]

Sarrazin, J.P.R. and Bider, J.R. (1973). Activity, a neglected parameter in population estimates–The development of a new technique. *J. Mammal.* **54**, 369–82. [448, 504]

Sauder, D.W., Linder, R.L., Dahlgren, R.B. and Tucker, W.L. (1971). An evaluation of the roadside technique for censusing breeding waterfowl. *J. Wildl. Manag.* **35**, 538–43. [451]

Scattergood, L.W. (1954). Estimating fish and wildlife populations: a survey of methods. *In* O. Kempthorne (ed.), *Statistics and Mathematics in Biology,* Iowa State College Press: Ames, 273–85. [2, 19, 54]

Schaefer, M.B. (1951). Estimation of the size of animal populations by marking experiments. *U.S. Fish and Wildlife Service Fisheries Bulletin* **69**, 191–203. [64, 431]

Schevill, W.E. (ed.). (1974). *The Whale Problem: a Status Report.* Harvard University Press: Cambridge, Massachusetts. [459]

Schnabel, Z.E. (1938). The estimation of the total fish population of a lake. *Amer. Math. Mon.* **45**, 348–52. [130, 139]

Schnell, J.H. (1968). The limiting effects of natural predation on experimental cotton rat populations. *J. Wildl. Manag.* **32**, 698–711. [400]

Schroder, G.D. and Geluso, K.N. (1975). Spatial distribution of *Dipodomys spectabilis* mounds. *J. Mammal.* **56**, 363–8. [480, 484]

Schuerholz, G. (1974). Quantitative evaluation of edge from aerial photographs. *J. Wildl. Manag.* **38**, 913–20. [459]

Schultz, V. (1959). A contribution toward a bibliography on game kill and creel census procedures. *Misc. Publ. No. 359, Md. Agric. Exp. Sta.* [304]

Schultz, V. (1961). *An Annotated Bibliography of the Uses of Statistics in Ecology – a Search of 31 Periodicals.* Off. Tech. Ser., Dept. Comm., TID—3908, 1–315. [1]

Schultz, V. and Brooks, S.H. (1958). Some statistical aspects of the relationship of quail density to farm composition. *J. Wildl. Manag.* **22**, 283–91. [53]

Schultz, V. and Byrd, M.A. (1957). An analysis of covariance of cottontail rabbit population data. *J. Wildl. Manag.* **21**, 315–19. [53]

Schultz, V., Eberhardt, L.L., Thomas, J.M. and Cochran, M.I. (1976). *A Bibliography of Quantitative Ecology.* Dowden, Hutchinson and Ross, Stroudsburg, Pennsylvania. [1, 28]

Schultz, V. and Muncy, R.J. (1957). An analysis of variance applicable to transect population data. *J. Wildl. Manag.* **21**, 274–8. [53]

Schumacher, F.X. and Eschmeyer, R.W. (1943). The estimation of fish populations in lakes and ponds. *J. Tennessee Acad. Sci.* **18**, 228–49. [141–2]

Schwartz, C.C. (1974). Analysis of survey data collected on bobwhite in Iowa. *J. Wildl. Manag.* **38**, 674–8. [452]

Seal, H.L. (1954). The estimation of mortality and other decremental probabilities. *Skand. Aktuarietidskrift* **37**, 137–62. [411]

Sealander, J.A., Griffin, D.N., De Costa, J.J. and Jester, D.B. (1958). A technique for studying behavioural responses of small mammals to traps. *Ecology* **39**, 541–2. [84]

Sebens, K.P. (1976). Individual marking of soft-bodied intertidal invertebrates *in situ:* a vital stain technique applied to the sea anemone, *Anthopleura xanthogrammica. J. Fish. Res. Bd. Canada* **33**, 1407–10. [488]

Seber, G.A.F. (1962). The multi-sample single recapture census. *Biometrika* **49**, 330–49. [194–5, 212, 273, 511, 514, 530]

Seber, G.A.F. (1965). A note on the multiple recapture census. *Biometrika* **52**, 249–259. [169, 199]

Seber, G.A.F. (1967). Asymptotic variances in multinomial distributions. *New Zealand Statistician* **2**(1), 23–4. [9]

Seber, G.A.F. (1970a). The effects of trap resonse on tag–recapture estimates. *Biometrika* **26**, 13–22. [60, 85, 151]

Seber, G.A.F. (1970b). Estimating time-specific survival and reporting rates for adult birds from band returns. *Biometrika* **57**, 313–18. [213, 241–5, 526]

Seber, G.A.F. (1971). Estimating age-specific survival rates from bird-band returns when the reporting rate is constant. *Biometrika* **58**, 491–7. [252, 527–8]

Seber, G.A.F. (1972). Estimating survival rates from bird-band returns. *J. Wildl. Manag.* **36**, 405–13. [245, 255, 528]

Seber, G.A.F. (1977). *Linear Regression Analysis.* Wiley: New York. [10–11, 453, 465, 523]

Seber, G.A.F. (1979). Transects of random length. *In* R.M. Cormack, G.P. Patil and D.S. Robson (eds.), *Sampling Biological Populations*, Statistical Ecology Series, Vol. 5, 183–92. International Co-operative Publishing House, Burtonsville, Maryland. [460, 467]

Seber, G.A.F. and Le Cren, E.D. (1967). Estimating population parameters from catches large relative to the population. *J. Animal Ecol.* **36**, 631–43. [318]

Seber, G.A.F. and Pemberton, J.R. (1979). The line intercept method for studying plant cuticles from rumen and faecal analyses. *J. Wildl. Manag.* **43**, 916–25. [18, 460]

Seber, G.A.F. and Whale, J.F. (1970). The removal method for two and three samples. *Biometrics* **26**, 393–400. [312, 315]

Selleck, D.M. and Hart, C.M. (1957). Calculating the percentage of kill from sex and age ratios. *Calif. Fish and Game* **43**(4), 309–16. [380]

Sen, A.R. (1970a). Relative efficiency of sampling systems in the Canadian waterfowl harvest survey, 1967–68. *Biometrics* **26**, 315–26. [28, 53]

Sen, A.R. (1970b). On the bias in estimation due to imperfect frame in the Canadian waterfowl surveys. *J. Wildl. Manag.* **34**, 703–6. [489]

Sen, A.R. (1971a). Sampling techniques for estimation of mite incidence in the field with special reference to the tea crop. *In* G.P. Patil, E.C. Pielou and W.E. Waters (eds.), *Statistical Ecology*, Vol. 2, 253–65. Penn. State Univ. Press, University Park, Penn. [453]

Sen, A.R. (1971b). Successive sampling with two auxilary variables. *Sankhyā B* **33**, 371–8. [489]

Sen, A.R. (1971c). Increased precision in Canadian waterfowl harvest survey through successive sampling. *J. Wildl. Manag.* **35**, 664–8 [360, 489]

Sen, A.R. (1971d). Some recent developments in waterfowl sample survey techniques. *Applied Statistics* **20**, 139–47, [360]

Sen, A.R. (1972). Some nonsampling errors in the Canadian waterfowl mail survey. *J. Wildl. Manag.* **36**, 951–4. [489]

Sen, A.R. (1973a). Response errors in Canadian waterfowl surveys. *J. Wildl. Manag.* **37**, 485–91. [489]

Sen, A.R. (1973b). Theory and application of sampling on repeated occasions with several auxiliary variables. *Biometrics* **29**, 381–5. [489]

Sen, A.R. (1973c). Estimation of memory bias in wildlife mail surveys. *Bull. Intern. Stat. Inst.* **45**, 401–11. [489]

Sen. A.R. (1976). Developments in migratory game bird surveys. *J. Amer. Statist. Assoc.* **71**, 43–8. [489]

Sen, A.R., Sellers, S. and Smith, G.E.J. (1975). The use of a ratio estimate in successive sampling. *Biometrics* **31**, 673–83. [489]

Sen, A.R. and Southward, G.M. (1977). Sampling techniques in fisheries and wildlife. *Bull. Internat. Stat. Inst.* **47**, 516–34. [454, 548]

Sen, A.R., Tourigny, J. and Smith, G.E.J. (1974). On the line transect sampling method. *Biometrics* **30**, 329–40. [462, 471]

Sen, A.R., Smith, G.E.J. and Butler, G. (1978a). On a basic assumption in the line transect method. *Biometrical Journal* **20**, 363–9. [462]

Sen, A.R., Tourigny, J. and Smith, G.E.J. (1978b). Corrections to "On the line transect sampling method". *Biometrics* **34**, 328–9. [462]

Seshadri, V., Csorgo, M. and Stevens, M.A. (1969). Tests for the exponential distribution using Kolmogorov-type statistics. *J. Roy. Statist. Soc., B*, **31**, 499–509. [46]

Severinghaus, C.W. and Maguire, H.F. (1955). Use of age composition data for determining sex ratios among adult deer. *N.Y. Fish and Game J.* **2**(2), 242–6. [375]

Sharot, T. (1976a). Sharpening the jackknife. *Biometrika* **63**, 315–21. [18]

Sharot, T. (1976b). The generalized jackknife: finite samples and subsample sizes. *J. Amer. Statist. Assoc.* **71**, 451–4. [18]

Sheppe, W. (1965). Characteristics and uses of *Peromyscus* tracking data. *Ecology* **46**, 630–4. [20, 54]

Shetter, D.S. (1967). Effect of jaw tags and fin excision upon the growth, survival and exploitation of hatchery rainbow trout fingerlings in Michigan. *Trans. Amer. Fish. Soc.* **96**, 394–9. [83]

Shibata, K. (1971). Studies on echo counting for estimation of fish stocks. I. Overlap counting and reading of S-type echo counter. *Bull. Japanese Soc. Sci. Fish.* **37**, 711–9. [453]

Shiyomi, M. and Nakamura, K. and Vemura, M. (1976). A rapid estimation of animal population density on the assumption of negative binomial distribution. *Res. Popul. Ecol.* **18**, 28–38. [451]

Sichel, H.S. (1973). On a significance test for two Poisson variables. *Applied Statistics* **22**, 50–57. [123]

Sinclair, A.R.E. (1969). Aerial photographic methods in census of small mammals. *E. Afr. Agric. For. J.* **34**, 87–93. [456]

Sinclair, A.R.E. (1972). Long term monitoring of mammal populations in the Serengeti: census of non-migratory ungulates, 1971. *E. Afr. Wildl. J.* **10**, 287–298. [455–8]

Sinclair, A.R.E. (1973). Population increases of buffalo and wildebeest in the Serengeti. *E. Afr. Wildl. J.* **11**, 93–107. [456–7]

Sindermann, C.J. (1961). Parasite tags for marine fish. *J. Wildl. Manag.* **25**, 41–7. [94]

Siniff, D.B., Cline, D.R. and Erickson, A.W. (1970). Population densities of seals in the Weddell Sea, Antarctica, in 1968. *In* M.W. Holdgate (ed.), *Antarctic Ecology*, Vol. 1, 378–94. Academic Press: London and New York. [25]

Siniff, D.B., De Master, D.P., Hofman, R.J. and Eberhardt, L.L. (1977). An analysis of the dynamics of a Weddell seal population. *Ecol. Monogr.* **47**, 319–35. [459, 505]

Siniff, D.B. and Jessen, C.R. (1969). A simulation model of animal movement patterns. *Adv. Ecol. Res.* **6**, 185–219. [446]

Siniff, D.B. and Skoog, R.O. (1964). Aerial censusing of caribou using stratified random sampling. *J. Wildl. Manag.* **28**, 391–401. [28, 454]

Skalski, J.R. and Robson, D.S. (1979). Tests of homogeneity and goodness-of-fit to a truncated geometric model for removal sampling. *In* R.M. Cormack, G.P. Patil and D.S. Robson (eds.), *Sampling Biological Populations*, Vol. 5, 283–314. International Co-operative Publishing House, Burtonsville, Maryland. [503]

Skellam, J.G. (1948). A probability distribution derived from the binomial distribution by regarding the probability of success as variable between the sets of trials. *J. Roy. Statist. Soc. B.* **10**, 257–61. [62, 88, 177]

Skellam, J.G. (1952). Studies in statistical ecology. I: Spatial pattern. *Biometrika* **39**, 346–62. [43]

Skellam, J.G. (1958). The mathematical foundations underlying the use of line transects in animal ecology. *Biometrics* **14**, 385–400. [40]

Skutt, H.R., Bock, F.M., Haugstad, P., Holter, J.B., Hayes, H.H. and Silver, H. (1973). Low-power implantable transmitters for telemetry of heart rate and temperature from white-tailed deer. *J. Wildl. Manag.* **37**, 413–7. [448]

Slade, N.A. (1976). Analysis of social structure from multiple capture data. *J. Mammal.* **57**, 790–5. [487]

Slater, L.E. (ed.), (1963). Bio-telemetry: the Use of Telemetry in Animal Behaviour and *Physiology in Relation to Ecological Problems.* Macmillan: New York. [94]

Slobodkin, L.B. (1962). *Growth and Regulation of Animal Populations.* Holt, Rinehart and Winston: New York. [395, 403]

Smirnov, V.S. (1967). The estimation of animal numbers based on the analysis of population structure. *In* K. Petrusewicz (ed.), *Secondary Productivity of Terrestrial Ecosystems.* vol. 1, 199–223. Institute of Ecology, Polish Academy of Sciences: Warsaw. [353]

Smith, A.D. (1964). Defecation rates of mule deer. *J. Wildl. Manag.* **28**, 435–44. [54]

Smith, G.E.J. (1979). Some aspects of line transect sampling when the target population moves. *Biometrics* **35**, 323–9. [468]

Smith, H.D., Jorgensen, C.D. and Tolley, H.D. (1972). Estimation of small mammals using recapture methods: partitioning of estimator variables. *Acta theriologica* **17**, 57–66. [449]

Smith, M.H. (1968). A comparison of different methods of capturing and estimating numbers of mice. *J. Mammal:* **49**, 455–62. [565]

Smith, M.H., Blessing, R., Chelton, J.G., Gentry, J.B., Golley, F.B. and McGinnis, J.T. (1971). Determining density for small mammal populations using a grid and assessment lines. *Acta theriologica* **16**, 105–25. [448, 504]

Smith, M.H., Boize, B.J. and Gentry, J.B. (1973). Validity of the center of activity concept. *J. Mammal.* **54**, 747–9. [446]

Smith, M.H., Gardner, R.H., Gentry, J.B., Kaufman, D.W. and O'Farrel, M.H. (1975). Density estimation of small mammal populations. *In* F.B. Golley, K. Petrusewicz and L. Ryskowski (eds.), *Small mammals: Their Productivity and Population Dynamics*, 25–53. Intern. Biol. Prog. 5., Cambridge Univ. Press, Great Britain. [84, 449, 487]

Smith, T.D. (1977). A matrix model of sperm whale populations. *Rep. int. Whal. Commn* **27**, 337–42. [550]

Smith, T.G. (1975). Parameters and dynamics of ringed seal populations in the Canadian eastern Arctic. *Rapp. P.-v. Réun. Cons. int. Explor. Mer* **169**, 281–295. [550]

Snedecor, G.W. (1946). *Statistical Methods*, 4th edn. Iowa State Univ. Press.: Ames (Iowa). [263]

Sonleitner, F.J. (1973). Mark–recapture estimates of over-wintering survival of the Queensland fruit fly, *Dacus trioni,* in field cages. *Res. Popul. Ecol.* **14**, 188–208. [228, 505]

Sonleitner, F.J. and Bateman, M.A. (1963). Mark–recapture analysis of a population of Queensland fruit-fly, *Dacus tryoni* (Frogg.), in an orchard. *J. Animal Ecol.* **32**, 259–69. [205]

South, A. (1965). Biology and ecology of *Agriolamax reticulatus* (Müll.) and other slugs: spatial distribution. *J. Animal Ecol.* **34**, 403–17. [25]

Southern, H.N. (1965). The trap-line index to small mammal populations. *J. Zool., Lond.* **147**, 217–21. [564]

Southward, G.M. (1976). Sampling landings of halibut for age composition. *International Pacific Halibut Commission Sci. Rep. No. 58,* 31 pp. [548]

Southwick, C.H. and Siddiqi, M.R. (1968). Population trends of Rhesus monkeys in villages and towns of Northern India, 1959–65. *J. Animal Ecol.* **37**, 199–204. [53]

Southwood, T.R.E. (1966). *Ecological Methods.* Methuen: London. [20, 21, 25–6, 52, 84–5, 93–4, 96, 327, 393, 397, 399, 480, 562, 572]

Spenceley, G.W., Spenceley, R.M. and Epperson, E.R. (1952). *Smithsonian Logarithmic Tables to Base e and Base 10.* Smithsonian Institute: Washington. [136]

Spigarelli, S.A., Romberg, G.P. and Thorne, R.E. (1973). A technique for simultaneous echo location of fish and thermal plume mapping. *Trans. Amer. Fish. Soc.* **102**, 462–6. [453]

Spinage, C.A. (1970). Population dynamics of the Uganda defassa waterbuck (*Kobus defassa Ugandae* Neumann) in the Queen Elizabeth Park, Uganda. *J. Animal Ecol.* **39**, 51–78. [547]

Spinage, C.A. (1972). African ungulate life tables. *Ecology* **53**, 645–52. [546]

Spitz, F. (1963). Les techniques d'échantillonage utilisées dans l'étude des populations de petits mammifères. *La Terre et la Vie* **2**, 203–37. [20]

Ssentongo, G.W. and Larkin, P.A. (1973). Some simple methods of estimating mortality rates of exploited fish populations. *J. Fish. Res. Bd. Canada* **30**, 695–8. [273]

Stair, J. (1957). Dove investigations. *Arizona Game and Fish Dept. Job Completion Rept., P.-R. Proj. W–53–R–7, WP3, J–9,* 1–32. [251]

Standen, V. and Latter, P.M. (1977). Distribution of a population of *Cognettia sphagnetorum* (Enchytraeidae) in relation to microhabitats in a blanket fog. *J. Animal Ecol.* **46**, 213–29. [480]

Stasko, A.B. and Pincock, D.G. (1977). Review of underwater biotelemetry, with emphasis on ultrasonic techniques. *J. Fish. Res. Bd. Canada* **34**, 1261–85. [448]

Stickel, L.F. (1948). The trap line as a measure of small mammal populations. *J. Wildl. Manag.* **12**, 153–61. [564]

Stiteler, W.M. and Patil, G.P. (1971a). Variance-to-mean ratio and Morisita's index as measures of spatial patterns in ecological populations. *In* G.P. Patil, E.C. Pielou and W.E. Waters (eds.), *Spatial Patterns and Statistical Distributions,* Statistical Ecology Series, Vol. 1, 423–59. Penn. State Univ. Press, University Park, Penn. [479]

Stiteler, W.M. and Patil, G.P. (1971b). Statistical measurement of spatial patterns. *Bull. Intern. Stat. Inst.* **44**, 57–85. [479]

Stoddart, L.C. (1970). A telemetric method for detecting jackrabbit mortality. *J. Wildl. Manag.* **34**, 501–7. [94]

Storm, G.L., Andrews, R.D., Phillips, R.L., Bishop, R.A., Siniff, D.B. and Tester, J.R. (1976). Morphology, reproduction, dispersal, and mortality of Midwestern red fox populations. *Wildl. Monogr. No. 49*, 82 pp. [527, 547]

Stormer, F.A., Hoekstra, T.W., White, C.M. and Kirkpatrick, C.M. (1977). Frequency distribution of deer pellet groups in Southern Indiana. *J. Wildl. Manag.* **41**, 779–82. [451]

Stott, B. (1968). Marking and tagging. *In* W.E. Ricker (ed.), *Methods for Assessment of Fish Production in Fresh Waters,* IBP Handbook No. 3, 78–92. Blackwell Scientific Publications: Oxford. [93]

Stott, R.S. and Olson, D.P. (1972). An evaluation of waterfowl surveys on the New Hampshire coastline. *J. Wildl. Manag.* **36**, 468–77. [456]

Stradling, D.J. (1970). The estimation of worker ant populations by the mark–release–recapture method: an improved marking technique. *J. Animal Ecol.* **39**, 575–91. [106]

Strandgaard, H. (1967). Reliability of the Petersen method tested on a roe-deer population. *J. Wildl. Manag.* **31**, 643–51. [82, 84, 111, 565]

Suzuki, O. (1973). A rational sampling method for estimation of demersal fish abundance. *Bull. Japanese Soc. Sci. Fish,* **39**, 1013–9. [451]

Swed, F.S. and Eisenhart, C. (1943). Tables for testing randomness of grouping in a sequence of alternatives. *Ann. Math. Statist.* **14**, 66–87. [186]

Swift, D.M. and Steinhorst, R.K. (1976). A technique for estimating small mammal population densities using a grid and assessment lines. *Acta theriologica* **21**, 471–80. [448]

Swinebroad, J. (1964). Net-shyness and wood thrush populations. *Bird-banding* **35**, 196–202. [180]

Taber, R.D. (1956). Uses of marking animals in ecological studies: marking of mammals; standard methods and new developments. *Ecology* **37**, 681–5. [93]

Taber, R.D. (1969). Criteria of sex and age. *In* R.H. Giles, Jr. (ed.), *Wildlife Management Techniques,* 3rd edn, 325–402. The Wildlife Society: Washington. [393]

Taber, R.D. and Cowan, I.M. (1969). Capturing and marking wild animals. *In* R.H. Giles, Jr (ed.), *Wildlife Management Techniques,* 3rd edn, 277–317. The Wildlife Society Washington. [85, 93]

Takahasi, K. (1961). Model for the estimation of the size of a population by using the capture–recapture method. *Ann. Inst. Statist. Math.* **12**, 237–48. [160]

Tanaka, R. (1951). Estimation of vole and mouse populations on Mount Ishizuchi and on the uplands of Southern Shikoku. *J. Mammal.* **32**, 450–8. [84, 145–8]

Tanaka, R. (1952). Theoretical justification of the mark-release index for small mammals *Bull. Kochi Women's College* **1**, 38–47. [145]

Tanaka, R. (1956). On differential response to live traps of marked and unmarked small mammals. *Annot. Zool. Jap.* **29**, 44–51. [84]

Tanaka, R. (1963a). On the problem of trap-response types of small mammal populations. *Res. Popul. Ecol.* **5**, 139–46. [84]

Tanaka, R. (1963b). Examination of the routine census equation by considering multiple collisions with a single-catch trap in small mammals. *Jap. J. Ecol.* **13**, 16–21. [569]

Tanaka, R. (1967). New regression formula to estimate the whole population for recapture-addicted small mammals. *Res. Popul. Ecol.* **9**, 83–94. [503]

Tanaka, R. (1970). A field study on the effect of prebaiting on censusing by the capture–recapture method in a vole population. *Res. Popul. Ecol.* **12**, 111–25. [84, 503]

Tanaka, R. (1972). Investigation into the edge effect by use of capture–recapture data in a vole population. *Res. Popul. Ecol.* **13**, 127–51. [446–9]

Tanaka, R. (1974). An approach to the edge effect in proof of the validity of Dice's assessment lines in small-mammal censusing. *Res. Popul. Ecol.* **15**, 121–37. [447, 449]

Tanaka, R. and Kanamori, M. (1967). New regression formula to estimate the whole population for recapture-addicted small mammals. *Res. Popul. Ecol.* **9**, 83–94. [568]

Tanaka, R. and Murakami, O. (1977). Progress in the study of census methods for estimating population density of small rodents in Japan. *In* M. Morisita (ed.), *Studies on Methods. of Estimating Population* Density, Biomass and Productivity in Terrestrial Animals. *JIBP Synthesis* **17**, 112–22. University of Tokyo Press. [84, 449, 503]

Tanaka, R. and Teramura, S. (1953). A population of the Japanese field vole infested with Tsutsugamuchi disease. *J. Mammal.* **34**, 345–52. [46]

Tanaka, S. (1953). Precision of age-composition of fish estimated by double sampling method using the length for stratification. *Bull. Jap. Soc. Fish.* **19**(5), 657–70. [33]

Tanaka, S. (1975). Some considerations on the methods for calculating survival rate from catch per unit effort data. *Bull Japanese Soc. Sci. Fish* **41**, 121–8. [330]

Tanton, M.T. (1965). Problems of live-trapping and population estimation for the wood mouse, *Apodemus sylvaticus* (L.). *J. Animal Ecol.* **34**, 1–22. [174, 176–7]

Tanton, M.T. (1969). The estimation and biology of populations of the bank vole and wood mouse. *J. Animal Ecol.* **38**, 511–29. [174, 184]

Tayler, C.K. and Saayman, G.S. (1974). A method for determining the composition, deployment and stability of groups of free-ranging dolphins. *Zeits. Säugetierkunde* **37**, 116–9. [459]

Taylor, L.R. (1961). Aggregation, variance and the mean. *Nature* **189**, 732–5. [479]

Taylor, L.R. (1971). Aggregation as a species characteristic. *In* G.P. Patil, E.C. Pielou and W.E. Waters (eds.), *Spatial Patterns and Statistical Distributions,* Statistical Ecology Series, Vol. 1, 357–77. Penn. State Univ. Press, University Park, Penn. [480]

Taylor, L.R., Woiwod, I.P. and Perry, J.N. (1978). The density-dependence of spatial behaviour and the rarity of randomness. *J. Animal Ecol.* **47**, 383–406. [480]

Taylor, R.H. and Williams, R.M. (1956). The use of pellet counts for estimating the density of populations of the wild rabbit, *Oryctolagus cuniculus* (L.). *N.Z. J. Sci. and Technol.* **38B**, 236–56. [54]

Taylor, S.M. (1966). Recent quantitative work on British Bird populations: a review. *The Statistician* **16**, 119–70. [84, 174]

Tesch, F.W. (1968). Age and growth. *In* W.E. Ricker (ed.), *Methods for Assessment of Fish Production in Fresh Waters,* IBP Handbook No. 3, 93–123. Blackwell Scientific Publications: Oxford. [393]

Tester, A.L. (1955). Estimation of recruitment and natural mortality rate from age-composition and catch data in British Columbia herring populations. *J. Fish. Res. Bd. Canada* **12,** 649–81. [426–7]

Tester, J.R. and Siniff, D.B. (1965). Aspects of animal movement and home range data obtained by telemetry. *Trans. Nth Amer. Wildl. Conf.* **30,** 379–92. [174]

Thomasson, R.L. and Kapadia, C.H. (1968). On estimating the parameter of a truncated geometric distribution. *Ann. Inst. Statist. Math.* **20,** 519–23. [171, 582]

Thompson, H.R. (1956). Distribution of distance to *n*th neighbour in a population of randomly distributed individuals. *Ecology* **37,** 391–4. [42, 44–5]

Thompson, R.L. and Gidden, C.S. (1972). Territorial basking counts to estimate alligator populations. *J. Wildl. Manag.* **36,** 1081–8. [487]

Thorburn, D. (1977). On the asymptotic normality of the jackknife. *Scand. J. Statist.* **4,** 113–8. [18]

Thorne, R.E. (1971). Investigations into the relation between integrated echo voltage and fish density. *J. Fish. Res. Bd. Canada* **28,** 1269–23. [453]

Thorne, R.E. (1977). A new digital hydroacoustic data processor and some observations on herring in Alaska. *J. Fish. Res. Bd. Canada* **34,** 2288–94. [454]

Thorne, R.E. and Dawson, J.J. (1974). An acoustic estimate of the escapement of sockeye salmon (*Oncorhynchus nerka*) into Lake Washington in 1971. *J. Fish. Res. Bd. Canada* **31,** 222–5. [453]

Thorne, R.E., Reeves, J.E. and Millikan, A.E. (1971). Estimation of the Pacific Lake (*Merluccins productus*) population in Port Susan, Washington, using an echo integrator. *J. Fish. Res. Bd. Candda* **28,** 1275–84. [453]

Tomlinson, P.K. (1971). Some sampling problems in fishery work. *Biometrics* **27,** 631–41. [548]

Tomlinson, R.E. (1968). Reward banding to determine reporting rate of recovered mourning dove bands. *J. Wildl. Manag.* **32,** 6–11. [250, 360]

Trent, T.T. and Rongstad, O.J. (1974). Home range and survival of cottontail rabbits in southwestern Wisconsin. *J. Wildl. Manag.* **38,** 459–72. [447, 546, 561]

Trippensee, R.E. (1948). *Wildlife Management, Upland Game and General Principles.* McGraw-Hill: New York. [21]

Tucker, C.J., Miller, L.D. and Pearson, R.L. (1975). Shortgrass Prairie spectral measurements. *Photogramm. Eng.* **41,** 1157–62. [459]

Tukey, J.W. (1958). Bias and confidence in not quite large samples (abstract). *Ann. Math. Statist.* **29,** 614. [18, 509]

Turček, F.J. (1958). Zonologische Arbeitsmethoden für Wirbeltiere. *In* J. Balogh (ed.), *Lebensgemeinschaften der Landtiere,* 415–50. [21]

Turner, F.B. (1960a). Size and dispersion of a Louisiana population of the cricket frog. *Agris gryllus. Ecology* **41,** 258–68. [41, 134, 144, 160, 205]

Turner, F.B. (1960b). Tests of randomness in recaptures of *Rana pipretiosa. Ecology* **41,** 237–9. [228]

Turner, F.B., Hoddenbach, G.A., Medica, P.A. and Lannom, J.R. (1970). The demography of the lizard, *Uta stansburiana* Baird and Girard, in Southern Nevada. *J. Animal Ecol.* **39,** 505–20. [397, 548]

Usher, M.B. (1969). The relation between mean square and block size in the analysis of similar patterns. *J. Ecol.* **57,** 505–14. [482]

Usher, M.B. (1972). Developments in the Leslie matrix model. *In* J.N.R. Jeffers (ed.), *Mathematical Models in Ecology,* 12th Symposium of the British Ecology Society, 29–60. Blackwell, Oxford and Edinburgh. [551]

Usher, M.B. (1975). Analysis of patterns in real and artificial plant populations. *J. Ecol.* **63,** 569–86. [482]

Van Den Avyle, M.J. (1976). Analysis of seasonal distribution patterns of young largemouth bass (*Micropterus salmoides*) by use of frequency-of-capture data. *J. Fish. Res. Bd. Canada* **33,** 2427–32. [184]

Van Etten, R.C., Switzenberg, D.F. and Eberhardt, L. (1965). Controlled deer hunting in a square-mile enclosure. *J. Wildl. Manag.* **29,** 59–73. [299]

Van Noordwijk, M. (1978). A mark–recapture study of coexisting zygopteran populations *Odonatologica* **7,** 353–74. [238, 505]

Van Sickle, J. (1977). Mortality rates from size distributions: the application of a conservation law. *Oecologia* **27,** 311–81. [549]

Van Winkle, W. (1975). Comparison of several probabilistic home-range models. *J. Wildl. Manag.* **39,** 118–23. [447]

Van Winkel, W., Jr, Martin, D.C. and Sebetich, M.J. (1973). A home-range model for animals inhabiting an ecotone. *Ecology* **54,** 205–9. [447]

Varley, G.C. and Gradwell, G.R. (1970). Recent advances in insect population dynamics. *Ann. Rev. Ent.* **15,** 1–24. [393, 572]

Vaughan, D.S. and Saila, S.B. (1976). A method for determining mortality rates using the Leslie matrix. *Trans. Amer. Fish. Soc.* **105,** 380–3. [550]

Wagner, F.H. and Stoddart, L.C. (1972). Influence of coyote predation on black-tailed jackrabbit populations in Utah. *J. Wildl. Manag.* **36,** 329–42. [452]

Wakeley, J.S. and Mendall, H.L. (1976). Migrational homing and survival of adult female eiders in Maine. *J. Wildl. Manag.* **40,** 15–21. [529]

Wallin, L. (1971). Spatial pattern of trappability of two populations of small mammals. *Oikos* **22,** 221–4. [447]

Waloff, N. and Blackith, R.E. (1962). The growth and distribution of the mounds of *Lasius flavus* (Fabricius) (Hym.: Formicidae) in Silwood Park, Berkshire. *J. Animal Ecol.* **31,** 421–37. [21, 41]

Walter, G.G. (1975). Graphical methods for estimating parameters in simple models of fisheries. *J. Fish. Res. Bd. Canada* **32,** 2163–8. [542]

Walters, C.J. (1969). A generalised computer simulation model for fish population studies. *Trans. Amer. Fish. Soc.* **98,** 505–12. [572]

Warner, K. (1971). Effects of jaw tagging on growth and scale characteristics of landlocked Atlantic salmon, *Salmo salar. J. Fish. Res. Bd. Canada* **28,** 537–42. [488]

Warren, W.G. (1971). The centre-satellite concept as a basis for ecological sampling. *In* G.P. Patil, E.C. Pielou and W.E. Waters (eds), *Sampling and Modeling Biological Populations and Population Dynamics,* Statistical Ecology Series, Vol. 2. 87–118. Penn. State Univ. Press, University Park, Penn. [483]

Warren, W.G. (1972). Point processes in forestry. *In* P.A.W. Lewis (ed.), *Stochastic Point Processes,* 801–16. Wiley: New York. [484]

Warren, W.G. (1979). Trends in the sampling of forest populations. *In* R.M. Cormack, G.P. Patil and D.S. Robson (eds.), *Sampling Biological Populations,* Statistical Ecology Series Vol. 5, 329–54. International Co-operative Publishing House, Burtonsville Maryland. [460]

Waters, T.F. (1960). The development of population estimate procedures in small trout lakes. *Trans. Amer. Fish. Soc.* **89,** 287–94. [86]

Watson, R.M. (1969). Aerial photographic methods in census of animals. *E. Afr. Agric. For. J.* **34**, 32–7. [456]

Watson, R.M., Parker, I.S.C. and Allan, T. (1969a). A census of elephant and other large mammals in the Mkomazi region of northern Tanzania and southern Kenya. *E. Afr. Wildl. J.* **7**, 11–26. [455–6]

Watson, R.M., Graham, A.D. and Parker, I.S.C. (1969b). A census of the large mammals of Loliondo Controlled Area, Northern Tanzania. *E. Afr. Wildl. J.* **7**, 43–59. [455–6]

Watson, R.M., Freeman, G.M. and Jolly, G.M. (1969c). Some indoor experiments to simulate problems in aerial censusing. *E. Afr. Agric. For. J.* **34** (special issue). 56–62. [456]

Watson, R.M., Jolly, G.M. and Graham, A.D. (1969d). Two experimental censuses. *E. Afr. Agric. For. J.* **34**, (special issue), 60–2.

Watt, K. (ed.), (1966). *Systems Analysis in Ecology.* Academic Press: New York. [572]

Webb, W.L. (1942). Notes on a method of censusing snowshoe hare populations. *J. Wildl. Manag.* **6**, 67–9. [36]

Webb, W.L. (1965). Small mammal populations on islands. *Ecology* **46**, 479–88. [327]

Welch, H.E. (1960). Two applications of a method of determining the error of population estimates of mosquito larvae by the mark and recapture technique. *Ecology* **41**, 228–9. [112]

Westerskov, K. (1963). Superior survival of black-necked over ring-necked pheasants in New Zealand. *J. Wildl. Manag.* **27**, 239–45. [253]

Westman, W.E. (1971). Mathematical models of contagion and their relation to density and basal area sampling techniques. *In* G.P. Patil, E.C. Pielou and W.E. Waters (eds.), *Spatial Patterns and Statistical Distributions,* Stastical Ecology Series Vol. 1, 515–36. Penn. State Univ. Press, University Park, Penn. [478]

Wetherill, G.B. (1967). *Elementary Statistical Methods.* Methuen. London. [308]

Wheeler, G.G. and Calhoun, J.B. (1968). Manual for conducting ICSM census category 04 (octagon census and assessment trap lines). *ICSM manual series No. 4, Parts 1 and 2, Edition 1:* 1–50. [504]

White, E.G. (1970). A self-checking coding technique for mark–recapture studies. *Bull. Ent. Res.,* **60**, 303–7. [93, 96]

White, E.G. (1971a). A versatile FORTRAN computer program for the capture–recapture stochastic model fo G.M. Jolly. *J. Fish. Res. Bd. Canada* **28**, 443–6. [205, 506]

White, E.G. (1971b). A computer program for capture–recapture studies of animal populations: a Fortran listing for the stochastic model of G.M. Jolly. *New Zealand Tussock Grasslands and Mountain Lands Institute Spec. Publ. No. 8,* 33 pp. [506]

White, E.G. (1975). Identifying population units that comply with capture–recapture assumptions in an open community of alpine grasshoppers. *Res. Popul. Ecol.* **16**, 153–87. [505]

Whitehouse, S. and Steven, D. (1977). A technique for aerial radio tracking. *J. Wildl. Manag.* **41**, 771–5. [448]

Whitlock, S.C. and Eberhardt, L.L. (1956). Large-scale dead deer surveys: methods, results and management implications. *Trans. Nth Amer. Wildl. Conf.* **21**, 555–66. [361]

Widrig, T.M. (1954). Method of estimating fish populations with application to Pacific sardine. *U.S. Fish and Wildl. Serv., Fish. Bull.* **56**, 141–66. [329]

Wiegert, R.G. (1964). Population energetics of meadow spittlebugs (*Philaenus spumarius* L.) as affected by migration and habitat. *Ecol. Monogr.* **34**, 217–41. [28]

Wierzbowska, T. (1972). Statistical estimation of home range size of small rodents. *Ekol. Polska A* **20**, (49), 781–831. [446]

Wierzbowska, T. (1975). Review of methods for estimating the parameters of the home range of small forest rodents from the aspect of sample size. *Acta theriologia* **20**, 3–22. [446–7]

Wierzbowska, T. and Petrusewicz, K. (1963). Residency and rate of disappearance of two free-living populations of the house mouse (*Mus musculus* L.). *Ekol. Pol.* (*A*), **11**, 557–74. [181]

Wight, H. (1959). Eleven years of rabbit-population data in Missouri. *J. Wildl. Manag.* **23**, 34–9. [53]

Wilbur, H.M. (1975). The evolutionary and mathematical demography of the turtle *Chrysemys picta*. *Ecology* **56**, 64–77. [546]

Wilbur, H.M. and Landwehr, J.M. (1974). The estimation of population size with equal and unequal risks of capture. *Ecology* **55**, 1339–48. [496]

Williams, C.B. (1964). *Patterns in the Balance of Nature*. Academic Press: London. [572]

Williams, W.P. (1965). The population density of four species of freshwater fish, roach (*Rutilus rutilus* (L.)), bleak (*Alburnus alburnus* (L.)), dace (*Leuciscus leuciscus* (L.)), and perch (*Perca fluviatilis* (L.)) in the River Thames at Reading. *J. Animal Ecol.* **34**, 173–85. [567]

Williamson, M.H. (1967). Introducing students to the concepts of population dynamics. *In* J.M. Lambert (ed.), *The Teaching of Ecology*, 169–175. Blackwell, Oxford and Cambridge. [551]

Wilson, L.F. and Gerrard, D.J. (1971). A new procedure for rapidly estimating European pine sawfly (*Hymenoptera: diprionidae*). *Canadian Entomologist* **103**, 1315–22. [451]

Winsor, C.P. and Clark, G.L. (1940). A statistical study of variation in the catch of plankton nets. *J. Mar. Res.* **3**, 1–34. [276, 426]

Wise, J.P. (1972). Yield-per-recruit estimates for eastern tropical Atlantic yellowfin tuna. *Trans. Amer. Fish. Soc.* **101**, 75–9. [542]

Wittes, J.T. (1972). On the bias and estimated variance of Chapman's two-sample capture–recapture population estimate. *Biometrics* **28**, 592–7. [60]

Wittes, J.T. (1974). Applications of a multinomial capture–recapture model to epidemiological data. *J. Amer. Statist. Assoc.* **69**, 93–7. [491]

Wittes, J.T., Colton, T. and Sidel, V.W. (1974). Capture–recapture methods for assessing the completeness of case ascertainment when using multiple information sources. *Journal of Chronic Diseases* **27**, 25–36. [491]

Wittes, J.T. and Sidel, V.W. (1968). A generalization of the simple capture–recapture model with applications to epidemiological research. *Journal of Chronic Diseases* **21**, 287–301. [491]

Wolcott, T.G. (1977). Optical tracking and telemetry for nocturnal field studies. *J. Wildl. Manag.* **41**, 309–12. [448]

Wood, G.W. (1963). The capture–recapture technique as a means of estimating populations of climbing cutworms. *Canadian J. Zool.* **41**, 47–50. [108]

Wood, J.E. (1959). Relative estimates of fox population levels. *J. Wildl. Manag.* **23**, 53–63. [56, 560]

Woodbury, A.M. (1956). Uses of marking animals in ecological studies: marking amphibians and reptiles. *Ecology* **37**, 670–4. [93]

Yapp, W.B. (1956). The theory of line transects. *Bird Study* **3**, 93–104. [40]

Young, H. (1958). Some repeat data on the Cardinal. *Bird-Banding* **29**, 219–23. [180]

Young, H. (1961). A test for randomness in trapping. *Bird-Banding* **32**, 160–2. [184–6]

Young, H., Nees, J. and Emlen, J.T. Jr. (1952). Heterogeneity of trap response in a population of house mice. *J. Wildl. Manag.* **16**, 169–80. [84, 160, 165–6, 179]

Young, H., Strecker, R.L. and Emlen, J.T. Jr. (1950). Localisation of activity in two indoor populations of house mice. *Mus musculus. J. Mammal.* **31**, 403–10. [166]

Youngs, W.D. (1972). Estimation of natural and fishing mortality rates from tag recaptures. *Trans. Amer. Fish. Soc.* **101**, 542–5. [523]

Youngs, W.D. (1974). Estimation of the fraction of anglers returning tags. *Trans. Amer. Fish. Soc.* **103**, 616–8. [523]

Youngs, W.D. (1976). An analysis of the effect of seasonal variability of harvest on the estimate of exploitation rate. *Trans. Amer. Fish. Soc.* **105**, 45–7. [523]

Youngs, W.D. and Robson, D.S. (1975). Estimating survival rate from tag returns: model tests and sample size determination. *J. Fish. Res. Bd. Canada* **32**, 2365–71. [518–22]

Zahl, S. (1974). Application of the *S*-method to the analysis of spatial pattern. *Biometrics* **30**, 513–24. [483]

Zahl, S. (1977a). A comparison of three methods for the analysis of spatial pattern. *Biometrics* **33**, 681–92. [482]

Zahl, S. (1977b). Jacknifing an index of diversity. *Ecology* **58**, 907–13. [18]

Zippin, C. (1956). An evaluation of the removal method of estimating animal populations. *Biometrics* **12**, 163–9. [312–15]

Zippin, C. (1958). The removal method of population estimation. *J. Wildl. Manag.* **22**, 82–90. [312–14]

Zubrzycki, S. (1963). O minimaksowym szacowaniu licznosci populacji. *Zastosow. Mat.* **7**, 183–94. [121]

Zubrzycki, S. (1966). Explicit formulas for minimax admissible estimators in some cases of restrictions imposed on the parameter. *Zastosow. Mat.* **9**, 31–52. [121]

ADDITIONAL REFERENCES

Cook, P. (1979). Statistical inference for bounds of random variables. *Biometrika* **66**, 367–74. [58]

Cook, R.D. and Jacobson, J.O. (1979). A design for estimating visibility bias in aerial surveys. *Biometrics* **35**, 735–42. [458]

Cormack, R.M. (1979). Spatial aspects of competition between individuals. *In* R.M. Cormack and J.K. Ord (eds.), *Spatial and Temporal Analysis in Ecology*, Statistical Ecology Series, vol. 8, 151–211. International Co-operative Publishing House, Burtonsville, Maryland. [478]

Cormack, R.M. (1981). Loglinear models for capture–recapture experiments on open populations. *In* R.W. Hiorns and D. Cooke (eds.), *The Mathematical Theory of the Dynamics of Biological Populations* (to appear). Academic Press, London. [541]

Darroch, J.N. and Ratcliff, D. (1980). A note on capture–recapture estimation. *Biometrics* **36**, 149–53. [136]

De Vries, P.G. (1979). A generalization of the Hayne-type estimator as an application of line intersect sampling. *Biometrics* **35**, 743–8. [470]

Gates, C.E. and Smith, P.W. (1980). An implementation of the Burnham–Anderson distribution-free method of estimating wildlife densities from line transect data. *Biometrics* **36**, 155–60. [462]

Good, I.J., Lewis, B.C., Gaskins, R.A. and Howell, L.W. (1979). Population estimation by the removal method assuming proportional trapping. *Biometrika* **66**, 485–94. [504]

Iwao, S. (1979). The $\overset{*}{m}$–m method for analyzing the distribution patterns of single- and mixed-species populations. *In* G.P. Patil and M. Rosenzweig (eds.), *Contemporary Qualitative Ecology and Related Ecometrics*, Statistical Ecology Series, vol. 12, 215–28. International Co-operative Publishing House, Burtonsville, Maryland. [480]

Nelson, L.J., Anderson, D.R. and Burnham, K.P. (1980). The effect of band loss on estimates of annual survival. *J. Field Ornithology* **51**, 30–8. [251, 525]

North, P.M. (1977). A novel clustering method for estimating numbers of bird territories. *Applied Statistics* **26**, 149–55. [20]

North, P.M. (1979). Relating Grey Heron survival rates to winter weather conditions. *Bird Study* **26**, 23–8. [527]

North, P.M. and Cormack, R.M. (1981). On Seber's method of estimating age-specific bird survival rates from ringing recoveries. *Biometrics* **37**, 103–12. [528]

North, P.M. and Morgan, B.J.T. (1979). Modelling heron survival using weather data. *Biometrics* **35**, 667–81. [528]

Ord, J.K. (1979). Time-series and spatial patterns in ecology. *In* R.M. Cormack and J.K. Ord (eds.), *Spatial and Temporal Analysis in Ecology*, Statistical Ecology Series, vol. 8, 1–94. International Co-operative Publishing House, Burtonsville, Maryland. [478]

Otis, D.L. (1980). An extension of the change-in-ratio method. *Biometrics* **36**, 141–7. [355]

Perry, J.N. and Mead, R. (1979). On the power of the index of dispersion test to detect spatial pattern. *Biometrics* **35**, 613–22. [14]

Quinn II, T.J. (1979). The effects of school structure on line transect estimators of abundance. *In* G.P. Patil and M.L. Rosenzweig (eds.), *Contemporary Quantitative Ecology and Related Ecometrics*, Statistical Ecology Series, vol. 12, 473–91. International Co-operative Publishing House, Burtonsville, Maryland. [467]

Ramsey, F.L. (1979). Parametric models for line transect surveys. *Biometrika* **66**, 505–12. [462]

Ripley, B.D. (1979). Tests of "randomness" for spatial point patterns. *J. Roy. Statist. Soc. B.* **41**, 368–74. [485]

INDEX

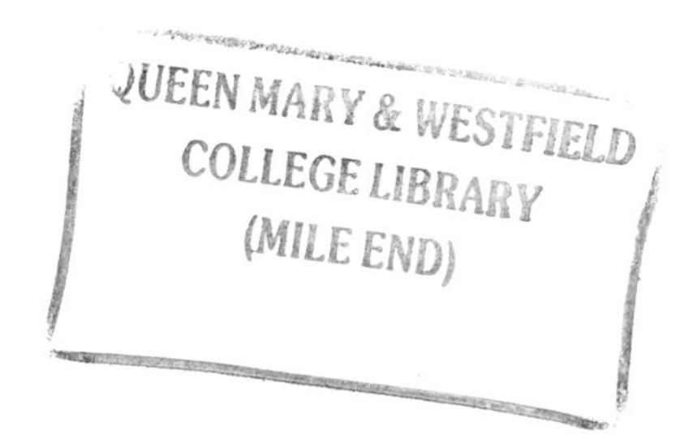